TODAY'S TECHNICIAN ™

SHOP MANUAL

For Automotive Electricity and Electronics

EIGHTH EDITION

Barry Hollembeak

✷ Cengage

Australia • Brazil • Canada • Mexico • Singapore • United Kingdom • United States

Today's Technician™: Shop Manual for Automotive Electricity and Electronics, Eighth Edition

Barry Hollembeak

SVP, Product: Erin Joyner

VP, Product: Thais Alencar

Portfolio Product Director: Jason Fremder

Portfolio Product Manager: Emily Olsen

Learning Designer: Mary Clyne

Content Manager: Sangeetha Vijay, Lumina Datamatics Ltd.

Digital Project Manager: Elizabeth Cranston

VP, Product Marketing: Jason Sakos

Director, Product Marketing: Neena Bali

Content Acquisition Analyst: Erin McCullough

Production Service: Lumina Datamatics Ltd.

Designer: Felicia Bennett

Cover Image Source: IRINA SHI/Shutterstock.com.

Last 3 editions: © 2019, © 2015, © 2011

For product information and technology assistance, contact us at
Cengage Customer & Sales Support, 1-800-354-9706 or support.cengage.com.

For permission to use material from this text or product, submit all requests online at **www.copyright.com.**

Library of Congress Control Number: 2022922242

Shop Manual ISBN: 978-0-357-76640-8
Package ISBN: 978-0-357-76638-5

Cengage
200 Pier 4 Boulevard
Boston, MA 02210
USA

Cengage is a leading provider of customized learning solutions. Our employees reside in nearly 40 different countries and serve digital learners in 165 countries around the world. Find your local representative at **www.cengage.com.**

To learn more about Cengage platforms and services, register or access your online learning solution, or purchase materials for your course, visit **www.cengage.com.**

Notice to the Reader

Printed in the United States of America
Print Number: 05   Print Year: 2024

Contents

Photo Sequence

Job Sheets

Preface

Thanks to the support the *Today's Technician*™ series has received from those who teach automotive technology, Cengage, the leader in automotive-related textbooks and learning solutions, is able to live up to its promise to provide new editions of the series every few years. We have listened and responded to our critics and fans and have presented this new updated and revised eighth edition. By revising this series on a regular basis, we can respond to changes in the industry, in technology, in the certification process, and the ever-changing needs of those who teach automotive technology.

We have also listened to instructors when they said that something was missing or incomplete in the previous edition. We have responded to those and have included the results in this eighth edition.

The *Today's Technician*™ series features textbooks and digital learning solutions that cover all mechanical and electrical systems of automobiles and light trucks. The individual titles correspond to the ASE (National Institute for Automotive Service Excellence) certification areas and are specifically correlated to the 2022 standards for Automotive Service Technicians, Master Service Technicians, as well as to the standards for Maintenance and Light Repair.

Additional titles include remedial skills and theories common to all of the certification areas and advanced or specific subject areas that reflect the latest technological trends. *Today's Technician*™: *Automotive Electricity and Electronics, 8e* is designed to give students a chance to develop the same skills and gain the same knowledge that today's successful technician has. This edition also reflects the most recent changes in the guidelines established by the ASE Education Foundation.

The purpose of the ASE Education Foundation is to evaluate technician training programs against standards developed by the automotive industry and recommend qualifying programs for accreditation by ASE. Programs can earn ASE accreditation upon the recommendation of the Education Foundation. The ASE Education Foundation's national standards reflect the skills that students must master. ASE accreditation through the Education Foundation's evaluation ensures that accredited training programs meet or exceed industry-recognized, uniform standards of excellence.

The technician of today and the future must know the underlying theory of all automotive systems, and be able to service and maintain those systems. Dividing the material into two volumes, a Classroom Manual and a Shop Manual, provides the reader with the information needed to begin a successful career as an automotive technician without interrupting the learning process by mixing cognitive and performance learning objectives into one volume.

The design of Cengage's *Today's Technician*™ series was based on features that are known to promote improved student learning. The design was further enhanced by a careful study of survey results, in which respondents were asked to value particular features. Some of these features can be found in other textbooks, while others are unique to this series.

Each Classroom Manual contains the principles of operation for each system and subsystem. The Classroom Manual also contains discussions on design variations of key components used by different vehicle manufacturers. It also looks into emerging technologies that will be standard or optional features in the near future. This volume is organized to build upon basic facts and theories. The primary objective of this volume is to allow the reader to gain an understanding of how each system and subsystem operates. This understanding is necessary to diagnose the complex automobiles of today and tomorrow. Although the basics contained in the Classroom Manual provide the knowledge needed for diagnostics, diagnostic procedures appear only in the Shop Manual. An understanding of the underlying theories is also a requirement for competence in the skill areas covered in the Shop Manual.

A spiral-bound Shop Manual delivers hands-on learning experiences with step-by-step instructions for diagnostic and repair procedures. Photo Sequences are used to illustrate some of the common service procedures. Other common procedures are listed and are accompanied with fine line drawings and color photos that allow the reader to visualize and conceptualize the finest details of the procedure. This volume also contains the reasons for performing the procedures, as well as when that particular service is appropriate.

The two volumes are designed to be used together and are arranged in corresponding chapters. Not only are the chapters in the volumes linked together, but the contents of the chapters are also linked. The linked content is indicated by marginal callouts that refer the reader to the chapter and page where the same topic is addressed in the companion volume. This valuable feature saves users the time and trouble of searching the index or table of contents to locate supporting information in the other volume. Instructors will find this feature especially helpful when planning the presentation of material and when making reading assignments.

Both volumes contain clear and thoughtfully selected illustrations, many of which are original drawings or photos specially prepared for inclusion in this series. This means that art is a vital part of each textbook and not merely inserted to increase the number of illustrations.

Highlights of the New Edition—Classroom Manual

The text, photos, and illustrations in the eight edition have been updated throughout to highlight the latest developments in automotive technology. New subject areas include smart charging systems, vehicle network security, LED headlights, and connected vehicle technology. Former Chapter 15 was split to focus all coverage on passive restraint and occupant safety systems. Vehicle safety systems was pulled from this chapter and now has its own chapter that includes coverage of ADAS. The last chapter still discusses electric vehicle operation, but has expanded coverage of EV, HEV, and PHEV components and operation. Although Chapter 17 covers details associated with alternative powered vehicles, all pertinent information about hybrid and electric vehicles is included in the main text that concerns relative topics. For example, the discussion of batteries in Chapter 5 includes coverage of HEV and EV batteries and ultra-capacitors. Chapter 6 includes AC motor principles and the operation of the integrated starter/generator. Chapter 7 includes the HEV charging system, including regenerative braking and the DC/DC converter.

The flow of basic electrical to more complex electronic systems has been maintained. Chapters are arranged to enhance this flow and reduce redundancy.

Chapter 1 introduces the student to the automotive electrical and electronic systems with a general overview. This chapter emphasizes the interconnectivity of systems in today's vehicles, and describes the purpose and location of the subsystems, as well as the major components of the system and subsystems. The goal of this chapter is to establish a basic understanding for students to build their learning on. All systems and subsystems that are discussed in detail later in the text are introduced, and their primary purpose is described. Chapter 2 covers the underlying basic theories of electricity and includes discussion of Ohm's and Kirchhoff's laws. This is valuable to the student and the instructor because it covers the theories that other textbooks assume the reader knows. All related basic electrical theories are covered in this chapter.

Chapter 3 applies those theories to the operation of electrical and electronic components, and Chapter 4 covers wiring and the proper use of wiring diagrams. Emphasis is on using the diagrams to determine how the system works and how to use the diagram to isolate the problem.

The chapters that follow cover the major components of automotive electrical and electronic systems, such as batteries, starting systems and motor designs, charging systems, and basic lighting systems. This is followed by chapters that detail the functions of the body computer, input components, and vehicle communication networks. From here the student is guided into specific systems that utilize computer functions.

Current electrical and electronic systems are used as examples throughout the text. Most of these systems are discussed in detail. This includes computer-controlled interior and exterior lighting, night vision, adaptive lights, instrumentation, and electrical/electronic accessories. Coverage includes intelligent wiper, immobilizer, adaptive cruise control systems, and ADAS. Chapter 15 details the passive restraint systems currently used. Chapter 16 focuses on vehicle safety systems such as Park Assist, Lane Departure Warning, Lane Centering, Collision Avoidance, and other ADASs including autonomous systems.

Highlights of the New Edition—Shop Manual

Like the Classroom Manual, updates to the Shop Manual match current trends. This updated manual includes service information related to the new topics covered in the Classroom Manual. In addition, new Photo Sequences that accompany the new material were added. The purpose of these detailed photos is to show students what to expect when they perform the same procedure. They also help familiarize students with a system or type of equipment they may not encounter at school. Although the main purpose of the textbook is not to prepare someone to successfully pass an ASE exam, this textbook includes all the information required to do so.

To stress the importance of safe work habits, Chapter 1 is dedicated to safety, and includes general HEV safety. As with the Classroom Manual, HEV system diagnosis is included within the main text. This provides the student with knowledge of safe system diagnosing procedures so they know what to expect as they further their training in this area. Included in this chapter are common shop hazards, safe shop practices, safety equipment, and the legislation concerning and the safe handling of hazardous materials and wastes.

Chapter 2 covers special tools and procedures including the use of isolation meters and expanded coverage of scan tools. Included is a section on what it entails to be an electrical systems technician. This section covers relationships, completing the work order, and ASE certification. Another section emphasizes the importance of proper diagnostic procedures.

Chapter 3 leads the student through basic troubleshooting and service. This includes the use of various test equipment to locate circuit defects and how to test electrical and electronic components. Chapter 4 provides experience with wiring repairs along with extended coverage and exercises on using the wiring diagrams.

The remaining chapters have been thoroughly updated. The Shop Manual cross-references to the Classroom Manual using marginal notes. This provides students the benefit of being able to quickly reference the theory of the component or system that they are now working with.

The textbook uses currently accepted service procedures as examples throughout the text. These procedures also serve as the basis for new job sheets inserted in the Shop Manual chapters.

Classroom Manual

Features of this manual include the following:

Cognitive Objectives

These objectives outline the chapter's contents and identify what students should know and be able to do upon completion of the chapter. *Each topic is divided into small units to promote easier understanding and learning.*

Terms to Know List

A list of key terms appears immediately after the Objectives. Students will see these terms discussed in the chapter. Definitions are also found in the Glossary at the end of the manual.

A Bit of History

This feature gives the student a sense of the evolution of the automobile. This feature not only contains nice-to-know information but also should spark some interest in the subject matter.

Margin Notes

The most important Terms to Know are highlighted and defined in the margins. Common trade jargon also appears in the margins and gives some of the common terms used for components. This helps students understand and speak the language of the trade, especially when conversing with an experienced technician.

Author's Notes

This feature includes simple explanations, stories, or examples of complex topics. These are included to help students understand difficult concepts.

Cross-References to the Shop Manual

When necessary, references to the appropriate page in the Shop Manual are given. Although the chapters of the two manuals are synchronized, material covered in other chapters of the Shop Manual may be fundamental to the topic discussed in the Classroom Manual.

Summary

Each chapter concludes with a summary of key points from the chapter. The key points help the reader review the chapter's contents.

Review Questions

Each chapters have short-answer essay, fill in the blanks, and multiple-choice questions. These questions accurately assess the student's competence in the objectives stated at the beginning of the chapter.

Instrumentation and Warning Lamps   371

returns the display to zero. The trip odometer continues to store trip mileage even if this function is not selected for display.

If the IC chip fails, some manufacturers provide for the replacement of the chip. Depending on the manufacturer, the new chip may be programmed to display the last odometer reading. Most replacement chips display an X, S, or * to indicate the odometer chip is not the original. If the odometer IC chip cannot be programmed, a door sticker is fixed to the vehicle to indicate that the odometer has been replaced.

AUTHOR'S NOTE Federal Motor Vehicle Safety Standards require the odometer be capable of storing up to 500,000 miles in nonvolatile memory. Most odometer readouts are up to 199,999.9 miles.

If an error occurs in the odometer circuit, the display changes to notify the driver. The form of error message differs among manufacturers. Some systems display "ERROR," while others may use dashed lines.

Federal and state laws prohibit tampering with the correct mileage as indicated on the odometer. If the odometer requires replacement, follow all applicable laws.

Tachometers

The electric tachometer receives voltage pulses from the ignition system, usually the ignition coil (**Figure 12-25**). The negative (−) side of the coil provides the tachometer signal as the switching unit opens the primary circuit. Each voltage pulse represents the generation of one spark at the spark plug. The rate of spark plug firing is in direct relationship to the speed of the engine. A circuit within the tachometer converts the ignition pulse signal into a varying voltage that rotates the needle.

The digital tachometer can be a continuously displayed gauge, or be a part of a multiguage display. The digital tachometer receives its voltage signals from the ignition module or PCM via the bus network and displays the readout in a bar graph (**Figure 12-26**) or quartz analog gauge. The PCM or ignition module typically receives the engine speed data from the Crankshaft Position (CKP) sensor.

The multiguage system has a built-in power supply that provides a 5-volt reference signal to the other monitored systems for the gauge. Also, the gauge has a **watchdog circuit** incorporated into it. The power on/off watchdog circuit supplies a reset voltage to the microprocessor in the event that pulsating output signals from the microprocessor are interrupted.

Shop Manual
Chapter 12, page 630

Figure 12-25 Electrical tachometer wired into the ignition system.

Instrumentation and Warning Lamps   381

Summary

- Through the use of gauges and indicator lights, the driver is capable of monitoring several engine and vehicle operating systems.
- The gauges include a speedometer, odometer, tachometer, oil pressure, charging indicator, fuel level, and coolant temperature.
- The most common types of electromechanical gauges are the d'Arsonval, three-coil, two-coil, and air-core.
- Computer-driven quartz swing needle displays are similar in design to the air-core electromagnetic gauges used in conventional analog instrument panels.
- All gauges require the use of a variable resistance sending unit. Styles of sending units include thermistors, piezoresistive sensors, and mechanical variable resistors.
- Digital instrument clusters use digital and linear displays to notify the driver of monitored system conditions.
- The most common types of displays used on electronic instrument panels are: light-emitting diodes (LEDs), liquid crystal displays (LCDs), vacuum fluorescent displays (VFDs), and a phosphorescent screen that is the anode.
- A head-up display system displays visual images onto the inside of the windshield in the driver's field of vision.
- In the absence of gauges, warning lamps indicate alters about important engine and vehicle functions.

Review Questions

Short-Answer Essays

1. What are the most common types of electromagnetic gauges?
2. Describe the operation of the piezoresistive sensor.
3. What is a thermistor used for?
4. What is meant by *electromechanical*?
5. Describe the operation of the air-coil gauge.
6. What is the basic difference between conventional analog and computer-driven analog instrument clusters?
7. Describe the operating principles of the digital speedometer.
8. Explain the operation of IC chip–type odometers.
9. Describe the operation of the electronic fuel gauge.
10. Describe the operation of quartz analog speedometers.

Fill in the Blanks

1. The purpose of the tachometer is to indicate _____.
2. A piezoresistive sensor monitors _____ changes.
3. The most common style of fuel level sending unit is a _____ variable resistor.
4. The combination valve activates the brake warning light by _____ pressure in the brake hydraulic system.
5. In a three-coil gauge, the _____ produces a magnetic field that bucks or opposes the low-reading coil. The _____ coil and the bucking coil are wound together, but in opposite directions. The _____ coil is positioned at 90° angle to the low-reading and bucking coils.
6. A _____ circuit completes the warning light circuit to ground through the ignition switch when it's in the START position.
7. Digital instrument clusters use _____ and _____ displays to notify the driver of monitored system conditions.
8. _____ is a calculation using the final drive ratio and the tire circumference to obtain accurate vehicle speed signals.

Shop Manual

To stress the importance of safe work habits, the Shop Manual dedicates one full chapter to safety. Other important features of this manual include the following:

Performance-Based Objectives

These objectives define the contents of the chapter and define what the student should have learned upon completion of the chapter. These objectives also correspond with the list of required tasks for ASE certification. *Each ASE task is addressed.*

Although not designed to simply prepare someone for the ASE certification exams, the ASE task lists are one of the focuses of this textbook. These tasks are defined generically when the procedure is commonly followed and specifically when the procedure is unique for specific vehicle models. Procedures for imported- and domestic-model automobiles and light trucks are included.

Terms to Know List

Terms in this list are also defined in the Glossary at the end of the manual.

Special Tools List

Whenever a special tool is required to complete a task, it is listed in the margin next to the procedure.

Cautions and Warnings

Throughout the text, warnings alert the reader to potentially hazardous materials or unsafe conditions. Cautions advise the student of things that can go wrong if instructions are not followed or if an unacceptable part or tool is used.

Basic Tools List

Each chapter begins with a list of the basic tools needed to perform the tasks included in the chapter.

Margin Notes

The most important Terms to Know are highlighted and defined in the margins. Common trade jargon also appears in the margins and gives some of the common terms used for components. This feature helps students understand and speak the language of the trade, especially when conversing with an experienced technician.

Photo Sequences

Detailed Photo Sequences illustrate many service procedures. These detailed photographs show students what to expect when they perform particular procedures. They can also provide the student a familiarity with a system or type of equipment, which the school may not have.

Service Tips

Whenever a shortcut or special procedure is appropriate, it is described in the text. These tips generally describe common procedures used by experienced technicians.

Cross-References to the Classroom Manual

Reference to the appropriate page in the Classroom Manual is given whenever necessary. Although the chapters of the two manuals are synchronized, material covered in other chapters of the Classroom Manual may be fundamental to the topic discussed in the Shop Manual.

Customer Care

This feature highlights those little things a technician can do or say to enhance customer relations.

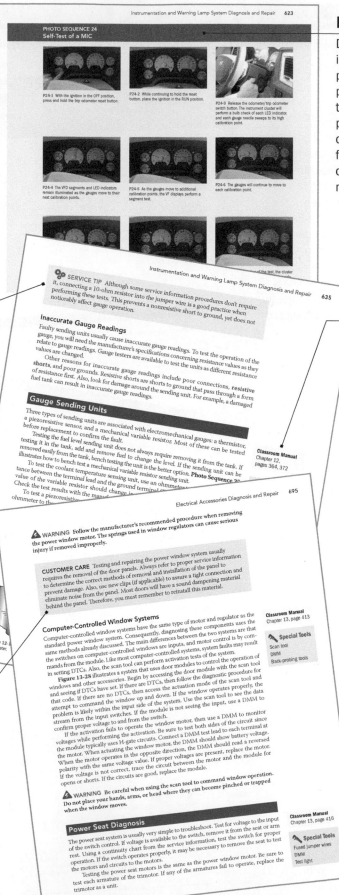

Case Studies

Beginning with Chapter 3, each chapter ends with a case study describing a particular vehicle problem and the logical steps a technician might use to solve the problem.

ASE-Style Review Questions

Each chapter contains ASE-style review questions that reflect the performance-based objectives listed at the beginning of the chapter. These questions can be used to review the chapter as well as to prepare for the ASE certification exam.

ASE Challenge Questions

Each technical chapter ends with five ASE challenge questions. These are not mere review questions; rather, they test students' ability to apply general knowledge to the contents of the chapter. To answer these questions the student may need to do some additional research.

CASE STUDY

The owner of a 2016 Charger says the automatic door lock feature fails to operate. The technician connects a scan tool to the diagnostic link connector (DLC) and initiates the actuator output test. The door locks operate properly while in this test mode. The technician then initiates the switch test. The scan tool indicates that the driver's side door ajar switch is constantly reading open. The service information instructs the technician to disconnect the door ajar switch connector. When disconnected, the scan tool indicates that the switch is open. Next, the technician connects a fused jumper wire across the harness side connector terminals. The scan tool still indicates that the switch is open. According to the service information, the input state should indicate closed when the jumper wire is connected.

The technician tests for voltage from the cabin compartment node (CCN) to the switch connector. The voltage should be 12 volts, but 0 volt is measured. Inspection of the wire between the CCN and the connector is good. According to the service information, the CCN must be replaced since it's not sending out the 12-volt reference signal to the left-front door switch.

The resistance of the terminal lead from the CCN to the switch is measured. Per the service manual specifications, the resistance value is too high, and the CCN requires replacement.

ASE-Style Review Questions

1. All of the following can cause the horn to sound continuously **EXCEPT:**
   A. A short to power in the circuit between the relay and horn.
   B. A short to ground between relay and horn.
   C. The horn switch contacts are stuck closed.
   D. A short to ground between the relay coil and the horn switch.

2. The two-speed windshield wiper operates in HIGH position only.
   Technician A says the l...
   worn...

4. Aligning the ACC sensor is being discussed.
   Technician A says the sensor's height determines the screen's height.
   Technician B says the screen is positioned directly in front of the sensor.
   Who is correct?
   A. A only
   B. B only
   C. Both A and B
   ...er A nor B
   ...rm will trip
   ...entry.
   ...ose lock
   ...witch may
   ...d B
   ...nor B
   ...igh-
   ...block.

---

580   Chapter 11

ASE Challenge Questions

1. The high-beam headlamps of a vehicle equipped with a Twilight Sentinel automatic headlamp system are inoperative; the low beams are working correctly.
   A voltmeter connected across the high-beam contacts of the dimmer switch indicates 0.15 volt when the high beams are "on."
   Technician A says the circuit from the Sentinel amplifier to the dimmer switch has excessive resistance.
   Technician B says the dimmer switch is faulty.
   Who is correct?
   C. Both A and B
   D. Neither A nor B

4. The left turn signals of a vehicle are flashing very slowly; the right turn signals are operating correctly.
   Technician A says this may be caused by excessive circuit resistance on the left turn signal circuit.
   Technician B says this may be occurring because someone may have installed bulbs with higher-than-normal wattage ratings on the left side of the vehicle.
   Who is correct?
   A. A only
   B. B only
   C. Both A and B
   D. Neither A nor B

5. All of the following could cause premature failure of a composite bulb **EXCEPT:**
   A. High charging system voltage.
   ...essive bulb circuit ground resistance.
   ...dling.

Name _____   Date _____

SHOP SAFETY SURVEY

Safety   41

Objective: Upon completion of this job sheet you should be able to evaluate the safety of your workplace.

JOB SHEET
1

ASE Education Foundation Correlation

This job sheet addresses the following Foundational and Workplace tasks:

Shop and Personal Safety

1.   Identify general shop safety rules and...

Instrumentation and Warning Lamp System Diagnosis and Repair   661

DIAGNOSTIC CHART 12-1

PROBLEM AREA:	Instrument cluster gauges.
SYMPTOMS:	One or all gauges fluctuate from low or high to normal readings.
POSSIBLE CAUSES:	1. Poor ground connection.
2. Excessive resistance.
3. Poor connections.
4. Faulty sending unit.
5. Defective printed circuit.

DIAGNOSTIC CHART 12-2

PROBLEM AREA:	Instrument cluster gauges.
SYMPTOMS:	One or all gauges read high.
POSSIBLE CAUSES:	1. Faulty instrument voltage regulator.
2. Shorted printed circuit.
3. Faulty sending unit.
4. Short to ground in sending unit circuit.
5. Faulty gauge.
6. Poor sending unit ground connection.

DIAGNOSTIC CHART 12-3

PROBLEM AREA:	Instrument cluster gauges.
SYMPTOMS:	One or all gauges read low.
POSSIBLE CAUSES:	1. Faulty instrument voltage regulator.
2. Poor sending unit ground connection.
3. Improper bulb application.
4. Improper connections..

DIAGNOSTIC CHART 12-4

PROBLEM AREA:	Instrument cluster gauges.
SYMPTOMS:	One or all gauges fail to read.
POSSIBLE CAUSES:	1. Blown fuse.
2. Open in the printed circuit.
3. Faulty gauge.
4. Poor common ground connection.
5. Open in the sending unit circuit.
6. Faulty sending unit.
7. Faulty instrument voltage regulator.
8. Poor electrical connection to the cluster.
9. Shorted sending unit circuit.

We Support
Education Foundation

Job Sheets

Located at the end of each chapter, job sheets provide a format for students to perform procedures covered in the chapter. A reference to the ASE task addressed by the procedure is included on the Job Sheet.

Diagnostic Charts

Some chapters include detailed diagnostic charts that list common problems and most probable causes. They also list a page reference in the Classroom Manual for better understanding of the system's operation and a page reference in the Shop Manual for details on the procedure necessary for correcting the problem.

Supplements

Instructor Resources

The *Today's Technician*™ series offers a robust set of instructor resources, available online at Cengage's Instructor Companion Site. The following tools have been provided to meet any instructor's classroom preparation needs:

- An Instructor's Manual including lecture outlines, teaching tips, and complete answers to end-of-chapter questions.
- PowerPoint® presentations with images that coincide with each chapter's content coverage.
- Cengage Learning Testing Powered by Cognero® provides hundreds of test questions in a flexible, online system. You can choose to author, edit, and manage test bank content from multiple Cengage Learning solutions and deliver tests from your LMS, or you can simply download editable Word documents from the Instructor Companion Site.
- An Image Gallery includes photos and illustrations from the text.
- The Job Sheets from the Shop Manual are provided in Word format.
- End-of-Chapter Review Questions are provided in Word format, with a separate set of text rejoinders available for instructors' reference.
- To complete this powerful suite of planning tools, correlation guides are provided to the ASE Education Foundation tasks and to the previous edition.
- An Educator's Guide.

Sign up or sign in at www.cengage.com to search for and access this product and its online resources.

Mindtap for Automotive Electricity and Electronics, 8E

Today's leading online learning platform, MindTap for *Today's Technician*™: *Automotive Electricity and Electronics, 8e*, gives you complete control of your course to craft a personalized, engaging learning experience that challenges students, builds confidence, and elevates performance. MindTap introduces students to core concepts from the beginning of your course using a simplified learning path that progresses from understanding to application and delivers access to an enhanced ebook with videos, animations, simulations, learning activities, auto-graded quizzes, and job sheets. Use MindTap for MindTap for *Today's Technician*™: *Automotive Electricity and Electronics, 8e*, as is, or personalize it to meet your specific course needs. You can also easily integrate MindTap into your learning management system (LMS).

Reviewers

The author and publisher would like to extend special thanks to the following instructors for their contributions to this text:

Donal Howell
College of the Sequoias

Randall Wesenick
Automotive Instructor

Rick Cartile
Automotive Instructor, Meridian Technology

Derrek Keesling
Department Chair Automotive Technology College of Lake County

David Shields
Meridian Technology CenterMeridian Technology Center

Joseph Gumina
Professor of Automotive City College of San Francisco

Orlando Grijalva
Automotive Instructor Tarrant County College

Tony DeVillier
Associate Professor, Delgado Community College

Eric Gomez
Transportation Education Director Eastern New Mexico University Roswell

CHAPTER 1
SAFETY

Upon completion and review of this chapter, you should be able to:

- Explain how safety is a part of professionalism.
- Describe personal safety responsibilities.
- Explain the proper application of each type of eye protection device.
- Properly lift heavy objects.
- Inspect power tools before use.
- Raise a vehicle using a floor jack and safety stands.
- Raise a vehicle using a hoist.
- Demonstrate the ability to properly run the engine in the shop.
- Classify fires and fire extinguishers.
- Locate, identify, and inspect fire extinguishers in the shop.
- Explain the proper use of the fire extinguisher.
- Define hazardous materials.
- Explain the right-to-know law or workplace hazardous materials information systems (WHMIS).

- Describe the responsibilities of the employer and the employee concerning hazardous materials.
- Describe what constitutes hazardous waste.
- Explain the proper procedure for disposing of hazardous waste.
- Describe the basic safety rules of servicing electrical systems.
- Work around batteries safely.
- Explain the safety precautions associated with charging and starting systems.
- List the safety precautions associated with servicing the air bag system.
- Explain the safety precautions that are necessary when servicing the antilock brake system.
- Explain the safety precautions necessary when servicing electric vehicles (EVs) and hybrid electric vehicles (HEVs).

Terms to Know

Air bag module
Air bag system
Antilock brake systems (ABS)
Asbestos
BAT
Carbon monoxide (CO)
Caustic
Conductors
Face shields
Fire blanket
Fire extinguishers
Flammable

Floor jacks
Hand tools
Hazard Communication Standard
Hazardous materials
Hazardous waste
High-efficiency particulate-arresting (HEPA)
High-voltage service plug
Hoists
Material safety data sheets (MSDS)
Occupational safety glasses

One-hand rule
Pneumatic tools
Power tools
Resource Conservation and Recovery Act (RCRA)
Right-to-know laws
Safety goggles
Safety stands
Vehicle lift point
Volatile
Workplace hazardous materials information systems (WHMIS)

Introduction

Being a professional technician is more than being knowledgeable about automotive systems; it's also an attitude. Being a professional technician includes having an understanding of all the hazards that may exist in the workplace. One of the most obvious traits of a professional is the ability to work productively and safely. This is where knowledge becomes very important. You need it to be productive and you need it to ensure your own safety and the safety of others. This chapter discusses the safety concerns associated with working in an automotive repair shop and the safety concerns associated with the vehicle's fuel and emission systems. In addition to basic shop safety, working on the vehicle's fuel and emission systems presents many special concerns.

Safety is everyone's concern. However, never assume the person working next to you is as conscientious as you are. You must be aware of what is going on around you at all times. As a professional technician, you must perform your work in a manner that protects not only you, but others in the workplace as well.

Personal Safety

Personal safety encompasses all aspects of preventing injury, including awareness, attitudes, and dress. All three of these are manifested through neat work habits. Cleaning up spills, keeping tools clean, and organizing the tools and materials in the shop help prevent accidents. Rushing to complete a job may result in a lack of consideration for personal safety and may ultimately cause an accident. Taking time to be neat and safe is rewarded by fewer accidents, higher customer satisfaction, and better pay.

Dress and Appearance

Nothing displays professional pride and a positive attitude more than the way you dress (**Figure 1-1**). Customers demand a professional atmosphere in the service shop. Your appearance instills customer confidence, as well as expresses your attitude toward safety and your chosen profession. Wearing proper and neat clothing can prevent injuries.

Loose-fitting clothing, or clothing that hangs out freely, can cause serious injury. Long-sleeve shirts should have their cuffs buttoned or rolled up tightly. Shirttails should

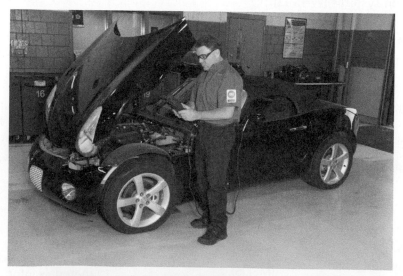

Figure 1-1 Wearing clean and properly fitting clothes along with proper safety equipment is an indication of how serious you are about safety.

be tucked in at all times. Some job positions within an automotive repair facility may require the employee to wear a necktie. If a necktie is worn in the shop area, it should be tucked inside your shirt. Clip-on ties are recommended if you must wear a tie.

Long hair is a serious safety concern. Very serious injury can result if hair gets caught in rotating machinery, fan belts, or fans. If your hair is long enough to touch the bottom of your shirt collar, it should be tied back and tucked under a hat.

Jewelry has no place in the automotive shop. Rings, watches, bracelets, necklaces, earrings, and so forth can cause serious injury. The gold, silver, and other metals used in jewelry are excellent **conductors** of electricity. Your body is also a good conductor. When electrical current flows through a conductor, it generates heat. The heat can be great enough to cause severe burns. Jewelry can also get caught in moving parts, causing serious cuts. Necklaces can cause serious injury or even death if they get caught in moving equipment.

You should wear shoes or boots that will protect your feet in the event something falls, or you stumble into something. It's a good idea to wear safety shoes or boots with steel toes and shanks. Most safety shoes also have slip-resistant soles. Tennis and jogging shoes provide little protection and are not satisfactory footwear in the automotive shop. Never wear any type of open toe or open top shoes, such as sandals.

Smoking, Alcohol, and Drugs in the Shop

Due to the potential hazards, never smoke when working in the shop. A spark from a cigarette or lighter may ignite flammable materials in the workplace. If the shop has designated smoking areas, smoke only in those areas. As a courtesy to your customers, do not smoke in their vehicles. Nonsmokers may not appreciate cigarette odor in their vehicles.

The use of drugs or alcohol must be avoided while working in the shop. Even a small amount of drugs or alcohol affects reaction time. In an emergency, slow reaction time may cause personal injury. If a heavy object falls off the workbench and your reaction time is slowed by drugs or alcohol, you may not get out of the way quickly enough to prevent injury. Also, you are a hazard to your coworkers if you are not performing at your best. Penalties for failure to adhere to your employer's policies concerning smoking, alcohol, and drugs may include termination of employment and possible legal actions.

Eye Protection

The importance of wearing proper eye protection cannot be overemphasized. Every working day there are more than 1,000 eye injuries in the United States. Many of these injuries result in blindness. Sadly, almost all of these are preventable. The safest and surest way to protect your eyes is to wear proper eye protection any time you enter the shop. At a minimum, wear eye protection when grinding, using power tools, hammering, cutting, chiseling, or performing service under the vehicle. In addition, wear eye protection when doing any work that can cause sparks, dirt, or rust to enter your eyes, and when you are working around chemicals. Remember, just because you are not doing the work yourself does not mean you cannot suffer an eye injury. Many eye injuries are caused by a coworker. Wear eye protection any time you are near an eye hazard.

The key to protecting your eyes is the use of *proper* eye protection. All eye protection must meet the newest requirements of the American National Standards Institute (ANSI). Currently this standard is identified as ANSI Z87.1. Regular prescription glasses do not provide adequate protection. Regular glasses are designed to impact standards that are far below those that are required in the workplace. The lens may stop a flying object, but the frame may allow the lens to pop out and hit your face, causing injury. In addition, regular glasses do not provide side protection. Prescription glasses that meet ANSI Z87.1 are available.

Figure 1-2 Examples of some of the different types of eye protection: (A) occupational safety glasses with side shields; (B) safety goggles that may be worn over prescription glasses; and (C) a face shield that is worn over safety glasses or goggles and protects the face.

There are many types of eye protection (**Figure 1-2**). One of the best ways to protect your eyes is to wear **occupational safety glasses**. These glasses are light and comfortable. They are constructed of tempered glass or safety plastic lens and have frames that prevent the lens from being pushed out upon impact. They have side shields to prevent the entry of objects from the side. Occupational safety glasses are available in prescription lens so they can be worn instead of regular corrective lens glasses.

Safety goggles fit snugly around the area of your eyes to prevent the entry of objects and to provide protection from liquid splashes. The force of impact on the lens is distributed throughout the entire area where the safety goggles are in contact with your face and forehead. Safety goggles are designed to fit over regular glasses.

Face shields are clear plastic shields that protect the entire face. They are used when there is potential for sparks, flying objects, or splashed liquids, which can cause neck, facial, and eye injuries. The plastic is not as strong or impact resistant as occupational safety glasses or safety goggles. If there is a danger of high-impact objects hitting the face shield, it's a good practice to wear safety glasses under the face shield.

Safety glasses provide little or no protection against chemicals. When working with chemicals, such as battery acid, refrigerants, and cleaning solutions, safety goggles should be worn. Full-face shields are not intended to provide primary protection for your eyes. They are designed to provide primary protection for your face and neck and should be worn in addition to eye protection.

Before removing your eye protection, close your eyes and lean forward. Pieces of metal, dirt, or other foreign material may have accumulated on the outside. These could fall into your eyes when you remove your glasses or shield.

Eyewash Fountains

Eye injuries may occur in various ways in an automotive shop. The following are some common types of eye injuries:

1. Thermal burns from excessive heat.
2. Irradiation burns from excessive light, such as from an arc welder.
3. Chemical burns from strong liquids, such as gasoline or battery electrolyte.

Figure 1-3 An eyewash fountain is used to remove chemicals and dirt from the eyes.

4. Foreign material in the eye.
5. Penetration of the eye by a sharp object.
6. A blow from a blunt object.

Wearing safety glasses and observing shop safety rules will prevent most eye accidents. If a chemical gets into your eyes, it must be washed out immediately to prevent a chemical burn. An eyewash fountain is the most effective way to wash the eyes, and every shop should be equipped with some eyewash facility (**Figure 1-3**). Be sure that you know the location of the eyewash fountain in the shop.

Hearing Protection

When performing any work or machining operation that will generate loud noise, wear some form of hearing protection (**Figure 1-4**). Ear plugs work well for providing up to 30 dB reduction in noise levels. Another option is to use specially designed hearing protection earmuffs.

Figure 1-4 Ear plugs or earmuffs provide the necessary hearing protection.

Figure 1-5 First-aid kit and typical contents.

First-Aid Kits

First-aid kits should be clearly identified and conveniently located (**Figure 1-5**). These kits contain such items as bandages and ointments required for minor cuts. All shop personnel must be familiar with the location of first-aid kits. At least one of the shop personnel should have basic first-aid training, and this person should oversee administering first aid and keeping the first-aid kits filled.

Hand Protection

Good hand protection is often overlooked (**Figure 1-6**). A scrape, cut, or burn can seriously impair your ability to work for many days. A well-fitting pair of heavy work gloves should be worn while grinding, welding, or when handling high-temperature components. Special rubber gloves are recommended for handling **caustic** chemicals. Caustic chemicals can destroy or eat through something and are considered extremely corrosive.

Many technicians wear latex, vinyl, or nitrile gloves to help protect their hands and to keep them clean. These are similar to the type of gloves worn by doctors and dentists during examinations. Latex gloves are inexpensive and provide good hand protection; however, some people are allergic to latex. If you wear latex gloves and develop a rash or redness on your hands, discontinue use. Vinyl gloves are also available and provide good resistance to tears and many nonaggressive liquids. Also, vinyl gloves are latex-free, so

Figure 1-6 Different types of gloves that are typically worn in the automotive service shop.

those who are allergic to latex can wear them. At a higher cost, nitrile gloves are latex-free synthetic rubber gloves that are superior to latex or vinyl in puncture resistance. In addition, nitrile gloves resist a wide range of chemicals that are harmful to either latex or vinyl.

Latex, vinyl, or nitrile gloves should be worn if you have an open cut or other injury on your hand, to prevent infection and the spread of diseases. In addition, these gloves should be worn if you are required to render first aid or medical assistance to someone who is injured. Because of the serious nature of blood-borne pathogens (disease- and infection-causing microorganisms carried by blood and other potentially infectious materials), it's important that you take every precaution to protect yourself regardless of the perceived status of the individual you are assisting. In other words, whether or not you think the blood/body fluid is infected with blood-borne pathogens, you treat it as if it is.

Rotating Belts and Pulleys

Many times the technician must work around rotating parts such as fans, generators, power steering pumps, air pumps, water pumps, and air conditioner compressors (**Figure 1-7**). Other rotating equipment or components of concern include tire changers, spin balancers, drills, bench grinders, and drive shafts. Always think before acting. Be aware of where you are placing your hands and fingers at all times. Do not place rags, tools, or test equipment near moving parts. In addition, make sure you are not wearing any loose clothing or jewelry that can get caught.

Electric Cooling Fans

Be very cautious around electric cooling fans. Some of these fans will operate even if the ignition switch is turned off. They are controlled by a temperature-sensing unit in the engine block or radiator and may turn on any time the coolant temperature reaches a

Figure 1-7 The accessory drive system on an engine can present serious safety concerns.

Figure 1-8 When lifting a heavy object, keep your back straight and lift with your legs.

certain temperature. Before working on or around an electric cooling fan, you should become familiar with its operation, and, if necessary, you should disconnect the electrical connector to the fan motor or the negative battery cable.

Lifting

Back injuries are one of the most crippling injuries in the industry, yet most of them are preventable. Most occupational back injuries are caused by improper lifting practices. These injuries can be avoided by following a few simple lifting guidelines:

1. Do not lift a heavy object by yourself. Seek help from someone else.
2. Do not lift more than you can handle. If the object is too heavy, use proper equipment to lift it.
3. Do not attempt to lift an object if there is not a good way to hold onto it. Study the object to determine the best balance and grip points.
4. Do not lift with your back. Your legs have some of the strongest muscles in your body. Use them.
5. Place your body close to the object. Keep your back and elbows straight (**Figure 1-8**).
6. Make sure you have a good grip on the object. Do not attempt to readjust the load once you have lifted it. If you are not comfortable with your balance and grip, lower the object and reposition yourself.
7. When lifting, keep the object as close to your body as possible. Keep your back straight and lift with your legs.
8. While carrying the object, do not twist your body to change directions. Use your feet to turn your whole body in the new direction.
9. To set the load down, keep the object close to your body. Bend at the knees and keep your back straight. Do not bend forward or twist.
10. If you need to place the object onto a shelf or bench top, place an edge of the object on the surface and slide it into place. Do not lean forward.

Tool and Equipment Safety

Technicians would not be able to do their job without tools and equipment. Most injuries caused by tools and equipment are the result of improper use, improper maintenance, and/or carelessness.

Figure 1-9 Assortment of hand tools.

Hand Tools

Hand tools use only the force generated from the body to operate (**Figure 1-9**). They multiply the force received through leverage to accomplish the work. Here are some very simple steps that you can take to help assure safe hand tool use:

1. Do not use tools that are worn out or broken.
2. Do not use a tool to do something that it was not designed to do. Use the proper tool for the job.
3. Keep your tools clean and in good condition.
4. Point sharp edges of tools away from you.
5. Do not hold small components in your hands while using tools such as screwdrivers. The tool may slip and cause injury to your hand.
6. Examine your work area for things that can cause injury if a tool slips or a fastener breaks loose quickly. Readjust yourself or the tool to avoid injuries.
7. Do not put sharp tools in your pockets.

Power Tool Safety

 WARNING **Always wear safety glasses when operating a power tool.**

Power tools use forces other than those generated by the body (**Figure 1-10**). They can use compressed air, electricity, or hydraulic pressure to generate and multiply force. Many times a technician will be required to use power tools when performing electrical service.

Figure 1-10 Examples of power tools.

Drills and hole saws will be used to install new accessories onto the vehicle or to drill holes for wiring to pass through. Grinders, drill presses, and hydraulic presses may be used to help fabricate or modify components. **Pneumatic tools** are powered by compressed air and are used to remove or fasten components. All of these tools can cause injury if not used properly. Use the following guidelines when working with power tools:

> **Pneumatic tools** are often called *air tools*.

1. Ask your instructor if you are not sure of the correct operation of a tool.
2. Always wear proper eye protection when using power tools.
3. Check that all safety guards and safety equipment are installed on the tool.
4. Before using an electrical tool, check the condition of the plug and cord. The plug should be a three-pronged plug. Never cut off the grounding prong. Do not use the tool if the wires are frayed or broken. Plug the tool only into a grounded receptacle.
5. Before using an air tool, check the condition of the air hose. Do not use the tool if the hose shows signs of weakness such as bulges or fraying. Also, the tool should be properly oiled.
6. Before using a hydraulic tool, check the condition of all hoses and gauges. Do not use the tool if any of these are defective.
7. Make sure other people are not in the area when you turn the tool on.
8. Do not leave the area with the tool still running. Stay with the tool until it stops. Then disconnect it.
9. Make all adjustments to the tool before turning it on.
10. If the tool is defective or does not pass your safety inspection, put a sign on it and report it to your supervisor or instructor.

Inspecting Power Tools. Improper use of power tools is the cause of many accidents each year. Many of these accidents could have been avoided if the tool was checked before it was used.

Use this checklist for the electric power tool(s) that you are inspecting. Do not plug the electrical cord into the socket until the inspection is completed. Consider the following:

1. Name of tool.
2. Is the tool clean?
3. Are safety guards in place?
4. Does the tool appear to be in good condition?
5. Is there a ground terminal on the plug?
6. Is the electrical cord in good condition?

If you answered *no* to any of the previous questions, tag the tool and report the defect to your instructor.

Use this checklist for the pneumatic tool(s) that you are inspecting. Do not connect the tool to the air outlet until the inspection is completed. Consider the following:

1. Name of tool.
2. Is the tool clean?
3. Are safety guards in place?
4. Does the tool appear to be in good condition?
5. Is the air hose in good condition?

If you answered *no* to any of the previous questions, tag the tool and report the defect to your instructor.

Compressed Air Safety

Compressed air is used in an automotive shop to do many things. However, cleaning off your clothes is not one of them. Dirt and other objects blown off your clothes can cause serious injury to you and others. There may be dirt in the nozzle or hose that will be

Figure 1-11 Only approved safety air nozzles should be used.

ejected at a high rate of speed and can be forced into someone's skin or eyes. In addition, most shops have compressed air systems equipped with automatic oilers. The pressure can push the oil and air bubbles through your skin and into your blood.

Use only approved safety nozzles when using compressed air to dry parts that have been cleaned (**Figure 1-11**). Safety nozzles have a relief passage that prevents high pressures from being expelled directly out the front. It's best not to use compressed air to dry parts; however, there are instances where air must be used to dry small passages. In these cases, use only air pressure that has been regulated to about 25 psi.

Check the air hoses for signs of wear. Do not use them if they are bulging and frayed or if the couplers are damaged.

Lifting the Vehicle

Many service procedures require lifting the vehicle. There are two basic methods of lifting the vehicle from the floor: floor jack and safety stands, and hoists. Each requires the technician to follow certain safety rules to prevent injury and vehicle damage.

Floor Jack and Safety Stand Use

Floor jacks are used to lift a vehicle a short distance off the floor or when only a portion of the vehicle needs to be raised (**Figure 1-12**). Before using a floor jack, check it for signs of hydraulic fluid leaks and for damage that would compromise its safe use. Before lifting the vehicle, place wheel blocks in front of and behind one of the tires that will remain on the ground.

Many jack manufacturers and service information provide illustrations for the proper **vehicle lift points** on a vehicle (**Figure 1-13**). Vehicle lift points are the areas the

Special Tools
Floor jack
Safety stands
Wheel blocks
Service information

Figure 1-12 Floor jacks are used to raise a vehicle a short distance off the floor.

19 inches
(483 mm)

30 inches
(762 mm)

☒ Drive on hoist   ▥ Frame contact hoist

▢ Floor jack   ▤ Outboard twin
                            post hoist

Figure 1-13 Lift point illustrations are usually provided in the service information.

manufacturer recommends for safe vehicle lifting. These areas are structurally strong enough to sustain the stress of lifting. If this information is not available, always place the floor jack on major strength parts. These areas include the frame, cross member, and differential. If you are in doubt about the proper lift point, ask your instructor. Never lift on sheet metal or plastic parts.

The floor jack is to be used only to lift the vehicle off the floor. It's not intended to support the vehicle while someone is under it. Use **safety stands** to support the vehicle (**Figure 1-14**). Use one safety stand for each quarter of the vehicle that is lifted (**Figure 1-15**). Place the safety stand under the frame or a major support component of the vehicle. When the vehicle is lowered onto the stands, make sure that they do not tilt.

Safety stands are also referred to as floor stands or jack stands.

Figure 1-14 Safety stands should be used to support the vehicle after it has been lifted by a floor jack.

Figure 1-15 Support the vehicle by safety stands located at each corner that is lifted. Do not rely on the floor jack to support the vehicle.

Before using the floor jack, make sure that it has a sufficient rating to lift and sustain the weight of the vehicle. Next, inspect it for proper lubrication and hydraulic fluid level. Check the operation of the jack while looking for signs of hydraulic fluid leaks. If the jack does not pass any one of these inspections, tag it and notify your instructor immediately.

To lift the entire vehicle, begin by placing the vehicle's transmission into PARK. Place it in first gear if the vehicle has a manual transmission. Set the parking brake and place wheel blocks around the rear wheels (**Figure 1-16**). Position the floor jack under the front of the vehicle at a location strong enough to support the weight. The jack should be centered between the front tires and positioned so that the lift will be straight up and down (**Figure 1-17**).

Figure 1-16 Before lifting the vehicle with floor jacks, block the wheels to prevent them from rolling.

⚠ **WARNING** If you are lifting only one wheel of the vehicle, be careful not to lift it so high that it can slip off the jack saddle.

Operate the jack until the jack saddle contacts the vehicle lift point. Check for good contact. If things look good, lift the front of the vehicle a couple of inches off the floor. Recheck the position of the jack. Continue to check the jack position throughout the lifting procedure. If the vehicle or jack begins to lean, lower the jack and reset it. Lift the vehicle to the required height. Do not lift higher than is necessary.

Do not get under a vehicle that is supported only by a floor jack. Place safety stands under the vehicle in locations that will support the weight. Use two safety stands to support the front of the vehicle. Make sure that the safety stands are located where they will not lean or slip. Slowly lower the vehicle onto the stands.

Place the floor jack under the rear of the vehicle (**Figure 1-18**). Follow the same procedure to raise the rear of the vehicle. Use two safety stands to support the rear of the vehicle.

Front cross member

Figure 1-17 Example of correct lifting location for raising the front of the vehicle.

Figure 1-18 Example of proper lift procedure for raising the rear of the vehicle.

When the vehicle is properly lifted and supported by safety stands, it is safe to work under the vehicle.

Use the same lift points to lower the vehicle. Once one end of the vehicle is on the floor, place wheel blocks around those wheels. Then lower the other end.

Hoist Safety

Hoists are used when the entire vehicle needs to be raised, usually high enough for the technician to stand underneath the vehicle (**Figure 1-19**). When a vehicle is placed on the hoist, it must be centered. The balance of the vehicle must be taken into consideration, as well as the effects on balance if a heavy component is removed from the vehicle.

Locate the correct lift points in the service information for the vehicle you are working on. Other sources of lift point location information include the Automotive Lift Institute and lift manufacturers. Pay close attention to any warnings or special considerations listed there.

Center the vehicle over the hoist; keep in mind the vehicle's center of gravity and balance point. Locate the hoist pads under the lift points (**Figure 1-20**). Adjust the pads so that the vehicle will be lifted level. Vehicles that have had their suspensions modified to raise or lower the vehicle's height may require special lifting considerations. In addition, vehicle accessories (such as running boards) may present obstacles to safe

Special Tools
Frame contact hoist
Service information

Figure 1-19 One type of hoist that may be used in the automotive shop. The hoist is used to raise the entire vehicle.

Figure 1-20 Locate the hoist pad at the correct lift point for the vehicle you are lifting.

lifting of the vehicle. In these instances, use of special lift arm adapters may be required. If the vehicle is not level on the pads, or the pads are not in the proper position, lower the vehicle and readjust as needed. Never get under a vehicle that is not sitting properly on the hoist.

Lift the vehicle a few inches off the floor. Confirm good pad contact by shaking the vehicle while observing for signs of any movement. If the vehicle is not secure on the hoist, or unusual noises are heard while lifting, lower it to the floor and reset the pads.

With the vehicle at the desired height, lock the hoist (**Figure 1-21**). Do not get under the vehicle until the hoist locks have been set.

To lower the vehicle, release the locks and put the control valve into the lower position (**Figure 1-22**). Once it is returned to the floor, push the contact pads out of the path of the tires.

Figure 1-21 Make sure the hoist's locks are properly engaged before continuing to do any other tasks on or around the vehicle.

Figure 1-22 Release the hoist locks and lower the vehicle.

Figure 1-23 Testing the ventilation system before using it.

Figure 1-24 Connect the ventilation system hose to the vehicle's exhaust before starting the engine. Make sure the hose is secure and there are no kinks in the hose.

Running the Vehicle While in the Shop

Many times it will be necessary to run the engine while the vehicle is in the shop. This presents the possibility of **carbon monoxide (CO)** poisoning if it's not done safely. Carbon monoxide is an odorless, colorless, and toxic gas produced as a result of the combustion process. All shops should be equipped with a ventilation system that will remove the vehicle exhaust. If the shop is not equipped with a ventilation system, a hose must be routed from the vehicle's exhaust to outdoors.

⚡ **WARNING Some of the early warning signs of carbon monoxide poisoning are headaches, dizziness, and blurred vision. If you experience any of these while in the shop, notify your instructor immediately and get some fresh air. If symptoms persist, seek medical attention.**

Check the operation of the shop's ventilation system before connecting to the vehicle. Turn on the ventilation motor and place your hand over the hose (**Figure 1-23**). You should feel a strong, consistent vacuum. A weak vacuum indicates a restriction or a leak in the ventilation system. If you encounter a problem, notify your instructor.

Connect the ventilation hose to the vehicle's exhaust (**Figure 1-24**). Place wheel blocks around a tire and place the transmission in PARK or NEUTRAL. Set the parking brake. The engine is now ready to be started.

Be aware of the warning signs of carbon monoxide poisoning. Report any symptoms to your instructor immediately.

⚡ **WARNING Be careful when working around electric engine cooling fans. These fans are controlled by a thermostat and can come on without warning, even when the engine is not running. Whenever you must work around these fans, disconnect the electrical connector to the fan motor before reaching into the area around the fan. Make sure you reconnect the connector before you return the car to the customer.**

Fire Hazards and Prevention

Fires are classified by the types of materials that are involved (**Table 1-1**). Technicians should be able to locate the correct fire extinguisher to control all the types of fires they are likely to experience (**Figure 1-25**). Technicians must also be able to fight a fire in an emergency. However, do not risk your life fighting a fire. When fighting a fire, be sure that you can escape by backing out of an exit. Do not attempt to fight a fire that has spread beyond the capabilities of the fire fighting equipment you have.

Table 1-1   A Guide to Fire Classification and Fire Extinguisher Types

	Class of Fire	Typical Fuel Involved	Type of Extinguisher
Class ▲**A** **Fires** (green)	For Ordinary Combustibles Put out a class A fire by lowering its temperature or by coating the burning combustibles.	Wood Paper Cloth Rubber Plastics Rubbish Upholstery	Multipurpose dry chemical
Class ■**B** **Fires** (red)	For Flammable Liquids Put out a class B fire by smothering it. Use an extinguisher that gives a blanketing, flame-interrupting effect; cover whole flaming liquid surface.	Gasoline Oil Grease Paint Lighter fluid	Carbon dioxide Halogenated agent Standard dry chemical Purple K dry chemical Multipurpose dry chemical
Class ●**C** **Fires** (blue)	For Electrical Equipment Put out a class C fire by shutting off power as quickly as possible and by always using a nonconducting extinguishing agent to prevent electric shock.	Motors Appliances Wiring Fuse boxes Switchboards	Carbon dioxide Halogenated agent Standard dry chemical Purple K dry chemical Multipurpose dry chemical
Class ★**D** **Fires** (yellow)	For Combustible Metals Put out a class D fire of metal chips, turnings, or shavings by smothering or coating with a specially designed extinguishing agent.	Aluminum Magnesium Potassium Sodium Titanium Zirconium	Dry power extinguishers and agents only

Figure 1-25 Know the location and types of fire extinguishers that are available in the shop.

Figure 1-26 Be sure to know what type of fires the extinguisher will put out.

Labels on the fire extinguisher will indicate the types of fires that it will put out (**Figure 1-26**). Become familiar with the use of a fire extinguisher.

Fire Extinguisher Use

Fire extinguishers are portable apparatuses containing chemicals, water, foam, or special gas that can be discharged to extinguish a small fire. Tour the shop area and become familiar with the location of the fire extinguishers. Use a report sheet to record the locations. Also indicate the type of each extinguisher and what kinds of fires it will extinguish. Check the gauge and record the state of charge for each extinguisher.

⚠ **WARNING** **The following is not intended to be used as a lab exercise unless expressly directed by your instructor. Use the photo sequence as a guide to become familiar with fire extinguisher use. You must be willing to fight a fire in the shop, if the occasion arises.**

⚠ **WARNING** **Do not risk your life fighting a fire. If it's evident that the fire is out of control, get out. Always be aware of where you are and the location of the nearest exit. Do not open garage doors in the event of a fire because the extra oxygen will intensify the flames.**

The proper use of a fire extinguisher is very important. It's possible to deplete an extinguisher and still not put out even the smallest of fires if the extinguisher is used improperly. Procedures vary depending on the agents used. Technicians must become familiar with all the extinguishers equipped in the shop. **Photo Sequence 1** illustrates the proper use of a multipurpose dry chemical extinguisher. This type of extinguisher is the most widely used in the automotive shop.

Fire Blankets

A **fire blanket** is a safety device that can be used to extinguish small fires (**Figure 1-27**). These blankets are made of nonflammable materials such as fiberglass or aramid fibers. Fires that can be put out by a fire blanket include grease/oil fires and electrical fires. Also, fire blankets are useful for putting out clothing fires since they do not stick to fire-damaged skin.

Fire blankets work by preventing the supply of oxygen to a fire. Wrapping something that is burning in a blanket smothers the flames. To get the benefits of your fire blanket, you must use it properly. It's important to read the instructions and become familiar with the proper procedures prior to having to use them in an actual emergency.

When using a fire blanket, you must protect your hands from the fire. Wrap your hands into the top edge of the blanket as you put the blanket on the flame.

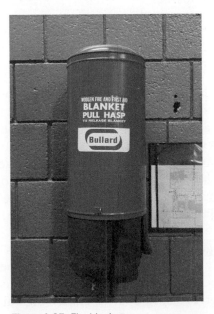

Figure 1-27 Fire blanket.

PHOTO SEQUENCE 1
Using a Dry Chemical Fire Extinguisher

P1-1 Multipurpose dry chemical fire extinguisher.

P1-2 Hold the fire extinguisher in an upright position.

P1-3 Pull the safety pin from the handle.

P1-4 Free the hose from its retainer and aim it at the base of the fire.

P1-5 Stand 8 feet from the fire; do not approach any closer to the fire than this distance.

P1-6 Squeeze the lever while sweeping the hose from side to side. Keep the hose aimed at the base of the fire.

Gasoline

Gasoline is so commonly found in automotive repair shops that its dangers are often forgotten. A slight spark or an increase in heat can cause a fire or explosion. Gasoline is a very explosive liquid and is very powerful. One exploding gallon of gasoline has a force equal to 14 sticks of dynamite. The expanding vapors from gasoline are extremely dangerous, and these vapors are present even in cold temperatures. Gasoline vapors are heavier than air; therefore, when an open container of gasoline is sitting about, the vapors spill out over the sides of the container onto the floor. These fumes are more **flammable** (support combustion) than liquid gasoline and can easily explode.

Never smoke around gasoline since even the droppings of hot ashes can ignite the gasoline. If an engine has a gasoline leak or you have caused a leak by disconnecting a fuel line, stop the leak and clean up the spilled gasoline immediately. While stopping the leak, be extra careful not to cause sparks. If any rags are used in the cleanup of the gasoline, they must be placed in an approved container (**Figure 1-28**). Due to

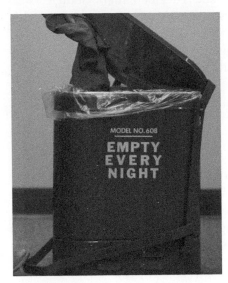

Figure 1-28 Dirty rags and towels must be stored in approved containers.

Figure 1-29 Gasoline must be stored in approved containers. Store any type of combustible materials in an approved safety cabinet.

extreme fire hazards, it's important to immediately clean up any gasoline spilled on the floor. Also, many of the compounds in petroleum are toxic, especially if they are in high concentrations.

The chemicals in petroleum that do not evaporate quickly are biodegradable. Optimum degradation occurs if the gasoline is diluted and there is enough air, water, and nutrients for the microbes to eat up the chemicals. These properties of gasoline are an advantage in the cleanup and disposal of small spills. Spreading absorbent material such as kitty litter, sand, ground corncobs, straw, sawdust, woodchips, peat, synthetic absorbent pads, or even dirt can stop the flow and soak up the gasoline. Keep in mind that the absorbent does not make gasoline nonflammable.

Brooms can be used to sweep up the absorbent material and put it into buckets, garbage cans, or barrels. Remember to control ignition sources. Be aware of local laws concerning gasoline spills. Some states or municipalities require notification of any gasoline spill larger than 5 gallons.

Gasoline should always be stored in approved containers (**Figure 1-29**) and never in glass containers. If the glass container is knocked over or dropped, a terrible explosion can occur. Approved gasoline storage cans have a flash-arresting screen at the outlet. These screens prevent external ignition sources from igniting the gasoline within the can while the gasoline is being poured.

Follow these safety precautions regarding gasoline containers:

1. Always use approved gasoline containers that are painted red for proper identification.
2. Do not fill gasoline containers completely. Always leave the level of gasoline at least 1 inch (25 mm) from the top of the container. This action allows expansion of the gasoline at higher temperatures. If gasoline containers are completely full, the gasoline will expand when the temperature increases. This expansion forces gasoline from the can and creates a dangerous spill.

3. If gasoline containers must be stored, place them in a well-ventilated area such as a storage shed. Filled gasoline containers must be stored within an approved safety cabinet.
4. When a gasoline container must be transported, be sure it's secured against upsets. Do not transport or fill gasoline containers on plastic truck bed liners. Static electricity can be generated and ignite the vapors.
5. Do not store a partially filled gasoline container for long periods of time because it may give off vapors and produce a potential danger.
6. Never leave gasoline containers open except while filling or pouring gasoline from the container.
7. Do not prime an engine with gasoline while cranking the engine.
8. Never use gasoline as a cleaning agent.

Diesel Fuel

A **volatile** substance easily vaporizes or explodes. Although diesel fuel is not as volatile as gasoline, it should still be stored and handled in the same way as gasoline. It's also not as refined as gasoline and tends to be a "dirty" fuel. It normally contains impurities, including active microscopic organisms that can be highly infectious. If diesel fuel happens to enter an open cut or sore, thoroughly wash it immediately. If it gets into your eyes, flush them immediately and seek medical help.

Solvents

Cleaning solvents are not as volatile as gasoline, but they are still flammable. They should be treated and stored in the same way as gasoline. Whenever using solvents, wear eye and hand protection.

Rags

Oily and greasy rags can also cause fires. Used rags should be stored in an approved container and never thrown out with normal trash. Like gasoline, oil is a hydrocarbon and can ignite with or without a spark or flame.

EV and HEV Battery Fires

It is highly unlikely that a high-voltage (HV) lithium-ion battery fire will be extinguished using a conventional fire extinguisher. These batteries are commonly used in Electric Vehicles (EVs) and Hybrid Electric Vehicles (HEVs) When these batteries catch fire, they are susceptible to a self-destructive chain reaction known as "thermal runaway," resulting in a feedback loop of rising temperatures. A HV battery fire can exceed 5,000°F (2,760°C). Also, the battery can re-ignite several times after the initial fire is extinguished.

Water is probably the only element that is available to you for fighting this type of fire. However, safety experts say it takes thousands of gallons of water directed and focused in a stream onto the battery to extinguish the fire. Most EVs and HEVs hide their batteries or locate them under the floor pan, making it very difficult to get water directly onto the battery. Water also presents another issue; it may cause a violent flare-up as the water molecules separate into explosive hydrogen and oxygen gases.

The best thing to do is to make sure someone has called 911 and has notified the fire department that the vehicle involved is an EV or HEV. If the HV battery appears to be the source of the fire, everyone must leave the building immediately and get a safe distance away from it.

Hazardous Materials

Hazardous materials are materials that can cause illness, injury, or death or pollute water, air, or land. Many solvents and other chemicals used in an automotive shop have warning and caution labels that should be read and understood by everyone who uses them. Many service procedures generate what are known as hazardous wastes. Examples of hazardous waste are used or dirty cleaning solvents and other liquid cleaners.

Right-to-Know Laws

In the United States, **right-to-know laws** concerning hazardous materials and wastes protect every employee in a workplace. The general intent of these laws is for employers to provide a safe working place as it relates to hazardous materials. The right-to-know laws state that employees have a right to know when the materials they use at work are hazardous. The right-to-know laws started with the **Hazard Communication Standard** published by the Occupational Safety and Health Administration (OSHA) in 1983. This document was originally intended for chemical companies and manufacturers that required employees to handle hazardous materials in their work situation. At the present time, most states have established their own right-to-know laws. Meanwhile, the federal courts have decided to apply these laws to all companies, including automotive service shops.

Under the right-to-know laws, employers have three responsibilities regarding their employees' handling of hazardous materials. The first responsibility concerns employee training and providing information. All employees must be trained about the types of hazardous materials they will encounter in the workplace. All employees must be informed about their rights under legislation regarding the handling of hazardous materials. In addition, information about each hazardous material must be posted on **material safety data sheets (MSDS)** available from the manufacturer (**Figure 1-30**). These sheets contain extensive information and facts about hazardous materials. In Canada, MSDS are called **workplace hazardous materials information systems (WHMIS)**.

The employer has a responsibility to place MSDS where they are easily accessible by all employees. The MSDS provide extensive information about the hazardous material such as the following:

1. Chemical name.
2. Physical characteristics.
3. Protective equipment required for handling.
4. Explosion and fire hazards.
5. Other incompatible materials.
6. Health hazards such as signs and symptoms of exposure, medical conditions aggravated by exposure, and emergency and first-aid procedures.
7. Safe handling precautions.
8. Spill and leak procedures.

The second responsibility of the employer is to make sure that all hazardous materials are properly labeled. The label information must include health, fire, and reactivity hazards posed by the material, as well as the protective equipment necessary to handle the material. The manufacturer must supply all warning and precautionary information about hazardous materials, and this information must be read and understood by the employee

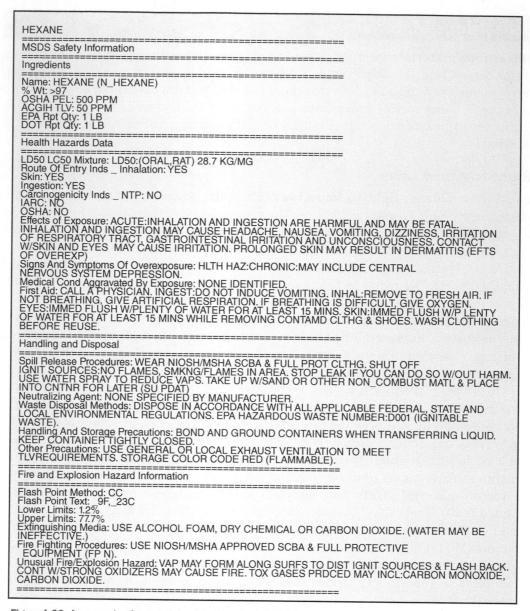

```
HEXANE
===============================================================
MSDS Safety Information
===============================================================
Ingredients
===============================================================
Name: HEXANE (N_HEXANE)
% Wt: >97
OSHA PEL: 500 PPM
ACGIH TLV: 50 PPM
EPA Rpt Qty: 1 LB
DOT Rpt Qty: 1 LB
===============================================================
Health Hazards Data
===============================================================
LD50 LC50 Mixture: LD50:(ORAL,RAT) 28.7 KG/MG
Route Of Entry Inds _ Inhalation: YES
Skin: YES
Ingestion: YES
Carcinogenicity Inds _ NTP: NO
IARC: NO
OSHA: NO
Effects of Exposure: ACUTE:INHALATION AND INGESTION ARE HARMFUL AND MAY BE FATAL.
INHALATION AND INGESTION MAY CAUSE HEADACHE, NAUSEA, VOMITING, DIZZINESS, IRRITATION
OF RESPIRATORY TRACT, GASTROINTESTINAL IRRITATION AND UNCONSCIOUSNESS. CONTACT
W/SKIN AND EYES  MAY CAUSE IRRITATION. PROLONGED SKIN MAY RESULT IN DERMATITIS (EFTS
OF OVEREXP)
Signs And Symptoms Of Overexposure: HLTH HAZ:CHRONIC:MAY INCLUDE CENTRAL
NERVOUS SYSTEM DEPRESSION.
Medical Cond Aggravated By Exposure: NONE IDENTIFIED.
First Aid: CALL A PHYSICIAN. INGEST:DO NOT INDUCE VOMITING. INHAL:REMOVE TO FRESH AIR. IF
NOT BREATHING, GIVE ARTIFICIAL RESPIRATION. IF BREATHING IS DIFFICULT, GIVE OXYGEN.
EYES:IMMED FLUSH W/PLENTY OF WATER FOR AT LEAST 15 MINS. SKIN:IMMED FLUSH W/P LENTY
OF WATER FOR AT LEAST 15 MINS WHILE REMOVING CONTAMD CLTHG & SHOES. WASH CLOTHING
BEFORE REUSE.
===============================================================
Handling and Disposal
===============================================================
Spill Release Procedures: WEAR NIOSH/MSHA SCBA & FULL PROT CLTHG. SHUT OFF
IGNIT SOURCES:NO FLAMES, SMKNG/FLAMES IN AREA. STOP LEAK IF YOU CAN DO SO W/OUT HARM.
USE WATER SPRAY TO REDUCE VAPS. TAKE UP W/SAND OR OTHER NON_COMBUST MATL & PLACE
INTO CNTNR FOR LATER (SU PDAT)
Neutralizing Agent: NONE SPECIFIED BY MANUFACTURER.
Waste Disposal Methods: DISPOSE IN ACCORDANCE WITH ALL APPLICABLE FEDERAL, STATE AND
LOCAL ENVIRONMENTAL REGULATIONS. EPA HAZARDOUS WASTE NUMBER:D001 (IGNITABLE
WASTE).
Handling And Storage Precautions: BOND AND GROUND CONTAINERS WHEN TRANSFERRING LIQUID.
KEEP CONTAINER TIGHTLY CLOSED.
Other Precautions: USE GENERAL OR LOCAL EXHAUST VENTILATION TO MEET
TLVREQUIREMENTS. STORAGE COLOR CODE RED (FLAMMABLE).
===============================================================
Fire and Explosion Hazard Information
===============================================================
Flash Point Method: CC
Flash Point Text:  9F,_23C
Lower Limits: 1.2%
Upper Limits: 77.7%
Extinguishing Media: USE ALCOHOL FOAM, DRY CHEMICAL OR CARBON DIOXIDE. (WATER MAY BE
INEFFECTIVE.)
Fire Fighting Procedures: USE NIOSH/MSHA APPROVED SCBA & FULL PROTECTIVE
  EQUIPMENT (FP N).
Unusual Fire/Explosion Hazard: VAP MAY FORM ALONG SURFS TO DIST IGNIT SOURCES & FLASH BACK.
CONT W/STRONG OXIDIZERS MAY CAUSE FIRE. TOX GASES PRDCED MAY INCL:CARBON MONOXIDE,
CARBON DIOXIDE.
===============================================================
```

Figure 1-30 An example of a material safety data sheet (MSDS).

before handling the material. Pay great attention to the information on the label. By doing so, you will use the substance in the proper and safe way, thereby preventing hazardous conditions.

The third responsibility of the employer is for maintaining permanent files regarding hazardous materials. These files must include information on hazardous materials in the shop, proof of employee training programs, and information about accidents such as spills or leaks of hazardous materials. The employer's files must also include proof that employees' requests for hazardous material information, such as MSDS, have been met. The employer must maintain a general right-to-know compliance procedure manual.

There are responsibilities for the employees as well. Employees must be familiar with the intended purpose of the substance, the recommended protective equipment, accident and spill procedures, and any other information regarding the safe handling of hazardous materials. This training must be given annually to employees and provided to new employees as part of their job orientation.

A BIT OF HISTORY

During the 1960s, disabling injuries increased 20% and 14,000 workers were dying on the job each year. In pressing for prompt passage of workplace safety and health legislation, Senator Harrison A. Williams Jr. called attention to the need to protect workers against such hazards as noise, cotton dust, and asbestos. Representative William A. Steiger also worked for passage of a bill to protect workers. On December 29, 1970, President Richard M. Nixon signed The Occupational Safety and Health Act of 1970, also known as the Williams-Steiger Act.

Hazardous Waste

Waste is considered a **hazardous waste** if it's on the Environmental Protection Agency (EPA) list of known and harmful materials or if it has one or more of the following characteristics:

1. Any material that reacts violently with water or other chemicals is considered hazardous.
2. If a material releases cyanide gas, hydrogen sulfide gas, or similar gases when exposed to low-pH acid solutions, it's hazardous.
3. If a material burns the skin or dissolves metals and other materials, it's considered hazardous.
4. Materials are hazardous if they leach one or more of eight heavy metals in concentrations greater than 100 times the primary drinking water standard. These materials are considered toxic.
5. A liquid is hazardous if the temperature at which the vapors on the surface of the fuel will ignite when exposed to an open flame is below 140°F (60°C), and a solid is hazardous if it ignites spontaneously.

A complete list of EPA hazardous wastes can be found in the Code of Federal Regulations. It should be noted that no material is considered hazardous waste until the shop is finished using it and is ready to dispose of it. New oil is not a hazardous waste; however, used oil is. Once you drain oil from an engine, you have generated hazardous waste and now become responsible for its proper disposal. There are many other wastes that need to be handled properly after you have removed them, such as batteries, brake fluid, transmission fluid, and engine coolant.

No fluids drained from a vehicle should be allowed to enter sewage drains. Some fluids, such as coolant, can be captured and recycled in the shop with special equipment. Filters for fluids (transmission, fuel, and oil filters) also need to be handled in designated ways. Used filters need to be drained and then crushed or disposed of in a special shipping barrel. Most regulations demand that oil filters be drained for at least 24 hours before they are disposed of or crushed.

Federal and state laws control the disposal of hazardous waste materials. It's the responsibility of the employer and the employee to ensure that everyone in the shop is familiar with these laws. Hazardous waste disposal laws include the **Resource Conservation and Recovery Act (RCRA)**. This law basically states that hazardous waste generators are responsible for the waste from the time it becomes a waste material until the proper waste disposal is completed. Therefore, the user must store hazardous waste material properly and safely and be responsible for the transportation of this material until it arrives at an approved hazardous waste disposal site where it's processed according to the law. A licensed waste management firm normally does the disposal. The hazardous waste coordinator for the shop should have a written contract with the hazardous waste hauler.

The RCRA controls these types of automotive waste:

1. Paint and body repair products waste.
2. Solvents for parts and equipment cleaning.
3. Batteries and battery acid.
4. Mild acids used for metal cleaning and preparation.
5. Waste oil, engine coolants, or hydraulic fluids.
6. Air-conditioning refrigerants.
7. Engine oil, transmission, and fuel filters.

NEVER, under any circumstances, use these methods to dispose of hazardous waste material:

1. Pour hazardous wastes on weeds to kill them.
2. Pour hazardous wastes on gravel streets to prevent dust.
3. Throw hazardous wastes in a dumpster.
4. Dispose of hazardous wastes anywhere but an approved disposal site.
5. Pour hazardous wastes down sewers, toilets, sinks, or floor drains.
6. Bury hazardous wastes in the ground.

Handling and Disposing Shop Wastes

To assure that you are following the required procedures concerning the handling and disposal of hazardous waste, the following sections are provided as guidelines. Be sure to familiarize yourself with federal, state, and local regulations in your area.

Oil Disposal. Used oils can be recycled, properly disposed, or used to fuel waste oil burners. In all cases, proper collection of used oil is essential. A drip table or screen table is commonly used to collect the oil into a designated collection bucket. These tables will collect oil as it drips off parts. Be sure not to mix oils unless approved by the recycle company or allowed by the waste oil furnace.

Oil filters must be drained for at least 24 hours before disposal. Once properly drained, they are crushed and then set to be properly disposed of.

Any hazardous waste that is stored on the premises must be properly labeled. Also, be sure to adhere to the regulations as to the maximum quantity that may be stored on site.

Solvents. If possible, reduce the use of solvents by utilizing less toxic alternatives. For example, water-based cleaning solvents should be used instead of petroleum-based solvents. Used solvents that have become too dirty to continue their use must be disposed of properly. Usually, this requires a hazardous waste management company to come to the shop on a regular basis and service the equipment. However, some equipment can be serviced in-house. When storing solvents, be sure they are in approved containers and are tightly sealed. Evaporation of the solvent may have an environmental impact.

Properly label used solvents and store them in containment areas. Store solvents with other compatible wastes only.

Asbestos Exposure

Asbestos describes a number of naturally occurring fibrous materials and is classified as a carcinogen. Asbestos has been identified as a health hazard and has been shown to cause diseases that result in lung cancers known as mesothelioma. When asbestos fibers are breathed in, they cause scarring of the lungs and damage the air passages. The scars result in holding locations for the asbestos fibers that the body is unable to expel. Asbestos has been/is used for clutch discs, brake pads and shoes, and gaskets. Because asbestos is a health hazard, all OSHA and EPA regulations must be adhered to when disposing of parts that contain this material.

The EPA does not regulate the removal of brake pads and clutches unless more than 50% of the shop's work involves grinding or debonding of the asbestos material. If the shop is above this 50% requirement, then the asbestos materials are regulated as a hazardous waste and must be handled according to the set regulations.

Even if your shop is below the 50% rule, it's very important that as a technician you protect yourself from the dangers associated with inhaling asbestos dust. The EPA recommends that the shop capture asbestos dust into a separate container. The OSHA-preferred method of servicing components consisting of asbestos dust is to use a low-pressure/wet-cleaning method. Never use compressed air to blow off the dust from the parts.

The wet-cleaning method is done by mixing water with an organic solvent. The solvent is flowed over the brake parts prior to removal of the brake drum or around the brake disc prior to brake disassembly. A catch container is positioned to trap the contaminated solvent. During disassembly of the brakes, the solvent is reapplied to assure all asbestos dust is removed.

Another asbestos cleaning method is to use a **high-efficiency particulate-arresting (HEPA)** vacuum cleaner (**Figure 1-31**). The HEPA vacuum cleaner captures the asbestos dust in a special filter. The brake or clutch parts are covered with a special tent. Built-in gloves in the tent allow the technician to clean the parts while viewing through a window. This procedure prevents direct contact with the asbestos and draws the dust particles into the container.

When the HEPA filter is full, it's wetted with a mist of water and then removed. The filters must be placed into an impermeable container, properly labeled, and properly disposed of in accordance with all laws.

Another procedure that is allowed only if the shop performs less than five brake or clutch services a week is the wet wipe method. The technician uses a spray bottle or low-pressure air-charged container to spray a mist of water and detergent onto all brake and clutch components. The components are then wiped with a clean cloth. To reduce the risks of asbestos poisoning, the following guidelines should be followed:

- Never smoke while working around asbestos.
- Wash your hands, arms, and face before eating.
- Take showers at the completion of your work shift.
- Do not wear your work clothes home. Change into your work clothes when you arrive at work and change out of them when leaving.

Glovebag collection system

HEPA vacuum cleaner

Figure 1-31 A high-efficiency particulate-arresting (HEPA) vacuum cleaner used to capture asbestos fiber dust.

AUTHOR'S NOTE More information concerning work environment safety can be found by contacting the United States EPA Office of Compliance at http:// es.inel.gov or the Coordinating Committee for Automotive Repair (CCAR)-Greenlink at http://www.ccar-greenlink.org.

Electrical System Safety

There are many safety requirements that must be followed when working on the vehicle's electrical system. In addition to personal safety, there is the concern of damaging the electrical system with improper service techniques. The following are a few of the safety rules.

Battery Safety

Before attempting to do any type of work on or around the battery, you must be aware of certain precautions. To avoid personal injury or property damage, follow these precautions:

1. Battery acid is very corrosive. Do not allow it to come into contact with your skin, eyes, or clothing. If battery acid should get into your eyes, rinse them thoroughly with clean water and seek immediate medical attention. If battery acid comes into contact with your skin, wash with clean water. Baking soda added to the water will neutralize the acid. If the acid is swallowed, drink large quantities of water or milk followed by milk of magnesia and a beaten egg or vegetable oil.
2. When making connections to a battery, be careful to observe polarity, positive to positive and negative to negative.
3. When disconnecting battery cables, always disconnect the negative (ground) cable first.
4. When connecting battery cables, always connect the negative cable last.
5. Avoid any arcing or open flames near a battery. The vapors produced by the cycling of a battery are very explosive. Do not smoke around a battery.
6. Follow the manufacturer's instructions when charging a battery. Charge the battery in a well-ventilated area. Do not connect or disconnect the charger leads while the charger is turned on.
7. Do not add electrolyte to the battery if it is low. Add only distilled water.
8. Do not wear any jewelry while servicing the battery. These items are excellent conductors of electricity. Severe burns may result if current flows through them by accidental contact with the battery positive terminal and a ground.
9. Never lay tools across the battery. They may come into contact with both terminals, shorting out the battery and causing it to explode.
10. Wear safety glasses and/or a face shield when servicing the battery.

Batteries are very dangerous components of the vehicle. It's important that you be able to demonstrate the ability to work around the battery in a safe manner. Throughout this manual there will be many instances where you will be required to perform a task involving the battery. Chapter 5 covers the subject of removing and testing the battery. Do not perform any tests or disconnect the battery until you have completed that chapter. The purpose of this section is to assist you in becoming more familiar with the battery to allow you to work safely around it.

Point out the following components of the battery to your instructor:

1. Negative terminal.
2. Positive terminal.
3. Vents.

Answer the following questions concerning battery safety (answer written or orally per your instructor):

1. Why be concerned about battery acid?
2. What should be done if battery acid splashes into your eyes?
3. What should you do if battery acid gets onto your skin?
4. What is meant by polarity? Why is it a concern when connecting a battery?
5. Which terminal must be disconnected first?
6. When connecting battery cables, which cable is to be connected last?
7. Why is wearing jewelry discouraged around the battery?
8. Why is smoking not allowed around the battery?
9. Why are tools not to be laid across the top of the battery?
10. What safety protection should be worn while servicing or working around the battery?
11. What other safety precautions must be observed?

If you do not understand any of the safety precautions associated with working on or around the battery, ask your instructor.

Starting System Service Safety

Before testing or servicing the starting system, become familiar with these precautions that should be observed:

1. Refer to the recommendations given in the service information for correct procedures for disconnecting a battery. Some vehicles with on-board computers must be supplied with an auxiliary power source to maintain computer memories.
2. Disconnect the battery ground cable before disconnecting any of the starter circuit's wires or removing the starter motor.
3. Be sure the vehicle is properly positioned on the hoist or on safety stands.
4. Before performing any cranking test, be sure the vehicle transmission is in PARK or NEUTRAL and the parking brakes are applied. Put wheel blocks in front of and behind one tire.
5. Follow the service information procedures for disabling the ignition system.
6. Be sure all test leads are clear of any moving engine components.
7. Never clean any electrical components in solvent or gasoline. Clean with denatured alcohol, or wipe only with clean rags.

Charging System Service Safety

The following are some general rules for servicing the generator and the charging system:

1. Do not run the vehicle with the battery disconnected. The battery acts as a buffer and stabilizes any voltage spikes that may cause damage to the vehicle's electronics.
2. When performing charging system tests, do not allow output voltage to increase over 16 volts (V).
3. If the battery needs to be recharged, disconnect the battery cables while charging.
4. Do not attempt to remove electrical components from the vehicle with the battery connected.
5. Before connecting or disconnecting any electrical connections, the ignition switch must be in the OFF position, unless directed otherwise in the service information.
6. Avoid contact with the **BAT** terminal of the generator while the battery is connected. BAT is the terminal identifier for the conductor from the generator to the battery positive terminal. Battery voltage is always present at this terminal.

⚠ **Caution**
Always double-check the polarity of the battery charger's connections and leads before turning the charger on. Incorrect polarity can damage the battery or cause it to explode.

Air Bag Safety

The **air bag system** is designed as a supplemental restraint that, in the case of an accident, will deploy a bag out of the steering wheel or passenger-side dash panel to provide additional protection against head and face injuries. An air bag system demands that the technician pay close attention to safety warnings and precautions when working on or around it. Most air bags are deployed by an explosive charge. Accidental deployment of the air bag can result in serious injury. When working on or around the steering wheel or **air bag module**, be aware of your hands and arms. The air bag module is the air bag and inflator assembly together in a single package (**Figure 1-32**). Do not place your arm over the module. If the air bag deploys, injury can result.

Air bag systems contain a means of deploying the bag even if the battery is disconnected. This system is needed in the event the battery is damaged or disconnected during an accident. The reserve energy can be stored for over 30 minutes after the battery is disconnected. Follow the service information procedures for disabling the system.

When carrying the air bag module, carry it so that the bag and trim are facing up and away from your body (**Figure 1-33**). In the event of accidental deployment, the charge will be away from you. Do not face the module toward any other people.

When you place the module on the bench, face the bag and trim up (**Figure 1-34**). This provides a free space for the bag to expand if it deploys. If the module will be stored for any period of time, it must be stored in a cool dry place. Store the module with the trim up and do not place anything on top of the module.

Figure 1-32 Typical air bag module.

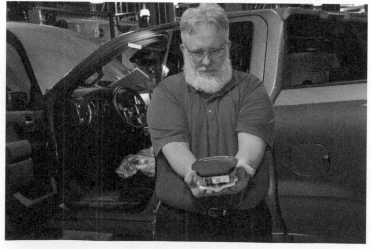

Figure 1-33 Proper method of carrying an air bag module.

Figure 1-34 Place the air bag module on the bench with trim facing up.

While troubleshooting the air bag system, do not use electrical testers such as battery-powered or AC-powered voltmeters, ohmmeters, and so on, or any other equipment except those specified in the service information. Do not use a test light to troubleshoot the system.

When it's necessary to make a repair or replace a component in the air bag system, always disconnect the negative battery cable before making the repair. It's a good practice to insulate the terminal with tape or a rubber hose to prevent it from coming into contact with the battery post. Some manufacturers recommend that the air bag inflator module(s) be disconnected in addition to the negative battery cable.

AUTHOR'S NOTE Be sure to follow all manufacturer's warnings, cautions, and special service notes when working on or around the air bag system.

Although it's unlikely that an air bag module will inflate on its own, it is possible to ignite it while performing service. To prevent injury, be aware of where the module(s) is located in the vehicle. The most common location is in the center of the steering wheel. However, all late-model vehicles also have a passenger-side air bag (**Figure 1-35**). Some manufacturers are installing air bag systems for the back-seat passengers as well.

Passenger side
(dash mounted)

Driver side
(steering wheel
mounted)

Figure 1-35 An air bag system with both a driver and a passenger-side air bag module.

Figure 1-36 The air bag wiring is usually identified with a bright yellow harness and/or connector.

Figure 1-37 Some manufacturers identify air bag circuits in their service information with a warning symbol.

⚠ WARNING Obey all of the warnings in the service information when working on or around the air bag system. Failure to follow these warnings may result in air bag deployment and injury.

To prevent accidental deployment of the air bag system, disconnect the negative battery cable before disconnecting or connecting any electrical connectors in the system. It's important to be able to recognize the components of the air bag system. Most manufacturers place the wiring of the air bag system into a bright yellow harness tube or use bright yellow insulation or connectors (**Figure 1-36**). The wires are usually tagged to alert the technician. Walk around an air bag–equipped vehicle with your instructor. Your instructor will point out the components of the system and review the necessary safety precautions.

Some manufacturers denote air bag–related circuitry in their service information with a warning symbol (**Figure 1-37**).

Antilock Brake System Safety

The **Antilock brake system (ABS)** automatically pulsate the brakes to prevent wheel lockup under panic stop and poor traction conditions. ABS is available on most of today's vehicles. There are many different systems used, and each has its own safety requirements regarding servicing the system. Become familiar with the warnings and cautions associated with the system you are working on by studying the service information before performing any service.

Certain components of the ABS are not intended to be serviced individually. Do not attempt to remove or disconnect these components. Only those components with approved removal and installation procedures in the service information should be serviced.

Some operations require that the tubes, hoses, and fittings be disconnected. Many earlier ABS used high hydraulic pressures (up to 2,800 psi [19,305 kPa]) and an accumulator to store this pressurized fluid. Before disconnecting any lines or fittings, the accumulator must be fully depressurized. Many late-model ABS do not use an accumulator; therefore, these systems do not require depressurizing. However, always refer to the correct service information before servicing a brake system. The following is a common method of depressurizing the ABS. However, follow the service information procedures for the vehicle you are working on:

1. Place the ignition switch in the OFF position.
2. Pump the brake pedal a minimum of 20 times.
3. There should be noticeable change in pedal feel when the accumulator is discharged.

General Electric and Hybrid Electric Vehicle Safety

Since the EV and HEV system can use voltages in excess of 650 volts (both DC and AC), it's vital that the service technician be familiar with, and follow, all safety precautions. Failure to perform the correct procedures can result in electrical shock, battery leakage, explosion, or even death. The following are some general service precautions to be aware of:

Special Tools

Class III DMM capable of reading 400 VDC

Insulating gloves

Insulating tape

- Test the integrity of the insulating gloves prior to use.
- Wear high-voltage insulating gloves when disconnecting the service plug.
- Do not attempt to test or service the system for 5 minutes after the high-voltage service plug is removed. At least 5 minutes is required to discharge the high-voltage capacitors inside the inverter module.
- Never cut the high-voltage power cables (**Figure 1-38**). The wire harnesses, terminals, and connectors of the high-voltage system are identified by orange or blue. In addition, high-voltage components may have a "High Voltage" caution label attached to them (**Figure 1-39**).
- Never open high-voltage components.
- Use Class III 1,000 volt rated DMM and test leads.
- Use insulated tools when required.
- Do not wear metallic objects that may cause electrical shorts.
- Follow the service information diagnostic procedures.
- Wear protective safety goggles when inspecting the high-voltage (HV) battery.
- Before touching any of the high-voltage system wires or components, wear insulating gloves, make sure the high-voltage service plug is removed, and disconnect the auxiliary battery.

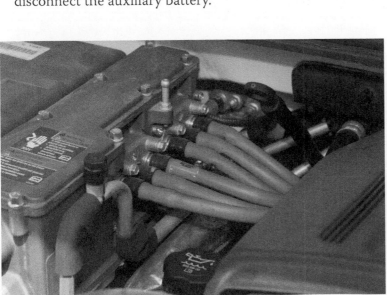

Figure 1-38 High-voltage cables are identified with bright orange or blue conduit or insulation.

Figure 1-39 High-voltage warning label. Be sure to familiarize yourself with all warnings associated with service on EV and HEV vehicles.

- Remove the service disconnect plug prior to performing a resistance check.
- Remove the service plug prior to disconnecting or reconnecting any connectors or components.
- Isolate any high-voltage wires that have been removed with insulation tape.
- Properly torque the high-voltage terminals.

⚠️ **WARNING** **The technician must verify that the system remains powered down when performing any repairs that involve contact with high-voltage or hybrid components or systems.**

⚠️ **WARNING** **The high-voltage checkout procedure must be performed to ensure that the vehicle is properly powered down.**

⚠️ **WARNING** **Whenever the vehicle has been left unattended, recheck that the service disconnect has not been reinstalled.**

⚠️ **WARNING** **Prior to performing any diagnostic or service procedure, you must thoroughly read and follow all applicable high-voltage safety procedures.**

⚠️ **WARNING** **Be sure to utilize proper safety equipment when working on any high-voltage system. Failure to do so may result in serious injury or death.**

⚠️ **WARNING** **Wait a minimum of 5 minutes after performing the high-voltage battery disconnect procedure before accessing the high-voltage system. Failure to do so may result in serious injury or death.**

High-Voltage Service Plug

The HEV is equipped with a **high-voltage service plug** that disconnects the HV battery from the system. Usually, this plug is located at the battery (**Figure 1-40**). Prior to disconnecting the high-voltage service plug, the vehicle must be turned off. Some manufacturers also require that the negative terminal of the auxiliary battery be disconnected. Once the high-voltage service plug is removed, the high-voltage circuit is shut off at the intermediate position of the HV battery and there is no high voltage in the vehicle's systems.

Figure 1-40 The high-voltage service plug is usually located at the HV battery.

Figure 1-41 To install the service plug, make sure that the lever is down and then fully locked.

> ⚙ **SERVICE TIP** Diagnostic trouble codes will be erased once the batteries are disconnected. Prior to disconnecting the system, be sure to check and record any DTCs.

The high-voltage service plug assembly contains a safety interlock reed switch. The reed switch is opened when the clip on the high-voltage service plug is lifted. The open reed switch turns off power to the service main relay (SMR). The main fuse for the high-voltage circuit is inside the high-voltage service plug assembly.

However, never assume that the high-voltage circuits are off. The removal of the high-voltage service plug does not disable the individual HV batteries. Use a Class III DMM to verify that 0 volts is in the system before beginning service. When testing the circuit for voltage, set the voltmeter to the 400 VDC scale.

After the high-voltage service plug is removed, a minimum of 5 minutes must pass before beginning service on the system. This is required to discharge the high voltage from the condenser in the inverter circuit.

To install the high-voltage service plug, make sure the lever is locked in the DOWN position (**Figure 1-41**). Slide the plug into the receptacle, and lock it in place by lifting the lever upward. Once it's locked in place, it closes the reed switch.

Preparing HEV for Service

When working on an HEV, always assume the HV system is live until you have proven otherwise. If the vehicle has been driven into the service department, you know that the HV system was energized since most HEVs do not move without the HV system operating.

It's critical that proper tools be used when working on the HV system. These include protective hand tools and a Class III DMM with an insulation test function. The meter must be capable of checking for isolation up to 1,000 volts and measuring resistance at over 1.1 megaohms. In addition, the DMM insulation test function is used to confirm proper isolation of the HV system components after a repair is performed.

Always remove the service disconnect plug and perform the HV checkout procedure to prove that the HV system has been powered down. Never perform repairs until you have performed this procedure to ensure the system is safe to work on. Follow **Photo Sequence 2** as a guide to the safety procedures that must be followed whenever servicing the HEV's high-voltage systems.

Special Tools
Safety goggles
Insulating gloves
Class III DMM

PHOTO SEQUENCE 2
HEV High-Voltage Isolation

P2-1 Tools required to perform this procedure include safety goggles, 1,000-volt-rated insulating gloves, and a Class III digital multimeter (DMM).

P2-2 Remove the key from the ignition.

P2-3 Test the integrity of the insulating gloves prior to use. Wear the gloves until the high-voltage isolation procedure is complete.

P2-4 Put on your eye protection.

P2-5 If equipped, disconnect the 12-volt inline connector to isolate the 12-volt battery. If the vehicle does not have the in-line connector, disconnect the negative (−) terminal of the auxiliary (12-volt) battery. Always disconnect the auxiliary battery prior to removing the high-voltage service plug.

P2-6 Remove the high-voltage service plug and put it into your tool box or somewhere it cannot be accidentally reinstalled by someone else.

P2-7 Cover the high-voltage service plug receptacle with insulation tape.

P2-8 After waiting at least 5 minutes for the high-voltage condenser inside the inverter to discharge, remove any covers necessary to access the isolation test locations.

P2-9 Use a Class III DMM to confirm that high-voltage circuits have 0 volts before performing any service procedure.

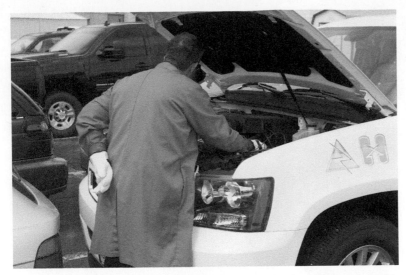

Figure 1-42 Follow the one-hand rule while testing the high-voltage system of an HEV.

Whenever possible use the **one-hand rule** when servicing the HV system (**Figure 1-42**). The one-hand rule means working with only one hand while servicing the HV system so that in the event of an electric shock the high voltage will not pass through your body. It's important to follow this rule when performing the HV checkout procedure since confirmation of HV system power down has not been proven yet.

Insulated Glove Integrity Test

The insulating gloves that the technician must wear for protection while servicing the HV system must be tested for integrity before they are used. If there is a leak in the gloves, high-voltage electricity can travel through the hole to the technician's body. The insulating gloves must meet Class "0" requirements of a rating of 1,000 volts (**Figure 1-43**). In addition, the technician should wear rubber-soled shoes, cotton clothing, and safety glasses with side shields. Remove all jewelry and make sure metal zippers are not exposed. Always have a second set of insulation gloves available and let someone in the shop know their location. When preparing to work on an HEV, let others know in the event they must come to your aid.

Figure 1-43 Class "0" insulation gloves must be worn when working on high-voltage systems.

Figure 1-44 Test the insulating gloves for leaks before each use.

To test the insulating gloves:

1. Remove the gloves from leather protectors and inspect for any tears or worn spots.
2. Blow air into the gloves to inflate them and seal the opening by folding the base of the glove.
3. Slowly rotate the base of the glove toward the fingers to increase the pressure (**Figure 1-44**).
4. Look and feel for pinholes.

Repeat the steps for each glove. This procedure must be performed for each new set of gloves and before each use.

 WARNING **Do not use the gloves if they fail the leak test or are damaged.**

⚡ Caution

Once the high-voltage service plug is removed, do not operate the power switch. Doing so may damage the hybrid vehicle control ECU.

Summary

- Being a professional technician means more than having knowledge of vehicle systems. It also requires an understanding of all the hazards in the work area.
- As a professional technician, you should work responsibly to protect yourself and the people around you.
- Technicians must be aware that it's their responsibility to prevent injuries in the shop, and that their actions and attitudes reflect how seriously they accept that responsibility.
- Long-sleeve shirts should have their cuffs buttoned or rolled up tightly, and shirttails should be tucked in at all times. Neckties should be tucked inside your shirt; ideally, only clip-on ties should be worn.
- Long hair should be tied back and tucked under a hat.

- Jewelry has no place in the automotive shop.
- The safest and surest method of protecting your eyes is to wear proper eye protection anytime you enter the shop.
- When working around rotating pulleys and belts, be aware of where you are placing your hands and fingers at all times.
- Most occupational back injuries are caused by improper lifting practices.
- Most injuries caused by tools and equipment are the result of improper use, improper maintenance, and carelessness.
- Never use compressed air for cleaning off your clothes.
- The floor jack is to be used only to lift the vehicle off the floor. Use safety stands to support the vehicle after it's lifted.

- Fires are classified by the types of materials involved. Fire extinguishers are classified by the type of fire they will extinguish.
- Batteries can cause serious injury if all safety rules are not followed when working on or around them.
- The air bag system demands that the technician pay close attention to safety warnings and precautions when working on or around them.
- Air bag systems contain a means of deploying the bag even if the battery is disconnected. The reserve energy can be stored for over 30 minutes after the battery is disconnected.
- Since the hybrid system can use voltages in excess of 650 volts (both DC and AC), it's vital that the service technician be familiar with, and follow, all safety precautions.
- Remove the high-voltage service plug and put it where it cannot be accidentally reinstalled by someone else.

- Do not attempt to test or service the high-voltage system of an HEV for 5 minutes after the high-voltage service plug is removed.
- Use a Class III DMM and test leads to confirm that high-voltage circuits have 0 volts before performing any service procedure on the high-voltage system of an HEV.
- The insulating gloves that are worn to protect the technician while servicing the high-voltage system of an HEV must be tested for integrity before use.
- Do not use the insulating gloves if they fail the leak test or are otherwise damaged.
- The HEV is equipped with a high-voltage service plug that disconnects the HV battery from the system.
- Never assume that the high-voltage circuits of an HEV are off.

ASE-Style Review Questions

1. Which of the following is *not* included in proper gasoline spill cleanup?
   A. Spread absorbent material to stop the flow and soak up the gasoline.
   B. Immediate containment of the spill.
   C. Use a soap and water solution to wash the gasoline into a container.
   D. Be aware that the absorbent does not reduce flammability concerns.

2. *Technician A* says the right-to-know laws require employers to train employees regarding hazardous waste materials.
   Technician B says the right-to-know laws have no provisions requiring proper labeling of hazardous materials.
   Who is correct?
   A. A only
   B. B only
   C. Both A and B
   D. Neither A nor B

3. All of the following statements concerning hybrid high-voltage system safety are true, EXCEPT:
   A. Disconnect the motor generators prior to turning the ignition off.
   B. Test the insulating gloves for leaks prior to use.
   C. Do not attempt to test or service the system for 5 minutes after the high-voltage service plug is removed.
   D. Turn the power switch to the off position prior to performing a resistance check.

4. *Technician A* says electrical fires are extinguished with Class A fire extinguishers.
   Technician B says gasoline is extinguished with Class B fire extinguishers.
   Who is correct?
   A. A only
   B. B only
   C. Both A and B
   D. Neither A nor B

5. *Technician A* says material safety data sheets (MSDS) explain employers' and employees' responsibilities regarding handling and disposal of hazardous materials.

   Technician B says material safety data sheets (MSDS) contain specific information about hazardous materials.

   Who is correct?

   A. A only
   B. B only
   C. Both A and B
   D. Neither A nor B

6. *Technician A* says a solid that ignites spontaneously is considered a hazardous material.

   Technician B says a liquid is considered a hazardous material if the vapors on the surface will ignite when exposed to an open flame whose temperature is below 140°F (60°C).

   Who is correct?

   A. A only
   B. B only
   C. Both A and B
   D. Neither A nor B

7. *Technician A* says safety glasses are sufficient when working with battery acid and refrigerants.

   Technician B says full-face shields are designed to provide protection for the face, neck, and eyes of the technician.

   Who is correct?

   A. A only
   B. B only
   C. Both A and B
   D. Neither A nor B

8. Which of the following statements on HEV safety is correct?

   A. Always place the high-voltage service plug where someone will not accidentally reinstall it.
   B. Use a Class III DMM set to 400 VDC to determine if the high-voltage system voltage is at 0 volts before servicing.
   C. Test the integrity of the insulating gloves prior to use.
   D. All of these choices.

9. Which of the following statements concerning gasoline storage is true?

   A. Always use approved gasoline containers that are painted blue for proper identification.
   B. Prevent air from entering the container by filling gasoline containers completely full.
   C. Always transport gasoline containers on plastic bed liners.
   D. Do not store a partially filled gasoline container for long periods of time.

10. *Technician A* says once the service plug is disconnected there is no high voltage in the vehicle systems.

    Technician B says prior to disconnecting the high-voltage service plug, the vehicle must be turned off and the auxiliary battery may need to be disconnected.

    Who is correct?

    A. A only
    B. B only
    C. Both A and B
    D. Neither A nor B

Name _____   Date _____

SHOP SAFETY SURVEY

Objective: Upon completion of this job sheet you should be able to evaluate the safety of your workplace.

ASE Education Foundation Correlation ————————————————————

This job sheet addresses the following Foundational and Workplace tasks:

Shop and Personal Safety

1.   Identify general shop safety rules and procedures.

2.   Utilize safe procedures for handling of tools and equipment.

3.   Identify and use proper placement of floor jacks and jack stands.

4.   Identify and use proper procedures for safe lift operation.

5.   Utilize proper ventilation procedures for working within the lab/shop area.

6.   Identify marked safety areas.

8.   Identify the location and use of eye wash stations.

9.   Identify the location of the posted evacuation routes.

10.   Comply with the required use of safety glasses, ear protection, gloves, and shoes during lab/shop activities.

11.   Identify and wear appropriate clothing for lab/shop activities.

12.   Secure hair and jewelry for lab/shop activities.

We Support
ASE | **Education Foundation**

Procedure

1. Before you begin to evaluate your workplace, evaluate yourself. Are you dressed for work?   ☐ Yes   ☐ No

   If yes, why? _____

   If no, what must you correct to be properly dressed? _____

   _____

2. Do your safety glasses meet all required standards (ANSI)?   ☐ Yes   ☐ No

   Do they have side shields? _____

3. Carefully inspect your shop, noting any potential hazards._____

NOTE: A hazard is not necessarily a safety violation but is an area of which you must be aware.

4. Are there safety areas marked around grinders and other machinery?

   ☐ Yes ☐ No

5. What is the air pressure regulator set to? _____

6. Where are the tools stored in your shop? _____

   Are they clean and neatly stored?   ☐ Yes   ☐ No

7. If you could, how would you improve the tool storage?

   _____

   _____

8. What kind of hoist is used in your shop?

_____

9. Ask your instructor to demonstrate hoist usage.

10. Where is the first-aid kit in your shop?

_____

11. Where is the main power shutoff or emergency shutoff controls located?

_____

12. List the location of the exits.

_____

_____

13. Describe the emergency evacuation procedures.

_____

_____

14. Where are the hazardous materials stored?

_____

_____

15. What is the procedure for handling hazardous waste?

_____

_____

16. What is the procedure to be followed in your shop in case of an accident?

_____

_____

Instructor's Evaluation:

Grade	Rubric
0	No exposure due to: ☐ Absence ☐ Other:
1	Student unsuccessfully attempted the task
2	Student completed the task with significant instructor assistance.
3	Satisfactory performance of task. The student was able to perform elements of the task with minimal instructor assistance.
4	Competent/Proficient. Able to perform all elements of the task successfully and independently.

Instructor's Notes:

Name _____   Date _____

FIRE HAZARD INSPECTION

Objective: Upon completion of this job sheet, you should be able to identify workplace fire hazards.

ASE Education Foundation Correlation

This job sheet addresses the following Foundational and Workplace tasks:

Shop and Personal Safety

7.   Identify the location and the types of fire extinguishers and other fire safety equipment; demonstrate knowledge of the procedures for using fire extinguishers and other fire safety equipment.

We Support

ASE | Education Foundation

Procedure

1. Are there any flammable liquids stored in your shop?

    Are they stored properly?   ☐ Yes ☐ No

    Why or why not? _____

    _____

2. Where are the fire extinguishers located in your shop?

    _____

    _____

    What class fires can each extinguisher be used on?

    _____

    _____

3. Explain to your instructor how to use each fire extinguisher in your shop.

4. Does your shop have a fire blanket?   ☐ Yes ☐ No

    If so, where is it kept?

    _____

    _____

5. Where are the fire alarms located?

    _____

    _____

6. Where are the fire exits located?

    _____

    _____

7. Where are the fire escape routes posted?

_____

_____

8. Where are the fireproof cabinets located?

_____

_____

9. Where are dirty shop towels to be disposed?

_____

_____

Instructor's Evaluation:

Grade	Rubric
0	No exposure due to: ☐ Absence ☐ Other:
1	Student unsuccessfully attempted the task
2	Student completed the task with significant instructor assistance.
3	Satisfactory performance of task. The student was able to perform elements of the task with minimal instructor assistance.
4	Competent/Proficient. Able to perform all elements of the task successfully and independently.

Instructor's Notes:

Name _____   Date _____

HYBRID SAFETY

Objective: Upon completion of this job sheet, you should be familiar with the critical safety procedures involved in servicing a high-voltage hybrid system.

ASE Education Foundation Correlation

This job sheet addresses the following Foundational and Workplace tasks:

Shop and Personal Safety

14.   Identify vehicle systems which pose a safety hazard during service due to high voltage such as: hybrid/electric drivetrain, lighting systems, ignition systems, A/C systems, injection systems, etc.

We Support
ASE | Education Foundation

Tools and Materials

• HEV
• Service information

Safety Considerations and PPE

• Insulating gloves
• Eye protection

⚡ **WARNING** **You will be working around high voltage while doing this job sheet. Follow all warnings and cautions in the manufacturer's service procedures. Even though the service disconnect is removed, the HV battery is still "live."**

Procedure

1. Use the service information data and determine the location on the vehicle for the high-voltage service plug.

   _____

   _____

   _____

2. What must be done prior to disconnecting the service plug?

   _____

   _____

3. How long must you wait after the plug is disconnected before servicing the system?

   _____

   _____

4. Access the 12-volt auxiliary battery and remove the negative terminal.

5. Test the insulating gloves for leaks. Are the gloves safe to use?   ☐ Yes   ☐ No

   If no, inform your instructor.

6. Put on the insulating gloves and eye protection.

7. Remove the service plug. What device(s) are integrated into the service plug assembly?

_____

_____

8. Reinstall the service plug.

9. Review your observations with your instructor.

Instructor's Evaluation:

Grade	Rubric
0	No exposure due to: ☐ Absence ☐ Other:
1	Student unsuccessfully attempted the task
2	Student completed the task with significant instructor assistance.
3	Satisfactory performance of task. The student was able to perform elements of the task with minimal instructor assistance.
4	Competent/Proficient. Able to perform all elements of the task successfully and independently.

Instructor's Notes:

CHAPTER 2
SPECIAL TOOLS AND PROCEDURES

Upon completion and review of this chapter, you should be able to:

- Explain the proper use of jumper wires.
- Demonstrate the proper use of a test light.
- Explain the proper use of a logic probe.
- Explain the proper use of analog volt/amp/ohmmeters.
- Demonstrate the proper use of a digital voltmeter.
- Demonstrate the proper use of a digital ohmmeter.

- Describe when to use the different types of multimeters.
- Correctly set up a lab scope.
- Describe the purpose of the scan tool.
- Use Ohm's law to determine electrical values in different types of circuits.
- Locate service information.
- Explain the concepts of working as an electrical systems technician.

Basic Tool

Basic mechanic's tool set

Terms to Know

Ammeter
Average responding
Backprobe
Captured signal
Cycle
D'Arsonval movement
Diagnostic trouble codes (DTCs)
Digital multimeter (DMM)
Duty cycle
Flat rate

Frequency
Glitches
Hertz
Jumper wire
Logic probe
Multimeter
Noise
Ohmmeter
Open circuit testing
Pulse width
Root mean square (RMS)

Scan tool
Sine wave
Sinusoidal
Square wave
Test light
Vehicle identification number (VIN)
Voltage drop
Voltmeter
Work order

Introduction

This chapter covers many of the typical shop procedures that the electrical systems technician may encounter. This includes proper troubleshooting procedures, the use of special test equipment, the use of service information, and workplace practices.

Also, this chapter covers what it means to work as an electrical systems technician. This includes compensation programs, the importance of workplace and customer relations, communication, and certification.

Basic Electrical Troubleshooting

Classroom Manual
Chapter 2, page 19

To diagnose electrical components and circuits properly, you must be able to use many different types of electrical test equipment. In this chapter, you will learn when and how to use the most common types of test equipment. You will also learn which test instrument is best to use to identify the cause of the various types of electrical problems.

AUTHOR'S NOTE In the past voltmeters, ammeters, and ohmmeters were often separate pieces of test equipment. Today these meters are integrated into a single digital multimeter (DMM), also called a digital volt/ohmmeter (DVOM). The DMM is discussed later in this chapter.

Classroom Manual
Chapter 2, page 21

Electrical current is a term used to describe the movement or flow of electricity. The greater the number of electrons flowing past a given point in a given amount of time, the more current the circuit has. This current, like the flow of water or any other substance, can be measured. The unit for measuring electrical current is the ampere. An **ammeter** is used to measure electrical current flow in a circuit.

Classroom Manual
Chapter 2, page 22

When any substance flows, it meets resistance. As current flows through a resistance, there is less voltage left over. The amount of voltage drop is a direct result of the amount of resistance. An **ohmmeter** can measure the resistance to electrical flow. The ohmmeter sends a small current through the leads and the component being tested. It then measures the voltage drop and calculates the resistance value in ohms.

Voltage is electrical pressure. Voltage is the force developed by the attraction of electrons to protons. The more positive one side of the circuit is, the more voltage is present in the circuit. Voltage does not flow; rather, it's the pressure that causes current to flow.

Classroom Manual
Chapter 2, page 19

To have electricity, a force is needed to move the electrons between atoms. This force is the pressure between a positive and negative point within an electrical circuit. This force, also called electromotive force (EMF), is measured in units called volts. One volt is the amount of pressure (force) required to move one ampere of current through a resistance of one ohm. A **voltmeter** is used to measure voltage.

The amount of current that flows in a circuit is determined by the resistance in that circuit. As resistance goes up, the current goes down. The energy used by a load is measured in volts. Amperage stays constant in a circuit, but the voltage is dropped as it powers a load. Measuring voltage drop determines the amount of electrical energy changed to another form of energy by the load.

Troubleshooting electrical problems involves using meters, test lights, and jumper wires to determine if any part of the circuit is open, shorted, or if there is unwanted resistance.

To troubleshoot a problem, always begin by verifying the customer's concern. Operate the affected system along with other systems to get a complete understanding of the problem. Often there are other problems, which are not as evident or bothersome to the customer, that provide helpful information for diagnostics. Obtain the correct wiring diagram for the vehicle and study the affected circuit. From the diagram, you should be able to identify testing points and probable problem areas. Then test and use logic to identify the cause of the problem.

An ammeter and a voltmeter connected to a circuit at the different locations shown in **Figure 2-1** should give readings as indicated when there are no problems in the circuit. An open exists whenever there is not a complete path for current flow. If there is an open anywhere in the circuit, the ammeter will read zero current. If the open is in the 1-ohm

Figure 2-1 A basic circuit being tested with an ammeter and a voltmeter.

resistor, a voltmeter connected from C to ground will read zero. However, if the resistor is open and the voltmeter is connected to points B and C, the reading will be 12 volts. The reason is that the battery, ammeter, voltmeter, 2-ohm resistor, and 3-ohm resistor are all connected together to form a series circuit. Because of the open in the circuit, there is only current flow in the circuit through the meter, not the rest of the circuit. This current flow is very low because the meter has such high resistance. Therefore, the voltmeter will show a reading of 12 volts, indicating no voltage drop across the resistors.

To help you understand this concept, look at what happens if the 2-ohm resistor is open instead of the 1-ohm resistor. A voltmeter connected from point C to ground would indicate 12 volts. The 1-ohm resistor in series to the high resistance of the voltmeter would have little effect on the circuit. If an open occurred between point E and ground, a voltmeter connected from point B, C, D, or E to ground would read 12 volts. A voltmeter connected across any one of the resistors, from B to C, C to D, or D to E, would read 0 volts because there will be no voltage drops if there is no current flow.

A short would be indicated by excessive current and/or abnormal voltage drops. These examples illustrate how a voltmeter and ammeter can be used to check for problems in a circuit. An ohmmeter also can be used to measure the values of each component and compare these measurements to specifications. If there is no continuity across a part, it is open. If there is more resistance than called for, there is high internal resistance. If there is less resistance than specified, the part is shorted.

⚡ **WARNING Because the human body is a conductor of electricity, observe all safety rules when working with electricity.**

⚡ **Caution**
Any broken, frayed, or damaged insulation material requires replacement or repair to the wire. Exposed conductor material from damaged insulation can result in a safety hazard and damage to circuit components.

Test Equipment

Since electricity is an invisible force, the proper use of test tools will permit the technician to "see" the flow of electrons. Knowing what is being looked at and being able to interpret various meter types will assist in electrical system diagnosis. A number of common tools

Figure 2-2 Using a fused jumper wire to bypass the switch.

and instruments are used to diagnose and repair electrical circuits correctly. The most common tools are jumper wires, test lights, voltmeters, ammeters, and ohmmeters.

Jumper Wires

One of the simplest types of test equipment is the **jumper wire**. Connecting one end of the jumper wire to battery positive will provide an excellent 12-volt power supply for testing a component. If the component does not operate when it's in its own circuit but does operate when battery voltage is jumped to it, the component is good and the fault is within the circuit (**Figure 2-2**).

> **AUTHOR'S NOTE** The term *battery positive* refers to any portion of the circuit from the battery positive terminal to the load (insulated side) that has battery voltage.

Jumper wires can be used to bypass individual wires, connectors, components, or switches (**Figure 2-3**). Bypassing a component or wire helps to determine if that part is faulty. If the problem is no longer evident after installing the jumper wire, the part being bypassed is usually faulty.

To protect the circuit being tested, the jumper wire should be fitted with an in-line fuse holder or circuit breaker. This will allow the quick changing of fuses to protect the circuit and help prevent damage to the circuit if the jumper wire is connected improperly.

⚡ **WARNING** **Never connect a jumper wire across the terminals of the battery. The battery could explode, causing serious injury.**

Test Lights

There are two types of test lights commonly used in diagnosing electrical problems: non-powered and self-powered. A **test light** is used when the technician needs to look for electrical power in the circuit. A typical nonpowered test light has a transparent handle that contains a light bulb. A sharp probe extends from one end of the handle while a wire with a clamp extends from the other end (**Figure 2-4**). If the circuit is operating properly, clamping the lead of the test light to ground and probing the circuit at a point of voltage should light the lamp (**Figure 2-5**).

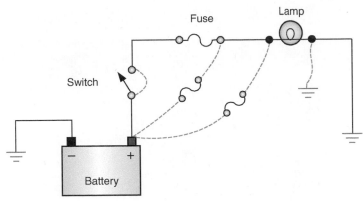

Figure 2-3 Examples of locations a fused jumper wire can be used to bypass a portion of the circuit, and to test the ground circuit. Anytime you bypass the circuit fuse, use a jumper wire that is fitted with a fuse. Never bypass the load component.

Figure 2-4 A typical 12-volt test light is used to probe for voltage in a circuit.

Figure 2-5 If voltage is present, the test light will illuminate.

A test light is limited because it does not display how much voltage is at the point of the circuit being tested. However, by understanding the effects of voltage drop, the technician will be able to interpret the brightness of the test light and relate the results to the expectations of a good circuit. Connecting the test lamp before the voltage drop should light the lamp brightly. If the lamp is connected after a voltage drop, the lamp will light dimly. The light should not illuminate if it's probing for voltage after the last resistance.

> ⚙ **SERVICE TIP** A headlight bulb fitted with wires and clips will make a good test light for checking circuits that carry 5 amps or more of current. Do not use this lamp to test circuits that carry less than 5 amps of current.

It must be understood that just because the test light lights brightly, that does not mean the circuit may not have problems. If the light is bright, that means the circuit can handle the amount of current needed to light the test light. Most test lights require very little amperage to turn on (e.g., 250 mA). If testing a circuit that needs to handle 8 amps of current, you should use a test light that will require that amount of current.

⚠ **Caution**

Do not use a test light to test for power in an electronic circuit, such as a computer or printed circuit board. The increased current draw of the test light may damage the system components.

Figure 2-6 Typical self-powered continuity tester.

Another type of circuit tester is the self-powered continuity tester (**Figure 2-6**). The continuity tester has an internal battery that powers a light bulb. With the power in the circuit turned off or disconnected, the ground clip is connected to the ground terminal of the load component. Probing the feed wire will illuminate the light if the circuit is complete (has continuity). If there is an open in the circuit, the lamp will not illuminate.

⚡ **WARNING** **Never use a self-powered test light to test the air bag system. The battery in the tester can cause the air bag to deploy.**

Logic Probes

Computer-controlled systems use a pulsed voltage to transmit messages or to operate a component. A standard or self-powered test light should not be used to test these circuits since they may damage the computer. However, a **logic probe** (**Figure 2-7**) can be used. A logic probe looks like a test light, except it has three different-colored LEDs. The red LED will light if a high voltage is present at the point in the circuit being tested. The green LED will light to indicate the presence of low voltage. The yellow LED lights to indicate the presence of voltage pulses. If the voltage is pulsed from a high level to a low level, the yellow LED will be on, and the red and green LEDs will cycle, indicating the change in voltage. Logic probes can be used only on digital circuits. Analog circuits change voltage over time; the logic probe works on only pulsed voltages.

Figure 2-7 A typical logic probe.

Multimeters

The **multimeter** is one of the most versatile tools used to diagnose electrical systems. It can be used to measure voltage, resistance, and amperage. In addition, some types of multimeters are designed to test diodes and measure frequency, duty cycle, temperature, and rotation speed. Multimeters are available in analog (swing needle) and digital display.

Analog Meters

Analog meters use a sweeping needle and a scale to display test values (**Figure 2-8**). All analog meters use a **D'Arsonval movement** (**Figure 2-9**). A D'Arsonval movement is a small coil of wire mounted in the center of a permanent horseshoe-type magnet. A pointer or needle is mounted to the coil. When taking a measurement, current flows through the coil and creates a magnetic field around the coil. As the coil's magnetic field interacts with the magnetic field of the permanent magnet, the coil rotates within the permanent magnet. The amount of rotation is determined by the strength of the magnetic field around the coil. Since the needle moves with the coil, it reflects the amount of coil movement and its direction.

Digital Meters

With modern vehicles incorporating computer-controlled systems, today's technician must know how to properly use a **digital multimeter (DMM)** (**Figure 2-10**). Digital multimeters display values using liquid crystal displays instead of a swinging needle. Basically, they are computers that determine the measured value and display it for the technician. Computer systems have integrated circuits (ICs) that use very low amounts of current. Analog meters can overload computer circuits and burn out the IC chips since they allow a large amount of current to flow through the circuit.

On the other hand, most digital multimeters have very high input resistance (impedance), which prevents the meter from drawing current when connected to a circuit. Most DMMs have at least 10 megohms (10 million ohms) impedance. This high impedance reduces the risk of damaging computer circuits and components.

Digital meters rely on electronic circuitry to measure electrical values. The measurements are displayed with LEDs or on a liquid crystal display (LCD). Digital meters tend to give more accurate readings and are certainly much easier to read. Digital meters display the measurement in a numerical value rather than reading a scale at the point where the needle lines up. This also eliminates the almost certain error caused by viewing an analog meter at an angle.

Special Tools

Analog volt or ohmmeter
DMM
Backprobing tools

⚠ Caution

Do not use an analog meter on a computer-controlled circuit unless expressly directed to do so in the service information. Damage to the circuit or computer may result.

Digital multimeters (DMM) are also referred to as DVOMs (digital volt/ohmmeters).

Figure 2-9 D'Arsonval movement is the basis for the movement of an analog meter.

Figure 2-8 An analog meter.

Figure 2-10 Digital multimeter.

Special Tool

DMM

Classroom Manual
Chapter 2, page 19

Classroom Manual
Chapter 2, page 22

Voltmeters

A voltmeter can be used to measure the voltage available at the battery. It can also test the voltage available at the terminals of any component or connector. In addition, a voltmeter tests voltage drop across an electrical circuit, component, switch, or connector.

A voltmeter has two leads: a red positive lead and a black negative lead. The red lead should be connected to the positive side of the circuit or component. The black lead should be connected to ground or to the negative side of the component. A voltmeter is connected in parallel with a circuit (**Figure 2-11**).

Figure 2-12 shows how to check for voltage in a closed circuit. The voltage at point A is 12 volts positive. There is a drop of 6 volts over the 1-ohm resistor, and the reading is 6 volts positive at point B. The remaining voltage drops in the motor load, and the voltmeter reads 0 at point C, indicating normal motor circuit operation.

When reading voltage in the same circuit that has an open (**Figure 2-13A**), 12 volts will be indicated at any point ahead of the open. In this case, 12 volts are read at points A, B, and C, but not X. Since the circuit is open and no current flows, no voltage is dropped across the resistor or motor.

The loss of voltage due to resistance in wires, connectors, and loads is called **voltage drop**. Voltage drop is the amount of electrical energy converted to another form of energy. For example, electrical energy is converted to heat energy to make a lamp light. It's the heat that makes the lamp light.

Testing for voltage by disconnecting a connector and then measuring for applied voltage is called **open circuit testing**. This test can be misleading since virtually no current is flowing through the voltmeter, and the circuit is not loaded. Thus, the meter may display full battery voltage although there is excessive resistance in the circuit.

An ohmmeter is not reliable for testing a circuit under load. Since the ohmmeter uses an internal battery to measure resistance, any component being tested with an ohmmeter must be isolated from vehicle voltage. Even then, the ohmmeter may not read low resistances accurately. For example, consider the starter motor circuit with the heavy battery cable that powers the starter motor. If the motor was drawing 150 amps and there were 0.040 ohms of resistance in that cable, the voltage drop would be 6 volts! This illustrates why voltage drop testing under load is the preferred method for diagnosing excessive resistance.

To measure voltage drop across each portion of the circuit, determine what point is the most positive and what point is the most negative. A point in the circuit can be either positive or negative, depending on what is being measured. Referring to **Figure 2-13B**, if the

Figure 2-11 Connecting a voltmeter in parallel to the circuit.

Figure 2-12 Checking voltage in a closed circuit.

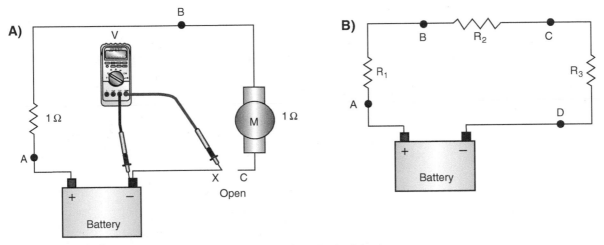

Figure 2-13 Using a voltmeter to (A) find an open circuit and (B) to check voltage drop.

voltage drop across R_1 is being measured, the positive meter lead is placed at point A and the negative lead at point B. The voltmeter will measure the difference in volts between these points. However, to measure the voltage drop across R_2 the polarity of point B is positive and point C is negative. The reason is that point B is the most negative for R_1, yet it's the most positive point for R_2. To measure the voltage drop across R_3, point C is positive, and point D is negative. The positive lead of the voltmeter should be placed as close as possible to the positive side of the battery in the circuit. The procedure for measuring voltage drop of the left side headlamp circuit is shown in **Photo Sequence 3**.

> ⚙️ **SERVICE TIP** Using a voltmeter to measure voltage drop is one of the easiest ways of testing a circuit's ability to carry current under load.

Often, voltage drop testing requires the technician to **backprobe** a connector to access the terminal. Backprobing requires using a backprobing tool (**Figure 2-14**) that will slide into the back side of a connector. The tool should be inserted between the wire's insulation and the connector's seal until it contacts the terminal (**Figure 2-15**). It's important to use as small a backprobing tool as possible. The wire, connector, or terminal will not be damaged when performed properly.

Figure 2-14 Backprobing tools.

Figure 2-15 Backprobing tool properly inserted into the back of a connector.

PHOTO SEQUENCE 3
Performing a Voltage Drop Test

P3-1 The tools required to perform this task include a voltmeter, backprobing tools, and fender covers.

P3-2 Set the voltmeter to its lowest DC voltage scale.

P3-3 To test the voltage drop of the whole insulated side of the left headlight circuit, connect the red (positive) voltmeter test lead to the battery positive (+) terminal.

P3-4 Use the backprobing tool to connect the black (negative) voltmeter test lead into the low-beam terminal of the headlight socket. Make sure that you are connected to the input side of the headlight.

P3-5 Turn on the headlights (low beam) and observe the voltmeter readings. The voltmeter will indicate the amount of voltage dropped between the battery and the headlight.

In the past, the main concern was that all wiring had resistance values low enough to allow enough voltage to the load for proper operation. Today, communication signals are sent over wiring using very low current and cannot operate properly if circuit resistance causes voltage drops. Today's vehicles, with low-current computer circuits, cannot withstand voltage drops. Engineers determine conductor and component sizing to limit voltage drop to 3% of the source voltage. However, what constitutes an acceptable voltage drop depends on the amount of current. A starter motor circuit with 150 amps current draw can still operate with a voltage drop of .5 volt, but that same voltage drop on an injector circuit will have detrimental effects. For low-current circuits, 100 mV per connection is the maximum amount of voltage drop. Computer control circuits should not exceed 100 mV total.

Figure 2-16 shows two headlights connected to a 12-volt battery by two wires. Each wire has a resistance of only 0.05 ohm. Each headlight has a resistance of 2 ohms. The two headlights are wired in parallel and have a total resistance of 1 ohm. Total circuit resistance is 1.1 ohms (1 ohm + 0.05 ohm + 0.05 ohm). Therefore, the circuit current is 10.9 amps (I = E/R, I = 12V/1.1 ohms). The voltage drop across the bulbs is 10.9 volts (E = I × R, E = 10.9 amps × 1 ohm). Although it may not be very noticeable, the light bulbs will not be as bright as they would be if the wires had 0 resistance and the bulbs had 12 amps of current.

A voltmeter can also be used to check for proper circuit grounding. If the voltmeter reading indicates full battery voltage to the lights, but the bulbs are not illuminated, the bulbs or sockets could be bad, or the ground connection is faulty. An easy way to check for a defective bulb is to replace it with a known good bulb.

If the bulbs are good, the problem lies in either the light sockets or the ground wires. Connect the voltmeter between the bulb socket's ground wire and a good ground. If the light socket is defective, the voltmeter will read 0 volts. If the socket is not defective, but the ground wire is broken or disconnected, the voltmeter will read very close to battery voltage. In fact, any voltage reading would indicate a bad or poor ground circuit. The higher the voltage, the greater the problem.

Figure 2-16 Measuring the voltage drop across a lamp. Notice the wires have some resistance.

Special Tool

DMM

Classroom Manual
Chapter 2, page 22

 Caution

Since the ohmmeter
is self-powered, never
use an ohmmeter on
a powered circuit.

Thus far, we have discussed using a voltmeter on a direct current (DC) circuit. The voltmeter can also measure alternating current (AC). There are two methods used to display AC voltage: **root mean square (RMS)** and **average responding**. If the AC voltage signal is a true **sine wave**, both methods will display the same value. However, most automotive sensors do not produce a pure sine wave signal. The technician must know how the different meters will display the AC voltage reading under these circumstances. The type of display the voltmeter uses can be found in the meter's operating manual.

Ohmmeters

In contrast to the voltmeter, which uses the voltage available in the circuit, an ohmmeter is battery powered. The circuit being tested must be electrically open or isolated. If the power is on in the circuit, the ohmmeter may be damaged.

The two ohmmeter leads are placed across (in parallel) with the circuit or component being tested. The red lead is placed on the positive side of the circuit, and the black lead is placed on the negative side of the circuit (**Figure 2-17**). The meter sends current through the component and determines the amount of resistance based on the voltage dropped across the load. The scale of an ohmmeter reads from 0 to infinity (∞). A 0 reading means there is no resistance in the circuit and may indicate a short in a component that should show a specific resistance. For example, a coil winding should have a high resistance value; a 0-ohm reading would indicate the coil windings are being bypassed. An infinity reading indicates a number higher than the meter can measure. This usually is an indication of an open circuit.

AUTHOR'S NOTE There is no correlation between the ohmmeter reading and continuity. The ohmmeter reads the resistance value determined during the test. For example, a reading of 0 ohms does not mean there is 0 continuity. It means there is no resistance between the test points. An ohmmeter reading of 250 ohms does not mean there are 250 units of continuity but 250 ohms of resistance.

The test chart, shown in **Table 2-1**, illustrates the readings that may be expected from an ohmmeter or voltmeter under different conditions. It's important to become familiar with these examples in order to analyze circuits.

Figure 2-17 Measuring resistance with an ohmmeter. The meter is connected in parallel with the component being tested after power is removed from the circuit.

Table 2-1 Circuit Test Chart

Type of Defect	Test Unit	Expected Results
Open	Ohmmeter	∞ infinite resistance between conductor ends
	Test light	No light after open
	Voltmeter	Ø volts at end of conductor after the open
Short to ground	Ohmmeter	Ø resistance to ground
	Test light	Lights, if connected across the fuse
	Voltmeter	Generally not used to test for ground
Short	Ohmmeter	Lower than specified resistance through load component Ø resistance to adjacent conductor
	Test light	Light will illuminate on both conductors
	Voltmeter	A voltage will be read on both conductors
Excessive resistance	Ohmmeter	Higher than specified resistance through the circuit
	Test light	Light illuminates dimly
	Voltmeter	Voltage will be read when connected in parallel over resistance

Ohmmeters are also used to trace and check wires or cables. Assume that one wire of a four-wire cable is to be found. Connect one ohmmeter probe to the known wire at one end of the cable and touch the other probe to each wire at the other end of the cable. Any evidence of resistance indicates the correct wire. Using this same method, you can check a suspected defective wire. If low resistance is shown on the meter, the wire is intact. However, this does not mean that testing the wire under load is not required. If infinite resistance is measured, the wire is defective (open). If the wire is okay, continue checking by connecting the probe to other leads. Any indication of resistance indicates that the wire is shorted to one of the other wires and that the harness is defective.

Ammeters

Circuit problems can also be identified by using an ammeter. An ammeter measures current flow in a circuit. An ammeter must be placed in series with the circuit being tested (**Figure 2-18**). Normally, this requires disconnecting a wire or connector from a component and connecting the ammeter between the wire or connector and the component. The red lead of the ammeter should always be connected to the most positive side of the connector, and the black lead should be connected to the least positive side.

Most handheld multimeters have a 10-ampere protection device. This is the highest amount of current flow the meter can read. When using the ammeter, start on a high scale and work down to obtain the most accurate readings.

Classroom Manual
Chapter 2, page 21

 Special Tool
DMM

⚠ **Caution**
Never place the leads of an ammeter across the battery or a load. This puts the meter in parallel with the circuit and will blow the fuse in the ammeter or possibly destroy the meter.

Figure 2-18 Measuring current flow with an ammeter. The meter must be connected in series with the circuit.

Figure 2-19 An ammeter with an inductive pickup. The inductive pickup eliminates the need to connect the meter in series with the circuit.

⚠ Caution

When testing for a short, always use a fused jumper wire. Never bypass the circuit fuse with a non-fused jumper wire. The jumper wire fuse should be rated at no more than 50% higher capacity than specifications. This offers circuit protection and provides enough amperage for testing. After the problem is found and corrected, be sure to install the specified rating of fuse for circuit protection.

It's much easier to test current using an ammeter with an inductive pickup (**Figure 2-19**). The pickup clamps around the wire or cable being tested. The inductive clamp has an arrow on it to indicate proper attachment to the conductor. The arrow indicates current flow direction using the conventional flow theory of positive to negative. These ammeters measure amperage based on the magnetic field created by the current flowing through the wire. This type of pickup eliminates the need to separate the circuit to insert the meter leads.

Because ammeters are built with very low internal resistance, connecting them in series does not add any appreciable resistance to the circuit. Therefore, an accurate measurement of the current flow can be taken.

For example, assume that a circuit shown in Figure 2-18 normally draws 5 amps and is protected by a 10-amp fuse. If the circuit constantly blows the fuse, a short exists somewhere in the circuit. Mathematically, each light should draw 1.25 amperes (5/4 = 1.25). To find the short, disconnect all lights by removing them from their sockets. Then, close the switch and read the ammeter. With the load disconnected, the meter should read 0 amperes. If there is any reading, the wire between the fuse block and the socket is shorted to ground.

If 0 amps were measured, reconnect each light in sequence; the reading should increase 1.25 amperes with each bulb. If, when making any connection, the reading is higher than expected, the problem is in that part of the light circuit.

⚙ **SERVICE TIP** Most handheld DMMs have the option for an inductive pickup that will measure amperage between 10 and 20 amps (depending on the model). This makes measuring amperage on some circuits easier since it eliminates the need to connect the test leads in series.

Additional DMM Functions

Some multimeters feature additional functions besides measuring AC or DC voltage, amperage, and ohms. Many multimeters can measure engine revolutions per minute (rpm), ignition dwell, diode condition, distributor condition, frequency, and temperature.

Some meters have a MIN/MAX function that displays the minimum, maximum, and average values received by the meter during the time the test was being recorded. This feature is valuable when checking sensors, output commands, or circuits for electrical **noise**. Noise is usually the result of radio frequency interference (RFI) or electromagnetic induction (EMI). Noise is an unwanted voltage signal that rides on a signal. The noise causes slight increases and decreases in the voltage signal to, or from, the computer. Another definition of noise is an AC signal riding on a DC voltage. The computer may attempt to react to the small changes in the signal due to the noise. This means the computer responds to the noise and not the voltage signal, resulting in incorrect component operation.

Classroom Manual
Chapter 2, page 46

Also, some multimeters may have the capabilities to measure **duty cycle**, pulse width, and frequency. Duty cycle is the measurement of the amount of "on" time as compared to the time of a **cycle** (**Figure 2-20**). A cycle is one set of changes in a signal that repeats itself several times. The duty cycle is displayed in a percentage. For example, a 60% duty cycle means the device is on 60% of the time and off 40% of the time of one cycle.

Pulse width is similar to duty cycle except that it's a measurement of time the device is turned on within a cycle (**Figure 2-21**). Pulse width is usually measured in milliseconds.

As previously mentioned, some multimeters can measure **frequency** (**Figure 2-22**). Frequency is a measure of the number of cycles that occur in one second. The higher the frequency, the more cycles that occur in a second. Frequency is measured in **hertz**. If the cycle occurs once per second, the frequency is 1 hertz. If the cycle is 300 times per second, the frequency is 300 hertz.

The meter's trigger level must be set to accurately measure duty cycle, pulse width, and frequency. The trigger level tells the meter when to start counting. Trigger levels can be set at certain voltage levels or at a rise or fall in the voltage. Normally, meters have a

Figure 2-20 Duty cycle.

Figure 2-21 Pulse width.

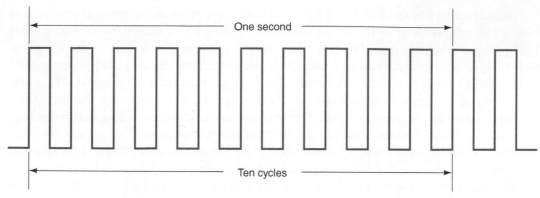

Figure 2-22 Frequency signal.

built-in trigger level that corresponds with the voltage setting. If the voltage does not reach the trigger level, the meter will not begin to recognize a cycle. On some meters, you can select between a rise or fall in voltage to trigger the cycle count. A rise in voltage is a positive increase in voltage. This setting is used to monitor the activity of devices whose power feed is controlled by a computer. A fall in voltage is negative voltage. This setting is used to monitor ground-controlled devices.

Reading the DMM

With the increased use of the DMM, the technician needs to be able to read the meter accurately. By becoming proficient in the use of the DMM, technicians will have confidence in their conclusions and recommended repairs. This will eliminate the replacement of parts that were not faulty. There are deviations between DMM manufacturers as to how the display is presented, but most follow the method described here.

Multimeters either have an "auto range" feature, in which the meter automatically selects the appropriate scale, or must be manually set to a particular scale. Either way, to designate particular ranges and readings, meters display a prefix before the reading or range. Meters use the prefix because they cannot display long numbers. Values such as 20,200 Ω cannot be displayed as a whole number. As a result, scales are expressed as a multiple of tens or use the prefix K, M, m, and μ (**Table 2-2**). The prefix K stands for kilo and represents 1,000 units. For example, a reading of 10K equals 10,000. An M stands for mega and represents 1,000,000 units. A reading of 10M would represent 10,000,000. An m stands for milli and represents 0.001 of a unit. A reading of 10m would be 0.010. The symbol μ stands for micro and represents 0.000001 of a unit. In this case, a reading of 125.0 μ would represent 0.000125.

Table 2-2   Common Prefixes Used on Digital Multimeters

Prefix	Symbol	Relation to Basic Unit
Mega	M	1,000,000
Kilo	K	1,000
Milli	m	$0.001 \text{ or } \dfrac{1}{1,000}$
Micro	μ	$0.000001 \text{ or } \dfrac{1}{1,000,000}$
Nano	n	0.000000001
Pico	p	0.000000000001

If the display has no prefix before the unit being measured (V, A, Ω), the reading displayed is read directly. For example, if the reading was 1.243 V, the actual voltage value is 1.243. However, if a prefix is displayed, then the decimal point will need to be floated to determine actual readings. If the prefix is M (mega), the decimal is floated six places to the right. For example, a reading display of 2.50 MΩ is actually 2,500,000 ohms. A reading display of 0.250 MΩ is actually 250,000 ohms.

A prefix of K (kilo) means the decimal point needs to move three places to the right. For example, a display reading of 56.4 KΩ is actually 56,400 ohms. A reading of 1.264 KΩ is actually 1,264 ohms.

If the prefix is m (milli), the decimal is floated three places to the left. For example, a reading of 25.4 mA represents 0.0254 amperes. A display of 165.0 mA is actually 0.165 amperes.

Finally, if the prefix is a μ (micro), the decimal is floated six points to the left. A reading displayed as 125.3μ would represent 0.0001253 amperes while a reading of 4.6 μA is actually 0.0000046 amperes.

When using the ohmmeter function of the DMM, make sure power to the circuit being tested is turned off. Also, be sure to calibrate the meter before taking measurements. Meter calibration is done by holding the two test leads together. Most DMMs will self-calibrate, while others will need to be adjusted by turning a knob until the meter reads zero. Some meters do not have a means of calibrating, so the value read while the leads are touching will need to be subtracted from the test readings. Connect the DMM in parallel to the portion of the circuit being tested. The DMM will read zero (or close to zero) even on the lowest scale if continuity is good. The DMM will display an infinite reading if the continuity is very poor. This reading is usually shown as a "1.000," a "1," or an "OL."

OL stands for *over limit.*

Insulation Tester

An insulation tester can be a stand-alone meter or an additional function of a DMM (**Figure 2-23**). It measures the insulation resistance on the powered-down HV system when servicing an EV or HEV. Isolation tests are performed only on systems that have the power removed.

To test for insulation leakage of the high-voltage system, begin by inserting the insulation test probe into the + terminal and the ground lead into the − terminal. Do not connect the leads to the volt-ohm terminals. Turn the knob to the "INSULATION"

Figure 2-23 An insulation meter is used on many EVs and HEVs to determine insulation leaks in the high-voltage system.

position. In this position, a battery level check is performed. Confirm that the battery level test has passed. Press the "RANGE" button to select the desired voltage range. If performing an isolation test on an HEV, this should be set to 500 volts. Next, connect the leads to the circuit you are testing. The meter will automatically detect if the circuit is powered. The meter display should indicate a series of dashed lines. If the meter displays >30 V, then there is a voltage greater than 30 volts in the circuit, and the test meter will not perform the insulation test. Push and hold the "TEST" button on the red insulation test lead. While the test is in progress, the applied voltage will be displayed in the lower right corner of the screen, and the resistance in M ohms or G ohms will be displayed in the center of the screen. Leave the test leads attached to the circuit and release the "TEST" button. The meter will continue to display the resistance reading while the circuit is discharged through the meter.

Lab Scopes

A lab scope is a visual DMM (**Figure 2-24**). Technicians have used scopes to diagnose ignition, fuel injection, and charging systems for many years. These "tune-up scopes" were usually part of a large diagnostic machine, although some were stand-alone units. In recent years, the lab scope has become the diagnostic tool of choice for many good technicians.

Special Tool

Lab scope

The lab scope is advantageous in diagnosing many electrical problems quickly and accurately. Digital and analog volt/ohm/ammeters do not react fast enough to read systems that cycle quickly.

Although the lab scope can measure more than just voltage, we will focus on its function as a very fast-reacting voltmeter. The scope allows the technician to view voltage over time (**Figure 2-25**). These voltage readings appear as a voltage trace on the lab scope's screen.

An upward movement of the trace indicates an increase in voltage, and a downward movement of this trace represents a decrease in voltage. As the voltage trace moves across the screen, it represents a specific length of time.

Figure 2-24 The grids of a lab scope.

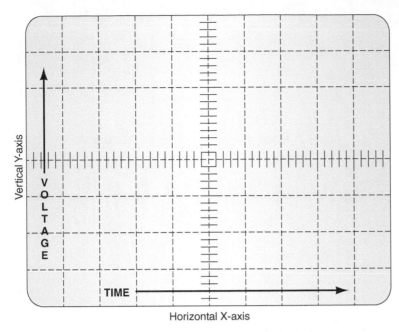

Vertical Y-axis

V O L T A G E

TIME

Horizontal X-axis

Figure 2-25 Grids on the screen serve as a time and voltage reference.

The divisions of the scope screen are set up as a grid pattern (called the graticule). The horizontal movement of the waveform represents time. Voltage is measured with the vertical position of the waveform. Since the scope displays voltage over time, the waveform moves from the left (the beginning of measured time) to the right (the end of measured time). The value of the divisions can be adjusted to improve the view of the voltage waveform. For example, the vertical scale can be adjusted so that each division represents 0.5 volt, and the horizontal scale can be adjusted so that each division equals 0.005 (5 milliseconds). These adjustments allow you to view small changes in voltage that occur in a very short period of time. The grid serves as a reference for the measurements.

The lab scope is the best method to verify an input or output signal. Monitoring voltage or amperage offers the ability to measure frequency, pulse width (time), and duty cycle (time ON versus time OFF).

Since a scope displays actual voltage, it will display any electrical noise or disturbances accompanying the voltage signal (**Figure 2-26**). Electrical noise can cause intermittent problems with unpredictable results. When a computer receives a voltage signal with noise, it will try to react to the minute changes. As a result, the computer responds to the noise rather than the voltage signal.

Electrical disturbances or **glitches** are momentary changes in the signal. Intermittent shorts to ground or power can cause glitches. These problems can occur for only a moment or may last for some time. A lab scope is handy for identifying these and other causes of intermittent problems. Changes in the voltage signal while wiggling or pulling on a wiring harness can detect any terminal looseness. This type of testing is commonly referred to as a "wiggle test."

The lab scope can be analog or digital. The trace of an analog scope is produced by a beam of light moving across a cathode ray tube. The advantage of this type of lab scope is the display is in real time. The digital scope, commonly called a digital storage oscilloscope or DSO, converts the voltage signal into digital information and stores it in its memory. To help with diagnosis, you can freeze the **captured signal** for closer analysis. DSOs also can capture low-frequency signals. Low-frequency signals tend to flicker when displayed on an analog screen. To have a clean waveform on an analog scope, the signal must be

The squares formed by the grid are called major divisions. These are further defined by smaller divisions called minor divisions.

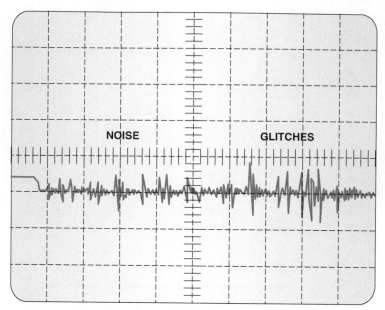

Figure 2-26 RFI noise and glitches showing as voltage signals.

repetitive and occurring in real time. The signal on a DSO is not quite real time. Rather, it displays the signal as it occurred a short time before.

The delay of a DSO is very slight. Most DSOs have a sampling rate of one million samples per second. This is quick enough to serve as an excellent diagnostic tool and allows slight changes in voltage to be observed. Slight and quick voltage changes cannot be detected on an analog scope.

Both an analog and a digital scope can be dual-trace scopes (**Figure 2-27**). This means they both can display two traces at the same time. By watching two traces simultaneously, you can watch the cause and effect of a sensor and compare a good or normal waveform to the one being displayed.

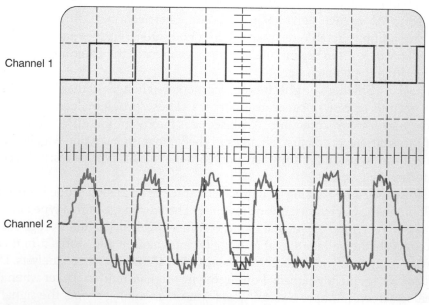

Figure 2-27 A dual-trace scope can show two patterns simultaneously.

Waveforms

A waveform represents voltage over time. Any change in the amplitude of the trace indicates a change in the voltage. When the trace is a straight horizontal line, the voltage is constant (**Figure 2-28**). A diagonal line up or down represents a gradual increase or decrease in voltage. A rapid rise or fall in the trace indicates a sudden change in voltage.

Scopes can display AC and DC, one at a time or both, as in the case of noise caused by RFI. The consistent change of polarity and amplitude of the AC signal causes slight changes in the DC voltage signal. A normal AC signal changes its polarity and amplitude over a period of time. The waveform created by AC voltage is typically a sine wave (**Figure 2-29**). One complete sine wave shows the voltage moving from zero to its positive peak, moving down through zero to its negative peak, and returning to zero. If the rise and fall from positive and negative are the same, the wave is said to be **sinusoidal**. If the rise and fall are not the same, the wave is nonsinusoidal.

The waveform on a scope is commonly called a trace.

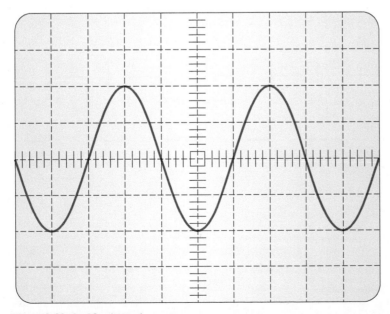

Figure 2-28 A waveform showing a constant voltage.

Figure 2-29 An AC voltage sine wave.

One complete sine wave is a cycle. The number of cycles that occur per second is the frequency of the signal. Checking frequency or cycle time is one way of checking the operation of some electrical components. Input sensors and permanent magnet voltage generators are the most common automotive components that produce AC voltage that can be checked on a scope (**Figure 2-30**). AC voltage waveforms should also be checked for noise and glitches. These may send false information to the computer.

DC voltage waveforms may appear as a straight line or a line showing a change in voltage. Sometimes a DC voltage waveform will appear as a **square wave**, which shows voltage making an immediate change (**Figure 2-31**). This type of wave represents voltage being applied (circuit being turned on), voltage being maintained (circuit remaining on), and no voltage applied (the circuit is turned off). Of course, a DC voltage waveform may also show gradual voltage changes.

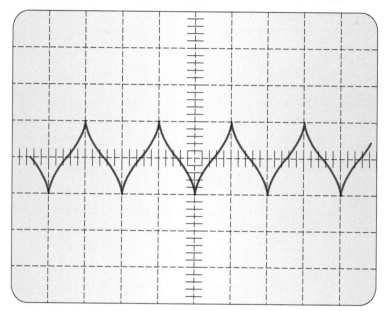

Figure 2-30 An AC voltage trace from a typical permanent magnetic generator-type sensor.

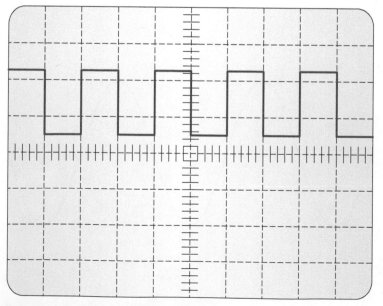

Figure 2-31 Typical square wave.

Scope Controls

Depending on the manufacturer and the model of the scope, the type and number of its controls will vary. However, most scopes have controls for intensity, vertical (Y-axis) adjustments, horizontal (X-axis) adjustments, and trigger adjustments. The intensity control is used to adjust the brightness of the trace. This allows for clear viewing regardless of the light around the scope screen.

The vertical adjustment (Volts/Div) controls the voltage value shown per division (**Figure 2-32**). If the scope is set at 0.5 (500 millivolts) volt/div, a 5-volt signal will need 10 divisions. Likewise, if the scope is set to 1 Volt/Div, 5 volts will need only five divisions. While using a scope, it's important to set the vertical so that voltage can be accurately read. Setting the voltage too low may cause the waveform to move off the screen, while setting it too high may cause the trace to be flat and unreadable. While reading the trace, adjust the voltage value to get the best resolution. The vertical position control allows the vertical position of the trace to be moved anywhere on the screen.

The horizontal control is the trace's time control (Time/Div) and sets the time base of the scope's sweep rate (**Figure 2-33**). The time per division can be set from very short periods (millionths of a second) to full seconds. If the time per division is set too low, the complete trace may not show across the screen. Setting the time per division too high can crowd the trace and inhibit detailed observation. Also, horizontal control allows the horizontal position of the trace to be set on the screen.

Trigger controls tell the scope when to begin a trace across the screen. Setting the trigger is important when trying to observe the timing of something. Proper triggering will allow the trace to repeatedly begin and end at the same points on the screen. There are usually numerous trigger controls on a scope. The trigger mode selector has a NORM and AUTO position. NORM means no trace will appear on the screen until a voltage signal occurs within the set time base. The AUTO setting will display a trace regardless of the time base.

Slope and level controls are used to define the actual trigger voltage. The slope switch determines whether the trace will begin on a rising or falling of the voltage signal (**Figure 2-34**). The level control determines where the time base will be triggered according to a certain point on the slope.

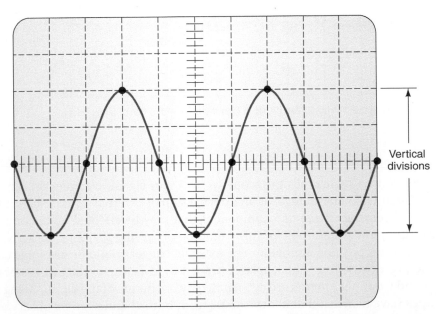

Figure 2-32 Vertical divisions represent voltage.

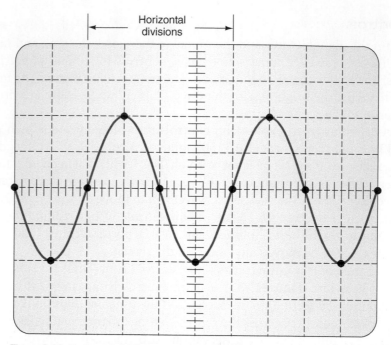

Figure 2-33 Horizontal divisions represent time.

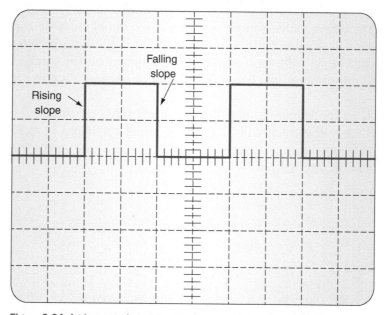

Figure 2-34 A trigger can be set to start the trace with a rise or fall of voltage.

A trigger source switch tells the scope which input signal to trigger on. This can be Channel 1, Channel 2, line voltage, or an external signal. External signal triggering is very useful when observing a trace of a component that may be affected by the operation of another component. An example of this would be observing fuel injector activity when changes in throttle position are made. The external trigger would be a voltage change at the throttle position sensor. The displayed trace would be the cycling of a fuel injector. Channel 1 and Channel 2 inputs are determined by the points of the circuit being probed. Some scopes have a switch that allows inputs from both channels to be observed simultaneously or alternately.

Lab scopes manufacturers offer many accessories that make the scope even more helpful. Current probes and clamps, pressure transducers, and frequency generators are just some examples. In addition, software overlays can be used with recorded waveforms to determine ignition timing, valve opening and closing events, and the engine's volumetric efficiency.

Scan Tools

The introduction of powertrain computer controls brought with it the need for tools capable of troubleshooting electronic control systems. There are a variety of computer **scan tools** available today that do just that (**Figure 2-35**). When connected to the computer through diagnostic connectors, a scan tool can access **diagnostic trouble codes (DTCs)**, run tests to check system operations, monitor system activity, and set system parameters. Trouble codes and test results are displayed on an LED screen or printed on the scan tool's printer.

A scan tool receives its testing information from one of several sources. Some scan tools have a programmable read-only memory (PROM) chip that contains all the information needed to diagnose specific model lines. This chip usually requires updating at least once a year to keep the tool compatible with new vehicles.

The scan tool can be designed to perform specific functions on certain vehicles. For example, some may only read DTCs from the PCM on domestic vehicles. Professional grade scan tools can test many on-board computer systems such as engine, antilock brake, air bags, and active suspension systems, depending on the year and make of vehicle. In addition, they can be used for programming immobilizer keys, resetting memories, service minder resets, sensor calibration, and bi-directional controls.

The scan tool plugs into the data link connector (DLC) (**Figure 2-36**). The DLC is a multi-pin connector that houses the data communication lines to the vehicle's computers for on-board diagnostics (OBD) and gathering live data streams with a scan tool. Before 1996, there was no standard DLC location, configuration, or terminal arrangement. Scan tools used on pre-1996 (known as OBD-I) vehicles required adapters to connect the scan

Special Tool
Scan Tool

The scan tool is also called a *scanner* and a *scan tester*.

Figure 2-35 An example of a handheld scan tool.

Figure 2-36 The scan tool connects to the diagnostic link connector (DLC).

Figure 2-37 Scan tool adapter kit.

tool to all of the different configurations. After 1996, OBD II mandates (ASE J-1962) required that the DLC be of a standardized configuration, shape, and terminal arrangement. The mandate only applies to the DLC used for emission-related functions and data, so manufacturers can continue to use their proprietary connectors for access to other systems. Today, most manufacturers utilize the J-1962 DLC to access all vehicle systems. However, adapters for connecting the scan tool to a specific DLC are still needed on some vehicles **(Figure 2-37)**.

The OBD-II DLC is usually located under the instrument panel on the driver's side, within 2 feet (0.61 m) of the steering wheel. Also, J1962 allows for two standardized interfaces, called type A and type B. Both DLC types have the same 16-pin, D-shaped female connector. However, the groove that separates the rows of terminals is different (**Figure 2-38**). The groove of the type A DLC is a single groove across the length of the terminal rows. The groove of the type B DLC is interrupted in the middle. These configurations make it impossible for a type A scan tool male connector to fit a type B DLC. But the type B scan tool connector can plug into a type A DLC. The type A connector is used for 12-volt supply vehicles, whereas type B is used for 24-volt vehicles.

When the scan tool is powered up, you may be given the option to manually input the vehicle data or to have the scan tool perform a VIN scan to retrieve the vehicle data. Typically, the information needed to start a diagnostic session includes the model year, make of vehicle, type of engine, and equipped options.

As automotive computer systems become more complex, the diagnostic capabilities of scan tools continue to expand. Most scan tools can store or freeze data into memory during a road test and then play the data back when the vehicle is returned to the shop.

Newer scan tools connect to the internet so they can display diagnostic information based on the fault code. They may also be able to reference Service Bulletins and display sensor specifications for the vehicle being tested.

Scan Tool Features

Scan tools vary depending on the manufacturer, but many have the following features:

1. A display window to show data and messages to the technician. Messages are displayed from left to right. Most scan tools display at least four readings at the same time.
2. An adapter cord that plugs into the scan tool and connects to the DLC.
3. A serial interface for connecting optional devices, such as a printer, terminal, or personal computer.
4. A keypad that allows the technician to enter data and reply to tool messages.

Typical keys that may be on a scan tool include the following:

1. Numbered keys covering digits 0 through 9.
2. Horizontal or vertical arrow keys that allow the technician to move backward and forward through test modes and menus.
3. ENTER keys to enter information into the tester.
4. PAGE BACK key that allows the technician to interrupt the current procedure and go back to the previous modes.
5. MORE key that allows the technician to obtain additional diagnostic information from the scan tool software.
6. YES and NO keys to allow the technician to select or reject specific procedures.

 Caution
Always keep scan tool leads away from rotating parts such as belts and fan blades. Personal injury or property damage may result if scan tool leads become tangled in rotating parts.

 WARNING **The scan tool and the vehicle manufacturer's recommended diagnostic procedures must be followed while diagnosing computer systems. Improper test procedures may result in scan tool and computer system damage.**

Figure 2-38 The type A DLC (left) is used on 12-volt supply systems while type B (right) is used on 24-volt supply systems.

The advantages of using a scan tool include:

1. Quick access to data from various on-board computers. Accessing this computer data significantly reduces diagnostic time.
2. The ability to be used on several different vehicle manufacturer products.
3. Some have the ability to access service bulletin information.
4. The ability for the technician to observe computer data during various operating conditions when a specific problem may occur. Most scan tools have a snapshot capability that freezes computer data into the scan tool memory for a specific period of time. This data may be played back after the technician returns to the shop.
5. Most can be connected to a printer, and a copy of the scan tool data can be printed. This can help improve communications between the customer, service writer, and the technician.
6. Can be connected to a personal computer (PC). This connection allows data to be transferred from the scan tool to the PC. This data may be saved and recalled at a future time. This information may be transferred to an off-site diagnostic center for analysis.

Static Strap

You are probably familiar with static charges in one form or another. The most common experience with static electricity is when you slide your feet across a carpeted floor and then touch something. You might feel and see a slight static discharge. The action of sliding your feet across the carpet placed a slight electrical charge on you. A change in the number of electrons on you puts you at a different charge level than the objects around you. When you touch one of the objects, there is a discharge between you and the object. Although this discharge generally does nothing to you, other than perhaps surprise you, it can potentially damage electronic circuitry. You need to discharge the static electricity before you begin working on an electronic component or processor.

To effectively work on computer circuits, some precautions are necessary. Generally, these can be summarized by the statement that you must be at the same electrical potential as the component you are working on and the vehicle you are working in. Many manufacturers suggest using a static strap that connects you, the component, and the ground system of the vehicle together (**Figure 2-39**). The theory behind this is to place all things

Wrist strap

Work surface mat

Ground cord assembly

Alligator clip to body ground

Figure 2-39 Static straps are needed to protect electronic devices.

that will touch at the same electrical potential so that a discharge will not occur. Even if you do not have special static straps, run jumper ground wires between the components and the vehicle, and ground yourself to the vehicle before you begin working.

Memory Keepers

Memory keepers, or battery maintainers, supply voltage to the vehicle's computers so they retain the values stored in their memory chips when the battery is disconnected. Radio presets, clock settings, memory seat positions, and computer functions can be lost when the battery is disconnected. A simple memory keeper (**Figure 2-40**) plugs into the power outlet socket, and a 9-volt battery will keep the settings for up to 4 hours. These devices should not be used if service is being performed to the air bag system.

⚠ **Caution**
Reduce electrical load before connecting the memory keeper. Close all doors and make sure the ignition is in the OFF position.

Service Information

With today's complex electrical systems, it would be impossible to repair every customer concern brought into the service bay without having the correct service information. The service information (in either paper form or electronic format) is one of the most important and valuable tools for today's technician. It provides information concerning system description, service procedures, specifications, diagnostic information, and testing procedures. In addition, the service information provides information concerning wiring harness connections and routing, component location, and fluid capacities. Service information can be acquired from the vehicle manufacturer or through aftermarket suppliers.

It's also important to stay current with updated service information. This is usually published as Technical Service Bulletins (TSBs). These documents may provide information concerning fixes for a problem, new part numbers to replace a defective unit, corrections to service information, and general system operation information.

To obtain the correct information, you must be able to identify the vehicle you are working on. This may involve using the **vehicle identification number (VIN)**. The VIN has codes for model year and engine. Which numbers are used varies between manufacturers, but the service information will provide instructions for proper VIN decoding.

Figure 2-40 A memory keeper will keep the computer memories operational while the battery is disconnected.

Procedural information provides the steps necessary to perform the task. Most service manuals provide illustrations to guide the technician through the task. To get the most out of the service manual, you must use the correct manual for the vehicle and system being worked on and follow each step in order. Some technicians lead themselves down the wrong trail by making assumptions and skipping steps.

Torque, end play, and clearance specifications may be located within the text of the procedural information. In addition, specifications may be provided in a series of tables. The heading above the table provides a quick reference to the specification information provided.

Diagnostic procedures are often presented in a chart form (**Figure 2-41**) or a tree. The procedure guides you through the process as system tests are performed. The result of a test then directs you to different steps. Keep following the steps until the problem is isolated.

A1 CHECK THE DTCs FROM BOTH THE CONTINUOUS AND ON-DEMAND PAM SELF-TESTS

Check the PAM DTCs from the continuous and on-demand self-tests.

Are any PAM DTCs recorded?

Yes : REFER to DTC CHARTS .

No : GO to A2.

A2 CHECK THE MESSAGE CENTER FOR CORRECT OPERATION

While observing the message center, disable and enable the parking aid system.

Does the message center display REAR PARK AID OFF when the parking aid system is disabled, and display REAR PARK AID ON when the parking aid system is enabled?

Yes : GO to A3.

No : REFER to INFORMATION CENTER - INSTRUMENTATION, MESSAGE CENTER, AND WARNING CHIMES .

A3 CHECK FOR CORRECT REVERSE GEAR INPUT

Apply the parking brake.

Ignition ON.

Select REVERSE.

Enter the following diagnostic mode on the scan tool: PAM DataLogger.

Monitor the PAM TRANSGR PID.

Does the PID read REVERSE (transmission in reverse)?

Yes : GO to A4.

No : GO to A5.

A4 CHECK THE PARKING AID SENSORS FOR CORRECT ALIGNMENT

Figure 2-41 A diagnostic chart leads the technician to the cause of the fault.

Figure 2-42 Computerized service information and web-based materials retrieval systems have replaced paper service information.

Since the service manual is divided into several major component areas, a table of contents is provided for easy access to the information. Each component area of the vehicle is covered under a section in the service information. Using the table of contents identifies the section to turn to. Once in the appropriate section, a smaller, more specific table of contents will direct you to the information.

Today, most service and parts information is provided through computer services (**Figure 2-42**). The computer system helps the technician find the required information quicker and easier than in a book-type manual. Using the computer keyboard, light pen, touch-sensitive screen, or mouse, the technician makes choices from a series of menus on the monitor. Computer systems are popular since libraries of printed service information can take large areas of space. If needed, the information can be printed to paper.

Another tool that can be of assistance to the technician is hotline services. There are many companies and organizations that provide online assistance to technicians. The hotline assistants use database information and factory service manuals, TSBs, and other sources to help the technician repair the vehicle.

Also, some companies and organizations provide a service of tracking the service history of vehicles. Manufacturers will generally do this for any of their vehicles that are serviced in their dealerships. Subscription services are available that will provide similar information to aftermarket service centers. This information lets the technician know if service has been performed on the vehicle that may be related to the present problem. Also, some Web site organizations will track if the vehicle has ever been involved in an accident, stolen, or caught in a flood.

Working as an Electrical Systems Technician

To be a successful automotive technician, you need to have good training, a desire to succeed, and a commitment to becoming a good technician and a good employee. A good employee works well with others and strives to make the business successful. The required training is not just in the automotive field. Good technicians need to have good reading, writing, and math skills. These skills will allow you to better understand and use the material found in service information and textbooks and provide you with the basics for good communication with customers and others.

Compensation

Technicians are typically paid according to their abilities. Most often, new or apprentice technicians are paid by the hour. While being paid, they are learning the trade and the business. Time is usually spent working with master technicians or doing low-skill jobs. As apprentices learn more, they can earn more and take on more complex jobs. Once technicians have demonstrated a satisfactory level of skills, they can go on **flat rate**.

Flat rate is a pay system in which technicians are paid for the amount of work they do. Each job has a flat rate time. Pay is based on that time, regardless of how long it took to complete the job. To explain how this system works, let's look at a technician who is paid $30.00 per flat rate hour. If a job has a flat rate time of 3 hours, the technician will be paid $90.00 for the job, regardless of how long it takes to complete the job. Experienced technicians beat the flat rate time most of the time. Their weekly pay is based on the time "turned in," not on the time spent. If the technician turns in 60 hours of work in a 40-hour workweek, they actually earned $45.00 each hour worked. However, if they turned in only 30 hours in the 40-hour week, the hourly pay is $22.50.

The flat rate system favors good technicians who work in a shop that has a large volume of work. Flat rate times make it easier to provide customers with accurate repair estimates. It also rewards skilled and productive technicians.

Workplace Relationships

You enter into a business agreement with your employer when you begin a job. As an employee, you are selling your time, skills, and efforts. In return, your employer pays you for these resources. As part of the employment agreement, your employer also has responsibilities to you:

- Instruction and Supervision—You should be told what is expected of you. A supervisor should observe your work, tell you if it's satisfactory, and offer ways to improve your performance.
- A Clean, Safe Place to Work—An employer should provide a clean and safe work area and a place for personal cleanup.
- Wages—You should know how much you are to be paid, what your pay will be based on, and when you will be paid before accepting a job.
- Fringe Benefits—When hired, you should be told what benefits to expect, such as paid vacations and employer contributions to health insurance and retirement plans.
- Opportunity—You should be given a chance to succeed and possibly advance within the company.
- Fair Treatment—All employees should be treated equally, without prejudice or favoritism.

On the other side of this business transaction, employees have responsibilities to their employers. Your obligations as an employee to the employer include the following:

- Regular Attendance—A good employee is reliable.
- Following Directions—As an employee, you are part of a team, and doing things your way may not serve the company's best interests.
- Responsibility—You must be willing to answer for your behavior and work. You need to realize that you are legally responsible for your work.
- Productivity—Remember, you are paid for your time, skills, and effort.
- Loyalty—Loyalty is expected. Being loyal means you will act in your employer's best interests, both on and off the job.

Customer Relations

Another responsibility you have is good customer relations. Learn to listen and communicate clearly. Be polite and organized, particularly when dealing with customers. Always be as honest as you possibly can.

Look like and present yourself as a professional, which is what automotive technicians are. Professionals are proud of what they do, and they show it. Always dress and act appropriately and watch your language, even when you think no one is nearby.

Respect the vehicles you work on. They are important to the lives of your customers. Always return the vehicle to the owner in a clean, undamaged condition. Remember, a vehicle is one of the largest expenses a customer has. Treat it that way. It doesn't matter if you like the car. It belongs to the customer; treat it respectfully.

Explain the repair process to the customer in understandable terms. Whenever you are explaining something to a customer, make sure you do this in a simple way without making the customer feel stupid. Always show the customer respect and be courteous. Not only is this the right thing to do, but it also leads to loyal customers.

Communicating with the Customer

Depending on the size of the service center that you are working in, you may or may not talk directly with the customer. A service advisor or manager might be in between the consumer and you, the technician. In either case, getting the correct information from the customer cannot be overemphasized. The customer is likely the person who was driving the vehicle when the problem showed up. The conversation that someone has with the customer can be extremely useful and save hours of fruitless work. A repair ticket that states "drivability problem" will require the technician to figure out the driving conditions that were present when the problem occurred. It's possible that the technician will not be able to duplicate the conditions and not observe the problem. The vehicle is returned to the consumer with the note of "no problem found." Think about how frustrating this could be, especially if the customer experiences the problem the very next day. The conversation with the customer should have revealed important information for the technician. When did the problem occur? What specifically did the vehicle do or not do? What was the outside temperature? Was the engine warm or cold? What driving conditions produced the problem? Can you duplicate the problem with the technician in the vehicle?

Think about how much more information the technician has in the second example. Less time will likely be necessary to fix the vehicle because the technician has a starting point. Once you get out in the field, try to develop your communication skills and especially your ability to listen to the customer. By really listening to your customer, you can save yourself hours of frustration and increase your chances of a successful repair. Better repairs bring customers back to the service center the next time they need service.

Completing a Work Order

The **work order** is a contract between the customer and the shop to perform specified services. In larger shops, a service writer is employed to interact with the customer and begin the work order process. In smaller shops, the technician may be required to perform the duties of a service writer. Regardless of who is responsible for the write-up, a work order must be written for every vehicle brought to the shop for repairs.

The work order must contain certain information (**Figure 2-43**). This includes information about the customer, the customer's vehicle, and the customer's concern or service request. In addition, estimates of the cost to perform the service are included on the work order and the estimated date and time the work will be completed.

AUTHOR'S NOTE The work order is also referred to as a Repair Order.

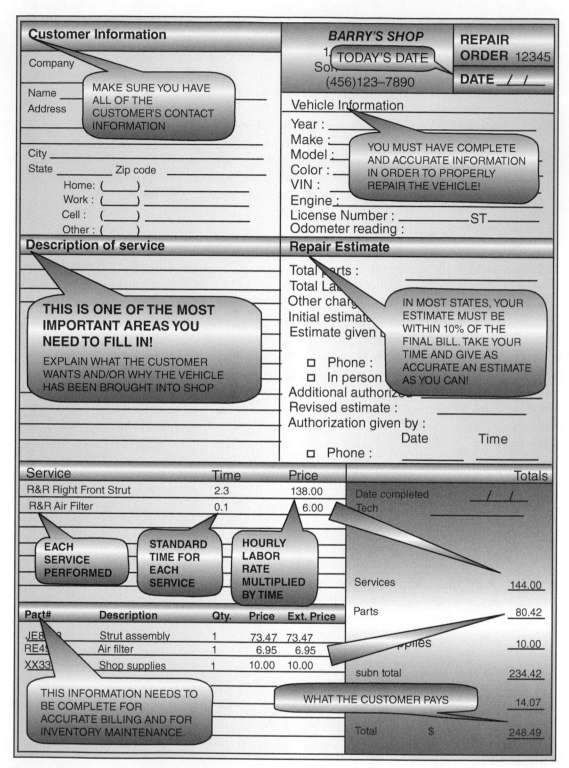

Figure 2-43 The work order must be completed properly.

The work order is a legal document that is an agreement to provide services and legally protects both the customer and the shop. The work order is signed by the customer who authorizes the shop to perform the service and accepts all terms that are noted on the work order. Also, the customer is protected against being charged more than the estimate given by the shop. If the costs exceed the original estimate, the customer must authorize the higher amount. Some states allow shops to be within 10% of the estimate, while others hold the shop to the estimated amount. Other functions of the work order are for use by payroll and general record keeping.

Today, most service facilities use a form of shop management software. The information to complete the work order is inputted using the computer's keyboard. The software also takes information from the work order and saves it in various files, each defined by its purpose. The software package also helps in the estimation of repair costs.

One of the responsibilities of the technician is to document the cause and correction of the concern. It may not be possible to provide an estimate to the customer at the time of the original write-up since the problem has not been diagnosed yet. In this case, the customer agrees to pay the shop to perform the diagnostic routine. Often a limit is agreed upon by the parties. Once the problem has been identified, and the root cause is determined, the cost of the repair must be estimated, and the customer gives approval. The cause of the complaint needs to be documented in such a manner that the customer and service writer understand what the issue is. The correction actions are documented for purposes of warranty. Whatever the repair entailed is to be recorded. For example, if a wire was repaired, document this repair and the location of the repair. If parts are installed, list the parts.

ASE Certification

An obvious indicator of your knowledge and abilities, as well as your dedication to the trade, is ASE certification. The National Institute for Automotive Service Excellence (ASE) has established a voluntary certification program for automotive, heavy-duty truck, collision repair technicians. In addition to these programs, ASE also offers individual testing in many specialty areas. Currently there are 57 ASE certification tests.

ASE certification combines voluntary testing with on-the-job experience to confirm that technicians have the skills needed to work on today's vehicles (**Figure 2-44**). ASE recognizes two distinct levels of service capability—the automotive technician and the master automotive technician.

After passing at least one exam and providing proof of two years of hands-on work experience, you become ASE certified. A technician who passes one examination receives an automotive technician shoulder patch. The Master Automotive Technician patch is awarded to technicians who pass all eight of the basic automotive certification exams. To

Figure 2-44 ASE certification communicates your pride in your profession to the customers.

keep your Master Technician certification, you must renew each of the certifications every five years. Recertification requires you take the recertification test for the expiring certifications. The recertification tests have about half of the question as the certification test.

You may receive credit for one of the two years of work experience by substituting relevant formal training in one of the following:

- Complete two years of an ASE accredited program at a high school, trade school, technical institute, or community college.
- Short courses. For shorter courses of training, you may substitute two months of training for one month of work experience.
- Apprenticeship programs. You may receive full credit for the experience requirement by satisfactorily completing a three- or four-year ASE accredited OEM Sponsored Co-op program (apprenticeship).
- Passing three of more ASE Entry Level Certifications Program Tests may substitute for up to one year of work experience.

Each certification test consists of 40 to 80 multiple-choice questions. A panel of technical service experts write the questions, including domestic and import vehicle manufacturers, repair and test equipment and parts manufacturers, working automotive technicians, and automotive instructors. All questions are pretested, and quality checked on a national sample of technicians before they are included in the actual test. Many test questions force the test taker to choose between two different repair or diagnostic methods. The test questions focus on basic technical knowledge, repair knowledge and skill, and testing and diagnostic knowledge and skill.

Diagnostics

The true measure of a good technician is their ability to find and correct the cause of problems. Service manuals and other information sources will guide you through the diagnosis and repair procedures, but these guidelines will not always lead you to the exact cause of the problem. To do this, you must use your knowledge and take a logical approach to troubleshooting. Diagnosis is not guessing, and it's more than following a series of interrelated steps in order to find the solution to a specific problem. Diagnosis is a way of looking at systems that are not functioning as they should and finding out why. It's knowing how the system should work and deciding if it's working correctly. By understanding the purpose and operation of the system, you can accurately diagnose problems.

Most good technicians use the same basic diagnostic approach. Simply because this is a logical approach, it can quickly lead to the cause of a problem. Logical diagnosis follows these steps:

1. Gather information about the problem.
2. Verify that the problem exists.
3. Thoroughly define what the problem is and when it occurs.
4. Determine if there are any other related issues.
5. Research all available information to determine the possible causes of the problem.
6. Isolate the problem by testing.
7. Repair the problem.
8. Verify the repair.

It's often best to reference the wiring diagrams to get the whole picture when diagnosing electrical system issues. Often, electrical systems are interconnected together. By recognizing this interconnection, you can determine what portions of the circuit are working. By eliminating what is working, you are left with what is not.

Summary

- A test light is used when the technician needs to "look" for electrical power in the circuit. The test light allows the technician to see if current is at a point in the circuit by lighting the light.
- A logic probe provides a means for testing voltages on electronic circuits without damaging the circuit.
- Digital multimeters (DMM) display values using liquid crystal displays instead of a swinging needle. They are computers that determine the measured value and display it for the technician.
- A voltmeter measures the voltage potential between two points in a circuit.
- An ohmmeter is used to measure the resistance of a circuit or part of a circuit.
- An ammeter is a special meter used to measure current flow in a circuit.
- Electrical noise is an unwanted voltage signal that rides on a signal. Noise is usually the result of radio frequency interference (RFI) or electromagnetic induction (EMI).
- Duty cycle is the percentage of on time the circuit component is turned on compared to the total time of the cycle.
- A cycle is one set of changes in a signal that repeats itself several times.
- Pulse width is the amount of time, measured in milliseconds, that a component is turned on.
- Frequency is a measure of the number of cycles that occur in one second.
- Hertz is the measurement of frequency.
- A lab scope provides a visual display of electrical waves.
- Glitches may result from momentary shorts to ground, shorts to power, or opens in the circuit.
- A sine wave is a waveform that shows voltage-changing polarity from positive to negative.
- Having straight vertical sides and a flat top indicating a fast-acting on-off voltage state identifies a square wave pattern.
- Scan tools interface with the vehicle's computer system to allow the technician to "talk" with the computers.
- The service manual (in either paper form or electronic format) is one of today's technician's most important and valuable tools.
- Technical Service Bulletins (TSBs) may provide information concerning fixes for a problem, new part numbers to replace a defective unit, corrections to service manual information, and general system operation information.
- Service manual procedural information provides the steps necessary to perform the task.
- Since the service manual is divided into several major component areas, a table of contents is provided for easy access to the information.
- Service and parts information can also be provided through computer services.
- Technicians are typically paid according to their abilities. New or apprentice technicians are paid by the hour. Once technicians have demonstrated a satisfactory level of skills, they can go on flat rate.
- You enter into a business agreement with your employer when you begin a job. You sell your time, skills, and efforts when you become an employee. In return, your employer pays you for these resources.
- As part of the employment agreement, your employer also has responsibilities to you: instruction and supervision; a clean, safe place to work; wages; fringe benefits; opportunity; and fair treatment.
- As an employee, your obligations to your employer include regular attendance, following directions, responsibility, productivity, and loyalty.
- Be polite, respectful, organized, and honest when communicating with customers.
- An obvious sign of your knowledge and abilities, as well as your dedication to the trade, is ASE certification.
- The work order is a contract between the customer and the shop to perform specified services.
- The true measure of a good technician is an ability to find and correct the cause of problems.
- Diagnosis is not guessing, and it's more than following a series of interrelated steps in order to find the solution to a specific problem.

ASE-Style Review Questions

1. *Technician A* says a test light is ideal for checking for voltage on a low-current, low-power circuit.
   Technician B says to use an analog multimeter to test computer-controlled circuits.
   Who is correct?
   A. A only
   B. B only
   C. Both A and B
   D. Neither A nor B

2. The use of an ammeter is being discussed.
   Technician A says the ammeter is used to measure current flow.
   Technician B says the ammeter must be connected in parallel to the circuit being tested.
   Who is correct?
   A. A only
   B. B only
   C. Both A and B
   D. Neither A nor B

3. A DMM is being used to measure current flow. The meter is displaying 85.5 mA.
   Technician A says this represents 0.0855 amperes.
   Technician B says the decimal point needs to be moved six points to the left.
   Who is correct?
   A. A only
   B. B only
   C. Both A and B
   D. Neither A nor B

4. *Technician A* says a 10% duty cycle indicates that the load device is turned on most of the time.
   Technician B says the pulse width is measured in degrees.
   Who is correct?
   A. A only
   B. B only
   C. Both A and B
   D. Neither A nor B

5. A vehicle is being tested for a draw against the battery with the ignition switch in the OFF position. The specifications state the draw should be between 10 and 30 milliamps. The DMM reads 0.251 amps.
   Technician A says this draw is within the specification range.
   Technician B says the draw is too high.
   Who is correct?
   A. A only
   B. B only
   C. Both A and B
   D. Neither A nor B

6. *Technician A* says an ohmmeter that reads 0 ohms means there is no continuity of the circuit or component being tested.
   Technician B says an infinite reading on the ohmmeter means there is good continuity.
   Who is correct?
   A. A only
   B. B only
   C. Both A and B
   D. Neither A nor B

7. *Technician A* says a voltmeter connected in parallel to the load device will indicate the voltage drop across the device.
   Technician B says an ohmmeter reading of 0.00 Ω when connected in parallel to the coil of an A/C compressor indicates the coil is shorted.
   Who is correct?
   A. A only
   B. B only
   C. Both A and B
   D. Neither A nor B

8. *Technician A* says the upward voltage traces on a lab scope screen indicates a specific length of time.
   Technician B says the lab scope provides accurate voltage over time readings.
   Who is correct?
   A. A only
   B. B only
   C. Both A and B
   D. Neither A nor B

9. Testing for a faulty switch is being discussed.
   Technician A says to use a jumper wire to bypass the switch and connect the various circuits controlled by the switch.
   Technician B says to use an ammeter to test continuity through the switch.
   Who is correct?
   A. A only
   B. B only
   C. Both A and B
   D. Neither A nor B

10. *Technician A* says the service manual (in either paper form or electronic format) is one of the most important and valuable tools for today's technician.
    Technician B says Technical Service Bulletins (TSBs) provide information concerning fixes for a problem, new part numbers to replace a defective unit, corrections to service manual information, and general system operation information.
    Who is correct?
    A. A only
    B. B only
    C. Both A and B
    D. Neither A nor B

ASE Challenge Questions

1. *Technician A* says a voltmeter measures the electrical potential between two points of the circuit.
   Technician B says an ammeter reading of 0.00 when connected in series to a circuit indicates a short to ground.
   Who is correct?
   A. A only
   B. B only
   C. Both A and B
   D. Neither A nor B

2. A scan tool allows a technician to:
   A. measure circuit resistance.
   B. determine load voltage drop.
   C. view computer inputs.
   D. load test the battery.

3. All of the following are true concerning the use of an ohmmeter EXCEPT:
   A. Connect the ohmmeter with the circuit power off.
   B. An "OL" indicates that the reading is over limits.
   C. 0.00 indicates no resistance.
   D. Connect the test leads of the meter in series to the load.

4. The signal output of a 5-volt throttle position sensor is being checked (from idle to the wide-open throttle position) with a lab scope. The test lead is set to the 10 × position. Which of the following represents the correct position of the vertical adjustment selector?
   A. 0.1 volt
   B. 0.5 volt
   C. 2 volts
   D. 5 volts

5. *Technician A* says logic probes can be very helpful in testing analog signals.
   Technician B says the nonpowered test lights should not be used to test most computer circuits.
   Who is correct?
   A. A only
   B. B only
   C. Both A and B
   D. Neither A nor B

6. The insulation test function is being discussed.
   Technician A says the test is performed with the circuit powered.
   Technician B says the meter displays the circuit's voltage.
   Who is correct?
   A. A only
   B. B only
   C. Both A and B
   D. Neither A nor B

7. All of the following concerning the use of jumper wires are true EXCEPT:
   A. The jumper wire should be fused.
   B. The jumper wire can be used to bypass a switch.
   C. The jumper wire can be used to bypass the load device.
   D. The jumper wire can be used to supply an alternate ground circuit.

8. When a connector is unplugged, and applied voltage is measured, the reading is battery voltage.
   Technician A says this means the circuit is good.
   Technician B says an open circuit voltage test does not indicate if the circuit has resistance.
   Who is correct?
   A. A only
   B. B only
   C. Both A and B
   D. Neither A nor B

9. While using a test light to check for voltage, the light comes on bright.
   Technician A says this indicates that the circuit can handle the load that is required by the test light.
   Technician B says the amperage to turn the light on depends on the bulb's type and rating.
   Who is correct?
   A. A only
   B. B only
   C. Both A and B
   D. Neither A nor B

10. Which of the following statement is most correct?
    A. Voltage drop testing tests the circuit under load.
    B. An ammeter must be connected in parallel to the load.
    C. An ohmmeter must be connected in series to a circuit.
    D. An insulation meter is used to measure high amperage.

Name _____   Date _____

USING OHM'S LAW TO CALCULATE ELECTRICAL PROPERTIES

Objective: Upon completion of this job sheet, you should be able to calculate circuit values using Ohm's law.

ASE Education Foundation Correlation ─────────────────────

This job sheet addresses the following **MLR/AST/MAST** task:

A.4.   Demonstrate knowledge of electrical/electronic series, parallel, and series-parallel circuits using principles of electricity (Ohm's Law). **(P-1)**

We Support
|Education Foundation

Tools and Materials

• A calculator

Procedure

Using Ohm's law, solve the following problems:

Exercise 1—Series Circuit

Refer to the following circuit.

Use Ohm's law to calculate the following values, when $R_1 = 2$ ohms and $R_2 = 4$ ohms

    Total circuit resistance = _____ ohms

    Circuit current = _____ amps

    Current through R_1 = _____ amps

    Current through R_2 = _____ amps

    Voltage drop across R_1 = _____ volts

    Voltage drop across R_2 = _____ volts

If the resistance of R_1 increases to 8 ohms, what are the new values?

    Total circuit resistance = _____ ohms

    Circuit current = _____ amps

    Current through R_1 = _____ amps

    Current through R_2 = _____ amps

    Voltage drop across R_1 = _____ volts

    Voltage drop across R_2 = _____ volts

Exercise 2—Series Circuit

Refer to the following circuit.

Use Ohm's law to calculate the following values, when $R_1 = 3$ ohms and $R_2 = 6$ ohms:

Total circuit resistance = _____ ohms

Circuit current = _____ amps

Current through R_1 = _____ amps

Current through R_2 = _____ amps

Voltage drop across R_1 = _____ volts

Voltage drop across R_2 = _____ volts

12 V

Exercise 3—Parallel Circuit

Refer to the following circuit.

Use Ohm's law to calculate the following values, when $R_1 = 12$ ohms and $R_2 = 12$ ohms:

Total circuit resistance = _____ ohms

Circuit current = _____ amps

Current through R_1 = _____ amps

Current through R_2 = _____ amps

Voltage drop across R_1 = _____ volts

Voltage drop across R_2 = _____ volts

12 V

Exercise 4—Parallel-Series Circuit

Refer to the following circuit.

Use Ohm's law to calculate the following values, when $R_1 = 1$ ohm, $R_2 = 3$ ohms, $R_3 = 2$ ohms, and $R_4 = 2$ ohms:

Total circuit resistance = _____ ohms

Circuit current = _____ amps

Current through R_1 = _____ amps

Current through R_2 = _____ amps

Current through R_3 = _____ amps

Current through R_4 = _____ amps

Voltage drop across R_1 = _____ volts

Voltage drop across R_2 = _____ volts

Voltage drop across R_3 = _____ volts

Voltage drop across R_4 = _____ volts

Exercise 5—Parallel-Series Circuit

Refer to the following circuit.

Use Ohm's law to calculate the following values, when R_1 = 1 ohm, R_2 = 3 ohms, R_3 = 2 ohms, and R_4 = 10 ohms:

Total circuit resistance = _____ ohms

Circuit current = _____ amps

Current through R_1 = _____ amps

Current through R_2 = _____ amps

Current through R_3 = _____ amps

Current through R_4 = _____ amps

Voltage drop across R_1 = _____ volts

Voltage drop across R_2 = _____ volts

Voltage drop across R_3 = _____ volts

Voltage drop across R_4 = _____ volts

Exercise 6—Parallel Circuit

Refer to the following circuit.

Use Ohm's law to calculate the following values, when $R_1 = 2$ ohms, $R_2 = 3$ ohms, and $R_3 = 6$ ohms:

Total circuit resistance = _____ ohms

Circuit current = _____ amps

Current through R_1 = _____ amps

Current through R_2 = _____ amps

Current through R_2 = _____ amps

Voltage drop across R_1 = _____ volts

Voltage drop across R_2 = _____ volts

Voltage drop across R_3 = _____ volts

Exercise 7—Series-Parallel Circuit

Refer to the following circuit.

Use Ohm's law to calculate the following values, when $R_1 = 1$ ohm, $R_2 = 2$ ohms, $R_3 = 3$ ohms, and $R_4 = 6$ ohms:

Total circuit resistance = _____ ohms

Circuit current = _____ amps

Current through R_1 = _____ amps

Current through R_2 = _____ amps

Current through R_3 = _____ amps

Current through R_4 = _____ amps

Voltage drop across R_1 = _____ volts

Voltage drop across R_2 = _____ volts

Voltage drop across R_3 = _____ volts

Voltage drop across R_4 = _____ volts

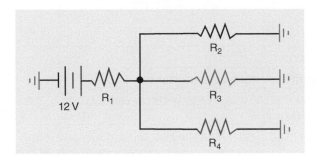

Exercise 8—Series-Parallel Circuit

Refer to the following circuit that has resistance on the ground side.

Use Ohm's law to calculate the following values, when R_1 = 6 ohms, R_2 = 3 ohms, and R_3 = 2 ohms:

Total circuit resistance = _____ ohms

Circuit current = _____ amps

Current through R_1 = _____ amps

Current through R_2 = _____ amps

Current through R_3 = _____ amps

Voltage drop across R_1 = _____ volts

Voltage drop across R_2 = _____ volts

Voltage drop across R_3 = _____ volts

Instructor's Evaluation:

Grade	Rubric
0	No exposure due to: ☐ Absence ☐ Other:
1	Student unsuccessfully attempted the task
2	Student completed the task with significant instructor assistance.
3	Satisfactory performance of task. The student was able to perform elements of the task with minimal instructor assistance.
4	Competent/Proficient. Able to perform all elements of the task successfully and independently.

Instructor's Notes:

Name _____   Date _____

METER SYMBOL INTERPRETATION

Objective: Upon completion of this job sheet, you should be able to convert commonly found symbols into numeric values.

ASE Education Foundation Correlation ────────────────────────────────

This job sheet addresses the following **MLR/AST/MAST** task:

A.5.   Demonstrate proper use of a digital multimeter (DMM) when measuring source voltage, voltage drop (including grounds), current flow, and resistance. **(P-1)**

We Support

 Education Foundation

Procedure

Convert the following values into the electrical units noted:

1. 2.4 KΩ = _____ Ω

2. 954 mV = _____ V

3. 5. 76 Ω = _____ Ω

4. 2 mA = _____ A

5. 22 KΩ = _____ Ω

6. 4.5 mA = _____ A

7. 456 mA = _____ A

8. 1,024 mV = _____ V

9. 0.786 KΩ = _____ Ω

10. 32 KΩ + 112Ω = _____ Ω

11. 1,400Ω = _____ KΩ

12. 0.000235 A = _____ mA

13. 0.987 V = _____ mV

14. 5 KV = _____ mV

15. 123,955 Ω = _____ KΩ

16. 144,000 mA = _____ A

17. 126 mV + 11.874 V = _____ V

18. 320,000 Ω = _____ Ω

19. 0.000045 A = _____ mA

20. 12,600 mV = _____ V

Instructor's Evaluation:

Grade	Rubric
0	No exposure due to: ☐ Absence ☐ Other:
1	Student unsuccessfully attempted the task
2	Student completed the task with significant instructor assistance.
3	Satisfactory performance of task. The student was able to perform elements of the task with minimal instructor assistance.
4	Competent/Proficient. Able to perform all elements of the task successfully and independently.

Instructor's Notes:

Name _____   Date _____

USING A TEST LIGHT

Objective: Upon completion of this job sheet, you should be able to properly use a test light to check continuity in electrical circuits.

ASE Education Foundation Correlation

This job sheet addresses the following **MLR** tasks:

A.1.   Research vehicle service information such as fluid type, vehicle service history, service precautions, technical service bulletins, and recalls including vehicles with advanced driver assistance systems (ADAS). **(P-1)**

A.7.   Describe types of test lights; use appropriate test light to check operation of electrical circuits per service information. **(P-2)**

A.9.   Use wiring diagrams to trace electrical/electronic circuits. **(P-1)**

This job sheet addresses the following **AST/MAST** tasks:

A.7.   Describe types of test lights; use appropriate test light to check operation of electrical circuits per service information. **(P-1)**

A.9.   Use wiring diagrams during the diagnosis (troubleshooting) of electrical/electronic circuit problems. **(P-1)**

We Support

ASE | Education Foundation

Tools and Materials

- A vehicle
- A test light
- Wiring diagrams for the vehicle
- Fender covers

Safety Considerations and PPE

- Safety glasses

Procedure

1. Using the proper wiring diagram and your instructor's assistance, determine which wire is the positive feed and which wire is the ground for one of the low-beam headlights on your assigned vehicle. Identify the color of each wire:

   Positive _____ Negative _____

2. Test proper operation of the test light by connecting the negative lead to the negative post of the battery and touching the probe tip to the positive lead of the battery. If the test light is working properly, the light should illuminate. If the light does not turn on, inform your instructor. Did the light work properly?   ☐ Yes   ☐ No

3. With the headlight switch in the OFF position, disconnect the headlight connector.

4. Connect the negative lead of the test light to a good ground. Locate the probe of the test light into the connector cavity identified as positive feed. Did the test light come on?   ☐ Yes   ☐ No

   Explain your results: _____

   _____

   _____

5. Turn the headlight switch into the ON position and retest the circuit again. Did the test light come on?   ☐ Yes ☐ No

6. Turn the headlight switch OFF.

7. Move the wire clip of the test light to the positive post of the battery and locate the probe into the cavity identified as the ground. Does the test light come on?   ☐ Yes ☐ No

   Explain the reason: _____

   _____

   _____

Instructor's Evaluation:

Grade	Rubric
0	No exposure due to: ☐ Absence ☐ Other:
1	Student unsuccessfully attempted the task
2	Student completed the task with significant instructor assistance.
3	Satisfactory performance of task. The student was able to perform elements of the task with minimal instructor assistance.
4	Competent/Proficient. Able to perform all elements of the task successfully and independently.

Instructor's Notes:

Name _____   Date _____

USE OF A VOLTMETER

Objective: Upon completion of this job sheet, you should be able to measure available voltage and voltage drop.

ASE Education Foundation Correlation ⎯⎯⎯⎯⎯⎯⎯⎯⎯⎯⎯⎯⎯⎯⎯⎯⎯⎯⎯⎯⎯

This job sheet addresses the following **MLR/AST/MAST** tasks:

A.1.   Research vehicle service information such as fluid type, vehicle service history, service precautions, technical service bulletins, and recalls including vehicles with advanced driver assistance systems (ADAS). **(P-1)**

A.2.   Identify electrical/electronic system components and configuration. **(P-1)**

A.5.   Demonstrate proper use of a digital multimeter (DMM) when measuring source voltage, voltage drop (including grounds), current flow, and resistance. **(P-1)**

A.9.   Use wiring diagrams to trace electrical/electronic circuits. **(P-1)**

We Support
Education Foundation

Tools and Materials

- A vehicle
- A DMM
- Fender covers
- Wiring diagram for the vehicle
- Basic hand tools

Safety Considerations and PPE

- Safety glasses

Procedure

1. Set the DMM to the appropriate scale to read 12 volts DC.

2. Connect the meter across the battery (positive to positive and negative to negative). What is your reading on the meter? _____ volts.

3. With the meter still connected across the battery, turn on the headlights of the vehicle. What is your reading on the meter? _____volts.

4. Keep the headlights on. Connect the positive lead of the meter to the point on the vehicle where the battery's ground cable attaches to the frame. Keep the negative lead where it is.

   What is your reading on the meter? _____volts

   What is being measured? _____

5. Disconnect the meter from the battery and turn off the headlights.

6. Refer to the correct wiring diagram and determine what wire at the right headlight delivers current to the lamp when the headlights are on, and low beams selected. Color of the wire.

   _____

7. From the wiring diagram, identify where the headlight is grounded.

   Ground location _____

8. Connect the negative lead of the meter to the point where the headlight is grounded.

9. Connect the positive lead of the meter to the power input of the headlight.

10. Turn on the headlights.

What is your reading on the meter? _____volts

What is being measured? _____

11. What is the difference between the reading here and the battery's voltage?
_____volts

12. Explain why there is a difference.

_____

_____

_____

_____

Instructor's Evaluation:

Grade	Rubric
0	No exposure due to: ☐ Absence ☐ Other:
1	Student unsuccessfully attempted the task
2	Student completed the task with significant instructor assistance.
3	Satisfactory performance of task. The student was able to perform elements of the task with minimal instructor assistance.
4	Competent/Proficient. Able to perform all elements of the task successfully and independently.

Instructor's Notes:

Name _____   Date _____

USE OF AN OHMMETER

Objective: Upon completion of this job sheet, you should be able to check the continuity of a circuit and measure resistance on a variety of components.

ASE Education Foundation Correlation

This job sheet addresses the following **MLR/AST/MAST** tasks:

A.1.   Research vehicle service information such as fluid type, vehicle service history, service precautions, technical service bulletins, and recalls including vehicles with advanced driver assistance systems (ADAS). **(P-1)**

A.2.   Identify electrical/electronic system components and configuration. **(P-1)**

A.5.   Demonstrate proper use of a digital multimeter (DMM) when measuring source voltage, voltage drop (including grounds), current flow, and resistance. **(P-1)**

A.9.   Use wiring diagrams to trace electrical/electronic circuits. **(P-1)**

We Support

ASE | **Education Foundation**

Tools and Materials

- A vehicle
- A DMM
- Wiring diagrams for the vehicle
- Fender covers

Safety Considerations and PPE

- Safety glasses

Note: An ohmmeter works by sending a small amount of current through the path to be measured. Because of this, all circuits and components being tested must be disconnected from power. An ohmmeter must never be connected to an energized circuit; doing so may damage the meter. The safest way to measure ohms is to disconnect the negative battery cable before taking resistance readings.

Procedure

1. Where is the fuse panel or power distribution box located?

   _____

2. With no power to the fuses, check the resistance of each fuse.

   Summarize your findings.

   _____

   _____

3. Disconnect the primary wires leading to the ignition coil.

   Now connect the leads of the digital meter across the terminals of the coil.

   Your reading is: _____ ohms

4. Reconnect the wires to the coil.

   Carefully remove one spark plug wire from the spark plug and ignition coil or distributor cap.

   Now connect the leads of the digital meter across the wire.

   Your reading is: _____ ohms

5. Carefully reinstall the spark plug wire.

   Locate the cigar lighter inside the vehicle.

   Now connect the leads of the digital meter from the heating coil to its case.

   Your reading is: _____ ohms

6. Refer to the service manual and find out how to remove the bulb in the dome light. Remove it.

   Now connect the leads of the digital meter across the bulb.

   Your reading is: _____ ohms

7. Reinstall the bulb.

   Remove the rear brake light bulb.

   Now connect the leads of the digital meter across the bulb.

   Your reading is: _____ ohms

8. Reinstall the bulb.

   Disconnect the wire connector to one of the headlights.

   From the wiring diagram, identify which terminals are for low-beam operation.

   Now connect the leads of the digital meter across the low-beam terminals.

   Your reading is: _____ ohms

9. From the wiring diagram, identify which terminals are for high-beam operation. Now connect the leads of the digital meter across the high-beam terminals.

   Your reading is: _____ ohms

10. You measured the resistance across several different light bulbs. On each, you should have read a different amount of resistance. Based on your findings, which light bulb would be the brightest and which would be the dimmest? Explain why.

   _____

   _____

   _____

Instructor's Evaluation:

Grade	Rubric
0	No exposure due to: ☐ Absence ☐ Other:
1	Student unsuccessfully attempted the task
2	Student completed the task with significant instructor assistance.
3	Satisfactory performance of task. The student was able to perform elements of the task with minimal instructor assistance.
4	Competent/Proficient. Able to perform all elements of the task successfully and independently.

Instructor's Notes:

Name _____   Date _____

LOCATING SERVICE INFORMATION

Objective: Upon completion of this job sheet, you should be able to locate the following service information; service procedures, technical service bulletins, and repair history.

ASE Education Foundation Correlation

This job sheet addresses the following **MLR/AST/MAST** task:

A.1.   Research vehicle service information, such as fluid type, vehicle service history, service precautions, technical service bulletins, and recalls including vehicles equipped with advanced driver assistance systems (ADAS). **(P-1)**

We Support
ASE | Education Foundation

Tools and Materials

• Applicable service information

Procedure

1. What system has your instructor assigned you?

_____

2. What procedure has your instructor assigned?

_____

3. Reference the service procedures assigned. Confirm proper selection with your instructor.

Instructor's notes: _____

4. Are there any precautions or warnings list in the procedure?

☐ Yes   ☐ No

If yes, what does the first one relate to?

_____

_____

5. Are there any TSBs associated with the system you are assigned?

☐ Yes   ☐ No

If yes, list them by TSB number:

_____

_____

6. Look up any Repair History associated with the vehicle.

Is there any Repair History?   ☐ Yes   ☐ No

If yes, what is the most recent repair?

_____

_____

_____

Instructor's Evaluation:

Grade	Rubric
0	No exposure due to: ☐ Absence ☐ Other:
1	Student unsuccessfully attempted the task
2	Student completed the task with significant instructor assistance.
3	Satisfactory performance of task. The student was able to perform elements of the task with minimal instructor assistance.
4	Competent/Proficient. Able to perform all elements of the task successfully and independently.

Instructor's Notes:

Name _____   Date _____

LAB SCOPE SETUP

Objective: Upon completion of this job sheet, you should be able to setup a lab scope to test an input sensor.

ASE Education Foundation Correlation

This job sheet addresses the following **AST** tasks:

A.9.   Use wiring diagrams to trace electrical/electronic circuits. **(P-1)**

A.13.   Test and measure circuit using an oscilloscope and/or graphing multimeter (GMM); interpret results; determine needed action. **(P-2)**

This job sheet addresses the following **MAST** tasks:

A.9.   Use wiring diagrams during the diagnosis (troubleshooting) of electrical/ electronic circuit problems. **(P-1)**

A.13.   Test and measure circuit using an oscilloscope and/or graphing multimeter (GMM); interpret results; determine needed action. **(P-1)**

We Support

ASE | **Education Foundation**

Tools and Materials

- A vehicle
- Lab scope
- Wiring diagram for the vehicle
- Lift or jacks with stands

Safety Considerations and PPE

- Safety glasses

Note: The intent of this job sheet is to acquaint you with basic setup procedures of the lab scope to obtain the proper waveform. The best reference is the lab scope manufacturer's User Guide. Also, use your instructor as a resource during this activity.

Procedure

1. What input speed sensor was assigned to you by your instructor?

   _____

2. Referring to the service information, identify the location of the input speed sensor connector. Where is it located?

   _____

   _____

3. Backprobe the signal circuit of the sensor connector.

4. Is the expected voltage trace going to be AC or DC?

   ☐ AC   ☐ DC

5. What is the expected voltage reading from the sensor?

   _____

6. What Volt/Div are you starting with?

   _____

7. What Time/Div are you starting with?

_____

8. What trigger are you going to use for the setup?

_____

_____

9. Connect the lab scope to the backprobe tools and observe the trace while the sensor is providing an input signal.
General results:

_____

_____

_____

10. Change the time and voltage levels to get the clearest waveform. At what setting is this accomplished?

_____

_____

_____

11. Capture the waveform and play it back. Does the waveform indicate any concerns?

_____

_____

_____

12. What is the highest voltage value of the waveform?

_____

_____

_____

Instructor's Evaluation:

Grade	Rubric
0	No exposure due to: ☐ Absence ☐ Other:
1	Student unsuccessfully attempted the task
2	Student completed the task with significant instructor assistance.
3	Satisfactory performance of task. The student was able to perform elements of the task with minimal instructor assistance.
4	Competent/Proficient. Able to perform all elements of the task successfully and independently.

Instructor's Notes:

_____

CHAPTER 3
BASIC ELECTRICAL TROUBLESHOOTING AND SERVICE

Upon completion and review of this chapter, you should be able to:

- Describe how different electrical problems affect the circuit.
- Diagnose and repair circuit protection devices.
- Test switches with a variety of test instruments and determine needed action.
- Test relays and relay circuits for proper operation and determine needed action.
- Test variable and fixed resistors and determine needed action.
- Test capacitors and determine needed action.
- Diagnose diodes for opens, shorts, and other defects.
- Diagnose transistors for opens, shorts, and other defects.
- Locate and repair opens in a circuit.
- Locate and repair shorts in a circuit.
- Locate and repair the cause of unwanted high resistance in a circuit.
- Set up a digital storage oscilloscope (DSO) and identify waveforms

⚒ **Basic Tools**

Basic mechanic's tool set

Service information

Terms to Know

Analog signal	Fusible link	Pulse width modulation (PWM)
Capacitor	Gauss gauge	Relay
Circuit	Normally open (NO)	Rheostat
Darlington	Open circuit	Shorted circuit
Digital signal	Overload	Short to ground
Diode	Peak	Stepped resistor
Feedback	Peak-to-peak voltage	Variable resistor
Fuse	Potentiometer	

Introduction

Troubleshooting electrical issues involve the same tools and methods, regardless of which **circuit** has the problem. All electrical circuits must have proper voltage, current, and resistance. Testing for the presence of these elements, measuring them, and comparing your measurements to specifications are the key to effective diagnosis. You must have a solid understanding of the basic electrical properties to do this.

Voltage is the electrical pressure that causes electrons to move, provided there is a complete path for them to do so. *Current* is the aggregate flow of electrons through a circuit and is defined as the rate of electron flow. *Resistance* is the opposition to current flow. An electrical circuit must have resistance to change electrical energy to light, heat, or movement.

The term **circuit** means a circle and is the path of electron flow consisting of the voltage source, conductors, load component, and return path to the voltage source.

An electrical circuit may develop an open, short, or excessive voltage drop that will cause it to operate improperly. An **open circuit** is a circuit that has is a break in continuity. The open can be on either the insulated side or the ground side. A **shorted circuit** decreases the resistance of the circuit. A short can be across to another circuit or to ground. When there is a circuit-to-circuit (coper-to-copper) short, one of the circuits is no longer controlled only by its switch. Since the shorted circuit becomes a new parallel leg, the entire parallel circuit will turn on and off, with the switch controlling the other circuit. With this type of problem, many strange things can happen. A short to ground also creates a new parallel leg. This new leg has very low resistance and causes the current in the circuit to increase drastically.

High-resistance problems can occur anywhere in the circuit. However, the effect of high resistance is the same regardless of where it is. Additional or unwanted resistance in series with a circuit will always reduce the current in the circuit. It also reduces the amount of voltage drop by the component in the circuit.

Testing for Circuit Defects

Electrical circuits may develop an open, a short, a ground, or an excessive voltage drop that will cause the circuit to operate improperly.

Testing for Opens

Classroom Manual
Chapter 3, page 79

It's possible to test for opens using a digital multimeter (DMM), a DSO, a test light, a self-powered test light, or a fused jumper wire. The test equipment used will depend on the circuit and the accessibility of the components.

You need to determine the correct operation of the circuit before attempting to determine what is wrong. **Figure 3-1** shows the voltmeter readings that a properly operating parallel circuit should have at different locations.

Special Tools
DMM
Test light
Self-powered test light
Fused jumper wire
DSO

Figure 3-1 Expected voltmeter readings in a properly operating parallel circuit.

Figure 3-2 Knowing the expected voltage at different locations in the circuit will assist in locating an open.

The easiest method of testing a circuit is to start at the most accessible place and work from there. If the load component is easily accessible, test for voltage at the input to the load (**Figure 3-2**). Remember, to get reliable test results, the load component must remain in the circuit. Don't do open circuit testing unless directed to do so in the service information. When making a test connection, think ahead of what the expected results should be. Use the following procedure for locating the open:

An open circuit will not blow a fuse. However, a blown fuse opens the circuit.

1. With the switch in the ON position, measure the voltage at point A. If the voltage is 10.5 volts or higher, check the ground side (point B). If the voltage at point B is less than 10.5 volts, there is excessive resistance or an open in the ground circuit. If the voltage at point A is less than 10.5 volts, continue testing.
2. Work toward the battery. Test all connections for voltage. If voltage is present at a connection, the open is between that connection and the previously tested location (**Figure 3-3**). Use a fused jumper wire to bypass that section of the circuit to confirm the location of the open.

Figure 3-3 Working from the battery positive terminal toward ground, an open is located between the points where voltage was last measured and where it was not.

Figure 3-4 Properly operating complex parallel circuit.

3. If battery voltage is present at point B, the open is in the ground circuit. Use a fused jumper wire to connect the ground circuit. Then retest the component.

In more complex circuits, the open may have very different results. In a normally operating circuit, the voltmeter readings would be as indicated in **Figure 3-4**. If an open occurs in the ground side of the circuit, the circuit converts to a series circuit (**Figure 3-5**). This type of circuit defect is a form of **feedback** that results in lamps coming on when not intended to. If the electrons cannot find a path to ground through the intended circuit, they will attempt to find an alternate path. This may result in turning on components that are in the path. Lamp 3 has the normally applied voltage, but lamps 1, 2, and 4 are in series and illuminate dimmer than normal. The voltmeter will read 12 volts at the locations illustrated in Figure 3-5. However, the voltmeter will not indicate 0 volt on the ground side of bulb 1. Using a fused jumper wire from the ground terminal of the bulb to chassis ground would confirm the open circuit.

Testing for Shorts

Classroom Manual
Chapter 3, page 79

Locating a copper-to-copper short can be one of the most challenging tasks for a technician. If the short is within a component, the component will operate at less than optimum or not at all. An ohmmeter will measure the component's resistance. If there is a short, then the amount of resistance will be lower than specified. If specifications for the component are not available, it may be necessary to replace the component with a known good unit. Do this only after first determining that the component's insulated and ground side circuits are in good condition.

Figure 3-5 An open in the ground circuit can convert the circuit to a series circuit. The dashed line represents the resulting current path to ground.

If the short is between circuits, the result will be components operating when not intended (**Figure 3-6**). Visually check the wiring for signs of chafing, burned insulation, and melted conductors that will indicate a short. Also, check shared connectors of the two affected circuits. Corrosion can form between two connector terminals and result in the short.

If the visual inspection does not isolate the cause of the copper-to-copper short, remove one of the fuses for the affected circuits. (If the affected circuits share a common fuse, remove it.) Install a buzzer (fitted with flat blade terminals) across the fuse holder terminals (**Figure 3-7**). Activate the circuit that has the buzzer connected to it. In Figure 3-6, if the buzzer is connected to fuse B, then turn on switch 1. Disconnect the

Special Tools

DMM
Buzzer fitted with terminals and in-line fuse

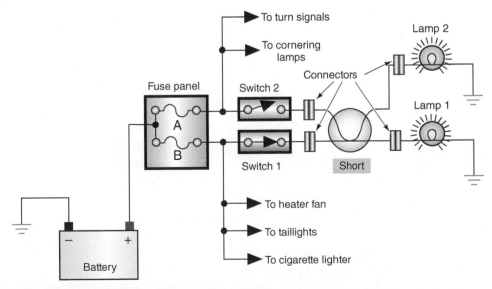

Figure 3-6 A copper-to-copper short between two circuits.

In-line
20-ampere
fuse

Seat belt buzzer

Figure 3-7 The buzzer will sound until the cause of the short is found and corrected.

loads that are supposed to be activated by this switch (lamp 1). Disconnect the wire connectors from the load back to the switch. If the buzzer stops when a connector is disconnected, the short is in that portion of the circuit.

Testing for a Short to Ground

Classroom Manual
Chapter 3, page 81

A short may not blow
a fuse, depending on
the amount of current
flowing.

A fuse that blows immediately upon installation indicates a **short to ground**. This condition allows current to return to ground before it has reached the intended load component. If the circuit is unfused, the insulation and conductor will melt. Not all shorts will blow a fuse, however. If the short to ground is on the ground side of the load component, but before a grounding switch, the component will not turn off (**Figure 3-8**). If the short to ground is after the load and grounding switch (if applicable), circuit operation may not be affected.

To confirm that the circuit has a short to ground before the load, remove the fuse and connect a test light in series across the fuse connections. If the test light illuminates, the circuit has a short to ground. An ohmmeter can be used by first removing the fuse and then testing for continuity between the output side of the fuse box and chassis ground. With the switch open, any reading other than infinite or "OL" may indicate a short to ground. Always check the service information since a very high resistance value

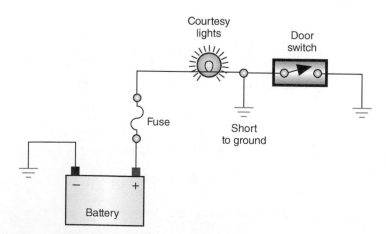

Courtesy
lights

Door
switch

Fuse

Short
to ground

Battery

Figure 3-8 A ground in this location will cause the lamp to remain on.

may be normal for that circuit, especially if a computer or electronic components are involved. If the switch is closed, the ohmmeter should read the resistance value of the load(s).

> ⚙️ **SERVICE TIP** By carefully studying the wiring diagram of the circuit, the short's general location may be determined. With the test light or ohmmeter connected at the fuse box as discussed, pay attention to the light or readings while turning the switch on and off. Also, disconnect the load device and other in-line connectors while observing the test equipment.

When attempting to pinpoint the location of the short to ground, it may be challenging to use a test light or voltmeter because the fuse blows before any testing can be conducted. To prevent this, connect a cycling circuit breaker (fitted with alligator clips) across the fuse holder (**Figure 3-9**). The circuit breaker will continue to cycle open and closed, allowing you to test for voltage.

Testing for shorts may be complicated if a single fuse protects several circuits and if the ground location isn't accessible. A couple of methods can be used to locate the fault.

One method is to connect a test light, in series with a cycling circuit breaker, across the fuse holder (**Figure 3-10**). While observing the test light, disconnect individual circuits one at a time until the light goes out. The fault is in the circuit that was disconnected when the light went off.

🔦 **Special Tools**

Test light

Circuit breaker fitted with terminals

Gauss gauge or compass

DMM

⚠️ **Caution**

Use a circuit breaker rated between 25 and 30 amperes. The use of a circuit breaker rated too high will damage the circuit.

Figure 3-9 Use a circuit breaker to protect the circuit while checking for the short to ground.

Figure 3-10 The test light will allow you to see when you have located the faulty circuit.

Figure 3-11 The needle of a compass or Gauss gauge will fluctuate over the portion of the circuit that has current flowing through it. The needle will stop fluctuating at the location of the short to ground.

A second method uses a **Gauss gauge** or a compass to locate the short to ground. A Gauss gauge is a meter that is sensitive to the magnetic field surrounding a wire conducting current. The gauge or compass works on the principle that a conductor carrying current develops a magnetic field around it. With a cycling circuit breaker in place of the blown fuse, trace the circuit's path with the gauge or compass. The needle will fluctuate as long as the gauge is over the conductor and the circuit breaker is cycling. The needle will stop fluctuating when the point of the short to ground is passed (**Figure 3-11**). This method will work even through the vehicle's trim. It will be necessary to follow all circuits protected by the fuse. Consult the wiring diagram for this information.

Commercial short finders are available to assist in locating a short to ground. First, connect the transmitter in series to the circuit. Do this by installing it at the fuse, circuit breaker, relay, or disconnecting a connector. Next, follow the wire harness holding the receiver a little above it. The illumination of a LED or an audible sound alerts you that the receiver is over the location of the short. Some short finders may identify the direction and distance to the short.

Testing for Voltage Drop

Voltage drop (when considered a defect) defines the portion of applied voltage used up in other points of the circuit rather than by the load component. It is a resistance in the circuit that reduces the amount of electrical pressure available beyond the resistance. Excessive voltage drop may appear on either the insulated or the ground return side of a circuit, and you should test both. The circuit must be active with current flowing to test for voltage drop. The source voltage must be as specified before voltage drop readings can be valid.

Excessive voltage drop caused by high resistance can cause dim or flickering lamps, inoperative load components, or slower-than-normal electrical motor speeds. Excessive resistance will not cause the fuse to blow.

For testing for circuit resistance, voltage drop testing is the preferred method over using an ohmmeter to measure resistance. Remember that an ohmmeter works by sending a low voltage through the conductors to calculate resistance. This low voltage does not provide enough current to see how the circuit will behave under a full electrical load. It's possible to have a circuit or component pass the specifications for an ohmmeter test yet fail the voltage drop test. The voltage drop test will test the circuit under load and provide a more accurate representation of the circuit's integrity.

A buzzer can be substituted for the test light.

Classroom Manual
Chapter 3, page 82

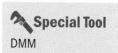

Special Tool

DMM

To perform a voltage drop test on any circuit, connect the positive voltmeter lead to the most positive portion of the circuit and the negative lead to the most negative portion of the circuit. Follow the instructions shown in **Photo Sequence 4** to conduct a voltage drop test. Consult the service information for the maximum amount of voltage drop allowed. When the voltage drop decreases to within specifications, this indicates the location of the excessive resistance. In Photo Sequence 4, a faulty relay was the cause of the excessive resistance.

When testing the ground side of the circuit, the ground connection terminal of the load component is the most positive location, and the battery negative post is the most negative (**Figure 3-12**). Usually, more than 0.1 volt indicates excessive resistance in the ground circuit.

According to many manuals, the maximum allowable voltage drop for an entire circuit, except for the drop across the load, is 10% of the source voltage. Although 1.2 volts may be considered the maximum acceptable amount, it's still too much for low-current circuits. Many good technicians use 0.5 volt as the maximum allowable drop of a circuit. However, 0.1 volt is the maximum voltage drop of any one wire or connector. This is the most important specification to consider and remember.

If access to two points in a circuit using a single voltmeter is difficult, it's possible to calculate voltage drop by testing for available voltage. Use Ohm's law to determine the amount of voltage drop that should be across a component. Test for available voltage on both sides of the load component (**Figure 3-13**). Subtract the available voltage readings to obtain the amount of voltage drop across the component.

Voltage drop testing is also preferred over open voltage testing. Refer to the circuit illustrated in **Figure 3-14** as an example of how open circuit testing may lead to incorrect troubleshooting results. If the connector at the lamp is disconnected and you use a voltmeter to measure the applied voltage at the connector terminal with the switch closed, it will probably read full battery voltage (**Figure 3-14A**). This would be true even though the measurement is taken after the resistance. No voltage drop is measured over the resistance because the circuit isn't loaded, and the meter has high impedance. Doing the test by leaving the connector attached to the load and back probing the feed wire, the circuit is loaded (provided there are no opens). Connecting the voltmeter to measure voltage drop will now indicate a problem with the circuit (**Figure 3-14B**), and the voltmeter will read available voltage at that point in the circuit.

The higher the circuit current, the higher the allowable voltage drop. The lower the circuit current, the less the allowable voltage drop.

Figure 3-12 Testing the ground side of the starter motor circuit for high resistance by measuring voltage drop. Notice the voltmeter connections have the most positive portion of the circuit at the starter motor housing.

PHOTO SEQUENCE 4
Voltage Drop Test to Locate High Circuit Resistance

P4-1 Tools required to test for excessive resistance in a starting circuit include fender covers, a DMM, and a remote starter switch.

P4-2 Connect the positive lead of the meter to the positive battery post. If possible, don't connect the lead to the cable clamp.

P4-3 Connect the meter's negative lead to the main battery terminal on the starter motor.

P4-4 To conduct a voltage drop test, current must flow through the circuit. In this test, the ignition system is disabled, and the engine is cranked using a remote starter switch.

P4-5 With the engine cranking, read the voltmeter. The reading is the amount of voltage drop.

P4-6 If the reading is greater than specifications, test at the next connection toward the battery. In this instance, the next test point is the starter side of the solenoid.

P4-7 Crank the engine and touch the negative test lead to the starter side of the solenoid. Observe the voltmeter while the engine is cranking.

P4-8 Test in the same manner on the battery side of the solenoid. This is the voltage drop across the positive circuit from the battery to the solenoid.

Available voltage point A = 12.00 V
Minus available voltage point B = 6.00 V
Voltage drop across lamp 1 = 6.00 V

Available voltage point C = 6.00 V
Minus available voltage point D = 0.00 V
Voltage drop across lamp 2 = 6.00 V

Total voltage drop between points A and D = 12.00 V

Figure 3-13 Using available voltage to calculate voltage drop over a component.

Figure 3-14 Comparison between readings of an open circuit test and a loaded voltage drop test.

Testing Circuit Protection Devices

Classroom Manual
Chapter 3, page 74

Special Tools

DMM

Test light

⚠ **Caution**

Fuses and other protection devices don't wear out. They fail because something went wrong. Never replace a fuse or fusible link, or reset a circuit breaker without finding out why it failed.

A blade type fuse is called a spade fuse.

Classroom Manual
Chapter 3, page 76

Another name for a **fusible link** is fuse link.

⚡ **Caution**

Always disconnect the battery ground cable prior to servicing any fusible link.

Classroom Manual
Chapter 3, page 77

Circuit protection devices turn off the system whenever excessive current or an **overload** occurs. There are three basic types of **fuses** in automotive use: cartridge, blade, and ceramic. A fuse is a replaceable element that will melt should the current passing through it exceed the fuse rating. Most older domestic cars and a few imports use the cartridge fuse. To check this type of fuse, look for a break in the internal metal strip. Discoloration of the glass cover or glue bubbling around the metal caps is an indication of overheating. Late-model vehicles use blade or spade fuses. To check the fuse, pull it from the fuse panel and look at the fuse element through the transparent plastic housing. Look for internal breaks and discoloration. Many older European imports used ceramic fuses. To check this type of fuse, look for a break in the contact strip on the outside of the fuse. All fuse types can be tested using an ohmmeter or a test light. If the fuse is good, there will be continuity through it.

Fuses are rated by the current level at which they blow. A three-letter code identifies the type and size of the fuse. Blade fuses have codes ATC or ATO. All glass SFE fuses have the same diameter, but the length varies with the current rating. Ceramic fuses are available in two sizes, code GBF (small) and the more common code GBC (large). The insulator has the amperage rating embossed on it. Codes such as AGA, AGW, and AGC indicate the length and diameter of the fuse. Fuse lengths in each of these series are the same, but the current rating can vary. The identification code and current rating are usually stamped on the end cap.

The blade fuse's case color identifies its current rating (**Table 3-1**). In addition, manufacturers stamp the top of the fuse with the rating. The insulator on ceramic fuses is color coded to indicate different current ratings.

In most instances, a dedicated fuse box or panel hold the fuses. Typical locations for the fuse box include; under the dashboard, behind a panel in the foot well, or in the engine compartment. The fuse layout within the fuse box is generally identified by numbers or by abbreviated names of the protected circuits. On late-model vehicles, there may be icons or symbols indicating which circuits the fuse serves. The owner's and service information covers the fuse identification system in greater detail.

A **fusible link** is a conductor with a special heat-resistant insulation. When there is an overload in the circuit, the link melts and opens the circuit. Fusible links are often installed in the positive battery lead to the ignition switch and other circuits that have power with the key off.

A fusible link is a short length of small-gauge wire installed in a conductor. Since the fusible link is a lighter gauge of wire than the main conductor, it melts and opens the circuit before damage can occur in the rest of the circuit. Special insulation that bubbles when it overheats covers the fusible link. Bubbling of the insulation indicates that the internal link has melted. If the insulation appears good, pull lightly on the wire. If the link stretches, the wire has melted. Testing the condition of the link is also done using a test light, voltmeter, or ohmmeter.

To replace a fusible link, cut the protected wire where it's connected to the fusible link. Then, tightly crimp or solder a new fusible link of the same rating as the original link. Since the insulation on the manufacturer's fusible links is flameproof, never fabricate a fusible link from ordinary wire because the insulation may not be flameproof.

Many late-model vehicles use maxi-fuses instead of fusible links. Maxi-fuses look and operate like two-pronged blade or spade fuses, except they are much larger and can handle more current. (Typically, a maxi-fuse is four to five times larger.) Maxi-fuses are located in a fuse box in the engine compartment and/or passenger compartment, under the dash or the rear seat.

Table 3-1 Typical Color Coding of Protection Devices

Blade Fuse Color Coding	
Ampere Rating	**Housing Color**
4	pink
5	tan
10	red
15	light blue
20	yellow
25	natural
30	light green
Fuse Link Color Coding	
Wire Link Size	**Insulation Color**
20 GA	blue
18 GA	brown or red
16 GA	black or orange
14 GA	green
12 GA	gray
Maxi-Fuse Color Coding	
Ampere Rating	**Housing Color**
20	yellow
30	light green
40	amber
50	red
60	blue

SERVICE TIP To calculate the correct fuse rating, use Watt's law: watts divided by volts equals amperes. For example, if you install a 55-watt pair of fog lights, divide 55 by the battery voltage (12 volts) to determine how much current the circuit must carry. Since 55/12 = 4.58, the current is approximately 5 amperes. The correct fuse should be rated slightly higher than the normal current flow to allow for current surges. In this case, an 8- or 10-ampere fuse would do the job.

Maxi-fuses are easier to inspect and replace than are fuse links. Look at the fuse element through the transparent plastic housing to determine if the fuse has blown. If there is a break in the element, the maxi-fuse has blown. To replace it, pull it from its fuse box or panel. Always replace a blown maxi-fuse with a new one having the same ampere rating.

Some circuits are protected by circuit breakers. Like fuses, circuit breakers are rated in amperes. There are two types of circuit breakers: cycling or those that must be manually reset.

Classroom Manual
Chapter 3, page 77

A circuit breaker is typically abbreviated c.b. in a fuse chart in the service information.

In the cycling type, the bimetal arm will begin to cool once the current to it stops. Once it returns to its original shape, the contacts are closed, restoring circuit power. If the current is still too high, the circuit break repeats the cycle of opening the circuit.

There are two types of noncycling or resettable breakers. One resets by disconnecting power to the circuit. There is a coil wrapped around a bimetal arm (**Figure 3-15A**). When there is excessive current and the contacts open, a small current passes through the coil. This current through the coil isn't enough to operate a load, but it does heat both the coil and the bimetal arm. This keeps the arm in the open position until power is removed.

The second type resets by depressing a button. A spring pushes the bimetal arm down and holds the contacts together (**Figure 3-15B**). When an overcurrent condition exists, and the bimetal arm heats up, the arm bends enough to overcome the spring, and the contacts snap open. The contacts stay open until pushing the reset button snaps the contacts together again.

A visual inspection of a fuse or fusible link will not always determine if it has an open (**Figure 3-16**). To accurately test a circuit protection device, use an ohmmeter, voltmeter, or test light.

With the fuse or circuit breaker removed from the vehicle, connect the ohmmeter's test leads across the protection device's terminals (**Figure 3-17**). On its lowest scale, the ohmmeter should read 0 to 1 ohm. If it reads infinite, the protection device is open. Test a fusible link in the same way (**Figure 3-18**). Before connecting the ohmmeter across

SERVICE TIP Before using a test light, it's good practice to check the tester's lamp. To do this, simply connect the test light across the battery. The light should come on.

Figure 3-15 The two basic types of circuit breakers.

Figure 3-16 A fuse or fusible link can have a hidden fault that you cannot see.

Figure 3-17 A good fuse will have zero resistance when tested with an ohmmeter.

Figure 3-18 An ohmmeter can test a fusible link after disconnecting the link from power.

the fusible link, ensure there is no current flow through the circuit. To be safe, disconnect the negative cable of the battery.

To test a circuit protection device with a voltmeter, check for available voltage at both terminals (**Figure 3-19A**). If the device is good, voltage will be present on both sides. A test light can be used in place of a voltmeter (**Figure 3-19B**). When the test lamp's probe touches each test terminal, the lamp should illuminate.

Measuring the voltage drop across a fuse or other circuit protection device will tell you more about its condition than just if it's open. A voltage drop of near zero indicates the fuse, fuse link, or circuit breaker is in good condition. If the voltmeter reads 12 volts, the fuse is open. Any reading between 0 and 12 volts indicates some voltage drop. Make sure you check the fuse holder for resistance as well. Replace any circuit protection device that has excessive voltage drop across it.

CUSTOMER CARE Any time you install additional electrical accessories for customers, provide them with information concerning the type and size of fuses installed so they can put this information with their owner's manual.

Diagnosing PTC Circuit Protection Devices

PTCs are rated much like a fuse; so, if it experiences an excessive overload, it can be damaged. If the circuit experiences a current overload, the resistance of the PTC increases, which in turn reduces the amperage to a safe level. At this point, the PTC acts

⚠ Caution

Fuses are rated by amperage and voltage. Never install a larger rated fuse into a circuit than the one specified by the manufacturer. Doing so may damage or destroy the circuit. Also, don't replace a fusible link with a resistor wire or vice versa.

⚠ Caution

Do not use an unfused jumper wire to bypass the protection device. Circuit damage may result.

Classroom Manual
Chapter 3, page 78

Top view of mini-fuse

Window through which light can be seen

Figure 3-19 (A) Voltmeter test of a circuit protection device. Battery voltage should be present on both sides. (B) The test light will illuminate on both terminals if the fuse is good.

as the load for the circuit. Usually, the PTC will not go to a completely open state during this time.

Consider the circuit shown in **Figure 3-20** that uses a PTC to protect the power window motor. If the motor is inoperative and a check of voltage to the motor indicates that 0 volt is present, the problem may be the PTC. Often the PTC is installed into the fuse box. In many cases, it isn't serviceable as a separate unit, but the fuse box usually has test ports. Because disconnecting the battery to perform resistance tests with an ohmmeter results in computers resetting, loss of adaptive memories, and other set values, it's more desirable to use a voltmeter to test the PTC. Simply measure the voltage drop between the input side of the PTC and the output side while operating the accessory. Typically, there should be less than 0.050 voltage drop across these points. Because the PTC's resistance increases as the circuit's temperature increases, a high voltage drop indicates an increase in the PTC's resistance. This may be due to a short in the circuit. Confirm this is not the case by disconnecting in-line and component connectors while observing the voltmeter. If the voltmeter reading decreases toward 0 volt when a connector is unplugged, the circuit has a short.

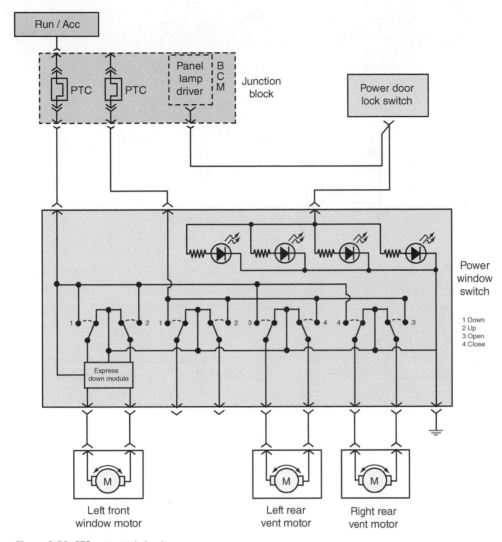

Figure 3-20 PTC-protected circuit.

Testing and Replacing Electrical Components

All electrical components can fail. Testing them is the best way of determining if they are good or bad. For the most part, the proper way to check electrical components is determined by what the component is supposed to do. If you think about what something is supposed to do and how it does it, you can figure out how to test it. Sometimes, removing the component and testing it on a bench is the best way to check it.

Switches

The easiest method of testing a **normally open (NO)** switch is to use a fused jumper wire to bypass the switch (**Figure 3-21**). An NO switch will not allow current flow when it's in its rest position. The contacts are open until they are acted on by an outside force that closes to complete the circuit. If the circuit operates with the switch bypassed, the switch is defective. Also, check the voltage drop across switches. Ideally, there should be no voltage drop when the switch is closed. Any voltage drop indicates resistance, and the switch requires replacement.

Classroom Manual
Chapter 3, page 51

Figure 3-21 Using a fused jumper wire to bypass the switch.

Special Tools
Jumper wires
Test light
DMM

A voltmeter or test light can be used to check for voltage on both sides of the switch (**Figure 3-22**). A faulty NO switch would have voltage present at the input side of the switch but not on the output side when in the ON position.

> **SERVICE TIP** Using an ohmmeter to test a switch is suitable for determining if the switch works and if the contacts move; however, it may not accurately indicate that a resistance is present. A voltage drop test is better for determining if there is resistance in the switch.

If the ignition switch provides power feed to the circuit being tested, it must be in the RUN position.

Removing the switch from the vehicle allows for bench testing it with an ohmmeter. With the switch contacts open, there should be no continuity between the terminals (**Figure 3-23**). There should be zero resistance through the switch contacts when the

Figure 3-22 (A) Using a voltmeter to test a switch, the voltage should be the same on both sides of the switch. (B) Using a test lamp to test a switch, the lamp should illuminate on both sides of the switch.

Open switch   Closed switch

Figure 3-23 Testing the continuity through a switch with an ohmmeter. With the switch open, there should be infinite resistance. With the switch closed, there should be zero resistance.

Headlamp switch connector
(connector end view)

Pin number	Circuit	Circuit function
B1	38 (BK/O)	Power supply to battery
B2	195 (T/W)	Tail lamp switch feed
I	19 (LB/R)	Instrument panel lamp feed
IGN		Not used
R	14 (BR)	Tail lamp and side marker lamps
H	15 (R/Y)	Headlamp dimmer switch feed
DN		Not used
D1	54 (LG/Y)	Interior lamp switch feed
D2	706 (GY)	Battery saver door switch feed

Figure 3-24 Example of headlight switch connector callouts used for continuity checks.

contacts are closed. On complex ganged-type switches, consult the service information for a continuity diagram (**Figure 3-24**). If there is no continuity chart, use the wiring diagram to make your own chart.

Relays

A **relay** is a device that uses a low current to control a high-current circuit. The relay can be either a normally open or a normally closed design. Checking the operation of a relay can be done using a jumper wire, voltmeter, ohmmeter, or test light. If the terminals are easily accessible, the jumper wire and test light may be the fastest method.

Classroom Manual
Chapter 3, page 54

Figure 3-25 A relay circuit with a ground control switch.

Special Tools

Fused jumper wires
Test light
DMM

To obtain accurate
voltmeter test results,
the battery must be
fully charged and in
good condition.

Check the wiring diagram for the relay to determine if the control is through an insulated or a ground switch. Use **Figure 3-25** as a guide for testing a ground-switch–controlled relay. Follow these steps:

1. Use a voltmeter to check for available voltage to the battery side of the relay (terminal A). If battery voltage is not present at this point, the fault is in the circuit from the battery to the relay. If battery voltage is present, continue testing.
2. Probe for battery voltage at control terminal B. If voltage is not present at this terminal, the fault is in the relay coil. If voltage is present, continue testing.
3. Use a fused jumper wire to connect terminal B to a good ground. If the horn sounds, the fault is in the control circuit from terminal B to the horn switch ground. If the horn does not sound, continue testing.
4. Connect the fused jumper wire from the battery positive to terminal C. If the horn does not sound; there is a fault in the circuit from the relay to the horn ground. If the horn sounds, the fault is in the relay.

> **SERVICE TIP** You can use the procedures presented here to test the relay and to determine the type of fault it has. However, the easiest way to test a relay is to substitute it with a *known good* relay of the same type. If the circuit operates with the substitute relay, the old relay is the faulty component.

Caution

It's not recommended
to use a test light to
probe for power in a
computer-controlled
circuit. The increased
draw of the test light
may damage the
system components.

If the computer controls the relay, a test light is not recommended. The test light may draw more current than the circuit can carry and damage the computer. Refer to **Figure 3-26** for procedures using a voltmeter to test a relay. Use a DMM to read DC voltage on a scale that will display 12 volts. Follow these steps:

1. Connect the negative voltmeter test lead to a good ground.
2. Connect the positive voltmeter test lead to the output wire (terminal B). Turn on the ignition switch. If no voltage is present at this terminal, go to step 3. If the voltmeter reads 10.5 volts or higher, turn off the control circuit. The voltmeter should then read 0 volt. If it does, then the relay is good. If the voltmeter still reads any voltage, the relay is not opening and requires replacement.
3. Connect the positive voltmeter test lead to the power input terminal (terminal A). The voltmeter should indicate battery voltage. If voltage is below this value, the circuit from the battery to the relay is faulty. If the voltage value is correct, continue testing.

Figure 3-26 Testing relay operation with a voltmeter.

4. Connect the positive voltmeter test lead to the control circuit terminal (terminal C). The voltage should read 10.5 volts or higher. If not, check the circuit from the battery to the relay (including the ignition switch). If the voltage is 10.5 volts or higher, continue testing.

5. Connect the positive voltmeter test lead to the relay ground terminal (terminal D). If the meter reads more than 0.1 volt, there is a poor ground connection. If the reading is less than 0.1 volt, replace the relay.

If the relay terminals are not accessible, remove the relay from its holding fixture and bench test it. Use an ohmmeter to test for continuity between the relay coil terminals (**Figure 3-27**). If the meter indicates an infinite reading, replace the relay. If there is continuity, use a pair of fused jumper wires to energize the coil (**Figure 3-28**). Check for continuity through the relay contacts. If the meter indicates an infinite reading, the relay is defective. If there is continuity, the relay is good, and the circuits will require further testing.

Be careful not to touch the coil terminals with the ohmmeter test leads while the coil is energized.

Figure 3-27 Testing the resistance of the relay coil.

Figure 3-28 Bench testing a relay.

Be sure to check your service information for resistance specifications and compare the relay to them. It's easy to check for an open coil. However, a shorted coil will also prevent the relay from working. Low resistance across a coil would indicate that it's shorted. Too low resistance may also damage the transistors and/or driver circuits because of the excessive current that would result.

Testing Stepped Resistors

A **stepped resistor** has two or more fixed resistor values (**Figure 3-29**). Using an ohmmeter is a good method of testing a stepped resistor. To obtain accurate test results, it's a good practice to remove the resistor and bench test it. Be sure to place the ohmmeter on the correct scale to read the anticipated amount of resistance. Connect the ohmmeter leads to the two ends of the resistor (**Figure 3-30**). Compare the results with the manufacturer's specifications. Repeat for each of the resistances.

You can also use a voltmeter or DSO to check a stepped resistor. By measuring the voltage after each part of the resistor block and comparing the readings to specifications, you can tell if the resistor is good or not.

Testing Variable Resistors

A **variable resistor** provides an infinite number of resistance values within a range. As with the stepped resistor, an ohmmeter is good for testing a variable resistor. However, it's also possible to use a voltmeter, DSO, or test light.

A **rheostat** is a two-terminal variable resistor used to regulate the current in an electrical circuit. To test a rheostat, locate the input and output terminals and connect the ohmmeter test leads to them. Rotate the resistor knob slowly while observing the ohmmeter. The resistance value should remain within the specification limits and change smoothly and constantly. If the resistance values are out of limits or the resistance value jumps as the knob is turned, replace the rheostat.

Classroom Manual
Chapter 3, page 58

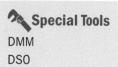
Special Tools
DMM
DSO

⚠ **Caution**
Operating a blower motor resistor out of the air flow of the heating, ventilation, and air-conditioning (HVAC) housing will cause the resistor to overheat and be damaged.

Classroom Manual
Chapter 3, page 58

Thermal limiter

Figure 3-29 A stepped resistor used in the heater blower motor circuit.

Figure 3-30 Ohmmeter testing of a stepped resistor.

Figure 3-31 Using an ohmmeter to test the continuity between terminals A and C of a potentiometer.

Figure 3-32 Testing continuity between terminals A and B of a potentiometer while moving the wiper.

If using a voltmeter, the voltage drop readings between the two terminals should be smooth and consistent. If using a test light, its brightness level should change when rotating the knob. The rheostat is defective if the light blinks at any point.

A **potentiometer** is a three-wire variable resistor that acts as a voltage divider to produce a continuously variable voltage output signal proportional to a mechanical position. To test a potentiometer, connect the ohmmeter test leads to terminals A and C (**Figure 3-31**). Check the results with specifications. Next, connect the ohmmeter test leads to terminals A and B (**Figure 3-32**). Check the resistance at the stop and observe the ohmmeter while moving the wiper to the other stop. The resistance values should be within specification and smooth and constant.

A voltmeter can be used in the same manner. However, jumper wires may need to gain access to the test points (**Figure 3-33**). Because potentiometers are primarily used in computer-controlled circuits, the use of a test light is not recommended. A DSO is an excellent tool to use for testing variable resistance circuits. The DSO allows for easy reading of the voltage trace as the wiper of the sensor is moved. The voltage indicated by the trace should be within the specified limit and should change smoothly as the wiper moves.

Special Tools
DMM
Test light
DSO

⚠ **Caution**
Do not pierce the insulation to test the potentiometer. These circuits usually operate between 5 and 9 volts. Piercing the insulation may break some of the wire strands, resulting in a voltage drop that will give errant information to the computer. Even if the conductor is not broken, moisture can enter and cause corrosion.

Figure 3-33 It may be necessary to use jumper wires to connect the wire connector to the sensor in order to measure voltage.

Figure 3-34 A capacitor with its rating marks.

Figure 3-35 Using the capacitor test function of the DMM, the reading should be within 6% of the capacitor's rating.

Testing Capacitors

Classroom Manual
Chapter 3, page 59

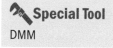

Special Tool
DMM

Capacitors store and release electrical energy. Because the capacitor stores voltage, it will also absorb voltage changes in the circuit. Control of damaging voltage spikes is possible due to the capacitor's ability to store voltage. In addition, this ability allows capacitors to reduce radio noise.

The capacitor can fail internally. To test for this, some DMMs have a capacitance test function. Before testing the capacitor, make sure to turn off the power to the capacitor. Next, you must dissipate the stored voltage in the capacitor. Connect a 12-volt test light across the terminals of the capacitor and leave connected until the light goes out. Remove or disconnect the capacitor from the circuit when the capacitor finishes discharging.

The capacitor will have its rating listed on its shell (**Figure 3-34**). Also, note the tolerance of the rating. If there is no tolerance range listed, use 6% as a general rule.

Based on the rating of the capacitor, select the correct micro-farad setting on the meter. If the meter does not have an auto-range function, select a range that is higher than the rating of the capacitor. Connect the black test lead of the DMM onto the ground terminal of the capacitor (or the case of the capacitor if it does not have a terminal). Connect the red test lead to the power terminal. The reading on the DMM should match the rating of the capacitor (**Figure 3-35**).

Testing Diodes

Classroom Manual
Chapter 3, page 63

Special Tools
Analog ohmmeter
DMM

A **diode** is an electrical one-way check valve that will allow current to flow in one direction only. Regardless of the bias of the diode, it should allow current flow in one direction only. A shorted diode allows current flow in both directions. An open diode does not allow current flow in either direction.

You may run into problems when checking a diode with a high-impedance digital ohmmeter. Since many diodes won't allow current flow through them unless the voltage is at least 0.6 volt, a digital meter may not be able to forward bias the diode. This will result in readings that indicate the diode is open, when in fact it may not be. Because of this problem, many digital multimeters have a special diode-testing feature. This feature allows for increased voltage at the test leads. Connect the meter's leads across the diode (**Figure 3-36**). Observe the reading on the meter. Then reverse the meter's leads and

Figure 3-36 Use the diode test function of the DMM to check a diode for an open and short.

observe the reading on the meter. The value displayed is the voltage required to forward bias the diode. A silicon diode should read between 400 and 800 mV in the forward direction and open in reverse. For a germanium diode, it's between 200 and 400 mV in the forward direction. If the diode is open, the meter will display "OL" or another reading to indicate infinity or out of range. Some meters during diode check will make a beeping noise when there is continuity.

A voltmeter can also be used to test diodes. Using the same logic as testing with an ohmmeter, test the voltage drop across the diode with the circuit turned on. The meter should read low voltage in one direction and near source voltage in the other direction.

The voltage displayed on the meter is the diode's turn-on voltage or diode drop.

Testing Zener Diodes

If the Zener diode is out of the circuit and you need to diagnose it for an open or short, then test the Zener as described for the standard diode. However, if you desire to measure its Zener voltage level, you will have to build a test circuit (**Figure 3-37**). This can be

Classroom Manual
Chapter 3, page 65

DC power
supply

Zener

Voltmeter

Figure 3-37 Making a test power supply to measure Zener voltage.

made using many styles of "project boxes" from most electronic stores. Set the power supply voltage to a value slightly higher than the Zener value. For example, for a 12-volt diode, the supply voltage should be about 15 volts. The value of the resistor (R) should limit the current to about 1 mA. For example, when using 15 volts with a 12-volt Zener, use a 3.3 K resistor.

Once the circuit is built, read the Zener voltage using a digital voltmeter. If the voltmeter indicates 600 mV, the diode is reverse biased and will need to be reinstalled into the circuit.

> ⚙ **SERVICE TIP** The avalanche diode is tested in the same manner as the Zener diode. However, most avalanche diodes used in the AC generator are rated between 20 and 30 volts.

Testing LEDs

Classroom Manual
Chapter 3, page 66

The turn-on voltage of an LED is usually between 1.5 and 2.5 volts. If your DMM has a diode test function, test the LED in the same manner as a standard diode. The difference will be that the meter will read between 1,600 and 2,500 mV when the diode conducts instead of the 600 mV you read on a standard diode.

It's possible to test an LED without the use of a DMM. First, you will need to build a test circuit like that shown (**Figure 3-38**). By plugging the LED into the circuit, the LED should light. If the LED doesn't light, reverse the diode's polarity. If it still doesn't light, then the LED is faulty.

Testing Zener Diodes and LEDs in a Circuit

It's not necessary to remove a Zener diode or the LED from the circuit to test it. To test a Zener, use a voltmeter and measure the voltage across it. Connect the negative lead to the anode and the positive lead to the cathode. The meter should read the Zener voltage. O volts indicate a shorted Zener. Confirm that the power and ground circuits to the Zener are good before replacing the diode. If the voltmeter reads a higher voltage than the Zener's rated voltage, the diode is open.

For an LED that is supposed to be lit but isn't, use a voltmeter to measure the voltage across it. If you measure more than 3 volts, the LED is open.

Figure 3-38 Testing an LED with a test circuit.

Testing Transistors

Although replacing a transistor that is part of an integrated circuit board is not often done in the automotive repair industry, some bipolar transistors can be easily tested and replaced. To test a bipolar transistor, you need to identify its type (NPN or PNP) and lead arrangement. There are several configurations of the transistor legs (**Figure 3-39**). Use a DMM to identify the different legs. Since transistors behave as back-to-back diodes, the collector and emitter are identifiable because the doping for the base-to-emitter junction is always much higher than that for the base-to-collector junction. Therefore, the forward voltage drop will be a couple of millivolts higher on the DMM reading when set on the diode test function. Follow **Photo Sequence 5** to identify the type of bipolar transistor.

If none of the six possible lead connection combinations indicates a pair of low readings, or if more than one combination results in a pair of low readings, the transistor is probably faulty. Keep in mind that the base-to-collector junction voltage drop is always slightly lower than the emitter-to-base junction drop.

With the type of transistor now known, use the diode function of the DMM to test it. Connect the meter's red lead to the base of the transistor and the black lead to the emitter. A good NPN transistor will read a voltage of between 450 and 900 mV. A good PNP transistor will read open. With the red lead still on the base, move the black lead to the collector. The reading should be the same as that in the previous test.

Next, reverse the meter leads and repeat the test. With the black lead connected to the base of the transistor and the red lead to the emitter, a good PNP transistor will read a voltage of between 450 and 900 mV. A good NPN transistor will read open. Leave the black lead on the base and move the red lead to the collector. The reading should be the same as that in the previous test.

Finally, place one meter lead on the collector and the other on the emitter. The meter should read open. Reverse your meter leads and the meter should read open. This is the same for both NPN and PNP transistors.

Testing Darlington Transistors

A **Darlington** is a special type of configuration usually consisting of two transistors fabricated on the same chip or mounted in the same package.

Classroom Manual
Chapter 3, page 67

Special Tool
DMM

⚠ Caution

Do not hold the transistor in your hand while testing it. For every degree the transistor increases in temperature, the base-to-emitter diode drop decreases by 2 mV. This is a significant amount when determining the base-to-emitter and base-to-collector junctions.

Classroom Manual
Chapter 3, page 71

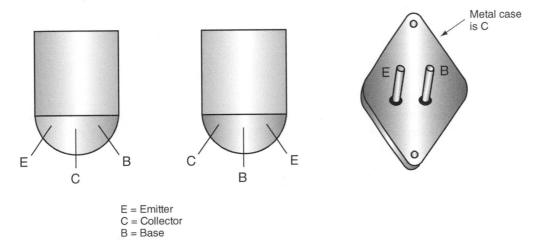

E = Emitter
C = Collector
B = Base

Figure 3-39 There are several configurations of the transistor legs.

PHOTO SEQUENCE 5
Identifying Bipolar Transistors

P5-1 Label the pins of the unknown device; 1, 2, and 3.

P5-2 Place the positive probe of the DMM on pin 1 and measure the diode drop to pins 2 and 3.

P5-3 If the positive probe is on the base of a good NPN transistor, you should read a low diode drop to pins 2 and 3. The base-to-collector diode drop will be slightly lower than the base-to-emitter reading.

P5-4 If one, or both, measurements to pins 2 and 3 are high, put the positive probe on pin 2 and retest. If still high, put the positive probe on pin 3 and retest.

P5-5 If all three pin readings are high, repeat the tests with the negative probe as the common pin. A pair of low readings now indicate a PNP transistor.

Some DMMs cannot read above 1.2 volts, resulting in a good Darlington testing as open. Confirm your DMM is able to read higher than 1.4 volts.

Special Tool

DSO

Testing is similar to that of normal bipolar transistors except that in the forward direction the base-to-emitter reading will be 0.2 to 1.4 volts when reading the DMM on the diode function. This higher voltage is due to the pair of junctions that are in series.

Digital Storage Oscilloscope (DSO)

The most significant advantage of a digital storage oscilloscope (DSO) is the speed at which it samples electrical signals. The mechanical switching speed of switches and relays is measured in thousands of a second or milliseconds. Electrical/electronic switching speed is measured in millionths of a second or microseconds. Radio frequency interference (RFI) is measured in billionths of a second. The DSO operates at 25 million samples per second. Another method of expressing this operating speed is to say the DSO can sample a signal in 40 billionths of a second.

⚙⚙ SERVICE TIP Some power transistors have built-in diodes that are reverse biased across the collector-to-emitter junction and resistors between the base-to-emitter junction. The resistance is usually 50 ohms. If not aware of this, you can be confused by the reading if testing the transistor as a standard bipolar transistor. You will need to know the specifications of the power transistor to test it properly. Power transistors without internal damper diodes test just about like bipolar transistors.

The high-speed sampling of the DSO allows this scope to display an extremely accurate, expanded graph of input sensor and output actuator operation compared to most multimeters. Such increased speed allows the DSO to display glitches or momentary defects in input sensors and output actuators. Some DSOs are capable of displaying as many as six traces at the same time (**Figure 3-40**).

DSO Screen

On the DSO screen, voltage and amperage are displayed vertically, and any changes are shown as vertical movement. Vertical grids on the screen provide a means of measuring the voltage or amperage. The level between the grids is adjustable (**Figure 3-41**). You must know the normal voltages in the circuit in order to select the proper voltage per division on the DSO that provides the most detail with the signal remaining on the screen.

Horizontal movement on the screen represents time. The milliseconds per division on the horizontal grid are adjustable with the time button on the DSO. If there are ten time divisions across the grid and the time set for each division is 5 ms, the time sweep will be 50 ms. Each time a DSO samples a voltage signal, it displays a dot on the screen. The DSO then connects the dots to provide a waveform. Reading a faster signal may require the selection of a shorter time base.

You must select the correct time base for the voltage signal being measured for the waveform to display properly on the screen. If the time base selected is too long and the voltage too high, the waveform is too small to read. Conversely, if the time base is too short and the voltage scale too low, the waveform is too large for the screen.

⚠ Caution

While diagnosing computer systems, always place test equipment, such as DSOs or scan testers, in a secure position where they will not fall onto the floor or into rotating components. Dropping the DSO or scan tool can cause severe meter damage.

Figure 3-40 Digital storage oscilloscope capable of tracing multiple waveforms.

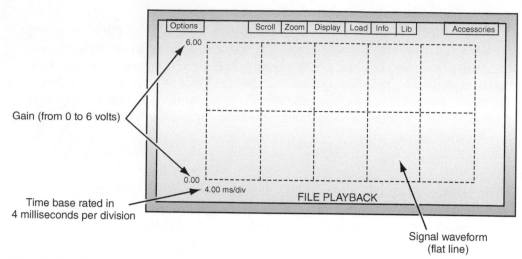

Gain (from 0 to 6 volts)

Time base rated in
4 milliseconds per division

Signal waveform
(flat line)

Figure 3-41 DSO vertical and horizontal screen grids.

Peak, Average, and Root Mean Square Related to an AC Voltage Waveform

The term **peak** represents the highest point in *one* cycle of an AC voltage waveform. When considering both the highest and lowest peaks in an AC voltage waveform, the term **peak-to-peak voltage** is the total voltage measured between these peaks. For example, an AC voltage waveform with a 60-volt peak would have a 120-volt peak-to-peak. Calculate the average AC voltage by multiplying 0.637 × peak voltage. The average voltage on a 60-volt peak would be 0.637 × 60 = 38.2 volts. volts.

In many cases, root mean square (RMS) describes AC voltage. For example, by dividing one cycle of an AC voltage waveform from a 120-volt household electrical outlet into four parts at 90° intervals, the instantaneous voltage and current are recorded for each degree in a 90° interval and then averaged. Multiplying 0.7071 × the peak voltage calculates the square root of the average. The peak voltage for the average 120-volt household outlet is about 170 volts at 60 hertz (Hz). Therefore, 0.7071 × 170 = 120.207 RMS. If this same voltage were measured using the average voltage method, it would be 108.29 volts.

Measuring AC Voltage and Time Base

To display a waveform for a 120-volt household electrical outlet, round off the peak voltage of 170 to 200 volts. If there are eight vertical voltage divisions on the DSO screen with four divisions above and below the center-line, select 50 volts per division to display the high and low peaks on the waveform. Assuming the 0-volt position in the waveform is positioned in the center of the screen, the 50 volts per division selection provides 200 volts above and below the screen centerline to display the 170-volt peak voltage above and below the center-line. Increasing the volts per division setting makes the peaks appear shorter on the screen.

In a 60-hertz AC voltage, one cycle occurs in approximately 18 milliseconds (ms). Displaying one complete AC cycle requires about 20 ms. The average DSO has 10 horizontal divisions. Since 20/10 = 2 ms per division, this time base selection displays one AC voltage waveform. If 4 ms per division is selected, 2 AC voltage waveforms are displayed. Increasing the time base displays more AC voltage cycles on the screen.

0.637 represents 63.7% of the peak voltage. Fifty percent is not used because the rate of change in voltage is slower toward the peak than it is nearer the 0 value. The voltage is above 50% for a longer time than it's below it.

The example of scope setup used here references tracing household voltage since the expected voltage values and cycle timing are known to most.

0.7071 is one divided by the square of two.

Conversely, decreasing the time base displays fewer AC voltage cycles on the screen, and the waveform appears expanded.

Setting the millisecond (ms) time base too low expands the waveform horizontally, resulting in an altered and incomplete waveform display. Adjust the timing so the screen displays only one to three cycles for the best display.

Some DSOs have a button selection that automatically selects baseline volts per division and ms per division. If you select the improper volts per division and ms per division for the voltage signal, this baseline should provide settings to display a waveform on the screen. If needed, you can adjust the volts per division and ms per division to provide the desired display.

Trigger and Trigger Slope

The trigger selection tells the DSO when to begin displaying a waveform. Until the DSO has a trigger level, it doesn't know when to begin the waveform display. When testing an input sensor that operates in a 0-volt to 5-volt range, select a trigger level of one-half this range.

Trigger slope informs the DSO whether the voltage signal is moving upward or downward when it crosses the trigger level. When a negative trigger slope is selected, the voltage signal is moving downward as it crosses the trigger level. Selecting a positive trigger level results in an upward voltage signal trace when it crosses the trigger level. A marker on the left side of the screen indicates the 0-volt or ground voltage position. DSO controls allow for moving the placement of the marker (**Figure 3-42**). A second marker at the top of the screen indicates the trigger location. Since the control buttons on DSOs vary, the technician must spend some time to become familiar with a particular DSO.

Types of Voltage Signals

An **analog signal** is a varying voltage within a specific range over a period of time (**Figure 3-43**). A throttle position sensor (TPS) produces an analog voltage signal each time the throttle is opened (**Figure 3-44**).

A **digital signal** is either on or off. It may be described as one that is always high or low. The leading edge of a digital signal represents an increasing voltage, while the trailing edge of the signal represents a decreasing voltage. If the component uses

Figure 3-42 Ground-level reference marker on the left side of the screen.

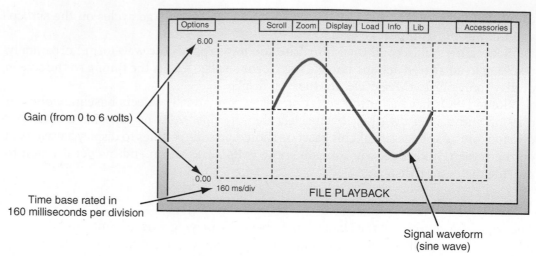

Gain (from 0 to 6 volts)

Time base rated in
160 milliseconds per division

Signal waveform
(sine wave)

Figure 3-43 Analog voltage signal.

Signal waveform

Figure 3-44 TPS analog voltage signal.

insulated side switching to turn on and off, the line across the top of the leading and trailing edge signals represents the length of the component's on time, called pulse width (**Figure 3-45**).

The computer measures the distance between the leading edge of a digital signal and the leading edge of the following signal to determine the waveform's frequency. The distance between the leading edge of one digital signal and the leading edge of the following digital signal is one cycle (**Figure 3-46**). The computer counts the number of cycles over a period of time to establish the frequency. For example, if 92 cycles occur per second, the frequency is 92 hertz.

The relationship between the on time and off time in a digital signal is called duty cycle. For example, if the component on time and off time are equal, the component has a 50% duty cycle. When the component has a 90% duty cycle, the component is on for 90% of the time in a cycle and off for 10% of the time in a cycle. The computer controls some outputs, such as a carburetor mixture control solenoid, by varying the pulse width while the frequency remains constant. This type of computer control is referred to as **pulse width modulation (PWM)**. Other outputs, such as fuel injectors, are controlled by varying the frequency and the pulse width.

Figure 3-45 The component on time in a cycle is called pulse width.

Figure 3-46 The distance between leading edges is one cycle.

Coupling

Based on the input voltage type (DC or AC) selected, you are telling the scope how to interpret the incoming signal. Basically, you are selecting the amplitude of the input as compared with what the scope knows as ground. Thus, a 12-volt square wave appears as a signal moving from 0 V to 12 V.

If you choose to set your input to AC coupling, the input signal is passed through a capacitor then to the processing circuit. The capacitor filters the signal, and the scope averages the input voltage level around zero. For example, if the same signal of the 12 V square wave just discussed were cycling at a duty cycle of 50%, the average would be 6 V. The signal would display as an AC sine wave with 3 volts above and 3 volts below zero.

The capacitor allows a cycling DC voltage to pass but stops any voltage that is not cycling (or is cycling at a very low frequency). There will be times you may wish to select AC coupling although you will be looking at a DC circuit since this will amplify the waveform.

User-Friendly DSOs

DSOs with simplified, user-friendly controls offer auto-ranging that automatically selects the proper voltage and time base for the signal received. The technician may use the DSO controls to turn off the auto-range function and manually select the voltage range and time base.

Pressing the menu key brings up various menus options. You can scroll through the menu to select a specific test. The DSO may display digital readings (number values) on the screen along with the waveforms. For example, the DSO may display the minimum, average, and maximum millivolt (mV) readings along with an O_2 sensor waveform.

Some user-friendly DSOs have a removable application module to help prevent scope obsolescence. A software program card plugs into the bottom of the DSO.

Case Study

A customer brings their vehicle to the shop because the dash lights are not illuminating. The technician checks the fuses and all are good. The technician then substitutes a bulb, known to be good, in the printed circuit, but it still does not illuminate. Next, they use a voltmeter to check for applied voltage to the panel light circuit. The test indicates that 12.6 volts are present.

The technician then performs a voltage drop test on the ground side of the circuit and the voltmeter indicates 12.6 volts. Testing indicates that the printed circuit has an open in the ground side of its circuit. Upon receiving written approval from the customer to perform the repairs, the technician replaces the printed circuit. Verification of the repair confirms the diagnosis.

ASE-Style Review Questions

1. Circuit defects are being discussed.

   Technician A says an open can be only on the ground side of the circuit.

   Technician B says an unwanted resistance can result from a corroded connector.

   Who is correct?

   A. A only
   B. B only
   C. Both A and B
   D. Neither A nor B

2. Testing the fuse is being discussed.

   Technician A says sometimes a visual inspection of a fuse or fusible link does not reveal that it's open.

   Technician B says to use a jumper wire to bypass the fuse in order to test the circuit.

   Who is correct?

   A. A only
   B. B only
   C. Both A and B
   D. Neither A nor B

3. Testing of a switch is being discussed.

   Technician A says a switch can be tested with a voltmeter.

   Technician B says to use a fused jumper wire to see if the switch is operating.

   Who is correct?

   A. A only
   B. B only
   C. Both A and B
   D. Neither A nor B

4. The voltmeter reading illustrated in Figure 3-47 indicates:

   A. An open in the circuit between the control module and the motor

   B. A properly operating PTC

   C. An open PTC

   D. None of these choices

Figure 3-47

5. An LED does not light. A voltmeter indicates 0 volt across the LED. All of the following can be the cause **EXCEPT**:

   A. The current-limiting resistor is open.

   B. The LED is open.

   C. Short to ground between the current-limiting resistor and the LED.

   D. Open ground circuit.

6. Voltage drop testing is being discussed.

   Technician A says it's possible to calculate voltage drop by testing for available voltage on both sides of the component.

   Technician B says excessive voltage drop can be on either the power side or the ground side of the circuit.

   Who is correct?

   A. A only   C. Both A and B

   B. B only   D. Neither A nor B

7. The results of conductor-to-conductor shorts are being discussed.

   Technician A says if the windings of an electric motor are shorted together, then the amount of resistance will be higher than specified.

   Technician B says a short between circuits can result in both circuits operating by closing one switch.

   Who is correct?

   A. A only   C. Both A and B

   B. B only   D. Neither A nor B

8. The testing of a shorted circuit is being discussed.

   Technician A says if a fuse blows as soon as it's installed, this indicates a short to ground.

   Technician B says if the short to ground is on the ground side of the load component, but before a grounding switch, the component will not turn off.

   Who is correct?

   A. A only   C. Both A and B

   B. B only   D. Neither A nor B

9. Testing of a transistor is being discussed.

   Technician A says to test the base-collector junction and the base-emitter junction as if they were standard diodes.

   Technician B says the resistance between the collector and emitter should read open circuit.

   Who is correct?

   A. A only   C. Both A and B

   B. B only   D. Neither A nor B

10. Testing of a potentiometer is being discussed.

   Technician A says a voltmeter can be used if jumper wires are connected to the wire connector and sensor to gain access to the test points.

   Technician B says the wires can be pierced to test for voltage.

   Who is correct?

   A. A only   C. Both A and B

   B. B only   D. Neither A nor B

ASE Challenge Questions

1. The fuse for an A/C blower motor circuit fails after a short period of time.

   Technician A says that the blower motor may be binding internally.

   Technician B says that the blower motor ground circuit may have excessive resistance.

   Who is correct?

   A. A only   C. Both A and B

   B. B only   D. Neither A nor B

2. Relay testing is being discussed.

   Technician A says that it's acceptable to bypass the relay control coil terminals with a fused jumper wire during the test sequence.

   Technician B says to check the voltage drop across the relay load contact terminals with the relay control circuit de-energized.

   Who is correct?

   A. A only   C. Both A and B

   B. B only   D. Neither A nor B

3. The troubleshooting of a parallel circuit that contains three dimly lit bulbs is being discussed. A voltmeter that is placed across each of the bulbs indicates 7.2 volts.

   Technician A says that the power supply that is common to all three bulbs may be faulty.

   Technician B says that the ground terminal that is common to all three bulbs may have excessive resistance.

   Who is correct?

   A. A only   C. Both A and B

   B. B only   D. Neither A nor B

4. An A/C compressor clutch coil spike suppression diode is being tested with ohmmeter DMM set on diode test mode. The meter reads infinite in both directions.

   Technician A says that the diode is electrically open.

   Technician B says that using this diode would result in the failure of the circuit fuse.

   Who is correct?

   A. A only   C. Both A and B

   B. B only   D. Neither A nor B

5. How much total resistance is in a 12-volt circuit that is drawing 4 amperes?

   A. 0.333 ohm   C. 4.8 ohms

   B. 3 ohms   D. 48 ohms

6. A voltmeter connected across the input and output terminals of an instrument cluster illumination lamp rheostat indicates 12.6 volts with the switch in the maximum brightness position and the engine off. Which of the following statements is true?

   A. The voltage available to the lamps will be 12.6 volts.

   B. The voltage available to the lamps will be 0 volt.

   C. The rheostat is operating correctly.

   D. More information is required to determine if the lamps will operate correctly.

7. The left-rear and right-rear taillights and the left-rear brake light of a vehicle illuminate dimly whenever the brake pedal is pressed; however, the right-rear brake light operates properly.

   Technician A says the left-rear taillight and brake light may have a poor ground connection.

   Technician B says the brake light switch may have excessive resistance.

   Who is correct?

   A. A only   C. Both A and B

   B. B only   D. Neither A nor B

8. The horn of a vehicle equipped with a horn relay sounds weak and distorted. Which of the following is the least likely cause of this problem?

   A. High resistance in the relay load circuit.

   B. High resistance in the horn ground circuit.

   C. Excessive voltage drop between the relay load contact and the horn.

   D. Excessive voltage drop across the relay coil winding.

9. The power seats don't move fore or aft. The technician notices that the motor attempts to move the seat, then stops. The LEAST likely cause of this is:

   A. A binding track

   B. A short in the motor armature

   C. A faulty PTC

   D. An obstruction under the seat

10. The current draw of a window motor is being measured.

    Technician A says the ammeter can be connected to the power side of the motor.

    Technician B says the ammeter can be connected on the ground side of the motor.

    Who is correct?

    A. A only   C. Both A and B

    B. B only   D. Neither A nor B

Name _____   Date _____

VOLTAGE DROP TESTING VS OPEN CIRCUIT TESTING

Upon completion of this job sheet, you should be able to perform a voltage drop test and an open circuit test and determine the benefits and limitations of each to locating high-resistance faults.

ASE Education Foundation Correlation

This job sheet addresses the following **MLR** tasks:

A.5.   Demonstrate proper use of a digital multimeter (DMM) when measuring source voltage, voltage drop (including grounds), current flow, and resistance. **(P-1)**

A.6.   Demonstrate knowledge of the causes and effects from shorts, grounds, opens, and resistance problems in electrical/electronic circuits. **(P-1)**

A.7.   Describe types of test lights; use appropriate test light to check operation of electrical circuits per service information. **(P-2)**

A.8.   Use fused jumper wires to check operation of electrical circuits per service information. **(P-2)**

This job sheet addresses the following **AST/MAST** tasks:

A.5.   Demonstrate proper use of a digital multimeter (DMM) when measuring source voltage, voltage drop (including grounds), current flow, and resistance. **(P-1)**

A.6.   Demonstrate knowledge of the causes and effects from shorts, grounds, opens, and resistance problems in electrical/electronic circuits. **(P-1)**

A.7.   Describe types of test lights; use appropriate test light to check operation of electrical circuits per service information. **(P-1)**

A.8.   Use fused jumper wires to check operation of electrical circuits per service information. **(P-2)**

We Support
ASE | **Education Foundation**

Tools and Materials

- A vehicle
- Fender covers
- DMM
- 35-ohm resistor attached to fused jumper wires
- Test light
- Fused jumper wire

Safety Considerations and PPE

- Safety glasses

Procedure

1. Measure the voltage across the battery terminals and record your results.

   _____

   _____

2. Connect one end of the fused jumper wire with the 35-ohm resistor to the battery positive (+) terminal. CAUTION: Be sure the other end of the jumper wire does not come into contact with ground.

3. With the black lead of the DMM connected to the negative (–) terminal of the battery, connect the red lead to the free end of the fused jumper wire. Record the voltage reading.

_____

_____

4. Steps 1 through 3 represented open circuit testing. This would be the same as disconnecting a connector and testing for applied voltage on the feed terminal. The free end of the installed jumper wire represents the disconnected connector. Is this reading in step 3 the same as in step 1?   ☐ Yes ☐ No

   Explain why: _____

_____

5. Measure the voltage across the battery terminals and record.

_____

_____

6. Connect the terminal of the test light to the negative (–) post of the battery or a good ground. Connect the probe of the test light to the positive (+) terminal of the battery. Explain the behavior of the test light.

_____

_____

7. Move the probe of the test light to the free end of the jumper wire with the resistor installed. How does the test light behave?

_____

_____

8. With the test light still connected, connect the black lead of the DMM to ground and the red lead to the tip of the probe. What is the voltage reading?

_____

_____

9. Steps 5 through 8 show testing the circuit under load. The test light represents the load. Why is the voltage different in Step 8 than the reading you had in Step 3?

_____

_____

10. With the test light still connected as in Steps 6 and 7, place the red DMM test lead at the positive (+) terminal of the battery and the black lead at the probe of the test light. What is the voltage reading?

_____

_____

11. What does this voltage represent?

_____

_____

12. Explain to your instructor how you would perform the load and voltage drop tests on the ground side of a circuit. Demonstrate this using the jumper wire with the resistor installed, the test light, and the DMM.

_____

_____

_____

Instructor's Evaluation:

Grade	Rubric
0	No exposure due to: ☐ Absence ☐ Other:
1	Student unsuccessfully attempted the task
2	Student completed the task with significant instructor assistance.
3	Satisfactory performance of task. The student was able to perform elements of the task with minimal instructor assistance.
4	Competent/Proficient. Able to perform all elements of the task successfully and independently.

Instructor's Notes:

Name _____   Date _____

TESTING FOR AN OPEN CIRCUIT

Upon completion of this job sheet, you should be able to test a circuit and locate the open.

ASE Education Foundation Correlation

This job sheet addresses the following **MLR** tasks:

A.1.   Research vehicle service information such as fluid type, vehicle service history, service precautions, technical service bulletins, and recalls including vehicles with advanced driver assistance systems (ADAS). **(P-1)**

A.2.   Identify electrical/electronic system components and configuration. **(P-1)**

A.5.   Demonstrate proper use of a digital multimeter (DMM) when measuring source voltage, voltage drop (including grounds), current flow, and resistance. **(P-1)**

A.6.   Demonstrate knowledge of the causes and effects from shorts, grounds, opens, and resistance problems in electrical/electronic circuits. **(P-1)**

A.7.   Describe types of test lights; use appropriate test light to check operation of electrical circuits per service information. **(P-2)**

A.8.   Use fused jumper wires to check operation of electrical circuits per service information. **(P-2)**

A.9.   Use wiring diagrams to trace electrical/electronic circuits. **(P-1)**

This job sheet addresses the following **AST/MAST** tasks:

A.1.   Research vehicle service information such as fluid type, vehicle service history, service precautions, technical service bulletins, and recalls including vehicles with advanced driver assistance systems (ADAS). **(P-1)**

A.2.   Identify electrical/electronic system components and configuration. **(P-1)**

A.5.   Demonstrate proper use of a digital multimeter (DMM) when measuring source voltage, voltage drop (including grounds), current flow, and resistance. **(P-1)**

A.6.   Demonstrate knowledge of the causes and effects from shorts, grounds, opens, and resistance problems in electrical/electronic circuits. **(P-1)**

A.7.   Describe types of test lights; use appropriate test light on an electrical circuit per service information. **(P-1)**

A.8.   Use fused jumper wires to check operation of electrical circuits per service information. **(P-1)**

A.9.   Use wiring diagrams during the diagnosis of electrical/electronic circuit problems. **(P-1)**

We Support
ASE | Education Foundation

Tools and Materials
- A vehicle
- Fender covers
- DMM
- Test light
- Fused jumper wire
- Wiring diagram for the vehicle

Safety Considerations and PPE
- Safety glasses

Procedure

1. What is the customer's complaint?

   _____

   _____

2. Can the complaint be verified?   ☐ Yes   ☐ No

   (If no, consult your instructor.)

3. Are there any other related symptoms?   ☐ Yes   ☐ No

   If yes, describe the symptom.

   _____

   _____

4. Following the wiring diagram, use a voltmeter, fused jumper wire, or test light to trace the circuit to locate the open.

5. Describe the location of the open.

   _____

   _____

   _____

Instructor's Evaluation:

Grade	Rubric
0	No exposure due to: ☐ Absence ☐ Other:
1	Student unsuccessfully attempted the task
2	Student completed the task with significant instructor assistance.
3	Satisfactory performance of task. The student was able to perform elements of the task with minimal instructor assistance.
4	Competent/Proficient. Able to perform all elements of the task successfully and independently.

Instructor's Notes:

Name _____   Date _____

TESTING FOR A SHORT TO GROUND

Upon completion of this job sheet, you should be able to test a circuit and locate the short to ground.

ASE Education Foundation Correlation

This job sheet addresses the following **MLR** tasks:

A.1.   Research vehicle service information such as fluid type, vehicle service history, service precautions, technical service bulletins, and recalls including vehicles with advanced driver assistance systems (ADAS). **(P-1)**

A.2.   Identify electrical/electronic system components and configuration. **(P-1)**

A.5.   Demonstrate proper use of a digital multimeter (DMM) when measuring source voltage, voltage drop (including grounds), current flow, and resistance. **(P-1)**

A.6.   Demonstrate knowledge of the causes and effects from shorts, grounds, opens, and resistance problems in electrical/electronic circuits. **(P-1)**

A.7.   Describe types of test lights; use appropriate test light to check operation of electrical circuits per service information. **(P-2)**

A.8.   Use fused jumper wires to check operation of electrical circuits per service information. **(P-2)**

A.9.   Use wiring diagrams to trace electrical/electronic circuits. **(P-1)**

This job sheet addresses the following **AST/MAST** tasks:

A.1.   Research vehicle service information such as fluid type, vehicle service history, service precautions, technical service bulletins, and recalls including vehicles with advanced driver assistance systems (ADAS). **(P-1)**

A.2.   Identify electrical/electronic system components and configuration. **(P-1)**

A.5.   Demonstrate proper use of a digital multimeter (DMM) when measuring source voltage, voltage drop (including grounds), current flow, and resistance. **(P-1)**

A.6.   Demonstrate knowledge of the causes and effects from shorts, grounds, opens, and resistance problems in electrical/electronic circuits. **(P-1)**

A.7.   Describe types of test lights; use appropriate test light on an electrical circuit per service information. **(P-1)**

A.8.   Use fused jumper wires to check operation of electrical circuits per service information. **(P-1)**

A.9.   Use wiring diagrams during the diagnosis of electrical/electronic circuit problems. **(P-1)**

We Support
ASE | Education Foundation

Tools and Materials

- A vehicle
- Fender covers
- DMM
- Test light
- Fused jumper wires

- Circuit breaker fitted with alligator clips
- Gauss gauge
- Wiring diagram for the vehicle

Safety Considerations and PPE

- Safety glasses

Procedure

1. Which circuit is affected by the short to ground?

   _____

2. Are there any other related symptoms? ☐ Yes ☐ No

   If yes, describe the symptom(s).

   _____

3. Pull the fuse or circuit protection for the affected circuit from the fuse box.

4. With ignition switch in the RUN position, connect the test light across the fuse terminals in the fuse box. Does the test light illuminate?   ☐ Yes   ☐ No

5. What does this test indicate?

   _____

   _____

   _____

6. With the test light connected across the fuse terminals of the fuse box, disconnect components and connectors in the affected circuit that are identified in the wiring diagram.

7. Did the test light go out when a component was disconnected?   ☐ Yes   ☐ No

8. What can be concluded thus far?

   _____

   _____

9. Connect the test circuit breaker across the fuse terminals of the fuse box and use the Gauss gauge to find the location of the short to ground.

10. Describe the location of the short to ground.

    _____

    _____

    _____

    _____

Instructor's Evaluation:

Grade	Rubric
0	No exposure due to: ☐ Absence ☐ Other:
1	Student unsuccessfully attempted the task
2	Student completed the task with significant instructor assistance.
3	Satisfactory performance of task. The student was able to perform elements of the task with minimal instructor assistance.
4	Competent/Proficient. Able to perform all elements of the task successfully and independently.

Instructor's Notes:

Name _____   Date _____

TESTING CIRCUIT PROTECTION DEVICES

Upon completion of this job sheet, you should be able to test circuit protection devices for opens.

ASE Education Foundation Correlation

This job sheet addresses the following **MLR** tasks:

A.1. Research vehicle service information such as fluid type, vehicle service history, service precautions, technical service bulletins, and recalls including vehicles with advanced driver assistance systems (ADAS). **(P-1)**

A.2. Identify electrical/electronic system components and configuration. **(P-1)**

A.5. Demonstrate proper use of a digital multimeter (DMM) when measuring source voltage, voltage drop (including grounds), current flow, and resistance. **(P-1)**

A.7. Describe types of test lights; use appropriate test light on an electrical circuit per service information. **(P-2)**

A.11. Inspect and test fusible links, circuit breakers, and fuses. **(P-1)**

This job sheet addresses the following **AST/MAST** tasks:

A.1. Research vehicle service information such as fluid type, vehicle service history, service precautions, technical service bulletins, and recalls including vehicles with advanced driver assistance systems (ADAS).

A.2. Identify electrical/electronic system components and configuration. **(P-1)**

A.5. Demonstrate proper use of a digital multimeter (DMM) when measuring source voltage, voltage drop (including grounds), current flow, and resistance. **(P-1)**

A.7. Describe types of test lights; use appropriate test light on an electrical circuit per service information. **(P-1)**

A.11. Inspect and test fusible links, circuit breakers, and fuses; determine needed action. **(P-1)**

We Support
ASE | Education Foundation

Tools and Materials
- A vehicle equipped with fusible links and PTC circuit protection
- Fender covers
- A DMM
- Test light

Safety Considerations and PPE
- Safety glasses

Procedure
1. Locate the fuse panel or power distribution center.
2. Check that the test light is working properly by connecting it across the battery.
3. Connect the negative lead of a test light to a good ground.
4. Turn the ignition switch to the RUN position.
5. Touch the probe of the test light onto the metal test tabs on each side of the fuse.

6. Did the test light illuminate on each side of the fuse? ☐ Yes ☐ No Why or why not?

_____

_____

_____

7. Repeat for all fuses in the fuse box. Record your findings.

_____

_____

_____

8. Remove any fuses that failed the test.

9. Visually inspect the fuse. Record your findings.

_____

_____

_____

10. Use an ohmmeter and test the fuses. Record your findings.

_____

_____

_____

11. Locate a fusible link on the vehicle.

12. Use a voltmeter and measure the voltage drop over the link. Record your results.

_____

_____

_____

13. Disconnect power to the fusible link and use an ohmmeter to test the link. Record your results.

_____

_____

_____

14. Locate a PTC on the vehicle.

15. Use service information to identify the PTC and its location from the circuit assigned to you by your instructor.

_____

_____

16. While operating the accessory, measure the voltage on the input side of the PTC.

_____

17. While operating the accessory, ensure the voltage on the output side of the PTC.

_____

18. Connect the red test lead to the input side of the PTC and the black lead to the output side of the PTC and operate the system the PTC protects. Record the voltmeter reading.

_____

19. Based on your observations, what can you conclude about the functionality of the PTC?

_____

_____

Instructor's Evaluation:

Grade	Rubric
0	No exposure due to: ☐ Absence ☐ Other:
1	Student unsuccessfully attempted the task
2	Student completed the task with significant instructor assistance.
3	Satisfactory performance of task. The student was able to perform elements of the task with minimal instructor assistance.
4	Competent/Proficient. Able to perform all elements of the task successfully and independently.

Instructor's Notes:

Name _____   Date _____

TESTING SWITCHES

Upon completion of this job sheet, you should be able to test a switch and properly determine needed repairs.

ASE Education Foundation Correlation

This job sheet addresses the following **MLR** tasks:

A.1. Research vehicle service information such as fluid type, vehicle service history, service precautions, technical service bulletins, and recalls including vehicles with advanced driver assistance systems (ADAS). **(P-1)**

A.2. Identify electrical/electronic system components and configuration. **(P-1)**

A.5. Demonstrate proper use of a digital multimeter (DMM) when measuring source voltage, voltage drop (including grounds), current flow, and resistance. **(P-1)**

A.7. Describe types of test lights; use appropriate test light on an electrical circuit per service information. **(P-2)**

A.8. Use fused jumper wires to check operation of electrical circuits per service information. **(P-2)**

A.9. Use wiring diagrams to trace electrical/electronic circuits. **(P-1)**

A.12. Inspect, test, repair, and/or replace components, connectors, terminals, harnesses, and wiring in electrical/electronic systems (including solder repairs). **(P-2)**

This job sheet addresses the following **AST/MAST** tasks:

A.1. Research vehicle service information such as fluid type, vehicle service history, service precautions, technical service bulletins, and recalls including vehicles with advanced driver assistance systems (ADAS). **(P-1)**

A.2. Identify electrical/electronic system components and configuration. **(P-1)**

A.5. Demonstrate proper use of a digital multimeter (DMM) when measuring source voltage, voltage drop (including grounds), current flow, and resistance. **(P-1)**

A.7. Describe types of test lights; use appropriate test light on an electrical circuit per service information. **(P-1)**

A.8. Use fused jumper wires to check operation of electrical circuits per service information. **(P-1)**

A.9. Use wiring diagrams during the diagnosis of electrical/electronic circuit problems. **(P-1)**

A.12. Inspect, test, repair, and/or replace components, connectors, terminals, harnesses, and wiring in electrical/electronic systems (including solder repairs); determine needed action. **(P-1)**

We Support ASE | Education Foundation

Tools and Materials
- A vehicle
- Fender covers
- DMM
- Test light
- Fused jumper wires
- Wiring diagram for the vehicle

Safety Considerations and PPE
- Safety glasses

Procedure

1. Locate the brake light switch (or another switch as directed by your instructor).

2. Disconnect the switch from the wire harness.

3. Use an ohmmeter to measure the resistance of the switch with the brake pedal released. Record your results.

   _____

4. With the ohmmeter still connected across the switch terminal, press the brake pedal and record the ohmmeter reading:

   _____

5. Based on your results, is the switch operating properly?   ☐ Yes   ☐ No

   Why? _____

   _____

6. With the electrical connector to the switch still unplugged, connect a fused jumper wire across the battery feed and brake light circuits.

7. Do the brake lights come on?   ☐ Yes   ☐ No

   If Yes, what is the faulty component?

   _____

   _____

Instructor's Evaluation:

Grade	Rubric
0	No exposure due to: ☐ Absence ☐ Other:
1	Student unsuccessfully attempted the task
2	Student completed the task with significant instructor assistance.
3	Satisfactory performance of task. The student was able to perform elements of the task with minimal instructor assistance.
4	Competent/Proficient. Able to perform all elements of the task successfully and independently.

Instructor's Notes:

Name _____   Date _____

TESTING RELAYS

Upon completion of this job sheet, you should be able to test a relay and properly determine needed repairs.

ASE Education Foundation Correlation ────────────────────────

This job sheet addresses the following **MLR** tasks:

A.1.   Research vehicle service information such as fluid type, vehicle service history, service precautions, technical service bulletins, and recalls including vehicles with advanced driver assistance systems (ADAS). **(P-1)**

A.2.   Identify electrical/electronic system components and configuration. **(P-1)**

A.5.   Demonstrate proper use of a digital multimeter (DMM) when measuring source voltage, voltage drop (including grounds), current flow, and resistance. **(P-1)**

A.7.   Describe types of test lights; use appropriate test light to check operation of electrical circuits per service information. **(P-2)**

A.8.   Use fused jumper wires to check operation of electrical circuits per service information. **(P-2)**

A.9.   Use wiring diagrams to trace electrical/electronic circuits. **(P-1)**

A.12.   Inspect, test, repair, and/or replace components, connectors, terminals, harnesses, and wiring in electrical/electronic systems (including solder repairs). **(P-2)**

This job sheet addresses the following **AST/MAST** tasks:

A.1.   Research vehicle service information such as fluid type, vehicle service history, service precautions, technical service bulletins, and recalls including vehicles with advanced driver assistance systems (ADAS). **(P-1)**

A.2.   Identify electrical/electronic system components and configuration. **(P-1)**

A.5.   Demonstrate proper use of a digital multimeter (DMM) when measuring source voltage, voltage drop (including grounds), current flow, and resistance. **(P-1)**

A.7.   Describe types of test lights; use appropriate test light to check operation of electrical circuits per service information. **(P-1)**

A.8.   Use fused jumper wires to check operation of electrical circuits per service information. **(P-1)**

A.9.   Use wiring diagrams during the diagnosis of electrical/electronic circuit problems. **(P-1)**

A.12.   Inspect, test, repair, and/or replace components, connectors, terminals, harnesses, and wiring in electrical/electronic systems (including solder repairs); determine needed action. **(P-1)**

We Support

ASE | Education Foundation

Tools and Materials

- A vehicle
- Fender covers
- DMM
- Fused jumper wires
- Test light
- Wiring diagram for the vehicle

Safety Considerations and PPE

- Safety glasses

Procedure

1. Locate the low-speed fan relay (or another relay as directed by your instructor).

2. Disconnect the relay from the relay box.

3. Check the wiring diagram for the relay being tested.

    Identify the terminal callouts and their purpose.

    _____

    _____

    _____

    _____

    _____

4. Is the control circuit of the relay through an insulated or a ground switch?

    _____

    _____

5. Use a voltmeter to check for available voltage to the two terminals of the relay box that connect to the relay terminals.

    Is voltage available at both terminals? ☐ Yes ☐ No

    If no, what would be your next step?

    _____

    _____

6. Connect a test light or DMM across the two terminals of the relay box that connects to the control side of the relay.

7. Use the switch(es) that normally control the relay and turn them ON. Did the test light illuminate or the voltmeter read close to battery voltage? ☐ Yes ☐ No

    If No, what does this indicate?

    _____

    _____

    If Yes, what does this indicate?

    _____

    _____

8. Use a fused jumper wire to connect the high-current input terminal of the relay box to the high-current output terminal. Did the fan come on? ☐ Yes ☐ No

    If No, what does this indicate?

    _____

    _____

    _____

If Yes, what does this indicate?

_____

_____

_____

9. Use the service information to locate the specifications for the resistance of the relay coil. Record your findings.

_____

10. Use an ohmmeter to measure the resistance of the relay's coil. Is the reading within specifications?   ☐ Yes   ☐ No

If No, what does the reading indicate the problem is?

_____

_____

11. Use a pair of fused jumper wires to energize the coil. Check for continuity through the relay contacts. What is your reading?

_____

_____

12. Based on your results of this series of relay tests, what are your conclusions?

_____

_____

_____

_____

_____

Instructor's Evaluation:

Grade	Rubric
0	No exposure due to: ☐ Absence ☐ Other:
1	Student unsuccessfully attempted the task
2	Student completed the task with significant instructor assistance.
3	Satisfactory performance of task. The student was able to perform elements of the task with minimal instructor assistance.
4	Competent/Proficient. Able to perform all elements of the task successfully and independently.

Instructor's Notes:

_____

Name _____   Date _____

TESTING A DIODE

Upon completion of this job sheet, you should be able to test a diode for a short or open.

ASE Education Foundation Correlation

This job sheet addresses the following **MLR** tasks:

A.1.   Research vehicle service information such as fluid type, vehicle service history, service precautions, technical service bulletins, and recalls including vehicles with advanced driver assistance systems (ADAS). **(P-1)**

A.2.   Identify electrical/electronic system components and configuration. **(P-1)**

A.5.   Demonstrate proper use of a digital multimeter (DMM) when measuring source voltage, voltage drop (including grounds), current flow, and resistance. **(P-1)**

A.12.   Inspect, test, repair, and/or replace components, connectors, terminals, harnesses, and wiring in electrical/electronic systems (including solder repairs). **(P-2)**

This job sheet addresses the following **AST/MAST** tasks:

A.1.   Research vehicle service information such as fluid type, vehicle service history, service precautions, technical service bulletins, and recalls including vehicles with advanced driver assistance systems (ADAS). **(P-1)**

A.2.   Identify electrical/electronic system components and configuration. **(P-1)**

A.5.   Demonstrate proper use of a digital multimeter (DMM) when measuring source voltage, voltage drop (including grounds), current flow, and resistance. **(P-1)**

A.12.   Inspect, test, repair, and/or replace components, connectors, terminals, harnesses, and wiring in electrical/electronic systems (including solder repairs); determine needed action. **(P-1)**

We Support
ASE | **Education Foundation**

Tools and Materials
- An assortment of diodes
- DMM

Procedure

1. Which side is the stripe around the diode on?   ☐ Anode   ☐ Cathode

2. Use the ohmmeter function of the DMM and measure the resistance in both directions through the diode. Record your results.

   Forward biased _____

   Reverse biased _____

3. Explain why a DMM ohmmeter is not recommended for testing a diode.

   _____

   _____

   _____

4. Use a DMM to test a second diode using the diode test function. Connect the red test lead to the anode and the black lead to the cathode side of the diode. Record the meter readings.

   _____

   _____

5. What does this reading represent?

_____

_____

6. Reverse the test leads across the diode. What is the reading?

_____

_____

7. What does this reading represent?

_____

_____

8. What is your conclusion concerning this diode?

_____

_____

Instructor's Evaluation:

Grade	Rubric
0	No exposure due to: ☐ Absence ☐ Other:
1	Student unsuccessfully attempted the task
2	Student completed the task with significant instructor assistance.
3	Satisfactory performance of task. The student was able to perform elements of the task with minimal instructor assistance.
4	Competent/Proficient. Able to perform all elements of the task successfully and independently.

Instructor's Notes:

_____

Name _____   Date _____

USING A LAB SCOPE ON SENSORS AND SWITCHES

Upon completion of this job sheet, you should be able to connect a lab scope and observe the activity of various sensors and switches.

ASE Education Foundation Correlation ─────────────────────────

This job sheet addresses the following **MLR** tasks:

A.1.	Research vehicle service information such as fluid type, internal combustion engine operation, vehicle service history, service precautions, technical service bulletins, and recalls including vehicles equipped with advanced driver assistance systems (ADAS).**(P-1)**
A.5.	Demonstrate proper use of a digital multimeter (DMM) when measuring source voltage, voltage drop (including grounds), current flow, and resistance. **(P-1)**
A.9.	Use wiring diagrams to trace electrical/electronic circuits. **(P-1)**
A.12.	Inspect, test, repair, and/or replace components, connectors, terminals, harnesses, and wiring in electrical/electronic systems (including solder repairs). **(P-2)**

This job sheet addresses the following **AST/MAST** tasks:

A.1.	Research vehicle service information such as fluid type, internal combustion engine operation, vehicle service history, service precautions, technical service bulletins, and recalls including vehicles equipped with advanced driver assistance systems (ADAS).**(P-1)**
A.5.	Demonstrate proper use of a digital multimeter (DMM) when measuring source voltage, voltage drop (including grounds), current flow, and resistance. **(P-1)**
A.9.	Use wiring diagrams during the diagnosis of electrical/electronic circuit problems. **(P-1)**
A.12.	Inspect, test, repair, and/or replace components, connectors, terminals, harnesses, and wiring in electrical/electronic systems (including solder repairs); determine needed action. **(P-1)**

This job sheet addresses the following **AST** task:

A.13.	Test and measure circuit using an oscilloscope and/or graphing multimeter (GMM); interpret results; determine needed action. **(P-2)**

This job sheet addresses the following **MAST** task:

A.13.	Test and measure circuit using an oscilloscope and/or graphing multimeter (GMM); interpret results; determine needed action. **(P-1)**

We Support
Education Foundation

Tools and Materials

- A vehicle with accessible sensors and switches
- Service information for the assigned vehicle
- Component locator manual for the assigned vehicle
- Fender covers
- Lab scope
- DMM

Procedure

1. Connect the lab scope test leads across the battery. Make sure the scope is properly set. Observe the trace on the scope. Is there evidence of noise? Explain.

   _____

   _____

   _____

   _____

2. Locate the A/C compressor clutch control wires. Connect the DMM to read available voltage. Start and run the engine. Observe the meter, and then turn the compressor on. What happened on the meter?

   _____

   _____

   Now connect the lab scope to the same point with the compressor turned off. Observe the waveform and then turn the compressor on. What happened to the trace?

   _____

   _____

3. Turn off the engine but keep the ignition on. Locate the throttle position (TP) sensor and identify the purpose of each wire to it. List each wire and describe the purpose of each.

   _____

   _____

4. Connect the DMM to read reference voltage at the TP sensor. What do you read?

   _____

   _____

   Now move the leads to read the output of the sensor. Starting with the throttle closed, slowly open the throttle until it's wide open. Watch the voltmeter while doing this. Describe your readings.

   _____

   _____

5. Now connect the lab scope to read reference voltage at the TP sensor. Describe what you see on the trace (is the voltage steady, evidence of noise, EMI, RFI?).

   _____

   _____

   Now move the leads to read the output of the sensor. Starting with the throttle closed, slowly open the throttle until it's wide open. Watch the trace while doing this. Describe your readings.

   _____

   _____

   _____

6. Locate the oxygen sensor and identify the purpose of each wire to it. Connect the DMM to read voltage generated by the sensor. (To do this, you may use an electrical connector for the O_2 sensor that is positioned away from the hot exhaust manifold.) Now run the engine. Watch the meter as the O_2 sensor warms and describe what happened below.

_____

_____

_____

7. Now connect the lab scope to read voltage output from the sensor. Watch the trace and describe what happened below.

_____

_____

_____

8. Explain what you observed as the differences between testing with a DMM and a lab scope.

_____

_____

_____

Instructor's Evaluation:

Grade	Rubric
0	No exposure due to: ☐ Absence ☐ Other:
1	Student unsuccessfully attempted the task
2	Student completed the task with significant instructor assistance.
3	Satisfactory performance of task. The student was able to perform elements of the task with minimal instructor assistance.
4	Competent/Proficient. Able to perform all elements of the task successfully and independently.

Instructor's Notes:

_____

DIAGNOSTIC CHART 3-1

PROBLEM AREA:	Electrical opens.
SYMPTOMS:	Electrical component will not operate.
POSSIBLE CAUSES:	**1.** Broken conductor. **2.** Defective switch. **3.** Defective relay. **4.** Blown fuses. **5.** Burned fusible links. **6.** Burned or defective circuit breakers.

DIAGNOSTIC CHART 3-2

PROBLEM AREA:	Excessive resistance resulting in lowered electrical output.
SYMPTOMS:	Electrical components fail to operate or operate at reduced efficiency.
POSSIBLE CAUSES:	**1.** Excessive resistance. **2.** Corroded or damaged conductor. **3.** Excessive resistance in the switch. **4.** Excessive resistance in the relay. **5.** Improper stepped resistor values **6.** Defective variable resistor.

DIAGNOSTIC CHART 3-3

PROBLEM AREA:	Copper-to-copper short or short to ground.
SYMPTOMS:	No electrical component operation or operation when another control switch is activated.
POSSIBLE CAUSES:	**1.** Broken or burned insulation and/or connectors causing copper-to-copper short. **2.** Broken or burned insulation and/or connectors causing a short to ground.

DIAGNOSTIC CHART 3-4

PROBLEM AREA:	Switch.
SYMPTOMS:	Component fails to turn on.
POSSIBLE CAUSES:	Faulty terminals to switch. Burned switch contacts. Internal open in switch.

DIAGNOSTIC CHART 3-5

PROBLEM AREA:	Switch.
SYMPTOMS:	Component fails to turn off.
POSSIBLE CAUSES:	Short across the terminals to switch. Stuck switch contacts. Internal short in switch.

DIAGNOSTIC CHART 3-6

PROBLEM AREA:	Relays.
SYMPTOMS:	Component fails to turn on.
POSSIBLE CAUSES:	Faulty terminals to relay.
	Open in relay coil.
	Shorted relay coil.
	Burned relay high-current contacts.
	Internal open in relay high-current circuit.

DIAGNOSTIC CHART 3-7

PROBLEM AREA:	Relay.
SYMPTOMS:	Component fails to turn off.
POSSIBLE CAUSES:	Short across the relay terminals.
	Stuck relay high-current contacts.

DIAGNOSTIC CHART 3-8

PROBLEM AREA:	Stepped resistors.
SYMPTOMS:	Component fails to turn on.
POSSIBLE CAUSES:	Faulty terminal connections to the stepped resistor.
	Open in the input side of the resistor.
	Open in the output side of the resistor.

DIAGNOSTIC CHART 3-9

PROBLEM AREA:	Stepped resistors.
SYMPTOMS:	Component fails to operate at certain speeds or brightness.
POSSIBLE CAUSES:	Excessive resistance at the terminal connections to the stepped resistor.
	Open in one or more of the resistor's circuit.
	Short across one or more of the resistor's circuit.

DIAGNOSTIC CHART 3-10

PROBLEM AREA:	Variable resistors.
SYMPTOMS:	Component fails to turn on.
POSSIBLE CAUSES:	Faulty terminal connections to the variable resistor.
	Open in the input side of the resistor.
	Open in the output side of the resistor.
	Open in the return circuit of the resistor.
	Short to ground in the output circuit.
	Short to ground in the input circuit.

DIAGNOSTIC CHART 3-11

PROBLEM AREA:	Variable resistors.
SYMPTOMS:	Component fails to operate at certain speeds or brightness.
POSSIBLE CAUSES:	Excessive resistance at the terminal connections to the variable resistor. Excessive internal resistance. Intermittent opens in the resistor.

DIAGNOSTIC CHART 3-12

PROBLEM AREA:	Circuit protection devices.
SYMPTOMS:	Component fails to turn on.
POSSIBLE CAUSES:	Faulty terminal connections to the protection device. Excessive current draw due to electrical short. Excessive current draw due to mechanical restrictions. Incorrect amperage rating of fuse for the circuit.

CHAPTER 4

WIRING REPAIR AND READING WIRING DIAGRAMS

Upon completion and review of this chapter, you should be able to:

- Perform repairs to copper wire using solderless connections.
- Solder splices to copper wire.
- Repair aluminum wire according to manufacturer's requirements.
- Repair twisted/shielded wire.
- Replace fusible links.

- Repair and/or replace the terminals of a hard-shell connector.
- Repair and/or replace the terminals of weather and metri-pack connectors.
- Read a wiring diagram to correctly determine the operation of the circuit.
- Use the wiring diagram to diagnose possible causes for the system fault.

Terms to Know

Color codes	Pull to seat	Splice clip
Crimping	Push to seat	Troubleshooting
Ground straps	Solderless connectors	
Heat shrink tube	Splice	

Introduction

Many electrical repairs will involve the replacement or repairing of a damaged conductor. To locate the problem area, today's technician must be capable of reading and understanding electrical diagrams and schematics. The first step in electrical diagnostics is to confirm the battery's condition. Once the battery is determined to be operating correctly, perform a visual inspection. At some point, you will probably need to reference the wiring diagrams to complete the electrical system troubleshooting process. Troubleshooting is the diagnostic procedure of locating and identifying the root cause of the fault. It's a systematic process of elimination by using cause and effect. By using the wiring diagrams, you should understand how the circuit is supposed to work. This is essential before attempting to determine why it does not work.

The component locator will assist the technician in finding the location of the electrical components and harnesses shown in the schematic. Usually, the component locator will also list the locations of connectors, grounds, and splices. Most electronic generated wiring diagrams will provide a hyperlink that will jump to the component locator section.

The process of troubleshooting an electrical concern is as follows:

1. Confirm the concern. Perform a check of the system to understand what is wrong. If the on-board computer monitors the faulty system, enter diagnostics to retrieve any diagnostic trouble codes (DTCs). DTCs are the output of the self-diagnostics program in the form of alpha/numeric codes that indicate faulty circuits or components.
2. Study the electrical schematic. Determine if there are any shared circuits. Try to operate the shared circuits. If the shared circuits operate correctly, the problem is isolated to the wiring or components of the problem system. If the shared circuits don't operate, the problem is usually in the power or ground circuit.
3. Locate and repair the fault. The fault is located by narrowing down the possible causes and taking measurements as required. Before replacing any components, check the ground and power leads. If these are good, then the component is bad.
4. Test the repair. Repeat a check of the system to confirm that it's operating properly.

Classroom Manual
Chapter 4, page 86

 Caution

Manufacturers may use solid copper wire in low-voltage, low-current circuits where flexibility is not required. Do not use solid wire where high voltage, high current, or flexibility is required unless the manufacturer used solid wire originally.

Wire Repair

Not all electrical repairs involve removing and replacing a faulty component. Many times the cause of the malfunction is a damaged conductor. The repair to the circuit must not increase the resistance of the circuit. Also, The repair should be a permanent repair. There are many methods to repair a damaged wire. The type of repair used will depend on factors such as:

1. Type of repair required.
2. Ease of access to the damaged area.
3. Type of conductor.
4. Size of wire.
5. Circuit requirements.
6. Manufacturer's recommendations.

The most common methods of wire repair include wrapping damaged insulation with electrical tape or tubing, crimping the connections with solderless connectors, and soldering splices.

Copper Wire Repairs

Copper wire is the most commonly used primary wire in the automobile. The insulation may break down, or the wire may break due to stress or excessive motion. Also, damage to the wire may be due to excessive current flow. Repair or replace the wire for any of these conditions.

⚡ WARNING **Repair of air bag wiring must be done to the manufacturer's specifications.**

In some instances, bypassing a length of wire that is not accessible may be necessary. In this case, cut the wire before it enters the inaccessible portion and at the other end where it leaves the area. Install a replacement wire and reroute it to the load component (**Figure 4-1**). Be sure to protect the wire by using conduit, straps, hangers, and grommets as needed.

The two most common methods of splicing copper wire are with solderless connectors or by soldering.

Crimping. **Crimping** of **solderless connectors** is an acceptable method to **splice** wires that are not subjected to weather elements, dirt, corrosion, or excessive movement. Also, do not use crimped connections in electronic circuits. A poor connection, or corrosion over time, can result in the improper electronic control operation of the system. Do the following to make a splice using solderless connections:

1. The crimping tool has different areas for performing several functions (**Figure 4-3**). This single tool will cut the wire, strip the insulation, and crimp the connector. Use the correct size of stripping opening on the crimping tool to remove enough insulation to allow the wire to penetrate the connector completely.
2. Place the wire into the connector and crimp the connector (**Figure 4-4**). To get a proper crimp, place the open area of the connector facing toward the anvil. Be sure to compress the wire under the crimp.

Crimping means bending or deforming by pinching a connector to hold the wire connection securely in place.

Solderless connectors are hollow metal tubes covered with insulating plastic. They can be butt connectors or terminal ends (**Figure 4-2**).

Splice is a term used to mean joining the ends of two or more electrical conductors at a single point.

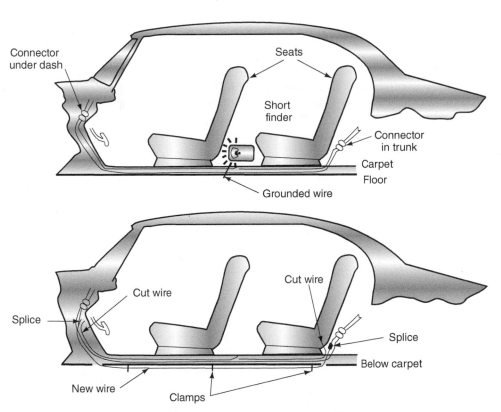

Figure 4-1 Routing a new wire to replace a damaged wire in an inaccessible location.

Figure 4-2 Types of solderless connectors and terminals sometimes used for temporary wire repair.

Figure 4-3 A typical crimping tool used for making electrical repairs.

3. Insert the stripped end of the other wire into the connector and crimp in the same manner.

4. Use electrical tape or a piece of **heat shrink tube** to provide additional protection. Heat shrink tube is plastic tubing that shrinks in diameter when exposed to heat. Use a controlled heat source such as a heat gun to shrink the tubing. Never expose the tube to open flames.

Adhesive splices provide moisture-proof connections. The adhesive splice has a heat-shrinkable adhesive sleeve that—when heated—melts and flows to fill any voids and create a permanent seal. These are far more reliable than standard solderless connectors.

Another type of crimping connector is the tap splice connector. This type of connector allows for adding circuits in parallel with an existing circuit without stripping the wires (**Figure 4-5**). Although tap connectors make connecting wires easy, do not use them to make a permanent repair. They are unreliable for making and maintaining good connections.

Adding parallel circuits decreases the circuit's total resistance and increases the circuit's amperage. The wire size of the existing circuit(s) must be capable of carrying the additional amperage. Anytime an additional circuit is attached to an existing circuit, ensure the fuse and conductor of the circuit you are tapping into have a large enough capacity.

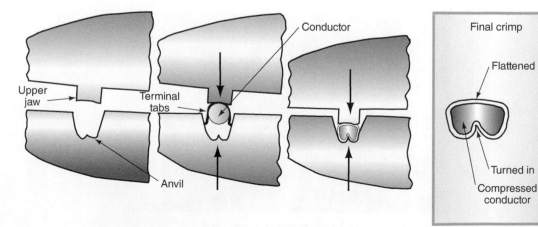

Figure 4-4 Properly crimping a connector.

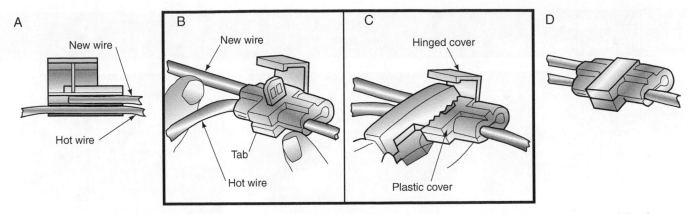

Figure 4-5 Using the tap connector to splice in another wire. (A) Place wires in position in the connector, (B) close the connector around the wires, (C) use pliers to force the tab into the conductors, and (D) close the hinged cover.

> **CUSTOMER CARE** As a professional technician, you should never have a vehicle leave the shop with a temporary repair. Almost all crimp-type repairs are temporary. A temporary repair is permissible when on a road call, and you need to get the vehicle drivable again to bring it to the shop.

Soldering. Soldering is the best way to splice copper wires. Solder is an alloy of tin and lead. It is melted over a splice to hold the wire ends together. Soldering may be a splicing procedure, but it's also an art that takes much practice. **Photo Sequence 6** illustrates the soldering process when using a **splice clip**. A splice clip is a special connector used along with solder to assure a good connection. The splice clip is different from a solderless connection in that it does not have insulation. Some splice clips have a hole provided for applying solder (**Figure 4-6**).

Figure 4-6 Splice clip. Some splice clips have a hole for applying solder.

PHOTO SEQUENCE 6
Soldering Copper Wire

P6-1 Tools required to solder copper wire: 100-watt soldering iron, rosin core solder, crimping tool, splice clip, heat shrink tube, heat gun, safety glasses, sewing seam ripper, electrical tape, and fender covers.

P6-2 Place the fender covers over the vehicle fenders.

P6-3 Disconnect the fuse that protects the circuit being repaired. Note: if the circuit is not protected by a fuse, then disconnect the battery.

P6-4 If the wiring harness is taped, use a seam ripper to open the wiring harness.

P6-5 Cut out the damaged wire using the wire cutters on the crimping tool.

P6-6 Using the correct size stripper, remove about 1/2 inch (12 mm) of the insulation from both wires. Be careful not to nick or cut any of the wires.

P6-7 Determine the correct gauge and length of replacement wire.

P6-8 Using the correct size stripper, remove 1/2 inch (12 mm) of insulation from each end of the replacement wire.

P6-9 Select the proper-size splice clip to hold the splice. Don't install the clip at this time.

P6-10 Place the correct length and size of heat shrink tube over one of the wires. Slide the tube far enough away from the splice to keep it from being exposed to the heat of the soldering iron.

P6-11 Overlap the two splice ends and hold them in place with your thumb and forefinger.

P6-12 Center the splice clip around the wires and crimp in place. Ensure that the copper wires extend beyond the splice clip in both directions. Crimp the clip on both ends.

P6-13 Heat the splice clip with the pre-tinned soldering iron while applying solder to the opening in the back of the clip. Don't apply solder to the iron; the iron should be 180° away from the clip's opening.

P6-14 After the solder cools, slide the heat shrink tube over the splice.

P6-15 Heat the tube with the hot air gun until it shrinks around the splice. Don't overheat the tube.

P6-16 Re-tape the wiring harness.

The process of applying solder to the tip of the iron is called *tinning*.

If you are not using a splice clip, braid the wire ends together tightly using one of the methods illustrated (**Figure 4-7**). Remove about 1 inch (25 mm) of the insulation from the wires and join the wires. Then the splice should be heated with the soldering gun. It's important to note that when soldering, the solder should melt by the heat of the wire splice, not the heat of the soldering tool. The solder should melt and flow evenly among all of the wire strands (**Figure 4-8**). Insulate the splice with heat shrink tube or good-quality electrical tape.

Always use rosin core solder when making electrical repairs. Acid core solder is used for purposes other than electrical repairs and can cause the wire to corrode, which would lead to high resistance.

SERVICE TIP It's easier to heat the wire with the soldering iron if the solder is melted onto the tip of the iron first. This will transfer the heat from the iron to the wire more quickly.

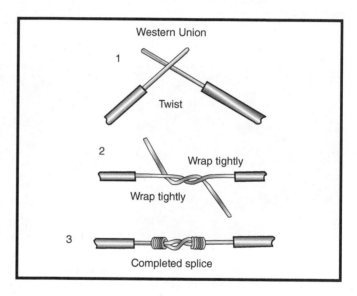

Figure 4-7 Methods for joining wires together.

Figure 4-8 When soldering, apply the solder to the joint, not to the tip of the soldering iron.

WARNING Before cutting into a wire to make a splice, look for other splices or connections. Never have two or more splices within 1.5 inches (40 mm) of each other. Also, always use wire of the same size or larger than the wire you are replacing.

Repairing Aluminum Wire

 WARNING Attempting to solder aluminum wire will damage the conductor.

General Motors has used single-stranded aluminum wire in limited applications where wire flexing will not occur.

    After cutting away the damaged wire, strip all wire of the last 1/4 inch (6 mm) of insulation. Be careful not to nick or damage the conductor. Apply a generous coating of petroleum jelly to the wire and connector (**Figure 4-9**). The petroleum jelly will prevent corrosion from developing in the core.

    Crimp the connector in the usual manner. Insulate the splice with heat shrink tube. Do not use electrical tape since it will not stay in place due to the petroleum jelly.

 WARNING Do not connect copper wire to aluminum wire. The two different metals will conduct differently and have different expansion rates that may cause the connection to break and present a fire hazard.

> **Special Tools**
> Crimping tool
> Petroleum jelly
> Safety glasses
> Electrical tape or heat shrink tube

Splicing Twisted/Shielded Wire

Many computer communication circuits use twisted/shielded wire. It protects the circuit from electrical noise that would interfere with the operation of the computer controls (**Figure 4-10**). These wires may carry as low as 0.1 amp (A) of current. It's important that the splice made in these wires does not have any resistance. The added resistance may give false signals to the computer or actuator. **Photo Sequence 7** illustrates the procedure for splicing shielded wire.

> **Special Tools**
> Crimping tool
> Splice clip
> Wire strippers
> 100-watt soldering gun
> Fender covers
> Safety glasses
> 60/40 rosin core solder
> Head shrink tube
> Head Gun

Figure 4-9 Apply petroleum jelly to the areas shown.

Figure 4-10 Twisted/shielded wire used in computer circuits.

PHOTO SEQUENCE 7
Splicing Shielded Wires

P7-1 Tools required to perform this task include a crimping tool, splice clip, wire strippers, soldering gun, rosin core solder, heat shrink tube, heat gun, and safety glasses.

P7-2 Locate and cut out the damaged section of wire.

P7-3 Being careful not to cut into the Mylar tape or to cut the drain wire, remove about 1 inch (25 mm) of the outer jacket from the ends of each cable.

P7-4 Unwrap the Mylar tape. Do not remove the tape from the cable.

P7-5 Untwist the conductors.

P7-6 Remove a small amount of the insulation from the ends of the conductors. The conductor ends should be staggered so that the splices will not be on top of each other. Also, the heat shrink tube should be located over the cable, but away from the work area.

P7-7 Locate a splice clip over the two ends of the conductors.

P7-8 Solder the splice.

P7-9 Repeat the splice and solder procedure for the second conductor. Make sure that the two splices will not touch each other.

PHOTO SEQUENCE 7 (CONTINUED)

P7-10 Wrap the conductors with the Mylar tape. Do not wrap the drain wire in the tape.

P7-11 Wrap the drain wire around the outside of the Mylar tape.

P7-12 Splice the drain wire with a clip and solder.

P7-13 Locate the heat shrink tube over the splice area and use the heat gun to shrink the tube over the cable.

Replacing Fusible Links

Not all opens within a fusible link are detectable by visual inspection only. Test for battery voltage on both sides of the fusible link to confirm its condition. If you need to replace the fusible link, cut it out of the circuit and solder a new fusible link in place.

There are two types of insulation used on fusible links: Hypalon and Silicone/GXL. Hypalon can be used to replace either type of link. However, do not use Silicone/GXL to replace Hypalon. To identify the type of insulation, cut the blown link's insulation back. The insulation of the Hypalon link is a solid color all the way through. The insulation of Silicon/GXL will have a white inner core.

When cutting off the damaged fusible link from the feed wire, cut it beyond the splice (**Figure 4-11**). When making the repair, don't use a fusible link longer than 9 inches (228 mm). This length will not provide sufficient overload protection.

If the damaged fusible link feeds two harness wires, use two fusible links. Splice one link to each of the harness wires (**Figure 4-12**).

Classroom Manual
Chapter 3, page 76

 Special Tools

Crimping tool
100-watt soldering iron
60/40 rosin core solder
Safety glasses
Fender covers

Figure 4-11 Replacing a fusible link.

Figure 4-12 Replacing a fusible link that feeds two circuits.

Caution

Disconnect the battery negative cable before performing any repairs to the fusible link.

Caution

Use the fusible link gauge required by the manufacturer.

CUSTOMER CARE If the fusible link, or any fuses, are blown, it's necessary to locate the cause. Fuses don't wear out. If they blow, it's due to an overload of current in the circuit. The fault can be identified and repaired the first time the vehicle is in the shop by using a cause-and-effect diagnosis approach. The following illustrates how to use this method:

Effect	Cause
Turn signals not operating	Blown fuse
Blown fuse	Short to ground
Short to ground	Broken wire in trunk
Broken wire	Loose jack handle

This method leads the technician to fix the fault and find the cause.

Classroom Manual
Chapter 4, page 91

Special Tools
Pick
Crimping tool
Safety glasses
Fender covers

Repairing Connector Terminals

The connector terminal will require repairs due to abuse, improper disconnecting procedures, and exposure potential to the elements. The method of repair depends on the type of connector.

AUTHOR'S NOTE There are many differing connector service procedures, even within the same family of connector types. The following are examples of typical procedures for the different types of connectors. Always refer to the correct service information for the procedure for the connector you are servicing.

Molded Connectors

You cannot disassemble one-piece, molded-type connectors for repairs (**Figure 4-13**). Molded in-line connectors allow the male and female connector halves to be separated; however, you cannot disassemble the connector itself. If the connector is damaged, cut it off and splice in a new connector.

Classroom Manual
Chapter 4, page 91

Hard-Shell Connectors

⚠ **WARNING** **Do not place your fingers or body next to the connector. If excessive force is needed to depress the tang, the pick may be pushed out the back and cause injury.**

Classroom Manual
Chapter 4, page 91

Hard-shell connectors usually provide a means of removing the terminals for repair. Use the special tool, or a proper size pick, to depress the locking tang of the connector (**Figure 4-14**). Pull the lead back far enough to release the locking tang from the connector. Remove the pick, and then pull the lead completely out of the connector. Make the repair to the terminal using the same procedures for repairing copper or aluminum wire.

Re-form the terminal locking tang to assure a good lock in the connector (**Figure 4-15**). Use the pick to bend the lock tang back into its original shape. Insert the lead into the back of the connector. A noticeable "catch" should be felt when the lead is halfway through the connector. Gently push back and forth on the lead to confirm that the terminal is locked in place.

Weather-Pack Connectors

Weather-pack connectors secure the terminals with a hinged secondary lock or a plastic terminal retainer. To perform repairs, first disconnect the two halves by pulling up on the primary lock while pulling the two halves apart (**Figure 4-16**). Unlock the secondary locks and swing them open (**Figure 4-17**).

Classroom Manual
Chapter 4, page 93

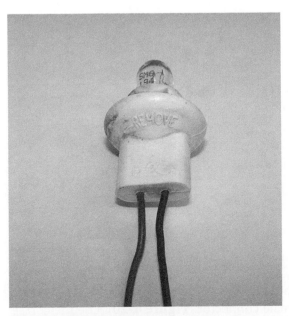

Figure 4-13 Molded connector halves are a one-piece design that cannot be disassembled for repairs.

Push narrow pick between terminal and connector body.

Figure 4-14 Depress the locking tang to remove the terminal from the connector.

Figure 4-15 Re-form the locking tang to its original position before inserting the terminal back into the connector.

Figure 4-16 The weather-pack connector has two locks. Use the primary lock to separate the halves.

Figure 4-17 Unlock the secondary lock to remove the terminals from the connector.

Figure 4-18 Use the recommended special tool to unlock the tang on the terminal.

Figure 4-19 After the lock tang has been depressed, remove the lead from the back of the connector.

Figure 4-20 Male and female connectors.

Special Tools

Weather-pack tool
Crimping tool
Safety glasses
Fender cover

Depress the terminal locking tangs using the special weather-pack tool. Push the cylinder of the tool into the terminal cavity from the front until it stops (**Figure 4-18**). Pull the tool out, and then gently pull the lead out of the back of the connector (**Figure 4-19**).

The terminal is either a male or a female connector (**Figure 4-20**). Use the correct terminal for the repair. Feed the wire through the seal and connect the repair lead in the normal manner of crimping and soldering (**Figure 4-21**). Re-form the terminal lock tang by bending it back into its original position (**Figure 4-22**).

Figure 4-22 Re-forming the locking tangs of the terminal.

Seal

Seal crimp   Crimp and solder

Core crimp

Figure 4-21 Crimp and solder the terminal to the lead.

> **SERVICE TIP** Use a weather-pack repair kit that provides new seals for a complete repair.

Insert the lead from the back of the connector until a noticeable "catch" is felt. Gently push and pull on the lead to confirm that it's locked to the connector. Close the secondary locks and reconnect the connector halves.

Metri-Pack Connectors

There are two types of metri-pack connectors: **pull to seat** and **push to seat**. The push-to-seat terminal removal is illustrated in **Figures 4-23** and **4-24**.

Remove the pull-to-seat terminal by inserting a pick into the connector and under the lock tang (**Figure 4-25**). Gently pull back on the lead while prying up on the lock tang. When the lock tang is free of the tab in the connector, push the lead through the *front* of the connector.

To repair the terminal, insert the stripped wire through the seal and the connector body (**Figure 4-26**). Crimp and solder the terminal to the wire. Pull the wire lead and terminal back into the connector body until the terminal is locked (**Figure 4-27**).

Ground Straps

Ground straps between the powertrain components and the vehicle's chassis complete the return path to the battery. Ground straps also connect sheet metal parts, even though there is no electrical circuit involved. In this instance, the purpose of the strap is to suppress electromagnetic induction (EMI). Sheet metal can behave like a large capacitor, and the resulting electrostatic field can interfere with computer-controlled circuits.

Classroom Manual
Chapter 4, page 93

> **Special Tools**
> Pick
> Crimping tool
> Safety glasses
> Fender covers

The terms "**pull to seat**" and "**push to seat**" depict the method used to install the terminals into the connector.

Classroom Manual
Chapter 4, page 89

> **Special Tools**
> DMM
> Lab scope

Connector

Wire

Locking tab

Weather seal

Push-to-seat terminal (enters from rear of connectors)

Figure 4-23 Use a pick to unlock the male terminal locking tang.

Front-entry release for push-to-seat terminal

Figure 4-24 Use a pick to unlock the nib of the terminal retainer for female terminals.

Lock tab

Rear-entry release for pull-to-seat terminal

Figure 4-25 Pull up on the lock tang to release the terminal from the connector.

Stripped wire

Weather seal

Connector body

Figure 4-26 You must install the wire lead into the seal and connector before attaching the terminal.

Connector

Wire

Weather seal

Terminal

Locking tang

Figure 4-27 Make sure that the terminal locks into the connector body.

Like any other conductor, resistance or opens in the ground strap, or its connections, can cause multiple issues. For example, the engine starter may not engage or crank slow, fuel injectors may not properly energize, and ignition coils may not function.

If electron flow cannot return through the designed ground path, it will attempt to find an alternate path. Sometimes this may be evidenced by pitting of the slip yoke at the rear of the transmission or by pitting of wheel bearings. This pitting results from arcing as the current jumps from one component to another and wears away the metal.

Ground straps used for electrical circuits can be tested using the voltage drop test. Each connection of the ground strap should have a voltage drop of 0 volt (V). In addition, the ground strap itself should have no voltage drop between its ends. Be sure to confirm that the connection terminals are tight and clean. It's a good practice to use the voltage drop test to confirm proper connections anytime the ground strap is removed and reinstalled.

Ground straps used to control EMI generally don't have electrical circuits involved, so a voltage drop test would not be possible. However, a voltmeter or lab scope set on the lowest scale may pick up any voltage that is passing through the strap. Also, you can use an ohmmeter to measure the resistance between connector ends.

> **SERVICE TIP** When troubleshooting an intermittent electrical problem, be sure to check the condition of *all* ground straps on the vehicle.

Reading Wiring Diagrams

When attempting to locate probable causes for system malfunction, it's important to have the correct wiring diagram for the vehicle you are working on. There may be different wiring diagrams for each model and even for the same models but equipped with different options. In some cases, it may be necessary to use the date of manufacture and/or the VIN to determine the correct diagram to use. The VIN plate is usually located on the cowl, next to the left upper instrument panel (**Figure 4-28**). It's visible from the outside of the vehicle. Other locations for the VIN plate are in the door opening, glove box, and engine compartment.

Classroom Manual
Chapter 4, page 95

Electronic wiring diagrams may also have a feature that performs plate to VIN transfer. By entering the license plate number and the issuing state, the software performs a data base look up to pull the matching VIN. Some use a barcode scanner to read the VIN. These methods are faster and more accurate than manually entering the VIN number.

Study the method used to identify circuits and **color codes**. Also, become familiar with the electrical symbols used by the manufacturer. Usually, this information is provided in the "How to use" section of the service information. Since displaying symbols on

Color codes are used to assist in tracing the wires. Usually, the first group of letters designates the base color of the insulation and the second group indicates the color of the tracer.

Figure 4-28 VIN plate location.

monitors depends on the quality of the monitor, some electronic platforms limit the use of symbols. Instead, components are drawn as boxes and labeled.

Before you begin to use a wiring diagram, you must first have an idea of what you are looking for. It may be a particular component, circuit, or connector. The best way to start the process is by identifying the component, or one of the components, which doesn't work correctly. Then look in the index for the wiring diagram and find where that component is shown. Computer-driven wiring diagrams may provide a search feature that simplifies locating the component within the wiring diagrams.

Once you have found the component or part of the circuit you were looking for, start at that component and trace through the circuit to identify all related components, connectors, and wires. When using the electronic version of the wiring diagram, the program may assist in this effort by highlighting circuits as you click on them **(Figure 4-29)**. If your system does not have the highlighting feature, print off the diagram(s) and use colored pencils or markers to trace the circuits. You can also draw your own simplified diagram onto a piece of paper. It doesn't have to be pretty to work; it just needs to be accurate.

AUTHOR'S NOTE Wiring diagrams are drawn with the components in their normal key off state.

For tracing to have any value, you need to identify the power source for the component and/or for the circuit, all related loads (sometimes this involves tracing the circuit back through other pages of the wiring diagram), and the ground connection for the component and for all of the related loads. During this step you should be able to answer the following questions concerning the circuit:

1. Where does the voltage come from?
2. When should voltage be present at any point in the circuit?
3. Are there any other events that must happen in order for the circuit to operate?
4. What is the expected voltage and amperage at any point in the circuit?

Figure 4-29 Electronic wiring diagrams may offer a highlighting feature that make tracing circuits easier. In this case the Z92 circuit is highlighted.

After you have studied the circuit, you should know how the circuit is supposed to work. Now describe the problem you are hoping to solve. Ask yourself, "What could cause this?" Limit your answers to those items in your traced wiring diagram, and to the description of the problem. It's wise to make a list of all probable causes of the problem; then number them according to probability. For example, if no dash lights come on, it's possible that all of the bulbs burned out. However, this is not as probable as a blown fuse. After you have listed the probable causes in order of probability, look at the wiring diagram to identify how you can quickly test to find out which is the cause. Diagnostics becomes easier as your knowledge of electricity grows. It also becomes easier with a good understanding of how the circuit works.

Figure 4-30 is a schematic for a blower control circuit of a heater system. By tracing through the circuit, you should be able to determine the correct operation of the system. It cannot be overemphasized that you should not attempt to figure out why the circuit is not working until you understand how it is supposed to work.

In this circuit, battery voltage is applied to the fuse box when the ignition switch is in the RUN position. The 30-amp fuse protects circuit 181 to the motor. Notice that circuit 181 connects to the blower motor resistor block through connector C606. When it leaves the resistor block (after flowing through the thermal limiter), it goes through connector C001 and attaches to the motor through connector C002. Details of these connectors are shown with the schematic.

Figure 4-30 Heater system wiring diagram.

The circuit then leaves the motor and enters the resistor block through the same connectors. There is a splice in the resistor block that connects this wire to a series of resistors. Connector C606 directs the various circuits out of the resistor block to the blower switch and ground through connector C613. If the switch is placed in the OFF position, the circuit to ground is opened and the blower motor should not operate. If the switch is placed in the LOW position, the current will flow through all of the resistors and the switch through circuit 260. The switch completes the circuit to ground. As the switch is placed in different speed positions, the amount of resistance and the circuit number change. The motor speed should increase as the amount of resistance decreases.

If the customer states that the heater motor does not work at all, in any speed position, consider the following possible causes:

1. Open fuse.
2. Open in the lead from the ignition switch to the fuse box.
3. Bad ground connection after the switch.
4. Inoperative blower motor.
5. Open in circuit 181.
6. Open thermal limiter.
7. Open in the orange wire between connectors C001 and C002.
8. Open in the black wire between connectors C002 and C001.
9. Disconnected, damaged, or corroded connector C606 or C613.
10. Faulty switch.

However, if the customer states that the motor does not operate only in the low position, the problem is limited to three possibilities:

1. The third resistor in the series is open.
2. An open in circuit 260 from the resistor block to the blower switch.
3. A faulty switch.

By understanding the way the system is supposed to work, the problem of determining where to look for the problems is simplified. Practice in reading wiring schematics is the only good teacher.

Once the potential problem areas are determined, use the wire color codes to locate the exact wires. Test the leads for expected voltages at different locations in the circuit.

The example just discussed hopefully has given an understanding of the advantages to using a wiring diagram before actually performing diagnostic tests. Today's vehicles have an ever-increasing number of electrical accessories and include the use of computer modules to perform many tasks. When the manufacturers design the vehicle, many of the electrical circuits are constructed in a parallel arrangement. All of this combines to make the electrical circuits more challenging to diagnose. Proper use of the wiring diagram can help you get the diagnosis accomplished intelligently.

To illustrate again how the proper use of the wiring diagram helps to make intelligent diagnostics, consider a simple lighting circuit (**Figure 4-31**). In this circuit, the fuse is wired in series to the three parallel lamps. The common components of this circuit include the fuse and the conductors connecting the battery to the point identified as S1. After the S1 splice location, the three lamps branch off. If lamp A is the only nonfunctioning component of the circuit, by using the wiring diagram you should be able to determine possible fault areas very quickly. If lamps B and C function, the open must be beyond common-point S1. However, if none of the lamps function, the open is probably before the S1. By using the wiring diagram, diagnostic time has been saved by not having to test half of the circuit.

In the previous examples of wire diagram usage, we focused mainly on locating opens or faulty components. The wiring diagram is also an excellent tool for locating shorts to ground. Consider the circuit shown in **Figure 4-32** and assume that fuse 4 is blown.

Knowing that the fuse blows as a result of excessive current flow in the circuit that can result from a short to ground condition, you need to isolate the location of the short.

> **AUTHOR'S NOTE** This type of diagnostic procedure of using the wiring diagram to isolate possible fault locations is sometimes referred to as "common-point diagnostics."

Fuse 4 in this circuit protects many parallel circuits that encompass a wiring harness that goes from the front to the rear of the vehicle. To trace out the circuits within the harness would be very time consuming. Using the wiring diagram and common-point diagnostics will identify the most likely location of the short.

Figure 4-31 Lighting circuit used as an example of using the wiring diagram to locate an open.

Figure 4-32 Example of using the wiring diagram to locate a short.

The first step is to isolate the faulty circuit. Turn off all loads that are protected by fuse 4. With all circuits turned off, there should be no current flow through the circuit. Remove the fuse and insert a test light across the fuse box terminals. This installs the test light in series with the circuit. With all loads still off, turn the ignition switch to the RUN position and observe the test light. If the test light is illuminated, the short is in the common circuit between the fuse and the switches. In this instance, you would then be looking for a short in the dashboard area after the fuse and before the individual circuit switches. The next step would be to separate connectors C203 and C201 one at a time to see if the test light goes out. This will isolate the circuit further.

If the test light did not illuminate when the ignition switch was placed in the RUN position, the location of the short is after the switches. Replace the fuse with a new properly rated fuse and turn on each circuit separately until the fuse blows. The short would be in the section of the circuit that, when energized, caused the fuse to blow.

Keep in mind that the customer may have brought this vehicle in because the A/C was not working. If all that you did was replace the fuse and it did not blow when the A/C was turned on, you missed the fact that the fuse may blow when the backup lamp switch is closed. Without referencing the wiring diagram, it's possible to make the assumption that there is an intermittent short in the A/C circuit or within the compressor clutch. Remember, one of your diagnostic steps is to determine related symptoms and that the last step is to verify your repair as successful.

Case Study

The vehicle owner states that the brake lights don't light. They also say the dome light is not working. The technician verifies the problem, checks the battery for good connections and tests the fusible links. All are in good condition.

A study of the wiring diagram indicates that the brake light and dome light circuits share the same fuse. The wiring diagrams also indicate that these two circuits share the ignition switch illumination circuit. A check of the ignition switch illumination light shows that it's not operating either. The technician checks the fuse identified in the wiring diagram. It is blown. Installing a replacement fuse and retesting resulted in the dome and brake lights working properly for three tests, and then the fuse blew again.

Upon further testing, an intermittent short to ground was found in the ignition switch illumination circuit in the steering column. The technician soldered in a repair wire to replace the damaged section. After completing all repairs, a verification test indicated proper operation of all circuits.

ASE-Style Review Questions

1. Splicing copper wire is being discussed.
   Technician A says it's acceptable to use solderless connections.
   Technician B says acid core solder should not be used on copper wires.
   Who is correct?
   A. A only
   B. B only
   C. Both A and B
   D. Neither A nor B

2. Use of wiring diagrams is being discussed.
   Technician A says a wiring diagram can help determine related circuits.
   Technician B says the wiring diagram will give the exact location of the components in the car.
   Who is correct?
   A. A only
   B. B only
   C. Both A and B
   D. Neither A nor B

3. The fuse for the parking lights is open.
   Technician A says the blown fuse may be due to high resistance in one of the parking light bulb connectors.
   Technician B says the fuse probably wore out due to age.
   Who is correct?
   A. A only
   B. B only
   C. Both A and B
   D. Neither A nor B

4. Repairs to a twisted/shielded wire are being discussed.
   Technician A says a twisted/shielded wire carries high current.
   Technician B says because a twisted/shielded wire carries low current, any repairs to the wire must not increase the resistance of the circuit.
   Who is correct?
   A. A only
   B. B only
   C. Both A and B
   D. Neither A nor B

5. Replacement of fusible links is being discussed.
   Technician A says not all open fusible links are detectable by visual inspection.
   Technician B says to test for battery voltage on both sides of the fusible link to confirm its condition.
   Who is correct?
   A. A only
   B. B only
   C. Both A and B
   D. Neither A nor B

6. *Technician A* says troubleshooting is the diagnostic procedure of locating and identifying the cause of the fault.
   Technician B says troubleshooting is a step-by-step process of elimination by the use of cause and effect.
   Who is correct?
   A. A only
   B. B only
   C. Both A and B
   D. Neither A nor B

7. *Technician A* says a replacement fusible link should be at least 9 inches (25 mm) long.
   Technician B says if the damaged fusible link feeds two harness wires, use one replacement fusible link.
   Who is correct?
   A. A only
   B. B only
   C. Both A and B
   D. Neither A nor B

8. Repairing connectors is being discussed.
   Technician A says molded connectors are a one-piece design and cannot be separated for repairs.
   Technician B says to replace the seals when repairing the weather-pack connector.
   Who is correct?
   A. A only
   B. B only
   C. Both A and B
   D. Neither A nor B

9. *Technician A* says all metri-pack connectors use male terminals.
   Technician B says metri-pack connectors are called pull to seat or push to seat to depict the method used to install the terminals into the connector.
   Who is correct?
   A. A only
   B. B only
   C. Both A and B
   D. Neither A nor B

10. *Technician A* says there may be a different wiring diagram for each model of a vehicle.
    Technician B says it may be necessary to use the VIN to get the correct wiring diagram for the vehicle.
    Who is correct?
    A. A only
    B. B only
    C. Both A and B
    D. Neither A nor B

ASE Challenge Questions

1. *Technician A* says to use acid core solder whenever soldering aluminum wires.
   Technician B says do not use solderless connectors if a weather-resistant connection is desired.
   Who is correct?
   A. A only
   B. B only
   C. Both A and B
   D. Neither A nor B

2. Which of the following electrical troubleshooting routines is in the correct sequence?
   A. Study the electrical diagram, confirm the concern, locate and repair the fault, and test the repair.
   B. Study the electrical diagram, locate and repair the fault, confirm the concern, and test the repair.

C. Confirm the concern, study the electrical diagram, test the repair, and locate and repair the fault.

D. Confirm the concern, study the electrical diagram, locate and repair the fault, and test the repair.

3. Wire repair procedures are being discussed.

Technician A says aluminum wire should never be soldered; it should only be crimped whenever repairs are necessary.

Technician B says damaged twisted/ shielded wires should be connected with splice clips and then the connections should be soldered.

Who is correct?

A. A only
B. B only
C. Both A and B
D. Neither A nor B

4. *Technician A* says that when replacing a fusible link, the gauge size of the replacement fusible link can be decreased but never increased.

Technician B says that a 14-gauge fusible link can be replaced with an equivalent length of 14-gauge stranded wire.

Who is correct?

A. A only
B. B only
C. Both A and B
D. Neither A nor B

5. *Technician A* says that a wire identified as "754 LG/W (H)" on a wiring diagram refers to a wire located in circuit 754 that is light green with white hash marks.

Technician B says that a wire that is labeled 181 BR/O refers to an 18-gauge wire that is brown with an orange stripe.

Who is correct?

A. A only
B. Both A and B
C. B only
D. Neither A nor B

6. Using **Figure 4-33**, what would be the **LEAST LIKELY** cause for fuse 16 blowing?

A. A short to ground between terminal 2 of the fuse and terminal 86 of the relay.

B. A short to ground between terminal 85 of the relay and terminal 47 of the IMP.

C. A short to ground at terminal 30 of the relay.

D. A short to ground in circuit X2.

7. A customer states that their vehicle's tail lamps are not functioning. When testing the lamp operation, it's noted that the tail/stop lamps and the turn signals don't operate. During the diagnostic routine, the technician also notices that the backup lamps are also not functioning. All fuses test good. Using all of the wiring diagrams with indicated test voltages in **Figure 4-34a** through **d**, what is the most likely cause of this condition?

A. G301 is open.

B. G302 is open.

C. Faulty integrated power module.

D. Open in the Z918 circuit.

Figure 4-33

Figure 4-34a

Figure 4-34b

Figure 4-34c

Figure 4-34d

8. Referring to **Figure 4-35**, how is the blower speed controlled?

A. The A/C–heater controller uses a high-side driver for each speed circuit.

B. The totally integrated power module uses PWM to control the ground circuit of the motor.

C. A/C–heater controller uses a low-side driver for each speed circuit.

D. All of these choices.

9. Referring to the wiring diagram illustrated in **Figure 4-36a** through **d**, the fog lights don't work when the headlight switch is in the AUTO position, but work in the HEAD and PARKLAMP positions. What is the most likely cause?

A. Open in the headlamp switch resistor between the 3 and 6 positions.

B. Open in the resistor below the 3 position of the headlamp switch.

C. Short to ground in circuit L80.

D. Short to voltage at terminal 4 of headlamp switch.

Figure 4-35

Figure 4-36a

Figure 4-36b

Figure 4-36c

Figure 4-36d

10. Referring to **Figure 4-37a** through **d**, the coil for cylinder 4 and the injector for cylinder 3 both fail to function. What is the possible cause?

   A. Poor connection at splice S134.

   B. Open in the K13 circuit.

   C. Poor connection in the S185 splice.

   D. Open in circuit between S184 and PCM C1/38 terminal.

Figure 4-37a

Figure 4-37b

Figure 4-37c

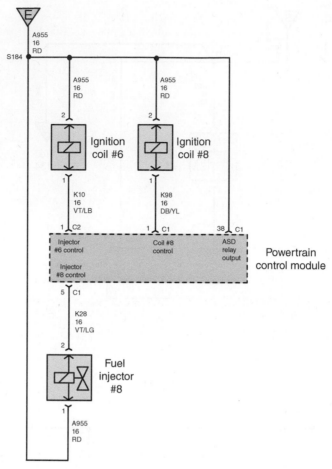

Figure 4-37d

Name _____   Date _____

PART IDENTIFICATION ON A WIRING DIAGRAM

Objective: Upon completion of this job sheet, you should be able to locate different parts on a wiring diagram.

ASE Education Foundation Correlation

This job sheet addresses the following **MLR/AST/MAST** tasks:

A.2.   Identify electrical/electronic system components and configuration. **(P-1)**

A.9.   Use wiring diagrams to trace electrical/electronic circuits. **(P-1)**

We Support
ASE | Education Foundation

Tools and Materials

• A wiring diagram in a service manual or from electronic service information system for the system and vehicle assigned by your instructor.

Procedure

Study the wiring diagram; then answer the following questions.

1. Are all circuit grounds clearly marked in the wiring diagrams?

   _____
   _____

2. What is represented by lines that cross each other?

   _____
   _____

3. Are most switches shown in their normally open or normally closed position?

   _____
   _____

4. Do all wires have a color code listed by them?

   _____

5. Is the internal circuitry of all components shown in their schematic drawing?

   _____

6. Locate the following electrical components for the assigned vehicle

Component	Location	Drawing
Windshield wiper motor	_____	_____
Dome (courtesy) light	_____	_____
A/C compressor clutch	_____	_____
Turn signal flasher unit	_____	_____
Fuse	_____	_____
Fuel gauge sending unit	_____	_____

7. Attach print out of the wiring diagram (if using electronic retrieval system) for the fuel gauge sending unit.

Instructor's Evaluation:

Grade	Rubric
0	No exposure due to: ☐ Absence ☐ Other:
1	Student unsuccessfully attempted the task
2	Student completed the task with significant instructor assistance.
3	Satisfactory performance of task. The student was able to perform elements of the task with minimal instructor assistance.
4	Competent/Proficient. Able to perform all elements of the task successfully and independently.

Instructor's Notes:

Name _____   Date _____

USING A COMPONENT LOCATOR

Objective: Upon completion of this job sheet, you should be able to identify components and their locations indicated on the wiring diagram.

ASE Education Foundation Correlation ——————————————————————

This job sheet addresses the following **MLR/AST/MAST** tasks:

A.2.   Identify electrical/electronic system components and configuration. **(P-1)**

A.9.   Use wiring diagrams to trace electrical/electronic circuits. **(P-1)**

We Support
ASE | Education Foundation

Tools and Materials

• Service information with component locator capabilities.

Procedure

Using the component locator for the vehicle that has been assigned, the student will find the locations for the taillight, headlight, and stoplight circuits, connectors, connections, plugs, and switches.

1. **Taillight circuit:**

   _____

   Location of source of power: _____

   Location of the switch: _____

   _____

   Does the switch control the operation of the taillights directly, or is it an input to a module? DIRECTLY INPUT

   If a module, which one and where is it located?

   _____

   _____

   List the ID numbers of any in-line connectors between the switch (or module) and the taillights: _____

   List the location(s) of the in-line connectors: _____

   _____

   _____

   Where is the location of the ground connection for the left taillight?

   _____

   _____

2. **Headlight circuit:**

   Location of source of power: _____

   _____

Location of switch: _____

_____

Does the switch control the operation of the headlights directly, or is it an input to a module? DIRECTLY INPUT

If a module, which one and where is it located?

_____

_____

List the ID numbers of any circuit splices between the switch (or module) and the left headlight: _____

List the location(s) of the splices: _____

_____

_____

Where is the location of the ground connection for the left headlight?

_____

_____

3. **Stoplight circuit:**

Location of source of power: _____

Location of switch: _____

Does the switch control the operation of the stoplights directly, or is it an input to a module? DIRECTLY INPUT

If a module, which one and where is it located?

_____

_____

On which side of the vehicle does the wiring harness run between the switch (or module) and the stoplights? _____

Instructor's Evaluation:

Grade	Rubric
0	No exposure due to: ☐ Absence ☐ Other:
1	Student unsuccessfully attempted the task
2	Student completed the task with significant instructor assistance.
3	Satisfactory performance of task. The student was able to perform elements of the task with minimal instructor assistance.
4	Competent/Proficient. Able to perform all elements of the task successfully and independently.

Instructor's Notes:

Name _____   Date _____

INTRODUCTION TO WIRING DIAGRAMS

Objective: Upon completion of this job sheet, you should be able to find the power source, ground connection, and controls for electrical circuits using a wiring diagram.

ASE Education Foundation Correlation _____

This job sheet addresses the following **MLR/AST/MAST** tasks:

A.2.   Identify electrical/electronic system components and configuration. **(P-1)**

A.9.   Use wiring diagrams to trace electrical/electronic circuits. **(P-1)**

We Support
Education Foundation

Tools and Materials

Vehicle service information

Procedure

Use the following wiring diagrams to complete each exercise.

Option: Your instructor will assign you a vehicle to look up similar circuits using an electronic wiring diagram format.

Exercise 1

Using the diagram in **Figure 4-38**, answer the following (list wire colors):

1. How is the windshield wiper motor circuit powered? _____

    _____

    _____

    _____

    _____

2. How is the circuit grounded?

    _____

    _____

    _____

    _____

Figure 4-38

3. How is the circuit controlled?

   a. If it's controlled by a switch, is the switch normally open or closed?

   b. Is the circuit power switched or ground switched?

   c. Is the circuit controlled by a variable resistor?

   d. Is the circuit controlled by a mechanical or vacuum-operated device?

   _____
   _____
   _____
   _____

4. How is the circuit protected (identify all fuses)?

   _____
   _____
   _____

Exercise 2

Using the wiring diagram in **Figure 4-39**, answer the following:

1. How is the circuit powered? _____

   _____
   _____
   _____
   _____
   _____

Figure 4-39

2. How is the circuit grounded?

_____

_____

_____

_____

3. How is the circuit controlled?

   a. If it's controlled by a switch, is the switch normally open or closed?

   b. Is the circuit power switched or ground switched?

   c. Is the circuit controlled by a variable resistor?

   d. Is the circuit controlled by a mechanical or vacuum-operated device?

_____

_____

_____

_____

4. How is the circuit protected?

_____

_____

_____

_____

Exercise 3

Using the wiring diagram in **Figure 4-40**, answer the following:

1. How is the circuit powered?

_____

_____

_____

_____

_____

2. How is the circuit grounded?

_____

_____

_____

_____

3. How is the circuit controlled?

   a. If it's controlled by a switch, is the switch normally open or closed?

   b. Is the circuit power switched or ground switched?

   c. Is the circuit controlled by a variable resistor?

   d. Is the circuit controlled by a mechanical or vacuum-operated device?

_____

_____

_____

_____

4. How is the circuit protected?

_____

_____

_____

_____

Figure 4-40

Exercise 4

Using the wiring diagram in **Figure 4-41**, answer the following:

1. How is the circuit powered?

_____

_____

_____

_____

2. How is the circuit grounded?

_____

_____

_____

_____

3. How is the circuit controlled?

    a. If it's controlled by a switch, is the switch normally open or closed?

    b. Is the circuit power switched or ground switched?

    c. Is the circuit controlled by a variable resistor?

    d. Is the circuit controlled by a mechanical or vacuum-operated device?

_____

_____

_____

_____

Figure 4-41

4. How is the circuit protected?

_____

_____

_____

_____

Exercise 5

Using the wiring diagram in **Figure 4-42**, answer the following:

1. How is the circuit powered?

_____

_____

_____

_____

2. How is the circuit grounded?

_____

_____

_____

_____

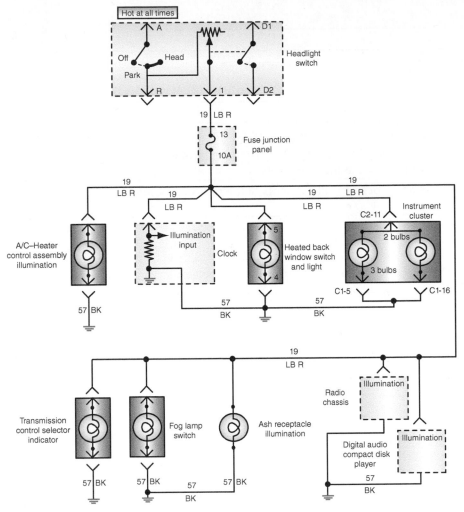

Figure 4-42

3. How is the circuit controlled?

   a. If it's controlled by a switch, is the switch normally open or closed?

   b. Is the circuit power switched or ground switched?

   c. Is the circuit controlled by a variable resistor?

   d. Is the circuit controlled by a mechanical or vacuum-operated device?

_____

_____

_____

_____

4. How is the circuit protected?

_____

_____

_____

_____

Instructor's Evaluation:

Grade	Rubric
0	No exposure due to: ☐ Absence ☐ Other:
1	Student unsuccessfully attempted the task
2	Student completed the task with significant instructor assistance.
3	Satisfactory performance of task. The student was able to perform elements of the task with minimal instructor assistance.
4	Competent/Proficient. Able to perform all elements of the task successfully and independently.

Instructor's Notes:

Name _____   Date _____

SOLDERING COPPER WIRES

Objective: Upon completion of this job sheet, you should be able to properly solder two copper wires.

ASE Education Foundation Correlation

This job sheet addresses the following **MLR** tasks:

A.12. Repair and/or replace connectors, terminal ends, and wiring of electrical/electronic systems (including solder repair). **(P-2)**

This job sheet addresses the following **AST/MAST** tasks:

A.12. Inspect, test, repair, and/or replace components, connectors, terminals, harnesses, and wiring in electrical/electronic systems (including solder repairs); determine needed action. **(P-1)**

We Support

ASE | Education Foundation

Tools and Materials

- Two pieces of copper wire
- 100-watt soldering iron
- 60/40 rosin core solder
- Crimping tool
- Splice clip
- Heat shrink tube
- Heating gun

Safety Considerations and PPE

- Safety glasses

Procedure:

1. Using the correct size stripper, remove about 1/2 inch (12 mm) of the insulation from both wires. Be careful not to nick or cut any of the wires.

2. Select the proper size splice clip to hold the splice.

3. Place the correct length and size of heat shrink tube over the wire. Slide the tube far enough away so the wires are not exposed to the heat of the soldering iron.

4. Overlap the two splice ends and hold in place with thumb and forefinger. While the wire ends are overlapped, center the splice clip around the wires and crimp into place.

5. Heat the splice clip with the soldering iron while applying solder to the opening in the back of the clip. Apply only enough solder to make a good connection. The solder should travel through the wire.

6. After the solder cools, slide the heat shrink tube over the splice and heat the tube with the hot air gun until it shrinks around the splice. Do not overheat the tube.

Instructor's Evaluation:

Grade	Rubric
0	No exposure due to: ☐ Absence ☐ Other:
1	Student unsuccessfully attempted the task
2	Student completed the task with significant instructor assistance.
3	Satisfactory performance of task. The student was able to perform elements of the task with minimal instructor assistance.
4	Competent/Proficient. Able to perform all elements of the task successfully and independently.

Instructor's Notes:

DIAGNOSTIC CHART 4-1

PROBLEM AREA	Burned fusible link.
SYMPTOMS	Several electrical components fail to operate.
POSSIBLE CAUSES	1. Short to ground in circuit. 2. Excessive circuit current flow due to high mechanical resistance of actuators.

DIAGNOSTIC CHART 4-2

PROBLEM AREA	Burned, broken, or defective connectors, insulation, or conductors.
SYMPTOMS	Open or short circuits that prevent component operation.
POSSIBLE CAUSES	1. Corroded wiring. 2. Improper or no wire protectors. 3. Corroded connectors. 4. Damaged conductor. 5. Damaged connectors.

CHAPTER 5

BATTERY DIAGNOSIS AND SERVICE

Basic Tools

Basic mechanic's tool set
Service information

Upon completion and review of this chapter, you should be able to:

- Demonstrate all safety precautions and rules associated with servicing the battery.
- Perform a visual inspection of the battery, cables, and terminals.
- Correctly slow and fast charge a battery, in or out of the vehicle.
- Describe the differences between slow and fast charging.
- Perform a battery terminal test and accurately interpret the results.
- Perform a battery leakage test and determine the needed corrections.
- Test a conventional battery's specific gravity.

- Perform an open circuit test and accurately interpret the results.
- Perform a conductance test of the battery and accurately interpret the results.
- Perform a battery drain test and accurately determine the causes of battery drains.
- Remove, clean, and reinstall the battery properly.
- Jump-start a vehicle by use of a booster battery and jumper cables.
- Determine the cause of HV battery system failures.
- Measure HV battery module voltages with a DMM.

Terms to Know

Battery ECU	Conductance	Refractometer
Battery leakage test	Fast charging	Slow charging
Battery terminal test	Hydrometer	Stabilize
Capacity test	Jump assist	State of charge
Charge	Open circuit voltage test	
Charge rate	Parasitic drains	

Introduction

Classroom Manual
Chapter 5, page 113

The battery is the heart of the electrical system of the vehicle. It's important not to overlook it when servicing most electrical problems. A discharged or weak battery can affect more than just the starting of the engine. Because of its importance, the battery should be checked whenever the vehicle is in the shop for service. A battery test series will show the state of charge and output voltage of the battery, which determines if it's good, needs recharging or must be replaced.

There are many different manufacturers of battery test equipment. Most modern testers are computer based and conduct the tests automatically after selection of a particular test (**Figure 5-1**). Always follow the procedures given by the tester's manufacturer.

Figure 5-1 A computer-based generator, regulator, battery, and starter tester.

General Precautions

Before attempting to do any type of work on or around the battery, the technician must be aware of certain precautions. To avoid personal injury or property damage, take the following precautions:

1. Battery acid is very corrosive. Do not allow it to come in contact with skin, eyes, or clothing. If battery acid gets into your eyes, rinse them thoroughly with clean water and receive immediate medical attention. If battery acid comes in contact with skin, wash with clean water. Baking soda added to the water will help neutralize the acid. If you swallow acid, drink large quantities of water or milk followed by milk of magnesia and a beaten egg or vegetable oil.
2. When making connections to a battery, be careful to observe polarity, positive to positive and negative to negative.
3. When disconnecting battery cables, always disconnect the negative (ground) cable first.
4. When connecting battery cables, always connect the negative cable last.
5. Avoid any arcing or open flames near a battery. The vapors produced by the battery cycling are very explosive. Do not smoke around a battery.
6. Follow manufacturer's instructions when charging a battery. Charge the battery in a well-ventilated area. Do not connect or disconnect the charger leads with the charger turned on.

7. Do not add additional electrolyte to the battery if it's low. Add only distilled water.
8. Do not wear any jewelry or watches while servicing the battery. These items are excellent conductors of electricity. They can cause severe burns if current flows through them by accidental contact with the battery positive terminal and ground.
9. Never lay tools across the battery. They may come into contact with both terminals, shorting out the battery and causing it to explode.
10. Wear safety glasses or face shield when servicing the battery.
11. If the battery's electrolyte is frozen, allow it to defrost before doing any service or testing of the battery. While it's defrosting, look for leaks in the case. Leakage means the battery is cracked and needs replacement.

Battery Inspection

Classroom Manual
Chapter 5, page 115

 Special Tools
Fender covers
Safety glasses
Battery filler bottle

Before performing any electrical tests, inspect the battery cables, and terminals. A complete visual inspection of the battery should include the following items:

1. Battery date code: This provides information as to the age of the battery (**Figure 5-2**).
2. Condition of battery case: Check for dirt, grease, and electrolyte condensation. These contaminants can create an electrical path between the terminals and cause the battery to drain. Also, check for damaged or missing vent caps and cracks in the case. Excessive tightening of the hold-down fixture, excessive under-hood temperatures, buckled plates from extended undercharged conditions, freezing, or excessive charge rate can crack or buckled the case.
3. Electrolyte level, color, and odor: Discoloration of electrolyte and the presence of a rotten egg odor indicate an excessive charge rate, excessive deep cycling, impurities in the electrolyte solution, or an old battery. If necessary, add distilled water until the level covers the top of the plates by 1/2 inch (12 mm). After adding water, charge the battery before performing any tests.

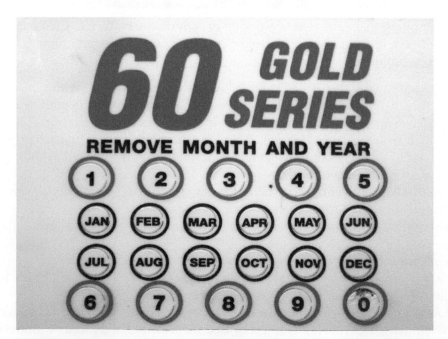

Figure 5-2 The battery sticker will usually have the purchase date, plus additional information.

Figure 5-3 Inspect the condition of the battery cables and terminals.

4. Condition of battery cables and terminals: Check for corrosion, broken clamps, frayed cables, and loose terminals (**Figure 5-3**). These conditions result in voltage drop between the battery terminal and the end of the cable. The sulfuric acid that vents out with the battery gases attacks the battery terminals and cables. Sulfuric acid reacting with lead and copper creates lead sulfate and copper sulfate deposits. These deposits are resistive to electron flow and limit the amount of current supplied to the electrical and starting systems. If the deposits are bad enough, the resistance can increase to a level that prevents the starter from cranking the engine.

5. Battery abuse: This includes using bungee cords and 2 × 4s for hold-down fixtures, too small of a battery rating for the application, and obvious neglect to periodic maintenance. In addition, inspect the terminals for indications of being hit with a hammer (an attempt to tighten the clamp) and for improper cable removal procedures. Finally, check for proper cable length.

6. Battery tray and hold-down fixture: Check for proper tightness. Also, check for signs of acid corrosion of the tray and hold-down unit. Replace as needed.

7. If the battery has a built-in hydrometer, check its color indicator (**Figure 5-4**).

Classroom Manual
Chapter 5, page 127

Classroom Manual
Chapter 5, page 122

> ⚙ **SERVICE TIP** Grid growth can cause the battery plate to short out the cell. If there is normal electrolyte level in all cells but one, that cell is probably shorted and the electrolyte has converted to hydrogen gas.

One common cause of early battery failure is overcharging. If the charging system supplies a voltage level over 15.5 volts, the plates may become warped due to the excess heat generated during overcharging. Overcharging also causes the active material to disintegrate and shed off the plates.

If the charging system does not produce enough current to keep the battery charged, the lead sulfate can become crystallized on the plates. If this happens, the sulfate is difficult to remove, and the battery will resist recharging. The recharging process converts the sulfate on the plates. Undercharging does not convert the sulfate and it will harden on the plates.

Sulfate that does not convert back to H_2SO_4 hardens on the plates. This results in battery sulfation, which permanently damages the battery.

Green dot

Dark

Clear

Top of battery

Sight glass

Clear plastic rod

Green ball

65% or above state of charge

Below 65% state of charge

Low level electrolyte

Figure 5-4 Built-in hydrometer used to indicate a battery's state of charge.

Some technicians believe the internal resistance of a fully charged battery is high; however, this not true. Keep in mind that the coating of lead sulfate on the battery plates acts as an insulator. The more sulfate on the plates, the higher the battery's internal resistance. The increased resistance of a discharged battery allows it to accept a higher rate of charge without gassing or overheating than a battery near a full charge.

Charging the battery removes the sulfate and reduces the amount left to sustain the reverse chemical reaction. The battery's "natural absorption rate" is the level of charge current that can be applied without overheating the battery and breaking down the electrolyte into hydrogen and oxygen. Overcharging occurs when the charge current is greater than the natural absorption rate.

Vibration is another common reason for battery failure. If the battery is not secure, the plates will shed the active material due to excessive vibration. Over time the shed material accumulates at the bottom of the battery and can create an electrical connection between the plates. The shorted cell will not produce voltage, resulting in a battery with only 10.5 volts across the terminals. With this reduced amount of voltage, the starter usually will not be capable of starting the engine. Installing the proper hold-down fixtures will prevent this problem.

Even during normal battery operation, the active materials on the plates will shed over time. The negative plate also becomes soft. Both of these events will reduce the effectiveness of the battery.

⚠ WARNING **There are many safety precautions associated with charging the battery. The hydrogen gases produced by a charging battery are very explosive. Exploding batteries are responsible for over 15,000 injuries per year that are severe enough to require hospital treatment. Keep sparks, flames, and lighted cigarettes away from the battery. Also, do not use the battery to lay tools on. They may short across the terminals and result in the battery exploding. Always wear eye protection and proper clothing when working near the battery. Also, most jewelry is an excellent conductor of electricity. Do not wear any jewelry when performing work on or near the battery. Do not remove the vent caps while charging. Turn the charger off before connecting or disconnecting the charger leads.**

Charging the Battery

To **charge** the battery means to pass an electric current through the battery in an opposite direction than during discharge. You can charge the battery in or out of the vehicle. The safest method is to remove the battery before charging it. If charging the battery with it still in the vehicle, it's important to protect any vehicle computers by removing the negative battery cable.

Turn the charger off before connecting or disconnecting the charger leads. Connect the cable leads to the battery terminals, observing polarity. Attempting to charge the battery with reversed cable leads will result in battery damage and possible damage to vehicle's onboard computers. For this reason, many battery chargers have a warning system to alert the technician that they connected the cables in reverse polarity. Rotate the clamps slightly on the terminals to assure a good connection.

Depending on the requirements and amount of time available, the battery can be slow or fast charged. Each method of charging has its advantages and disadvantages.

 WARNING Before charging a battery that has been in cold weather, check the electrolyte for ice crystals. A discharged battery will freeze at a higher temperature than a fully charged battery. Do not attempt to charge a frozen battery. Forcing current through a frozen battery may cause it to explode. Allow it to warm at room temperature for a few hours before charging.

Slow Charging

Slow charging the battery has two advantages:

- It's the only way to restore the battery to a fully charged state.
- It minimizes the chances of overcharging the battery.

Slow charging means holding the charge rate between 3 and 15 amps for a long period of time. Slow charging the battery causes the lead sulfate on the plates to convert to lead dioxide and sponge lead throughout the thickness of the plate.

Fast Charging

Fast charging uses a high current for a short period of time to boost the battery. Fast charging the battery will bring the state of charge up high enough to crank the engine. However, fast charging cannot recharge the battery as effectively as slow charging. Fast charging the battery converts only the lead sulfate on the outside of the plates. The

Classroom Manual
Chapter 5, page 118

Special Tools

Safety glasses
Battery charger
Voltmeter
Fender covers

 Caution

If removing the battery from the vehicle, disconnect the negative battery cable first. Use a battery carrying tool to lift out the battery (**Figure 5-5**).

Slow charging is also called "trickle charging."

Carrying tool   Carrying strap

Figure 5-5 Always use a battery carrier to lift the battery.

conversion does not go through the plates. After the battery has fast charged to a level that it will crank the engine, it should be slow charged to a full state.

Charge Rate

The **charge rate** is the speed at which the battery can safely be recharged at a set amperage. The charge rate required to recharge a battery depends on several factors:

1. Battery capacity. High-capacity batteries require longer charging time.
2. State of charge.
3. Battery temperature.
4. Battery condition.
5. Battery type.

Slow charging is the easiest on the battery. However, slow charging requires a long period of time. The basic rule of thumb for slow charging the battery is 1 amp for each positive plate in one cell.

Slow charging of the battery may not always be practical due to the time involved. In these cases, fast charging is the only alternative. To determine the charging time for a full charge based on charge rate amperes, use **Table 5-1**.

An alternative method is to connect a voltmeter across the battery terminals while it's charging. If the voltmeter reads fewer than 15 volts, the charging rate is low enough. If the voltmeter reads over 15 volts, reduce the charging rate until voltage reads below 15 volts. Keeping the voltage at 15 volts will ensure the quickest charge and a safe rate for the battery.

There are three methods of determining a fully charged battery:

1. Specific gravity holds at 1.264 or higher after stabilizing the battery.
2. An open circuit voltage test indicates 12.68 or higher after stabilizing the battery.
3. The ammeter on the battery charger falls to approximately 3 amps or less and remains at that level for 1 hour.

Recharging Recombination Batteries

Classroom Manual
Chapter 5, page 123

Most recombination batteries (including gel cell, valve-regulated, and absorbed glass mat) batteries will accept recharging very well. This is due, in part, to their low internal

Table 5-1 Table Showing the Rate and Time of Charging a Battery. Electrolyte Temperatures Should Not Exceed 125°F (51.7°C) During Charging

Open Circuit Voltage	Battery Specific Gravity	State of Charge (%)	Charging Time of Full Charge at 80°F (27°C)					
			At 60 A	At 50 A	At 40 A	At 30 A	At 20 A	At 10 A
12.6	1.265	100	Full charge					
12.4	1.225	75	15 min.	20 min.	27 min.	35 min.	48 min.	90 min.
12.2	1.190	50	35 min.	45 min.	55 min.	75 min.	95 min.	180 min.
12.0	1.155	25	50 min.	65 min.	85 min.	115 min.	145 min.	280 min.
11.8	1.120	0	65 min.	85 min.	110 min.	150 min.	150 min.	370 min.

⚙ **SERVICE TIP** If a severely discharged battery will not take a slow charge, connect a good battery in parallel (with jumper cables). Fast charge for 30 minutes; then disconnect the good battery and slow charge the discharged battery.

resistance. However, overcharging is very harmful to recombination batteries. Since recombination batteries use a special sealing design, overcharging will dry out the electrolyte by forcing the oxygen and hydrogen from the battery through the safety valves.

A continually undercharged battery will build up a layer of sulfate on the positive plate. The sulfate then acts to resist the flow of electrons. This may also result in plate shedding which reduces battery performance and shortens battery life.

When recharging the recombination battery, disconnect the negative battery cable. Charge the battery directly at the battery terminals (do not use any jumper cables or chassis ground locations). The charger must properly limit the voltage to no more than 14.4 volts and no less than 13.8 volts at 68°F (20°C). This requires charging equipment designed to recharge recombination batteries. Older battery chargers that use higher voltages and charging rates usually cannot recharge recombination batteries. Charging a recombination battery at too high a voltage may experience shorter battery life because of damage resulting from the battery temperature increasing to over 100°F (37.8°C).

Most modern battery chargers are compatible with recombination-type batteries and have a switch to cycle between flooded and recombination batteries. Check the documentation supplied with the charger you are using to recharge a recombination cell battery.

It's a good practice to measure the voltage across the recombination battery before recharging it. If the voltage is less than 10 volts, it may take several hours to recharge the battery since the charger will apply very low current until the battery voltage increases.

Hybrid vehicles with 12-volt auxiliary batteries are recharged similarly to other recombination batteries. However, there may be special service procedures that must be performed in preparation of isolating and charging the battery.

Maintaining Memory Functions

Today's vehicles have several computer control modules that use memories to store values and presets. These include radio station selection, memory seat positions, transmission shift schedule learning, and adaptive fuel strategies. These memories may require battery voltage to maintain them and reset when battery voltage is disconnected. If the battery is disconnected to perform battery tests, battery service, or recharged, using a memory keeper will retain the memory values. However, if the battery is disconnected to perform electrical circuit tests or repairs, do not use a memory keeper. In this case you will need to restore the memory functions.

Classroom Manual
Chapter 5, page 115

Battery Test Series

After the battery and cables have been completely inspected and any problems have been corrected, the battery is ready to be tested further. Most test procedures require a fully charged battery.

Battery Terminal Test

The **battery terminal test** checks for poor electrical connections between the battery cables and terminals. Use a voltmeter to measure voltage drop across the cables and terminals. It's a good practice to perform the battery terminal test anytime the battery cable is disconnected and reconnected to the terminals. By performing this test, comebacks, due to loose or faulty connections, can be reduced.

Classroom Manual
Chapter 5, page 127

Figure 5-6 Voltmeter connections for the battery terminal test.

Figure 5-7 Use battery clamp pullers to remove the cable end from the terminal. Do not pry the clamp off.

Figure 5-8 A terminal cleaning tool used to clean the cable's clamps and the battery's terminals.

Special Tools

Safety glasses
Voltmeter
Terminal pliers
Terminal puller
Terminal and clamp cleaner
Fender covers

Classroom Manual
Chapter 5, page 118

The battery must be fully charged to perform this test. To test the negative terminal connection, connect the positive voltmeter test lead to the cable clamp and connect the negative meter lead to the battery terminal (**Figure 5-6**). Follow the vehicle manufacturer's recommended method to disable the ignition system to prevent the vehicle from starting. This may require the removal of the fuel pump or electronic fuel injection (EFI) relay.

Crank the engine and observe the voltmeter reading. If the voltmeter shows over 0.3 volt, there is a high resistance at the cable connection. Remove the battery cable using the clamp puller (**Figure 5-7**). Clean the cable ends and battery terminals (**Figure 5-8**).

Perform the test for the positive terminal in the same manner. Connect the voltmeter leads across the terminal and clamp, observing polarity.

Battery Leakage Test

A dirty battery can cause battery drain. The dirt can conduct current flow over the battery case and drain a battery as quickly as leaving a light on. A **battery leakage test** determines if current is flowing across the battery case. Set a voltmeter to a low DC volt scale and connect the positive test lead to the battery's positive terminal. Move the negative

Figure 5-9 Using a voltmeter to perform the battery leakage test.

test lead across the top and sides of the battery case (**Figure 5-9**). If the meter reads voltage, the dirt is completing a current path from the battery's negative terminal to its positive terminal. Keep in mind that you should not measure voltage anywhere on the case of the battery. If voltage is present, remove the battery. Then use a baking soda and water mixture to clean the case of the battery. When cleaning the battery, don't allow the baking soda and water solution to enter its cells. After the case is clean, rinse it off with clean water.

State of Charge Test

Measuring the **state of charge** is a check of the battery's electrolyte and plates. The state of charge test uses the measurement of the specific gravity to determine the charge of the battery.

To use the hydrometer to test the battery's state of charge:

1. Remove all battery vent caps.
2. Check the electrolyte level. It must be high enough to withdraw the correct amount of solution into the **hydrometer**. A hydrometer measures the specific gravity of a liquid (**Figure 5-10**).
3. Squeeze the bulb and place the pickup tube into the electrolyte of a cell.
4. Slowly release the bulb. Draw in enough solution to suspend the float in the barrel. Hold the hydrometer in a vertical position.

Read the specific gravity where the float scale intersects the top of the solution (**Figure 5-11**). Correct the reading by compensating for temperatures (**Figure 5-12**).

If the electrolyte level is too low to perform the test, add distilled water to the cell. Mix the water and electrolyte by charging the battery. Do not take hydrometer readings until the water and electrolyte are mixed.

Test Results. As a battery becomes discharged, its electrolyte has a larger percentage of water. Thus, a discharged battery's electrolyte will have a lower specific gravity number than that of a fully charged battery.

A fully charged battery will have a hydrometer reading near 1.265. Remember, to correct the reading based on the temperature of the electrolyte. If the corrected hydrometer reading is below 1.265, the battery needs recharging, or it may be defective.

Special Tools
Voltmeter
Fender covers
Safety glasses

Classroom Manual
Chapter 5, pages
118, 122

Special Tool
Hydrometer

Figure 5-10 A temperature correction hydrometer measures the specific gravity of the electrolyte solution, indicating the battery's state of charge.

Figure 5-11 Read the specific gravity of the electrolyte at the point where the electrolyte intersects the float. (A) A low reading; (B) a high reading.

Figure 5-12 Correct the specific gravity reading according to the temperature of the electrolyte.

To determine if the battery is defective using a hydrometer, checking every cell. If the specific gravity has a 0.050-point variation between the highest and lowest cell readings, the battery is defective (**Figure 5-13**). If the cells have an equal gravity (even if all are low) regenerating the battery is possible by recharging.

Specific gravity testing is not the sole determinant of battery condition. If the battery cells do not have equal specific gravity, replace the battery. When the specific gravity of all the cells is good or bad, consider the battery's voltage before making a conclusion about

Water level

Acid content

| 1.260 | 1.260 | 1.195 | 1.260 | 1.250 | 1.250 |

Figure 5-13 A defective cell can be determined by the specific gravity readings.

the battery's condition. A battery with low specific gravity and acceptable voltage is normally only discharged, perhaps due to a charging system problem. However, a battery with good specific gravity readings but low voltage readings requires replacement.

Optical Refractometer

A **refractometer** uses light refractions to determine and display a very accurate specific gravity reading. Place a couple of drops of electrolyte on the glass slide of the refractometer (**Figure 5-14**). While holding the meter's lens up to your eye, the light will refract through the sample and display the specific gravity in the window (**Figure 5-15**).

Special Tool
Refractometer

> **SERVICE TIP** If the vehicle has many circuits that place a constant drain on the battery (computer, clock, memory radios, etc.), you should disconnect the negative battery cable before taking a voltmeter reading.

Figure 5-14 Placing an electrolyte sample onto the refractometer's slide.

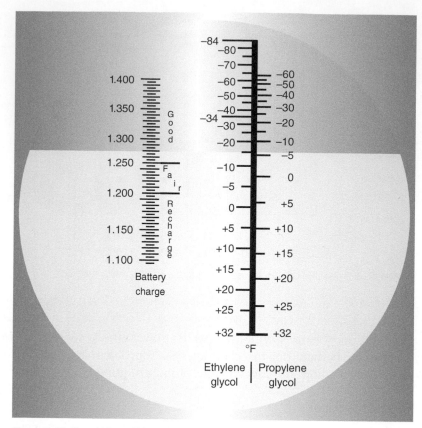

Figure 5-15 The window will indicate the specific gravity.

Open Circuit Voltage Test

The battery **open circuit voltage test** determines the battery's state of charge. Do this test if a hydrometer is unavailable or when testing a sealed battery. Before performing the open circuit voltage test, **stabilize** the battery. Stabilizing the battery means to remove the surface charge by placing a large load on the battery for 15 seconds. Connect a voltmeter across the battery terminals, observing polarity (**Figure 5-16**). Measure the open circuit voltage. Take the reading to the 1/10 volt. A good battery will have an open circuit voltage of 12.6 volts.

Figure 5-16 Open circuit voltage test using a voltmeter.

Open Circuit Voltage Table	
Open Circuit Voltage	Charge Percentage
11.7 volts or less	0%
12.0 volts	25%
12.2 volts	50%
12.4 volts	75%
12.6 volts or more	100%

Figure 5-17 Open circuit voltage test results relate to the specific gravity of the battery's cells.

To analyze the open circuit voltage test results, consider that a battery at a temperature of 80°F (26.7°C), in good condition, should show about 12.4 volts. The battery is considered charged if the state of charge is 75% or more. **Figure 5-17** illustrates the relationship between open circuit voltage test results and charge percentage.

CUSTOMER CARE One of the best things you can do for your customers is to assist them in choosing the correct battery. Battery selection needs to be based on the make of the vehicle, electrical options on the vehicle, driving habits, and climatic conditions. The largest current capacity rating that can be achieved in a given battery group may benefit some customers but may be a waste of money for others.

Classroom Manual
Chapter 5, page 125

Battery Conductance Testing

Conductance is a measurement of the battery's plate surface that's available for chemical reaction. This determines how much power the battery can supply and describes the ability of a battery to conduct current. Conductance testing of the battery has proven to be a reliable test of the battery's capacity (**Figure 5-18**). The higher the conductance value (or lower internal resistance), the better the performance potential of the battery.

Conductance testing sends a low-frequency AC signal through the battery. The tester capturers a portion of the returning AC pulse and calculates a conductance measurement. Conductance battery testers can accurately test batteries even if they are not fully charged. Internal battery damage is also detectable without charging the battery. An on-screen display directs the technician through the steps required to perform the battery test.

A fully charged battery will have a conductance reading between 110% and 140% of its CCA rating. As the plate surfaces sulfate, the cell will lose its conductance. If the battery loses its cranking ability, the conductance will drop below the CCA rating. At this point, battery replacement is necessary. **Photo Sequence 8** illustrates a typical procedure for using the conductance tester for determining a battery's condition.

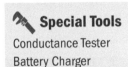

Special Tools
Conductance Tester
Battery Charger

Figure 5-18 Handheld conductance tester.

PHOTO SEQUENCE 8
Using a Conductance Battery Tester

P8-1 The GR8 is a common battery/starter/charging system tester that uses conductance.

P8-2 With the tester connected to the power supply, use the arrow keys to select the System Test icon in the Main Menu.

P8-3 Connect the small clamps from the tester to the battery terminals.

P8-4 Select the POST TYPE based on battery design.

P8-5 Press the NEXT soft key to continue.

P8-6 Determine the battery type and make the appropriate selection.

P8-7 Select the battery's rating units.

P8-8 Use the UP or DOWN arrow keys to select the battery rating.

P8-9 The load test will begin.

P8-10 After the test, the tool will display a state of health decision and analyze the battery's state of charge. Use the UP or DOWN arrow keys to scroll through the screens.

P8-11 The results of the battery test series can be printed out.

P8-12 Pressing the End key will return to the Main Menu.

Battery Drain Test

Classroom Manual
Chapter 5, page 115

If a customer has an issue with their battery being dead after not driving the vehicle for a short while, the problem may be a current drain from one of the electrical systems. The most common cause for this type of drain is a light that's not turning off—such as glove box, trunk, or engine compartment illumination lights.

These **parasitic drains** on the battery can cause various driveability problems. With low battery voltage, several problems can result; for example:

1. The computer may go into backup mode or "limp-in" mode of operation.
2. The computer may set false trouble codes.
3. To compensate for the low battery voltage, the computer may raise the engine speed.

The procedure for performing the battery drain test may vary according to the manufacturer. However, battery drain can often be observed by connecting an ammeter in series with the negative battery cable or by placing an inductive ammeter pickup lead around the negative cable. If the meter reads 30 mA or more, with the ignition OFF, all doors closed, all lights off, and computers timed out there is excessive drain. Visually check the trunk, glove box, and under-hood lights to see if they are on. If they are, remove the bulb and watch the battery drain. If the drain is now within specifications, find out why the circuit is staying on and repair the problem. If the cause of the drain is not the lights, go to the fuse panel or distribution center and remove one fuse at a time while watching the ammeter. When the drain decreases, the circuit protected by the fuse you removed last is the source of the problem.

The following is a typical procedure for determining and locating parasitic drains against the battery. First, it's necessary to determine if a draw is occurring by connecting an inductive ammeter clamp around the negative battery cable. The meter must be capable of reading less than 1 amp accurately. With all accessories turned off, the ammeter reading indicates the current draw against the battery. If the ammeter reads higher than allowed current, the cause of the drain must be determined. It's normal for some vehicles to have higher than 30 mA draw for up to an hour after turning off the ignition switch to allow the computers to perform their "administrative" tasks. Be sure to confirm the normal time-out period with the proper service information.

If an inductive clamp is not used, then the ammeter needs to be connected in series with the negative battery cable and the battery terminal. It may be advisable to use a test light in series with the negative battery cable and the battery terminal first (**Figure 5-19**). This prevents the vehicle's computers from powering down and then powering back up again when the ammeter is connected. The additional amperage of the computers powering back up may blow the fuse in the ammeter. Before disconnecting the battery cable from the terminal, connect the test light's clip to the cable clamp and hold the test light's

Special Tools

Test light
DMM
Terminal pliers
Terminal puller
Cable clamp spreader
Safety glasses
Fender covers

The open circuit voltage reading must be 11.5 volts or higher to perform the battery drain test.

If the vehicle is equipped with computer-controlled air suspension systems, it may be necessary to disconnect the module to eliminate it from the test.

Figure 5-19 Using a test light to prevent the vehicle's computers from powering down.

probe onto the terminal so that it will not interfere with the removal of the clamp. While removing the clamp maintain connection of the lamp with the clamp and the terminal. If the test light is on, there is a drain against the battery that's sufficient to light the lamp. After the time period has expired, the test light should go out. If not, connect the ammeter leads in series between the cable and terminal (with the test light still connected). Once the ammeter is connected, remove the test light and read the amount of current draw.

An alternate method of isolating the circuit that has the excessive drain is to measure the voltage drop across the fuse. Since a fuse may protect several circuits, it's possible to mask the problem by pulling fuses. For example, if a fuse protects the power supply circuits to three different computers, pulling that fuse will cause them all to totally turn off. Usually, a computer does not turn off completely when it powers down, but the draw is very slight. If one of the three computers was failing to power down, then excessive drain would be against the battery. Pulling the fuse powers down all three computers and may reset the defective computer, eliminating the cause of the parasitic load. After reinstalling the fuse there no longer will be a parasitic load. This did identify the since pulling the fuse caused the load reading on the ammeter to drop. However, with the fuse reinstalled, the load is no longer indicated and isolating the cause is made more difficult. It would be advantageous to identify the circuit first, and then unplug the computers one at a time until the circuit with the draw is identified.

A fuse has resistance so that it can get hot and burn through if excessive current passes through it; this resistance will provide a voltage drop. To identify the circuit with a parasitic load, use the test tabs on the top of the fuse (maxi-fuses will require removing the lens). Connect a voltmeter set on the millivolt scale across the test points and observe the meter's reading (**Figure 5-20**). If there is no (or very little) current flowing through the

Figure 5-20 Using a voltmeter across a fuse to determine if the circuit as current flowing.

Fuse Value	Fuse Type	Divide by
5	Mini	16.5
10	Mini	7.5
15	Mini	4.5
20	Mini	3.5
25	Mini	2.5
30	Mini	2.0
20	Cartridge	1.0
30	Cartridge	1.5
40	Cartridge	1.0
50	Cartridge	0.5

Figure 5-21 Determining how much current is flowing through a fuse.

circuit, the voltmeter will read 0 volt. However, current flow through the fuse will cause the voltmeter to produce a reading. The reading indicates how much current is flowing in the circuit, which is a function of the rating of the fuse (**Figure 5-21**). For example, if the voltmeter reads 20 mV when connected across a 15-amp fuse, the current draw is 20/4.5 = 4.4 amps.

> **AUTHOR'S NOTE** It's not important to determine the amount of current flow. You are not looking for 1 or 2 mA but more than 30 mA; therefore, any reading displayed on the voltmeter indicates excessive current flow.

> **AUTHOR'S NOTE** Using an ammeter to test for battery drain will not indicate an internal short in the battery.

Battery Removal and Cleaning

It's natural for dirt and grease to collect on the top of the battery. If allowed to accumulate, the dirt and grease can form a conductive path between the battery terminals. This may result in a drain on the battery. Also, normal battery gassing will deposit sulfuric acid as the vapors condense. Over time, the sulfuric acid will corrode the battery terminals, cable clamps, and hold-down fixtures. As the corrosion builds on the terminals, it adds resistance to the entire electrical system.

Periodic battery cleaning will eliminate these problems. To clean the battery correctly, it's best to remove it from the vehicle. Removing the battery protects the vehicle's finish and other under-hood components. Follow **Photo Sequence 9** for the procedure for removing the battery from the vehicle and cleaning it. Consult the manufacturer's service information for precautions concerning the vehicle's computer controls.

When replacing the battery into the vehicle, make sure it's properly seated in the tray. Connect the hold-down fixture and secure the battery. Do not overtighten the hold-down fixture. Install the positive cable and tighten; then install the negative cable. Be sure to observe polarity. Perform a battery terminal test to confirm good connections.

Classroom Manual
Chapter 5, page 128

Special Tools

Baking soda
Cleaning brushes
Terminal pliers
   (Figure 5-22)
Cable clamp spreader
Terminal puller
Terminal and clamp
   cleaning tool
Battery-lifting strap
Putty knife
Protective terminal
   coating
Safety glasses
Heavy rubber gloves
Fender covers

⚠ **Caution**

Do not allow the baking soda solution to enter the battery cells. This will neutralize the acid in the electrolyte and destroy the battery.

Figure 5-22 Battery terminal pliers.

The baking soda and water solution consists of one tablespoon per quart of water.

CUSTOMER CARE Before installing the battery back into the vehicle, it's a good practice to clean the battery tray. First scrape away any heavy corrosion with a putty knife. Next, clean the tray with a baking soda and water solution (**Figure 5-23**). Flush with water and allow to dry. After the tray has dried, paint it with rust-resistant paint. After the paint has completely dried, coat the tray with silicone base spray. These extra steps will protect the tray from corrosion.

Brush

Battery tray

Figure 5-23 After scraping the battery tray clean with a putty knife, use the baking soda mixture to remove all electrolyte residues.

PHOTO SEQUENCE 9
Removing and Cleaning the Battery

P9-1 Tools needed to remove the battery from the vehicle include rags, baking soda, pan, terminal pliers, cable clamp spreader, terminal puller, assorted wrenches, terminal and clamp cleaning tool, battery lifting strap, safety glasses, heavy rubber gloves, rubber apron, and fender covers.

P9-2 Place the fender covers on the vehicle to protect the finish.

P9-3 Loosen the clamp bolt for the negative cable using terminal pliers and wrench of correct size. Be careful not to put excessive force against the terminal.

PHOTO SEQUENCE 9 (CONTINUED)

P9-4 Use the terminal puller to remove the cable from the terminal. Do not pry the cable off of the terminal.

P9-5 Locate the negative cable away from the battery.

P9-6 Loosen the clamp bolt for the positive cable and use the terminal puller to remove the cable. If the battery has a heat shield, remove it.

P9-7 Disconnect the hold-down fixture.

P9-8 Using the battery-lifting clamp, remove the battery out of the tray. Keep the battery away from your body. Wear protective clothing to prevent acid spills onto your hands.

P9-9 Transport the battery to the bench. Keep it away from your clothes.

P9-10 Mix a solution of baking soda and water.

P9-11 Brush the solution over the battery case. Be careful not to allow the solution to enter the battery's cells.

P9-12 Flush the solution off with water.

PHOTO SEQUENCE 9 (CONTINUED)

P9-13 Use a wire brush to remove corrosion from the hold-down brackets.

P9-14 Use the terminal cleaning tool to clean the cable.

P9-15 Use the cleaning tool to clean the battery posts.

P9-16 Install the battery into the tray and install the hold-down hardware. Next, install the cables, positive first followed by the negative cable.

P9-17 Coat the battery terminals with corrosion-preventive spray. Perform a terminal test to assure tight connections.

⚠ **WARNING** **Be careful not to touch the positive terminal with the wrench when tightening the negative cable clamp.**

Spray the cable clamps with a protective coating to prevent corrosion. A little grease or petroleum jelly will prevent corrosion as well. Also, protective pads that go under the clamp and around the terminal are available (**Figure 5-24**).

Figure 5-24 A protective pad under the battery clamp prevents corrosion of the clamp and terminal.

Jumping the Battery

There will be times when you will have to use a boost battery and jumper cables to jump-start a vehicle (**Figure 5-25**). It's important to follow all safety precautions. Jump-starting a dead battery can be dangerous if not done correctly. Follow these steps to safely jump-start most vehicles:

1. Make sure the two vehicles are not touching each other. The excessive current flow through the vehicles' bodies can damage the small ground straps that attach the engine block to the frame. These small wires are designed to carry only 30 amps. Up to 400 amps may attempt to flow through the straps if the vehicles are touching.
2. For each vehicle, engage the parking brake and put the transmission in neutral or park.
3. Turn off the ignition switch and all accessories, on both vehicles.
4. Attach one end of the positive jumper cable to the disabled battery's positive terminal.
5. Connect the other end of the positive jumper cable to the booster battery's positive terminal.
6. Attach one end of the negative jumper cable to the booster battery's negative terminal.
7. Attach the other end of the negative jumper cable to an engine ground on the disabled vehicle.
8. Attempt to start the disabled vehicle. If the disabled vehicle does not start readily, start the jumper vehicle and run at fast idle to prevent excessive current draw.
9. Once the disabled vehicle starts, disconnect the ground-connected negative jumper cable from its engine block.
10. Disconnect the negative jumper cable from the booster battery.
11. Disconnect the positive jumper cable from the booster battery and then from the other battery.

Special Tools

Booster battery
Jumper cables
Jumper box
Fender covers
Safety glasses

Caution

Do not use more than 16 volts to jump-start a vehicle equipped with an engine control module. The excess voltage may damage the electronic components.

Figure 5-25 Proper jumper cable connections and sequence for jump-starting a vehicle.

Figure 5-26 A typical jumper box.

⚡ **WARNING** **Do not connect this cable end to the battery negative terminal. Doing so may create a spark that will cause the battery to explode.**

⚡ **WARNING** **A rapidly discharged battery will create hydrogen gas. Do not attach jumper cables to a weak battery if starting the vehicle has been attempted. Wait for at least 10 minutes before connecting the jumper cable and attempting to start the vehicle.**

A jumper box is often used instead of jumping from another vehicle's battery (**Figure 5-26**). The jumper box contains a small 12-volt battery that can supply enough current to start the vehicle. The box is connected to the vehicle battery by clamps similar to that of a battery charger. When the jumper box is not in use, it's plugged into a 120-volt wall socket to keep it charged. Observe polarity when connecting the jumper box.

HV Battery Service

Classroom Manual
Chapter 5, page 129

It's important to remember that most hybrid vehicles have two separate batteries. One is the high-voltage (HV) battery pack; the other is the conventional 12-volt battery. In most hybrids, the HV battery pack provides the electrical power to start the engine and powers the electric motors. The 12-volt battery is used to power the basic electrical system, such as the lights, accessories, and power equipment. When diagnosing a problem with the HV battery, it's important to consider that the 12-volt battery also supplies the power for the electronic controls that monitor and regulate the hybrid system's operation. If the 12-volt battery source is not operating properly, the hybrid system will not either. Consequently, never ignore the 12-volt system when working on an HV system.

Most hybrid vehicles have a continuous drain on the 12-volt battery system. In addition, a HV battery pack that does not see regular use will drain. Due to the battery drain when the vehicle is not in use, most manufacturers recommend that HEVs be started and run for at least 10 minutes every month. This keeps the HV battery charged enough to operate the vehicle. However, it may not be enough to keep the 12-volt battery properly charged. When servicing the batteries on a HEV, remember that if the battery's voltage

Special Tools
DMM
Scan tool

drops below a specific level, the emissions malfunction indicator lamp (MIL) and/or hybrid warning lights may illuminate.

Before working on or around the HV battery and energy systems, review all safety warnings, precautions, and procedures outlined in Chapter 1. Remember that these are guidelines; the manufacturer's service literature provides specific instructions and safety warnings. Following these safety precautions is not limited to servicing of the electrical system; all services including air-conditioning, engine, transmission, and bodywork may require services completed around and/or with HV systems. If there is any doubt as to whether something has high voltage or not, or if the circuit is sufficiently isolated, *test it!!*

A battery electronic control unit (ECU) monitors the condition of the HV battery assembly. The **battery ECU** determines the state of charge (SOC) of the HV battery by monitoring voltage, current, and temperature. The battery ECU collects data and transmits it to the HV ECU to be used for proper charge and discharge control.

The battery ECU also controls the operation of the battery blower motor to maintain proper HV battery temperature.

The HV battery stores power generated by MG1 and recovered by MG2 during regenerative braking (**Figure 5-27**). The HV battery must also supply power to the electric motor when the vehicle is first started from a stop or when additional power is needed. A typical HV battery uses several nickel-metal-hydride modules and can provide over 270 volts (**Figure 5-28**).

When the vehicle is moving, the HV battery is subjected to repetitive charge and discharge cycles. The HV battery is discharged by MG2 during acceleration mode and then is recharged by regenerative braking. An amperage sensor (**Figure 5-29**) is used so the battery ECU can transmit requests to the HV ECU to maintain the SOC of the HV battery. The battery ECU attempts to keep the SOC at 60%. The battery ECU also monitors delta SOC to determine if it's capable of maintaining acceptable levels of charge. The normal, low-to-high SOC delta is 20%.

If the battery ECU sends abnormal messages to the HV ECU, the HV ECU illuminates the warning light and enters fail-safe control. DTCs and informational codes set along with freeze frame data. This will occur if the battery ECU determines the insulation resistance of the power cable to be 100k ohms or less.

MG1 functions as the control element for the planetary gear set. It recharges the HV battery and supplies electrical power to drive MG2. MG1 also functions as the starter for the engine. MG2 is used for power at low speeds and for supplemental power when needed at higher speeds.

Figure 5-27 Layout of the generating HEV transaxle.

Figure 5-28 The modules of the HV battery.

Figure 5-29 The amperage sensor.

⚙️ **SERVICE TIP** If inspection and testing fail to locate the leak, then it's possible that water entered into the battery assembly or into the converter/inverter assembly.

Whenever an HV battery malfunction occurs, use the scan tool to view the "HV Battery Data List." This provides all HV battery system information. Since the HV battery is usually constructed of a series of modules, the scan tool may display the voltage for each module. Usually, a module will have 7.2 volts so the scan tool reading should be between 7 and 8 volts for each module. Some manufacturers will combine two modules to make a block with a normal voltage of 14.4 volts. The scan tool reading for each block should be between 14 and 16 volts (**Figure 5-30**). Regardless of how the scan tool displays the voltage readings, compare each reading with the others. They should be within 0.3 volt of each other when the contactors are open. If any voltage readings are higher or lower than the others, that module or block may be damaged.

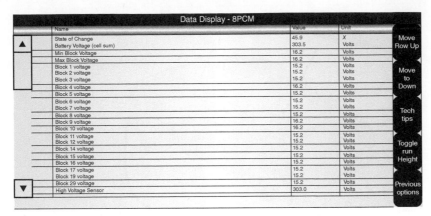

Figure 5-30 Scan tool reading of HV battery voltages.

The scan tool may also display the HV battery SOC. Based on the SOC of the HV battery, the hybrid controller will perform the following:

- 6% SOC all vehicle operation shut down
- 35% SOC e-motors are not used (drive is disabled)
- 40% SOC e-motors are de-rated
- 50% SOC engine restarts to charge HV battery
- Approximately, 52% to 68% SOC normal operating range
- 80% SOC regenerative braking is disabled
- 94% SOC all vehicle operation shut down

If the scan tool does not display the HV battery voltages, then it will be necessary to test the battery with a digital multimeter (DMM). This may require removal of the HV battery pack from the vehicle and the removal of the battery's case.

> **SERVICE TIP** Some battery blocks may read 0 volt when the service plug is removed. Be sure to confirm proper voltages with the service information.

High-Voltage Battery Charging

Driving the vehicle or running the engine charges the HV battery. These are the best methods of charging the HV battery pack. If the SOC of the HV battery is too low to allow the engine to run, the HV battery will need to be recharged. This may require the use of a special high-voltage battery charger. In addition, some manufacturers will allow only specially trained people to recharge the battery.

To recharge the HV battery, a special Hybrid/EV battery service and de-power tool is required. The de-power function discharges the battery pack or battery sections to make them safe for transportation. Section balance equalizes the sections of the battery pack to ensure optimal battery pack operation. Sections are charged or discharged individually to bring the pack into alignment. The view pack function generates a detailed status report of the HV battery, including temperature sensors and individual cell voltages.

Most manufacturers recommend that HV battery recharging be performed outside. When using the charger, the immediate area should be secured and marked with warning tape. The correct cable is connected between the vehicle and the charger (**Figure 5-31**). It will require about 3 hours to recharge the battery to an SOC of about 50%.

Some manufacturers provide a **jump assist** feature so the HV battery can be charged without special charging equipment. The process can be initiated by pushing a button or using a scan tool. The purpose of jump assist is to transfer enough energy from the 12-volt

Special Tools

High-voltage battery charger
Warning tape

Caution

Do not activate the jump assist mode for an extended period. The 12-volt battery will be damaged because of excessive current transfer.

Jump assist is also referred to as "Charge assist."

Figure 5-31 HV battery charger connection.

battery to the HV battery to allow the engine to be started. Jump assist does not fully charge the HV battery; it works to start the engine so the system can recharge the HV battery. The HV system needs to recharge the HV battery after a jump assist is performed.

When jump assist is initiated, one of the hybrid system modules (depending on manufacturers and system) converts the 12 volts into high-voltage DC. The HV current is routed to the HV battery pack until a calibrated amount of charge (amp hrs) has been delivered and the HV battery is above a specified voltage (e.g., 300 volts). At this time, the charging process is complete, and the electric motors should start the engine.

Depending on the manufacturer, HV battery condition, and temperature, the jump assist function may take up to an hour to complete.

CUSTOMER CARE It's unusual for the HV battery to lose enough charge to not start the engine; anytime the HV battery requires recharging or jump assist, inspect the HV system for failures. The HV battery will drain to a low SOC if the vehicle is not driven for an extended amount of time. Most manufacturers require the vehicle to be used at least 10 minutes every month to keep the HV battery charged.

On some vehicles, jump assist is automatically initiated after a failed crank has occurred. This will happen when the driver requests an engine crank and the HV system does not have enough energy to start the engine. For automatic jump assist to occur, the HV battery SOC must be at least 40% and the HV battery voltage must be between 200 and 358 volts. A message on the HEV monitor screen notifies the driver of activation of the jump assist mode (**Figure 5-32**).

Some manufacturers provide scan tool initiation of the jump assist. Scan tool initiation of jump assist mode bypasses any voltage requirements that occur with a failed crank.

Figure 5-32 Driver notification that automatic jump assist has started after a failed crank.

This mode usually requires applying a 30-amp battery charge to the 12-volt battery during the process. To assure a successful jump assist, note the following tips:

- The charger must have an output of at least 30 amps.
- Do not charge the battery with the charger set on "High" or "Jump."
- Make sure to leave the ignition key in the OFF position to provide the HV battery with all available current.
- To prevent damage to the 12-volt battery, do not charge the HV battery through the 12-volt battery for an extended period.
- Be sure to disconnect the battery charger as soon as the jump assist completes.
- A low 12-volt system does not require a jump assist procedure. If the 12-volt battery is low, perform a traditional jump start.

Following a successful jump assist, the HEV monitor screen and the scan tool will confirm the completion of the procedure.

The jump assist feature can be aborted if any of the following occurs:

- Turning the key to the START position when the system is being charged.
- Incorrect installation of the 12-volt battery charger (reversed cables).
- The 12-volt battery charger is removed.
- The control module determines conditions are not suitable for charging.

A button is provided to initiate the charge process if a Ford Escape or Mariner Hybrid fails to start. First, make sure the ignition is in the OFF position. Next, open the access panel on the driver's side foot well. Located behind this access is the jump-start button (**Figure 5-33**). Press the button and then wait at least 8 minutes before continuing. If you attempt to continue before the 8 minutes have elapsed, the energy from the 12-volt battery will not be able to supply enough power to start the engine by the battery pack.

Pressing the jump-start button sends a request that the system send energy from the 12-volt battery to the HV battery pack. If the 12-volt battery has ample energy, it will be enough to start the engine. However, if the auxiliary battery is weak, it should be jump-started rather than the high-voltage battery pack.

After the 8-minute wait time, the HEV system warning lamp will start to blink. After the warning lamp stops blinking (this may take up to 2 minutes), attempt to start the engine. If the engine still does not start, wait a couple of minutes and repeat the procedure again. If the engine still does not start, the low-voltage battery must be recharged, or the vehicle jump-started through the 12-volt battery.

Jump-start
button

Figure 5-33 Jump-start button located in kick panel.

Recycling Batteries

The materials used to make lead-acid batteries are recyclable. This is the best practice for the removal of used batteries. Do not discard batteries with regular trash, as they contain metals and chemicals that are hazardous to the environment. The Rechargeable Battery Recycling Corporation (RBRC) promotes recycling of rechargeable batteries in North America. RBRC collects batteries from consumers and businesses and sends them to recycling companies. Ninety-eight percent of all lead-acid batteries are recycled. During the recycling process, the lead, plastic, and acid are separated. All of these compounds may be used in manufacturing new batteries.

High-voltage NiMH batteries contain several components, including nickel, copper, and steel, which have value as recycled materials. Properly sealed NiMH batteries that are not leaking are considered dry cell batteries and are not hazardous waste. However, always refer to federal, state, local, and provincial laws and regulations governing the recycling of these batteries. If NiMH batteries are found to be leaking, they are regulated as hazardous waste under federal and state regulations.

AUTHOR'S NOTE In California, NiMH batteries must be managed under California Universal Waste Rules.

CASE STUDY

A customer states that their vehicle does not start without having to jump the battery. The technician learns that this happens every time the customer attempts to start the vehicle. The customer also says the voltmeter in the dash has been reading higher than normal.

The technician verifies the concern. The engine turns over very slowly for a few seconds and then does not turn. After jumping the battery to get the vehicle into the shop, the technician makes a visual inspection of the battery and cables. The open circuit voltage test shows a voltage of 12.5 volts across the terminals. When the battery is subjected to the capacity test, the voltage drops to 7.8 volts at 80°F. After 10 minutes, the open circuit voltage is back up to 12.5 volts. The technician determines that the battery is sulfated. The battery can't handle the load of cranking the engine, which is why it always needs to be jumped. The higher-than-normal voltmeter readings indicate the charging system is trying to keep the battery charged. The technician calls the customer with a price quote. The customer agrees to have the battery replaced. While replacing the battery, the technician cleans the battery tray, the cable clamps, and sprays the clamps with a corrosion protector.

ASE-Style Review Questions

1. Battery terminal connections are being discussed.
   Technician A says when disconnecting battery cables, always disconnect the negative cable first.
   Technician B says when connecting battery cables, always connect the negative cable first.
   Who is correct?
   A. A only
   B. B only
   C. Both A and B
   D. Neither A nor B

2. A customer's battery is always dead when they attempt to start their car in the morning. After jumping the battery one time in the morning, the car will start throughout the day with no problems. All of the following can be the cause **EXCEPT:**
   A. The starter motor is drawing too much current.
   B. The glove box light is staying on.
   C. A computer is not powering down.
   D. A relay contact is stuck closed.

3. The specific gravity of a battery has been tested. All cells have a corrected reading of about 1.200.
   Technician A says the battery needs to be recharged before further testing.
   Technician B says the battery is sulfated and needs to be replaced.
   Who is correct?
   A. A only
   B. B only
   C. Both A and B
   D. Neither A nor B

4. When charging a battery:
   A. Connect a voltmeter across the battery terminals while the battery is charging and keep the charge rate at fewer than 10 volts.
   B. Disconnect the negative battery cable before charging to prevent damage to the alternator and computers.
   C. Charge at an amperage rate of 50% of the CCA rating.
   D. All of these choices.

5. Which statement concerning the battery leakage test is correct?
   A. The test is used to determine if the battery can provide current and voltage when loaded.
   B. A voltmeter reading of 0.05 when performing the test is acceptable.
   C. Both A and B.
   D. Neither A nor B.

6. The battery open circuit test is being discussed.
   Technician A says the battery must be stabilized before the open circuit voltage test is performed.
   Technician B says a test result of 12.4 volts is acceptable.
   Who is correct?
   A. A only
   B. B only
   C. Both A and B
   D. Neither A nor B

7. *Technician A* says a fully charged battery will have a conductance reading between 110% and 140% of its CCA rating.
   Technician B says as cell plate surfaces sulfate, the conductance of the cell will increase.
   Who is correct?
   A. A only
   B. B only
   C. Both A and B
   D. Neither A nor B

8. *Technician A* says the best way to charge the HV battery is by allowing the engine to run.
   Technician B says an on-board jump assist method may be available to start the engine if the HV battery SOC is too low.
   Who is correct?
   A. A only
   B. B only
   C. Both A and B
   D. Neither A nor B

9. While jump-starting a vehicle with a second vehicle, a puff of smoke is observed and the engine ground cable is burned.
   Technician A says this happened because the two vehicles were touching.
   Technician B says this was caused by connecting the negative jumper cable to the disabled vehicle's engine ground.
   Who is correct?
   A. A only
   B. B only
   C. Both A and B
   D. Neither A nor B

10. Ice crystals are found in the electrolyte. This can be caused by:
    A. A discharged battery.
    B. Use of tap water.
    C. Reversed battery connections.
    D. Improper hold-downs.

ASE Challenge Questions

1. Charging a battery with a battery charger is being discussed.

   Technician A says that battery-charging voltage should never exceed 15 volts.

   Technician B says that a battery can be considered fully charged when its open circuit voltage exceeds 12.1 volts after it has been stabilized.

   Who is correct?

   A. Technician A   C. Both A and B
   B. Technician B   D. Neither A nor B

2. *Technician A* says that a battery terminal test is performed by placing voltmeter leads between the positive battery post and the negative battery terminal.

   Technician B says that the total amount of voltage drop between the negative battery post and the negative battery terminal should not exceed 300 mV.

   Who is correct?

   A. Technician A   C. Both A and B
   B. Technician B   D. Neither A nor B

3. *Technician A* says that an ammeter is used when performing a battery leakage test.

   Technician B says that a fully charged battery will have a specific gravity of at least 1.265 at 80°F (27°C).

   Who is correct?

   A. Technician A   C. Both A and B
   B. Technician B   D. Neither A nor B

4. *Technician A* says that slow charging the battery starts the charge at about 3 amps and slowly ramps it up to 60 amps.

   Technician B says that fast charging is the only way to restore the battery to a fully charged state.

   Who is correct?

   A. Technician A   C. Both A and B
   B. Technician B   D. Neither A nor B

5. The battery drain test is being discussed.

   Technician A says that an ammeter that can accurately read less than 1 amp should be used.

   Technician B says that a drain caused by an internally shorted battery will not be indicated with an ammeter.

   Who is correct?

   A. Technician A   C. Both A and B
   B. Technician B   D. Neither A nor B

Name _____   Date _____

INSPECTING A BATTERY

Objective: Upon completion of this job sheet, you should be able to visually inspect a battery.

ASE Education Foundation Correlation —————————————————————

This job sheet addresses the following **MLR/AST/MAST** tasks:

A.5.   Demonstrate proper use of a digital multimeter (DMM) when measuring source voltage, voltage drop (including grounds), current flow, and resistance. **(P-1)**

B.4.   Inspect and clean battery; fill battery cells (if applicable); check battery cables, connectors, clamps, and hold-downs. **(P-1)**

B.7.   Identify electrical/electronic modules, security systems, radios, and other accessories that require reinitialization or code entry after reconnecting vehicle battery. **(P-2)**

We Support

ASE | **Education Foundation**

Tools and Materials

- A vehicle with a 12-volt battery
- A DMM
- Basic tool kit
- Baking soda
- Terminal cleaning brush
- Battery cable puller
- Wire brush

Safety Considerations and PPE

- Safety glasses
- Rubber gloves
- Rubber or acid proof apron

Procedure

1. Describe the general appearance of the battery.

   _____

   _____

   _____

2. Describe the general appearance of the cables and terminals.

   _____

   _____

   _____

3. Check the tightness of the cables at both ends. Describe their condition.

   _____

   _____

   _____

4. Set the DMM to read DC voltage. Connect the positive lead of the DMM to the positive terminal of the battery. Put the negative lead of the DMM on the battery case, and move it all around the top and sides of the case. What readings do you get on the voltmeter?

_____

_____

_____

_____

5. What is indicated by the readings?

_____

_____

_____

6. Measure the voltage of the battery. Your reading was: _____ volts.

7. Perform a battery terminal test for each cable and record your results:

Positive Cable _____

Negative Cable _____

8. What do you know about the condition of the battery based on the visual inspection and the tests that you did?

_____

_____

_____

_____

Instructor's Evaluation:

Grade	Rubric
0	No exposure due to: ☐ Absence ☐ Other:
1	Student unsuccessfully attempted the task
2	Student completed the task with significant instructor assistance.
3	Satisfactory performance of task. The student was able to perform elements of the task with minimal instructor assistance.
4	Competent/Proficient. Able to perform all elements of the task successfully and independently.

Instructor's Notes:

Name _____   Date _____

REMOVING, CLEANING, AND INSTALLING A BATTERY

Objective: Upon completion of this job sheet, you should be able to remove, clean, and reinstall a battery. Also, you should be able to identify any modules that require memory functions or have reinitialization procedures.

ASE Education Foundation Correlation

This job sheet addresses the following **MLR/AST/MAST** tasks:

B.2.   Confirm proper battery capacity, size, type, and application for vehicle; perform battery capacity and load test. **(P-1)**

B.3.   Maintain or restore electronic memory functions as recommended by manufacturer. **(P-2)**

B.4.   Inspect and clean battery; fill battery cells (if applicable); check battery cables, connectors, clamps, and hold-downs. **(P-1)**

B.7.   Identify electrical/electronic modules, security systems, radios, and other accessories that require reinitialization or code entry after reconnecting vehicle battery. **(P-2)**

We Support
ASE | Education Foundation

Tools and Materials

- A vehicle with a 12-volt battery
- A wash pan
- A box of clean baking soda
- Water
- Cleaning brush
- Hand tools
- Battery cable puller
- Battery terminal cleaner and brushes
- Cable end spreader
- Memory keeper
- Service information
- Distilled water

Safety Considerations and PPE

- Safety glasses
- Rubber gloves
- Rubber apron

Procedure

NOTE: Before disconnecting the battery, record the stations of all presets on the radio. Also, use the service information and note any computer modules that require memory restore procedures, reinitialization, or code entry after reconnecting the battery.

_____

_____

_____

_____

_____

_____

1. Identify the battery type: ☐ Flooded   ☐ Gel cell   ☐ AGM

2. What is the rating of the battery?

_____

3. Is this the correct battery for the vehicle?   ☐ Yes   ☐ No

   If *no*, consult with your instructor.

4. Connect the memory keeper.

5. Disconnect the negative cable of the battery. Use the cable puller if the cable is difficult to remove.

6. Carefully remove the battery hold-down strap. Describe the condition of the hold-down.

_____

_____

_____

7. Disconnect the positive cable of the battery. Use the cable puller if the cable is difficult to remove. Move anything that may interfere with the removal of the battery.

8. With a terminal cleaner, clean off the terminals of the battery.

9. With a battery-carrying strap, remove the battery from the vehicle and place it in a wash pan, or on the floor close to a drain.

10. Inspect the battery tray. Describe its condition and your recommendations.

_____

_____

_____

11. Make a mixture of baking soda and water. The mixture should be like a paste.

12. With the brush, scrub the top, sides, and bottom of the battery with the baking soda paste. Be careful not to allow the paste to get inside the battery.

13. After all of the battery has been scrubbed, wash the paste off with clean water.

14. Allow the battery to drip dry; then wipe the water off the battery with a clean rag.

15. Use the leftover paste to clean the battery tray and the hold-down assembly.

16. Rinse these off with water.

17. Use a clean, dry cloth to wipe the tray dry. Then place the battery onto the tray.

18. Check the electrolyte level in the cells. Are any low? _____

_____

19. Clean the battery cable terminals and spread them slightly with the spreader tool.

20. Install the battery hold-down assembly; make sure it's tight and that the battery is positioned properly.

21. Connect the positive cable to the battery.

22. Connect the negative cable to the battery.

23. Remove the memory keeper.

24. Reset all preset stations on the radio and other accessories with memory.

_____

_____

_____

Instructor's Evaluation:

Grade	Rubric
0	No exposure due to: ☐ Absence ☐ Other:
1	Student unsuccessfully attempted the task
2	Student completed the task with significant instructor assistance.
3	Satisfactory performance of task. The student was able to perform elements of the task with minimal instructor assistance.
4	Competent/Proficient. Able to perform all elements of the task successfully and independently.

Instructor's Notes:

Name _____   Date _____

PERFORM AN OPEN CIRCUIT VOLTAGE TEST (STATE-OF-CHARGE)

Objective: Upon completion of this job sheet, you should be able to determine the state of charge of a battery using the open circuit voltage test.

ASE Education Foundation Correlation

This job sheet addresses the following **MLR/AST/MAST** tasks:

A.1.   Research vehicle service information such as fluid type, vehicle service history, service precautions, technical service bulletins, and recalls including vehicles with advanced driver assistance systems (ADAS). **(P-1)**

A.5.   Demonstrate proper use of a digital multimeter (DMM) when measuring source voltage, voltage drop (including grounds), current flow, and resistance. **(P-1)**

B.1.   Perform battery state-of-charge test; determined needed action. **(P-1)**

We Support

ASE | Education Foundation

Tools and Materials

- A vehicle with a 12-volt battery
- Hydrometer or refractometer
- Service information
- DMM
- Fender covers

Safety Considerations and PPE

- Safety glasses

Procedure

1. Perform a battery state of charge test:

   a. Record the specific gravity readings for each cell:

   (1) _____ (2) _____ (3) _____ (4) _____ (5) _____ (6) _____

   b. If the battery is a sealed battery, what is the open circuit voltage?
   _____ volts

2. Summarize the battery's state of charge from the above.

   _____

   _____

   For this section, you will use the service information to identify service procedures to test the hybrid vehicle auxiliary battery.

1. What is the specified rating for the battery?

   _____

2. Record any warnings or cautions posted in the service information concerning working on or around the auxiliary battery:

_____

_____

_____

_____

3. Record any procedures that must be performed prior to testing the battery:

_____

_____

_____

_____

4. Describe the procedure for testing the battery:

_____

_____

_____

_____

5. Perform the state-of-charge test and record your results:

_____

_____

_____

_____

Instructor's Evaluation:

Grade	Rubric
0	No exposure due to: ☐ Absence ☐ Other:
1	Student unsuccessfully attempted the task
2	Student completed the task with significant instructor assistance.
3	Satisfactory performance of task. The student was able to perform elements of the task with minimal instructor assistance.
4	Competent/Proficient. Able to perform all elements of the task successfully and independently.

Instructor's Notes:

_____

Name _____   Date _____

USING A BATTERY CONDUCTANCE TESTER

Objective: Upon completion of this job sheet, you should be able to test a battery using a conductance tester and determine needed service action.

ASE Education Foundation Correlation ——————————————————————

This job sheet addresses the following **MLR/AST/MAST** tasks:

A.1.   Research vehicle service information such as fluid type, vehicle service history, service precautions, technical service bulletins, and recalls including vehicles with advanced driver assistance systems (ADAS). **(P-1)**

B.2.   Confirm proper battery capacity for vehicle application; perform battery capacity test and load test. **(P-1)**

We Support
ASE | **Education Foundation**

Tools and Materials

- A vehicle with a 12-volt battery
- Service information
- Conductance tester
- Fender covers

Safety Considerations and PPE

- Safety glasses

Procedure

1. The preferred test position is at the battery posts. Where is the battery located?

    _____

    _____

2. Inspect the battery and terminals per Job Sheets 23 and 24.

3. Check the electrolyte level. If low, add distilled water to fill to 1/2 inch (12 mm) above the top of the plates.

4. What is the CCA rating of the battery you are working on?

    _____

5. What is the battery type?

    Flooded _____ Gel Cell _____ AGM _____ Other _____

6. If you must test at a remote-post location, it should have both a positive and a negative post. Otherwise, you must remove the battery.

7. If needed, plug the tester into the correct electrical supply.

    Connect the test leads to the battery terminals observing polarity. What is the voltage displayed by the tester?

    _____

8. Follow the on-screen instructions to select "Battery Test" and select if the battery is in the vehicle or removed.

9. Which cables need to be connected to the battery or jump posts?

_____

10. Continue to follow the on-screen instruction; then press ENTER to start the test.

11. Record the battery test results.

_____

_____

Instructor's Evaluation:

Grade	Rubric
0	No exposure due to: ☐ Absence ☐ Other:
1	Student unsuccessfully attempted the task
2	Student completed the task with significant instructor assistance.
3	Satisfactory performance of task. The student was able to perform elements of the task with minimal instructor assistance.
4	Competent/Proficient. Able to perform all elements of the task successfully and independently.

Instructor's Notes:

_____

Name _____   Date _____

PERFORMING A BATTERY DRAIN TEST

Objective: Upon completion of this job sheet, you should be able to test parasitic loads against battery and determine needed service action.

ASE Education Foundation Correlation

This job sheet addresses the following **MLR** task:

A.10.   Measure key-off battery drain (parasitic draw). **(P-2)**

This job sheet addresses the following **MLR/AST/MAST** tasks:

A.1.   Research vehicle service information such as fluid type, vehicle service history, service precautions, technical service bulletins, and recalls including vehicles with advanced driver assistance systems (ADAS). **(P-1)**

A.5.   Demonstrate proper use of a digital multimeter (DMM) when measuring source voltage, voltage drop (including grounds), current flow, and resistance. **(P-1)**

This job sheet addresses the following **AST/MAST** tasks:

A.10.   Diagnose the cause(s) of excessive key-off battery drain (parasitic draw); determine needed action. **(P-1)**

We Support

ASE | Education Foundation

Tools and Materials

- A vehicle with a 12-volt battery
- Service information
- DMM capable of accurately reading less than 1 amp
- Test light
- Battery terminal pullers
- Fender covers

Safety Considerations and PPE

- Safety glasses

Procedure

1. According to the service information, what is the maximum amount of key-off current draw?

   _____

   _____

2. According to the service information, how long after key off does it take for the systems to shut down before accurate test results can be obtained?

   _____

   _____

3. Make sure the ignition is in the OFF position, all accessories are turned off, and all doors are closed. Are there any obvious causes of battery drain such as lights that did not turn off?   ☐ Yes   ☐ No

4. If an amp clamp is not available, use the test light and connect it between the battery terminal and clamp. Being careful not to lose the connection of the test light, separate the cable clamp from the battery terminal. Is the test light illuminated?

   ☐ Yes   ☐ No

5. With the test light still connected between the battery terminal and the clamp, connect the DMM set to read amps in series between the clamp and battery terminal. Once the ammeter is connected, disconnect the test light. What is the indicated key-off draw?

   _____

   _____

6. Does the ammeter reading indicate a possible problem?

   ☐ Yes   ☐ No

7. If the answer to question 6 is *Yes*, reconnect the test light.

8. Making sure the test light connections stay intact, reconnect the cable to the battery terminal.

9. Using the DMM set to the DC mV scale, measure the voltage drop across each fuse. Did a fuse indicate that current was flowing through it?   ☐ Yes   ☐ No

10. If yes, use the wiring diagram for the vehicle and determine the affected circuit.

   _____

   _____

   Trace the circuit to locate the cause for the excessive draw. Record your results:

   _____

   _____

Instructor's Evaluation:

Grade	Rubric
0	No exposure due to: ☐ Absence ☐ Other:
1	Student unsuccessfully attempted the task
2	Student completed the task with significant instructor assistance.
3	Satisfactory performance of task. The student was able to perform elements of the task with minimal instructor assistance.
4	Competent/Proficient. Able to perform all elements of the task successfully and independently.

Instructor's Notes:

_____

Name _____   Date _____

HV BATTERY SERVICE AND DIAGNOSIS

Objective: Upon completion of this job sheet, you should be able to diagnose the condition of the HV battery.

ASE Education Foundation Correlation

This job sheet addresses the following **MLR/AST/MAST** task:

A.3.   Retrieve and record DTCs, OBD monitor status, and freeze frame data; clear codes and data when directed. **(P-1)**

We Support
ASE | Education Foundation

Tools and Materials

- Scan tool
- Isolation gloves
- Service information

Safety Considerations and PPE

- Safety glasses

⚠ WARNING You will be working around high voltage while doing this job sheet. Follow all warnings and cautions in the manufacturer's service procedures. Even though the service disconnect is removed, the HV battery is still "live."

Procedure

1. Use the scan to check for DTCs related to the HV battery and record your results:

   _____

   _____

2. Navigate to the battery control module and record the following information, if available (Note: you may need to modify the chart for the vehicle you are working on.):

PID	READING	UNIT
Hybrid Battery State of Charge		%
Hybrid Battery Current Sensor		Amps
Module Voltage (1)		Volts
Module Voltage (2)		Volts
Module Voltage (3)		Volts
Module Voltage (4)		Volts
Module Voltage (5)		Volts
Module Voltage (6)		Volts
Module Voltage (7)		Volts
Module Voltage (8)		Volts
Module Voltage (9)		Volts
Module Voltage (10)		Volts
Module Voltage (11)		Volts
Module Voltage (12)		Volts
Module Voltage (13)		Volts

PID	READING	UNIT
Module Voltage (14)		Volts
Module Voltage (15)		Volts
Module Voltage (16)		Volts
Module Voltage (17)		Volts
Module Voltage (18)		Volts
Module Voltage (19)		Volts
Module Voltage (20)		Volts
HV Battery Impedance		Ohms
BPCM Package Voltage		Volts
BPCM HV Bus Voltage		Volts

3. Are all module voltages displaying approximately the same voltage values?

☐ Yes   ☐ No

4. Remove the HV battery service disconnect plug and note if the voltage in any of the modules or blocks changes.

_____

If there was a change, why did it occur?

_____

_____

_____

Instructor's Evaluation:

Grade	Rubric
0	No exposure due to: ☐ Absence ☐ Other:
1	Student unsuccessfully attempted the task
2	Student completed the task with significant instructor assistance.
3	Satisfactory performance of task. The student was able to perform elements of the task with minimal instructor assistance.
4	Competent/Proficient. Able to perform all elements of the task successfully and independently.

Instructor's Notes:

Name _____   Date _____

BATTERY CHARGING

Objective: Upon completion of this job sheet, you should be able to properly recharge a 12-volt battery.

ASE Education Foundation Correlation

This job sheet addresses the following **MLR/AST/MAST** tasks:

A.1.   Research vehicle service information such as fluid type, vehicle service history, service precautions, technical service bulletins, and recalls including vehicles with advanced driver assistance systems (ADAS). **(P-1)**

A.5.   Demonstrate proper use of a digital multimeter (DMM) when measuring source voltage, voltage drop (including grounds), current flow, and resistance. **(P-1)**

B.5.   Perform battery charging according to manufacturer's recommendations. **(P-1)**

We Support
ASE | Education Foundation

Tools and Materials

- Battery charging station
- DMM
- Memory keeper
- Service information

Safety Considerations and PPE

- Safety glasses

⚠ **WARNING** **Be sure to have the proper charging equipment for the battery type. Also, the battery may release gasses that are flammable and harmful to breath. Perform the battery charging in an area that's well ventilated.**

Procedure

1. Use the DMM to measure the current voltage of the battery:

   _____

2. Connect the memory keeper.

3. Disconnect the negative battery cable at the battery.

4. Use the service information to determine the charge rate and time based on your findings in step 1:

   _____

5. Follow the User's Guide for the charging station and recharge the battery. At the completion of the charging procedure, reconnect the negative cable. Reset or reinitialize any modules that require this procedure.

Instructor's Evaluation:

Grade	Rubric
0	No exposure due to: ☐ Absence ☐ Other:
1	Student unsuccessfully attempted the task
2	Student completed the task with significant instructor assistance.
3	Satisfactory performance of task. The student was able to perform elements of the task with minimal instructor assistance.
4	Competent/Proficient. Able to perform all elements of the task successfully and independently.

Instructor's Notes:

Name _____   Date _____

JUMP STARTING A VEHICLE

Objective: Upon completion of this job sheet, you should be able to properly jump start a vehicle using a booster battery or an auxiliary power supply.

ASE Education Foundation Correlation

This job sheet addresses the following **MLR/AST/MAST** task:

B.6.   Jump-start vehicle using jumper cables and a booster battery or an auxiliary power supply. **(P-1)**

We Support
ASE | Education Foundation

Tools and Materials
• Battery jumper
• Jumper cables
• Booster battery
• DMM

Safety Considerations and PPE
• Safety glasses

⚠ **WARNING Be sure to follow all battery safety cautions and warnings.**

Procedure

1. What method of jump-starting the vehicle were you assigned?

   Booster battery and jumper cables _____ (Go to step 2)

   Battery jumper _____ (Go to step 13)

Procedure for Using Booster Battery and Jumper Cables

2. Check the state of charge of the booster battery.

   _____

3. Engage the parking brake and put the transmission in neutral or park.

4. Turn off the ignition switch and all accessories.

5. Attach one end of the positive jumper cable to the vehicle's battery's positive terminal.

6. Connect the other end of the positive jumper cable to the booster battery's positive terminal.

7. Attach one end of the negative jumper cable to the booster battery's negative terminal.

8. Attach the other end of the negative jumper cable to an engine ground on the disabled vehicle.

⚠ **WARNING Do not connect this cable end to the battery negative terminal. Doing so may create a spark that will cause the battery to explode.**

9. Attempt to start the disabled vehicle. Did the vehicle start? Yes _____ No _____

   If No, contact your instructor.

10. Once the disabled vehicle starts, disconnect the ground-connected negative jumper cable from its engine block.

11. Disconnect the negative jumper cable from the booster battery.

12. Disconnect the positive jumper cable from the booster battery and then from the other battery.

Procedure for Using Booster Battery and Jumper Cables

13. Check the state of charge of the battery jumper. _____

14. Connect the positive cable from the battery jumper to the battery's positive terminal.

15. Connect the negative cable from the battery jumper to an engine ground on the disabled vehicle.

⚠ **WARNING Do not connect this cable end to the battery negative terminal. Doing so may create a spark that will cause the battery to explode.**

16. Attempt to start the disabled vehicle. Did the vehicle start? Yes _____ No _____

   If No, contact your instructor.

17. Once the disabled vehicle starts, disconnect the ground-connected negative jumper cable from its engine block.

18. Disconnect the positive jumper cable from the battery jumper.

Instructor's Evaluation:

Grade	Rubric
0	No exposure due to: ☐ Absence ☐ Other:
1	Student unsuccessfully attempted the task
2	Student completed the task with significant instructor assistance.
3	Satisfactory performance of task. The student was able to perform elements of the task with minimal instructor assistance.
4	Competent/Proficient. Able to perform all elements of the task successfully and independently.

Instructor's Notes:

DIAGNOSTIC CHART 5-1

PROBLEM AREA:	Battery
SYMPTOMS:	No-start condition after vehicle is parked overnight.
POSSIBLE CAUSES:	1. Glove box, trunk, interior illumination lights not turning off due to defective switches. 2. Circuit short to ground. 3. Circuit copper-to-copper shorts. 4. Contaminated battery case.

DIAGNOSTIC CHART 5-2

PROBLEM AREA:	Battery
SYMPTOMS:	Battery requires jump-starting after engine shutdown.
POSSIBLE CAUSES:	1. Low specific gravity. 2. Defective battery cell(s).

DIAGNOSTIC CHART 5-3

PROBLEM AREA:	Battery
SYMPTOMS:	Inadequate current to start engine under heavy load conditions and battery will not accept a charge.
POSSIBLE CAUSES:	1. Contaminated battery case resulting in constant current draw. 2. Contaminated battery terminals. 3. Undercharged battery. 4. Defective battery. 5. Sulfated battery. 6. Damaged battery.

DIAGNOSTIC CHART 5-4

PROBLEM AREA:	Battery
SYMPTOMS:	Starter cranks engine slowly or fails to turn engine; low state of charge.
POSSIBLE CAUSES:	1. Discharged battery. 2. Contaminated terminal clamps. 3. Defective battery cables.

DIAGNOSTIC CHART 5-5

PROBLEM AREA:	HV battery
SYMPTOMS:	Low SOC.
	Warning lamp illumination.
POSSIBLE CAUSES:	1. Poor electrical connections. 2. Faulty motor generator or circuits. 3. Faulty HV ECU or circuits. 4. Faulty battery ECU or circuits. 5. Faulty amperage sensor or circuits. 6. Faulty HV battery.

CHAPTER 6

STARTING SYSTEM DIAGNOSIS AND SERVICE

Upon completion and review of this chapter, you should be able to:

- Perform a systematic diagnosis of the starting system.
- Determine the cause of slow- and no-crank conditions.
- Perform a quick check test series to determine the problem areas in the starting system.
- Perform and accurately interpret the results of a current-draw test.
- Perform and accurately interpret the results of an insulated circuit resistance test.

- Perform and accurately interpret the results of a ground circuit test.
- Diagnose the starter solenoid by performing the solenoid test series.
- Perform the no-crank test and recommend needed repairs as indicated.
- Remove and reinstall a starter motor.
- Disassemble, clean, inspect, repair, and reassemble a starter motor.
- Identify Stop/Start system inhibits.

Terms to Know

Crocus cloth

Current-draw test

Hydrostatic lock

No-crank

Quick test

Relative compression test

Slow-crank

Solenoid circuit resistance test

Introduction

Perhaps one of the most aggravating experiences a car owner may have is their engine failing to start. However, not all starting problems are the fault of the starting system. The ignition and fuel systems must also be in proper condition to perform their functions. In addition, the engine's internal condition must be such that compression, correct valve timing, and free rotation are all obtained.

AUTHOR'S NOTE Remember that the starting system operates in conjunction with the battery and charging systems. Keep the interdependency of these three systems in mind and that all three systems must perform properly. The starting, battery, and charging systems need to be diagnosed as a complete system.

In this chapter, you will perform the tests required to make an intelligent decision concerning the condition and operation of the components of the starting system. You will also remove, disassemble, reassemble, and reinstall the starter motor.

Starting System Service Cautions

Along with the precautions outlined in Chapter 1, follow these additional precautions when servicing the starter system:

1. Refer to the manufacturer's manuals for correct procedures for disconnecting a battery. Some vehicles with on-board computers require an auxiliary power source.
2. Disconnect the battery ground cable before disconnecting any of the starter circuit's wires or removing the starter motor.
3. Be sure the vehicle is positioned correctly on the hoist or on safety jack stands.
4. Before performing any cranking test, ensure the vehicle is in PARK or NEUTRAL, and the parking brakes are applied.
5. Follow the manufacturer's directions for disabling the ignition system.
6. Be sure the test leads are clear of any moving engine components.
7. Never clean any electrical components in solvent or gasoline. Clean with compressed air, denatured alcohol, or wipe with clean rags.

Starting System Troubleshooting

Customer concerns concerning the starting system generally fall into four categories: **no-crank**, **slow-cranking**, starter spins but does not turn the engine, and excessive noise. As with any electrical system concern a systematic approach to diagnosing the starting system will make the task easier. First, the battery must be in good condition and fully charged. Perform a complete battery test series to confirm the battery's condition. Many starting system failures are attributable to battery problems. Performing starting system tests with a weak battery may produce misleading results. The conclusions may be erroneous and costly.

Before performing any tests on the starting system, begin with a visual inspection of the circuit. The battery terminals must be clean, and the starter motor must be properly grounded. Repair or replace any corroded or loose connections, frayed wires, or other trouble sources.

Modern computer-controlled starting system diagnostics are DTC driven. The service information system will provide diagnostic charts that guide you through a logical sequence to follow whenever diagnosing a starting system concern (**Figure 6-1**). The DTCs and/or symptoms determine which diagnostic chart to follow.

If the customer complains of a no-crank situation, attempt to rotate the engine by hand. You may be able to use a socket and ratchet on the crankshaft pulley bolt to rotate the engine. Confirm that you have disabled the engine from starting before attempting to rotate the engine. Rotate the crankshaft two complete rotations in a clockwise direction. If the engine does not rotate, it may be seized because it was operated with no oil, broken engine components, or **hydrostatic lock**. Since liquid is noncompressible, a leak that allows coolant from the cooling system can fill a cylinder to such a level that the piston cannot move upward. This condition is referred to as hydrostatic lock.

Several potential trouble spots in the circuit can cause a slow-crank or no-crank concern (**Figure 6-2**). Excessive voltage drops in these areas will cause the starter motor to operate slower than required to start the engine. The speed the starter motor rotates the engine is essential to engine starting. If the speed is too slow, compression is lost, and the air–fuel

No-crank means that the starter does not turn the engine with the ignition switch in the START position. A buzzing noise that accompanies the no-crank indicates the starter motor drive has engaged the ring gear, but the engine does not rotate. Also, the no-crank condition may not produce any clicking sounds from the starter motor or solenoid.

Slow-crank means the starter drive engages the ring gear, but because the engine turns too slow, it cannot start. Some manufacturers provide specifications for engine cranking speed.

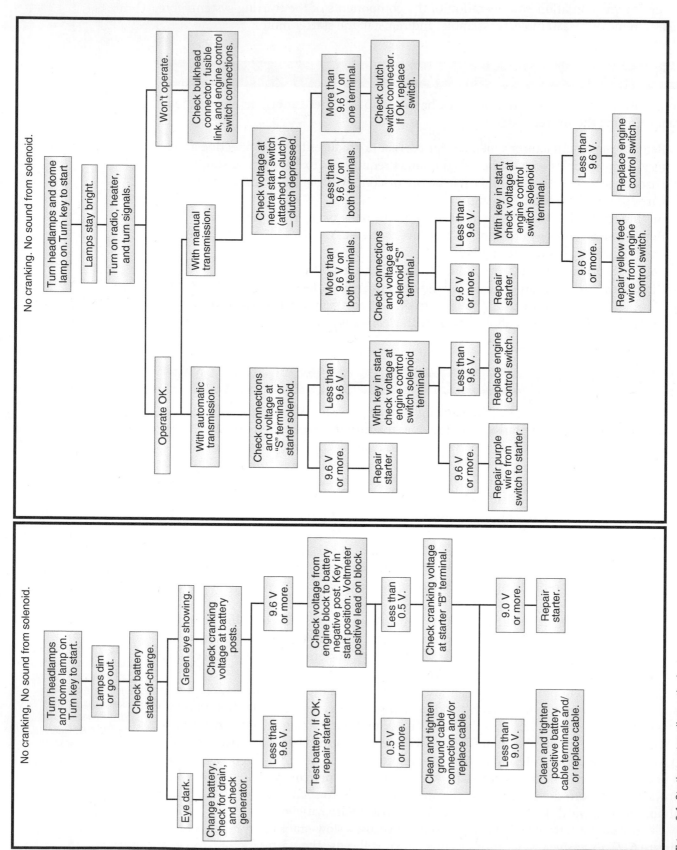

Figure 6-1 Starting system diagnostic chart.

Figure 6-1 (Continued)

Figure 6-1 (Continued)

Figure 6-2 Excessive wear, loose electrical connections, or excessive voltage drop in any of these areas can cause a slow- or no-crank condition.

mixture draw is impeded. Most manufacturers require a speed of approximately 250 rpm during engine cranking.

If the starter spins but the engine does not rotate, the most likely cause is a faulty starter drive. However, before faulting the starter drive, check the starter ring gear teeth for wear or breakage and for incorrect ring gear to pinion gear mesh. If the starter drive is at fault, remove and disassemble the starter motor and install a new drive mechanism.

A worn starter drive mechanism is the cause of most starter noises. Often, the starter drive can be replaced as a separate component of the starter.

CUSTOMER CARE Always treat the customer's car with respect. Place fender covers over the fenders when performing tasks under the hood. Do not lay tools on the vehicle's finish. Clean your hands before entering the vehicle. Place a seat protector over the seats and paper mats on the floorboards. Give the car back to the customer at least as clean as when you received it.

Testing the Starting System

As with the battery testing series, testing the starting system can be performed with a starting/charging system tester (**Figure 6-3**). Other diagnostic tools include the DMM and the lab scope.

The close relationship between starter and battery performance makes it essential to perform a complete battery test series before testing the starter system. If a fully charged battery fails the load test, replace it before continuing to the starting system tests.

Classroom Manual
Chapter 6, page 140

Quick Testing

The **quick test** will isolate the problem area and determine whether the starter motor, solenoid or control circuit is at fault. If the starter motor does not turn the engine at all

Figure 6-3 Most testers capable of measuring high current, such as the GR8, can be performed starter system tests.

and it's in good mechanical condition, performing the quick test can identify the problem area. To perform this test, make sure the transmission is in neutral and set the parking brake. Turn on the headlights. Next, turn the ignition switch to the START position while observing the headlights. Three things can happen to the headlights during this test:

1. They will go out.
2. They will dim.
3. They will remain at the same brightness level.

If the lights go out completely, the most likely cause is a poor battery cable connection. Test the connection using the battery terminal test. It will be necessary to remove the cable from the terminal and clean the cable clamp and battery terminals of all corrosion. Verify the connection quality again by performing the battery terminal test.

If the headlights dim when cranking the engine, the battery may be discharged. Check the battery condition. If it's good, then there may be a mechanical condition in the engine preventing it from rotating. If the engine rotates when turning it with a socket wrench on the pulley nut, the starter motor may have internal damage. A bent starter armature, worn bearings, thrown armature windings, loose pole shoe screws, or any other worn component in the starter motor that will allow the armature to drag can cause a high-current demand.

If the lights stay brightly lit and the starter makes no sound (listen for a deep clicking noise), there is an open in the circuit. The fault is in either the solenoid or the control circuit. To test the solenoid, bypass the solenoid by bridging the BAT and S terminals (**Figure 6-4**).

Classroom Manual
Chapter 6, page 141

> ⚙ SERVICE TIP Check the fusible link if the engine does not crank and the headlights do not come on.

Figure 6-4 Bypassing the solenoid to determine if the solenoid or the control circuit is faulty.

CUSTOMER CARE Thrown starter windings indicates several different problems. The most common is that the driver keeps the ignition switch in the START position too long after the engine has started. Other causes include the driver opening the throttle plates too wide while starting the engine, which results in excessive armature speeds when the engine does start. Also, excessive heat buildup in the motor can throw the windings. The starter motor is designed to operate for a very short time. Operating the motor for longer than 15 seconds generates excessive heat very quickly. If the engine does not start after a 15-second crank, the starter motor needs to cool for at least 30 seconds (up to 2 minutes in some cases) before the next attempt to start the engine.

⚡ **WARNING A starter can draw up to 400 amps (A). The tool used to jump the terminals must be able to carry this high current and must have an insulated handle.**

If the starter rotates with the solenoid bypassed, the control circuit is at fault. The solenoid is at fault if the starter does not rotate and the lights do not dim. Also, listen for the starter drive to engage. If the starter rotates slowly and the headlights dim, there is excessive current draw, and the system will need additional testing.

DMM Testing

Another quick test of the starting system (and battery) is to use the Min/Max function of the DMM. First, measure the static voltage across the battery terminals. Activate the Min/Max feature on the DMM and start the engine. The DMM will record the minimum and maximum voltage values. The minimum voltage represents cranking battery voltage. The maximum voltage represents the output voltage from the charging system after the engine starts. If the cranking voltage is below specifications, the battery needs to be tested before continuing diagnostics of the starting system. If the battery is good, perform the current-draw test.

Current-Draw Test

If the starter motor cranks the engine, you should perform the **current-draw test**. The current-draw test measures the amount of current that the starter draws when actuated.

Special Tools

Fender covers

Starting/charging system tester

DMM with high current inductive lead

Second DMM to read battery voltage

Lab scope with high current inductive lead

Fused jumper wires

⚠️ **Caution**

Do not operate the starter motor for longer than 15 seconds. Allow the motor to cool between cranking attempts.

The specification for current draw is the maximum allowable; the specification for cranking voltage is the minimum allowable.

Special Tools

Fender covers

GR8 system tester

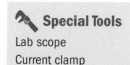
Special Tools

Lab scope

Current clamp

It determines the electrical and mechanical condition of the starting system. The following procedure is for a starting/charging system tester. A DMM or lab scope (both with the proper adapters) can also be used.

1. Connect the voltmeter test leads across the battery, observing polarity.
2. Connect the inductive amp probe around the battery ground cable. If there is more than one ground cable, clamp the probe around all of them.
3. Ensure all loads are turned off (lights, radio, etc.).
4. Follow the manufacturer's instructions to zero the ammeter.
5. If necessary, select STARTING TEST.
6. Follow the service information procedure to disable the ignition system to prevent the vehicle from starting. Some systems may require the removal of the fuel pump or electronic fuel injection relay or disconnecting the ignition module to prevent the engine from starting.
7. Crank the engine for 10–15 seconds and note the voltmeter and ammeter readings.

After recording the readings from the current-draw test, compare them with the manufacturer's specifications. If specifications are not available, correctly functioning systems, as a rule, will crank at 9.6 volts or higher. Current draw is dependent on engine size. Most V-8 engines will have a current draw of about 200 amps, six-cylinder engines about 150 amps, and four-cylinder engines about 125 amps.

If the readings obtained from the current-draw test are out of specifications, additional testing will isolate the problem. If the readings were on the borderline of the specifications, or there is an intermittent problem, then detailed testing for bad components will pinpoint potential failures.

The following provides an interpretation of the current-draw test:

- Voltage is 9.6 volts or more, and amperage is higher than specified—indicates impedance to starter motor rotation. This includes worn bushings, a mechanical blockage, and excessively advanced ignition timing. Shorted starter motor windings can cause high-current draw.
- Voltage is 9.6 volts or more, and amperage is lower than specified—indicates high electrical circuit resistance.
- Voltage is less than 9.6 volts, and current is higher than specified—indicates the amperage is draining the battery. Perform the battery test series to confirm the battery is good. If the battery passes this test series, see the first point above.
- Voltage is less than 9.6 volts, and current is lower than specified—indicates a faulty battery.

Obtaining the voltage reading at the battery during the current-draw test may not be an exact representation of the actual voltage delivered to the starter. Voltage losses due to corroded cables, poor connections, and faulty relays (or solenoids) may diminish the amount of voltage to the starter. Test these components before removing the starter from the vehicle.

Photo Sequence 10 illustrates the use of the GR8 tester as an example of using computer testing equipment. Different models of GR8 are available with different options. Always refer to the instruction manual for the type of tester you are using.

Lab Scope Testing

A lab scope or graphing multimeter is a good method of testing a starting system. Before disabling the ignition system to prevent the engine from starting, use the ammeter probe to view the starter signal. Most current probes convert amps to millivolts. Attach the amp probe over either the positive or negative battery cable. Be sure to orient the arrow on the current probe to the direction of current flow. Set the scope settings of 0.5 volt

PHOTO SEQUENCE 10
Using the Gr8 Starting System Test

P10-1 In the Main Menu, select the System Test icon.

P10-2 Follow any on-screen instructions for connecting the test leads to the vehicle.

P10-3 When prompted, start the vehicle's engine.

P10-4 A series of screens will display the complete results of the test. Use the UP or DOWN ARROW keys to scroll to each screen.

P10-5 Screen 1 of 2 shows the average cranking voltage, average cranking current (if using an amp clamp), and cranking time in seconds.

P10-6 Screen 2 of 2 shows the system performance. The Y-axis represents cranking voltage, and the X-axis represents time.

P10-7 Test results can be printed by using the PRINT soft key.

(V)/division and 100 ms/division to start with and adjust later as needed. Using the meter's record function will make interpreting the pattern results easier.

Initial engine rotation requires a high starter current. Once the engine is rotating, lower current is required. A typical pattern will indicate this higher current draw (rush-in current) at initial cranking (**Figure 6-5**). During this time, both the pull-in and hold-in windings of the starter solenoid are energized. Shortly after the high-current spike, the pull-in windings are de-energized, and current will drop to a lower value. After the engine starts, the current draw drops to zero as the ignition switch transitions to the RUN position. Excessively high current indicates a shorted winding or high mechanical resistance within the starter motor or engine. Low current indicates high electrical resistance in the circuit.

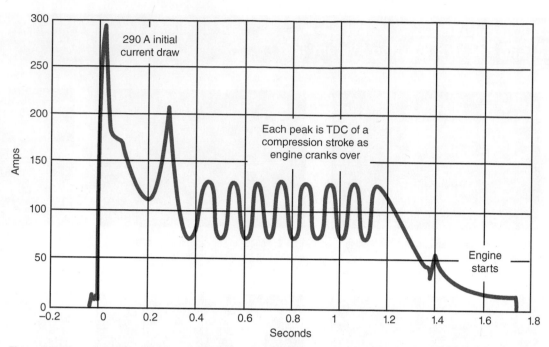

Figure 6-5 When the starter is energized, the current will spike and then drop as starter speed increases.

Relative Compression Testing

Although usually a test procedure for determining internal engine conditions, the **relative compression test** also identifies hard starting conditions. Good compression and a properly functioning starting system are critical for fast starts and good driveability. Due to the increased difficulty accessing spark plugs, the relative compression test has replaced the normal compression test procedure of removing the spark plugs and installing a compression gauge. Relative compression testing is a nonintrusive diagnostic routine that uses current draw during cranking to determine if a cylinder has lower compression relative to the other cylinders. This test also provides a good indication of the starting system.

The lab scope visualizes the effect that each cylinder's compression stroke has on starter current while the engine is cranking. Forcing the piston up during the compression stroke requires work. Since current represents work, the lab scope allows you to observe current on a small timeline scale.

To perform this test, clamp the lab scope's amp probe around the cable between the battery and the starter motor. With the ignition system disabled, crank the engine while monitoring the lab scope pattern. Initial current should go high and peak, followed by a ripple pattern (**Figure 6-6**). As the piston approaches top dead center of the compression stroke, the resistance increases and causes the current to increase. This should result in a trace with even peaks across the screen. If the compression of a cylinder is lower than that of other cylinders, then the current peak of that cylinder will also be lower (**Figure 6-7**).

Placing an external trigger placed around a spark plug makes it possible to identify which cylinder has the low compression. A adjust the scope to trigger off this signal. For example, if the lab scope is set to trigger off of the rpm pickup placed on the number one plug wire, then every time the screen is updated, the number one cylinder is the first one on the screen. Each cylinder number can be determined by following the engine's firing order.

Most engine analyzers have a variation of this test that will test the relative current comparison of each cylinder during cranking. These analyzers provide a readout of the results and may suggest the faulty cylinder.

Figure 6-6 Lab scope pattern showing good relative compression between cylinders.

Figure 6-7 Lab scope pattern indicating a cylinder with lower relative compression.

Insulated Circuit Resistance Test

Most manufacturers design their starting systems to have very little resistance to the flow of current to the starter motor. Most have less than 0.2 volt dropped on each side of the circuit. This means the voltage across the starter input terminal to the starter ground should be within 0.4 volt of battery voltage (**Figure 6-8**).

Voltage drops are measured by connecting a digital multimeter (DMM) set to the lowest voltmeter scale in parallel with the circuit section being tested. A load on the circuit must be applied to obtain a voltage drop reading. The following is a typical test procedure:

1. Connect the test leads as shown in **Figure 6-9**, depending on the circuit type. Connect the positive lead of the voltmeter to the positive battery post, and connect the negative lead to the starter battery (BATT) terminal.

Classroom Manual
Chapter 6, page 150

 Special Tools
DMM fender covers
Starting/charging system tester

SERVICE TIP As a general rule, allow up to 0.2 volt per cable and 0.1 volt per connection to be dropped in the high current circuit. Switches in the low current circuit can have a voltage drop as high as 0.3 volt. Most low current circuit connections should not exceed 0.1 volt. Use the wiring diagram for the vehicle to determine the number of conductors and connections used in the circuit. This will provide a specification for you if no other specifications are available.

Figure 6-8 Voltage drop testing to identify sources of excessive resistance.

2. Disable the ignition system as discussed in current-draw testing.
3. Crank the engine for 15 seconds and observe the voltmeter scale.

This tests for voltage drop in the entire circuit. If voltage drop is excessive, locate the cause of the excessive voltage drop by moving the negative voltmeter lead from the starter to the next connection toward the battery. With each move of the test lead, crank the engine while observing the voltmeter reading. Continue to test each connection toward the battery until a noticeable decrease in voltage drop is detected. The cause of the excessive voltage drop is between the current test location and the preceding point.

Ground Circuit Test

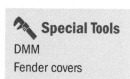

Special Tools
DMM
Fender covers

Classroom Manual
Chapter 6, pages
150, 153

A ground circuit test measures the voltage drop in the ground side of the circuit (**Figure 6-10**). An open starter motor connection to ground could cause the starter motor not to crank the engine. It can also cause an intermediate starter system problem. To perform the ground circuit test, connect the positive voltmeter lead to the starter motor case and the negative test lead to the battery's ground terminal. Remove any paint from the starter motor's case where you are connecting the test lead. Crank the engine while observing the voltmeter.

Less than 0.2 volt indicates the ground circuit is good. More than 0.2 volt indicates a poor ground circuit connection. A poor connection could result from loose starter mounting bolts, paint on the starter motor case, or a bad battery ground terminal post connection. Also, check the ground cable for high resistance or for being undersized.

Solenoid Circuit Resistance Testing

Special Tools
DMM fender covers
Starting/charging
system tester

High resistance in the solenoid will reduce the current flow through the solenoid windings and cause the solenoid to malfunction. If the solenoid has high resistance in the windings, it may result in the contacts burning and causing excessive resistance to the starter motor.

The **solenoid circuit resistance test** determines the electrical condition of the solenoid and the control circuit. To perform the solenoid circuit resistance test, first, disable the ignition system. Using a voltmeter or starting/charging system tester, connect the positive voltmeter lead to the BAT terminal of the solenoid. Connect the negative test lead

Figure 6-9 (A) Test lead connections for a starter-mounted solenoid. (B) Test lead connections for relay-controlled systems.

to the field coil terminal (M terminal). Crank the engine while observing the voltmeter reading. If the voltmeter reading indicates a voltage drop of greater than 0.2 volt, then the solenoid is defective. If the starter solenoid passes this test, follow these steps to test the starter solenoid switch circuit:

1. Disable the ignition system.
2. Connect the voltmeter leads to both solenoid switch terminals, observing polarity.
3. Crank the engine and observe the voltmeter reading.

Figure 6-10 Voltage drop testing of the ground-side circuit of the starting system.

Classroom Manual
Chapter 6,
pages 155, 157

Use the wiring diagram to follow the circuit and determine the number and location of any connection. If the indicated voltage drop is excessive, move the voltmeter leads up the circuit and test each component. The total voltage drop should be less than 0.1 volt per connection (with no more than 0.5 volt total).

Continue to move the voltmeter leads to test for voltage drop through the wires, starter relay, neutral safety switch, and ignition switch.

Open Circuit Test

Since voltage drop testing requires the circuit to be loaded, a no-crank complaint may prevent performing a voltage drop test. This is because most no-crank problems result from opens in the circuit. The easiest way to diagnose this problem is by using a test light. On a system that uses a starter motor-mounted solenoid, the M terminal is the end of the circuit (**Figure 6-11**). Connecting the positive lead of the test light to the M terminal and the negative lead to a good ground, the light should be on when placing the ignition switch in the START position (**Figure 6-12**). If the light comes on, there is no open in the insulated circuit (including the ignition switch, wires, neutral safety switch, solenoid, and all connections). The open is in either the starter motor or the ground circuit.

Special Tools

Test light
Fused jumper wires
Fender covers

Figure 6-11 Starting system using a solenoid shift.

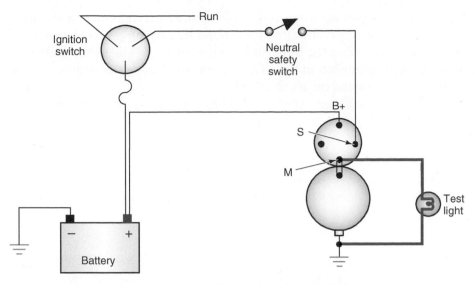

Figure 6-12 Test light connections for testing the solenoid and control circuit.

If the test light comes on very dim, there is very high resistance in the circuit. By working the test light through the circuit and back to the battery, the location of the high resistance should be found. If the test light does not come on, follow the same procedure of backtracking the circuit toward the battery until the open is found. Also, check for voltage at the B+ terminal (**Figure 6-13**). Connecting the test light as shown should light the bulb with the key off. The light should stay on when the starter motor is activated. If the light goes out in the START position, repair the cable or end connections.

You can use fused jumper wires to verify the location of the open by jumping across the defective component or connection. If jumping across the solenoid, for example, and the starter spins, there is an open in the solenoid. Use the same procedure to jump across the ignition switch, neutral safety switch, or open wires.

> ⚙ **SERVICE TIP** You can use a voltmeter instead of a test light. The voltage at the M terminal should be more than 9.6 volts.

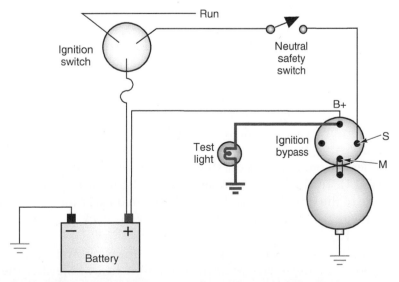

Figure 6-13 Test light connections for checking voltage at the BAT terminal.

If the test light comes on when connected to the M terminal, do a simple ground circuit test. This is done by connecting the ground lead of the test light onto the starter body and the positive lead to the M terminal of the solenoid. The light should illuminate bright with the ignition switch in the START position. If the ground circuit is good, the starter is suspect and should be bench tested.

Testing the Ford starting system is the same, except there is an additional battery cable to test.

> ⚙️ **SERVICE TIP** Most starter safety switches are adjustable. Sometimes adjusting the starter safety switch can fix a no-start problem.

Starter Motor Removal/Installation

If the tests indicate the necessity of removing the starter motor, the first step is disconnecting the battery from the system. Remove the negative battery cable. It's a good practice to wrap the cable clamp with tape or enclose it in a rubber hose to prevent accidental contact with the battery terminal.

It may be necessary to place the vehicle on a lift to access the starter motor. Before lifting the vehicle, disconnect all wires, fasteners, and so forth that are reachable from the top of the engine compartment.

⚡ **WARNING** **Check for proper pad-to-frame contact after the vehicle is a few inches above the ground. Shake the vehicle. If there are any unusual noises or movements of the vehicle, lower it and reset the pads.**

Special Tools
Fender covers
Battery cable puller

Disconnect the wires leading to the solenoid terminals. To prevent confusion during installation, use a piece of tape to identify the different wires or take a picture with your smartphone.

It may be necessary to disconnect and lower the exhaust system to provide space to remove the starter motor on some vehicles. Spray the exhaust system fasteners with penetrating oil to assist in removal.

Loosen the starter mounting bolts and remove all but one. Support the starter motor, remove the remaining bolt, and remove the starter motor.

⚡ **WARNING** **The starter motor is heavy; make sure it's secured before removing the last bolt.**

⚡ **Caution**

Take special care when handling the permanent magnet gear reduction starter. The permanent magnets are brittle and easily destroyed if the starter is dropped or struck by another object.

Reverse the procedure to install the starter motor. Be sure all electrical connections are tight. If you are installing a new or remanufactured starter, remove any paint that may prevent a good ground connection. Be careful not to drop the starter. Make sure it's properly supported.

Some General Motors starters use shims between the starter motor and the mounting pad (**Figure 6-14**). To check this clearance, insert a flat-blade screwdriver into the access slot on the side of the drive housing. Pry the drive pinion gear into the engaged position. Use a piece of wire 0.020 inch (0.5 mm) in diameter to check the clearance between the gears (**Figure 6-15**).

Excessive clearance produces a high-pitched whine while cranking the engine. If the clearance is too small, the starter will make a high-pitched whine after the engine starts and the ignition switch returns to the RUN position.

Shims adjust the proper pinion-turning clearance.

One shim will increase clearance by approximately 0.005".
More than one shim may be required.

Figure 6-14 Shimming the starter to obtain proper pinion-to-ring gear clearance.

Figure 6-15 Checking the clearance between the pinion gear and ring gear.

Starter Motor Disassembly

If the starter is defective, it can be disassembled, bench tested, and rebuilt. To reduce vehicle downtime to a minimum, many repair facilities do not rebuild starters. They replace them instead. However, many shops will replace the starter drive mechanism, which may require several of the following disassembling steps. The decision to rebuild or replace the starter motor is based on several factors:

1. What is best for the customer.
2. Shop policies.
3. Cost.
4. Time.
5. Type of starter.

Classroom Manual
Chapter 6,
pages 159, 160, 162

SERVICE TIP The major cause of drive housing breakage is too little clearance between the pinion and ring gears. It's always better to have a little more clearance than too small of a clearance.

⚡ **Caution**

Do not clean the starter motor components in solvent or gasoline. The residue left can ignite and destroy the starter. Wipe with clean rags, or use denatured alcohol to clean the starter components.

If you are rebuilding the stater, study the manufacturer's service information to become familiar with the disassembly procedures for the particular starter. **Figure 6-16** illustrates the disassembled view of a Delco Remy starter.

Photo Sequence 11 shows a typical procedure for disassembling a permanent magnet, gear reduction starter. Reference the specific manufacturer's service information for the starter motor you are working on.

Inspect the end frame and drive housing for cracks or broken ends. Check the frame assembly for loose pole shoes and broken or frayed wires. Inspect the drive gear for worn teeth and proper overrunning clutch operation. The commutator should be free of flat spots and should not be excessively burned. Check the brushes for wear. Replace them if worn past the manufacturer's specifications.

1. Lever
2. Plunger
3. Solenoid
4. Bushing
5. Spring
8. Coil
9. Armature
11. Grommet
31. Housing
32. Drive
33. Brushes
34. Washer
35. Bolt
36. Screw
37. Ring
38. Holder
39. Collar
40. Pin
41. Frame
44. Nut
45. Lead
46. Insulator
47. Shoe
48. Plate

Figure 6-16 Delco Remy 10MT starter.

PHOTO SEQUENCE 11
Typical Procedure for Disassembly of Gear Reduction Starter

P11-1 Disconnect the lead to the field coil and remove the solenoid retainer screws.

P11-2 Remove the solenoid housing while working the plunger off of the drive lever.

P11-3 Remove the frame through bolts.

P11-4 Separate the drive end frame from the body. Remove the seal also.

P11-5 Remove the O-ring from the end of the drive gear and then remove the retainer ring and C-clip.

P11-6 Remove the drive from the output shaft.

P11-7 Separate the output shaft and stationary gear assembly from the armature. Be sure to locate and retain the thrust ball located in a seat in the output shaft.

P11-8 Remove the lock ring from the output shaft and remove the stationary gear from the shaft.

P11-9 Remove the planetary gears from the output shaft.

P11-10 Remove the fasteners that attach the end plate to the brush plate.

P11-11 Remove the armature and brush assembly from the body.

P11-12 Separate the brush assembly from the armature.

Starter Motor Component Tests

Special Tools
DMM
Crocus cloth

Classroom Manual
Chapter 6,
pages 141, 144

With the starter motor disassembled and the components cleaned, you are ready to perform tests that will isolate the reason for the failure. The armature and field coils are checked for shorts and opens. If the armature or field coils are bad, it's best to replace the starter instead of rebuilding it.

Field Coil Testing

Test the field coil and frame assembly for opens and shorts to ground. In most cases, finding one of these conditions renders the starter unrebuildable in the field and will need to be replaced with a new unit.

Field coils are wired in many different ways and determine the testing procedure used. By looking at the wiring diagram, you will determine where the coils get their power and where they are grounded. Knowing these things is critical to testing the coils. The following procedure is valid for many, but not all, vehicles.

Using an ohmmeter, place one lead on the starter motor input terminal. Connect the other lead to the insulated brushes (**Figure 6-17**). The ohmmeter should indicate zero resistance. If there is resistance in the field coil, replace the starter.

Place one lead of the ohmmeter on the starter motor input terminal and the other lead on the starter frame (**Figure 6-18**). Since there should not be continuity, the ohmmeter

Figure 6-17 Testing the field coils for opens.

Figure 6-18 Testing the field coils for shorts to ground.

Figure 6-19 Testing the commutator for opens. There should be zero resistance between the segments.

should display an infinite reading. If the ohmmeter indicates continuity, there is a short to ground in the field coil, and you need to replace the starter.

Commutator Tests

Use an ohmmeter to test the armature commutator for opens and grounds. Clean the commutator with **crocus cloth**. Place the ohmmeter on the lowest scale and connect the test leads to any two commutator sectors (**Figure 6-19**). There should be 0 ohm of resistance. Replace the starter if there is resistance.

    To test the armature for short to ground, use an ohmmeter and connect one of the test leads to the armature shaft. Connect the other lead to the commutator segments (**Figure 6-20**). Check each commutator sector. There should be no continuity to ground. Replace the starter if any segment shows continuity to ground.

Brush Inspection

Use an ohmmeter to test continuity through the brush holder (**Figure 6-21**). Connect one test lead to the positive brush and the other lead to the negative brush. There should be no continuity between the brushes. If the ohmmeter indicated continuity, replace the brush holder.

    Another check of the brush assembly requires checking spring tension. To do this, install the brushes into the brush holder and slide the assembly over the commutator. Use a spring scale to measure the spring tension of the holders at the point where the spring

Classroom Manual
Chapter 6, page 142

Crocus cloth Polishes metals by removing very little metal.

Classroom Manual
Chapter 6, page 142

Figure 6-20 Testing the armature for short to ground. The meter should read infinite when placed on the shaft and the different segments of the commutator.

Figure 6-21 Typical brush holder.

lifts off the brush. If the spring tension is below specifications, replace the springs or the brush holder assembly.

Finally, measure the length of each brush (**Figure 6-22**). If they are shorter than the specifications, replace the brushes. Some starters may not have serviceable brushes and require replacing the brush holder assembly.

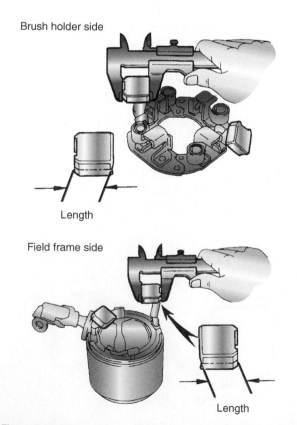

Brush holder side

Length

Field frame side

Length

Figure 6-22 Measure the length of the brushes to determine if they are worn.

Overrunning Clutch Inspection

The overrunning clutch may be inspected by sliding it onto the armature shaft. Attempt to rotate the clutch in both directions. If it's working properly, the clutch should rotate smoothly in one direction and lock in the other. If it fails to lock, or locks in both directions, replace the overrunning clutch.

Classroom Manual
Chapter 6, page 148

Starter Reassembly

Manufacturers use two methods of connecting the brushes; they are either soldered to the coil leads or screwed to terminals.

> ⚙ **SERVICE TIP** To seat the new brushes to the commutator, slide the brushes into their holders and then place the assembly onto the commutator. Slide a piece of fine sandpaper between the commutator and the brushes with the grain facing the brushes. Rotate the armature to sand down the face of the brushes so their contour matches the commutator.

Cut the old leads if the brushes are soldered to the coil leads (**Figure 6-23**). Place a piece of heat shrink tube over the brush connector. Crimp the new brush lead connector to the coil leads. Solder the brush connector to the coil lead with rosin core solder (**Figure 6-24**). Slide the heat shrink tube over the soldered connection and use a heating gun to shrink the tube.

To reassemble the starter motor, basically reverse the disassembly procedures. Additional steps are listed here:

1. Lubricate the splines on the armature shaft that the drive gear rides on with a high-temperature grease.
2. To install the snap ring onto the armature shaft, stand the commutator end of the armature on a block of wood. Position the snap ring onto the shaft and hold it in place with another block of wood. Hit the block of wood with a hammer to drive the snap ring onto the shaft (**Figure 6-25**).
3. Lubricate the bearings with high-temperature grease.
4. Apply sealing compound to the solenoid flange before installing the solenoid to the frame.

Special Tools

Feeler gauge set
100-watt soldering
    iron
Rosin core solder
Heat shrink tube
High-temperature
    grease
Two blocks of wood
Jumper cables
Fused jumper wire
Starting/charging
    system tester
Remote starter switch

Figure 6-23 Removing the worn brushes.

Figure 6-24 Soldering a new brush to the field coil lead.

Figure 6-25 With the snap ring centered on the shaft, use a hammer and block of wood to install it.

5. Use the scribe marks to locate the correct position of the frame-to-frame end and drive housing.

6. Check the pinion gear clearance. Disconnect the M terminal to the starter motor's field coils. Connect a jumper cable from the battery positive terminal to the S terminal of the solenoid. Connect the other jumper cable from the battery negative terminal to the starter frame (**Figure 6-26**). Connect a fused jumper wire from the M terminal and momentarily touch the other end of the jumper wire to the starter motor frame. This will shift the pinion gear into the cranking position and hold it there until the battery is disconnected. With the solenoid energized, push the pinion back toward the armature to remove any slack. Check the clearance with a feeler gauge (**Figure 6-27**). Compare clearance with specifications; normally, the clearance specifications are 0.010 to 0.140 inch (0.25 to 3.6 mm).

7. Perform the free spin test before installing the starter into the vehicle.

Figure 6-26 Jumper cable connections for checking the pinion gear clearance.

Figure 6-27 Checking the pinion gear to drive housing clearance.

> ⚙ **SERVICE TIP** Most starters have no provision to adjust the pinion clearance. However, if the clearance is excessive, it may indicate wear of the solenoid linkage or shift lever.

Stop/Start System Diagnostics

Diagnostics of the starter motor and the Stop/Start system circuits is very much like that discussed in this chapter. However, performing some tests will require disabling the Stop/Start feature. Most manufacturers provide a means of disabling the system by push button activation. In addition, it may be possible to disable Stop/Start by opening the hood since the system uses hood ajar inputs to allow the feature to operate only when the hood is closed. Opening the hood requires the engine to be started by the ignition start switch.

Classroom Manual
Chapter 6, page 174

> **AUTHOR'S NOTE** For the Stop/Start function to operate, the initial start must be by the regular ignition system.

A "SERVICE STOP/START SYSTEM" message or some other means will be used to alert the driver of a malfunction. If there is a malfunction in the Stop/Start system, the system will not shut off the engine (auto stop). If the Stop/Start system does not auto stop (engine does not shut off at stop), use a scan tool to monitor system inhibits (**Figure 6-28**). Typical auto stop inhibits include:

- The accelerator pedal is not in the idle position as reported by the throttle position sensor (TPS).
- The clutch is released (manual transmissions).
- The hood is open.
- The engine has not reached normal operating temperature.
- The driver's seat belt is not buckled.
- The outside temperature is below or above threshold.

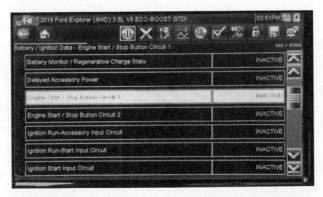

Figure 6-28 Use the scan tool to monitor all Stop/Start system inhibits.

- Cabin temperature is significantly different than temperature set on the temperature control.
- The temperature control system is set in the full defrost mode.
- The battery SOC is low.
- The transmission is in REVERSE.
- Transfer case is in 4LO mode.

There may be several other inhibitors that will prevent the engine from stopping. Reference the service information for a complete list. Confirm that all inhibitors are in their proper state. It may be necessary to diagnose the circuit of the offending inhibitor to locate the root cause of the issue.

CASE STUDY

A customer has a vehicle towed to the service center because it does not start. The technician verifies the concern and observes that the starter drive engages the flywheel, but the engine does not rotate. By turning the crankshaft, the technician confirms that the engine rotates. The engine turns freely through two complete revolutions.

Next, the technician performs a visual inspection of the starting system. All connections are cleaned and tightened. A complete battery test series is performed to confirm good battery condition. The battery passes all tests. The technician performs the solenoid circuit resistance test, which indicates an excessive voltage drop due to burned contacts. The extra resistance caused by the burned contacts prevented sufficient current flow to the starter motor. This did not allow enough torque to be generated to rotate the engine. The technician replaced the solenoid and verified the repair before returning the vehicle to the owner.

ASE-Style Review Questions

1. The starter circuit shown in **Figure 6-29** has a fully charged battery. The starter relay and solenoid do not operate with the ignition switch is in the START position.

   Technician A says this could be caused by a grounded circuit at terminal 85 of the starter relay.

   Technician B says this could be caused by an open to terminal 86 of the starter relay.

   Who is correct?

   A. A only

   B. B only

   C. Both A and B

   D. Neither A nor B

2. The following results are from a starter current-draw test:

   Voltage—9.4 volts.

   Current draw—363 amps.

   The specifications for the maximum current draw are 180 amps.

   Technician A says the test results indicate that the battery may be defective and needs testing.

   Technician B says the test results indicate a faulty natural safety start switch circuit.

   Who is correct?

   A. A only

   B. B only

   C. Both A and B

   D. Neither A nor B

Figure 6-29 Starter circuit.

3. While cranking the engine, a voltage drop of 600-mV is measured across the starter motor solenoid. What is the repair?

A. Replace the battery.

B. Replace the starter motor.

C. Confirm cable connections.

D. All of these choices.

4. In the starter system shown in **Figure 6-29**, a voltmeter indicates battery voltage to the input of the neutral safety switch and battery voltage on the output terminal with the transmission in neutral and the ignition switch in the START position. This indicates:

A. Normal operation.

B. The neutral safety switch is working properly.

C. A faulty starter relay coil winding.

D. An open neutral safety switch ground.

5. During a starter current-draw test, the voltage is 10.2 volts and amperage is above specifications. This could be caused by all of the following **EXCEPT**:

A. Excessive circuit resistance.

B. Shorted starter windings.

C. Worn starter motor bushings.

D. Internal engine failure.

6. Armature testing is being discussed.
Technician A says there should be no continuity between the commutator segments.
Technician B says there should be zero resistance between the commutator sectors and the armature shaft.
Who is correct?

A. A only   C. Both A and B

B. B only   D. Neither A nor B

7. *Technician A* says it's important to confirm battery condition before trying to test the starter system.
Technician B says the internal condition of the engine has little effect on the operation of the starting system.
Who is correct?

A. A only   C. Both A and B

B. B only   D. Neither A nor B

8. The Stop/Start feature does not shut the engine off when the vehicle is stopped.
Technician A says the feature may be inhibited due to a faulty seat belt buckle switch.
Technician B says the defrost mode may be selected on the HVAC system.
Who is correct?

A. A only   C. Both A and B

B. B only   D. Neither A nor B

9. Voltage drop testing of the solenoid control circuit is being discussed.

   Technician A says the maximum amount of voltage drop allowed is 0.9 volt.

   Technician B says the voltage drop across each connection should be less than 0.1 volt.

   Who is correct?

   A. A only

   B. B only

   C. Both A and B

   D. Neither A nor B

10. *Technician A* says most starter noises come from the armature.

    Technician B says if the engine does not rotate but the starter drive spins, the most likely cause is a shorted armature.

    Who is correct?

    A. A only

    B. B only

    C. Both A and B

    D. Neither A nor B

ASE Challenge Questions

1. A vehicle is towed into the shop due to a no-start (no-crank) condition. The headlights are turned on, and when the ignition key is placed in the START position, the lights remain bright but the starter does not crank the engine.

   Technician A says that the starter may be binding internally.

   Technician B says that the battery may be discharged.

   Who is correct?

   A. A only   C. Both A and B

   B. B only   D. Neither A nor B

2. A vehicle is being tested for a slow-crank condition. The starter current draw is 525 amps.

   Technician A says that the field windings in the starter may be shorted.

   Technician B says that the starter drive gear may be slipping.

   Who is correct?

   A. A only   C. Both A and B

   B. B only   D. Neither A nor B

3. The starter draw of a vehicle with a slow-crank condition is 75 amps. The battery has a good state of charge and it passed a capacity test.

   Technician A says that a starting circuit voltage drop test should be performed.

   Technician B says that the neutral safety switch may have excessive resistance.

   Who is correct?

   A. A only   C. Both A and B

   B. B only   D. Neither A nor B

4. A voltmeter that's connected across a starter solenoid's battery and motor terminals indicates 12 volts when the ignition key is turned to the START position. A distinct click is heard from the solenoid when this occurs, but the engine does not crank.

   Technician A says that there is excessive voltage drop in the circuit between the battery and the starter solenoid.

   Technician B says that the starter solenoid contacts are not contacting the terminal.

   Who is correct?

   A. A only

   B. B only

   C. Both A and B

   D. Neither A nor B

5. *Technician A* says incorrect pinion gear to ring gear clearance can result in noisy cranking and high starter draw.

   Technician B says starting circuit voltage drop should be checked before performing a starter draw test.

   Who is correct?

   A. A only

   B. B only

   C. Both A and B

   D. Neither A nor B

Name _____   Date _____

STARTER CURRENT DRAW TEST

Objective: Upon completion of this job sheet, you should be able to measure the current draw of a starter motor and interpret the results.

ASE Education Foundation Correlation

This job sheet addresses the following **MLR** task:

C.1.   Perform starter current-draw test. **(P-1)**

This job sheet addresses the following **AST/MAST** tasks:

C.1.   Perform starter current-draw test; determine needed action. **(P-1)**

C.7.   Differentiate between electrical and engine mechanical problems that cause a slow-crank or no-crank condition. **(P-2)**

We Support
ASE | **Education Foundation**

Tools and Materials

- A vehicle with a 12-volt battery
- Starting/charging system tester
- 2 DMMs with high current inductive pickup
- Lab scope with inductive pickup
- Service information

Safety Considerations and PPE

- Safety glasses

Procedure

1. Perform a battery test series to confirm the battery is good. If necessary, refer to the job sheets in Chapter 5. What was the result of the battery test?

   PASS _____ FAIL _____

   IF FAIL, WHY _____

   _____

2. Disable the ignition or fuel injection systems to prevent the engine from starting.

   a. How was this accomplished? _____

   _____

3. Expected starter current draw is: _____ amp(s).

   Voltage should not drop below: _____ volt(s).

4. Connect the tool(s) to the vehicle. What tool are you using to perform this test?

   _____

   _____

5. If applicable, zero the ammeter.

6. Be prepared to observe the amperage when the engine begins to crank and while it's cranking. Also, note the voltage when you stop cranking the engine.

The initial current draw was _____ amp(s)

After _____ seconds, the current draw was _____ amps and the voltage dropped to _____ volt(s).

7. Compare your measurements to the specifications. What is indicated by the test results?

_____

_____

_____

_____

_____

8. If the current draw was high, how would you differentiate between electrical and engine mechanical problems?

_____

_____

_____

_____

9. Reconnect the ignition or fuel injection system and start the engine.

Instructor's Evaluation:

Grade	Rubric
0	No exposure due to: ☐ Absence ☐ Other:
1	Student unsuccessfully attempted the task
2	Student completed the task with significant instructor assistance.
3	Satisfactory performance of task. The student was able to perform elements of the task with minimal instructor assistance.
4	Competent/Proficient. Able to perform all elements of the task successfully and independently.

Instructor's Notes:

Name _____   Date _____

STARTER SYSTEM TEST USING THE GR8

Objective: Upon completion of this job sheet, you should be able to measure the current draw of a starter motor and interpret the results.

ASE Education Foundation Correlation ——————————————

This job sheet addresses the following **MLR** task:

C.1.   Perform starter current-draw test. **(P-1)**

This job sheet addresses the following **AST/MAST** tasks:

C.1.   Perform starter current-draw test; determine needed action. **(P-1)**

C.7.   Differentiate between electrical and engine mechanical problems
                 that cause a slow-crank or no-crank condition. **(P-2)**

We Support
ASE | **Education Foundation**

Tools and Materials

- A vehicle with a 12-volt battery
- GR8 or similar starting system tester
- Service information

Safety Considerations and PPE

- Safety glasses

Procedure

1. Perform a battery test series to confirm the battery is good. If necessary, refer to the job sheets in Chapter 5. What was the result of the battery test?

   PASS _____   FAIL_____

   IF FAIL, WHY _____

   _____

2. Follow the on-screen prompts to select the Starter System Test.

3. Follow any on-screen instruction for connecting the tester to the vehicle.

4. When prompted, start the vehicle's engine.

5. What is the displayed test decision?

   _____

   _____

   _____

6. Are the complete results of the test able to be displayed?

   ☐ Yes   ☐ No

7. If yes, scroll to each screen and record the results here:

   Average cranking voltage: _____

   Average cranking current: _____

   Cranking time: _____

8. If possible, print off the test results and attach to this job sheet.

9. What is indicated by the test results? Compare your measurements to the specifications.

_____

_____

_____

_____

_____

Instructor's Evaluation:

Grade	Rubric
0	No exposure due to: ☐ Absence ☐ Other:
1	Student unsuccessfully attempted the task
2	Student completed the task with significant instructor assistance.
3	Satisfactory performance of task. The student was able to perform elements of the task with minimal instructor assistance.
4	Competent/Proficient. Able to perform all elements of the task successfully and independently.

Instructor's Notes:

Name _____   Date _____

TESTING THE STARTING SYSTEM CIRCUIT

Objective: Upon completion of this job sheet, you should be able to visually inspect the starting circuit and perform voltage drop testing for excessive resistance.

ASE Education Foundation Correlation

This job sheet addresses the following **MLR** tasks:

C.2.   Perform starter circuit voltage drop tests. **(P-1)**

C.3.   Inspect and test starter relays and solenoids. **(P-2)**

C.5.   Inspect and test switches, connectors, and wires of starter control circuits **(P-2)**

This job sheet addresses the following **AST/MAST** tasks:

C.2.   Perform starter circuit voltage drop tests; determine needed action. **(P-1)**

C.3.   Inspect and test starter relays and solenoids; determine needed action. **(P-2)**

C.5.   Inspect and test switches, connectors, and wires of starter control circuits; determine needed action. **(P-1)**

C.8.   Diagnose a no-crank condition using a wiring diagram and test equipment; determine needed action. **(P-1)**

We Support

ASE | **Education Foundation**

Tools and Materials
- A vehicle with a 12-volt battery
- Wiring diagram for the vehicle assigned
- A DMM
- Highlighter markers

Safety Considerations and PPE
- Safety glasses

Procedure

1. Using the service information, retrieve the wiring diagram for the starting system of the assigned vehicle and print it off. Highlight the positive side of the system in yellow. Highlight the negative side of the system in green. Identify the control circuit with orange.

2. Disable the ignition or fuel injection systems to prevent the engine from starting.

3. Connect the voltmeter to measure the voltage drop of the ground side of the starter motor circuit (from starter case to battery negative terminal).

   Crank the engine with the starter and observe the readings on the meter.

   Your reading was _____ volt(s).

4. What does this indicate?

_____

_____

_____

5. Connect the voltmeter across the battery's positive cable (from battery to starter motor). Crank the engine with the starter and observe the readings on the meter. Your reading was _____ volt(s).

6. This test measured the voltage drop across everything in the positive side of the starter motor circuit, including all relays and solenoids. How would you inspect a suspected faulty relay or solenoid?

_____

_____

_____

7. Use the wiring diagram and perform a voltage drop test on the control circuit.

8. What do the test results suggest?

_____

_____

9. What are your recommendations?

_____

_____

10. Reconnect the ignition or fuel injection systems to allow the engine to start.

_____

_____

Instructor's Evaluation:

Grade	Rubric
0	No exposure due to: ☐ Absence ☐ Other:
1	Student unsuccessfully attempted the task
2	Student completed the task with significant instructor assistance.
3	Satisfactory performance of task. The student was able to perform elements of the task with minimal instructor assistance.
4	Competent/Proficient. Able to perform all elements of the task successfully and independently.

Instructor's Notes:

_____

Name _____   Date _____

REMOVE AND INSTALL STARTER

Objective: Upon completion of this job sheet, you should be able to remove and install the starter.

ASE Education Foundation Correlation

This job sheet addresses the following **MLR** task:

C.4.   Remove and install starter in a vehicle. **(P-3)**

This job sheet addresses the following **AST** task:

C.4.   Remove and install starter in a vehicle. **(P-2)**

This job sheet addresses the following **MAST** task:

C.4.   Remove and install starter in a vehicle. **(P-1)**

We Support
ASE | Education Foundation

Tools and Materials

- A vehicle with a 12-volt battery
- Fender covers
- Battery cable puller
- Service information

Safety Considerations and PPE

- Safety glasses

Procedure

1. Position the vehicle over a hoist. Do not lift the vehicle at this time.

2. Disconnect the negative battery cable from the battery terminal. Isolate the cable clamp so it cannot touch the negative battery post.

3. Disconnect all wires, fasteners, and so on, which can be reached from the top of the engine compartment.

4. Disconnect the wires leading to the solenoid terminals. To prevent confusion, it's a good practice to use a piece of tape to identify the different wires.

5. On some vehicles, it may be necessary to disconnect the exhaust system to be able to remove the starter motor. Spray the exhaust system fasteners with a penetrating oil to assist in removal.

6. Loosen the starter mounting bolts and remove all but one.

7. Support the starter motor; remove the remaining bolt. Then remove the starter motor.

8. Identify the type of starter motor.

9. Does the motor require clearance shims?

Yes ☐   No ☐

If yes, describe the procedure for selecting the proper shim:

_____

_____

_____

_____

10. Reverse the procedure to install the starter motor. Be sure all electrical connections are tight. If you are installing a new or remanufactured starter, remove any paint that may prevent a good ground connection.

Instructor's Evaluation:

Grade	Rubric
0	No exposure due to: ☐ Absence ☐ Other:
1	Student unsuccessfully attempted the task
2	Student completed the task with significant instructor assistance.
3	Satisfactory performance of task. The student was able to perform elements of the task with minimal instructor assistance.
4	Competent/Proficient. Able to perform all elements of the task successfully and independently.

Instructor's Notes:

Name _____   Date _____

STOP/START FAMILIARIZATION

Objective: Upon completion of this job sheet, you should be able to identify Stop/Start features and inhibits.

ASE Education Foundation Correlation

This job sheet addresses the following **MLR** task:

C.6.   Demonstrate knowledge of an automatic idle-Stop/Start-stop system. **(P-2)**

This job sheet addresses the following **AST/MAST** task:

C.6.   Demonstrate knowledge of an automatic idle-Stop/Start-stop system. **(P-1)**

We Support
ASE | **Education Foundation**

Tools and Materials
- A vehicle Stop/Start with a 12-volt battery
- Scan tool
- Service information

Safety Considerations and PPE
- Safety glasses

Procedure

1. Based on the service information, describe how the Stop/Start feature is disabled for service:

    _____

    _____

    _____

2. Disable the Stop/Start feature.

    Was the process successful?   Yes ☐   No ☐

    If no, consult your instructor.

3. Enable the Stop/Start feature again.

    Was the process successful?   Yes ☐   No ☐

    If no, consult your instructor.

4. With the engine off, open the hood. Attempt to start the engine. What happened?

    _____

5. Shut the hood and assure the Stop/Start feature is enabled.

6. With the engine running, open the hood. What happened?

    _____

7. Use the scan tool to monitor the state of all inhibits listed in the service information. Are any input states preventing the Stop/Start system from functioning?

_____

_____

_____

_____

Instructor's Evaluation:

Grade	Rubric
0	No exposure due to: ☐ Absence ☐ Other:
1	Student unsuccessfully attempted the task
2	Student completed the task with significant instructor assistance.
3	Satisfactory performance of task. The student was able to perform elements of the task with minimal instructor assistance.
4	Competent/Proficient. Able to perform all elements of the task successfully and independently.

Instructor's Notes:

DIAGNOSTIC CHART 6-1

PROBLEM AREA:	Starting system operation
SYMPTOMS:	Starter fails to turn engine or operates at reduced efficiency. Excessive current draw.
POSSIBLE CAUSES:	1. Shorted armature. 2. Worn starter bushings. 3. Bent armature. 4. Thrown armature windings. 5. Loose pole shoes. 6. Grounded armature. 7. Shorted field windings.

DIAGNOSTIC CHART 6-2

PROBLEM AREA:	Starting system operation
SYMPTOMS:	No or reduced starter operation. Current draw too low.
POSSIBLE CAUSES:	1. Worn brushes. 2. Excessive circuit voltage drop.

DIAGNOSTIC CHART 6-3

PROBLEM AREA:	Starting system operation
SYMPTOMS:	Starter fails to rotate the engine or operates at reduced efficiency. Excessive starter circuit resistance.
POSSIBLE CAUSES:	1. Excessive starter circuit voltage drop. 2. Improper connections. 3. Corroded ground connections. 4. High resistance in solenoid or relay.

DIAGNOSTIC CHART 6-4

PROBLEM AREA:	Starting system operation
SYMPTOMS:	Starter fails to rotate engine or rotates too slowly. Failed starter test series.
POSSIBLE CAUSES:	1. Defective or worn starter. 2. Worn brushes. 3. Shorted field coils. 4. Open field coils. 5. Shorted armature. 6. Open armature.

DIAGNOSTIC CHART 6-5

PROBLEM AREA:	Starter control circuit
SYMPTOMS:	Starter does not operate when ignition switch is located in the START position. No sounds.
POSSIBLE CAUSES:	1. Open circuit. 2. Faulty ignition switch. 3. Park/neutral switch faulty or misadjusted. 4. Faulty starter relay/solenoid. 5. Faulty starter.

DIAGNOSTIC CHART 6-6

PROBLEM AREA:	Starter control circuit
SYMPTOMS:	Starter does not operate when ignition switch is located in the START position. Relay/solenoid clicks.
POSSIBLE CAUSES:	1. High resistance in starter control circuit. 2. High resistance in started circuit. 3. Faulty starter relay/solenoid. 4. Faulty starter.

DIAGNOSTIC CHART 6-7

PROBLEM AREA:	Starter drive
SYMPTOMS:	Starter spins but does not rotate the engine.
POSSIBLE CAUSES:	1. Defective one-way clutch. 2. Broken teeth on ring gear. 3. Faulty starter motor.

DIAGNOSTIC CHART 6-8

PROBLEM AREA:	Starter drive fails to disengage after engine starts.
SYMPTOMS:	Excessive noise after engine starts.
POSSIBLE CAUSES:	1. Faulty ignition switch. 2. Faulty relay/solenoid. 3. Faulty starter motor. 4. Improper starter mounting.

DIAGNOSTIC CHART 6-9

PROBLEM AREA:	Stop/Start
SYMPTOMS:	Engine will not shut off when vehicle is stationary.
POSSIBLE CAUSES:	1. The accelerator pedal is not in the idle position as reported by the throttle position sensor (TPS). 2. The clutch is released (manual transmissions) or input is missing. 3. The hood is open or faulty input switch. 4. The engine has not reached normal operating temperature or faulty engine temperature sensors. 5. The driver's seat belt is not buckled or faulty input. 6. The outside temperature is below or above threshold or faulty ambient temperature sensor. 7. Cabin temperature is significantly different than temperature set on the temperature control or faulty temperature sensors. 8. The temperature control system is set in the full defrost mode or faulty input. 9. The battery SOC is low. 10. The transmission is in REVERSE or faulty range sensor. 11. Transfer case is in 4LO mode or faulty input sensor.

CHAPTER 7

CHARGING SYSTEM TESTING AND SERVICE

Upon completion and review of this chapter, you should be able to:

- Diagnose charging system problems that cause an undercharge or no-charge condition.
- Diagnose charging system problems that cause an overcharge condition.
- Inspect, adjust, and replace generator drive belts, pulleys, and fans.
- Perform charging system output tests and determine needed repairs.
- Perform charging system circuit voltage drop tests and determine needed repairs.
- Perform voltage regulator tests and determine needed repairs.
- Remove and replace the AC generator.
- Disassemble, clean, and inspect AC generator components.
- Test and diagnose the rotor.
- Test and diagnose the stator.
- Perform diagnostic test routines on smart charging systems.
- Inspect the HV and LV circuits of the HEV inverter/converter module.

Terms To Know

Current output test	Full field test	Load shedding
Field current-draw test	Full fielding	Voltage output test

Introduction

WARNING Always wear safety glasses when performing charging system tests.

Whenever there is a charging system problem, conduct a complete battery test series before jumping in on the charging system. A faulty battery will not allow the charging system to work at its best. The AC generator is designed to maintain the battery's SOC, not to charge a dead battery. The battery must be in good condition and fully charged to obtain accurate charging system test results. It's also essential to perform a preliminary inspection of the charging system. Often, the visual inspection will reveal the source of the issue. Check the following items:

1. Condition of the drive belt (**Figure 7-1**). If the drive belt is worn or glazed, it will not allow enough rotor RPMs to produce sufficient current. Use a belt wear gauge to determine wear on ribbed serpentine belts (**Figure 7-2**). The belt is good if the gauge pin sits higher than the belt rib (**Figure 7-3**). If the gauge pin is lower than the belt rib, the belt requires replacement.
2. Drive belt tension (**Figure 7-4**).

Figure 7-1 Any of these conditions indicate the need to replace the generator drive belt.

Figure 7-2 A belt wear gauge for determining serpentine belt wear.

3. Drive belt tensioner condition.
4. Drive belt tensioner bearing failure.
5. Electrical connections to the AC generator.
6. Electrical connections to the regulator.
7. Ground connections at the engine and chassis.
8. Battery cables and terminals.
9. Fuses and fusible links.
10. Excessive current drain caused by an electrical load remaining powered up after turning off the ignition.
11. Symptoms of undercharging. These include slow-cranking, discharged battery, low instrument panel voltmeter readings, and the illumination of the charge indicator lamp.
12. Symptoms of overcharging. These include high voltmeter readings, battery boiling, and the illumination of the charge indicator lamp.

Figure 7-3 Using the belt wear gauge. If the gauge pin is above the rib, the belt is good.

Figure 7-4 Checking belt tension using a tension gauge.

The manufacturer of the vehicle you are working on may have several additional tests to perform. It's important to follow the procedures outlined by the manufacturer for the vehicle you are servicing.

SERVICE TIP To check the fusible link to the AC generator, use a voltmeter and measure the voltage drop across the link. This will identify any high resistance in the circuit.

Several types of testers are available for testing the charging system and AC generators. Some handheld multimeters have the ability to perform many tests. However, specialized battery/starting/charging system testers designed to test the entire system may be the fastest and easiest to use. Always follow the operating procedures for the specific tester you are using.

When performing the tests, be sure of the connections you are making. Connecting a test lead to the wrong terminal can result in AC generator damage, as well as damage to other electrical and electronic components. Refer to the service information to identify the various terminals for the AC generator and regulator.

⚡**WARNING Many charging system tests require operating the vehicle in the shop area. Always place wheel blocks against the drive wheels. Be sure there is proper ventilation of the vehicle's exhaust. Also, be aware of the drive belts and cooling fan. Be sure of where your hands and tools are at all times.**

Charging System Service Cautions

The following are some of the general rules when servicing the charging system:

1. Do not run the vehicle with the battery disconnected. The battery acts as a buffer and stabilizes any voltage spikes that may cause damage to the vehicle's electronics.
2. When performing charging system tests, do not allow the output voltage to increase over 16 volts (V).
3. Disconnect the battery cables when recharging the battery.
4. Do not attempt to remove electrical components from the vehicle with the battery connected.
5. Before connecting or disconnecting any electrical connections, the ignition switch must be in the OFF position.
6. Avoid contact with the BAT terminal of the AC generator while the battery is connected. Battery voltage is always present at this terminal.

AC Generator Noises

Noises that come from the AC generator can be from three sources. The causes of the noises are identifiable by the types of noises they make. A loose belt will make a squealing noise. Check the belt condition and tension. Replace the belt if necessary.

Faulty bearings can also cause a squealing noise. The bearings support the rotor in the housing halves. To test for bearing noises, use a length of hose, a long screwdriver, or a technician's stethoscope. By placing the end of the probe tool close to the bearings and listening on the other end, the tool transmits the bearing noise so you can hear it. Faulty bearings will require either AC generator replacement or disassembly of the AC generator to replace the bearings.

⚡ **WARNING This test is performed with the engine running. Use caution around the drive belts, fan, and other moving components.**

⚙ SERVICE TIP With the engine off, rub a piece of bar soap on the pulley surface of the drive belts. Do this one belt at a time until the noise stops. This way you will know which belt is the cause of the noise.

A whining noise can result from shorted diodes, a shorted stator, or worn rotor bearings. A quick way to test for the cause of a whining sound is to disconnect the wiring to the generator and run the engine. If there is no noise, the cause of the noise is a magnetic whine due to shorted diodes or stator windings. Use a lab scope to verify the condition of

Classroom Manual
Chapter 7, page 184

the diodes and stator. If the noise remains, the cause is mechanical and probably due to worn bearings.

Charging System Troubleshooting

Troubleshooting charts assist the technician in diagnosing the charging system. They give several possible causes for the customer complaint. They also instruct the technician as to what tests to perform or what service is required.

Voltage Output Testing

After completing the visual inspection and preliminary checks, the next step is to perform a **voltage output test**. This test quickly determines if the charging system is working properly. If the charging system is operating correctly, then check for battery drain. The following procedure is for performing the voltage output test:

1. Connect the voltmeter across the battery terminals, observing polarity.
2. Connect a scan tool to the DLC to monitor engine speed.
3. With the engine off, record the base voltage value across the battery.
4. Start the engine and bring the RPM up to between 1,500 and 2,000.
5. Observe the voltmeter reading. It should read between 13.5 and 14.5 volts.

If the charging voltage is too high, there may be a problem in the following areas:

1. Poor voltage regulator ground connection.
2. High resistance in the "sense" circuit between the battery and the PCM or voltage regulator.
3. Short to ground in the field coil control circuit, causing the AC generator to full field.
4. Loose or corroded battery cable terminals.
5. Defective voltage regulator or PCM.

If the charging voltage is too low, the fault might be:

1. Loose or glazed drive belt.
2. Discharged battery.
3. Loose or corroded battery cable terminals.
4. Defective AC generator.
5. Defective voltage regulator or PCM.

If the voltage reading is correct, perform a load test to check the voltage output under a load condition:

1. With the engine idling, turn on the headlights and the heater fan motor to high speed.
2. Increase the engine speed to approximately 2,000 RPM.
3. Check the voltmeter reading. It should increase a minimum of 0.5 volt over the previous base voltage reading. Some vehicle manufacturer specifications require a rise in voltage of 2.5 volts over the base voltage value.
4. If the voltage increases, the charging system is operating properly. If the voltage did not increase, perform the following test series to locate the fault.

Classroom Manual
Chapter 7, pages 183, 195

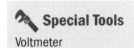 **Special Tools**
Voltmeter
Scan tool

> **SERVICE TIP** If the charging system passes the no-load test but fails the load test, check the condition and tension of the drive belt closely.

It's also possible to use the scan tool to diagnose the charging system's output voltage on most late-model vehicles. Most PCM-controlled charging systems use a battery voltage sense circuit to determine the state of charge for the battery and determine the AC generator's output voltage (**Figure 7-5**). If the PCM cannot operate the charging system correctly, a diagnostic trouble code (DTC) sets. In this case, follow the appropriate diagnostic routine for the set DTC to isolate the cause. However, not all faults result in a DTC. For example, if the sense circuit had excessive resistance due to a poor connection, the PCM would increase the generator output until it achieved the desired voltage level. Since the PCM is performing what is believed to be the correct function, no DTCs set. However, under this condition, the battery would be overcharging.

The PCM will use a temperature sensor to determine the battery temperature. This information, along with voltage sense, will determine a target voltage for the generator

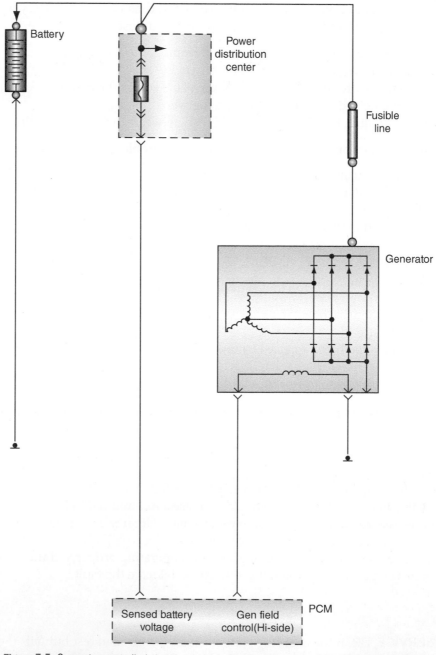

Figure 7-5 Computer-controlled charging system will use a sense circuit to monitor the state of charge (SOC) of the battery and the output of the generator.

Figure 7-6 The temperature of the battery is one factor that determines the charging system output voltage.

Figure 7-7 Scan tool graphing display of target and sensed voltage.

(**Figure 7-6**). In most cases, the PCM will energize the field circuit until it's about 0.50 volt above the target voltage and then turn off the field current until the voltage drops to 0.50 volt below the target voltage (**Figure 7-7**).

With a voltmeter connected across the battery terminals, compare the scan tool voltage display to the voltmeter reading. If there is a discrepancy between the two readings, test the voltage sense circuit.

> **SERVICE TIP** Keep in mind that a faulty temperature sensor, or a fault in its circuit, can cause the generator to undercharge or overcharge the battery.

Current Output Testing

Classroom Manual
Chapter 7, pages
183, 194

The AC generator outputs maximum current output only when the demands on the charging system require it. Using a carbon pile connected across the battery to create a load that draws and maintains battery voltage at 12.0 volts also reduces the regulator's signal voltage. The regulator will full field the winding to attempt to recharge the battery. This will produce the maximum current output to the battery. The **current output test** will determine the maximum output of the generator.

The following steps provide a typical procedure for performing the current output test using a DMM, an ammeter capable of reading the specified generator output, and a carbon pile.

Special Tools

Ammeter
DMM (2)
Scan tool
Carbon pile

> **SERVICE TIP** The DMM may be capable of reading the high amperage by setting the DMM to the DC Millivolts scale and connecting a current clamp to the "Volt" and "Common" inputs of the DMM. Place the current clamp around the alternator B+ wire.

1. Connect the carbon pile cables across the battery.
2. Prepare the DMM to read DC voltage and connect the test leads across the battery terminals.
3. Adjust the ammeter reading to zero.
4. Connect the ammeter's inductive pickup around all battery ground cables.
5. With the ignition switch in the RUN position (engine not running), observe the ammeter reading. This reading indicates how much current is required to operate any full-time accessories.
6. Start the engine and hold engine speed between 1,500 and 2,000 RPM.
7. Turn the load knob for the carbon pile slowly until the highest ammeter reading is obtained. Do not turn on any electrical loads at this time. Do not allow battery voltage to drop below 13 volts.
8. Return the load control knob to the OFF position.

Add the maximum output reading to the reading obtained in step 5. This total should be within 10% of the rated output of the AC generator.

If the ammeter reading indicates that output is 2 to 8 amps below the specification, an open diode or slipping belt may be the problem. An output reading of 10 to 15 amps below specifications indicates a possible shorted diode or slipping belt. If the AC generator output is below specifications, perform the full field test.

When testing the General Motors CS-130 or 144 AC generator, first use a voltmeter to test for voltage at the L and I terminals. Battery voltage should be indicated at both terminals with the ignition switch in the RUN position.

Circuit Voltage Drop Testing

Special Tools

DMM
Carbon pile

Excessive voltage drop is a primary cause of charging system problems. Testing all wires and connections should be a part of the diagnostic approach. Any individual wire or connection should not exceed 0.2 volt drop, while the total system drops should be less than 0.7 volt.

Figure 7-8 Voltage drop testing of the generator output circuit (A) and the generator ground circuit (B).

The ground side voltage drop should be less than 0.2 volt. The voltage drop test determines if the battery, regulator, and AC generator are all operating at the same potential.

When performing the voltage drop tests, load the system using a carbon pile or by turning on the headlights and the heater blower motor onto high. Try to load the system to between 10 and 20 amps with the engine running at about 2,000 RPM. Voltage drop test the circuit between the output terminal of the AC generator and the battery positive terminal (**Figure 7-8A**). Also, perform the voltage drop test between the battery negative terminal and the AC generator housing (**Figure 7-8B**).

Some manufacturers recommend measuring voltage drop when the generator is putting out its maximum. Always follow the recommendations of the manufacturer. Remember, if the AC generator is not putting out any current, there will be no voltage drop even if the circuit is very corroded.

 SERVICE TIP Turning on the headlights may be substituted for the carbon pile.

The general specifications for voltage drop testing are:

Insulated circuit: less than 0.7 volt.
Ground circuit: less than 0.2 volt.

If the circuit has excessive voltage drop, work up the circuit in sections to find the fault. Check every wire and connection.

Field Current-Draw Test

Because the production of a magnetic field requires field current, it's necessary to determine if current is flowing to the field coil. To perform the **field current-draw test** using the ammeter function of the DMM, follow these steps:

1. Confirm that the battery is fully charged.
2. Connect a carbon pile across the battery. Make sure the knob is in the OFF position before making the connection.

Classroom Manual
Chapter 7, pages
194, 199

Figure 7-9 Connect the amp probe around the field circuit wire to measure field current.

3. Place the DMM in the 10-amp mode and locate the induction pickup clamp around the field circuit wire at the AC generator (**Figure 7-9**). If a pickup clamp is not available, connect the DMM test leads in series to the field circuit.
4. Start the engine and run it at 2,000 RPM.
5. Using the carbon pile, load the generator to the rated output current.
6. Measure the amperage in the field circuit. Check results with the manufacturer's specifications. Readings typically range between 3 and 7 amps.

Another method of testing the field current is to use a lab scope (or DMM) with an inductive amp pickup. Monitor the waveform with the engine running (**Figure 7-10**). Use the scan tool's bi-directional controls to turn the field current on and off while observing the waveform. Look for spikes and drop off that indicate poor brush-to-slip ring contact.

 Special Tools

DMM
Lab Scope
Scan tool
Ammeter with inductive pickup
Multiplying coil
Carbon pile

> **SERVICE TIP** If no field current is measured with the engine running, substitute a headlamp at the generator field connector and see if the lamp illuminates with the engine idling. If it does, measure resistance at the field pins of the generator. An open indicates worn brushes or an open rotor.

For GM, A circuit systems with internal regulators, the test will require removing the field circuit connector from the AC generator and connecting a Y-type connector between terminals 1 and 2. Connect the common terminal of the Y connector to battery positive. With the amp pickup around the wire (or the test leads connected in series), run the engine at 2,000 RPM. The field current should be between 2 and 5 amps. Use the carbine pile to decrease battery voltage while observing the ammeter. As battery voltage decreases, the field current should increase. If the readings are within the specification limits, then the field circuit is good. If the readings are over specifications, a shorted field circuit or bad regulator may be the problem. If the readings are too low, then there is high electrical resistance that may be caused by worn brushes.

To test Ford's integral generator/regulator (IAR) system, use a voltmeter as follows:

1. With the ignition switch in the OFF position, connect the negative voltmeter lead to the generator housing.

Figure 7-10 Waveform of field current.

Regulator F terminal screw

Figure 7-11 Testing the IAR generator field circuit.

2. Connect the positive voltmeter lead to the F terminal screw of the regulator (**Figure 7-11**).
3. Check the voltmeter reading. It should indicate battery voltage. If it reads battery voltage, the field circuit is good.
4. If the voltmeter reading is less than battery voltage, disconnect the wiring plug from the regulator.
5. Connect the positive voltmeter lead to the I terminal of the plug.
6. Check the voltmeter reading. It should indicate 0 volt. If there is voltage present, repair the I lead from the ignition switch. The I lead is receiving voltage from another source.
7. If there was no voltage present in step 6, connect the positive voltmeter lead to the S terminal of the regulator wiring plug.
8. Check the voltmeter reading. If there is 0 volt, replace or service the regulator.
9. If there is a voltage reading, disconnect the wiring plug from the AC generator.
10. Check for voltage to the regulator wiring plug STA terminal.
11. If voltage is still present, repair the STA terminal wire lead to the AC generator. The STA terminal wire is receiving voltage from another source.
12. If no voltage is present, replace the rectifier bridge.

Full Field Test

The **full field test** will determine if the detected problem lies in the AC generator or the regulator. Perform the full field test only if the charging system fails the output test. This test is performed by manually **full fielding** the AC generator with the regulator bypassed. Full fielding supplies a constant full battery voltage to the field windings to produce maximum AC generator output. If this test still produces lower-than-specified output, the AC generator is the cause of the problem. If the output is within specifications with the regulator bypassed, then the problem is within the regulator.

When full fielding the system, the battery should be loaded to protect vehicle electronics and computers. Bypassing the voltage regulator takes away control of voltage output. The AC generator is capable of producing well over 30 volts. This increased voltage will damage the circuits not designed to handle that high of a voltage. A 20% increase in voltage output results in a 20% increase in current flow that can damage sensitive electronic circuitry.

Classroom Manual
Chapter 7, pages
195, 198

 Special Tools

Various jumper wires
Battery/starting/charging system tester
DMM
Scan tool

Figure 7-13 Quick check of the rotor and brushes.

Figure 7-12 Full fielding the GM 10SI AC generator by grounding the tab in the "D" test hole.

To perform the full field test on a General Motors' A circuit SI-type AC generator with an internal voltage regulator, insert a screwdriver into the D-shaped test hole (**Figure 7-12**). This test hole lines up with a small tab that's attached to the negative brush. The regulator is bypassed by inserting a screwdriver into the "D" hole about 1/2 inch (12.7 mm) and grounding it to the housing. Perform the output test again with the regulator bypassed. If the output is within specifications, the regulator is at fault.

A variation of the test calls for shorting the negative brush in the "D" hole while the ignition switch is in the RUN position. If the brushes and rotor are good, then the rear bearing should be magnetized and attract a metal screwdriver (**Figure 7-13**).

General Motors does not recommend a manual full field test on vehicles equipped with a CS or AD series generator. Instead, use the current output test to confirm the proper operation of the generator.

> ⚙ **SERVICE TIP** If the means of loading the battery is not available, do the full field test for a very short time. Do not allow voltage output to increase over 16 volts. Use the vehicle accessories to put a load on the battery.

> ⚠ **Caution**
>
> Not all AC generators can be manually full fielded. Check the manufacturer's procedures before attempting to full field an AC generator.

> ⚠ **Caution**
>
> Always refer to the manufacturer's service information for the correct location of the AC generator terminals. Incorrect test connections can damage the AC generator and the vehicle's electrical system components.

Ford Motor Company has utilized different designs of the integral regulator. The early design had one terminal, called the exciter, connected to the outside of the regulator (**Figure 7-14**). The wiring schematic for this design is illustrated in **Figure 7-15**. The field circuit can be grounded, and the regulator bypassed, by removing the protective cover from the field terminal (closest to the rear bearing).

Before full fielding the IAR AC generator, check the rotor and field circuit resistance by disconnecting the wiring plug to the regulator and connecting an ohmmeter between the regulator A and F terminals. Typically, the resistance should not be below 2.4 ohms (Ω).

If the resistance is less than 2.4 ohms, there is a short to ground somewhere in the circuit. Check for a failed regulator, shorted rotor, or shorted field circuits.

Figure 7-16 shows the wiring of the IAR system. To full field this system, disconnect the wiring connector to the AC generator and install a 12-gauge wire jumper between the B+ terminal blades (**Figure 7-17**). Connect another jumper wire from the regulator F

Figure 7-14 Integral regulator with exciter terminal.

Figure 7-15 Wiring schematic of an integral regulator with exciter terminal.

terminal screw to ground. Connect a voltmeter with the positive lead connected to one of the B+ jumper wire terminals, and the negative test lead to a good ground. Start the engine and perform the load output test. The regulator is faulty if the voltage rises to specifications. If the voltage does not rise to specifications, the AC generator needs to be serviced or replaced.

Many AC generators that use internal or computer-controlled regulators do not provide a means of doing manual full fielding. Typically, they may provide a method of full fielding the generator using a scan tool (**Figures 7-18** and **7-19**). The scan tool will direct the PCM to full field the generator and display the sensed battery voltage. While full fielding, the voltage should increase. However, if the voltage does not increase, you still must determine if the fault is in the generator, the insulated side of the field circuit, the field circuit between the generator and the PCM, or the PCM itself.

Classroom Manual
Chapter 7, page 203

Figure 7-16 Wiring diagram of Ford's IAR charging system.

Figure 7-17 Jumper wire connections between B+ terminals.

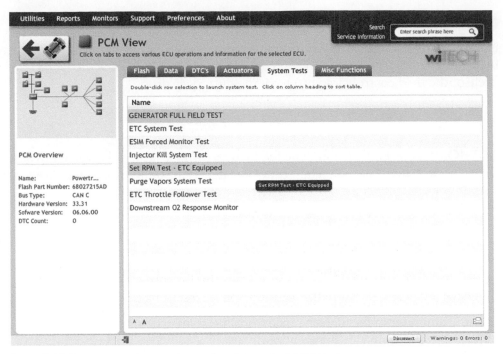

Figure 7-18 Some scan tools support an on-board generator full field test.

Figure 7-19 Running the full field test using a scan tool.

If the regulation is computer-controlled, study the wiring diagram for the charging system to determine if it's an A or a B circuit. By making this determination, you can use a jumper wire to connect the field circuit to either ground or voltage in place of the computer control. With a load against the battery, monitor the voltage output of the AC generator. If the voltage increases, the problem is not the AC generator but either the circuit between the AC generator and the computer or the computer itself. Be sure to remove the jumper wire as soon as the output voltage begins to increase. Never allow the voltage to increase over 16 volts.

⚠ **Caution**

Do not replace the regulator without first repairing any shorts in the rotor or field circuits. Doing so may damage the new regulator.

Regulator Test

Classroom Manual
Chapter 7,
pages 198, 200

The regulator test determines if the regulator maintains the correct voltage output under different load demands. To perform the regulator voltage test using a DMM, follow these steps:

1. Set the ammeter to the highest range (must be at least 10 amps).
2. Make sure the ammeter is reading zero.
3. Clamp the inductive pickup around the AC generator output wire.
4. Start the engine and hold between 1,500 and 2,000 RPM with all accessories turned off.
5. Allow the engine to run until the ammeter reads 10 amps or less. This indicates the battery is fully charged.
6. The voltmeter should read regulated voltage (13.5 to 14.5 volts).
7. Load the system between 10 and 20 amps by turning on the headlights or using a carbon pile.
8. The voltmeter should still read the specified regulated voltage.

Special Tools
Two DMMs
Inductive amp clamp
Scan tool
Carbon pile

Diode Pattern Testing

Classroom Manual
Chapter 7, page 189

An AC generator may have an open diode yet test close to manufacturer's specifications. Failure to detect an open diode during testing can cause a newly installed regulator to fail. In addition, an open diode can lead to the failure of other diodes.

A lab scope is an excellent diagnostic tool for testing the diodes. Set the lab scope on the lowest AC scale available. Connect the primary test leads onto the AC generator output terminal and ground. Start the engine and place a moderate load on the charging system (15 to 20 amps). Adjust the time scale so you can see at least 8 diode waveforms. Different patterns may appear. The amount of load placed on the system determines what is considered a normal waveform pattern.

The diode pattern (**Figure 7-20**) illustrates a good pattern if the AC generator is under a full load. The pattern shown in **Figure 7-21** is a good pattern for some AC generators under a light-load condition. The spikes are inductive voltages resulting from the fast cycling occurring during ignition and voltage regulator operation. This is one reason some manufacturers have started to use avalanche diodes in the output circuit of the generator.

Figure 7-20 Good AC level from the generator.

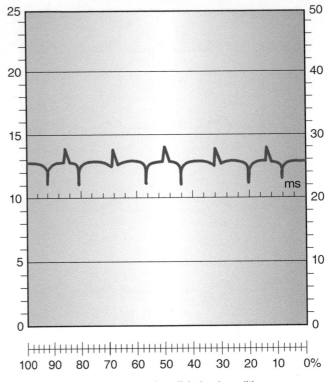

Figure 7-21 Normal pattern under a light-load condition.

Figures 7-22 through **7-25** illustrate patterns that have high-resistance, open, and shorted diodes. Remember to check the waveforms for electrical noise. If the diodes don't rectify all of the AC voltage, some of the AC voltage will ride on the DC output.

Figure 7-22 This waveform indicates an open diode.

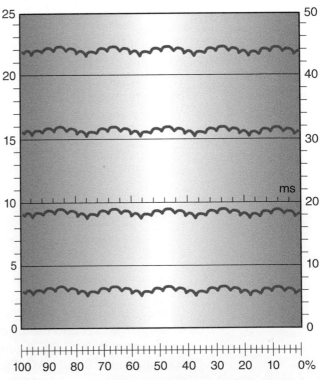

Figure 7-23 The uneven ripple of this pattern indicates a high resistance at the rectifier bridge and stator connection.

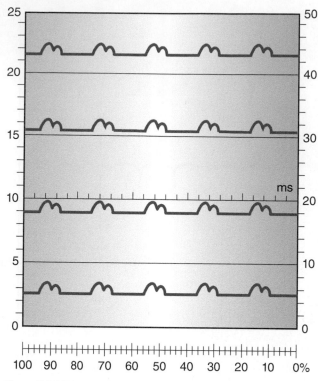

Figure 7-24 This uneven pattern indicates an open and a shorted diode.

Figure 7-25 A waveform indicating high resistance.

> ⚙️ **SERVICE TIP** Instead of using a carbon pile, it's possible to place a moderate load on the charging system by turning on the headlights and other electrical accessories.

A DMM can be used as an alternate method to test the action and functionality of the diodes by checking for AC voltage at the output terminal of the AC generator. With the DMM set on a low AC voltage scale, connect the positive test lead to the AC generator's output terminal (not at the battery), and the negative test lead to the case of the generator. With the engine running, turn on enough accessories to cause a 10- to 20-amp flow from the generator. Ideally, there should be 0 volt AC displayed on the meter. A voltage reading greater than 0.250 VAC indicates the diodes are not rectifying the AC output of the generator.

> ⚙️ **SERVICE TIP** In order to get accurate readings using a DMM, it must be capable of a root mean square reading.

The DMM testing just described looks at the average of the three stator windings. Consequently, if one leg of the stator is beginning to have a problem, it may not be indicated by the average reading on the DMM. A lab scope is a better choice to see what is occurring with each leg. A good pattern should not have a voltage over the peak-to-peak voltage limit of 0.250 VAC (**Figure 7-26**). Since each voltage pulse is an AC pattern from the generator, excessive AC voltage is easy to detect (**Figure 7-27**).

Figure 7-26 The waveform should not indicate a peak-to-peak voltage of over 0.250 volt AC.

Figure 7-27 Excessive AC voltage indicates an open diode in the diode trio.

You can also use a DMM to check for leaking diodes. Set the DMM to the milli-amp scale and connect the test leads in series with the AC generator output terminal and the wire (**Figure 7-28**). With the engine off, read the current flow. Current flow is typically 0.5 mA, but may be as high as 2 mA. If the current is above specifications, a diode is leaking.

⚙️ **SERVICE TIP** A leaking diode can result in a dead battery. Consider this when diagnosing a battery drain issue.

Figure 7-28 Using a DMM to test for faulty diodes.

Classroom Manual
Chapter 7, page 212

Special Tools

Ammeter

Inductive amp clamp

Charging System Requirement Test

It's possible to have a properly operating charging system, yet it does not meet the vehicle's electrical system requirements. If the installed AC generator does not supply sufficient current output to meet the demands of the vehicle, the customer may have complaints that are identical to those of a charging system that's not functioning at all. The actual AC generator output should be 10% to 20% greater than the load demand. The charging system requirement test determines the total electrical demand of the vehicle's electrical system.

To determine the vehicle's electrical requirement:

1. Make sure the ammeter reads zero.
2. Clamp the inductive pickup around all of the negative battery cables.
3. Turn the ignition switch to the RUN position. Do not start the engine.
4. Turn on all accessories to their highest positions.
5. Read the ammeter. The indicated amperage is the total load demand of the vehicle.

Using Computerized Charging System Testers

Due to all of the different methods manufacturers use for charging system operation and regulation, diagnostic equipment suppliers have developed several different computerized tools to assist the technician in properly diagnosing charging system issues. These testers perform the current output test as part of a test series of the system and then makes a determination of the results. Once set up, the tester will guide the technician through the required procedures. The tester may require periodic updating of its database so current procedures and specifications are available.

To use the GR8 tester, follow **Photo Sequence 12**. Be sure to always reference the tester's instruction manual for the tester model you are using.

PHOTO SEQUENCE 12
Using a Computerized Charging System Analyzer

P12-1 Power on the tester and wait for the selection screen to appear.

P12-2 Proceed to the Charging System Test.

P12-3 As the test progresses, the screen will notify you of what test leads need to be connected.

P12-4 During the "CHECKING FOR ALTERNATOR OUTPUT" test, the analyzer tests the output voltage of the AC generator.

P12-5 Directions are provided that guide you through the test. After completing a directive, press the NEXT key to continue.

P12-6 Continue to follow the instructions. In this step, gradually increase the RPM until the tester screen indicates to maintain the engine speed.

PHOTO SEQUENCE 12 (CONTINUED)

P12-7 While maintaining the engine speed, the tester records the measurements.

P12-8 If the test was successful, the tester will have you move into the next phase by pressing the NEXT key.

P12-9 The next test analyzes the AC generator functions at idle with no load. Also, during this test, the tester looks for diode ripple to determine the condition of the diodes. Data gathered during this time will be compared to data gathered in the following tests.

P12-10 You will be instructed to put a load on the system by turning on the headlights and heater blower motor to high speed while the engine is still idling. This test determines if the AC generator can meet the electrical demands.

P12-11 Next, the tester will test the AC generator output at higher engine speeds while still having the same loads as in the previous step. Gradually increase the engine speed until the screen displays that the RPM should be held steady.

P12-12 Once again the tester will gather data. Hold the engine speed until the tester finishes compiling the data.

P12-13 Finally, the tester will indicate it's analyzing the retrieved data.

P12-14 Once the analyzing process is complete, you are instructed to turn off all electrical loads and shut off the engine.

P12-15 The test result will be displayed on three different screens. The first screen shows load and no-load test results.

P12-16 The second screen shows a graph of the voltage range under load and no-load test conditions.

P12-17 The third screen will show the results of the diode test.

P12-18 The test results can be printed off using the PRINT button.

If the test results for the charging system indicate that there may be a problem, the tester can perform a voltage drop test. The cable drop test uses conductance to send current through the circuit. This allows for performing the test without starting the engine or loading the circuit. The voltage drop is calculated on the circuit's positive (+) and negative (−) sides, along with the total voltage drop. The amperage range for performing the cable drop test is 0 to 1,000 amps. The test requires two test lead connections. The tester's battery test leads are connected to the AC generator's output terminal and housing. The DMM test leads are connected across the battery.

To perform the cable drop test:

1. Select the Cable Drop Test icon in the Main Menu and follow the instructions on the display.
2. Select "ALT Circuit" function and press the NEXT key to continue.
3. Use the arrow keys to select the rated amperage of the AC generator and press the NEXT key to continue.
4. Connect the positive (+) clamp of the battery test leads to the AC generator output terminal.
5. Connect the negative (−) clamp to the AC generator housing.
6. Connect the positive (+) DMM clamp to the battery's positive (+) post.
7. Connect the negative clamp (−) to the battery's (−) negative post.

When the test starts, the GR8 screen will indicate that it's testing the circuit. The result screen indicates either PASS, CLEAN AND RETEST, or REPLACE.

Diagnosing the Smart Charging System

The smart charging system may set DTCs if there is a failure within the system. The system may default to a set output voltage depending on the failure.

A common occurrence is a customer bringing their vehicle to the repair facility with the charging system warning light illuminated. When measuring the generator output voltage, the reading is a steady 13.5 or 13.8 volts. This is a good indication that a charging system fault has occurred, and the charging system has gone into a default mode. Check for stored and active DTCs. If this is not possible, or if the fault codes do not isolate the cause of the problem, use systematic testing procedures to manually test the charging system.

Use the scan tool to monitor the charging system data PIDs **(Figure 7-29)**. Since many smart charging systems use more than one module, you will need to determine which module receives the information to make charging system decisions and which module is the driver. You may need to switch between these modules to gather all the necessary information.

Confirm battery voltage and goal voltage match each other. Although the actual voltage should switch over the desired voltage set point, if the actual voltage data PID fluctuates erratically, check the battery terminal connections for corrosion and tightness.

Also, confirm battery temperature input is correct. An inaccurate BTS reading can cause under or overcharging. You will also need to confirm the operation of the battery current sensor.

Observe the battery current sensor and battery voltage data PIDs **(Figure 7-30)**. A positive value from the current sensor indicates the generator is charging the battery. A negative value means the battery is supplying the electrical power for the vehicle's accessories. However, this does not automatically indicate an issue. The smart charging system can operate under different modes that turn off AC generator output. For example, the "Fuel Economy Mode" will turn off the generator. In this mode, generator output does not increase with load. If not aware of the operating conditions, you may mistakenly think the charging system is not operating properly.

Classroom Manual
Chapter 7, page 206

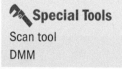
Special Tools
Scan tool
DMM

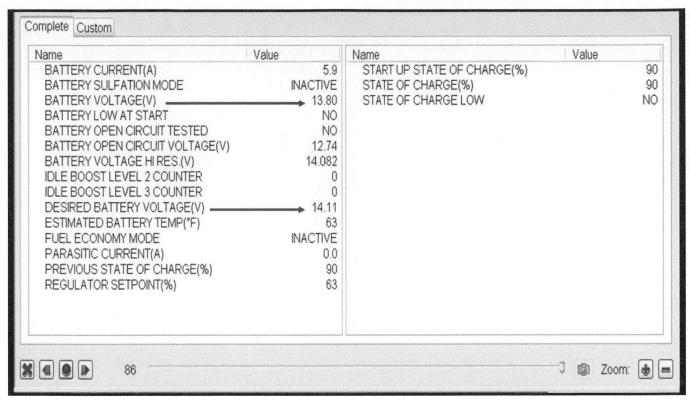

Figure 7-29 Scan tool data PIDs for the charging system.

Figure 7-30 Data PIDs showing the battery current sensor reading.

The battery current sensor may communicate its data over the LIN bus. Communication errors will result in the charging system going into default mode. The best way to test the LIN bus is to use a lab scope **(Figure 7-31)**. This test is also valid for the LIN bus used on interfaced generators.

When reading data PIDs, a pitfall to avoid is that the value displayed is what the module is attempting to do, not necessarily what it's accomplishing. Also, input values may be substituted in an attempt to achieve a desired outcome. For example, the data for a GM vehicle under full load may show the "F-terminal Signal" as 100% and the "L-terminal Command" as 65%. If the battery voltage is less than 12.6 volts, the charging system is not replenishing the battery. However, according to the data PIDs, it should be. This will require confirmation of charging system operation.

Figure 7-31 Lab scope waveform of the communication on the LIN bus.

First, you will need to reference the wiring diagrams to determine how the system works, terminal identification, and circuit identification. You can use a lab scope or a DMM to measure the duty cycle commands from the control module to the generator. If using a DMM, set it to the duty cycle function.

Select the proper trigger type for the circuit and connect the positive test lead to the "Command" or "Signal" circuits at the generator. Connect the negative test lead to the battery negative terminal.

While observing the duty cycle, increase the electrical loads by turning on accessories. Changes in electrical loads should change the duty cycle signals and commands at the generator. Lower electrical loads should have a lower duty cycle command. Higher electrical loads should have a higher duty cycle command.

In the previous example, the signal duty cycle reading with the lab scope or DMM was 0%. Normal signal duty cycle is about 35% at low load and increases as the load increases. In this example, the battery voltage failed to increase. This results in the substitution of the signal duty cycle value with progressively higher duty cycle values in an attempt to increase generator output.

Classroom Manual
Chapter 7, page 210

The smart charging system can determine the SOC of the battery and if an issue is preventing the charging system from maintaining an 80% SOC. If battery condition or charging system issues prevent the charging system from meeting the electrical demands, a **load shedding** request signal is broadcasted over the bus network, and certain electrical accessories will be turned off.

Load shedding events may be recorded, and associated information may be retrieved with a scan tool (**Figure 7-32**). The events may be stored in the control module of the affected accessory system(s), in one of the modules used for charging system control, or in a special scan tool menu. The information provided may help diagnose what the customer perceives as an intermittent issue with an accessory.

AUTHOR'S NOTE Although performing battery registration and programming is the final step in replacing some batteries, this procedure is included in this chapter since it's more a function of the charging system than the battery.

/c/Battery registration and programming. Replacing a battery may necessitate performing a battery registration and/or programming process. This procedure clears previously-stored battery statistics (battery charge level, current, voltage, temperature),

A/C DISENGAGE 1 HISTORY	RPM Unstable
A/C DISENGAGE 2 HISTORY	Battery Voltage
A/C DISENGAGE 3 HISTORY	RPM Unstable
A/C DISENGAGE 4 HISTORY	Battery Voltage
A/C DISENGAGE 5 HISTORY	RPM Unstable
A/C DISENGAGE 6 HISTORY	Battery Voltage
A/C DISENGAGE 7 HISTORY	RPM Unstable
A/C DISENGAGE 8 HISTORY	Battery Voltage

Figure 7-32 Systems affected due to load shedding will typically record the event.

resets the battery capacity to 80%, stores the current odometer reading, stores the readings of previous battery replacements, initializes power management functions, and allows battery capacity and type retrofit. Failure to perform the registration process may result in early battery failure. A properly equipped scan tool or battery tester can perform battery registration.

The charging system of some vehicles can adapt its strategies and self-learn that a new battery has been installed without performing the registration. This process works if the replacement battery is of the same type, rating, and size as the old battery. However, once a weak battery has been detected on some vehicles, many accessory systems will not clear the error messages and will not resume operation until completion of the battery registration process.

After installing a different battery type or rating, failure to complete the registration process may cause early battery failure. The modules that control the charging system must know the specifications of the new battery to modify the charging algorithm. For example, installing an AGM battery in place of the original flooded battery may lead to battery overcharging since the charging system will continue to use the flooded battery charging strategy. In this case, battery programming is necessary. This involves inputting a code on the battery sticker. Aftermarket batteries may not be coded, so a code may need to be generated with a scan tool or tester that supports this function. If you cannot perform this procedure, replace the battery with the same type as OEM.

AC Generator Removal and Replacement

AC generator removal varies according to the engine size, engine placement, and vehicle accessories (such as power steering and air-conditioning). The following is a typical procedure for removal and replacement of the AC generator:

1. Place fender covers over the fenders.
2. Disconnect the battery ground cable.
3. Disconnect the wiring harness connections to the AC generator.
4. Loosen the drive belt tensioner.
5. Remove the drive belt.
6. Remove the upper bolt that attaches the AC generator to the mounting bracket.
7. Remove the generator lower bolt while supporting the generator.
8. Remove the generator from the vehicle.

 Special Tools

Fender covers
Pry bar
Belt tension gauge
Battery terminal
  puller

⚠**WARNING** Never attempt to remove the AC generator or disconnect any wires to the generator without first disconnecting the battery's negative cable. Always wear safety glasses when working around the battery.

Reverse the removal procedure to install the AC generator. Slow charge the battery to ensure it's completely charged before starting the engine.

Belt Tension Adjustment

Classroom Manual
Chapter 7, page 184

🔧 **Special Tools**

Belt tension gauge
Belt tension tool kit

⚠ **Caution**

Do not increase the belt tension higher than specified by the manufacturer. Early AC generator bearing failure can occur if the belt tension is too tight.

Manufacturers use various methods to adjust the belt tension. However, regardless of the procedure, it's important to confirm you are installing the correct belt and that it's installed over the pulleys properly. The belt must be the proper length, width, depth, and type. There are two commonly used types of belts. The first is the *"V"* belt (**Figure 7-33**). This belt fits into a V-shaped groove in the pulleys. For proper operation, the belt must fit into the pulley so it makes full contact with the pulley (**Figure 7-34**).

Most manufacturers now use a serpentine belt (**Figure 7-35**). This belt has a series of ribs around the inside that matches grooves on the pulleys. Make sure the bibs fit properly to the grooves of the pulleys (**Figure 7-36**).

Belt tension adjustment on some engine packages is at the generator mounting bracket (**Figure 7-37**). To adjust the belt tension, leave the pivot and adjusting arm bolts loose. Look up the correct belt tension specification for the vehicle you are working on. Install

Figure 7-33 "V" belt.

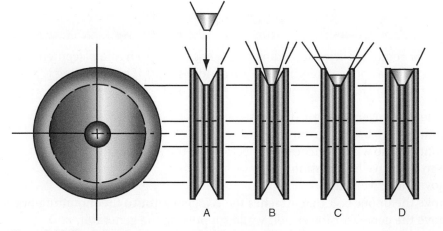

Figure 7-34 Proper installation and fit of the V belt are shown in D.

Figure 7-35 Serpentine belt.

Correct   Incorrect   Incorrect

Figure 7-36 Proper installation of the serpentine belt.

Square opening

AC generator

Figure 7-37 A typical method of adjusting belt tension at the brackets of the generator.

a belt tension gauge (refer to Figure 7-4) on the belt and use a square drive ratchet fitted into the square hole to rotate the generator until the belt tension gauge indicates the proper reading. If there is not a square hole or another method of moving the generator, apply pressure with a pry bar to the front housing of the AC generator only. Once the belt has the correct tension, tighten the bolts to the specified torque value. Manufacturers may provide different specifications for new and used belts. A belt is considered used if it has more than 15 minutes of run time.

Another method of adjusting the belt tension at the generator mounting brackets incorporates an adjusting screw (**Figure 7-38**). This eliminates the need to pry against the generator housing and provides an easier adjustment method. The procedure is the same as just described, but turn the adjusting screw until proper belt tension is obtained, and then tighten the pivot bolts to the correct torque.

⚠ Caution

Do not pry on the rear housing of the generator. It may crack the housing.

Adjusting screw for
belt tensioning

Figure 7-38 Some generators are fitted with an adjustment screw to
adjust the belt tension.

Many newer engines rigidly mount the generator to the bracket or engine block. Since the generator will not move, an idler pulley provides the means to adjust the belt tension. Many engines will have a pivot bolt and an adjusting slot (**Figure 7-39**). Some provide a fitting hole for a ratchet handle to fit into to provide an easy method of applying tension to the belt.

Most current engines use an automatic belt tensioner (**Figure 7-40**). This system has a tensioner that uses a spring or hydraulic pressure to maintain the proper belt tension automatically. The automatic tensioner maintains belt tension throughout the life of the belt. Using an automatic belt tensioner system increases belt life and prevents bearing damage to driven components. A tab or boss on the back of the canister fits into a hole in the bracket or engine block. This provides an anchor point. The spring

Pulley

Wrench
fittings

Pivot
bolt

Adjusting
slot

Lock bolt

Figure 7-39 The idler pulley may be the adjustment
point.

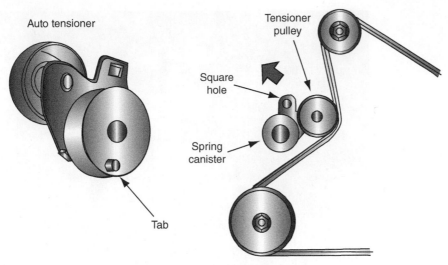

Figure 7-40 Most vehicles use an automatic belt tensioner.

Figure 7-41 Belt tensioner tools used to remove and install the belt on engines that use an automatic tensioner.

maintains a constant pressure that rotates the pulley assembly into the belt. The square hole is for using a break-over bar to rotate the tensioner away from the belt during belt removal and installation. Special tools are available for use on different types of automatic tensioners (**Figure 7-41**).

Some automatic tensioners have a built-in wear gauge (**Figure 7-42**). As the belt wears and stretches, the tensioner will continue to maintain the proper tension as it moves into the belt. As the belt wears, the tensioner will move, and the wear indicator marks align. The alignment of the marks indicates that the belt requires replacement.

Some manufacturers give belt tension specifications in frequency. A frequency-measuring tool equipped with a special microphone probe measures the frequency of the belt to determine proper tension. Place the microphone probe about 1 inch (25 mm) from the belt in the center of the span between the pulleys. Using your finger, pluck the belt a minimum of three times. The tool will display the frequency in hertz (Hz). Adjust the belt until the proper frequency is within specifications.

Photo Sequence 13 illustrates a typical procedure for inspecting and replacing a serpentine belt.

Another belt style is the "stretch belt." This style of belt does not have an automatic tensioner (or tensioner of any type). The design of the elastic core of the stretch belt keeps tension on the pulleys.

Figure 7-42 Some automatic tension adjusters have a wear indicator.

PHOTO SEQUENCE 13
Typical Procedure for Inspecting and Replacing a Serpentine Belt

P13-1 Visually inspect both sides of the belt.

P13-2 Replace the belt if it's glazed.

P13-3 Check for cracking and tearing of the belt ribs.

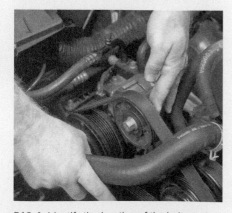

P13-4 Identify the location of the belt tension adjustment.

P13-5 If needed, loosen the mounting bracket bolt for the idler or generator.

P13-6 Pry the idler pulley or generator to release the tension on the belt and remove the belt.

PHOTO SEQUENCE 13 (CONTINUED)

P13-7 Confirm that the new belt is the same size as the old one.

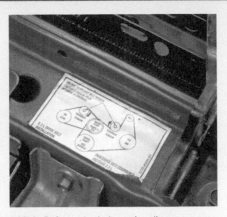

P13-8 Refer to the belt routing diagram located on a placard in the engine compartment or in the service information.

P13-9 Install the new belt over each of the pulleys following the recommended sequence.

P13-10 Pry out the idler or generator pulley to put tension against the belt.

P13-11 Confirm that the belt is properly located on each pulley.

P13-12 Measure the deflection of the belt or use a gauge as recommended by the manufacturer.

P13-13 Pry the idler pulley or generator to tighten the belt tension to specifications.

P13-14 Tighten the fasteners to the proper torque specifications.

P13-15 Start the engine and confirm proper belt operation and tracking.

Inspection of this type of belt is much like the serpentine belt, where the groove depth is the primary factor. The main difference is the procedures for removing and installing a new belt.

> **SERVICE TIP** If you are not going to reuse the old stretch belt, simply cut it to remove it. When installing a stretch belt, the belt is stretched over the largest pulley last.

⚠ **Caution**

Never use a screwdriver or other sharp object to install the belt. Cutting the belt or damaging the grooves will cause premature failure of the belt.

Many stretch belts can be removed and installed without special tools. If reusing the belt, the old belt may be "walked" off the smallest pulley by hand while the crankshaft is rotated clockwise **(Figure 7-43)**. During belt installation, the belt is held in place on the crankshaft pulley with a nylon wire tie **(Figure 7-44)**. Rotate the crankshaft pulley clockwise until the belt is positioned on the pulley. Continue to rotate the pulley until the wire tie breaks.

Some engine applications will require special tools to remove and install the belt **(Figure 7-45)**. **Photo Sequence 14** shows using a different set of tools to install the belt.

Figure 7-43 Some stretch belts are walked off the pulley while rotating the crankshaft.

Figure 7-44 Lock the belt to the pulley with a nylon wire tie.

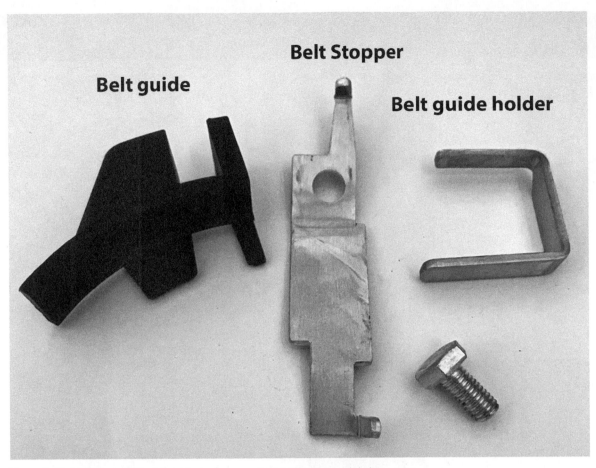

Figure 7-45 An example of the special tools necessary to install some stretch belts.

PHOTO SEQUENCE 14
Installing a Stretch Belt

P14-1 Remove any accessory belts necessary to access the stretch belt.

P14-2 Rotate the engine while feeding a tether strap between the pulley and the belt.

P14-3 With the strap fed ½ way around the pulley, pull on the end of the strap to remove the belt.

PHOTO SEQUENCE 14 (CONTINUED)

P14-4 Inspect the pulleys.

P14-5 Clean the pulleys to remove any debris in the grooves.

P14-6 Install the belt installation tool.

P14-7 Locate the belt around the crankshaft pulley, making sure it is fitted into the grooves. Locate the other end of the belt over the tool.

P14-8 Rotate the crankshaft counterclockwise to force the belt over the accessory's pulley.

P14-9 Continue to rotate the crankshaft pulley slowly while making sure the belt is sliding into place.

P14-10 Rotate the crankshaft pulley until the tool is free of the belt.

P14-11 Rotate the engine 360° and confirm the belt is properly seated in the accessory's pulley.

P14-12 Confirm the belt is properly seated into the crankshaft pulley.

AC Generator Disassembly

If the AC generator fails the previous tests, you must decide whether to rebuild or replace the AC generator. Several factors must be considered when making this decision:

1. What is best for the customer?
2. Shop policies.
3. Cost.
4. Time.
5. Type of AC generator.
6. Availability of parts

If the decision is made to disassemble the AC generator, study the manufacturer's service information and become familiar with the procedure for the particular AC generator being rebuilt. **Photo Sequence 15** shows the procedure for disassembling the Ford IAR AC generator.

With the AC generator disassembled, clean and inspect the components. Use a clean cloth to wipe the stator, rotor, and front bearing clean. Do not use solvent cleaners on these components. Inspect the front and rear bearings by rotating them on the rotor shaft. Check for noises, looseness, or roughness. Replace the defective bearing if any of these conditions are present.

Special Tool

Ohmmeter
100-watt soldering iron
Resin core solder
Arbor press
Soft jaw vise
Scribe
High-temperature bearing grease
400-grain emery cloth
Pulley puller
Clean rags
Heat sink grease

PHOTO SEQUENCE 15
Typical Procedure for IAR Generator Disassembly

P15-1 Always have a clean and organized work area. Tools required to disassemble the Ford IAR AC generator: rags, T20 TORX wrench, plastic hammer, arbor press, 100-watt soldering iron, soft jaw vise, safety glasses, and assorted nut drivers.

P15-2 Using a T20 TORX, remove the four attaching screws that hold the regulator to the AC generator rear housing.

P15-3 Remove the regulator and brush assembly as one unit.

P15-4 Using a T20 TORX, remove the two screws that attach the regulator to the brush holder. Separate the regulator from the brush holder. Remove the A terminal insulator from the regulator.

P15-5 Scribe or mark the two housing ends and the stator core for reference during assembly.

P15-6 Remove the three through bolts that attach the two housings.

PHOTO SEQUENCE 15 (CONTINUED)

P15-7 Separate the front housing from the rear housing. The rotor will come out with the front housing, while the stator will stay with the rear housing. NOTE: It may be necessary to tap the front housing with a plastic or dead weight hammer to get the halves to separate.

P15-8 Separate the three stator lead terminals from the rectifier bridge.

P15-9 Remove the stator coil from the housing.

P15-10 Using a T20 TORX, remove the four attaching bolts that hold the rectifier bridge.

P15-11 Remove the rectifier bridge from the housing.

P15-12 Use a socket to tap out the bearing from the housing.

P15-13 Clamp the rotor in a soft jaw vise.

P15-14 Remove the pulley-attaching nut, flat washer, drive pulley, fan, and fan spacer from the rotor shaft.

P15-15 Separate the front housing from the rotor. If the stop ring is damaged, remove it from the rotor; if not, leave it on the rotor shaft.

P15-16 Remove the three screws that hold the bearing retainer to the front housing.

P15-17 Remove the bearing retainer.

P15-18 Remove the front bearing from the housing. NOTE: It may be necessary to use an arbor press to remove the bearing if it does not slide out.

The rotor shaft rear bearing surface should be smooth. If the surface is not smooth, replace the rotor. Visual inspection of the rotor includes checking the slip rings for smoothness and roundness. Rings that are discolored, dirty, scratched, nicked, or have burrs, may be cleaned with fine-grit emery cloth. Be careful not to create any flat spots while polishing the slip rings.

Inspect the terminals and wire leads of the rotor and stator. Also, check both the rotor and the stator for signs of burnt insulation of the windings. If there is damage to the insulation, replace the component.

Inspect the housing halves for cracks. Also, check the fan and pulley for looseness on the rotor shaft and for cracks. Replace any part that does not pass inspection. Use a clean cloth to remove the heat transfer grease that's in the rectifier mounting area.

Inspect and test the AC generator's brushes anytime the unit is disassembled. Replace brushes worn shorter than 1/4 inch (6 mm) in length. Also, check the strength of the brush springs to assure they can maintain constant contact of the brushes with the slip rings. Test brush continuity with an ohmmeter. There should be zero resistance through the brush path. Replace the brushes if there is any resistance indicated.

Classroom Manual
Chapter 7, page 193

Classroom Manual
Chapter 7, page 186

Component Testing

Classroom Manual
Chapter 7, page 185

With the AC generator disassembled and cleaned, you can test the individual components. The chart in **Figure 7-46** illustrates the test connections and results for the major components.

Carefully inspect the rotor and stator windings for signs of discoloration or overheating. Often these components will look fine but will actually be damaged due to excessive heat. Use the blade of a knife and scrape the windings. If the coating or varnish flakes off, the windings have been overheated. Pay special attention to the connectors. If the rotor and stator pass the visual inspection, they need to be checked for opens and shorts to ground with an ohmmeter.

Use an ohmmeter to measure the resistance between the slip rings (**Figure 7-47**). Always check the manufacturer's service information for the correct specification for the unit you are working on.

 Special Tool
Ohmmeter

Component	Test Connection	Normal Reading	If Reading Was:	Trouble Is:
Rotor	Ohmmeter from slip ring to rotor shaft	Infinite resistance	Very low	Grounded
	Test lamp from slip ring to shaft	No light	Lamp light	Grounded
	Test lamp across slip rings	Lamp lights	No light	Open
Stator	Ohmmeter from any stator lead to frame	Infinite resistance	Very low	Grounded
	Test lamp from lead to frame	No light	Lamp lights	Grounded
	Ohmmeter across any pair of leads	Less than $1/2\Omega$	Any very high reading	Open
Diodes	Ohmmeter across diode, then reverse leads	Low reading one way; high reading other way	Both readings low Both readings high	Shorted Open
	12-V test lamp across diode, then reverse leads	Lamp lights one way, but not other way	No light either way Lamp lights both ways	Open Shorted

Figure 7-46 Guidelines for bench testing a generator.

Figure 7-47 Test connections for checking for opens in the rotor windings.

Figure 7-48 Testing the rotor for shorts to ground.

Next, connect the ohmmeter to measure continuity between each of the slip rings to the rotor shaft (**Figure 7-48**). The ohmmeter should show infinite resistance. If any of the ohmmeter test results are not within specifications, replace the rotor.

To test the stator for an open, connect the ohmmeter test leads to any pair of stator leads (**Figure 7-49**). Continue to test all possible combinations of pair connections to the stator. If the ohmmeter reads infinity between any two leads, a winding is open and the stator requires replacement.

To test the stator for a short to ground, connect the ohmmeter to the stator leads (one at a time) and the stator frame (**Figure 7-50**). The ohmmeter should read infinity on all three stator leads. If the reading is less than infinity, the stator is shorted to ground and must be replaced.

Classroom Manual
Chapter 7, page 187

Figure 7-49 Testing the stator for opens.

Figure 7-50 Testing the stator for shorts to ground.

Generator rectifier bridge

Generator housing

Insulated heat sink

Generator capacitor

Figure 7-51 Testing the rectifier bridge.

Test the diodes of the rectifier bridge using the diode test function of the DMM. Connect the test leads to the diode lead and the case (**Figure 7-51**). If the diode is good, it will show high resistance in one direction and about 600 mV in the opposite direction. Test all six diodes and replace any that are defective.

Remove the stator leads from the diodes before testing the stator windings.

Classroom Manual
Chapter 7, page 189

⚙ **SERVICE TIP** A shorted stator is difficult to test for because the resistance is very low for a normal stator. If all other components test okay, but the output was low, a shorted stator is the probable cause.

If the AC generator is equipped with a diode trio, test it for opens and shorts. The procedure is much the same as with the rectifier bridge test. With the DMM set to test diodes, connect one of the DMM test leads to the signal connector. Connect the other test

Classroom Manual
Chapter 7, page 194

Three connectors

Single connector

Figure 7-52 Diode trio test connections.

Generator brush

Brush holder

Brush retaining pin

Figure 7-53 When assembling a generator, use a pin to hold the brushes back so the rotor can fit into the brush holder.

lead to one of the three connectors (**Figure 7-52**). Reverse the test leads and retest. The DMM should read "OL" in one direction and above 600 mV in the other direction on all three connections. If not, replace the diode trio.

AC Generator Reassembly

The reassembly procedure is basically the reverse order of the disassembly. The following are suggestions to assist in the assembly process:

1. Always check the manufacturer's specifications for proper torque values of the attaching screws.
2. Use high-temperature grease on the bearings.
3. Check for free rotor rotation after installing and torquing the pulley to the rotor shaft.
4. Apply heat sink grease across the rectifier bridge base plate.
5. Protect the diodes from excess heat while soldering the connections. Use a pair of needle nose pliers as a heat sink.

When assembling a generator, always follow the recommendations of the manufacturer. It's critical that all screws and bolts that used insulating washers before disassembly have them installed during assembly. These insulators maintain proper circuit polarity. If a washer is left out, a short circuit will exist.

To aid in installing the rotor into the brushes, most generators have a hole that allows a pin or paper clip to be inserted into the brush holder to keep the brushes back and allow the rotor to fit into them (**Figure 7-53**). Before installing the pin, confirm the brush springs are in the proper position behind the brushes. After the rotor is in place, remove the pin, and the brushes will snap into place on the slip rings.

HEV Inverter/Converter

⚠️WARNING **Follow all safety precautions associated with working on or around the high-voltage system. Failure to follow all procedures may result in death.**

The HEV uses an inverter to invert the DC voltage from the HV battery into three-phase AC voltage for use by the motors. The inverter also rectifies the AC voltage from the motors when they are functioning as generators to charge the DC HV battery. The converter is a bidirectional DC/DC converter that changes the HV DC to LV DC to recharge the 12-volt battery.

Failures of these components to charge the batteries usually result in a DTC setting. This means that most of the diagnostics focus on the trouble code(s) retrieved and following the steps outlined by the manufacturer. Keep in mind that you need to confirm the condition of the HV cables, HV connections, LV connections, and circuits as in any system. In most cases, if there is no fault found in the cable or connections, the inverter or converter is replaced.

Classroom Manual
Chapter 7, page 217

Special Tools
Isolation meter
Class III DMM
Isolation gloves

CASE STUDY

A customer states that the vehicle's engine dies and requires a jump-start. The engine runs for a few minutes and then dies again. Turning on the headlights while the engine is running causes the engine to die immediately.

The technician boost starts the engine to confirm the customer's complaint. The engine dies just as reported. The technician slow charges the battery to full charge and then prepares to perform a battery test series. The battery passes all tests.

While performing the visual inspection of the charging system the technician finds the AC generator belt is worn and glazed. The technician does not know for certain yet if the belt is the only problem. However, the charging system tests will not be accurate if the belt is worn. The customer is informed of what the technician has found thus far, and that other tests will still need to be performed to confirm any other problems. The technician gives the customer an estimate for the belt and receives permission to replace it.

After installing the new belt, the technician performs the voltage and current output tests. The technician also checks the diode pattern. In addition, they perform the requirement test to confirm that the AC generator is the correct size. The charging system passed all tests.

The customer is notified that the car is ready to be picked up. When the customer arrives, the technician tactfully reminds the customer that all belts should be checked every six months and replaced per the manufacturer's maintenance schedule.

CASE STUDY

A customer states that the speedometer will read about 20 mph (32.2 kph) when the vehicle is stationary. Another shop has attempted to repair this condition by replacing the transmission output speed sensor, the PCM, and the underhood wiring harness. The technician tries to duplicate the issue and notices that the problem only occurs if the transmission is in any gear other than PARK or NEUTRAL.

Starting with the basics, the technician tests the battery and charging system. While monitoring the voltage across the battery terminals with their DMM, the charging system is loaded. Good AC generator voltage output is noted. The technician sets the DMM to read AC voltage and turns on the min/max function to capture the voltage readings. The DMM positive test lead is moved to the generator's output terminal. The low voltage is −2.87 volts, and the high voltage is 2.11 volts. Realizing that the voltage swing is too great, the technician determines that a diode is faulty.

With the customer's approval, the technician replaced the AC generator and set the belt to the proper tension. The retest shows that the new generator eliminated the voltage swing. Also, the speedometer works properly now. The vehicle was returned to a very happy customer.

The AC voltage was "tricking" the PCM into thinking the vehicle was going 20 mph (32.2 kph). By performing the basic tests first, the technician quickly identified the root cause of the issue.

ASE-Style Review Questions

1. Charging system testing is being discussed.
   Technician A says before attempting to test the charging system, the battery must be checked.
   Technician B says the battery's state of charge is not a concern to charging system testing.
   Who is correct?
   A. A only
   B. B only
   C. Both A and B
   D. Neither A nor B

2. AC generator noise complaints are being discussed.
   Technician A says a loose belt will make a grumbling noise.
   Technician B says a shorted diode can cause a whining noise.
   Who is correct?
   A. A only
   B. B only
   C. Both A and B
   D. Neither A nor B

3. *Technician A* says the voltage output test quickly determines whether the charging system is working properly.
   Technician B says when testing a charging system, the first step is to perform a visual inspection and preliminary checks of the charging system.
   Who is correct?
   A. A only
   B. B only
   C. Both A and B
   D. Neither A nor B

4. Test results of the voltage output test are being discussed.
   Technician A says if the charging voltage is too high, there may be a loose or glazed drive belt.
   Technician B says if the charging voltage is too low, the fault might be a grounded field wire from the regulator (full fielding).
   Who is correct?
   A. A only
   B. B only
   C. Both A and B
   D. Neither A nor B

5. Voltage drop testing is being discussed.
   Technician A says the total system drops should be less than 0.7 volt.
   Technician B says the ground side voltage drop should be less than 0.2 volt.
   Who is correct?
   A. A only
   B. B only
   C. Both A and B
   D. Neither A nor B

6. *Technician A* says the field current-draw test determines if current is available to the field windings.
   Technician B says a slipping belt can cause a low reading when performing the field current-draw test.
   Who is correct?
   A. A only
   B. B only
   C. Both A and B
   D. Neither A nor B

7. An ammeter connected in series between the AC generator output terminal and the output circuit wire reads about 4 mA with the engine off.
   Technician A says the field winding is shorted.
   Technician B says this indicates a leaking diode.
   Who is correct?
   A. A only
   B. B only
   C. Both A and B
   D. Neither A nor B

8. *Technician A* says full fielding means the field windings are constantly energized with full battery voltage.
   Technician B says if full fielding with the regulator bypassed produces lower-than-specified output, the regulator is the cause of the problem.
   Who is correct?
   A. A only
   B. B only
   C. Both A and B
   D. Neither A nor B

9. *Technician A* says the smart charging system may have a default generator output voltage if a fault is detected.
   Technician B says a DMM set to read AC voltage at the output terminal of the AC generator should read less than 0.250 VAC.
   Who is correct?
   A. A only
   B. B only
   C. Both A and B
   D. Neither A nor B

10. The results of current output testing are being discussed.
    Technician A says an open diode can result in an output that's 2 to 8 amps below specifications.
    Technician B says a current output reading of 10 to 15 amps below specifications indicates a possible slipping belt.
    Who is correct?
    A. A only
    B. B only
    C. Both A and B
    D. Neither A nor B

1. A vehicle's battery discharges in a very short period of time due to a shorted battery cell. Before replacing the battery, a charging system test is performed. The engine is started, and all accessories are turned off.

   Technician A says that the charging system's amperage output will be lower than normal.

   Technician B says that the generator's field current will be high when the engine is running.

   Who is correct?

   A. A only   C. Both A and B
   B. B only   D. Neither A nor B

2. The desired charging voltage PID shows 14.3 volts. The battery voltage PID is reading erratic voltages that exceed 16.2 volts.

   Technician A says the battery terminals may be corroded.

   Technician B says the drive belt may be glazed.

   Who is correct?

   A. A only   C. Both A and B
   B. B only   D. Neither A nor B

3. Charging system voltage is being measured at two places at the same time with the engine running. Connecting the voltmeter's leads across the battery terminals results in a 12.8-volt reading, while connecting the voltmeter to the AC generator output terminal and the AC generator case results in a 14.2-volt reading.

   Technician A says that there may be excessive resistance on the ground side of the charging system.

   Technician B says that there may be excessive resistance on the positive side of the charging system.

   Who is correct?

   A. A only   C. Both A and B
   B. B only   D. Neither A nor B

4. A customer says that their car's battery completely discharges every few days even though they drive the car about 50 miles (80.5 kilometers) every day. Testing with all accessories turned on showed the charging system voltage is 12.4 volts, and amperage is 60 amps at 1,300 RPM. A replacement generator produces the same measurements.

   Technician A says that there may be excessive electrical demand on the generator.

   Technician B says that the charging system output circuit may have excessive voltage drop.

   Who is correct?

   A. A only   C. Both A and B
   B. B only   D. Neither A nor B

5. Which of the following could result in battery overheating and eventual premature failure?

   A. Open field circuit.
   B. Shorted stator windings.
   C. Open diode.
   D. High resistance in the voltage regulator sensing circuit.

Name _____   Date _____

INSPECTING DRIVE BELTS

Upon completion of this job sheet, you should be able to visually inspect drive belts and check their tension.

ASE Education Foundation Correlation

This job sheet addresses the following **MLR** task:

D.2.   Inspect, adjust, and/or replace generator (alternator) drive belts; check pulleys and tensioners for wear; check pulley and belt alignment. **(P-1)**

This job sheet addresses the following **AST/MAST** task:

D.2.   Inspect, adjust, and/or replace generator (alternator) drive belts; check pulleys and tensioners for wear; check pulley and belt alignment; determine needed action. **(P-1)**

We Support
ASE | Education Foundation

Tools and Materials

- Two vehicles, one with a serpentine belt and one with V belts
- Service information for the vehicles assigned
- Belt wear gauge
- Belt tension gauge
- Belt tensioner tools

Safety Considerations and PPE

- Safety glasses

Procedure

1. On the vehicle with a serpentine belt, carefully inspect the belt and describe the general condition of the belt.

   _____

   _____

   _____

2. Use the belt wear gauge to determine if the belt is excessively worn. What is your determination?

   _____

   _____

   _____

3. Check the alignment of all pulleys that the belt uses. What are your recommendations?

   _____

   _____

   _____

4. Check the condition of all pulleys. What are your recommendations?

   _____

   _____

   _____

5. With the proper belt tension gauge, check the tension of the belt. Belt tension specifications _____ .

Actual _____

6. Based on the above, what are your recommendations?

_____

_____

_____

7. Describe the procedure for adjusting the tension of the belt.

_____

_____

_____

8. Set the belt tension to the proper specifications.

9. On the vehicle with V belts, you will find more than one drive belt. List the different belts by their purpose.

_____

_____

_____

10. Carefully inspect the belts and describe the general condition of each one.

_____

_____

_____

11. Check the alignment of all pulleys that the belt(s) uses. What are your recommendations?

_____

_____

_____

12. Check the condition of all pulleys. What are your recommendations?

_____

_____

_____

13. Check the tension of the AC generator drive belt. Belt tension should be _____. You found _____

14. Based on the above, what are your recommendations?

_____

_____

_____

15. Describe the procedure for adjusting the tension of the belt.

_____

_____

_____

16. Set the belt tension to the proper specification.

_____

_____

_____

Instructor's Evaluation:

Grade	Rubric
0	No exposure due to: ☐ Absence ☐ Other:
1	Student unsuccessfully attempted the task.
2	Student completed the task with significant instructor assistance.
3	Satisfactory performance of task. The student was able to perform elements of the task with minimal instructor assistance.
4	Competent/Proficient. Able to perform all elements of the task successfully and independently.

Instructor's Notes:

Name _____   Date _____

TESTING CHARGING SYSTEM OUTPUT

Upon completion of this job sheet, you should be able to measure the output of the charging system.

ASE Education Foundation Correlation

This job sheet addresses the following **MLR** task:

D.1.　　Perform charging system output test. **(P-1)**

This job sheet addresses the following **AST/MAST** tasks:

D.1.　　Perform charging system output test; determine needed action. **(P-1)**

D.5.　　Diagnose (troubleshoot) charging system for causes of undercharge, no-charge, or overcharge conditions; determine needed action. **(P-1)**

We Support
ASE | **Education Foundation**

Tools and Materials

- A vehicle
- Service information for the vehicle assigned
- DMM or lab scope
- Ammeter
- Inductive pickup
- Scan tool

Safety Considerations and PPE

- Safety glasses

Procedure

1. Identify the type and model of AC generator. What type is it?

   _____

   What are the output specifications for this AC generator?

   _____ amps and _____ volts at _____ RPM.

2. Make sure the knob is fully turned to the OFF position and connect the carbon pile across the battery terminals.

3. Set up the DMM to read battery voltage and connect the leads across the battery. Reading: _____

4. Attach the inductive pickup to the ammeter or DMM, zero the reading, and place the pickup around all ground cables from the battery.

   Ammeter reading with ignition off: _____

   Ammeter reading with ignition in RUN position: _____

5. Start the engine and run it at idle. With the headlights on and the blower motor fan on high, observe the output to the battery. The meter readings are:

   _____ amps

   _____ volts

6. Increase engine speed to the specified RPM while observing the meters. Record your readings:

_____ amps

_____ volts

7. Turn off the headlights and blower fan.

8. Increase engine speed to the specified RPM to measure maximum current output.

9. Slowly turn the knob of the carbon pile to increase the load until the highest ammeter reading possible is obtained without the battery voltage dropping below 13 volts. Turn the carbon pile OFF.

_____ amps

10. Compare the readings to specifications and give recommendations.

_____

_____

_____

_____

11. Are the readings below or above the specifications? _____

If below, go to step 12. If above, go to step 16.

12. Refer to the service information and list possible causes for the below specification results.

_____

_____

_____

13. If readings are below specifications, refer to the service information to find the proper way to full field the AC generator. Describe the method.

_____

_____

_____

_____

14. If possible, full field the generator and observe the output to the battery. The meter readings are:

_____ amps

_____ volts

15. Compare readings to specifications and give recommendations.

_____

_____

_____

16. If the readings were above specifications, refer to the service information and list what could be the causes:

_____

_____

_____

17. Describe what test procedure you would do to isolate the cause of the over specifications test results.

_____

_____

_____

Instructor's Evaluation:

Grade	Rubric
0	No exposure due to: ☐ Absence ☐ Other:
1	Student unsuccessfully attempted the task.
2	Student completed the task with significant instructor assistance.
3	Satisfactory performance of task. The student was able to perform elements of the task with minimal instructor assistance.
4	Competent/Proficient. Able to perform all elements of the task successfully and independently.

Instructor's Notes:

Name _____   Date _____

TESTING THE CHARGING SYSTEM CIRCUIT

Upon completion of this job sheet, you should be able to visually inspect and test the insulated and ground side charging system circuits.

ASE Education Foundation Correlation

This job sheet addresses the following **MLR** task:

D.4.   Perform charging circuit voltage drop tests. **(P-2)**

This job sheet addresses the following **AST/MAST** task:

D.4.   Perform charging circuit voltage drop tests; determine needed action. **(P-1)**

We Support
ASE | **Education Foundation**

Tools and Materials

• A vehicle
• Wiring diagram for the vehicle assigned
• A DMM

Safety Considerations and PPE

• Safety glasses

Procedure

1. Describe the general appearance of the AC generator and the wires that are attached to it.

   _____

   _____

   _____

2. Measure the open circuit voltage of the battery. Your measurement was
   _____ volts.

3. Using the wiring diagram, identify the output, input, and ground wires for the AC generator. Describe these wires by color and location.

   _____

   _____

   _____

   _____

4. Start the engine and allow it to run. Turn on the headlights in the high-beam mode.

5. Use the DMM to measure the voltage drop across the charging system's output wire. Record your readings.

   _____

   _____

6. Connect the DMM across the charging system's input wire. Measure the voltage drop and record your readings.

   _____

7. Connect the DMM across the charging system's ground wire. Measure the voltage drop and record your readings.

_____

8. What do the test results indicate?

_____

_____

_____

Instructor's Evaluation:

Grade	Rubric
0	No exposure due to: ☐ Absence ☐ Other:
1	Student unsuccessfully attempted the task.
2	Student completed the task with significant instructor assistance.
3	Satisfactory performance of task. The student was able to perform elements of the task with minimal instructor assistance.
4	Competent/Proficient. Able to perform all elements of the task successfully and independently.

Instructor's Notes:

_____

Name _____   Date _____

REMOVING AND INSTALLING THE AC GENERATOR

Upon completion of this job sheet, you should be able to properly remove, inspect, and replace the AC generator.

ASE Education Foundation Correlation

This job sheet addresses the following **MLR** task:

D.3.   Remove, inspect, and/or replace the generator (alternator). **(P-3)**

This job sheet addresses the following **AST** task:

D.3.   Remove, inspect, and/or replace the generator (alternator); determine needed action. **(P-2)**

This job sheet addresses the following **MAST** task:

D.3.   Remove, inspect, and/or replace the generator (alternator); determine needed action. **(P-1)**

We Support
ASE | Education Foundation

Tools and Materials
- A vehicle
- Service information
- Safety glasses
- Fender covers
- Belt wear gauge
- Belt tension gauge
- Belt tensioner tools

Safety Considerations and PPE
Safety glasses

⚠ **WARNING Never attempt to remove the AC generator or disconnect any wires to the generator without first disconnecting the battery's negative cable. Always wear safety glasses when working around the battery.**

Procedure

Note: It may be necessary to lift the vehicle on the hoist in order to access the ACV generator. Prior to raising the vehicle, disconnect/remove all components that can be accessed from the engine compartment.

_____

_____

_____

1. Place fender covers over the fenders.

2. Disconnect the battery ground cable.

3. Disconnect the wiring harness connections to the AC generator.

4. Describe the procedure for releasing drive belt tension.

_____

_____

_____

5. Remove the drive belt and inspect it as outlined in Job Sheet 36. Describe the condition of the belt:

_____

_____

_____

6. Remove the upper bolt that attaches the AC generator to the mounting bracket.

7. Remove the generator lower bolt while supporting the generator.

8. Remove the generator from the vehicle.

Inspect the AC generator for indications of overheating, damaged/loose terminals, worn bearings, damaged cooling fan, and damaged pulley. What are your findings?

_____

_____

_____

9. Check the condition of the belt drive, idler, and tension pulleys. Record your findings:

_____

_____

_____

10. Reverse the removal procedure to install the AC generator.

11. Locate the drive belt over the pulleys. Are the pulleys and belt properly aligned?   Yes ☐   No ☐

If no, consult with your instructor.

12. Adjust the belt tension as outlined in Job Sheet 36.

13. Connect all wires that attach to the AC generator.

14. Slow charge the battery to assure it's completely charged.

15. Connect the negative battery cable.

16. Check the output of the AC generator as outlined in Job Sheet 37. What are your results?

_____

_____

_____

_____

Instructor's Evaluation:

Grade	Rubric
0	No exposure due to: ☐ Absence ☐ Other:
1	Student unsuccessfully attempted the task.
2	Student completed the task with significant instructor assistance.
3	Satisfactory performance of task. The student was able to perform elements of the task with minimal instructor assistance.
4	Competent/Proficient. Able to perform all elements of the task successfully and independently.

Instructor's Notes:

DIAGNOSTIC CHART 7-1

PROBLEM AREA:	Charging system operation.
SYMPTOMS:	1. Battery is too low to start the engine. 2. Headlight illumination is dim. 3. Electrical accessories do not operate properly. 4. Charging indicator warning light illuminated.
POSSIBLE CAUSES:	1. Slipping or worn drive belt. 2. Poor battery cable connections. 3. Faulty voltage regulator. 4. Shorted stator. 5. Open stator. 6. Open diode. 7. Shorted diode. 8. Slipping or worn drive belt. 9. Worn brushes. 10. Open field coil circuit. 11. Shorted field coil. 12. Worn or slipping drive belt. 13. Worn brushes. 14. Faulty rectifier bridge. 15. Faulty diode trio. 16. Excessive resistance in system circuit. 17. Open in the circuit. 18. Improper ground connection. 19. Loose or corroded connection.

DIAGNOSTIC CHART 7-2

PROBLEM AREA:	Charging system operation (excessive charging).
SYMPTOMS:	1. Battery electrolyte level constantly low. 2. Bulbs burn out. 3. Brighter than normal bulb illumination. 4. Electrical system component failures.
POSSIBLE CAUSES:	1. Faulty voltage regulator. 2. Grounded field coil circuit. 3. Excessive resistance in sensing circuit. 4. Faulty generator.

DIAGNOSTIC CHART 7-3

PROBLEM AREA:	Voltage regulation.
SYMPTOMS:	1. Charging system overcharging the battery. 2. Charging system output below specifications.
POSSIBLE CAUSES:	1. Defective voltage regulator. 2. Open or short in sense circuit.

DIAGNOSTIC CHART 7-4

PROBLEM AREA:	Diodes.
SYMPTOMS:	**1.** Excessive noises. **2.** Low or no charging system output. **3.** AC generator fails output test.
POSSIBLE CAUSES:	**1.** Open diodes. **2.** Shorted diodes.

DIAGNOSTIC CHART 7-5

PROBLEM AREA:	AC generator pulleys.
SYMPTOMS:	**1.** Belts wear prematurely. **2.** Noises.
POSSIBLE CAUSES:	**1.** Pulley bent.

DIAGNOSTIC CHART 7-6

PROBLEM AREA:	AC generator.
SYMPTOMS:	**1.** Noises.
POSSIBLE CAUSES:	**1.** Bent fan blades. **2.** Worn or dry bearings.

DIAGNOSTIC CHART 7-7

PROBLEM AREA:	HEV 12-volt battery low SOC.
SYMPTOMS:	**1.** LV battery fails to charge. **2.** No HEV operation. **3.** MIL or HEV warning lamps.
POSSIBLE CAUSES:	**1.** 12-volt battery failure. **2.** Low SOC of 12-volt battery. **3.** High resistance in 12-volt battery cables or connections. **4.** Improper inverter/converter module ground connection. **5.** Poor or improper inverter/converter module LV connections. **6.** Excessive LV system loads. **7.** Faulty inverter/converter module.

DIAGNOSTIC CHART 7-8

PROBLEM AREA:	HEV 12-volt battery high SOC.
SYMPTOMS:	**1.** LV battery overcharged. **2.** No HEV operation. **3.** MIL or HEV warning lamps.
POSSIBLE CAUSES:	**1.** 12-volt battery failure. **2.** Improper jump-start procedures. **3.** Overcharging with a battery charger. **4.** High resistance in 12-volt battery cables or connections. **5.** Improper inverter/converter module ground connection. **6.** Poor or improper inverter/converter module LV connections. **7.** Faulty inverter/converter module.

CHAPTER 8

BODY COMPUTER SYSTEM DIAGNOSIS

Upon completion and review of this chapter, you should be able to:

- Describe the service precautions associated with servicing the BCM.
- Diagnose computer voltage supply and ground circuits.
- Explain the difference between hard and intermittent codes.
- Perform flash code retrieval on various vehicles.
- Erase fault codes.

- Perform a complete visual inspection of the problem system.
- Use a scan tool to enter BCM diagnostics.
- Perform basic actuator tests.
- Perform basic sensor tests.
- Properly flash the BCM.

Basic Tools

Basic mechanic's tool set

Service information

Terms to Know

Activation test

Breakout box

Diagnostic trouble code (DTC)

Electrostatic discharge (ESD) strap

Fail-safe

Flash

Flash codes

Hard codes

Impedance

Introduction

Because the body control module (BCM) operates many functions of the vehicle's electrical systems, it's important for today's technician to be able to properly diagnose problems with this system. Body computers have expanded to include climate control, lighting circuits, cruise control, antilock braking, electronic suspension systems, electronic shift transmissions, and alternator regulation. In some systems, the body controller function the direction lights, the rear window defogger, the illuminated entry, the intermittent wiper, and the antitheft systems (**Figure 8-1**).

Classroom Manual
Chapter 8, page 222

AUTHOR'S NOTE Many manufacturers are incorporating the functions of the typical BCM into other modules. For example, on many newer Chrysler vehicles the cabin compartment node (CCN) now performs the functions that were once the responsibility of the BCM. These vehicles no longer have a separate BCM. In this text, the term BCM references the stand-alone module and the functions contained within any other module.

Figure 8-1 The body controller operates many of the vehicle's electrical systems.

Classroom Manual
Chapter 8, page 223

A computer processes the physical conditions that represent information (data). The computer has four basic functions: input, processing, storage, and output. Understanding these four computer functions will help you organize the troubleshooting process. When testing a system you attempt to isolate a problem within one of these functions.

In the process of controlling the various electrical systems, the BCM continuously monitors operating conditions for possible system malfunctions. The computer compares system conditions against programmed parameters. The computer detects a malfunction if the conditions fall outside of these limits. A **diagnostic trouble code (DTC)** sets to indicate the portion of the system at fault. The technician can access this code for aid in troubleshooting.

If a malfunction results in improper system operation, the computer may minimize the effects by using **fail-safe** action. This provides limited system operation by substituting a fixed input value if a sensor circuit should fail. For example, if the automatic temperature control (ATC) system has a malfunction from the ambient temperature sensor, instead of shutting down the whole system, the computer will provide a fixed value as its input. This fixed value can be programmed into the computer's memory, or it can be the last received signal from the sensor prior to failure. This allows the system to operate on a limited basis instead of shutting down completely. Other faults may result in the ATC system switching to high fan speed, full heat, or defrost mode.

Some systems may substitute an invalid input value by referencing other inputs. The BCM uses these other inputs to calculate what the missing input value would be. For example, if the ambient air temperature sensor input is invalid, the BCM may look at the battery temperature sensor and the sun load sensor and calculate what the ambient temperature would be.

You need to know several things before accessing the computer's memory to gain information concerning system operation. You need to be familiar with what you're looking at and follow proper precautions when servicing these systems.

> **Fail-safe** action is commonly known as "limp-in mode" or "fail soft."

Electronic Service Precautions

The technician must take some precautions before servicing the body computer or any of its controlled systems. The BCM can withstand normal current draws associated with normal operation. However, overloading any of the system circuits will result in damage to the BCM. To prevent BCM and circuit damage, follow these service precautions:

1. Do not ground or apply voltage to any controlled circuits unless the service information instructs you to do so.
2. Use only a high-**impedance** multimeter (10 megaohms or greater) to test the circuits. Impedance is the combined opposition to current created by the meter's resistance, capacitance, and inductance. Never use a test light unless specifically instructed to do so in the service information.
3. Turn the ignition off before making or breaking electrical connections to the BCM.
4. Unless instructed otherwise in the service information, turn off the ignition switch before making or breaking any electrical connections to sensors or actuators.
5. Turn the ignition switch off whenever disconnecting or connecting the battery terminals. Also, turn it off when pulling and replacing the fuse.
6. Do not connect any other electrical accessories to the insulated or ground circuits of the computer-controlled systems.
7. Use only manufacturer's specific test and replacement procedures for the year and model of vehicle you are servicing.
8. When handling a computer, wear an **electrostatic discharge (ESD) strap** to ground your body to prevent static discharges that may damage electronic components.

> Static electricity can be 25,000 volts or higher.

By following these precautions, plus those listed in the service information, you can avoid having to replace expensive components.

Electrostatic Discharge

Some manufacturers mark certain components and circuits with a code or symbol to warn technicians that the units are sensitive to electrostatic discharge (**Figure 8-2**). Static electricity can destroy or render a component useless.

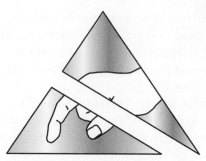

Figure 8-2 One type of Electrostatic Discharge (ESD) symbol used to warn the technician that the component or circuit is sensitive to static electricity.

When handling any electronic part, especially those that are static sensitive, follow the guidelines to reduce the possibility of electrostatic buildup on your body and the accidental discharge to the electronic part. If you are unsure if a part is sensitive to static, treat it as if it is.

1. Always touch a known good ground before handling the part. Repeat this step while handling the part and more frequently after sliding across a seat, sitting down from a standing position, or walking.
2. Avoid touching the electrical terminals of the part. It's good practice to keep your fingers off all electrical terminals as the oil from your skin can cause corrosion.
3. Do not remove a part from its protective package until it's time to install the part.
4. Before removing the part from its package, ground yourself and the package to a known good ground on the vehicle.

Electromagnetic Interference

Electromagnetic interference (EMI) or radio frequency interference (RFI) can cause problems with the vehicle's on-board computers. Unfortunately, an automobile's spark plug wires, ignition coil, and generator possess the ability to generate these interferences. Under the right conditions, EMI and RFI can trigger sensors or actuators. The result may be an intermittent problem with system operation.

To minimize the effects of EMI and RFI, ensure your visual inspection is thorough. Check to make sure that sensor wires running to the computer are routed away from potential EMI and RFI sources. Rerouting a wire by no more than an inch or two (25 to 50 mm) may keep EMI and RFI from falsely triggering or interfering with computer operation. EMI and RFI can be present on a voltage signal or on a ground.

Most manufacturers shield their BCM from EMI and RFI. However, this shielding will work only if the BCM is properly grounded. Always confirm a good ground before condemning the BCM. Some BCMs may have up to five grounds.

In addition, confirm the proper connection of all grounding straps. A loose or missing grounding strap can cause the computer to read erratic voltages and cause multiple issues.

Diagnosis of Computer Voltage Supply and Ground Circuits

Classroom Manual
Chapter 8, page 230

Like any other electrical or electronic component, a computer cannot operate properly unless it has satisfactory voltage supply at the required terminals and proper ground connections. A computer wiring diagram for the vehicle must be available for these tests.

First, measure the voltage across the battery terminals as a reference. Now, backprobe the battery feed terminal to the computer (if allowed by manufacturer) and connect the digital multimeter (DMM) leads to this terminal and to ground. The voltage at the battery feed terminal should be 0.5 volt (V) of battery voltage with the ignition switch off. Check the computer fuse and related circuit if the proper voltage is not present at this terminal. Turn the ignition switch to the RUN position and check applied voltage to *all* battery and ignition feed terminals of the computer. The voltage measured at these terminals should be within 0.5 volt of battery voltage. If the specified voltage is not available, test the voltage supply wires to these terminals. Voltage for these terminals may come through fuses, fuse links, or relays.

Special Tools

DMM

Lab scope

> 🛠 **SERVICE TIP** Never replace a computer unless you prove the ground and voltage supply circuits are in satisfactory condition.

Computer ground wires usually extend from the computer to a ground connection on the engine or battery. Often there is more than one ground wire. In addition, the fasteners used to attach the computer to the vehicle chassis may perform as a ground. With the ignition switch in the RUN position, perform a voltage drop test across the ground wires. Compare your results with specifications. A good circuit should not drop more than 0.2 volt on the ground circuit.

If the manufacturer does not allow for backprobing of the connector, use a **breakout box** or carefully forward probe the connector. In this case the voltage test is an open circuit. To achieve accurate results load the circuit. Use a test light that draws about 250 mA to load the circuit and then measure the voltage at the test lamp (**Figure 8-3**). When testing the battery and ignition feed circuits, the voltmeter should read within 500 mV of battery voltage. When reading the voltage of the ground circuit, the voltmeter should read less than 200 mV.

It's imperative to test all power and ground circuits. There may be multiple direct battery feeds and multiple grounds. Although these are often referred to as "redundant" powers and grounds, they don't actually become substitute circuits if the other drops off. For example, a module may have two direct power feeds and two ground circuits. If one of the power feeds should open, the other does not make up for it. In this case, each power feed may supply a microprocessor (μP), so if one is lost, one of the microprocessors fails to power up. The same may be true for the ground circuits. In fact, if a ground circuit is open, electrons will attempt to find alternate paths to grounds. The resulting feedback through the computer can result in very strange system operation.

A **breakout box** is a special tool that connects to the computer connectors, allowing you to test circuits, sensors, and actuators by providing test points.

Sensor Return Circuit

Many manufacturers use a single sensor return circuit for all BCM sensors. This makes a good ground critical since a sensor return circuit with high resistance will cause *all* of the sensors to indicate an erroneous voltage reading to the BCM. Consider a temperature sensing circuit as an example (**Figure 8-4**). (A) indicates the normal voltage sensed by the BCM with the NTC thermistor at 10 K ohms (O). (B) indicates the voltage sensed by the BCM with the NTC thermistor still at 10 K ohms but with an additional 200 ohms of resistance in the sensor's return circuit. The voltage is now off by 25 mV. The 200 ohms of resistance also affect all of the other sensors that share this return circuit and all of them will have a higher input voltage as a result. This may not seem like enough voltage to cause problems, but remember that most systems are very sensitive to small voltage changes.

Figure 8-3 Using a test light to load the circuit when the connector is unplugged.

> 🔧 **SERVICE TIP** When diagnosing computer problems, it's usually helpful to ask the customer about service work that has been performed lately on the vehicle. If service work has been performed, it's possible that a computer ground wire may be loose or disconnected.

In addition to causing some system operation problems, this type of fault may not set a fault code. The voltage values are within the normal operating parameters of the sensor, so continuity faults will not set. Also, since the resistance affects all sensors the computer may not detect rationality faults. System operation problems may occur because the computer is making bad decisions based on erroneous information.

Poor grounds can also allow EMI or noise to be present on the reference voltage signal. This noise causes minute changes in the voltage going to the sensor; therefore, the output signal from the sensor will also have these voltage changes. The computer will try to respond to these changes, which can cause a system operation problem. The best way to check for noise is to use a lab scope.

Figure 8-4 The effect of resistance on the sensor return circuit. (A) Normal sensor voltage. (B) Voltage reading with NTC thermistor at the same resistance, but sensor return circuit has additional resistance.

Connect the lab scope between the 5-volt reference signal to the sensor and ground. The trace on the scope should be flat (**Figure 8-5**). If noise is present, move the scope's negative probe to a known good ground. If the noise disappears, the sensor's ground circuit is faulty or has resistance. If the noise is still present, the voltage feed circuit is faulty or there is EMI in the circuit from another source, such as the AC generator. Find and repair the cause of the noise.

Circuit noise may be present in either the positive side or the negative side of a circuit. It may also be evident by a popping noise from the radio. However, noise can cause a variety of problems in any electrical circuit. The most common noise sources are electric motors, relays, solenoids, AC generators, ignition systems, switches, and air-conditioning (A/C) compressor clutches. Typically, noise results from an electrical device turning on and off. Sometimes the source of the noise is a defective suppression device. Some commonly used noise suppression devices are resistor-type secondary cables and spark plugs, shielded cables, capacitors, diodes, and resistors. If the source of the noise is not a poor ground or a defective component, check the suppression devices.

Resistors used for noise suppression don't eliminate the spikes but limit their intensity. When testing a circuit that uses a resistor to limit noise, large voltage spikes shown on the lab scope trace indicates the resistor may be bad. Clamping diodes suppress noise and induced voltages in circuits such as A/C compressor clutches and relays. If the diode is

Figure 8-5 (Top) Voltage signal trace of a good circuit. (Bottom) Trace of a circuit with excess noise.

bad, a negative spike will result (**Figure 8-6**). Capacitors or chokes control noise from motors and generators (**Figure 8-7**). To avoid much frustration during diagnosis of a computer system (especially the inputs), check the integrity of the ground as one of the first steps.

> ⚙ **SERVICE TIP** Considering the total number of vehicles produced that use computer control systems, very few customer complaints are actually due to the fault of the computer. Replace the computer only if all other possible causes have been checked and confirmed to operate properly. Always check the condition of the computer's battery feed and ground circuits before condemning the computer.

Diagnostic Trouble Codes

Once the computer performs a diagnostic test, or during the system's operation, the microprocessor software determines whether the system and its circuits performed properly. If the system or circuit fails, the computer sets a DTC.

Beginning in the 1996 model year, the Society of Automotive Engineers (SAE) published J2012 to describe industry-wide standards for a uniform DTC format. The trouble codes

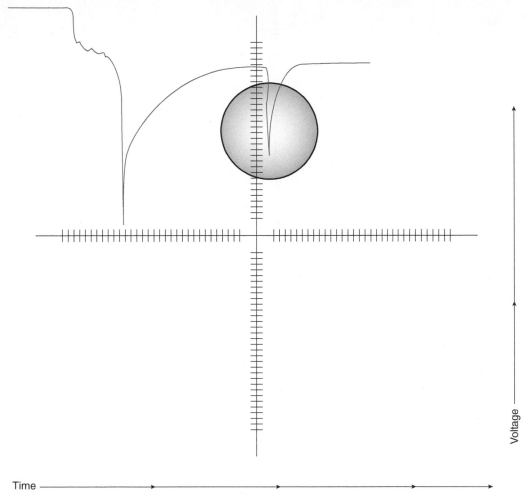

Figure 8-6 A voltage trace of an A/C circuit with a bad diode.

contain four characters, usually consisting of one letter and four numbers (**Figure 8-8**). The letter identifies the function of the device that has generated the fault code:

- P = Powertrain
- C = Chassis
- B = Body
- U = Network or DLC

The characters that follow identify the fault. These characters are usually numbers, but may also contain letters. Figure 8-8 is an example of a fault code that sets when the computer detects that the ignition switch input voltage is above a calibrated value. Using this code, you would follow a diagnostic routine in the service information to locate the cause.

Most BCMs can display the faults stored in memory by using a scan tool to retrieve the diagnostic trouble codes. Early systems used a method referred to as **flash codes**. Flash codes are DTCs displayed by flashing a lamp or light-emitting diode (LED).

AUTHOR'S NOTE Since OBD II's introduction in 1996, SAE no longer allows manufacturers to display emission-related DTCs by flash codes. Most manufactures have stopped using flash codes for DTC retrieval on most vehicle modules.

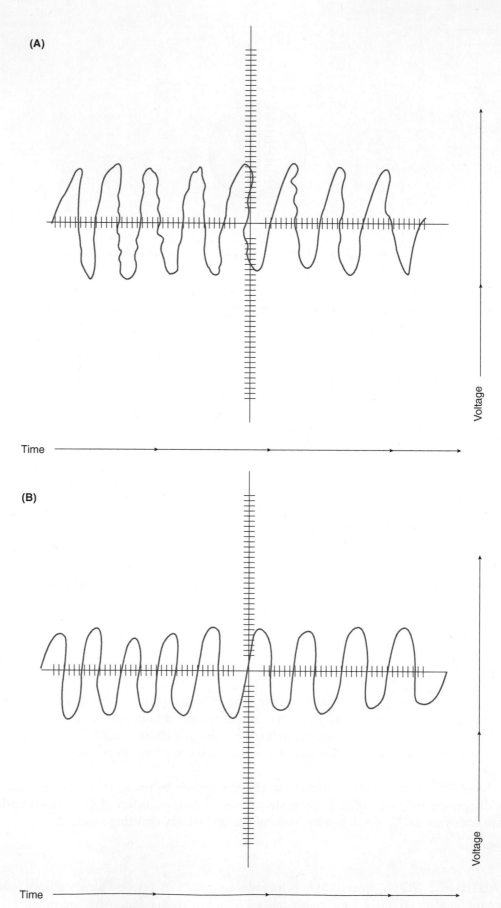

Figure 8-7 (A) Voltage trace of motor without a choke. (B) Voltage trace of a motor with a choke.

B2102

Figure 8-8 Example of a body function fault code.

Most computers used today can store codes for long periods of time; however, some older vehicles lose the code when the ignition is turned off. Systems that don't retain the code require that the technician test drive the vehicle and attempt to duplicate the fault. Once the computer detects a fault, the code must be retrieved before turning off the ignition again.

The trouble code does not necessarily indicate the faulty component; it only indicates that circuit of the system is not operating properly. For example, the code displayed may be B1312, indicating an A/C high-side temperature sensor problem. This does not mean the sensor is bad; the fault is in that circuit, which includes the wiring, connections, sensor, and BCM. To locate the problem, follow the diagnostic procedure in the service information for the code received.

> **AUTHOR'S NOTE** Usually "B" codes will not illuminate the "Check Engine" (MIL) lamp. Since 1996, the MIL only illuminates when there is an emissions related fault.

Diagnostic trouble codes are also called "fault codes."

Differences between Hard and Intermittent Codes

Some BCMs will store trouble codes in their memory until they are erased by the technician or until a set amount of engine starts have passed. Some computers will display two sets of fault codes. **Hard codes** are failures that were detected the last time the BCM tested the circuit. Intermittent codes are those that have occurred in the past but were not present during the last BCM test of the circuit. Depending on the scan tool or method of DTC retrieval, the hard codes and intermittent codes can be displayed together or separately.

Hard codes are also known as active codes. Intermittent codes are also called stored codes.

Most diagnostic charts are not much assistance in locating the causes of intermittent codes. This is because the testing at various points of the chart depends on the fault being present to locate the problem. If the fault is not present, you may be erroneously instructed to replace the BCM module, even though it's not defective.

Many intermittent problems are the result of poor electrical connections. Diagnosis should start with a good visual inspection of the connectors involved with the code. Even on hard codes, visually inspect the circuit before conducting any other tests.

Visual Inspection

A complete visual inspection is the most important check before diagnosing a BCM-controlled system. The visual inspection can identify faults that could cause you to waste time in diagnostics.

Inspect the following:

1. All sensors and actuators for physical damage.
2. Electrical connections into sensors, actuators, and control modules.
3. All ground connections.

4. Wiring for signs of burned or chafed spots, pinched wires, or contact with sharp edges or hot exhaust manifolds.
5. All vacuum hoses for pinches, cuts, or disconnects.

The time spent performing a visual inspection is worthwhile. Put forth the effort to check hidden wires and hoses under other components.

Entering Diagnostics

Caution

This section on entering BCM diagnostics is only a guide and intended to complement the service information. Improper methods of code retrieval may result in damage to the computer.

Special Tool

Scan tool

Caution

The procedures may change between models, years, and the type of instrument cluster installed. Refer to the correct service manual for the vehicle you are diagnosing.

There are as many methods of entering BCM diagnostics as there are vehicle manufacturers. Most require a scan tool (**Figure 8-9**). A scan tool is a microprocessor designed to communicate with the BCM. It will access trouble codes and run system operation, actuator, and sensor tests. The scan tool is plugged into the diagnostic link connector (DLC) and you choose the system requiring testing. Once trouble codes are retrieved consult the appropriate diagnostic chart for instructions on isolating the fault. It's also important to diagnose the codes in the order the manufacturer requires.

Using a Scan Tool

Connecting the scan tool into the DLC will allow the technician to access information concerning the operation of most vehicle systems. Some scan tools require adapters, cartridges, or electronic keys. Follow **Photo Sequence 16** as a typical method to enter body controller diagnostics.

When you have programmed the scan tool, by performing the initial entries, some entry options appear on the screen. These entry options vary depending on the scan tool and the vehicle. In the first selections, the tool is asking you to select the computer system to be tested. You make the desired selection to proceed with the desired test procedure. When you make a selection from the initial test selection menu, the scan tool moves on to the actual test selections.

Since the scan tool provides the voltage value that the BCM is sensing from an input circuit, diagnosing a circuit fault is simpler. For example, if you monitor the input voltage of a two-wire sensor, such as the ambient temperature sensor, and the scan tool indicates 5.0 volts on the circuit, the circuit has an open. The open could be anywhere in the circuit between the BCM and sensor ground (including the sensor). If the monitored voltage shows 0 volt, this usually indicates a short to ground. The short could be the sensor or in the circuit between the BCM and the sensor. A reading of 0 volt may also be caused by a faulty BCM not sending the reference voltage to the circuit.

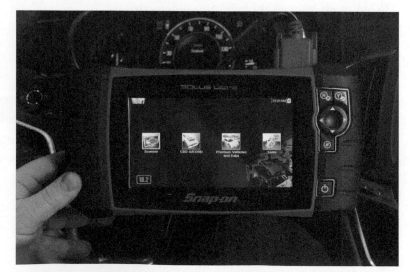

Figure 8-9 A scan tool is used to enter the computer's diagnostic capabilities.

PHOTO SEQUENCE 16
Typical Procedure for Scan Tool Diagnosis

P16-1 Turn the ignition switch to the OFF position.

P16-2 Select the proper scan tool data cable and adapters for the vehicle being tested. If required, connect the scan tool power cables to the vehicle's battery terminals (observing polarity) or to the accessory power outlet.

P16-3 Connect the scan tool cable to the DLC.

P16-4 Turn the ignition switch to the RUN (ON) position.

P16-5 Confirm the vehicle model year and engine size information.

P16-6 Select the system to be diagnosed.

P16-7 7 Operate the system and obtain the input sensor and output actuator data with the scan tool. It may be helpful to record and print the data. Compare the input sensor and output actuator data to the appropriate service information specifications. Identify any data that is not within specifications.

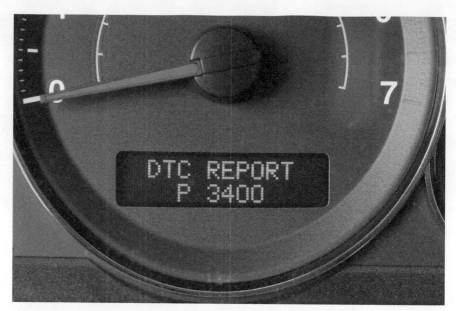

Figure 8-10 Some manufactures will display DTCs in the odometer.

If the scan tool cannot make a connection to the BCM, the following may be the cause:

- Loss of battery supply voltage to the BCM.
- Loss of BCM ground circuit.
- Defective communication circuit.
- Problem with the scan tool.
- Defective DLC.
- Defective BCM.

Alternate BCM Trouble Code Retrieval

Many manufacturers will provide a method of retrieving diagnostic fault codes without using a scan tool. The methods used to display DTCs vary between manufacturers and often between different lines of the same manufacturer. Always refer to the proper service information to determine the correct procedure to retrieve DTCs. Also, many manufacturers have discontinued displaying DTCs without a scan tool. Make sure that the vehicle you are working on supports this option.

Although flash codes are no longer allowed, some manufacturers provide a method of retrieving fault codes by displaying them in the odometer (**Figure 8-10**). For example, in select Chrysler vehicles cycling the ignition switch from OFF to RUN three times within 10 seconds without starting the engine displays all DTCs in the odometer. After all codes have cycled the odometer returns to the mileage reading.

AUTHOR'S NOTE Although other options for retrieving diagnostic trouble codes are sometimes available, it is usually best to use a scan tool since most will give the code and also supply a description of the code.

Testing Actuators

Classroom Manual
Chapter 8, page 238

Testing of actuators is included here to orient you to the basic procedures. Specific procedures are presented throughout this manual for individual systems and types of actuators.

Most computer-controlled actuators are electromechanical devices that convert the output commands from the computer into mechanical action. Uses of these actuators include; opening and closing switches, controlling vacuum flow to other components, and operating doors or valves, depending on the system's requirements.

Most systems allow for testing of the actuator through the scan tool or while in the correct mode. This method achieves a more accurate diagnostics of duty cycled. This will allow the technician to activate selected actuators to test their operation.

Special Tool
Scan tool DMM

⚡ **WARNING When performing actuation tests on systems such as power windows and power seats, be sure that you are not in a position that will allow you to be pinched or caught by moving components.**

If the actuator needs to be tested by means other than the scan tool, follow the manufacturer's procedures very carefully. Because some of the actuators used by the BCM operate with 5 to 7 volts, do not connect a jumper wire from a 12-volt source unless the service information directs this. Some actuators are easily tested with a voltmeter by testing for input voltage to the actuator. If the correct input voltage is observed with the circuit under an electrical load, check for a good ground connection. If both of these circuits are good, then the actuator is faulty. If using an ohmmeter to measure the resistance of an actuator, disconnect the actuator from the circuit first.

An ammeter measures the current of the actuator circuit. Simple on-off actuators should have the specified amperage anytime the actuator is energized. Higher than specified amperage reading would indicate a possible short in the actuator winding, mechanical binding of the actuator, or a short in the circuit. Lower than specified amperage indicates high circuit resistance, or a broken mechanical component of the actuator. The lower amperage is the result of no mechanical resistance to movement. A reading of 0 amp would indicate an open in the circuit. Pulse width modulated actuators have an ammeter reading that increases and decreases as the actuator pulse width increases and decreases, respectively.

Scan Tool Testing of Actuators

Most scan tools provide for monitoring actuator operation by watching the scan tool display of the computer controlling the actuator. As the computer turns the actuator on and off, the scan tool screen will indicate this command (**Figure 8-11**). Keep in

Figure 8-11 The scan tool may be capable of displaying the command of the computer to the actuator.

mind this indicates the computer's attempt to turn the actuator on and off, not necessarily that the actuator is working. While the actuator cycles from on and off, a voltmeter will indicate if the voltage levels are changing as the state of the actuator changes.

Some actuators will provide a feedback signal to the computer of correct operation. This feedback can be in the form of a potentiometer, commutator pulses, or a hard wire input back to the computer. Sometimes the feedback is by monitoring the voltage or the amperage of the control circuit. The scan tool may display these values. Consider the circuit of a BCM-controlled relay (**Figure 8-12**). With the relay de-energized the computer should see battery voltage on the control circuit. With the relay energized, the computer should see close to 0 volt. This may also be displayed as "High" or "Low" or as "True" or "False" on the scan tool (**Figure 8-13**).

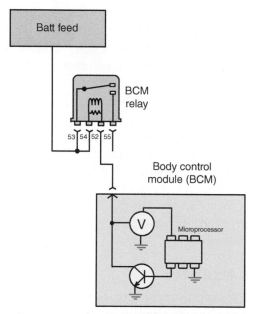

Figure 8-12 BCM relay driver control circuit.

Figure 8-13 Scan tool display of the operation for the control side of an actuator.

Figure 8-14 illustrates an activator circuit that monitors the current flow. In this case the scan tool may display the interrupted current (**Figure 8-15**). This value should match the reading of a connected ammeter. Keep in mind that the scan tool may display 0 amp anytime the computer detects a fault in the circuit. The fault may be an open, short to ground, or any high current draw condition that will cause the microprocessor to shut down the circuit driver to prevent damage.

Another function of the scan tool is the capacity to request an **activation test** (**Figure 8-16**). The scan tool activation tests places the scan tool into the function of the input to request the computer to follow through with the operation of the actuator. While performing the activation, a voltmeter or ammeter can confirm whether the request is being performed. Often, simply watching the actuator move confirms activation. If the activation test works, but normal operation fails, then the problem is likely in the inputs to the computer.

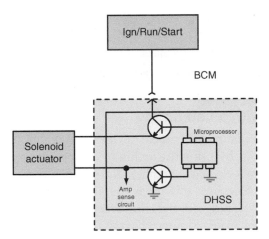

Figure 8-14 Dual high-side switch function that measures circuit current.

Figure 8-15 Scan tool displaying the current required for operating the actuator.

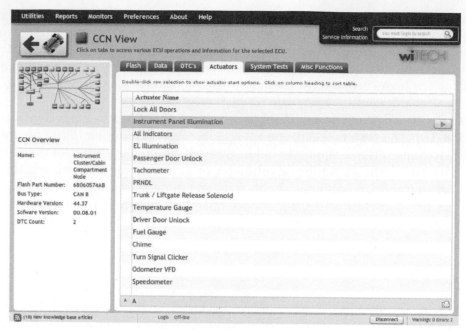

Figure 8-16 List of actuation tests available for this vehicle.

Testing Actuators with a Lab Scope

Special Tool

Lab scope

Since most actuators are electromechanical devices, they fail because they are electrically or mechanically faulty. By observing the action of an actuator on a lab scope, you will be able to watch an actuator's electrical activity. Normally, if there is a mechanical fault, it will affect the activator's electrical activity as well. Therefore, you get a good sense of the actuator's condition by watching the electrical trace on a lab scope.

To test an actuator, you need to know what type it is. Most actuators are solenoids. The computer controls the solenoid's action by controlling the control signal's pulse width. By watching the control signal, you can see the turning on and off of the solenoid (**Figure 8-17**). The voltage spikes result from the discharge energy from the coil in the solenoid.

Some actuators are controlled by pulse-width modulated signals (**Figure 8-18**). These signals show a changing pulse width. Control of these devices can be by varying the pulse width, signal frequency, and voltage levels.

Check both waveforms for amplitude, time, and shape. You should also observe changes to the pulse width as operating conditions change. A bad waveform will have

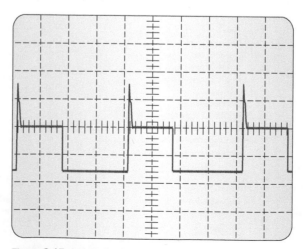

Figure 8-17 A typical solenoid control signal.

Figure 8-18 A typical pulse-width modulated solenoid control signal.

noise, glitches, or rounded corners. You should be able to see evidence that the actuator immediately turns on and off according to the computer's commands.

By monitoring injector type actuators, you can identify the BCM's driver's ability to complete the circuit to ground, compare commanded and actual pulse width, test the integrity of the actuators coil, confirm clamping diode function, and identify the pintle closing point (**Figure 8-19**). Obtain the waveform by backprobing the positive lead of the lab scope on the ground side of the actuator. Attach the negative test lead to battery ground. Set the voltage scale to read 100 volts with a 2ms time division. Start at this scale since the actuator may have a voltage spike of over 60 volts.

Full battery voltage should be indicated when the actuator is off. When the BCM activates the actuator, the voltage should be pulled to ground. This should be a straight downward trace indicating that the transistor is working properly. The length of time the voltage is low represents the pulse width. The actuator must stay on for the entire time of the pulse width as confirmed by comparing the time with the commanded time on the scan tool. During the pulse width time, the voltage should remain low. If the voltage begins to climb it may indicate a bad ground or a bad transistor. Turning off the actuator induces a voltage spike. If the coil is good, there should be an immediate increase in voltage. If the circuit uses a clamping diode, the voltage should reach the specified value. The flattening of the spike at the top indicates the use of a clamping diode.

Figure 8-19 An injector type actuator waveform. The slight bump in the falling edge indicates when the pintle closed.

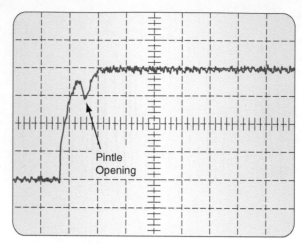

Figure 8-20 The amperage trace of the solenoid shows the pintle opening, and total amperage draw.

It is also possible to determine if the actuator pintle is working mechanically. A distinctive hump near the bottom of the trace as the voltage returns to battery voltage indicates where the pintle closed. If the pintle closed, it must have opened.

Using an amp clamp to monitor actuator current is ideal for identifying shorts, when the pintle opens, and determining if the coil is drawing the correct amount of current. (**Figure 8-20**).

A deflection in the ramp identifies pintle opening. Typically, the pintle should open at about two-thirds up the ramp. If the pintle is stuck there will be no deflection indicated. In addition, the falling edge of the waveform should indicate an immediate drop in current.

Flashing the BCM

Classroom Manual
Chapter 8, page 230

Most of the modules installed on today's vehicles utilize EEPROM chips that allow the basic programming of the computer to be altered or rewritten. This enables manufacturers to reduce costs by having the technician **flash** the computer instead of replacing it. To flash a computer means to remove the existing programming and overwrite it with new programming. The new program is downloaded into the scan tool and then into the BCM over a dedicated circuit.

When manufacturers first started developing reprogrammable ECUs it required factory level equipment and proprietary access to the updated software. As a result, independent repair facilities could not perform this function for their customers. To remedy this, the Society of Automotive Engineers (SAE) developed J2534 as a communication standard for ECU reprogramming. Its purpose is to create an application programming interface (API) for all manufacturers, allowing the aftermarket the ability to reprogram ECUs using a compliant pass-through device.

The standard J2534 applies to 2004 MY and later vehicles. The J2534 mandate requires vehicle manufactures to support aftermarket repair shops with flash reprogramming for any emissions-related computers installed on a vehicle that are reprogrammable by a manufacturer's dealership. Although this mandate effects 2004 and newer vehicles, some manufacturers provide J2534 support for older vehicles.

There are two subsets to the J2534 standard: J2534-1 and J2534-2. J2534-1 is primarily associated with emission modules. The J2534-1 reprogramming system consists of subscription software and a J2534-1 compliant pass-thru device. The subscription software is available directly from the manufacturer and operates on the shop's PC or laptop.

SAE J2534-2 defines optional features of the reprogramming device on non-emission ECUs. This standard allows the original equipment manufacturer (OEM) to reprogram all vehicle controllers if they release and publish the necessary protocol information about their non-emission ECUs.

Manufacturers will specify minimum equipment requirements to perform a flash. The shop must have a reliable high-speed internet connection (DSL or Cable Modem). A reliable connection is critical for any application that requires a live internet connection. Interruption of the connection may render the control module useless.

Prior to performing the reprogramming function, always check the condition of the vehicle's battery. A battery failure during the reprogramming process can result in a failure of the files to write correctly. Sometimes a failed operation can be recovered; however, there is a chance that failed reprogramming could permanently damage the control module. All manufacturers recommend using a battery charger to maintain a constant voltage to all modules during reprogramming.

Reprograming should be done using a dedicated shop computer restricted for this purpose only. The computer must meet the OEM requirements and have the supported operating system. Internet options such as enabling pop-ups and cookies, disabling firewalls, turning off all power saving modes, turning off screen savers, and so forth must also comply with stated requirements.

A subscription to the OEM websites is required to download the reprogramming files and programs. Some sites are free; others require the purchase of a subscription. Some manufacturers allow the purchase of temporary subscriptions that last for a few days.

Many manufacturers require downloading an API to the shop computer. The API is a library that includes specifications for routines, data structures, object classes, and variables.

The J2534-1 compliant pass-thru device (**Figure 8-21**) is a gateway between the vehicle's on-board computers and the shop's PC or laptop. The pass-thru device connects to the PC using either USB, serial cable, Ethernet cable, or wireless connections. The pass-thru device translates messages from the PC into the protocols used by the on-board computer and vice versa.

An important step in the reprogramming service process is determining if the control module requires a calibration update. Flashing a computer updates its part number or version number. Usually, a TSB is issued when a new calibration is released to address specific issues. The TSB will typically list the part number or version number the flash will apply for and may specify the new part number. In addition, the subscription software typically informs the technician of the need for an update. The scan tool may provide ECU data that will list the current part/version number. The update may apply if the current number does not match the latest version. However, it's not a good practice to arbitrarily update controllers just because one is available. Since reprogramming typically cannot be reversed, it's best to only update the controller if it addresses a specific customer concern. There is a chance the new calibration will make noticeable changes to the vehicle's operating characteristics. If the customer does not like any of the new settings there may not be a method to reload the old calibration.

Figure 8-21 The J2534-1 compliant pass-thru device is gateway between the on-board computer and the shop's computer.

> ⚙ **SERVICE TIP** If the reprogramming operation should fail, immediately retry the operation. Do not disconnect the interface tool.

Special Tools

Scan tool
Computer with
  internet access
Battery charger

Photo Sequence 17 shows a typical procedure for flashing a BCM. It's important to follow the service information procedures for the vehicle you are working on. Also, connecting a battery charger to keep the battery at 13.5 volts will prevent problems during the download resulting from low battery voltage. Check and clear any DTCs that may have set during the flash procedure.

PHOTO SEQUENCE 17
Flashing a BCM

P17-1 Use the scan tool to obtain the part number of the BCM and record the number.

P17-2 Connect a battery charger to the vehicle's battery. Maintain about 13.5 volts on the battery.

P17-3 Connect the scan tool to a PC that links the tool to the flash software. Some scan tools will connect directly to an internet site through a LAN or wireless connection.

P17-4 Follow the scan tool instructions to enter the computer's part number or the VIN in the field and select the NEXT button.

P17-5 Access the ECU requiring the flash update, select the desired flash file, and select the "Flash ECU" button to begin downloading the scan tool.

P17-6 Monitor the progress of the download to the scan tool.

P17-7 Follow any instructions concerning cycling of the ignition.

P17-8 Once the flash is complete, a confirmation screen will show if the file transfer was successful.

P17-9 DTCs may have set during the flash process. Erase any DTCs.

CASE STUDY

A customer is experiencing intermittent problems with several electrical options on their vehicle. The customer states that the systems will either operate erratically or stop operating completely. Retrieval of DTCs indicates several codes in many of the vehicle modules. Most of the DTCs relate to improper input values or feedback signals. The technician asks the customer if any services were performed on the vehicle lately and is informed that some body work was done a couple of weeks ago. Realizing that all of the DTCs are related to a possible single cause, the technician visually inspects all ground straps. During this inspection, a disconnected ground strap between the bulkhead and the engine hood is found. The connection is repaired and verification that all grounds are properly connected continues. The disconnected ground strap suppresses EMI that can result from the sheet metal behaving as a capacitor and the air space between the sheet metal forming an electrostatic field. Without proper grounding of the sheet metal, interference with computer-controlled circuits routed near the sheet metal can result. Upon verifying the repair, the vehicle was returned to the customer.

ASE-Style Review Questions

1. All of the following statements about servicing the BCM are true EXCEPT:
   A. Always check voltage supply circuits before replacing the BCM.
   B. Analog voltmeters must be used to test circuit voltage.
   C. Always check grounds before replacing the BCM.
   D. Turn the ignition switch to the OFF position before disconnecting or connecting components.

2. Diagnostic trouble codes are being discussed.
   Technician A says hard code failures are those that have occurred in the past but were not present during the last BCM test of the circuit.
   Technician B says intermittent codes are those that were detected the last time the BCM tested the circuit.
   Who is correct?
   A. A only
   B. B only
   C. Both A and B
   D. Neither A nor B

3. *Technician A* says DTCs will indicate the exact failure in the circuit.
   Technician B says DTCs will direct the technician to the circuit with a fault in it.
   Who is correct?
   A. A only
   B. B only
   C. Both A and B
   D. Neither A nor B

4. Accessing trouble codes is being discussed.
   Technician A says a scan tool is preferred since it may also provide a description of the code.
   Technician B says the body codes are prefixed with a "U."
   Who is correct?
   A. A only
   B. B only
   C. Both A and B
   D. Neither A nor B

5. When flashing a computer, all of the following are true EXCEPT:
   A. A battery charger should be connected to the vehicle's battery.
   B. Special instructions may be displayed on the scan tool screen during the flash.
   C. The flash process loads new software programming into the computer.
   D. The BCM must be removed from the vehicle before it's flashed.

6. *Technician A* says open circuit testing of the BCM power supplies will indicate if there is high resistance in the circuit.
   Technician B says that open circuit testing has the potential to lead to improper diagnostics.
   Who is correct?
   A. A only
   B. B only
   C. Both A and B
   D. Neither A nor B

7. If the scan tool fails to make connection with the BCM, all of the following could be the cause EXCEPT:

A. Loss of BCM power ground.

B. Open actuator circuit.

C. Open battery feed circuit to the BCM.

D. Faulty scan tool.

8. *Technician A* says a PWM actuator cannot be tested with a DMM.

Technician B says the PWM actuator may be tested with a lab scope.

Who is correct?

A. A only

B. B only

C. Both A and B

D. Neither A nor B

9. A high-side controlled actuator fails to operate. The scan tool displays 0 amp on the control circuit when the system is attempted to be turned on.

Technician A says this may indicate an open in the circuit.

Technician B says this may indicate a short to ground in the circuit.

Who is correct?

A. A only

B. B only

C. Both A and B

D. Neither A nor B

10. The relay coil circuit is controlled by a low-side driver. Direct battery feed is connected to the other terminal of the coil. If the BCM monitors voltage on the control circuit to determine proper operation, which of the following statements is **not** true?

A. The voltage reading by the BCM before the low-side driver should be battery voltage with the coil turned off.

B. The voltage reading by the BCM before the low-side driver should be close to 0 volt with the coil turned on.

C. A constant low-voltage reading by the BCM may be the result of an open in the circuit between the battery and the coil.

D. The BCM will not be capable of setting control circuit faults.

ASE Challenge Questions

1. The BCM-controlled ATC system is operating at full heat, high fan speed, with output from the defrost grids. This occurs whenever the vehicle is started and cannot be changed by the vehicle operator.

Technician A says a sensor circuit may have failed.

Technician B says for some reason the BCM may have initiated a fail-safe action.

Who is correct?

A. A only

B. B only

C. Both A and B

D. Neither A nor B

2. BCM system troubleshooting is being discussed.

Technician A says if the BCM "sees" a problem with any of its input sensors or output devices it will always command the "Check Engine" lamp to illuminate.

Technician B says most diagnostic charts are not much assistance for diagnosing intermittent problems.

Who is correct?

A. A only

B. B only

C. Both A and B

D. Neither A nor B

3. The in-car temperature sensor of a vehicle equipped with a BCM and ATC occasionally shorts internally, resulting in erratic operation of the ATC system. However, during diagnostic testing the ATC system performs correctly.

   Technician A says there may be an intermittent trouble code stored by the BCM.

   Technician B says the BCM will store an intermittent and a hard trouble code.

   Who is correct?

   A. A only   C. Both A and B

   B. B only   D. Neither A nor B

4. The diagnosis of BCM-controlled output actuators is being discussed.

   Technician A says if the actuator works during the activation test, the problem is probably associated with a faulty BCM.

   Technician B says some scan tools can control the operation of output actuators.

   Who is correct?

   A. A only   C. Both A and B

   B. B only   D. Neither A nor B

5. An actuator that receives direct battery feed and controlled by a low-side driver has a short to ground in the control circuit.

   Technician A says this may cause the battery to drain overnight.

   Technician B says the BCM will no longer have control of the system and may set a DTC.

   Who is correct?

   A. A only   C. Both A and B

   B. B only   D. Neither A nor B

6. Computer flashing performs which task?

   A. Repair hardware problems within the computer-controlled system.

   B. To update the software within the computer to correct or enhance operation.

   C. Determine if driver habits contribute to EEPROM issues.

   D. Correct for faulty input signals.

7. Which method may be used to check a computer-controlled actuator?

   A. Use a lab scope to monitor voltage levels.

   B. Use an ammeter to monitor current flow.

   C. Use a scan tool to activate the component.

   D. All of these choices.

8. When measuring the voltage on the power feed circuit to the computer, the voltmeter reads 12 volts with the connector unplugged. When load tested with a test light, the voltmeter reads 12 volts. When the connector is plugged back in and backprobed, the voltmeter reads 9.5 volts.

   Technician A says this may be due to an internal problem in the computer.

   Technician B says this may be due to an open computer ground circuit.

   Who is correct?

   A. A only   C. Both A and B

   B. B only   D. Neither A nor B

9. The scan tool indicates that an actuator is being turned on and off; however, the actuator does not operate. What is the *least* likely cause?

   A. Faulty computer.

   B. Open in the control circuit of the actuator.

   C. Open in the ground circuit of the actuator.

   D. Defective actuator device.

10. Using **Figure 8-22** of a BCM-controlled headlight system, what is the most likely cause that the headlights operate dimly on both sides?

    A. Poor battery feed connection to the BCM.

    B. Worn or corroded relay contacts.

    C. Faulty BCM.

    D. High resistance in the headlamp ground circuits.

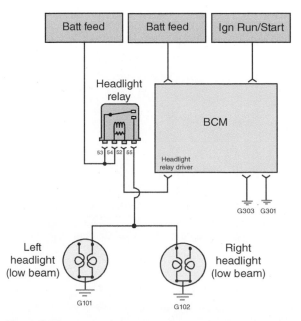

Figure 8-22

Name _____   Date _____

TESTING COMPUTER POWER AND GROUND CIRCUITS

Objective: Upon completion of this job sheet, you should be able to test the power and ground circuits of the BCM.

ASE Education Foundation Correlation

This job sheet addresses the following **MLR** tasks:

A.1.   Research vehicle service information such as fluid type, vehicle service history, service precautions, technical service bulletins, and recalls including vehicles with advanced driver assistance systems (ADAS). **(P-1)**

A.2.   Identify electrical/electronic system components and configuration. **(P-1)**

A.5.   Demonstrate proper use of a digital multimeter (DMM) when measuring source voltage, voltage drop (including grounds), current flow, and resistance. **(P-1)**

A.6.   Demonstrate knowledge of the causes and effects from shorts, grounds, opens, and resistance problems in electrical/electronic circuits. **(P-1)**

A.7.   Describe types of test lights; use appropriate test light to check operation of electrical circuits per service information. **(P-2)**

This job sheet addresses the following **AST/MAST** tasks:

A.1.   Research vehicle service information such as fluid type, vehicle service history, service precautions, technical service bulletins, and recalls including vehicles with advanced driver assistance systems (ADAS). **(P-1)**

A.2.   Identify electrical/electronic system components and configuration. **(P-1)**

A.5.   Demonstrate proper use of a digital multimeter (DMM) when measuring source voltage, voltage drop (including grounds), current flow, and resistance. **(P-1)**

A.6.   Demonstrate knowledge of the causes and effects from shorts, grounds, opens, and resistance problems in electrical/electronic circuits. **(P-1)**

A.7.   Describe types of test lights; use appropriate test light to check operation of electrical circuits per service information. **(P-1)**

We Support

ASE | Education Foundation

Tools and Materials

- DMM
- Test light
- Breakout box
- Backprobe tools
- Service information

Procedure

1. Using the proper service information, identify the following information for each direct battery feed circuit:

   BCM connector number and cavity: _____

   Circuit identification: _____

   Wire color code: _____

2. Identify the following information for each ignition feed or switched battery feed circuit:

BCM connector number and cavity: _____

Circuit identification: _____

Wire color code: _____

3. Identify the following information for each ground circuit:

BCM connector number and cavity: _____

Circuit identification: _____

Wire color code: _____

4. Identify the following information for each sensor return circuit:

BCM connector number and cavity: _____

Circuit identification: _____

Wire color code: _____

5. Disconnect the connector(s) that contain the direct battery feed circuits and measure the voltage at each of the battery feed terminals.

_____

_____

_____

6. Use a test light to load the circuit and retest the battery voltage at each terminal identified in step 1. Record your results.

_____

_____

_____

7. Were the readings different from those taken in step 5 and step 6? ☐ Yes   ☐ No Explain why or why not.

_____

_____

8. Reconnect the connector(s) and backprobe the terminals and measure the direct battery feed circuits identified in step 1. Record your results.

_____

_____

_____

9. If any of the direct battery feed circuits voltage values were below specifications, what would your next step be?

_____

_____

_____

10. Test each of the ignition or switched battery feed circuits identified in step 2 using open circuit, test light load, and backprobing methods and record your conclusions.

_____

_____

_____

_____

11. Disconnect the connector(s) that contain the circuits for the BCM grounds and measure the voltage at each of the ground terminals. Record your results.

_____

_____

12. Use a test light connected to battery positive to load the ground circuits and re-measure the voltage.

_____

_____

13. Reconnect the BCM connector and backprobe the ground circuits to measure voltage with the ignition in the RUN position. Record your results.

_____

_____

14. Disconnect the negative battery terminal and test the BCM ground circuits using an ohmmeter to test continuity from the disconnected BCM connector(s) and chassis ground. Record your results.

_____

_____

15. Comparing the tests performed on the ground circuit in steps 11 through 14, which do you believe to be the most accurate method of locating circuit faults and why?

_____

_____

_____

16. Describe how you would proceed to test the sensor return circuit.

_____

_____

_____

17. After reviewing step 16 with your instructor, perform the test you outlined and record your results.

_____

_____

_____

Instructor's Evaluation:

Grade	Rubric
0	No exposure due to: ☐ Absence ☐ Other:
1	Student unsuccessfully attempted the task
2	Student completed the task with significant instructor assistance.
3	Satisfactory performance of task. The student was able to perform elements of the task with minimal instructor assistance.
4	Competent/Proficient. Able to perform all elements of the task successfully and independently.

Instructor's Notes:

Name _____   Date _____

USING A SCAN TOOL

Upon completion of this job sheet, you should be able to hook up a scan tool and retrieve the codes from a computer.

ASE Education Foundation Correlation

This job sheet addresses the following **MLR** tasks:

A.1.	Research vehicle service information such as fluid type, vehicle service history, service precautions, technical service bulletins, and recalls including vehicles with advanced driver assistance systems (ADAS). **(P-1)**
A.2.	Identify electrical/electronic system components and configuration. **(P-1)**
A.3.	Retrieve and record DTCs, OBD monitor status, and freeze frame data; clear codes and data when directed. **(P-1)**
A.5.	Demonstrate proper use of a digital multimeter (DMM) when measuring source voltage, voltage drop (including grounds), current flow, and resistance. **(P-1)**

This job sheet addresses the following **AST** tasks:

A.1.	Research vehicle service information such as fluid type, vehicle service history, service precautions, technical service bulletins, and recalls including vehicles with advanced driver assistance systems (ADAS). **(P-1)**
A.2.	Identify electrical/electronic system components and configuration. **(P-1)**
A.3.	Retrieve and record DTCs, OBD monitor status, and freeze frame data; clear codes and data when directed. **(P-1)**
A.5.	Demonstrate proper use of a digital multimeter (DMM) when measuring source voltage, voltage drop (including grounds), current flow, and resistance. **(P-1)**
G.8.	Diagnose body electronic system circuits using a scan tool; check for module communication errors (data communication bus systems); determine needed action. **(P-2)**

This job sheet addresses the following **MAST** tasks:

A.1.	Research vehicle service information such as fluid type, vehicle service history, service precautions, technical service bulletins, and recalls including vehicles with advanced driver assistance systems (ADAS). **(P-1)**
A.2.	Identify electrical/electronic system components and configuration. **(P-1)**
A.3.	Retrieve and record DTCs, OBD monitor status, and freeze frame data; clear codes and data when directed. **(P-1)**
A.5.	Demonstrate proper use of a digital multimeter (DMM) when measuring source voltage, voltage drop (including grounds), current flow, and resistance. **(P-1)**
G.8.	Diagnose body electronic system circuits using a scan tool; check for module communication errors (data communication bus systems); determine needed action. **(P-1)**

We Support

 Education Foundation

Tools and Materials

- Vehicle equipped with a BCM
- Scan tool
- Service information
- DMM
- Fused jumper wire
- BOB

Procedure

1. Locate the data link connector for the BCM. Where is the connector located?

   _____

2. Connect the scan tool to the DLC and turn the ignition switch to the RUN position.

   What version is the scan tool software? _____

3. Select the body computer function. Record the following information about the BCM.

   Part number: _____

   Software version: _____

   Hardware version: _____

4. How is this information useful? _____

   _____

   _____

5. Retrieve any fault codes that may be in the BCM's memory. Record the codes below.

   _____

   _____

   _____

   _____

6. Are the fault codes hard codes or intermittent? _____

7. For the system with fault code retrieved in step 5, use the scan tool to monitor the desired state of the actuator while you attempt to operate the system inputs. Describe your results.

   _____

   _____

   _____

8. Identify the control and ground circuit for the actuator.

   _____

   _____

   _____

9. Following recommended practices of the vehicle manufacturer, either backprobe the actuator connector, use a jumper harness, or a breakout box to measure the voltage on each circuit with the ignition key off.

   _____

   _____

   _____

10. Repeat step 9 with the ignition switch in the RUN position.

    _____

    _____

    _____

11. Observe the voltmeter reading while attempting to operate the system and record your results.

_____

_____

_____

12. Based on your testing thus far, can you make any conclusions?

_____

_____

_____

13. Connect an ammeter to the control circuit of the actuator and attempt to operate the system again. Record your results.

_____

_____

_____

14. Use the scan tool's activation test function to command the operation of the actuator while observing the ammeter. Explain your results.

_____

_____

_____

15. Remove the ammeter from the circuit and replace it with a lab scope. Repeat step 14 and record your results.

_____

_____

_____

16. What are your conclusions?

_____

_____

_____

Instructor's Evaluation:

Grade	Rubric
0	No exposure due to: ☐ Absence ☐ Other:
1	Student unsuccessfully attempted the task
2	Student completed the task with significant instructor assistance.
3	Satisfactory performance of task. The student was able to perform elements of the task with minimal instructor assistance.
4	Competent/Proficient. Able to perform all elements of the task successfully and independently.

Instructor's Notes:

_____

Name _____   **Date** _____

FLASHING A BCM

Upon completion of this job sheet, you should be able to flash a BCM.

ASE Education Foundation Correlation

This job sheet addresses the following **MLR** task:

A.1.   Research vehicle service information such as fluid type, vehicle service history, service precautions, technical service bulletins, and recalls including vehicles with advanced driver assistance systems (ADAS). **(P-1)**

This job sheet addresses the following **AST** tasks:

A.1.   Research vehicle service information such as fluid type, vehicle service history, service precautions, technical service bulletins, and recalls including vehicles with advanced driver assistance systems (ADAS). **(P-1)**

G.9.   Describe the process for software transfer, software updates, or reprogramming of electronic modules. **(P-2)**

This job sheet addresses the following M**AST** tasks:

A.1.   Research vehicle service information such as fluid type, vehicle service history, service precautions, technical service bulletins, and recalls including vehicles with advanced driver assistance systems (ADAS). **(P-1)**

G.9.   Describe the process for software transfer, software updates, or reprogramming of electronic modules. **(P-1)**

We Support
ASE | Education Foundation

Tools and Materials
- Scan tool
- Access to flash software
- Battery charger
- Pass-thru device
- Shop computer
- Internet access
- Subscription

Procedure:

1. Is there a TSB that addresses flashing the BCM for your assigned vehicle?

   ☐ Yes   ☐ No

   If yes, what is the document number? _____

   If no, go to step 3.

2. According to the TSB, what is the purpose of the flash?

   _____
   _____
   _____

3. Connect the scan tool to the DLC and access the BCM. Obtain the BCM part number and record.

   _____

4. Describe the procedure used to obtain the flash download into your shop PC or scan tool.

_____

_____

5. Follow the procedure outlined in step 4 and download the flash file. Describe the navigation required to download the file to the scan tool.

_____

_____

6. How do you know the download was successful?

_____

_____

7. Return to the vehicle and connect a battery charger to the battery. Why is this step important?

_____

_____

8. Perform the flash of the BCM. Record any steps the scan tool directs you to perform during the flash.

_____

_____

_____

9. How do you know if the flash update was successful?

_____

10. After the flash is completed, check for DTCs. Were any DTCs set?   ☐ Yes   ☐ No

If yes, erase DTCs.

Instructor's Evaluation:

Grade	Rubric
0	No exposure due to: ☐ Absence ☐ Other:
1	Student unsuccessfully attempted the task
2	Student completed the task with significant instructor assistance.
3	Satisfactory performance of task. The student was able to perform elements of the task with minimal instructor assistance.
4	Competent/Proficient. Able to perform all elements of the task successfully and independently.

Instructor's Response

_____

DIAGNOSTIC CHART 8-1

PROBLEM AREA:	Scan tool does not connect to the BCM.
SYMPTOMS:	No response from the BCM.
POSSIBLE CAUSES:	1. Poor connection at the DLC. 2. Loss of battery feed to the DLC. 3. Loss of ground connection at the DLC. 4. Loss of BCM ground connection(s). 5. Loss of direct battery feed circuits to the BCM. 6. Loss of ignition switch or switched battery feed circuits to the BCM. 7. Faulty scan tool. 8. Defective BCM.

DIAGNOSTIC CHART 8-2

PROBLEM AREA:	BCM functions.
SYMPTOMS:	No BCM functions and safety systems are in default mode.
POSSIBLE CAUSES:	1. Loss of BCM ground connection(s). 2. Loss of direct battery feed circuits to the BCM. 3. Loss of ignition switch or switched battery feed circuits to the BCM. 4. Defective BCM.

DIAGNOSTIC CHART 8-3

PROBLEM AREA:	BCM software.
SYMPTOMS:	TSB update and/or improper system operation.
POSSIBLE CAUSES:	1. BCM requires flashing. 2. Defective BCM.

DIAGNOSTIC CHART 8-4

PROBLEM AREA:	Activators.
SYMPTOMS:	System fails to function.
POSSIBLE CAUSES:	1. Blown circuit fuse. 2. Open in the control circuit. 3. Open in the ground circuit. 4. Faulty actuator. 5. Open or poor BCM power and ground circuits. 6. Defective BCM.

CHAPTER 9
SENSOR DIAGNOSTIC ROUTINES

Basic Tool

Basic hand tools
Service information

Upon completion and review of this chapter, you should be able to:

- Describe the complexities of diagnosing intermittent faults.
- Properly utilize all tools and tool features to isolate the cause of an intermittent fault.
- Test thermistors used for temperature sensing for proper operation.
- Diagnose the thermistor circuit for proper operation.
- Perform proper diagnostics of pressure switch inputs.
- Properly diagnose Wheatstone bridge pressure sensors and their circuits.
- Diagnose frequency-generating capacitance discharge sensors.
- Determine proper pressure sensor operation by monitoring its signal with a lab scope.
- Use a digital multimeter (DMM) or lab scope to test and determine needed repairs to the potentiometer and its circuits.

- Use a DMM and a lab scope to diagnose the magnetic induction sensor.
- Use a DMM and a lab scope to diagnose the output of magnetic induction sensors.
- Properly diagnose the magnetic induction sensor circuits for opens, shorts, and short to ground conditions.
- Use a scan tool to diagnose the magnetic induction sensor.
- Use a DMM, scan tool, and lab scope to diagnose magnetoresistive (MR) sensors and circuits for opens, shorts, and short to ground conditions.
- Use a lab scope and DMM to determine proper operation of the Hall-effect sensor and its circuits.

Terms to Know

Air gap
Capacitance discharge sensor
Cold soak
Dual ramping
Environmental data
Freeze frame
Knock

Knock sensor
Magnetic induction sensors
Magnetoresistive (MR) sensors
Negative temperature coefficient (NTC)
Piezoelectric sensors

Piezoelectric transducer
Piezoresistive
Positive temperature coefficient (PTC)
Potentiometer
Sensor dropout
Thermistor
Wheatstone bridge

AUTHOR'S NOTE Although introduced in the Classroom Manual, photo cell service and diagnostics are covered in Chapter 11 where operation with advanced lighting systems is discussed.

Introduction

The automotive computer uses inputs as a means of gathering information. Missing or corrupted input data results in poor system performance. This chapter explores the methods used to diagnose temperature sensors, different types of pressure sensors, position sensors, motion detection sensors, and the related circuits for each.

All sensor diagnostics begin with a thorough inspection of the sensor and its related circuits. This includes looking for physical damage and proper electrical connection. If a sensor has a hose or other mechanical connection, inspect them as well. In addition, carefully examine the wiring harness and connectors between the sensor and the control module, including any dedicated ground circuits. Test all connectors and connector terminals for looseness, pullout, or damage. Investigate any other conditions that can cause the sensor not to perform properly.

This chapter provides examples of diagnostic routines. Although the diagnostics discussed may pertain to a specific function, the diagnostics of the circuits of most sensor types will be compatible.

Classroom Manual
Chapter 9, page 244

Intermittent Faults

One of the most aggravating repair situations for the vehicle owners is to have their vehicle returned to them with "No Problem Found" noted on the repair order. It's also very aggravating to the technicians when they cannot duplicate the problem the customer is experiencing. Intermittent problems challenge the technician to completely examine the system to determine possible causes for the fault. The answer is not to throw parts at it; instead, use a systematic diagnostic routine.

Intermittent faults with sensors usually set a diagnostic trouble code (DTC). However, the DTC is stored instead of being active when the technician retrieves them. This means the fault occurred at some point, but is not active right now and that the conditions necessary to set this DTC are not currently present.

Begin by reviewing any technical bulletins that are related to the problem the customer is experiencing. Perform any repairs directed by the bulletin. If there is not a bulletin that addresses the concern, continue with the diagnostics. This includes a thorough visual inspection of the sensor, sensor circuit wires, connectors, and terminals. Most intermittent conditions are the result of a poor connection somewhere in the circuit. It's a good practice to disconnect and clean all grounds involved in the system. If there are multiple stored DTCs, use the wiring information to determine common connections or grounds for all of the affected circuits. Perform additional testing if the visual inspection fails to isolate the cause.

If the scan tool provides **environmental data** information associated with the stored DTC, this may be helpful in duplicating the problem. Environmental data is a snap shot of conditions when the fault occurs. It also indicates how many ignition cycles since the detection of the fault and how long it was an active fault (**Figure 9-1**). In addition, OBD II faults may have an associated **freeze frame** with the DTC (**Figure 9-2**). Freeze frame is similar to environmental data, but contains OBD II–mandated information for when the fault first occurred. Both of these assist the technician in determining the operating conditions that set the code so they can attempt to duplicate the conditions.

Special Tools
Jumper wire
Scan tool
DMM
Backprobing tools

EV Data-PCM		
Names	Value	Units
Manifold Absolute Pressure Sensor Circuit High	P0108	
DTC Readiness Flag	Not Complete	
Odometer	1232.8	miles
Starts Since Set Counter	3	
Warning Indicator Request State	On	
Good Trip Counter	0	
DTC	01 08	
Number DTC	1	
Ignition Key Cycles	3	
Accumulation Timer	3	minutes
Warm-Up Cycles	0	
Key Cycles Since DTC Last Set Counter	4	
DTC Storage State	Stored	

Figure 9-1 Scan tool screen showing the details associated with the setting of a fault code.

CARB		
Name	Value	Units
PCM Mileage Since MIL On	0.0	miles
Freeze Frame Caused by DTC	01 08	
Freeze Frame DTC Priority	03	
PCM Odometer	1236.9921	miles
Open Loop due to Driving Conditions - Bank 2	False	
Open Loop - Bank 1	False	
Closed Loop - Bank 1	True	
Open Loop with DTC - Bank 1	False	
Closed Loop with DTC - Bank 1	False	
Open Loop due to Driving Conditions - Bank 1	False	
Open Loop - Bank 2	False	
Closed Loop - Bank 2	True	
Open Loop with DTC - Bank 2	False	
Closed Loop with DTC - Bank 2	False	
MAP	15	psi
MAP Voltage	4.94	volts
Calculated MAP	8.0	psi
Vacuum	16.17	in. Hg
Barometric Pressure	15	psi
Engine Load	58.8	%
TPS 2 Voltage	4.22	volts
APP Pedal Percent	0.0	%
APP 1 Voltage	0.4497	volts
APP 2 Voltage	0.22	volts

Figure 9-2 Partial screen capture of scan tool display for freeze frame.

⚙️ **SERVICE TIP** Do not erase the fault code until environmental or freeze frame data has been reviewed. Clearing the DTC may also clear this data.

Next, use the scan tool to clear the DTC. With the ignition in the RUN position, monitor the scan tool for at least 2 minutes while wiggling the harness and connectors for the sensor circuits. Monitor the scan tool while looking for the sensor data to change or for the DTC to reset. If this does not cause a change in the sensor data or the DTC to return, cycle the ignition key off and on several times, leaving the ignition on for at least 10 seconds at a time. If the DTC does not return after this test, start the engine and allow it to reach normal operating temperature.

If the code appears during any of the tests, then perform the diagnostic routines for an active code. If the code does not return, duplicate the conditions indicated by the environmental or freeze frame data. Using the data recording function of the scan tool may assist the technician in locating the fault. Also, a lab scope is a very useful tool to help diagnose intermittent conditions.

CUSTOMER CARE If you cannot duplicate an intermittent fault, don't attempt the repair by guessing. Some scan tool and OEM manufacturers provide a data recording tool that connects to the vehicle's bus network. When the problem the owner is experiencing occurs, they push a button to save the data recording. Upon returning to the shop, the technician is then able to retrieve the information and pinpoint the root cause.

Diagnosing Temperature Sensors and Their Circuits

Temperature sensors of various types are used in several different systems on today's vehicles. Due to their prominence, it's imperative that today's technician be able to diagnose these sensors and their associated circuits competently. This section details diagnostic routines for the most common forms of temperature-sensing systems.

Diagnosing Thermistor Circuits

A **thermistor** is a solid-state variable resistor made from a semiconductor material that changes resistance as a function of temperature. **Negative temperature coefficient (NTC)** thermistors reduce their resistance as the temperature increases. **Positive temperature coefficient (PTC)** thermistors increase their resistance as the temperature increases.

Regardless of the purpose or type, the thermistor circuits operate under the same principles (**Figure 9-3**). In most cases, the control module will detect a problem in the circuit and record a DTC. The DTC can relate to a circuit voltage too high, a circuit voltage too low, or a rationality fault.

Classroom Manual
Chapter 9, page 245

Special Tools
DMM
Scan tool
Lab scope
Jumper wires
Backprobing tools

Figure 9-3 Typical temperature sensor circuit.

Dual ramping is also referred to as dual resolution or dual range.

Classroom Manual
Chapter 9, pages 246

⚠️ **Caution**

Before disconnecting any computer system component, ensure the ignition switch is turned off. Disconnecting components may cause high induced voltages and computer damage.

Often, the temperature sensor can be tested while it is still installed in the system by back-probing the terminals and connecting a digital voltmeter or lab scope to the sensor terminals. The sensor should provide the specified voltage drop at any temperature. The scope trace of an NTC thermistor should indicate a smooth transition from a high voltage (about 4 volts) to a low voltage as the sensor warms.

If the system uses **dual ramping**, there should be a change in voltage to a higher reading at the specified voltage or temperature reading (**Figure 9-4**). If this switch does not occur, the fault is within the control module.

Consider Figure 9-3 as an example of circuit diagnosis. Start by disconnecting the sensor and observing the voltage on the scan tool. The displayed voltage should equal that used by the control module signal circuit (**Figure 9-5**). If the voltage displayed by the scan tool is 0, the problem is that the signal circuit is shorted either to chassis ground or the sensor ground circuit, or a faulty control module.

If the specified voltage is present with the sensor unplugged, use a jumper wire to connect the two terminals at the sensor harness connector. The scan tool should display 0 volt (V). If the voltage is still 5 volts, connect the signal circuit wire to chassis ground. If the reading is now 0 volt, the sensor ground circuit is open. While doing these steps, if the voltage is above 0 volt (but less than 5 volts), this indicates resistance in the circuit (**Figure 9-6**). If the voltage is high when connecting the signal circuit to

| COLD | | HOT | |
10 KΩ resistor		909 Ω resistor	
−20°F	4.7 V	110°F	4.2 V
0°F	4.4 V	130°F	3.7 V
20°F	4.1 V	150°F	3.4 V
40°F	3.6 V	170°F	3.0 V
60°F	3.0 V	180°F	2.8 V
80°F	2.4 V	200°F	2.4 V
100°F	1.8 V	220°F	2.0 V
120°F	1.2 V	240°F	1.6 V

Figure 9-4 Example of voltage specifications for a dual ramping temperature sensor circuit.

Figure 9-5 Disconnecting the sensor should cause the controller to register 5 volts. This will be the voltage displayed on the scan tool.

Data Display - PCM			
	Name	Value	Unit
▲	Fuel Level Percent	49.0	%
	Engine Coolant Temp	120.2	F
	Engine Coolant Temp Volt	0.0587	Volts
	Intake Air Temp Deg	93.2	F
	Intake Air Temp Volt	3.1821	Volts
	Ambient Temp	44.6	F
	Ambient Temp Voltage	2.69	Volts
▼	CAT Modeled Temp	−83.2	F

Figure 9-6 This voltage displayed on the scan tool indicates resistance in the circuit.

chassis ground, the resistance is in the signal circuit. If the voltage reading is above 0 volt when connecting the jumper wire across the two terminals, the resistance is in the sensor ground circuit.

> **SERVICE TIP** Because current flow through the sensor will affect the readings, don't leave the ohmmeter connected for longer than 15 seconds.

The circuits can also be tested using an ohmmeter. Disconnect the sensor and the control module connectors. Connect an ohmmeter from each sensor terminal to the control module terminal to which the wire is connected. Both sensor wires should indicate less resistance than specified by the vehicle manufacturer. If the wires have higher resistance than specified, repair the wires or wiring connectors. Also, test circuits for being shorted together.

If the control module sets a rationality-type DTC for the thermistor circuit, this means the sensed value does not agree with other inputs. To test for this type of fault, allow the vehicle to **cold soak** long enough to allow all the temperature sensors to be at room temperature. Use the scan tool to compare the different temperature values for such sensors as engine coolant temperature (ECT), intake air temperature (IAT), ambient temperature, and battery temperature. If a sensor value differs from the others, diagnose the sensor and circuit as described earlier.

A lab scope or the data-recording function of the scan tool can assist in locating intermittent faults. Since these instruments indicate a change in voltage over time, any change that occurs on the trace that cannot realistically happen in the time frame indicates a problem within the circuit. For example, engine temperature cannot increase 80 degrees in 2 seconds. Since body systems (such as air conditioning) use the ECT sensor, a fault with this sensor may result in the customer expressing many different concerns. **Figure 9-7** is a data recording of a normally operating ECT, and **Figure 9-8** is a data recording of an intermittent fault.

⚠ **Caution**
Never apply an open flame to a coolant temperature for test purposes. This action will damage the sensor.

Cold soak means allowing the vehicle to sit without the engine running long enough that the coolant temperature equalizes with ambient temperature.

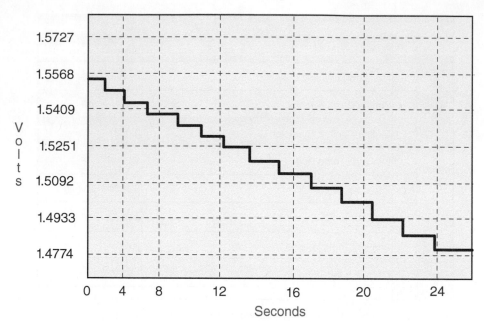

Figure 9-7 Normal NTC thermistor operation as temperature changes from cold to warm.

Figure 9-8 Data recording of ECT sensor circuit with an intermittent fault.

Diagnosing Pressure Sensors

Inputs used to determine pressures may be something as simple as a switch or can be a pressure transducer. This section discusses the routines used to diagnose pressure monitoring input circuits.

Pressure Switch Diagnosis

Pressure switch circuits are simple on-off type switches. They can control the operation of a warning lamp—for example, an oil pressure warning lamp. The pressure switch can turn on and off a system. An example of this would be a high-pressure cut-off switch used in some air-conditioning systems. Also, the pressure switch may be an input to a control module to indicate the proper operation of the system. This is the case of a pressure switch used in the hydraulic circuits of an electronically controlled automatic transmission.

Regardless of their function, pressure switches open and close at a certain pressure. Usually, this is done by a calibrated spring or disc. We will first look at the process of diagnosing an oil pressure warning lamp circuit.

Consider the simple circuit illustrated in **Figure 9-9**. This circuit uses a normally closed oil pressure switch that will open at a preset pressure. To test faulty warning lamp operation, first turn the ignition switch to the RUN position. The lamp should illuminate.

If the light does not come on during the prove-out, disconnect the oil pressure switch connector (**Figure 9-10**). Use a jumper wire to connect the switch circuit from the ignition switch to ground. The warning lamp should light with the ignition switch in the RUN position. If the light does not come on, either the bulb is burned out, or the wiring is faulty. If the lamp comes on, the problem is a faulty oil pressure switch or its ground.

If voltage is not present to the oil pressure switch, the bulb may be burned out. At this point, the instrument cluster will need to be removed. With the cluster removed, check for battery voltage to the panel connector. If voltage is present, substitute a known good bulb and test again.

If the customer states that the warning light stays on after starting the engine, test in the following manner. First, disconnect the lead to the sender switch. The light should go out with the ignition switch in the RUN position. If the light goes out, replace the oil pressure switch. If it does not, there is a short to ground in the wiring between the lamp and the oil pressure switch.

Computer-driven instrument cluster warning lamps may use normally open pressure switches that close when oil pressure increases. These systems use the switch as an

Classroom Manual
Chapter 9, pages
248, 266

⚒ Special Tools

Scan tool
DMM
Jumper wires
Backprobing tools

Figure 9-9 Normally closed oil pressure switch circuit.

Figure 9-10 Oil pressure switch and connector.

Figure 9-11 Computer-controlled warning lamp systems can use the pressure switch as an input.

If the customer states that the oil pressure light does not go out after starting the engine, first use a shop oil pressure gauge to confirm adequate oil pressure.

input to turn on or off the warning lamp; they don't directly operate it. Consider the system illustrated in **Figure 9-11**. The switch is a direct input to the powertrain control module (PCM). The PCM will send a sense voltage to the switch through a pull-up resistor. When oil pressure is below the threshold of the switch, the switch is open, and the voltage sense will be high. A data bus message will be sent to the instrument cluster requesting a light on activation. When the switch closes, the sense circuit is pulled low, and the light-off request is sent to the instrument cluster.

⚙ SERVICE TIP If there were a circuit problem in the data bus between the PCM and the instrument cluster, there would be multiple symptoms such as speedometer, tachometer, and engine temperature gauges not operating.

Preliminary testing of this circuit can be performed by simply unplugging the oil pressure switch. The lamp should be on with the ignition switch in the RUN position. Unplugging the oil pressure switch should turn off the lamp. If the results differ, perform additional testing to pinpoint the cause.

A scan tool accessing the instrument cluster module can command activation of the lamp. If the lamp does not come on when commanded, the problem is a faulty lamp, circuit board, or instrument cluster module. If the lamp does light when commanded, the fault is in the signal to the instrument cluster. Use the scan tool to access the data stream to the instrument cluster module. Confirm that the proper message is being received.

If unplugging the oil pressure switch does not turn the warning lamp off and the data bus message seen on the scan tool confirms that the requested state is to turn the lamp on, test the circuit from the PCM to the oil pressure switch for a short to ground. Use a voltmeter to test for the proper sense voltage level from the PCM. If the voltmeter reads 0 volt, either the circuit is shorted to ground, or the PCM is faulty. The circuit is not open because the symptom in the lamp is always off. Use an ohmmeter to test for continuity between the oil pressure switch connector and engine ground. The meter should have a very high-resistance reading. A low resistance indicates a short to ground in the circuit.

Unplug the PCM connector and test again between the circuit and engine ground. If it's now reading infinite, the PCM has an internal short.

If the lamp is always on (and the data bus indicates this is the requested state of the lamp), test the sense circuit between the PCM and the oil pressure switch for an open circuit. Unplug the oil pressure switch and use a jumper wire to connect the circuit to ground. The requested lamp state displayed on the scan tool should indicate a lamp-off command. If it does, replace the oil pressure switch. If the request does not change from on to off, backprobe the PCM connector for the sense circuit and use a jumper wire to connect the circuit to ground. If the lamp request changes to off, repair the open in the circuit between the PCM and the oil pressure switch. If the requested state does not change, the PCM is faulty.

Diagnosing Wheatstone Bridge Pressure Sensors

Pressure sensors are generally a form of a strain gauge that determines the amount of applied pressure by measuring the strain a material experiences when subjected to the pressure. Most strain gauge pressure sensors are a form of **piezoresistive** construction. A piezoresistive sensor changes resistance value as the pressure applied to the sensing material changes. A common piezoresistive pressure sensor is the **Wheatstone bridge**.

The Wheatstone bridge can measure small changes in resistance in a strain gauge. The manifold absolute pressure (MAP) sensor commonly uses the Wheatstone bridge as a sensing element. The MAP sensor is a primary sensor for determining the quantity of fuel to be injected into the combustion chamber. It determines the amount of air entering the combustion chamber at that instant. In most systems, it also serves as the means of determining the barometric pressure prior to starting the engine. Since the MAP sensor input indicates engine load, other body and chassis systems can use this information.

A defective MAP sensor may cause a rich or lean air-fuel ratio, excessive fuel consumption, and engine surging. It can also result in improper automatic transmission operation and/or improper air-conditioning system operation.

Diagnosis of MAP sensors differs between types. If the MAP sensor signal voltage is within the programmed value range of the PCM, a DTC may not set. For example, an open signal circuit will set a DTC, but resistance in the circuit that causes the voltage to be off by a few millivolts may not set a fault code. Also, a vacuum leak will cause the voltage values from the signal circuit to be different from the specifications.

A very simple test of the Wheatstone bridge MAP sensor is to monitor the MAP vacuum and voltage values displayed on the scan tool (**Figure 9-12**). With the ignition switch in the RUN position (engine off), record the MAP reading. At this time, the reading should be actual atmospheric pressure, and the voltage should confirm the reading. Compare this reading with the actual barometric pressure; the two should be equal. A MAP sensor that does not read correct barometric pressure cannot read correct vacuum. If the reading is incorrect, verify the condition of all the circuits to the MAP sensor, as discussed later. If the circuits are good, replace the sensor.

If the readings are within the normal ranges, start the engine and allow it to idle. The MAP vacuum and MAP voltage should indicate correct vacuum readings. Confirm this with a mechanical vacuum gauge connected to a port in the intake manifold.

Classroom Manual
Chapter 9, page 250

Special Tools
Scan tool
Vacuum gauge
Jumper wire
DMM

SERVICE TIP Anytime you use a scan tool to monitor a pressure reading, don't accept the displayed value until you confirm the voltage. A system substituting a pressure value may fool you into thinking the sensor is operating properly. A fault in the circuit can be determined by correlating the signal voltage to the pressure reading.

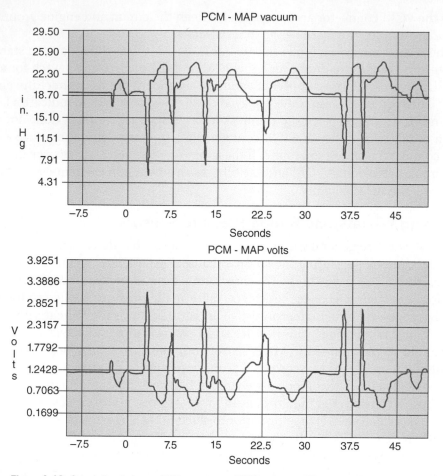

Figure 9-12 Correlation between MAP vacuum and MAP volts as different loads on the engine are applied.

To perform a quick test of the circuits, consider the schematic of a Wheatstone bridge sensor circuit (**Figure 9-13**). If the scan tool indicates a voltage reading of 5 volts, this would be caused by an open or short to voltage in the signal circuit, an open in the return circuit, or a faulty sensor. Unplug the sensor and use a jumper wire to jump the signal circuit to ground. The scan tool should now read 0 volt. If it still reads 5 volts, the circuit between the control module and the sensor connector is open. If the voltage is above 0 volt, there is excessive resistance in the signal circuit between the control module and the sensor connector. If the reading is 0 volt, move the jumper wire to short the signal circuit to the return circuit. The scan tool should now read 0 volt in a properly

Figure 9-13 MAP sensor circuit.

operating circuit. If the scan tool displays 5 volts, the return circuit is open. If the reading is above 0 volt, there is excessive resistance in the return circuit. If all of the readings are normal, the sensor is faulty.

If the scan tool reading with the ignition in the RUN position and sensor connected is 0 volt, the problem is either an open 5-volt supply circuit or a short to ground in the signal circuit. The sensor is faulty if the voltage changes to 5 volts when the sensor is unplugged. If the voltage remains 0 when the sensor is unplugged, use a voltmeter to check the supply circuit for 5 volts. If the voltage is low, there is either high resistance, an open, or a fault in the control module. Using an ohmmeter to test the continuity of the circuit can isolate these problems.

DMM Testing. To test the Wheatstone bridge sensor with a DMM, backprobe the sensor signal wire and turn the ignition switch to the RUN position (engine off). Connect the DMM between the signal wire and ground. The voltage reading indicates the barometric pressure signal from the sensor to the control module. Usually, this voltage is about 4 volts, depending on altitude. If the signal circuit voltage is not within the vehicle manufacturer's specifications compared with actual barometric pressure, confirm that the sensor's voltage supply and return circuits are good. If these circuits are good, replace the sensor.

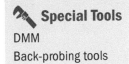

Special Tools
DMM
Back-probing tools

Pressure sensors installed in systems not exposed to atmospheric pressures (e.g., air-conditioning system high-side pressure sensor), mechanical gauges may need to be used to confirm proper voltage readings for the measured pressure. In some instances, it may be possible to remove the sensor from the system to expose it to atmospheric pressure.

If the voltage is acceptable with the ignition in the RUN position and the engine off, the next step is to confirm a change in voltage as the condition that the sensor is exposed to changes. For example, a MAP sensor is exposed to engine vacuum when the engine is running, and an air-conditioning (A/C) pressure transducer is exposed to high-side pressures as the A/C system cycles on and off, a boost pressure sensor is exposed to pressures greater than atmospheric as engine speed is increased. In the example of a MAP sensor, start the engine and observe the signal voltage. The voltage should have changed to a low value. If the voltage does not change, the sensor pressure chamber is obstructed or damaged, and the MAP sensor requires replacement. If the voltage changes, allow the engine to warm up to normal operating temperature while watching the voltage. Compare your voltage reading to specifications. If they are within specifications and the MAP sensor tests good, further testing of other components will be necessary.

For the Wheatstone bridge sensor circuit shown in Figure 9-13, if the voltage reading obtained in the initial test was 0 volt, unplug the MAP sensor connector and measure the voltage on each terminal with the ignition in the RUN position. On a three-wire connector, the voltage will usually be 5 volts on the supply circuit, 5 volts on the signal circuit, and 0 volt on the ground circuit. Obtaining these readings means the MAP sensor is suspect. If these voltages results differ, perform the same tests discussed earlier. In this case, confirm the changes in voltages as read on the DMM. If needed, use the ohmmeter function to test for continuity between the PCM connector and the MAP connector on the suspect circuit with the ignition in the OFF position. If there is continuity in the circuit, confirm PCM powers and grounds. If these are good, replace the PCM.

AUTHOR'S NOTE Some sensor signal circuits may have 0 volt when separating the connector. Always confirm the expected voltage values with the proper service information.

If the voltage recorded in the first step is inaccurate, backprobe the 5-volt supply circuit wire (connector plugged to the sensor) with a voltmeter with the ignition in the RUN position. If the reference wire is not supplying the specified voltage, check the voltage on this wire at the PCM connector. If the voltage is within specifications at the PCM, but low at the sensor, repair the 5-volt supply circuit wire for high resistance or an open. If this voltage is low at the control module, disconnect the control module's connector and use an ohmmeter to test for a short to ground. If a short to ground isn't indicated, check the voltage supply and ground circuits for the control module. If these circuits are satisfactory, replace the control module.

If the voltage supply circuit is good, test the sensor return (ground) circuit. With the ignition switch in the RUN position, backprobe the ground circuit terminal at the connector (with the connector plugged into the MAP sensor). Connect the voltmeter from the sensor ground wire to the battery ground. If the voltage drop across this circuit exceeds specifications, test the voltage drop on the PCM's ground circuits. If this is good, repair the ground wire from the MAP sensor to the control module.

Capacitance Discharge Sensor Diagnostics

Classroom Manual
Chapter 9, page 252

Special Tools

DMM
MAP sensor tester
Scan tool
Hand vacuum pump
Backprobing tools

Caution

Connecting any type of voltmeter directly to the voltage signal wire on a Ford MAP sensor may cause damage to the sensor.

Special Tools

Lab scope
Backprobing tools

Another variation of the piezo sensor uses capacitance discharge. Instead of using a silicon diaphragm, the **capacitance discharge sensor** uses a variable capacitor. Some of these sensors will use the discharged voltage as the signal and are diagnosed similarly to the Wheatstone bridge sensor. Other types of capacitance discharge sensors produce a digital voltage signal of varying frequency. On these sensors, use a voltmeter to check the 5-volt reference and ground circuits. However, testing the signal circuit of this type of sensor requires the use of a sensor tester that changes the sensor's varying frequency to a voltage reading or a scan tool that performs this function. **Photo Sequence 18** illustrates the use of the special sensor tester to diagnose the frequency-generating MAP sensor.

If the special sensor tester is unavailable, use a DMM that has a frequency measurement function. For example, to test a sensor used to measure vacuum, connect the positive meter lead to the sensor signal wire and the negative lead to ground. Set the meter to read hertz with a + trigger. With the ignition in the RUN position, measure the frequency on the sensor signal wire with no vacuum applied. Using a hand vacuum pump, increase the vacuum against the sensor in 5-in. Hg increments. Record the hertz reading at each step. **Figure 9-14** is an example of expected test results for a good sensor. Be sure to refer to the correct service information for the sensor you are testing. In this example, the hertz readings decrease as vacuum is applied.

Lab Scope Testing of the Pressure Sensor

Using a lab scope is a good method of testing the pressure sensor. As the exposed pressures change, the sensor voltage signal should increase and decrease (**Figure 9-15**). Lack of change, a slow response, or an erratic signal may be the result of the sensor or connecting wires being defective.

Vacuum Applied	Output Frequency
0 in. Hg	152–155 Hz
5 in. Hg	138–140 Hz
10 in. Hg	124–127 Hz
15 in. Hg	111–114 Hz
20 in. Hg	93–98 Hz

Figure 9-14 Examples of MAP sensor frequency with various vacuum values.

PHOTO SEQUENCE 18
Typical Procedure for Testing Frequency-Generating Sensors

P18-1 Remove the MAP sensor connector and vacuum hose.

P18-2 Connect the MAP sensor tester leads to the MAP sensor.

P18-3 Connect the MAP sensor connector to the harness connector.

P18-4 Connect the MAP sensor tester to the DMM.

P18-5 With the DMM set to DC voltage, observe the MAP sensor barometric pressure (BARO) reading on the voltmeter and compare this reading to specifications.

P18-6 Connect a hand vacuum pump to the MAP sensor and supply 5-in. Hg. Observe the MAP sensor reading and compare it to specifications.

P18-7 Increase the vacuum applied to the MAP sensor to 10-in. Hg and observe the reading. Compare this reading to specifications.

P18-8 Apply 15-in. Hg to the MAP sensor and compare the reading to specifications.

P18-9 Increase the vacuum 20-in. Hg to the MAP sensor and compare the reading to specifications. If any of the MAP sensor readings don't meet specifications, replace the MAP sensor.

Figure 9-15 Lab scope trace of normal MAP sensor operation.

Figure 9-16 Example of frequency output of the capacitance-type sensor.

A frequency-generating sensor produces a frequency that should change as the exposed pressures change. The upper voltage value of the trace should be 5 volts, while the lower voltage should be close to 0 volt (**Figure 9-16**). The sensor is defective if the frequency is erratic or does not change, or the voltage values are incorrect.

Diagnosing Piezoelectric Transducers

Classroom Manual
Chapter 9, page 251

Special Tools
DMM
Backprobing tools
Lab scope
Scan tool

Knock is the spontaneous auto-ignition of the remaining fuel-air mixture in the engine combustion chamber that occurs after normal combustion has started, causing the formation of standing ultrasonic waves.

Piezoelectric sensors produce a proportional voltage output resulting from the deformation of the element as pressure is applied. Since the **piezoelectric transducer** can measure the pressures associated with ultrasonic waves, it can perform many functions, including that of a **knock sensor** (detonation sensor). The knock sensor measures engine knock, or vibration, and converts the vibration into a voltage signal. Piezoelectric sensors are also used in passive restraint systems.

In this example, we will consider the knock sensor since it's one of the easiest piezoelectric-type sensors to understand and test. Diagnosing other piezoelectric sensors will be covered during the discussions on system diagnosis in later chapters.

Excessive spark advance for the given engine-operating conditions results in **knock**. The output voltage from the knock sensor circuit represents the strength of the engine knock and is read by the PCM. The knock sensor constantly produces an output voltage due to engine background noise, even when knock is not present (**Figure 9-17**). If knock occurs, the voltage output will increase (**Figure 9-18**). At this threshold, knock has occurred, and the PCM calculates the necessary amount of short-term spark retard to subtract from the spark advance based on the severity of the knock event. The amount of knock sensor voltage greater than the severity threshold voltage level determines the severity of the knock.

The procedure for checking a knock sensor varies depending upon vehicle make and year. Always follow the vehicle manufacturer's recommended test procedure and specifications. Some knock sensor circuits use a single wire and the sensor case to engine block

Figure 9-17 Normal knock sensor activity at idle.

Figure 9-18 Knock sensor activity when knock is occurring.

connection as ground. If there are two wires in the circuit, one is the signal circuit and the other is ground. Follow these steps for a typical knock sensor diagnosis:

1. Disconnect the knock sensor wiring connector, and turn on the ignition switch.
2. Connect a voltmeter from the disconnected knock sensor signal circuit to ground. The voltage should be 4 to 6 volts. If the voltage is within specifications, test the ground circuit as in step 3. If the voltage is about 12 volts, this indicates a short to a power circuit. If the specified voltage is not available at this signal circuit, backprobe the knock sensor wire at the PCM (or use a breakout box) and read the voltage at this terminal. If the voltage is satisfactory at this terminal, repair the knock sensor signal circuit for an open. If the voltage is not within specifications at the PCM terminal, replace the PCM.
3. If the voltage at the signal circuit connection to the sensor is within specifications, move the negative voltmeter lead to the ground circuit terminal of the sensor connector. If there is no voltage reading, the ground circuit is open. If the voltage reading is lower now than in the initial test, check for high resistance in the circuit.

4. If the above tests indicate there is not a problem with the signal or the ground circuits, then test the knock sensor. Use an ohmmeter connected between the knock sensor signal terminal and the ground terminal. If the knock sensor uses only one wire, the ground terminal is the case. Compare the ohmmeter reading with specifications, usually between 3,300 and 4,500 ohms (Ω). If the knock sensor does not have the specified resistance, replace the sensor.

Operation of the knock sensor can be observed with a lab scope. Some scan tools may also provide a graphed data display or recorded event display. Remember, the sensor is active at idle but when knock occurs the sensor should show increased voltage output.

Diagnosing Position and Motion Detection Sensors

Many automotive systems (such as electronic vehicle stability control, antilock brakes, airbags, and roll-over mitigation) require input data concerning position, motion, and speed. Today's technician will be called upon to diagnose these sensors. A common type of sensor to determine position and motion is the potentiometer. Motion and speed sensors that use magnetism include magnetoresistive (MR), inductive, variable reluctance (VR), and Hall-effect sensors. In addition, photoelectric sensors, solid-state accelerometers, axis rotation sensors, yaw sensors, and roll sensors are common components of many systems.

Testing the Potentiometer

Classroom Manual
Chapter 9, page 254

A **potentiometer** sensor can measure linear or rotary movement. Testing these sensors involves measuring the input voltage to the sensor and the feedback voltage to the computer. The feedback voltage to the computer should change smoothly as the resistance value of the sensor changes. To test these voltage signals, a series of jumper wires may be required (**Figure 9-19**). The jumper wires provide a method of gaining access to the terminals of weather-pack connectors without breaking the wire insulation.

Special Tools
DMM
Lab scope
Jumper wires

Figure 9-19 A jumper harness connected between the sensor and the wiring harness allows the technician to probe for voltage or test resistance without damaging the wiring.

Use an ohmmeter to measure changes in resistor values as the wiper moves across the internal fixed resistor. Disconnect the sensor from the system. Connect the ohmmeter leads to the reference and ground terminals (**Figure 9-20**). This measures the value of the fixed resistor. Check the results against specifications. If good, connect the ohmmeter test leads between the reference and the feedback terminals (**Figure 9-21**). Move the sensor's measurement arm or lever and observe the ohmmeter. The resistance should change smoothly and consistently as the wiper position is changed.

To illustrate a typical test procedure for a three-wire potentiometer circuit (**Figure 9-22**), begin by disconnecting the sensor connector and measuring the voltage on the three terminals with the ignition in the RUN position. Compare the results with specifications since some systems will have 5 volts, 5 volts, and 0 volt on the terminals (terminal A to terminal C), while others will have 5 volts, 0 volt, and 0 volt. If these voltages are within specifications, reconnect the sensor and backprobe the signal wire from the sensor.

With the ignition switch in the RUN position, connect a voltmeter between the signal wire and ground. Typical voltage readings are 0.5 to 1 volt, with the potentiometer in the "at home position." Always refer to the vehicle manufacturer's specifications.

⚠ **Caution**

Do not use a test light to test for voltage. This may damage the system. Also, do not probe for voltage by sticking the wire insulation.

The order of diagnostic testing is not set in stone. This presentation is an example of one testing method for locating the fault with a **throttle position sensor (TPS)** circuit. Not all of the tests are required to isolate the problem.

A – Reference
B – Feedback signal
C – Ground

Figure 9-20 Connecting an ohmmeter to test the potentiometer. This will give the fixed resistance value.

A – Reference
B – Feedback signal
C – Ground

Figure 9-21 Ohmmeter connection to test the wiper movement. The resistance should change smoothly as the wiper moves from one end to the other.

Potentiometer sensor

Figure 9-22 Potentiometer sensor circuit.

Observe the voltmeter reading while slowly moving the potentiometer wiper. Voltage should climb smoothly to the maximum specified level. The typical maximum voltage is between 3.5 and 4.5 volts. If the potentiometer does not reach the specified voltage or if the voltage signal is erratic, replace the sensor.

To test the sensor ground (return) circuit, do a voltage drop test. With the ignition switch in the RUN position, connect the voltmeter between the sensor ground wire at the sensor's harness connector and the battery ground. If the voltage drop across this circuit exceeds specifications (and the control module ground circuit is good), repair the ground wire from the sensor to the control module.

> **SERVICE TIP** While moving the potentiometer through its travel while observing the sensor voltage signal, tap the sensor lightly and watch for fluctuations on the voltmeter reading. Fluctuations indicate a defective sensor.

The lab scope is an excellent tool for testing the potentiometer since it displays every voltage value it sees. This means there is less chance of missing something while trying to use a voltmeter or an ohmmeter. Each time the wiper moves across the fixed resistor, the sensor should provide a smooth analog voltage signal (**Figure 9-23**). If the sensor is defective, glitches will appear in the sensor signal as the wiper is moved (**Figure 9-24**). Depending on circuit design, the glitch for an open can go either upward or downward. A glitch for a short will always go downward.

Evaluate the trace, asking yourself if what you see could actually occur with a good sensor. When looking at the scope trace, remember you are looking at voltage over time. For example, the voltages should still ramp when the potentiometer moves from one extreme position to the other. As the voltage moves from low to high and back again, a "rounded corner" should also be indicated on the trace. If the trace spikes straight up or down, this would indicate a problem with the sensor or circuit.

Figure 9-23 Normal TPS scope pattern as it's opened and closed again.

Figure 9-24 Faulty TPS waveform.

Figure 9-25 Magnetic inductive speed sensor circuit.

Diagnosing Magnetic Induction Sensors

Magnetic induction sensors use the principle of inducing a voltage into a winding by using a moving magnetic field (**Figure 9-25**). These sensors are commonly used to send data to the control module about the speed or position of the monitored component. For example, vehicle speed or wheel speed sensors (WSSs) can be of this design.

Improper timing pickup or rotational speed signals can result from circuit resistance. The zero cross characteristics of the magnetic induction sensor will accurately provide a timing reference provided that the target tooth width is close to the diameter of the sensor pole piece. However, the ideal timing signal occurs at the zero cross only if there is no electrical load on the sensor. A resistance load in the circuit will cause the inductance of the sensor coil to have a current that lags the open circuit generator voltage. This causes a phase shift in the output voltage.

> ⚙ SERVICE TIP In the case of the wheel speed sensor used on many antilock brake systems (ABS), loose wheel bearings or worn parts can alter the air gap.

Another factor that may result in improper sensor operation is an improper tooth **air gap**. The air gap is the distance between the sensor coil and the toothed ring. The signal's amplitude is directly related to the air gap. The space of the air gap is more critical at lower tone wheel speeds. Improper air gap can cause **sensor dropout**. This can occur when the sensor does not produce an output voltage at slower speeds.

> AUTHOR'S NOTE On some vehicles that use the ABS wheel speed sensor signal as an input to the speedometer, a customer may state that the speedometer drops off if they are moving at about 10 mph. Sensor dropout due to an excessive air gap can cause this.

To test the magnetic induction sensor, first check the resistance value of the coil. This will indicate if the coil is intact, open, or shorted. Disconnect the sensor from the system and use an ohmmeter to test the resistance value of the coil. Connect the ohmmeter across the coil terminals and record the reading. Compare the results with specifications. A lower than specified reading indicates a shorted winding. A higher than specified reading indicates resistance or an open.

The voltage generation of the sensor can be tested by connecting a voltmeter across the sensor terminals. The voltmeter must be in the alternating current (AC) position and

Classroom Manual
Chapter 9, page 254

🔧 **Special Tools**
DMM
Lab scope
Scan tool

Magnetic induction sensors are also called magnetic pulse generators.

on the lowest scale. Rotate the shaft while observing the voltage signal. It should increase and decrease with changes in shaft speed. Also, the frequency function of the DMM is useful for monitoring sensor output.

Lab scopes are ideal for testing magnetic induction sensors. Connect the lab scope leads across the sensor's terminals and rotate the shaft. The expected pattern is an AC signal that should be a perfect sine wave when the speed is constant (**Figure 9-26**). As the speed changes, the AC signal should change in amplitude and frequency.

Regardless of the instrument used, you need to look for intermittent dropout, erratic frequency changes, and unstable readings. If the signal output indicates a potential problem, be sure to inspect the tone wheel (**Figure 9-27**), and check and adjust the air gap before condemning the sensor (**Figure 9-28**).

If the sensor performs properly, test the circuits between the computer and the sensor. Disconnect both harness connectors and use an ohmmeter to check the integrity of the circuit. Check for opens and shorts to ground in both circuits. Also, test for a short across the two circuits. Another method is to reconnect the sensor and backprobe the harness connector at the computer and retest at this location. If there is a problem with the signal output now, the circuit wires are at fault.

Figure 9-26 WSS waveform pattern.

Figure 9-27 Inspect the tone wheel for damage.

Figure 9-28 Measuring the air gap.

Use a scan tool to monitor the sensor output. Usually, the scan tool will display the data as rpm or miles per hour. To accurately determine if a sensor is working properly, you must have a base to measure from. For example, on an ABS you can compare all four wheel sensor inputs to see if they all read the same value. When the torque converter is locked, the transmission input speed sensor should read the same speed as the engine crankshaft speed sensor. You can calculate the transmission output speed using the input speed and the current gear ratio.

Some magnetic induction sensor circuits use a bias voltage supplied from the system's computer for use in fault detection (**Figure 9-29**). In addition, the bias voltage also elevates the sensor signal off the common ground plane of the vehicle electrical system to reduce signal interference. The bias voltage varies from manufacturer to manufacturer. Typically, it's 5 volts; however, some manufacturers use a bias voltage of 1.5, 1.8, or 2.29 volts. **Figure 9-30** illustrates a signal voltage from a system that biases the voltage to 2.29 volts. Notice that the sensor voltage signal shifts to the positive. Always refer to the manufacturer's specifications to determine the required bias voltage when troubleshooting a magnetic induction sensor circuit.

The computer monitors the sensor signal at a point between the fixed pull-up resistor and the pickup coil. When power is applied to the circuit, current flows through the pull-up resistor and the pickup coil to ground. The voltage drop at the signal monitor point is a predetermined portion of the reference voltage and a known value that is part of the computer program.

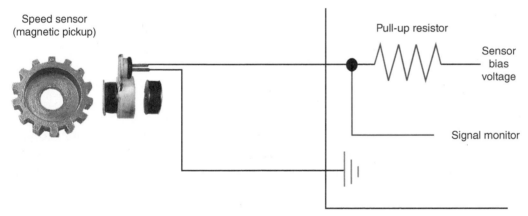

Figure 9-29 A speed sensor circuit that biases the signal circuit.

40 ms/Div

Figure 9-30 Output of a biased speed sensor. Notice the voltage is above 0 volt.

If an open circuit exists, no current flows through the circuit, and the pull-up resistor does not drop any voltage. The signal monitor voltage will be high. In this case, the computer will immediately set a trouble code for an open circuit fault.

If a shorted circuit exists, the pull-up resistor drops all (or nearly all) of the bias voltage. The signal monitor voltage will be lower than the programmed signal monitor voltage.

The simple voltage divider circuit shown in Figure 9-29 allows the computer to detect an electrical fault as soon as the ignition is turned on. The shaft does not need to rotate to detect these circuit defects. You can verify an open or short circuit fault by connecting a voltmeter between the high-voltage side of the pickup coil circuit and ground. Depending on the circuit fault, the meter should read close to full bias voltage or close to 0 volt with the ignition on.

Diagnosing the Magnetoresistive Sensor

Classroom Manual
Chapter 9, page 256

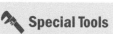
Special Tools

Scan tool
DMM
Lab scope
12-volt test light
Jumper wires

The alternating magnetic poles allow the computer to determine the direction of rotation.

Like the magnetic induction sensor, the air gap of the **magnetoresistive (MR) sensors** is an important factor in sensor operation. If the air gap is too wide, the current change in the circuit will cease to vary between 7 mA and 14 mA (**Figure 9-31**) and remain constant at one of these values. The sensor signal decreases as the air gap increases and will become too small to be recognized by the signal conditioning electronics in the computer.

The following describes the procedure for diagnosing the MR sensor that uses active target tone wheels with alternating magnetic poles (**Figure 9-32**). This type of sensor is common in ABS.

If there is an active ABS fault code for a WSS, visually inspect the wheel speed sensors, related wiring and electrical connections, and the ABS controller for obvious problems. If the inspection finds no problems, carefully backprobe the WSS harness connector and use the DMM to test for voltages on the supply circuit. With the ignition switch in the RUN position, the DMM should read above 10 volts. If the voltage is below 10 volts, check the circuit for a short to ground, high resistance, or an open.

To test for a short to ground, disconnect the WSS harness connector and the control module connector. Ensure the ignition is in the OFF position before disconnecting the components. Connect a 12-volt test light to the positive post of the battery and probe the supply circuit at the WSS connector. If the light illuminates, it indicates the circuit has a short to ground.

Use the ohmmeter function of the DMM to test the supply circuit for an open or high resistance. Another method to test the supply circuit for an open is to connect a jumper

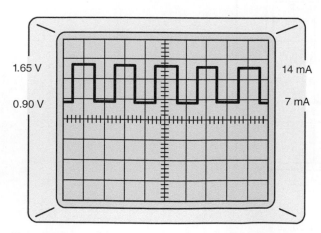

Figure 9-31 Output waveform of the magnetoresistive sensor.

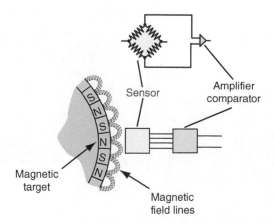

Figure 9-32 Active target with alternating magnetic poles.

wire between ground and the WSS supply circuit at the control module harness connector. Use a test light that is connected to a 12-volt source and probe the supply circuit at the WSS harness connector. If the light illuminates, test the power and ground circuit of the controller. If these are good, repair or replace the controller. If the light does not illuminate, the circuit is open.

If the supply circuit has greater than 10 volts, move the voltmeter test lead to measure the voltage on the signal circuit. The voltage here should be approximately 0.8 or 1.6 volts, depending on the position of the tone wheel. If the voltage reading is too high, test the circuit for a short to voltage condition. It's possible that the supply and the signal circuits are shorted together.

If the meter reads 0 volt during this test, test the circuit for a short to ground in the same manner as discussed for testing for short to ground in the supply circuit.

Test the signal circuit for an open condition by disconnecting the harness connector at the control module and connecting a jumper wire between ground and the signal circuit. Probe the signal circuit at the WSS harness connector using a 12-volt test light connected to a 12-volt source. If the light does not illuminate, the circuit is open.

Verify WSS operation by slowly rotating the tone wheel by hand while monitoring the voltage displayed on the DMM. Do this with the test lead still connected to the signal circuit. The sensor signal voltage should alternate between about 0.8 and 1.6 volts.

To test the circuit with a lab scope, connect the leads as you would for a voltmeter. Adjust the scope settings to read 0.5-volt divisions at a rate of about 20 ms. A good WSS scope waveform should have sharp square corners on the DC signal circuit to the control module (**Figure 9-33**).

> ✿ SERVICE TIP Erratic signal outputs can be caused by damage, missing teeth, cracks, corrosion, or looseness of the tone wheel. Also, wheel bearing failure can cause the WSS signal to be erratic.

For comparative reasons, use a scan tool to monitor the WSS inputs (**Figure 9-34**). While the vehicle is accelerating in a straight line, monitor the sensors while looking for an indication of one dropping out or reading different than the others. Although this identifies a problem, you will still need to use the DMM or lab scope to isolate the root cause.

20 ms/Div

Figure 9-33 Active target speed sensor output.

Data Display - ABS		
Name	Value	Unit
LF Wheel Speed	31.73	mph
RF Wheel Speed	31.89	mph
LR Wheel Speed	31.89	mph
RR Wheel Speed	31.89	mph
ABS Pump Feed	SNA/Not Programmed	Volts
Valve Feed	10.0	Volts
Brake Switch Status	Not Pressed	
Steering Angle Sensor Position	−92.96875	Degrees
Yaw Sensor	0.0	Degrees/sec
Pressure Sensor	16.24	psi
Lateral Acceleration	0.02	G
Rolls Complete	True	

Figure 9-34 The wheel speed sensor output can be read by a scan tool.

CUSTOMER CARE Uneven tire air pressures can cause the WSSs to read different speeds from each wheel position. This may cause the vehicle to enter ABS mode when not needed or cause the control module to set a DTC and inhibit ABS operation. Let your customers know that proper tire pressure maintenance is important for proper ABS operation.

Alternate Magnetoresistive Sensor Diagnosis. Another method for testing MR sensors is to design a tool that will visually display if the sensor is operating (**Figure 9-35**). Assemble the tool using a 9-volt battery, a 220-ohm resistor, and a light-emitting diode (LED). Job Sheet 48 steps you through the construction of this tool.

Figure 9-35 DYI special tool to test the active speed sensor.

Connect the tool to the sensor harness connector with the positive side of the 9-volt battery connected to the voltage input wire of the WSS. The signal side of the sensor connects to the 220-ohm resistor, an LED, and the negative post of the 9-volt battery. As the tone wheel is rotated, the changes in the magnetic poles cause the LED to blink if the WSS is operating properly.

Testing Hall-Effect Sensors

The best way to test the performance of Hall-effect inputs is to use a lab scope. With a lab scope, the unit can be checked while the monitored component is operating. However, a DMM can be used to confirm that the proper voltages and grounds are supplied to the sensor, but it will not be able to indicate the quality of the signal.

With the Hall-effect sensor connector plugged in, connect the positive lead of the lab scope to the signal wire by backprobing the connector. With the component operating, or rotating, the trace should show a clean 5-volt or 12-volt square wave pattern (based on design) that increases in frequency as shaft rpm increases (**Figure 9-36**). The voltage value should go to a full 5 or 12 volts. If it does not, there is resistance on the signal circuit between the Hall-effect sensor and the computer. The voltage should also return to 0 volt. If it does not, then there is resistance on the ground circuit. In addition, a sloping rising and falling line may indicate that the tone wheel is too far away from the magnet of the Hall-effect sensor or that the transistor is faulty (**Figure 9-37**). Check the trace for glitches and noise that may be the result of radio frequency interference (RFI) or electromagnetic interference (EMI) (**Figure 9-38**).

Classroom Manual
Chapter 9, page 258

Special Tools
Lab scope
DMM

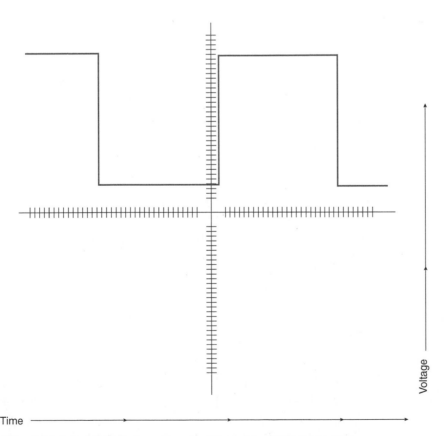

Figure 9-36 A good Hall-effect waveform should be a clean square wave pattern.

Time ⟶

Voltage ⟶

Figure 9-37 A sloping edge on either the rising or falling side may indicate an air gap that is too wide or the transistor in the Hall-effect sensor is faulty.

Time ⟶

Voltage ⟶

Figure 9-38 The Hall-effect trace pattern should be free of glitches and noise.

⚙ **SERVICE TIP** In many systems that use a Hall-effect sensor, it's possible to test the sensor by performing a scratch test. For example, if the fuel ignition system uses a Hall-effect crankshaft sensor, disconnect the sensor connector and use a terminal probe and a jumper wire to momentarily ground the signal wire to the PCM. If the circuit is good, the PCM should respond by activating the fuel pump and/or the shutdown relays. If the PCM responds now, the problem is the Hall-effect sensor.

CASE STUDY

A customer brings their minivan to the shop with a concern that the power liftgate does not open when pressing the key FOB or the inside button. The technician confirms the problem. The technician notes that during the initial testing of the power liftgate, the latch releases when using the outside handle. The technician accesses the liftgate module using their scan tool, but there are no DTCs recorded. While looking through the sensor data display on the scan tool, it is noticed that the door temperature sensor is reading 20°F (−6.7 °C). The temperature in the shop is 77°F (25°C). Knowing that the liftgate will not open if the temperature is too cold, the technician removes the sensor and tests its resistance. The sensor tested good. Inspection of the sensor connector does not indicate any problems. Next, the technician inspects the connector at the liftgate module. This inspection did not reveal any problems. After consulting the wiring diagram, the technician saw that there is an in-line connector in the D-pillar area. After de-trimming the pillar, the technician discovered that the connector terminals had corrosion. After cleaning the connector terminals, the sensor reading matched the temperature in the shop, and the lift gate operated properly. Prior to installing the D-pillar trim, the technician inspected for the cause of the corrosion and found signs of a small water leak from the taillight lens area. The leak was also repaired to assure the customer would not experience a repeat failure.

ASE-Style Review Questions

1. Using the circuit illustrated in Figure 9-3, the computer has set a fault code for the sense circuit being high.

   Technician A says this can be caused by a short to chassis ground in the return circuit.

   Technician B says the signal circuit may be shorted to the return circuit.

   Who is correct?

   A. A only
   B. B only
   C. Both A and B
   D. Neither A nor B

2. What is being tested in **Figure 9-39**?

   A. Open in the sensor signal circuit
   B. Open in the sensor return circuit
   C. Short to ground in the signal circuit
   D. Both A and B

3. The oil pressure light turns on while the engine is running. The oil pressure switch is a normally open switch. Unplugging the oil pressure switch does not turn the warning lamp off.

   Technician A says if the data bus message confirms that the requested state is to turn the lamp on, test the circuit from the PCM to the oil pressure switch for a short to ground.

   Technician B says to test the circuit between the switch and the PCM for an open.

   Who is correct?

   A. A only
   B. B only
   C. Both A and B
   D. Neither A nor B

Coolant
temperature
sensor

Computer

5 V

0 V

Fused jumper wire

Figure 9-39

4. Using a scan tool, the data PID shows a low voltage on the pressure sensor's supply circuit. This could be caused by any of the following EXCEPT:

   A. An open in the supply circuit wire
   B. A short to ground in the supply circuit wire
   C. High internal resistance in the sensor
   D. Excessive voltage drop in the supply circuit

5. Using a scan tool, the data PID shows a high voltage on the pressure sensor's signal circuit. The most likely cause would be:

   A. An open in the supply circuit
   B. An open in the signal circuit
   C. A short to ground in the supply circuit
   D. A short to ground in the signal circuit

6. *Technician A* says the knock sensor output voltage should increase as it detects an engine knock.
   Technician B says the knock sensor voltage should be a square wave pattern.
   Who is correct?

   A. A only   C. Both A and B
   B. B only   D. Neither A nor B

7. *Technician A* says a Hall-effect sensor that shows a trapezoidal-shaped waveform on the lab scope indicates that the tone wheel air gap is too wide.
   Technician B says the Hall-effect waveform should have round edges.
   Who is correct?

   A. A only   C. Both A and B
   B. B only   D. Neither A nor B

8. A potentiometer is suspected of being faulty. Voltmeter testing indicates 5 volts in the signal circuit when backprobed at the sensor, regardless of the wiper position.
   Technician A says an open sensor return circuit can cause this.
   Technician B says a short to ground in the supply circuit can cause this.
   Who is correct?

   A. A only   C. Both A and B
   B. B only   D. Neither A nor B

9. *Technician A* says a bent or damaged tone wheel could cause sensor dropout.
   Technician B says a damaged tone wheel could cause the output voltage to increase.
   Who is correct?

   A. A only   C. Both A and B
   B. B only   D. Neither A nor B

10. When inspecting magnetic reluctant type sensors, check all of the following EXCEPT:

    A. Bias voltage
    B. Proper contact between the sensor and the tone wheel
    C. Proper mounting of the sensor
    D. Tone wheel teeth condition

ASE Challenge Questions

1. *Technician A* says when diagnosing intermittent faults, it's good practice to substitute control modules to see if the problem goes away.
   Technician B says a circuit performance fault indicates that the continuity of the circuit is suspect.
   Who is correct?
   A. A only
   B. B only
   C. Both A and B
   D. Neither A nor B

2. The scan tool displays 5 volts for the ambient temperature sensor. This indicates:
   A. An open in the sensor return circuit
   B. An open in the signal circuit
   C. An open in the sensor
   D. All of these choices

3. A customer states that their air-conditioning system does not work. Testing reveals that the A/C clutch does not turn on when the A/C button is activated.
   Technician A says this can be caused by an open in the high-pressure cutout switch circuit.
   Technician B says a faulty A/C pressure transducer can cause this.
   Who is correct?
   A. A only
   B. B only
   C. Both A and B
   D. Neither A nor B

4. *Technician A* says the MAP sensor reading with the key on, engine off should equal barometric pressure.
   Technician B says the MAP sensor signal voltage should increase when the engine is started.
   Who is correct?
   A. A only
   B. B only
   C. Both A and B
   D. Neither A nor B

5. A vehicle with four-wheel ABS has a problem with the right rear wheel locking during heavy braking.
   Technician A says this could be caused by a bad speed sensor mounted at the wheel.
   Technician B says the speed sensor mounted at the differential could cause this problem.
   Who is correct?
   A. A only
   B. B only
   C. Both A and B
   D. Neither A nor B

Name _____   Date _____

TESTING AN ENGINE COOLANT TEMPERATURE SENSOR

Upon completion of this job sheet, you should be able to check the operation of an ECT sensor and its associated circuits.

ASE Education Foundation Correlation ————————————————————

This job sheet addresses the following **MLR** tasks:

A.1.   Research vehicle service information such as fluid type, vehicle service history, service precautions, technical service bulletins, and recalls including vehicles with advanced driver assistance systems (ADAS). **(P-1)**

A.2.   Identify electrical/electronic system components and configuration. **(P-1)**

A.3.   Retrieve and record DTCs, OBD monitor status, and freeze frame data; clear codes and data when directed. **(P-1)**

A.5.   Demonstrate proper use of a digital multimeter (DMM) when measuring source voltage, voltage drop (including grounds), current flow, and resistance. **(P-1)**

A.8.   Use fused jumper wires to check operation of electrical circuits per service information. **(P-2)**

A.9.   Use wiring diagrams to trace electrical/electronic circuits. **(P-1)**

This job sheet addresses the following **AST** tasks:

A.1.   Research vehicle service information such as fluid type, vehicle service history, service precautions, technical service bulletins, and recalls including vehicles with advanced driver assistance systems (ADAS). **(P-1)**

A.2.   Identify electrical/electronic system components and configuration. **(P-1)**

A.3.   Retrieve and record DTCs, OBD monitor status, and freeze frame data; clear codes and data when directed. **(P-1)**

A.5.   Demonstrate proper use of a digital multimeter (DMM) when measuring source voltage, voltage drop (including grounds), current flow, and resistance. **(P-1)**

A.8.   Use fused jumper wires to check operation of electrical circuits per service information. **(P-1)**

A.9.   Use wiring diagrams during the diagnosis (troubleshooting) of electrical/electronic circuit problems. **(P-1)**

G.8.   Diagnose body electronic system circuits using a scan tool; check for module communication errors (data communication bus systems); determine needed action. **(P-2)**

This job sheet addresses the following **MAST** tasks:

A.1.   Research vehicle service information such as fluid type, vehicle service history, service precautions, technical service bulletins, and recalls including vehicles with advanced driver assistance systems (ADAS). **(P-1)**

A.2.   Identify electrical/electronic system components and configuration. **(P-1)**

A.3.   Retrieve and record DTCs, OBD monitor status, and freeze frame data; clear codes and data when directed. **(P-1)**

A.5.   Demonstrate proper use of a digital multimeter (DMM) when measuring source voltage, voltage drop (including grounds), current flow, and resistance. **(P-1)**

A.8.   Use fused jumper wires to check operation of electrical circuits per service information. **(P-1)**

(Continued)

A.9.   Use wiring diagrams during the diagnosis (troubleshooting) of electrical/
electronic circuit problems. **(P-1)**

G.8.   Diagnose body electronic system circuits using a scan tool; check for module
communication errors (data communication bus systems); determine needed
action. **(P-1)**

We Support

AS≡ | **Education Foundation**

Tools and Materials

- Service information
- DMM
- Lab scope
- Scan tool
- Fused jumper wires
- Backprobing tools

TASK ONE—TESTING THE SENSOR

Procedure

1. Let the engine cool down completely.

2. Describe the location of the ECT sensor.

_____

3. What color of wires are connected to the sensor?

_____

4. Record the resistance specifications for a normal ETC sensor for this vehicle at shop
temperature.

_____

5. Disconnect the electrical connector to the sensor and measure the resistance of the
sensor.

Record your results: _____ ohms at approximately _____ degrees.

6. If not within specifications, what is your next step?

_____

_____

7. If the sensor resistance in step 6 is good, reconnect the sensor wires and backprobe
the signal circuit with a voltmeter or lab scope.

8. Start the engine and allow it to warm to normal operating temperature while
recording your voltage readings at different temperatures.

_____

_____

9. Is the voltage change smooth or erratic?

_____

10. Is the ETC a negative temperature coefficient (NTC) or positive temperature
coefficient (PTC) thermistor?

_____

11. Conclusions:

_____

_____

TASK TWO—CIRCUIT TESTING

Procedure

1. With the scan tool connected to the DLC, disconnect the sensor and observe the voltage on the scan tool with the ignition switch in the RUN position. Record your results.

   _____

   _____

2. If the voltage displayed is not correct, what would your next step be?

   _____

   _____

3. Use a fused jumper wire to short the two ECT sensor terminals together at the connector and observe the voltage displayed on the scan tool. Record your results.

   _____

4. If the voltage did not change, what would be your next step?

   _____

   _____

5. Describe what steps 1 and 3 have proven.

   _____

   _____

6. With the ignition switch in the OFF position, disconnect the PCM connector that houses the ECT sensor circuits.

7. Connect an ohmmeter test lead to each end of the ECT sensor sense circuit and record your results.

   _____

8. If the ohmmeter reading displayed is not correct, what would your next step be?

   _____

   _____

9. Connect an ohmmeter test lead to each end of the ECT sensor ground (return) circuit and record your results. _____

   _____

10. If the ohmmeter reading displayed is not correct, what would your next step be?

    _____

    _____

11. Connect one ohmmeter test lead to the ECT sensor ground (return) circuit and the other lead to the sense circuit at either connector. Record your results.

    _____

12. If the ohmmeter reading displayed is not correct, what would your next step be?

    _____

    _____

13. Conclusions:

    _____

    _____

Instructor's Evaluation:

Grade	Rubric
0	No exposure due to: ☐ Absence ☐ Other:
1	Student unsuccessfully attempted the task
2	Student completed the task with significant instructor assistance.
3	Satisfactory performance of task. The student was able to perform elements of the task with minimal instructor assistance.
4	Competent/Proficient. Able to perform all elements of the task successfully and independently.

Instructor's Notes:

Name _____   Date _____

PRESSURE SWITCH ANALYSIS

Upon completion of this job sheet, you will be able to test and determine the operation of a pressure switch.

ASE Education Foundation Correlation

This job sheet addresses the following **MLR** tasks:

A.1.   Research vehicle service information such as fluid type, vehicle service history, service precautions, technical service bulletins, and recalls including vehicles with advanced driver assistance systems (ADAS). **(P-1)**

A.2.   Identify electrical/electronic system components and configuration. **(P-1)**

A.5.   Demonstrate proper use of a digital multimeter (DMM) when measuring source voltage, voltage drop (including grounds), current flow, and resistance. **(P-1)**

A.8.   Use fused jumper wires to check operation of electrical circuits per service information. **(P-2)**

A.9.   Use wiring diagrams to trace electrical/electronic circuits. **(P-1)**

This job sheet addresses the following **AST** tasks:

A.1.   Research vehicle service information such as fluid type, vehicle service history, service precautions, technical service bulletins, and recalls including vehicles with advanced driver assistance systems (ADAS). **(P-1)**

A.2.   Identify electrical/electronic system components and configuration. **(P-1)**

A.5.   Demonstrate proper use of a digital multimeter (DMM) when measuring source voltage, voltage drop (including grounds), current flow, and resistance. **(P-1)**

A.8.   Use fused jumper wires to check operation of electrical circuits per service information. **(P-1)**

A.9.   Use wiring diagrams during the diagnosis (troubleshooting) of electrical/electronic circuit problems. **(P-1)**

G.8.   Diagnose body electronic system circuits using a scan tool; check for module communication errors (data communication bus systems); determine needed action. **(P-2)**

This job sheet addresses the following **MAST** tasks:

A.1.   Research vehicle service information such as fluid type, vehicle service history, service precautions, technical service bulletins, and recalls including vehicles with advanced driver assistance systems (ADAS). **(P-1)**

A.2.   Identify electrical/electronic system components and configuration. **(P-1)**

A.5.   Demonstrate proper use of a digital multimeter (DMM) when measuring source voltage, voltage drop (including grounds), current flow, and resistance. **(P-1)**

A.8.   Use fused jumper wires to check operation of electrical circuits per service information. **(P-1)**

A.9.   Use wiring diagrams during the diagnosis (troubleshooting) of electrical/electronic circuit problems. **(P-1)**

G.8.   Diagnose body electronic system circuits using a scan tool; check for module communication errors (data communication bus systems); determine needed action. **(P-1)**

Tools and Materials

- Service information
- DMM
- Fused jumper wires
- Any vehicle with a pressure switch included in system operation (oil pressure switch, A/C high-pressure cutout switch, transmission pressure switch, etc.)

Procedure

1. What is the function of the pressure switch for the vehicle and system assigned to you?

   _____

   _____

2. Identify the circuit(s) to the pressure switch.

   _____

   _____

3. Disconnect the switch connector and measure the voltage on the signal circuit with the ignition in the RUN position. Record your reading.

   _____

4. Is this within specifications?   ☐ Yes   ☐ No

   If no, list possible causes for the reading obtained.

   _____

   _____

5. Operate the system while the switch is disconnected. Record your results.

   _____

   _____

6. With the system still operating, use a fused jumper wire and short the signal circuit to the ground circuit. Record your results.

   _____

   _____

7. Based on your observations, is the switch a normally open or normally closed switch?

   _____

   _____

8. With the system turned off, measure the resistance of the switch. Record your readings.

   _____

9. With the system operating, measure the resistance of the switch. Record your readings.

   _____

10. Are the results those expected from the operation you determined in step 7?

    ☐ Yes   ☐ No

    If no, what could be the cause?

    _____

    _____

11. Based on your observations, describe the results of your evaluation of the knock sensor circuit.

_____

_____

Instructor's Evaluation:

Grade	Rubric
0	No exposure due to: ☐ Absence ☐ Other:
1	Student unsuccessfully attempted the task
2	Student completed the task with significant instructor assistance.
3	Satisfactory performance of task. The student was able to perform elements of the task with minimal instructor assistance.
4	Competent/Proficient. Able to perform all elements of the task successfully and independently.

Instructor's Notes:

Name _____   Date _____

TESTING THE POTENTIOMETER

Upon completion of this job sheet, you should be able to inspect and test a potentiometer.

ASE Education Foundation Correlation ──────────────────────────

This job sheet addresses the following **MLR** tasks:

A.1. Research vehicle service information such as fluid type, vehicle service history, service precautions, technical service bulletins, and recalls including vehicles with advanced driver assistance systems (ADAS). **(P-1)**

A.2. Identify electrical/electronic system components and configuration. **(P-1)**

A.5. Demonstrate proper use of a digital multimeter (DMM) when measuring source voltage, voltage drop (including grounds), current flow, and resistance. **(P-1)**

A.9. Use wiring diagrams to trace electrical/electronic circuits. **(P-1)**

This job sheet addresses the following **AST** tasks:

A.1. Research vehicle service information such as fluid type, vehicle service history, service precautions, technical service bulletins, and recalls including vehicles with advanced driver assistance systems (ADAS). **(P-1)**

A.2. Identify electrical/electronic system components and configuration. **(P-1)**

A.5. Demonstrate proper use of a digital multimeter (DMM) when measuring source voltage, voltage drop (including grounds), current flow, and resistance. **(P-1)**

A.9. Use wiring diagrams during the diagnosis (troubleshooting) of electrical/ electronic circuit problems. **(P-1)**

A.13. Test and measure circuit using an oscilloscope and/or graphing multimeter (GMM); interpret results; determine needed action. **(P-2)**

G.8. Diagnose body electronic system circuits using a scan tool; check for module communication errors (data communication bus systems); determine needed action. **(P-2)**

This job sheet addresses the following **MAST** tasks:

A.1. Research vehicle service information such as fluid type, vehicle service history, service precautions, technical service bulletins, and recalls including vehicles with advanced driver assistance systems (ADAS). **(P-1)**

A.2. Identify electrical/electronic system components and configuration. **(P-1)**

A.5. Demonstrate proper use of a digital multimeter (DMM) when measuring source voltage, voltage drop (including grounds), current flow, and resistance. **(P-1)**

A.9. Use wiring diagrams during the diagnosis (troubleshooting) of electrical/ electronic circuit problems. **(P-1)**

A.13. Test and measure circuit using an oscilloscope and/or graphing multimeter (GMM); interpret results; determine needed action. **(P-1)**

G.8. Diagnose body electronic system circuits using a scan tool; check for module communication errors (data communication bus systems); determine needed action. **(P-1)**

We Support

ASE | Education Foundation

Tools and Materials
- Service information
- Scan tool
- DMM
- Lab scope

Procedure

1. Identify the purpose of the potentiometer for your assigned task.

   _____

   _____

2. Identify the color code and purpose of the circuits to the potentiometer.

   _____

   _____

3. Unplug the harness connector at the potentiometer and measure the voltage at each terminal with the ignition switch in the RUN position.

   _____

   _____

4. Do the voltage readings match specifications?   ☐ Yes   ☐ No

   If no, what problem is indicated?

   _____

   _____

5. Measure the resistance across the potentiometer between the supply and ground terminals.

   _____

   Specification: _____

6. Measure the resistance between the signal and ground circuits as you move the wiper of the potentiometer. Describe your results.

   _____

   _____

7. Reconnect the harness connector to the potentiometer and backprobe the signal circuit terminal. With the ignition switch in the RUN position, what is the voltage on the signal circuit?

   _____

8. Move the potentiometer through its entire sweep while observing the voltmeter. Describe your observations.

   _____

   _____

9. What was the highest voltage observed? _____

   Specification: _____

10. Repeat steps 7 through 9 with a lab scope and record your observations.

_____

_____

Instructor's Evaluation:

Grade	Rubric
0	No exposure due to: ☐ Absence ☐ Other:
1	Student unsuccessfully attempted the task
2	Student completed the task with significant instructor assistance.
3	Satisfactory performance of task. The student was able to perform elements of the task with minimal instructor assistance.
4	Competent/Proficient. Able to perform all elements of the task successfully and independently.

Instructor's Notes:

Name _____   Date _____

DMM TESTING OF THE MAGNETIC INDUCTION SENSOR

Upon completion of this job sheet, you will have inspected and tested the magnetic induction speed sensor used to monitor wheel speed for the ABS system using a DMM.

ASE Education Foundation Correlation

This job sheet addresses the following **MLR** tasks:

A.1.   Research vehicle service information such as fluid type, vehicle service history, service precautions, technical service bulletins, and recalls including vehicles with advanced driver assistance systems (ADAS). **(P-1)**

A.2.   Identify electrical/electronic system components and configuration. **(P-1)**

A.5.   Demonstrate proper use of a digital multimeter (DMM) when measuring source voltage, voltage drop (including grounds), current flow, and resistance. **(P-1)**

A.9.   Use wiring diagrams to trace electrical/electronic circuits. **(P-1)**

This job sheet addresses the following **AST** tasks:

A.1.   Research vehicle service information such as fluid type, vehicle service history, service precautions, technical service bulletins, and recalls including vehicles with advanced driver assistance systems (ADAS). **(P-1)**

A.2.   Identify electrical/electronic system components and configuration. **(P-1)**

A.5.   Demonstrate proper use of a digital multimeter (DMM) when measuring source voltage, voltage drop (including grounds), current flow, and resistance. **(P-1)**

A.9.   Use wiring diagrams during the diagnosis (troubleshooting) of electrical/electronic circuit problems. **(P-1)**

This job sheet addresses the following **MAST** tasks:

A.1.   Research vehicle service information such as fluid type, vehicle service history, service precautions, technical service bulletins, and recalls including vehicles with advanced driver assistance systems (ADAS). **(P-1)**

A.2.   Identify electrical/electronic system components and configuration. **(P-1)**

A.5.   Demonstrate proper use of a digital multimeter (DMM) when measuring source voltage, voltage drop (including grounds), current flow, and resistance. **(P-1)**

A.9.   Use wiring diagrams during the diagnosis (troubleshooting) of electrical/electronic circuit problems. **(P-1)**

We Support
ASE | **Education Foundation**

Tools and Materials

- Service information
- DMM
- Lift or jacks with stands

Procedure

1. Safely lift the wheels of the vehicle from the ground.

2. Referring to the service information, identify the location of the RF wheel speed sensor connector. Where is it located?

   _____

   _____

3. Other than ABS, what other vehicle systems require information from the wheel speed sensors?

   _____

   _____

   _____

   _____

4. Locate and disconnect the speed sensor two-way connector harness for the assigned wheel location.

5. Connect your DMM to the sensor side of the connector and measure the resistance across the two terminals of the sensor.

   _____

   Specifications: _____

6. Set the DMM to read DC voltage and measure the voltage across the harness side of the connector with the ignition switch in the RUN position. _____

   Specifications: _____

7. What is the purpose of this voltage?

   _____

   _____

8. Reconnect the speed sensor and backprobe the (+) side of the connector. Read the voltage at this terminal _____

   Did the voltage drop? _____

9. Backprobe the (+) and (−) terminals of the speed sensor and connect the voltmeter across the two terminals. With the voltmeter set to read AC voltage, have an assistant start the vehicle and accelerate to 10 mph and maintain that speed. Record the voltmeter reading.

   _____

10. Accelerate to 20 mph and record the voltmeter reading.

    _____

11. Explain your observations.

    _____

    _____

    _____

12. Switch the DMM to read frequency and perform steps 8 and 9 again. Record your observations.

_____

_____

_____

Instructor's Evaluation:

Grade	Rubric
0	No exposure due to: ☐ Absence ☐ Other:
1	Student unsuccessfully attempted the task
2	Student completed the task with significant instructor assistance.
3	Satisfactory performance of task. The student was able to perform elements of the task with minimal instructor assistance.
4	Competent/Proficient. Able to perform all elements of the task successfully and independently.

Instructor's Notes:

Name _____   Date _____

LAB SCOPE TESTING A MAGNETIC INDUCTION SENSOR

Upon completion and review of this job sheet, you should be able to inspect and test an ABS wheel speed sensor with a lab scope.

ASE Education Foundation Correlation ————————————————

This job sheet addresses the following **MLR** tasks:

A.1. Research vehicle service information such as fluid type, vehicle service history, service precautions, technical service bulletins, and recalls including vehicles with advanced driver assistance systems (ADAS). **(P-1)**

A.2. Identify electrical/electronic system components and configuration. **(P-1)**

A.9. Use wiring diagrams to trace electrical/electronic circuits. **(P-1)**

This job sheet addresses the following **AST** tasks:

A.1. Research vehicle service information such as fluid type, vehicle service history, service precautions, technical service bulletins, and recalls including vehicles with advanced driver assistance systems (ADAS). **(P-1)**

A.2. Identify electrical/electronic system components and configuration. **(P-1)**

A.9. Use wiring diagrams during the diagnosis (troubleshooting) of electrical/electronic circuit problems. **(P-1)**

A.13. Test and measure circuit using an oscilloscope and/or graphing multimeter (GMM); interpret results; determine needed action. **(P-2)**

G.8. Diagnose body electronic system circuits using a scan tool; check for module communication errors (data communication bus systems); determine needed action. **(P-2)**

This job sheet addresses the following **MAST** tasks:

A.1. Research vehicle service information such as fluid type, vehicle service history, service precautions, technical service bulletins, and recalls including vehicles with advanced driver assistance systems (ADAS). **(P-1)**

A.2. Identify electrical/electronic system components and configuration. **(P-1)**

A.9. Use wiring diagrams during the diagnosis (troubleshooting) of electrical/electronic circuit problems. **(P-1)**

A.13. Test and measure circuit using an oscilloscope and/or graphing multimeter (GMM); interpret results; determine needed action. **(P-1)**

G.8. Diagnose body electronic system circuits using a scan tool; check for module communication errors (data communication bus systems); determine needed action. **(P-1)**

We Support
Education Foundation

Tools and Materials

- Wiring diagram
- Lab scope
- Lift or jacks with stands

Procedure

1. Safely lift the wheels of the vehicle from the ground.

2. Referring to the service information, identify the location of the assigned wheel speed sensor connector. Where is it located?

   _____

   _____

3. Other than ABS, what other vehicle systems require information from the wheel speed sensors?

   _____

   _____

   _____

   _____

4. Backprobe the speed sensor two-way connector harness for the RF wheel.

5. Connect the lab scope and observe the trace while the wheel is rotated at a constant speed.

   General results: _____

   _____

6. Increase the speed of the wheel while observing the graph. General results:

   _____

   _____

7. Record the operational action of the speed sensor and make any recommendation.

   _____

   _____

Instructor's Evaluation:

Grade	Rubric
0	No exposure due to: ☐ Absence ☐ Other:
1	Student unsuccessfully attempted the task
2	Student completed the task with significant instructor assistance.
3	Satisfactory performance of task. The student was able to perform elements of the task with minimal instructor assistance.
4	Competent/Proficient. Able to perform all elements of the task successfully and independently.

Instructor's Notes:

Name _____   Date _____

BUILDING A MAGNETORESISTIVE SENSOR TESTER

Tools and Materials

- 9-volt battery
- Project enclosure (Radio Shack #270-1801) 9-volt battery
- Snap connector (Radio Shack #270-324)
- Mini alligator clips (Radio Shack #270-1540)
- 2 feet of red wire
- 2 feet of black wire
- Green LED (Radio Shack #276-022)
- LED holder (Radio Shack #276-080)
- 220-ohm resistor (Radio Shack #271-1111)

Procedure

1. Use the schematic shown in **Figure 9-40** to construct the sensor tester.

2. Backprobe the terminals of an active wheel speed sensor and turn the ignition switch to the RUN position. Connect the tester leads to the sensor terminals, observing polarity, and rotate the wheel. Describe your results.

_____

_____

Figure 9-40 Magnetoresistive sensor tester schematic.

Instructor's Evaluation:

Grade	Rubric
0	No exposure due to: ☐ Absence ☐ Other:
1	Student unsuccessfully attempted the task
2	Student completed the task with significant instructor assistance.
3	Satisfactory performance of task. The student was able to perform elements of the task with minimal instructor assistance.
4	Competent/Proficient. Able to perform all elements of the task successfully and independently.

Instructor's Notes:

Name _____   Date _____

ACTIVE SPEED SENSOR CIRCUIT DIAGNOSIS

Upon completion of this job sheet, you will have inspected and tested the speed sensor circuit.

ASE Education Foundation Correlation

This job sheet addresses the following **MLR** tasks:

A.1.   Research vehicle service information such as fluid type, vehicle service history, service precautions, technical service bulletins, and recalls including vehicles with advanced driver assistance systems (ADAS). **(P-1)**

A.2.   Identify electrical/electronic system components and configuration. **(P-1)**

A.5.   Demonstrate proper use of a digital multimeter (DMM) when measuring source voltage, voltage drop (including grounds), current flow, and resistance. **(P-1)**

A.9.   Use wiring diagrams to trace electrical/electronic circuits. **(P-1)**

This job sheet addresses the following **AST** tasks:

A.1.   Research vehicle service information such as fluid type, vehicle service history, service precautions, technical service bulletins, and recalls including vehicles with advanced driver assistance systems (ADAS). **(P-1)**

A.2.   Identify electrical/electronic system components and configuration. **(P-1)**

A.5.   Demonstrate proper use of a digital multimeter (DMM) when measuring source voltage, voltage drop (including grounds), current flow, and resistance. **(P-1)**

A.9.   Use wiring diagrams during the diagnosis (troubleshooting) of electrical/electronic circuit problems. **(P-1)**

G.8.   Diagnose body electronic system circuits using a scan tool; check for module communication errors (data communication bus systems); determine needed action. **(P-2)**

This job sheet addresses the following **MAST** tasks:

A.1.   Research vehicle service information such as fluid type, vehicle service history, service precautions, technical service bulletins, and recalls including vehicles with advanced driver assistance systems (ADAS). **(P-1)**

A.2.   Identify electrical/electronic system components and configuration. **(P-1)**

A.5.   Demonstrate proper use of a digital multimeter (DMM) when measuring source voltage, voltage drop (including grounds), current flow, and resistance. **(P-1)**

A.9.   Use wiring diagrams during the diagnosis (troubleshooting) of electrical/electronic circuit problems. **(P-1)**

G.8.   Diagnose body electronic system circuits using a scan tool; check for module communication errors (data communication bus systems); determine needed action. **(P-1)**

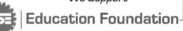

We Support
Education Foundation

Tools and Materials

- Service information
- Scan tool
- DMM

Procedure

1. Using the service information, determine what other vehicle systems require information from the wheel speed sensors (other than the ABS system)?

   _____

   _____

   _____

   _____

2. Establish communication with the assigned vehicle using the scan tool and check for DTCs in the ABS system. Are any DTCs present?   ☐ Yes   ☐ No

   If so, record them.

   _____

   _____

3. If applicable, list the possible causes for the DTC(s).

   _____

   _____

4. Visually inspect the sensor wiring. Check for any chafed, broken, or pierced wires. Also, visually inspect related harness connectors for broken, bent, spread, or pushed out terminals. Record your findings.

   _____

   _____

5. Inspect the tone wheel teeth for missing teeth, cracks, or looseness. Teeth should be perfectly square, not bent or nicked. Record your results.

   _____

   _____

6. Disconnect the WSS harness connector and use a DMM to measure the voltage of the supply circuit at the harness connector. What was the measured voltage?

   _____

7. Using a DMM, measure the voltage between the supply circuit and the signal circuit at the WSS harness connector. What were the results?

   _____

8. What do the results indicate?

   _____

   _____

9. Turn the ignition off and disconnect the ABM harness connector and the WSS connector.

10. Using a DMM, measure the resistance between the WSS signal circuit and the WSS supply circuit. What were the results?

    _____

    _____

11. What do the results indicate?

    _____

Instructor's Evaluation:

Grade	Rubric
0	No exposure due to: ☐ Absence ☐ Other:
1	Student unsuccessfully attempted the task
2	Student completed the task with significant instructor assistance.
3	Satisfactory performance of task. The student was able to perform elements of the task with minimal instructor assistance.
4	Competent/Proficient. Able to perform all elements of the task successfully and independently.

Instructor's Notes:

Name _____   **Date** _____

TESTING THE HALL-EFFECT SENSORS

Upon completion of this job sheet, you will be able to test a camshaft and crankshaft position sensor.

ASE Education Foundation Correlation —————————————————————

This job sheet addresses the following **MLR** tasks:

A.1.   Research vehicle service information such as fluid type, vehicle service history, service precautions, technical service bulletins, and recalls including vehicles with advanced driver assistance systems (ADAS). **(P-1)**

A.2.   Identify electrical/electronic system components and configuration. **(P-1)**

A.5.   Demonstrate proper use of a digital multimeter (DMM) when measuring source voltage, voltage drop (including grounds), current flow, and resistance. **(P-1)**

A.9.   Use wiring diagrams to trace electrical/electronic circuits. **(P-1)**

This job sheet addresses the following **AST** tasks:

A.1.   Research vehicle service information such as fluid type, vehicle service history, service precautions, technical service bulletins, and recalls including vehicles with advanced driver assistance systems (ADAS). **(P-1)**

A.2.   Identify electrical/electronic system components and configuration. **(P-1)**

A.5.   Demonstrate proper use of a digital multimeter (DMM) when measuring source voltage, voltage drop (including grounds), current flow, and resistance. **(P-1)**

A.9.   Use wiring diagrams during the diagnosis (troubleshooting) of electrical/ electronic circuit problems. **(P-1)**

A.13.   Test and measure circuit using an oscilloscope and/or graphing multimeter (GMM); interpret results; determine needed action. **(P-2)**

This job sheet addresses the following **MAST** tasks:

A.1.   Research vehicle service information such as fluid type, vehicle service history, service precautions, technical service bulletins, and recalls including vehicles with advanced driver assistance systems (ADAS). **(P-1)**

A.2.   Identify electrical/electronic system components and configuration. **(P-1)**

A.5.   Demonstrate proper use of a digital multimeter (DMM) when measuring source voltage, voltage drop (including grounds), current flow, and resistance. **(P-1)**

A.9.   Use wiring diagrams during the diagnosis (troubleshooting) of electrical/ electronic circuit problems. **(P-1)**

A.13.   Test and measure circuit using an oscilloscope and/or graphing multimeter (GMM); interpret results; determine needed action. **(P-1)**

We Support

ASE | **Education Foundation**

Tools and Materials

- A vehicle equipped with Hall-effect sensor
- Service information for the selected vehicle
- Lab scope
- DMM

Procedure

1. Describe the location on the vehicle for the Hall-effect sensor assigned to you.

   _____

   _____

2. What is the function of this sensor in the system?

   _____

   _____

3. According to the service information, what should be the voltages at each of the terminals of the sensor with the ignition switch in the RUN position?

   _____

   _____

4. Disconnect the sensor connector and measure the voltage to each terminal (harness side) with the ignition switch in the RUN position. Do the voltages agree with those found in step 3?   ☐ Yes   ☐ No

5. If the voltages don't agree, what is the likely cause?

   _____

   _____

6. Reconnect all disconnected connectors. Connect the lab scope to read the signal from the sensors to the control module. Select the appropriate voltage level for the sensor.

7. Operate the system while observing the sensor signals. Describe the signals received.

   _____

   _____

8. Do any of the patterns indicate a problem?   ☐ Yes   ☐ No

   If so, what is the likely cause?

   _____

   _____

Instructor's Evaluation:

Grade	Rubric
0	No exposure due to: ☐ Absence ☐ Other:
1	Student unsuccessfully attempted the task
2	Student completed the task with significant instructor assistance.
3	Satisfactory performance of task. The student was able to perform elements of the task with minimal instructor assistance.
4	Competent/Proficient. Able to perform all elements of the task successfully and independently.

Instructor's Notes:

_____

DIAGNOSTIC CHART 9-1

PROBLEM AREA:	Temperature sensor circuit performance.
SYMPTOMS:	Implausible voltage on the temperature sensor signal circuit; Improper system operation; Limp-in mode initiated.
POSSIBLE CAUSES:	**1.** Sensor signal circuit shorted to voltage. **2.** Sensor signal circuit shorted to ground. **3.** Sensor signal circuit shorted to sensor ground circuit. **4.** Signal circuit open. **5.** High resistance in signal circuit. **6.** Sensor ground circuit open. **7.** High resistance in sensor ground circuit. **8.** Faulty sensor. **9.** Internal controller fault.

DIAGNOSTIC CHART 9-2

PROBLEM AREA:	Temperature sensor circuit voltage low.
SYMPTOMS:	Improper system operation; Limp-in mode initiated.
POSSIBLE CAUSES:	**1.** Sensor signal circuit shorted to ground. **2.** Signal circuit open. **3.** High resistance in signal circuit. **4.** Sensor ground circuit open. **5.** High resistance in sensor ground circuit. **6.** Faulty sensor. **7.** Internal controller fault.

DIAGNOSTIC CHART 9-3

PROBLEM AREA:	Temperature sensor circuit voltage high.
SYMPTOMS:	Improper system operation; Limp-in mode initiated.
POSSIBLE CAUSES:	**1.** Sensor signal circuit shorted to voltage. **2.** Signal circuit open. **3.** High resistance in signal circuit. **4.** Sensor ground circuit open. **5.** High resistance in sensor ground circuit. **6.** Faulty sensor. **7.** Internal controller fault.

DIAGNOSTIC CHART 9-4

PROBLEM AREA:	Malfunction pressure switch or circuit.
SYMPTOMS:	Improper warning lamp operation; Improper system operation; Limp in mode initiated.
POSSIBLE CAUSES:	**1.** Open sense circuit. **2.** Open ground circuit. **3.** Sense circuit shorted to ground. **4.** Sense circuit shorted to voltage. **5.** Faulty pressure switch.

DIAGNOSTIC CHART 9-5

PROBLEM AREA:	Pressure sensor signal voltage high.
SYMPTOMS:	Improper system operation; Limp-in mode initiated.
POSSIBLE CAUSES:	**1.** Open signal circuit. **2.** Signal circuit shorted to voltage. **3.** Supply circuit shorted to battery voltage. **4.** Sensor ground circuit open. **5.** Faulty sensor. **6.** Internal controller fault.

DIAGNOSTIC CHART 9-6

PROBLEM AREA:	Pressure sensor signal voltage low.
SYMPTOMS:	Improper system operation; Limp-in mode initiated.
POSSIBLE CAUSES:	**1.** Signal circuit shorted to ground. **2.** Supply circuit shorted to ground. **3.** Open supply circuit. **4.** Faulty sensor. **5.** Internal controller fault.

DIAGNOSTIC CHART 9-7

PROBLEM AREA:	Piezoelectric sensor high-voltage fault.
SYMPTOMS:	System malfunction.
POSSIBLE CAUSES:	**1.** Sensor signal circuit shorted to battery voltage. **2.** Signal circuit open. **3.** Sensor ground circuit open. **4.** High resistance sensor ground circuit. **5.** Faulty sensor. **6.** Internal controller fault.

DIAGNOSTIC CHART 9-8

PROBLEM AREA:	Incorrect reading from potentiometer.
SYMPTOMS:	Sensor voltage at the controller is less than specifications.
POSSIBLE CAUSES:	**1.** Supply circuit open. **2.** Supply circuit shorted to ground. **3.** Signal circuit shorted to ground. **4.** Signal circuit shorted to the sensor ground circuit. **5.** Faulty sensor. **6.** Internal controller fault.

DIAGNOSTIC CHART 9-9

PROBLEM AREA:	Incorrect reading from potentiometer.
SYMPTOMS:	Sensor voltage at the controller is higher than specifications.
POSSIBLE CAUSES:	**1.** Signal circuit shorted to battery voltage. **2.** Signal circuit open. **3.** Signal circuit shorted to the supply circuit. **4.** Sensor ground circuit open. **5.** Faulty sensor. **6.** Internal controller fault.

DIAGNOSTIC CHART 9-10

PROBLEM AREA:	Incorrect reading from potentiometer.
SYMPTOMS:	Sensor voltage at the controller is not correct for position indicated by other inputs.
POSSIBLE CAUSES:	**1.** High resistance in the SIGNAL CIRCUIT. **2.** High resistance in the sensor ground circuit. **3.** High resistance in the voltage SUPPLY CIRCUIT. **4.** Faulty sensor. **5.** Internal controller fault.

DIAGNOSTIC CHART 9-11

PROBLEM AREA:	No signal from magnetic induction sensor.
SYMPTOMS:	No signal input detected by controller.
POSSIBLE CAUSES:	**1.** Sensor (+) circuit short to ground. **2.** Sensor (−) circuit short to ground. **3.** Sensor (+) circuit open. **4.** Sensor (−) circuit open. **5.** Sensor (+) circuit shorted to voltage. **6.** Sensor (−) circuit shorted to voltage. **7.** Sensor (−) circuit shorted to the sensor (+) circuit. **8.** Excessive sensor air gap. **9.** Faulty sensor. **10.** Internal controller fault.

DIAGNOSTIC CHART 9-12

PROBLEM AREA:	Intermittent signal from magnetic induction sensor.
SYMPTOMS:	Sudden change of output signal from the sensor.
POSSIBLE CAUSES:	**1.** Sensor (+) circuit short to ground. **2.** Sensor (−) circuit short to ground. **3.** Sensor (+) circuit open. **4.** Sensor (−) circuit open. **5.** Sensor (+) circuit shorted to voltage. **6.** Sensor (−) circuit shorted to voltage. **7.** Sensor (−) circuit shorted to the sensor (+) circuit. **8.** Excessive sensor air gap. **9.** Faulty sensor. **10.** Internal controller fault.

DIAGNOSTIC CHART 9-13

PROBLEM AREA:	Incorrect signal from magnetic induction sensor.
SYMPTOMS:	Incorrect reading from sensor as compared to other inputs.
POSSIBLE CAUSES:	**1.** High resistance in the sensor (+) circuit. **2.** High resistance in the sensor (−) circuit. **3.** Excessive sensor air gap. **4.** Faulty sensor. **5.** Internal controller fault.

DIAGNOSTIC CHART 9-14

PROBLEM AREA:	Incorrect speed input from active magnetoresistive sensor.
SYMPTOMS:	Incorrect reading from sensor as compared to other inputs.
POSSIBLE CAUSES:	**1.** Wiring harness, terminal, connector damage. **2.** Loose sensor mounting. **3.** Damaged tone wheel. **4.** Faulty sensor. **5.** Internal controller fault.

DIAGNOSTIC CHART 9-15

PROBLEM AREA:	Erratic performance from active magnetoresistive sensor.
SYMPTOMS:	Signal intermittently missing; Periodic drop off of signal.
POSSIBLE CAUSES:	**1.** Wiring harness, terminal, connector damage. **2.** Loose sensor mounting. **3.** Damaged tone wheel. **4.** Faulty sensor. **5.** Internal controller fault.

DIAGNOSTIC CHART 9-16

PROBLEM AREA:	Active magnetoresistive sensor circuit failure
SYMPTOMS:	**1.** No sensor output. **2.** Sensor circuit fails the diagnostic test. **3.** Sensor circuit low. **4.** Sensor circuit high.
POSSIBLE CAUSES:	**1.** Wiring harness, terminal, connector damage. **2.** 12-volt supply circuit shorted to ground. **3.** Signal circuit shorted to ground. **4.** 12-volt supply circuit shorted to voltage. **5.** 12-volt supply circuit open. **6.** Signal circuit shorted to voltage. **7.** Signal circuit open. **8.** Signal circuit shorted to 12-volt supply circuit. **9.** Faulty sensor. **10.** Internal controller fault.

DIAGNOSTIC CHART 9-17

PROBLEM AREA:	Hall-effect sensor input missing.
SYMPTOMS:	No signal is present during motion of monitored component.
POSSIBLE CAUSES:	1. Sensor voltage supply circuit shorted to battery voltage. 2. Sensor voltage supply circuit open. 3. Sensor voltage supply circuit shorted to ground. 4. Sensor signal circuit open. 5. Sensor signal circuit shorted to battery voltage. 6. Sensor signal circuit shorted ground. 7. Sensor signal circuit shorted to the voltage supply circuit. 8. Sensor ground circuit open. 9. Excessive sensor air gap. 10. Faulty sensor. 11. Internal controller fault.

DIAGNOSTIC CHART 9-18

PROBLEM AREA:	Intermittent Hall-effect sensor input.
SYMPTOMS:	Intermittent signal during motion of monitored component.
POSSIBLE CAUSES:	1. Sensor voltage supply circuit open. 2. Sensor voltage supply circuit shorted to ground. 3. Sensor signal circuit shorted to battery voltage. 4. Sensor signal circuit open. 5. Sensor signal circuit shorted ground. 6. Sensor signal circuit shorted to the voltage supply circuit. 7. Sensor ground circuit open. 8. Excessive sensor air gap. 9. Faulty sensor. 10. Damaged tone wheel. 11. Internal controller fault.

CHAPTER 10

VEHICLE MULTIPLEXING DIAGNOSTICS

Basic Tools

Hand tools
Fender covers
Service information

Upon completion and review of this chapter, you should be able to:

- Describe the purpose of U- and B-codes.
- Properly diagnose an ISO 9141-2 bus system and determine needed repairs.
- Properly diagnose an ISO-K bus system and determine needed repairs.
- Properly diagnose a class A bus system and determine needed repairs.
- Properly diagnose the J1850 bus system and determine needed repairs.
- Properly diagnose a Controller Area Network (CAN) bus system and determine needed repairs.
- Properly diagnose a Local Interconnect Network (LIN) bus system and determine needed repairs.
- Describe the necessary procedures to authenticate a scan tool and user.

Terms to Know

B-codes
J1962 breakout box (BOB)

U-codes
Vehicle module scan

Introduction

If a vehicle's multiplexing system should fail, symptoms can range from a single function (such as instrument gauges) not operating to multiple function failures, including engine no-start. Diagnosing the bus system isn't much different from diagnosing any other electrical system. Begin by verifying the customer's complaint; determine if there are any related symptoms, and then analyze the symptoms to develop a logical troubleshooting plan. It's important to understand how the bus system you are diagnosing should operate and the normal voltages and resistances on the system. Bus system failures include circuit opens, shorts, high resistance, and component failures. In addition, don't be quick to condemn the bus system if a module isn't communicating. It's possible that the module isn't powering up due to loss of battery voltage feed or loss of ground. This chapter discusses those items you must be aware of while diagnosing different bus networks. The most common bus systems are discussed here.

Since most bus systems communicate with the scan tool, they will have a point of connection at the data link connector (DLC). To assist in testing the data bus, a **J1962 breakout box (BOB)** is available (**Figure 10-1**). Since J1962 is the mandated DLC configuration for on-board diagnostics, second generation (OBD II), this tool will work on any OBD II–compliant vehicle. The J1962 BOB provides a pass-through test point that connects in series between the DLC and the scan tool. This provides easy testing of voltage and resistance of the DLC circuits without damaging the DLC.

Figure 10-1 The J1962 BOB makes pin out testing of the DLC easier.

Communication Fault Codes

Diagnostic trouble codes (DTCs) assigned to the vehicle communication network are called **U-codes**. These codes follow the same Society of Automotive Engineers (SAE) guideline as the P-codes used for powertrain faults. The prefix *U* indicates the fault is associated with network communications.

In addition, DTCs assigned to the vehicle's body systems and control modules are called **B-codes**. Typically, these codes refer to a failure of the system the module operates (such as a sensor failure). If a module relies on a bus message from another module but does not receive it, the first module may set both a U-code and a B-code. The U-code would be due to the loss of communication with the second module, and the B-code due to the system not able to perform a function. B-codes also follow the same SAE guideline as the P-codes.

Most modules on the bus network are capable of setting U-codes if they detect abnormal conditions. Most bus modules can detect loss of communication conditions with one or more modules or a bus failure. Some modules may also be able to monitor the actual voltage on the bus circuits and set additional trouble codes for conditions such as voltage high, low, shorted, or open bus circuits.

In addition, most modules report the status of the DTCs. If the conditions currently exist, the DTC is reported as being active. If the conditions no longer exist, then the DTC is reported as being stored.

AUTHOR'S NOTE It's important to remember that one fault can set multiple DTCs in many different modules. Just because a module recorded a fault does not mean the problem is in that module; it could be anywhere in the bus circuits.

Vehicle Module Scan

Not many vehicle systems are standalone and rely on inputs, outputs, and data from other systems. This involves using data bus networks, including different CAN bus networks, dedicated LIN buses, and MOST buses. In addition, manufacturer-specific network protocols may be used. It's not uncommon for one vehicle to use several different bus networks. Some scan tools provide a **vehicle module scan** function that will request all modules on the vehicle's bus networks to respond and then list those

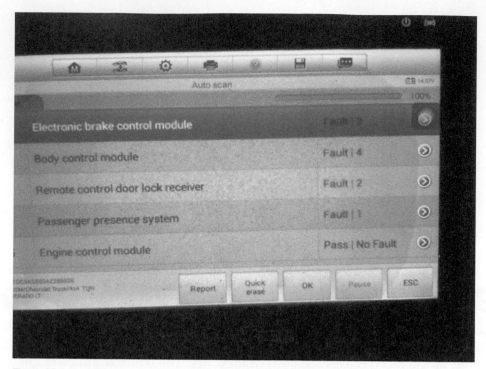

Figure 10-2 The vehicle module scan report is a quick method of determining if a module is communicating on the bus.

that replied **(Figure 10-2)**. In addition, the vehicle module scan report can indicate which modules have DTCs recorded.

Since most vehicle systems are interlinked, the vehicle module scan is very helpful. For example, a fault in the Antilock Brakes System (ABS) may prevent the Adaptive Cruise Control (ACC) system from operating. The vehicle module scan makes it simple to identify modules with DTCs in a single function. Without this feature, you must go into each module and read the DTCs.

Record all DTCs onto the repair order or into the shop management system. Note if the DTCs are current, pending, or historic. Some scan tools provide a record and print function for doing this. Some scan tools provide for uploading the data to a removable storage device from which the data can be printed. In addition, connecting the scan tool to the shop's Wi-Fi allows printing using a network printer. Regardless, you should record the DTCs and which modules stored them. Some DTCs will have Freeze Frame data associated with them. Freeze Frame records data associated with the running conditions of the vehicle at the time of the fault. Reference this information and use it to determine if multiple DTCs set at the same time.

Another useful function of the vehicle scan report is to use it to document the customer's concern and verify the repair. A pre-repair scan can be attached to the repair order to illustrate the issues identified during diagnostics to the customer. A post-vehicle scan shows that the repair was successful, and the DTCs did not return.

ISO 9141-2 Bus System Diagnostics

Classroom Manual
Chapter 10, page 274

The ISO 9141-2 standard bus system provides for communication links between the scan tool and the module. Some OBD II vehicles use this protocol for communication with the powertrain control module (PCM). In addition, some manufacturers use the system for communication between the scan tool and other modules on the vehicle **(Figure 10-3)**.

Figure 10-3 ISO 9142-2 bus system used to communicate between the scan tool and modules on the vehicle.

The ISO 9141-2 bus uses a K-line to transmit data from the module to the scan tool and an L-line for the module to receive data from the scan tool (**Figure 10-4**). The scan tool supplies bias to the module on the K-line, while the module supplies bias to the scan tool on the L-line. Communication occurs when the transmitting node pulls the voltage low. If a failure occurs in this bus system, communications between the scan

Figure 10-4 The K-line transmits data from the module to the scan tool, and an L-line receives data from the scan tool.

Special Tools

J1962 BOB

Scan tool

Digital multimeter
    (DMM)

Lab scope

tool and the module will not be possible. Since this bus system doesn't provide for communications between vehicle modules, the customer may not have any noticeable problems with vehicle or accessory operation.

When diagnosing a failure due to the scan tool not being able to communicate with the PCM, it's important to analyze the symptoms. Review the wiring diagram and bus system information to determine if other modules on the vehicle use the ISO 9141-2 bus system for diagnostics. If other modules are on the bus, then use the scan tool to attempt to connect with each of these modules. If the scan tool connects to any module using the ISO 9141-2 bus, then it's not a total bus failure, and you will need to diagnose for a partial bus failure. If no modules respond, you will need to diagnose for a total bus failure.

Referring to Figure 10-3, if node 1 responded but nodes 2 and 3 did not, then the first location to check would be the in-line connector. If only nodes 1 and 3 responded, the problem is in the bus circuit to node 2. Since this is a parallel circuit, the problem cannot be a short to ground or voltage. Either the bus has an open between the splice and the node, or the node isn't powering up due to faulty battery feed or ground circuits.

If no module responds, check voltages on the K-line and the L-line. Connect the J1962 BOB to the DLC and test for voltages at the proper terminals. Without the scan tool connected, there should be voltage only on the L-line. The module supplies voltage on the L-line. Zero voltage here can indicate an open circuit or short to ground. Disconnect the battery and use an ohmmeter to determine the type of fault. If this test indicates an open, the most likely location is between the DLC and the splice to the first node. A short to ground can be anywhere in the circuit. To locate the short, refer to the wiring diagram and determine if there is an in-line connector in the circuit. If the in-line connector is accessible, disconnect it and see if the resistance changes. If it does, the short is downstream of the in-line connector. If the ohmmeter reading does not change, the fault is between the DLC and the in-line connector. Next, disconnect the modules, one at a time, on the side of the in-line connector with the fault. After each module is disconnected, check the resistance reading. If disconnecting a module changes the reading, then that module has an internal fault and needs to be replaced. If the ohmmeter still indicates a short to ground after all of the modules are disconnected, the fault is in the wiring harness.

With the scan tool connected, voltage should be present on the K-line. Since the scan tool supplies voltage on the K-line, 0 volt (V) on this line means either the circuit has a short to ground or the scan tool is faulty. Use an ohmmeter to test for a short to ground. If the ohmmeter shows a short, follow the same procedure described for the L-line to locate the short.

ISO-K Bus System Diagnostics

Classroom Manual
Chapter 10, page 274

 Special Tools

J1962 BOB

Scan tool

DMM

Lab scope

Vehicles that use ISO-K for the communication connection between the module and the scan tool may have dedicated lines from individual terminals in the DLC to the module (**Figure 10-5**). The K-line from the DLC to the PCM will be from terminal 7 of the DLC. The ISO-K bus communicates between the scan tool and the module on a signal wire. This bus doesn't provide for inter-module communications. The scan tool supplies the voltage onto the K-line.

If communications between the scan tool and module are not possible, begin the diagnosis by connecting the J1962 BOB to the DLC. With the ignition key in the OFF position and the scan tool disconnected, there should be 0 volt on the K-line. If voltage is present on this circuit, there is a short to voltage between the DLC and the module, or internal to the module.

If there is no voltage, test for short to ground with an ohmmeter. A short to ground may be in the wiring between the DLC and the module, or internal to the module.

Figure 10-5 An ISO-K bus circuit connects several modules to the scan tool. Each module uses its own dedicated circuit from the DLC.

Next, connect the scan tool with the ignition switch in the OFF position. It would be best to use a lab scope to observe the voltages. Set the scope to read 10 ms per division on the time frame. At this time, the scope trace should show up to 12 volts. While observing the scope trace, turn the ignition key to the RUN position and attempt to establish communication between the scan tool and the module. The scope trace should indicate a digital signal as communication is established and as data is transmitted (**Figure 10-6**).

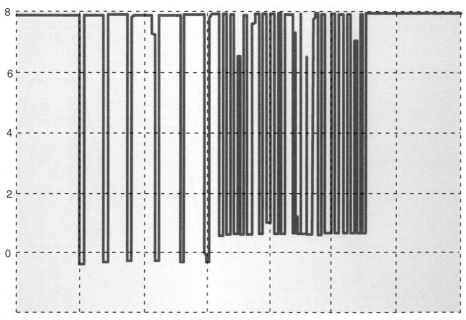

Figure 10-6 ISO-K transmission trace. The negative voltage is from the scan tool attempting to request data from the module. When the module is transmitting data to the scan tool, the voltage is pulled down to +0.5 volt.

Notice in the scope trace that the biased voltage from the scan tool is about 8.0 volts. When the scan tool is attempting to request data from the module, the voltage is pulled low, to −0.5 volt. Finally, when the module is transmitting data to the scan tool, the voltage is pulled to +0.5 volt.

Class A Bus System Diagnostics

Classroom Manual
Chapter 10, page 276

Special Tools

J1962 BOB
Scan tool
DMM
Lab scope

The class A bus system is a low-speed bus. This system has a master module and several slave modules. The master module will supply the bias voltage onto the bus system. If this module should fail, then no bus communications will be possible. In the Classroom Manual, the Chrysler Collision Detection (CCD) multiplexing system was presented as an example of how a class A bus operates. Here, the CCD bus system will be used to provide an example of failure modes and how to diagnose a typical class A bus system.

Depending on scan tool type, the scan tool may attempt to diagnose the bus system if communication fails to occur. If this happens, the scan tool will display the cause of the problem on the screen. The messages displayed can be as follows:

- Bus (−) open.
- Bus (+) and bus (−) open.
- Bus (+) open.
- Bus (+) shorted to bus (−).
- Bus bias level too high.
- Bus bias level too low.
- Bus shorted to 5 volts.
- Bus shorted to battery voltage.
- Bus shorted to ground.
- No bus bias.
- No bus termination.
- Not receiving bus messages correctly.

The CCD bus network uses the voltage difference between the two wires to transmit data (**Figure 10-7**). The bus bias on the two wires is about 2.50 volts. The bus can operate between 1.5 and 3.5 volts. Voltages that are outside of this range will not transmit data.

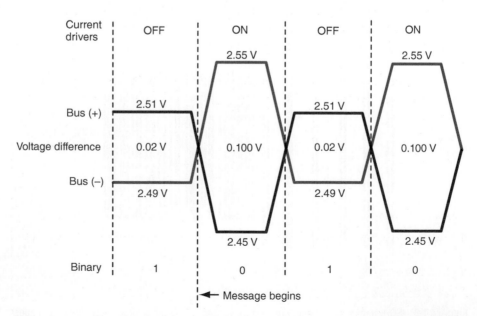

Figure 10-7 Normal CCD bus voltages during message transmission. The voltage difference transmits a bit 1 or a bit 0.

Since the CCD bus is a vehicle-wide bus system, failure of this bus will result in electrical systems not operating properly. First, determine if it's a total or partial bus failure by attempting to communicate with all of the modules on the vehicle. No modules communicating with the scan tool indicates a total bus failure. If it's possible to communicate with a module on the bus system, then approach the diagnostics as a partial bus failure.

The possible causes of a total bus failure include:

- A faulty master module.
- Faulty power or ground circuits to the master module.
- An open in one of the bus circuits from the master module.
- A short to ground in one of the bus circuits.
- A short to voltage in one of the bus circuits.
- The two bus wires shorted together.

> **⚙ SERVICE TIP** The PCM does not communicate with the scan tool on the CCD bus. It uses the K-line for this purpose. Don't be fooled into thinking that the problem is a partial CCD bus failure based on the response of the PCM.

To determine the type of fault causing total bus failure, use the J1962 BOB and connect it to the DLC. Measure the voltage on each bus circuit. The CCD system uses pin 3 of the DLC for CCD (+) and pin 11 for CCD (−). CCD (+) should have 2.49 volts with the ignition key in the RUN position. CCD (−) should have 2.51 volts.

A voltmeter reading of 12 volts on either circuit indicates a short to battery voltage in that circuit. To isolate the location of the short, turn the ignition switch to the OFF position and watch the voltmeter. If the voltage drops, then the short is to ignition-switched voltage. If the voltage is still reading 12 volts, the circuit is shorted to a direct battery feed. Remove fuses and relays one at a time while watching the voltmeter. If the voltage drops after removing a fuse or relay, note the identification of the fuse or relay and use the service information to determine what circuits and components are protected by that fuse or controlled by that relay. With this known, return to the vehicle, reinstall the fuse, and begin to unplug each component one at a time while observing the voltmeter. Once a component is unplugged and the voltage drops to normal, you have located the faulty component. If all components are unplugged and the voltage is still high, the fault is in the circuit's wiring for those components.

If the voltmeter indicates 0 volt on either or both of the bus circuits, the cause can be an open or a short to ground. To determine the type of fault, use an ohmmeter and measure the termination resistance of the bus system by disconnecting the battery and placing the test leads into pins 3 and 11. Most CCD bus systems will have 60 ohms (Ω) of termination resistance. However, some vehicles will have 120 ohms of termination resistance, so it's important to check the service information.

If the system should have 60 ohms and the ohmmeter reads 120 ohms, then one of the termination resistors is open, or there is an open in the bus circuit to a module that has one of the termination resistors. Usually, the termination resistors will reside in the PCM and the body control module (BCM). Each of these modules will have a 120-ohm termination resistor (**Figure 10-8**). To locate the open resistor, disconnect either the PCM or BCM while observing the ohmmeter. If the resistance does not change, then the open termination resistor is in the module that was just unplugged. If the resistance changes to infinite, this indicates that the remaining termination resistor in the circuit was just removed, so the fault is within the termination of the other module. Before replacing the

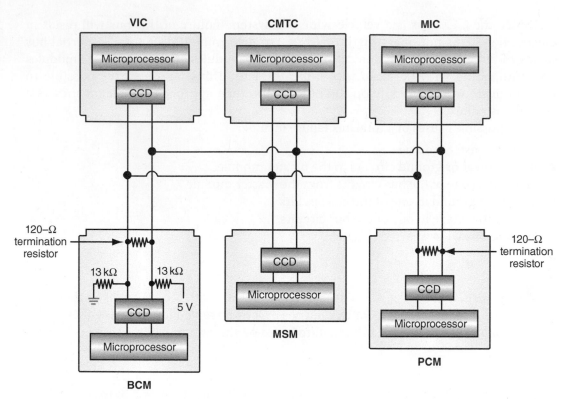

Figure 10-8 Usually, two modules will have termination resistors. Total circuit resistance will be 60 ohms.

module, check for an open in the bus circuits to that module. If the bus circuits are good, and module power and ground circuits are confirmed, replace the module.

If termination resistance is normal and there is no indication of a short to ground on any of the bus circuits, the fault is probably that the master module isn't powering up and not supplying the bias for the bus. Usually, the master module is the BCM in the CCD bus system. Confirm if the power and ground circuits are good before replacing the BCM.

If ohmmeter testing indicates a short to ground in one of the bus circuits, the fault may be a faulty module with an internal short, or the bus wire has a short to ground within the wiring harness. To isolate where the fault is, break the system circuit down into smaller sections. Examine the wiring diagram for the vehicle and determine if there is a connector that would separate the bus network. With the ohmmeter measuring the resistance between the affected circuit and chassis ground, unplug the connector. If the ohmmeter still reads low resistance, the fault is between the DLC and the connector. If the ohmmeter reads infinite, the fault is downstream of the connector. Now that this is known move the ohmmeter test leads (if necessary) to observe resistance readings, as modules on the appropriate side of the connector are unplugged. If the readings change after a module is unplugged, the fault is in the module. The fault is in the wiring harness if the reading is still low after all modules are unplugged.

SERVICE TIP If the scan tool cannot communicate with the PCM, and other modules on the CCD bus system have logged a loss of communication fault code against the PCM, check the common voltage supply circuits to the sensors. If these circuits short to ground, the PCM will turn off. This will also result in a no-start, and the PCM may not store any DTCs.

If the initial tests indicate a partial bus failure, try to determine which modules are not communicating. If there is more than one, use the wiring diagrams for the vehicle to determine if there is anything common between the affected modules. For example, look for a common connector, common splice, common voltage supply to the modules, or common ground for the modules. Locate this common component on the vehicle and confirm proper function and repair as needed.

When diagnosing the CCD bus system, remember the following:

1. Normal bus voltages with the ignition switch in the RUN position are approximately 2.5 volts.
2. Normal bus voltages with the ignition switch in the OFF position are 0 volt on both circuits.
3. Normal termination resistance is typically 60 ohms.
4. The wires must be twisted at the rate of one twist every 1¾ inches (44 mm).

J1850 Bus System Diagnostics

Classroom Manual
Chapter 10, pages
275, 278

The J1850 bus system can be either a 10.4 kb/s variable pulse width–modulated (VPWM) system or a 41.6 kb/s pulse width–modulated (PWM) system. We will use the Programmable Communication Interface (PCI) bus system to illustrate failure modes and how to diagnose a typical J1850 bus system. This system is similar to the class 2 bus system that General Motors uses.

Since the PCI bus is a vehicle bus system, a failure in it will cause electric systems not to operate properly. Like the other bus system diagnostics just discussed, begin by determining if the problem is a total bus or a partial bus failure.

Special Tools
J1962 BOB
Scan tool
DMM
Lab scope

If initial testing indicates a total bus failure, test the system for proper voltage and termination resistance. Use the J1962 BOB to provide a diagnostic test point to measure these values. Total bus failure can only result from a short to voltage or a short to ground in the circuit. Termination resistance will be different on every vehicle since each module provides its own termination. Since vehicles will have different options, the resistance will be different between vehicles. However, by measuring the resistance between pin 2 of the DLC (at the J1962 BOB) and chassis ground, a low reading will confirm a short to ground in the system.

The normal voltage on the PCI bus is 0 volt when the bus is at rest, and the voltage pulls up to near 7.5 volts when the bus is active (**Figure 10-9**). The best method of determining proper voltage on this system is with a lab scope. A DMM set to the voltmeter will not be able to give actual voltage values since the activity is too fast, so the meter will display the average. The MIN/MAX feature of the voltmeter will freeze the readings, but they still will not be actual voltages. However, the voltmeter will indicate if the bus is shorted to battery voltage (12 volts) or is a short to ground (0 volt). A reading of 0 volt may also indicate an open, but this type of fault will not result in total bus failure.

If initial testing indicates a short to battery voltage, isolate the location of the fault by breaking down the circuit into manageable pieces or smaller circuits. With the voltmeter still measuring PCI voltage, turn the ignition key to the OFF position. Wait for 20 seconds and read the voltmeter again. A voltmeter reading of 0 volt at this time indicates a short to ignition-switched voltage. If the voltmeter continues to display battery voltage, the bus is shorted to direct battery voltage. In either case, isolate the circuit shorted to the bus network by pulling fuses and relays one at a time while watching the voltmeter. If the voltage drops after pulling a fuse or relay, note the identification of the fuse or relay and use the service information to determine what circuits and components are protected by that fuse or controlled by that relay. Once this is known, return to the vehicle, reinstall

Figure 10-9 The normal voltage on the PCI bus is 0 volt when the bus is at rest. To transmit data, the voltage is pulled up to near 7.5 volts.

the fuse, and begin to unplug each component one at a time while observing the voltmeter. Once a component is unplugged, and the voltage drops to normal, that component is the fault. If all components are unplugged and the voltage is still high, the fault is in the circuit's wiring for those components.

If testing the termination resistance, as discussed earlier, indicates a short to ground, the fault may be an internal short within a module, or the PCI bus wire is shorted to ground within the wiring harness. Again, the trick is to break the system down into smaller circuits. Some vehicles use a common hub for all of the bus circuits. If this is the case, this location provides an excellent place to begin diagnostics. For example, **Figure 10-10** shows a PCI bus system that uses the BCM as a common hub for the bus. The BCM has three connectors attached to it that contain the different bus circuits from all of the modules. Unplugging these connectors (one at a time) while observing system operation breaks the system down into a few circuits. Once a connector is unplugged, and some of the bus systems become active, the fault is within the bus circuits of that connector.

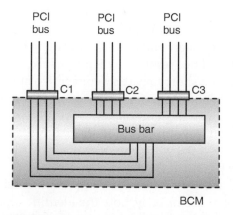

Figure 10-10 Common hub or bus bar within a controller helps make isolating the cause of bus failures easier.

It's possible that the connector has several bus wires in it. Use the ohmmeter to determine which bus wire is shorted to ground. Use the service information to determine what module is on that circuit and unplug the module. Again, measure resistance between the BCM connector and chassis ground. If the reading is infinite, the fault is in the module. A low reading indicates the wire has a short to ground.

If the bus system does not have a common hub, examine the wiring diagram to determine if there is a connector that would separate the bus network. Identify any connectors that can be used and their location. With the ohmmeter measuring the resistance between pin 2 of the DLC (using the J1962 BOB) and chassis ground, unplug the connector. If the ohmmeter still reads low resistance, the fault is between the DLC and the connector. If the ohmmeter reads any other value, the fault is downstream of the connector. Now that this is known move the ohmmeter test leads (if necessary) to observe resistance readings, as modules on the appropriate side of the connector are unplugged. As before, if the readings change after a module is unplugged, the fault is in the module. If the reading is still low after all modules are unplugged, the fault is in the wiring harness.

If the initial tests indicate a partial bus failure, try to determine which modules are not communicating. If there is more than one, use the wiring diagrams for the vehicle to determine if there is anything common between the affected modules. For example, look for a common connector, common splice, common voltage supply to the modules, or common ground for the modules. Locate this common component on the vehicle and confirm proper function and repair as needed.

Controller Area Network Bus Diagnostics

Vehicles that utilize controller area network (CAN) bus systems will generally have two or three different CAN systems. The vehicle body systems will communicate over the medium-speed CAN B network. Those systems, such as engine controls and antilock brakes that require data at a faster rate will communicate on the CAN C bus. Beginning in the 2005 model year, manufacturers started migrating to the new requirement for using a CAN bus network for diagnostics (often referred to as Diagnostic CAN C).

After verifying the customer's complaint, the next step is to determine which bus network is at fault. Usually, a scan tool can easily determine this. Use the scan tool to access the central gateway (CGW) module for the bus networks. Since this is the common component for all three CAN bus networks, it will usually provide a DTC that will indicate which bus network has the fault. If the scan tool cannot communicate to the central gateway module, the Diagnostic CAN C circuit is faulty.

With the faulty network identified, perform diagnostics of either CAN network using a DMM or a lab scope. However, methods are different based on the network and the type of fault. Follow this diagnostic procedure to isolate the cause of the fault:

1. If the fault code indicates an open circuit fault with either the CAN B (+) or the CAN B (−) circuit (**Figure 10-11**), then determine and locate an easily accessed module on the CAN B bus system. At the connector for this component, create a short to ground on the opposite CAN B circuit. For example, if the fault code indicates that the CAN B (−) is open, short the CAN B (+) circuit to ground. Since CAN B buses that operate at speeds up to 125 kb/s are fault tolerant and will operate in single-wire mode if an electrical potential exists between one of the circuits and ground, the open may not cause any noticeable symptoms. However, when the other circuit is shorted to ground, that module can no longer communicate. All other modules will communicate using the single-wire mode (**Figure 10-12**).

Classroom Manual
Chapter 10, pages
275, 281

Special Tools
J1962 BOB
Scan tool
DMM
Lab scope

Figure 10-11 An open bus (−) wire at CAN B bus Node 1 does not prevent the other nodes from being active on the bus.

With the wire shorted to ground, use the scan tool to determine which module isn't communicating on the bus. Locate this module on the vehicle. The fault will be between the splice from the affected bus circuit to the module.

2. If the fault code indicates that one of the CAN B circuits is high or low, then use a DMM to determine if the fault is a short to voltage (12 volts) or shorted to ground (0 volt). If the voltage on both circuits is between 500 mV and 1.0 volt, the two circuits are shorted together. Locating a shorted circuit will be similar to the procedure described earlier for the J1850 VPWM bus. Use the wiring to determine if any connectors can be used to break the circuits down into smaller sections.

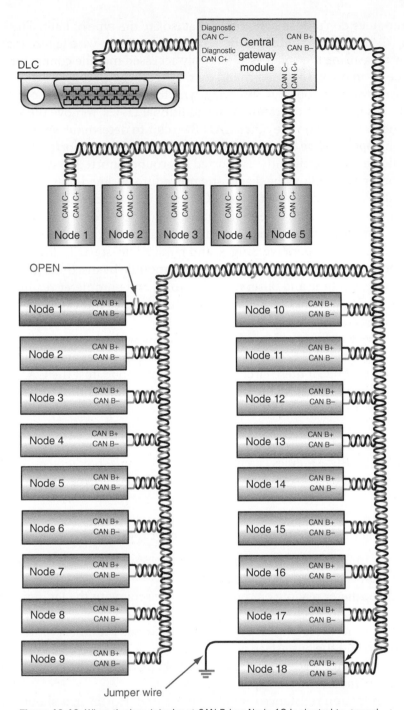

Figure 10-12 When the bus (+) wire at CAN B bus Node 18 is shorted to ground, all modules continue to communicate except Node 1. Node 1 cannot communicate since it no longer has a circuit.

Return to the vehicle and, with a voltmeter connected to the shorted circuit, observe the readings while opening the connectors. Once which side of the connector has the short is determined, disconnect modules or other connectors on that branch of the circuit to locate the root cause of the failure.

3. If the fault code indicates that both CAN B (+) and CAN B (−) circuits are low or high, a total bus failure has occurred. Symptoms may include headlamps on, warning indicator lamps on in the instrument panel, instrument panel backlighting at full intensity, gauges inoperative, and possible no-start. To locate the cause of this failure, use the diagnostic procedure for locating a short circuit, as described in step 2.

4. Some modules don't set specific codes that isolate the type of fault. They will set a generic "CAN B BUS" fault. First, determine if the single-wire failure is an open or short by measuring the voltages at an easily accessed module connector. If this fault is set due to an open, short to ground each CAN B circuit one at a time while monitoring the scan tool and observing symptoms. If shorting one of the circuits to ground does not affect operation or set additional DTCs, remove the short and repeat the procedure on the other CAN B circuit to determine where the open exists. The open will be on the CAN B circuit that was *not* shorted to ground. If the tests indicate that the circuit(s) is shorted to ground, shorted to voltage, or shorted to each other, then follow the diagnostic procedure outlined previously in step 2.

5. If the fault code indicated that a CAN C circuit is shorted, this would cause a total failure of the CAN C bus. The short can be to voltage, ground, or the two circuits together. Symptoms may include illuminated warning indicator lamps in the instrument panel, inoperative gauges, and possible no-start. To locate the short circuit, determine if it's caused by a defective module by disconnecting the modules one at a time and observing symptoms and ohmmeter readings on the CAN circuit. If disconnecting all of the modules fails to indicate a faulty module, the problem is in the wiring harness.

6. If the fault code indicates a loss of communication with a single module (on either the CAN B or the CAN C network), check for proper power and ground circuit at the affected module. If these circuits are confirmed good, test the CAN circuits for opens between the splice to the network and the module. If the circuits don't have an open, replace the module.

CUSTOMER CARE Modules can be very expensive and should be replaced only after all power feed and ground circuits are tested and confirmed good.

When diagnosing the CAN bus networks, remember these facts:

1. A normal voltmeter reading on the CAN B (+) circuit with the ignition key in the RUN position is between 280 and 920 mV.
2. A normal voltmeter reading on the CAN B (−) circuit with the ignition key in the RUN position is between 4.08 and 4.72 volts.
3. A normal voltmeter reading on the CAN B (+) circuit with the ignition key in the OFF position and the bus asleep is 0 volt.
4. A normal voltmeter reading on the CAN B (−) circuit with the ignition key in the OFF position and the bus asleep is battery voltage.
5. CAN B termination resistance cannot be measured. However, it should be infinite with the battery disconnected. A low reading indicates a short.
6. The CAN B bus can become active from inputs or from the ignition switch.
7. Normal CAN C bus termination is 60 ohms. However, if other modules on the CAN C bus network have termination resistance, the actual reading on an ohmmeter will be less than 60 ohms.
8. Normal Diagnostic CAN C termination is 60 ohms.
9. A normal voltmeter reading on the CAN C (+) circuit with the ignition key in the RUN position is about 2.60 volts.
10. A normal voltmeter reading on the CAN C (−) circuit with the ignition key in the RUN position is about 2.4 volts.
11. The CAN C bus may be event driven or be active only with the ignition in the RUN position.
12. The wires must be twisted at a rate of one twist every inch (25.4 mm).

Classroom Manual
Chapter 10, page 287

If the bus network is a star system, the diagnostics locations are the DLC and the common point connectors (star connectors). The star network architecture will have all bus circuits connected to this common point (**Figure 10-13**). Since these systems typically use CAN buses, voltage and termination resistance values are the same as for the stub architecture. In Figure 10-13, the vehicle does not use a Diagnostic CAN C bus; instead, both the CAN C and interior bus networks are in the DLC. This means the voltage and resistance of each bus can be measured at the DLC using a J1962. Each terminal of the star connector goes to a module, the DLC, or another star connector (**Figure 10-14**). The star connector provides an ideal location for isolating the circuits during diagnosis.

Figure 10-13 Star architecture typically does not use the Diagnostic CAN C bus since the vehicle's bus networks are all located in the DLC.

Figure 10-14 The star connectors are a common connection point that makes an ideal location for diagnosing bus network faults. Each terminal of the connector goes to one module, the DLC, or to the other connector.

For example, if the voltage and resistance readings at the DLC indicate that one of the CAN C bus circuits is shorted to ground, this would be the reason for the CAN C to be totally down. While leaving the ohmmeter connected to the terminal of the J1962 BOB that houses the faulty circuit, disconnect the star connector terminals one at a time until the ohmmeter no longer indicates a short to ground. The fault lies in the circuit just disconnected and is between the star connector and the module, or there may be an internal fault within the module.

> **SERVICE TIP** If there is more than one star connector in the bus network, voltage measurements at the DLC may look close to normal if one of the termination resistors was open. Use an ohmmeter to confirm that the proper termination resistance is on the circuit.

> **AUTHOR'S NOTE** The CAN C and the interior bus may each have more than one star connector. Since the star connector may contain termination resistance, if the circuit uses one star connector, it will have 60 ohms; if it uses two star connectors, then each will have 120 ohms.

Local Interconnect Network Bus Diagnostics

Classroom Manual
Chapter 10, page 288

Special Tools
Scan tool
DMM
Lab scope

The local interconnect network (LIN) bus is a supplemental bus network used along with the vehicle's main multiplexing system. The master module communicates data from the slave modules onto the main bus network. The scan tool will not have access to the slave module. Access to all LIN bus diagnostic procedures using a scan tool is through the master module.

Figure 10-15 is an illustration of a LIN bus system used for steering wheel–mounted radio controls. The steering column module (SCM) is the master module and connects to the CAN B bus network. The right steering wheel switch assembly is actually the slave module and is capable of sending data to the SCM and of receiving data from the SCM. The left steering wheel switch is a multiplex switch system that uses different resistance values for each switch position. The different resistances result in a unique voltage drop sensed by the slave module. The requested action of the driver that is indicated by pressing a switch is received by the slave module and sent to the master module. The master module then puts the data onto the CAN B bus network, which sends the message to the radio, which performs the requested action.

In this example, the scan tool will communicate with the SCM. By observing the data PIDs on the scan tool, the display should change values as input for each switch position is received. In addition, the master module (the SCM in this case) will store a DTC if a system indicates a fault. For example, if a switch in the left steering wheel switch assembly indicates that it's stuck in the pressed position for an excessive amount of time, the SCM will set a DTC for this condition.

If a customer states that the steering wheel radio controls fail to operate, use the scan tool to see if the master module receives *any* of the button inputs. If the master module indicates receiving any switch inputs, the LIN bus circuit is working. If the scan tool PIDs do not indicate a change in switch state, the LIN bus circuit, the slave module, or the master module has failed.

Figure 10-15 LIN bus system used for remote radio controls mounted on the steering wheel.

Figure 10-16 Normal LIN bus trace.

The most accurate way of determining LIN bus voltages is with a lab scope. Normal voltage on the LIN bus is up to 12 volts, which is pulled low during communications (**Figure 10-16**).

If using a voltmeter, remember it displays the average voltage reading. Disconnecting the slave module from the master module and testing the LIN bus circuit voltage with the ignition switch in the RUN position, the voltmeter reading will be about 9.5 volts. This is due to the master module pulse width modulating the signal in an attempt to communicate with the slave module. Reconnecting the slave module while reading the voltmeter will show about 8.3 volts since both modules are pulling the circuit low in an attempt to communicate.

If 0 volt is indicated by the voltmeter or lab scope when the slave module is disconnected, the cause could be a faulty master module, an open circuit between the master and slave module connector, or a short to ground in the LIN circuit. Use the voltmeter and ohmmeter to diagnose and locate the fault.

If normal voltage remains with the slave module disconnected, but then reads 0 volt when reconnected, the fault is in the slave module. In this case, the slave module has an internal short to ground.

> **CUSTOMER CARE** Many people don't realize the subjectivity of the communication networks to RFI and EMI. These signals may interfere with module communications. If the customer says they are experiencing intermittent problems with the vehicle, the problem may be due to an add-on feature such as a radio, cell phone, entertainment system, and so on. If diagnostics prove this to be the problem, inform the customer that these systems may require special shielding and location alterations to prevent future instances.

Enhanced Scan Tool Diagnostics

Some scan tools provide enhanced diagnostics of the CAN bus networks. In addition to performing the module scan, these units may provide advanced loss of communication screens, network topology screens, and so on. All of these additional features assist in diagnosing the CAN bus network. **Photo Sequence 19** provides an example of these features and how they are accessed. Since scan tool manufacturers differ in their approach to diagnostics and display of information, this photo sequence is provided only as a guide. Always refer to the instructions for the scan tool you are using.

PHOTO SEQUENCE 19
Advanced Scan Tool Functions

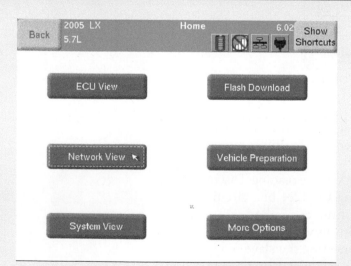

P19-1 The "Home" screen of the scan tool. Select "Network View."

P19-2 The network view provides an icon look of the CAN bus network from the scan tool to the different bus systems. Red-colored icons indicate a fault has been recorded in that system. From this screen select the "Advanced" button.

PHOTO SEQUENCE 19 (CONTINUED)

Description	Network	Active	Stored
▲ U0169 Lost Communication With Sunroof	⊗	6	0
U0186 Lost Communication With Audio	⊗	6	0
U0100 Lost Communication With	☑	0	2
U0121 Lost Communication With	☑	0	3
U0101 Lost Communication with TCM	☑	0	1
▼ U0151 Lost Communication with	☑	0	1

Enabling Conditions	
Ignition Status: RUN Position	Battery/Ignition DTC's: Yes
VIN Mismatch DTC: No	ECU Config DTC: No
CAN Electrical DTC's: Yes	CAN Signal DTC's: No

P19-3 This is the Loss of Communications Summary screen. A green check mark indicates that the module is communicating with the scan tool. The red Xs indicate that those modules are not communicating on the network bus. Under the "Active" column, the "6" indicates that six other modules have logged loss of communication fault codes against the module, and the faults are currently present. If any logged-against faults were stored, they would be under the "Stored" column. Press the "Network Topology" button.

P19-4 This is a live layout of the vehicle's communication system and is an actual representation of the vehicle's communication wiring. The color codes indicate if the module is active on the bus, if the vehicle isn't built with that module, and the different bus networks. The red icons indicate modules that are not communicating. In this case, a common connector or splice isn't the cause of the problem. The next logical diagnostic step would be to see if the two affected modules share a common power circuit or ground.

Description	Network	Active	Stored
▲ U0169 Lost Communication With Sunroof	⊗	6	0
U0186 Lost Communication With Audio	⊗	6	0
U0100 Lost Communication With	☑	0	2
U0121 Lost Communication With	☑	0	3
U0101 Lost Communication with TCM	☑	0	1
▼ U0151 Lost Communication with	☑	0	1

Enabling Conditions	
Ignition Status: RUN Position	Battery/Ignition DTC's: Yes
VIN Mismatch DTC: No	ECU Config DTC: No
CAN Electrical DTC's: Yes	CAN Signal DTC's: No

P19-5 Another option from the "Loss of Communication" screen is to press the "Advanced LoC" button. This advances to another loss of communications screen.

ECU Name	Network	E A	E S	C A	C S	S A	S S	L.A. A	L.A. S
▲ CCN	☑	0	0	0	0	0	0	0	0
FCM (CGW)	☑	0	0	2	3	0	0	0	0
TCM	☑	0	0	0	0	0	0	0	1
PCM	☑	0	0	0	0	0	0	0	2
ESM	☑	0	0	0	2	0	0	0	0
ABS	☑	0	0	0	0	0	0	0	3
SUNR	⊗	–	–	–	–	–	–	6	0
AMP	⊗	–	–	–	–	–	–	6	0
▼ HSM	☑	0	0	2	0	0	0	0	0

Enabling Conditions	
Ignition Status: RUN Position	Battery/Ignition DTC's: Yes
VIN Mismatch DTC: No	ECU Config DTC: No
CAN Electrical DTC's: Yes	CAN Signal DTC's: No

P19-6 This advanced loss of communications (LoC) screen indicates the type of fault code as "E" for electrical, "C" for communications, "S" for implausible signal, and "L.A." for logged against. Also, columns indicate the number of active or stored codes and which modules have logged the fault.

Data Bus Circuit Repair

Repairing the data bus circuit wires must be done in such a manner that the integrity of the system isn't compromised. Always repair the circuit by soldering, never use a crimp connector. Follow the repair procedure outlined in Chapter 4 for shielded circuits. When repairing the data bus circuit(s), consider the following:

1. Refer to the service information for the specification for the number of twists per meter or inch.
2. Any new wire(s) added must be the same length as that removed. The data bus + and − circuits must be the same length. Since the bus operates at a high speed, deviations in circuit length can disrupt the timing and result in improper data transmission.
3. If replacing both wires, twist the wires together prior to soldering them in place.
4. Use the same size wire as that removed.
5. If repairing both wires, stagger the splices.
6. Use heat shrink tube to provide a good insulating seal. Do not use electrical tape.

Scan Tool Authenticating

Classroom Manual
Chapter 10, page 294

Many automobile manufacturers require authenticated scan tool access to perform high-level diagnostics to combat unauthorized vehicle network manipulation threats. To use the scan tool for performing bi-directional control, module reprogramming, module calibration, and other high-level functions on modules on the private network, the vehicle manufacturer must have approved it as a secure scan tool. In addition, authentication of each user of the scan tool is necessary to access the private network diagnostic functions.

If the tool or the technician isn't authenticated with the vehicle's manufacturer, the scan tool will have limited "read-only" access. This access allows for reading and erasing DTCs, monitoring data PIDs, and other "low-level" functions. Functions for modules on the private network are more limited than functions for modules on the public network. The SGW controls the level of the user's access. Each user's level is determined based on the assigned role during the authentication process.

OEM scan tools have a verification tool that seamlessly performs code sharing in the background during the vehicle selection process. The aftermarket also has access to this verification tool. Compatible scan tools can connect to the secure gateway (SGW) module to open advanced functions using a secure connection between the scan tool manufacturer's server and the OEM's approved server. This process requires the shop to have an authorized OEM account and their technicians registered on the account.

Currently, www.AutoAuth .com is the authorized verification portal for FCA and other manufacturers. This is an approved "for a fee" site to provide secure gateway access. After establishing an account, the scan tool will automatically access the internet and communicate upstream to request and receive permission to bypass the SGW module (**Figure 10-17**).

Full diagnostic access requires the scan tool to have continued internet access throughout the diagnostic process. This includes during a test drive or anytime the vehicle is away from the shop. You may need to use a cell phone as a hot spot to maintain an internet connection.

Scan Tool

OEM SGW Authentication
System

Scan Tool
Manufacturer's Server

SGW Authentication
Bridge Server

Figure 10-17 The scan tool goes through the verification process automatically when connecting to a vehicle.

CASE STUDY

A customer has had their vehicle towed to the repair facility due to an intermittent no-start condition. Other shops have attempted to repair the problem, but the problem persists. The technician attempts to duplicate the problem, but the engine will start and run fine. The technician connects the scan tool and accesses the PCM to read fault codes. There are no active or stored DTCs. Realizing that the vehicle uses the CAN bus network, the technician does a vehicle scan report. The report shows the central gateway module has DTCs stored in it. The fault code is a stored code for "Loss of communication with PCM." There are no active faults. Since this is an intermittent problem, the technician checks the connectors at the PCM. Here a loose terminal is identified, and the technician makes the necessary repairs. The intermittent no-start was due to the vehicle's security system not communicating with the PCM; thus, the PCM would not allow the engine to start. The PCM did not set any fault codes because it was doing what it was supposed to. The CGW module did log the loss of communication fault against the PCM since it could not communicate with the PCM when the terminal connection came loose.

ASE-Style Review Questions

1. In a J1850 VPWM bus system, how can a short to ground be isolated?
   A. Use a fused jumper wire to supply battery voltage to pin 2 of the DLC while observing the DMM.
   B. Disconnect in-line connectors one at a time while observing the DMM.
   C. Use a fused jumper wire to short pin 15 of the DLC to ground while observing the DMM.
   D. All of these choices.

2. A short to ground in one of the CAN C bus circuits can cause all of the following to occur EXCEPT:
   A. No-start.
   B. No communication between the scan tool and any CAN C modules.
   C. CGW module logs a loss of communication fault code against CAN C modules.
   D. The bus will operate in single-wire mode.

3. All of the following can cause total bus failure of the J1850 VPWM bus system EXCEPT:

   A. Open bus wire to the BCM.

   B. Bus wire shorted to ground.

   C. Bus wire shorted to battery voltage.

   D. Internal short in a module.

4. The lab scope trace of an ISO-K bus system shows the voltage going to − 0.5 volt. This would indicate:

   A. The scan tool bias voltage.

   B. The module is attempting to transmit data to the scan tool.

   C. The scan tool is attempting to request data from the module.

   D. Excessive electrical noise on the circuit.

5. A class A bus that has 0 volt on both bus circuits may indicate:

   A. Both circuits are shorted to ground.

   B. The master module's power feed circuit is open.

   C. An open bus connector to the master module.

   D. All of these choices.

6. To isolate a bus circuit that is shorted to battery voltage, all of the following methods can be used EXCEPT:

   A. Use a jumper wire and connect the faulty circuit to ground.

   B. Turn the ignition switch off to see if it's switched voltage.

   C. Pull fuses one at a time to determine the shorted circuit.

   D. Pull relays one at a time to determine shorted circuits or components.

7. On a LIN bus, 0 volt is on the bus circuit with the slave module disconnected. What is the **LEAST LIKELY** cause?

   A. An open circuit between the master module and the slave module connector.

   B. A short circuit between the master module and the slave module connector.

   C. Faulty master module.

   D. Faulty slave module.

8. *Technician A* says on the LIN bus system, the scan tool accesses the slave modules to display their data.
   Technician B says the LIN slave module can send data only to the master module.
   Who is correct?

   A. A only   C. Both A and B

   B. B only   D. Neither A nor B

9. When the scan tool is connected, the voltage on the K-line reads zero.
   Technician A says this indicates a short to ground.
   Technician B says the scan tool may be faulty.
   Who is correct?

   A. A only   C. Both A and B

   B. B only   D. Neither A nor B

10. *Technician A* says normal voltages on the LIN bus is up to 12 volts.
    Technician B says 12 volts on the LIN bus indicates data transmission.
    Who is correct?

    A. A only   C. Both A and B

    B. B only   D. Neither A nor B

ASE Challenge Questions

1. *Technician A* says on a CAN B system with a short to ground in one of the bus circuits, it may be possible to locate the fault by jumping battery voltage to the faulty circuit.
   Technician B says to locate the short to ground, use a jumper wire and ground the opposite circuit.
   Who is correct?

   A. A only   C. Both A and B

   B. B only   D. Neither A nor B

2. *Technician A* says on an ISO 9141-2 bus system, without the scan tool connected, there should be voltage only on the L-line.
   Technician B says the module supplies the voltage on the L-line.
   Who is correct?

   A. A only   C. Both A and B

   B. B only   D. Neither A nor B

3. *Technician A* says on CAN B bus systems, it may be possible for the bus to operate if there is an electrical potential between one of the circuits and ground.

   Technician B says if the two CAN B circuits are shorted together, total bus failure will result.

   Who is correct?

   A. A only

   B. B only

   C. Both A and B

   D. Neither A nor B

4. One module fails to communicate on the bus.

   Technician A says this can be due to an internal short of one of the bus circuits.

   Technician B says this could be due to a faulty module ground circuit.

   Who is correct?

   A. A only

   B. B only

   C. Both A and B

   D. Neither A nor B

5. *Technician A* says U-codes identify a failure with a body system function.

   Technician B says B-codes indicate bus network communication failures.

   Who is correct?

   A. A only

   B. B only

   C. Both A and B

   D. Neither A nor B

Name _____   Date _____

RETRIEVING VEHICLE MODULE SCAN REPORT

Upon completion of this job sheet, you will be able to use a scan tool to retrieve a vehicle scan report.

ASE Education Foundation Correlation

This job sheet addresses the following **MLR** task:

A.3.   Retrieve and record DTCs, OBD monitor status, and freeze frame data; clear codes and data when directed. **(P-1)**

This job sheet addresses the following **AST** tasks:

A.3.   Retrieve and record DTCs, OBD monitor status, and freeze frame data; clear codes and data when directed. **(P-1)**

G.8.   Diagnose body electronic systems circuits using a scan tool; check for module communication errors (data communication bus systems); determine needed action. **(P-2)**

This job sheet addresses the following **MAST** tasks:

A.3.   Retrieve and record DTCs, OBD monitor status, and freeze frame data; clear codes and data when directed. **(P-1)**

G.8.   Diagnose body electronic systems circuits using a scan tool; check for module communication errors (data communication bus systems); determine needed action. **(P-1)**

We Support
ASE | **Education Foundation**

Tools and Materials

- Scan tool

Procedure

1. Connect the scan tool to the DLC and establish communications with the vehicle.

2. Enter the correct vehicle information or confirm the scan tool was able to pull the correct information.

3. Navigate the menu items to locate the Vehicle Module Scan function and perform the scan.

4. If possible, print off the Vehicle Scan Report.

5. How many modules reported to the scan tool? _____

6. Did any modules report they have recorded DTCs?   Yes _____   No _____

   If yes, record the modules: _____

   _____

   _____

   _____

7. Access the modules with recorded DTCs and print off any freeze frames associated with them.

8. Do the DTCs indicate a single fault or multiple faults? _____

9. Clear all DTCs and rerun the Vehicle Module Scan.

10. If possible, print off the Vehicle Scan Report.

Instructor's Evaluation:

Grade	Rubric
0	No exposure due to: ☐ Absence ☐ Other:
1	Student unsuccessfully attempted the task
2	Student completed the task with significant instructor assistance.
3	Satisfactory performance of task. The student was able to perform elements of the task with minimal instructor assistance.
4	Competent/Proficient. Able to perform all elements of the task successfully and independently.

Instructor's Notes:

Name _____   Date _____

J1850 BUS SYSTEM DIAGNOSIS

Upon completion of this job sheet, you will be able to use the scan tool and a DMM to monitor normal J1850 bus activity. You will also use the lab scope to observe normal bus operation. Finally, you will use the DMM to diagnose full and partial failures of the J1850 bus.

ASE Education Foundation Correlation

This job sheet addresses the following **MLR** tasks:

A.1.   Research vehicle service information such as fluid type, vehicle service history, service precautions, technical service bulletins, and recalls including vehicles with advanced driver assistance systems (ADAS). **(P-1)**

A.2.   Identify electrical/electronic system components and configuration. **(P-1)**

A.5.   Demonstrate proper use of a digital multimeter (DMM) when measuring voltage source, voltage drop (including grounds), current flow, and resistance. **(P-1)**

A.8.   Use fused jumper wires to check operation of electrical circuits per service information. **(P-2)**

A.9.   Use wiring diagrams to trace electrical/electronic circuits. **(P-1)**

This job sheet addresses the following **AST** tasks:

A.1.   Research vehicle service information such as fluid type, vehicle service history, service precautions, technical service bulletins, and recalls including vehicles with advanced driver assistance systems (ADAS). **(P-1)**

A.2.   Identify electrical/electronic system components and configuration. **(P-1)**

A.5.   Demonstrate proper use of a digital multimeter (DMM) when measuring voltage source, voltage drop (including grounds), current flow, and resistance. **(P-1)**

A.8.   Use fused jumper wires to check operation of electrical circuits per service information. **(P-1)**

A.9.   Use wiring diagrams during the diagnosis (troubleshooting) of electrical/electronic circuit problems. **(P-1)**

A.13.   Test and measure circuit using an oscilloscope and/or graphing multimeter (GMM); interpret results; determine needed action. **(P-2)**

G.8.   Diagnose body electronic systems circuits using a scan tool; check for module communication errors (data communication bus systems); determine needed action. **(P-2)**

This job sheet addresses the following **MAST** tasks:

A.1.   Research vehicle service information such as fluid type, vehicle service history, service precautions, technical service bulletins, and recalls including vehicles with advanced driver assistance systems (ADAS). **(P-1)**

A.2.   Identify electrical/electronic system components and configuration. **(P-1)**

A.5.   Demonstrate proper use of a digital multimeter (DMM) when measuring voltage source, voltage drop (including grounds), current flow, and resistance. **(P-1)**

A.8.   Use fused jumper wires to check operation of electrical circuits per service information. **(P-1)**

A.9.   Use wiring diagrams during the diagnosis (troubleshooting) of electrical/electronic circuit problems. **(P-1)**

A.13.   Test and measure circuit using an oscilloscope and/or graphing multimeter (GMM); interpret results; determine needed action. **(P-2)**

G.8.   Diagnose body electronic systems circuits using a scan tool; check for module communication errors (data communication bus systems); determine needed action. **(P-1)**

Tools and Materials

- J1962 BOB
- Scan tool
- DMM
- Lab scope
- Fused jumper wires
- Service information
- Vehicle with J1850 bus
- Instructor-installed bugs

Procedure

Task 1

1. Use the service information to identify the DLC cavity number(s), circuit identification, and color code(s). Record your findings.

    a. Cavity number(s) _____

    b. Circuit identification _____

    c. Color code(s) _____

2. Connect the J1962 BOB to the DLC, and connect the scan tool to the J1962 BOB. Turn the ignition switch to the RUN position.

3. If available, use the scan tool to perform an 1850 module scan. List all active modules on the J1850 bus.

    _____

    _____

    _____

4. Does the J1850 bus appear to be operating normally?   ☐ Yes   ☐ No

    Explain your answer.

    _____

    _____

5. Connect the DMM positive lead to the J1850 bus circuit of the J1962 BOB (a fused jumper wire may be required). Connect the negative lead to a good chassis ground. With the DMM set to read DC volts, observe and record the voltage range:

    MIN _____ MAX _____

6. Remove the DMM, and use a lab scope to observe the J1850 bus activity. If available, use the print function to print off the trace. If not available, describe the lab scope display.

    _____

    _____

    _____

7. Turn the ignition switch to the OFF position, and disconnect the vehicle battery.

8. Use the ohmmeter function of the DMM to measure the resistance between the J1962 BOB cavity for the J1850 bus and chassis ground. Observe and record the J1850 bus termination resistance. _____

9. Disconnect a regular node from the J1850 bus and record the bus termination resistance. _____

10. Disconnect the scan tool from the J1850 bus and record the bus termination resistance. _____

11. Disconnect the PCM from the J1850 bus and record the bus termination resistance. _____

12. Based on your observations, are the nodes wired in series or parallel? _____

13. Based on your observations, which module had the greatest impact on bus terminal resistance and why?

_____

_____

14. Reconnect all modules and then reconnect the vehicle battery.

Task 2

1. Start the engine, and use a fused jumper wire to short the J1850 bus circuit of the J1962 BOB to chassis ground. What are the observable conditions?

_____

_____

_____

2. Using the scan tool, perform a J1850 module scan and record the results.

_____

3. Use the voltmeter function of the DMM to observe and record the J1850 bus voltage.

_____

4. Does the measured voltage indicate normal operation?   ☐ Yes ☐ No

5. Turn the ignition switch to the OFF position, and disconnect the vehicle battery.

6. Use the ohmmeter function of the DMM; observe and record the J1850 bus termination resistance. _____

7. Explain the difference in this reading compared to Task 1, step 11.

_____

_____

_____

8. Remove the fused jumper wire used to short the bus to ground and reconnect the vehicle battery.

9. With the engine running, use a fused jumper wire to short the J1850 bus circuit of the J1962 BOB to battery voltage. What are the observable conditions?

_____

_____

_____

10. Use the scan tool to perform a J1850 module scan and record the results.

_____

_____

11. Use the voltmeter function of the DMM; observe and record the J1850 bus voltage.

_____

12. Turn the ignition switch to the OFF position. Observe and record the J1850 bus voltage. _____

13. If the voltage changed between step 11 and step 12, what does this indicate?

_____

_____

14. Based on your observations, can the J1850 bus be diagnosed with a DMM?
☐ Yes   ☐ No Explain your answer.

_____

_____

Task 3

1. A vehicle is brought to the shop with inoperative gauges and several electrical accessories not functioning properly. Can the condition be verified?   ☐ Yes   ☐ No

2. Use the scan tool to perform a J1850 module scan. Are there any modules present?   ☐ Yes   ☐ No

3. Disconnect the scan tool. With the ignition switch in the RUN position, use the J1962 BOB and the voltmeter function of the DMM to measure and record the J1850 bus voltage. _____

   Is the voltage within normal limits?   ☐ Yes   ☐ No

4. What does the voltmeter reading indicate?

_____

_____

5. Turn the ignition switch to the OFF position, and measure and record the J1850 bus voltage at the BOB.

_____

6. What conclusions, if any, can be made at this time?

_____

_____

_____

7. While the voltmeter is reading the indicated voltage of step 3, monitor the voltage while removing fuses from the power distribution center and junction box one at a time. Which fuses, when removed, caused a change in bus voltage?

_____

8. Use the service information to determine which circuits or systems are implicated by the fuses removed in step 7.

_____

9. At the vehicle, isolate the implicated components one at a time by disconnecting them and observing the voltmeter readings. Does isolating the component correct the problem?   ☐ Yes   ☐ No

If yes, identify the faulty component.

_____

If no, what is the next step?

_____

10. If the answer to step 9 was no, perform the tests you indicated that still need to be done. Was the root cause of the fault able to be determined?   ☐ Yes   ☐ No
If yes, describe the fault.

_____

If no, consult with your instructor.

_____

Task 4

1. Your instructor will inform you of the customer complaint. Record the problem.

_____

_____

2. Can the problem be verified?   ☐ Yes   ☐ No

3. Use the scan tool to communicate with the module(s) that operates the faulty system. Are communications with the module possible?   ☐ Yes   ☐ No

If yes, record any DTCs that are present.

_____

_____

If no, perform a J1850 module scan and record your results.

_____

_____

4. What problems are indicated?

_____

5. Based on these observations and your understanding of J1850 bus operation, what is the next logical step?

_____

_____

6. Based on your answer to step 5, assemble the necessary information needed to diagnose the fault. Record this information.

_____

_____

_____

7. Based on the information gathered, return to the vehicle and determine the cause of the fault. What is the necessary repair action?

_____

_____

_____

Instructor's Evaluation:

Grade	Rubric
0	No exposure due to: ☐ Absence ☐ Other:
1	Student unsuccessfully attempted the task
2	Student completed the task with significant instructor assistance.
3	Satisfactory performance of task. The student was able to perform elements of the task with minimal instructor assistance.
4	Competent/Proficient. Able to perform all elements of the task successfully and independently.

Instructor's Notes:

Name _____   Date _____

STUB NETWORK CAN BUS SYSTEM DIAGNOSIS

Upon completion of this job sheet, you will be able to identify the circuits that make up the stub-type network CAN bus and determine its operating characteristics. You will also see how circuit defects affect bus operation to aid in determining communication faults.

ASE Education Foundation Correlation

This job sheet addresses the following **MLR** tasks:

A.1.	Research vehicle service information such as fluid type, vehicle service history, service precautions, technical service bulletins, and recalls including vehicles with advanced driver assistance systems (ADAS). **(P-1)**
A.2.	Identify electrical/electronic system components and configuration. **(P-1)**
A.5.	Demonstrate proper use of a digital multimeter (DMM) when measuring voltage source, voltage drop (including grounds), current flow, and resistance. **(P-1)**
A.8.	Use fused jumper wires to check operation of electrical circuits per service information. **(P-2)**
A.9.	Use wiring diagrams to trace electrical/electronic circuits. **(P-1)**

This job sheet addresses the following **AST** tasks:

A.1.	Research vehicle service information such as fluid type, vehicle service history, service precautions, technical service bulletins, and recalls including vehicles with advanced driver assistance systems (ADAS). **(P-1)**
A.2.	Identify electrical/electronic system components and configuration. **(P-1)**
A.5.	Demonstrate proper use of a digital multimeter (DMM) when measuring voltage source, voltage drop (including grounds), current flow, and resistance. **(P-1)**
A.8.	Use fused jumper wires to check operation of electrical circuits per service information. **(P-1)**
A.9.	Use wiring diagrams during the diagnosis (troubleshooting) of electrical/electronic circuit problems. **(P-1)**
A.13.	Test and measure circuit using an oscilloscope and/or graphing multimeter (GMM); interpret results; determine needed action. **(P-2)**
G.8.	Diagnose body electronic systems circuits using a scan tool; check for module communication errors (data communication bus systems); determine needed action. **(P-2)**

This job sheet addresses the following **MAST** tasks:

A.1.	Research vehicle service information such as fluid type, vehicle service history, service precautions, technical service bulletins, and recalls including vehicles with advanced driver assistance systems (ADAS). **(P-1)**
A.2.	Identify electrical/electronic system components and configuration. **(P-1)**
A.5.	Demonstrate proper use of a digital multimeter (DMM) when measuring voltage source, voltage drop (including grounds), current flow, and resistance. **(P-1)**
A.8.	Use fused jumper wires to check operation of electrical circuits per service information. **(P-1)**
A.9.	Use wiring diagrams during the diagnosis (troubleshooting) of electrical/electronic circuit problems. **(P-1)**
A.13.	Test and measure circuit using an oscilloscope and/or graphing multimeter (GMM); interpret results; determine needed action. **(P-2)**
G.8.	Diagnose body electronic systems circuits using a scan tool; check for module communication errors (data communication bus systems); determine needed action. **(P-1)**

We Support

ASE | Education Foundation

Tools and Materials

- J1962 BOB
- Scan tool
- DMM
- Lab scope
- Fused jumper wires
- Back probing tools
- Service information
- Vehicle with full vehicle CAN bus system

Procedure

Task 1

1. Use the service information to identify the following CAN bus circuit numbers and color codes:

   NOTE: The circuits may be called by other names than those listed here.

   a. CAN B (+) _____

   b. CAN B (–) _____

   c. CAN C (+) _____

   d. CAN C (–) _____

   e. Diagnostic CAN C (+) _____

   f. Diagnostic CAN C (–) _____

2. Identify the cavity numbers the Diagnostic CAN bus uses at the DLC.

   a. Diagnostic CAN C (+) _____

   b. Diagnostic CAN C (–) _____

3. Identify the cavity numbers of a connector to a module on the CAN B bus network.

   a. Module _____

       i. CAN B (+) cavity _____

       ii. CAN B (–) cavity _____

4. Identify the cavity numbers for the CAN C bus network at the PCM connector.

   a. CAN B (+) _____

   b. CAN B (–) _____

5. Using the service information, determine the baud rate of each of the bus networks.

   a. CAN B (+) _____

   b. CAN B (–) _____

   c. CAN C (+) _____

   d. CAN C (–) _____

   e. Diagnostic CAN C (+) _____

   f. Diagnostic CAN C (–) _____

Task 2

1. Locate the CAN B module from Task 1, step 3, on the vehicle. With the ignition switch in the RUN position, use back probing tools to connect the voltmeter leads to the CAN B (+) circuit and chassis ground. Record your reading. _____

2. Read the voltage on the CAN B (–) circuit with the ignition switch in the RUN position. _____

3. Turn the ignition switch to the OFF position, and close all doors so the bus can power down. While the bus is powering down, record the voltages on each circuit.

    a. CAN B (+) _____

    b. CAN B (–) _____

4. Leave the ignition switch in the OFF position, and open a door. Record the bus voltages now.

    a. CAN B (+) _____

    b. CAN B (–) _____

5. Based on your observations, what is required to activate the CAN B bus network?

    _____

6. Turn the ignition switch to the OFF position and disconnect the vehicle battery.

7. Measure the resistance between the two CAN B bus circuits. _____

    Explain your results. _____

    _____

    _____

Task 3

1. Connect the J1962 BOB to the DLC, and turn the ignition switch to the RUN position. Use the scan tool to determine which modules on the vehicle are on the CAN B bus and record them.

    _____

    _____

    _____

2. Access the CAN B module connector from Task 1, step 3, and use a fused jumper wire to short one of the CAN B bus circuits to ground. Use the scan tool to determine which modules are active on the CAN B bus network.

    _____

    _____

    _____

3. Using a fused jumper wire and a back probing tool, short both CAN B bus circuits together. Use the scan tool to determine which modules are active on the CAN B bus network.

    _____

    _____

    _____

4. Using a fused jumper wire and a back probing tool, short both CAN B bus circuits to ground. Use the scan tool to determine which modules are active on the CAN B bus network.

_____

_____

_____

5. Use the scan tool and access the central gateway module. Record any fault codes listed.

_____

_____

_____

6. Remove the short from the bus circuit.

7. Based on your observations, is this bus network fault tolerant?   ☐ Yes   ☐ No
Explain your results.

_____

_____

Task 4

1. Use the scan tool to identify the modules that are on the CAN C bus network and record them.

_____

_____

_____

2. Locate the PCM connector identified in Task 1, and disconnect it from the PCM. Use a fused jumper wire and a back probing tool to short one of the CAN C bus circuits to chassis ground.

3. Use the scan tool and attempt to establish communication with any of the modules identified in step 1 (except the PCM). Is communication possible?

☐ Yes   ☐ No

Explain your results. _____

_____

4. Use a fused jumper wire and a back probing tool to short both of the CAN C bus circuits together. Use the scan tool and attempt to establish communication with any of the modules identified in step 1 (except the PCM). Is communication possible?

☐ Yes   ☐ No

Explain your results. _____

_____

5. Based on your observations, is this bus network fault tolerant?   ☐ Yes   ☐ No
Explain your results.

_____

_____

6. Remove all faults, but leave the PCM connector unplugged.

Task 5

1. Remove the scan tool from the DLC. Use a probing tool and the voltmeter function of the DMM to obtain and record the CAN C bus voltages during the following conditions.

   a. Ignition switch in RUN position.

   CAN C (+) _____

   CAN C (–) _____

   b. Ignition switch in OFF position.

   CAN C (+) _____

   CAN C (–) _____

   c. Ignition switch in OFF position and CAN B bus powered down.

   CAN C (+) _____

   CAN C (–) _____

   d. Ignition switch in OFF position; open a door.

   CAN C (+) _____

   CAN C (–) _____

2. Based on your observations, what is required to activate the CAN C bus network?

   _____

3. Turn the ignition switch to the OFF position and disconnect the vehicle battery.

4. Measure the resistance between the two CAN C bus circuits. _____

   Explain your results. _____

   _____

5. Reconnect all connectors.

Task 6

1. Connect the scan tool to the J1962 BOB and the DLC. Use a fused jumper wire to short one of the Diagnostic CAN C bus circuits at the BOB to ground. Attempt to communicate with the PCM. Was communication possible?   ☐ Yes   ☐ No

2. Remove the short, and move the fused jumper wire to short the other Diagnostic CAN C bus circuit to ground. Attempt to communicate with the PCM. Was communication possible?   ☐ Yes   ☐ No

3. Use a fused jumper wire to short the two Diagnostic CAN C bus circuits together. Attempt to communicate with the PCM. Was communication possible?
   ☐ Yes   ☐ No

4. Based on your observations, what is your conclusion?

   _____

   _____

   _____

5. Turn the ignition switch to the OFF position, and disconnect the vehicle battery.

6. Measure the resistance between the two Diagnostic CAN C bus circuits. _____

   Explain your results. _____

   _____

Instructor's Evaluation:

Grade	Rubric
0	No exposure due to: ☐ Absence ☐ Other:
1	Student unsuccessfully attempted the task
2	Student completed the task with significant instructor assistance.
3	Satisfactory performance of task. The student was able to perform elements of the task with minimal instructor assistance.
4	Competent/Proficient. Able to perform all elements of the task successfully and independently.

Instructor's Notes:

Name _____   Date _____

STAR NETWORK CAN BUS SYSTEM DIAGNOSIS

Upon completion of this job sheet, you will be able to identify the circuits that make up the star-type network CAN bus and determine their operating characteristics. You will also see how circuit defects affect bus operation to aid in determining communication faults.

ASE Education Foundation Correlation

This job sheet addresses the following **MLR** tasks:

A.1.	Research vehicle service information such as fluid type, vehicle service history, service precautions, technical service bulletins, and recalls including vehicles with advanced driver assistance systems (ADAS). **(P-1)**
A.2.	Identify electrical/electronic system components and configuration. **(P-1)**
A.5.	Demonstrate proper use of a digital multimeter (DMM) when measuring voltage source, voltage drop (including grounds), current flow, and resistance. **(P-1)**
A.9.	Use wiring diagrams to trace electrical/electronic circuits. **(P-1)**

This job sheet addresses the following **AST** tasks:

A.1.	Research vehicle service information such as fluid type, vehicle service history, service precautions, technical service bulletins, and recalls including vehicles with advanced driver assistance systems (ADAS). **(P-1)**
A.2.	Identify electrical/electronic system components and configuration. **(P-1)**
A.5.	Demonstrate proper use of a digital multimeter (DMM) when measuring voltage source, voltage drop (including grounds), current flow, and resistance. **(P-1)**
A.9.	Use wiring diagrams during the diagnosis (troubleshooting) of electrical/electronic circuit problems. **(P-1)**
A.13.	Test and measure circuit using an oscilloscope and/or graphing multimeter (GMM); interpret results; determine needed action. **(P-2)**
G.8.	Diagnose body electronic systems circuits using a scan tool; check for module communication errors (data communication bus systems); determine needed action. **(P-2)**

This job sheet addresses the following **MAST** tasks:

A.1.	Research vehicle service information such as fluid type, vehicle service history, service precautions, technical service bulletins, and recalls including vehicles with advanced driver assistance systems (ADAS). **(P-1)**
A.2.	Identify electrical/electronic system components and configuration. **(P-1)**
A.5.	Demonstrate proper use of a digital multimeter (DMM) when measuring voltage source, voltage drop (including grounds), current flow, and resistance. **(P-1)**
A.9.	Use wiring diagrams during the diagnosis (troubleshooting) of electrical/electronic circuit problems. **(P-1)**
A.13.	Test and measure circuit using an oscilloscope and/or graphing multimeter (GMM); interpret results; determine needed action. **(P-2)**
G.8.	Diagnose body electronic systems circuits using a scan tool; check for module communication errors (data communication bus systems); determine needed action. **(P-1)**

We Support

ASE | Education Foundation

Tools and Materials

- Scan tool
- J1962 BOB
- DMM (2)
- Lab scope
- Service information
- Vehicle with star CAN bus system

Procedure:

Task 1

Service Information

1. Using the service information, locate the following components on the vehicle and record their location:

   CAN C star connector(s) _____

   _____

   Interior bus star connector(s) _____

   _____

2. What components must be removed to access the star connectors?

   _____

   _____

   _____

3. List the circuit identification and wire colors for the CAN C and interior CAN circuits.

   NOTE: The circuits may be called by other names than those listed here.

Circuit Name	Circuit ID	Wire Color
CAN C (+)		
CAN C (−)		
Interior CAN (+)		
Interior CAN (−)		

4. Record the connector cavities where the bus circuits are located in the DLC:

Circuit Name	Cavity Number
CAN C (+)	
CAN C (−)	
Interior CAN (+)	
Interior CAN (−)	

Task 2

Voltage Measurements

This task uses the voltmeter function of the DMM to monitor normal bus activity.

1. Connect the J1962 BOB to the DLC.

2. Using two voltmeters, connect the red lead of one voltmeter to the CAN C bus (+) terminal on the BOB and the red lead of the other voltmeter to CAN C bus (−) terminal on the BOB. Connect both black leads to chassis ground.

3. Connect a scan tool to the J1962 BOB and start a diagnostic session. Measure the voltage levels on the CAN C bus with the ignition ON and ignition OFF.

    CAN C (+) Ignition ON _____ Ignition OFF _____

    CAN C (−) Ignition ON _____ Ignition OFF _____

4. Perform the same measurements for the interior CAN bus and record all voltage measurements:

    Scan tool disconnected

    Interior CAN (+) Ignition ON _____ Ignition OFF _____

    Interior CAN (−) Ignition ON _____ Ignition OFF _____

    Scan tool connected

    Interior CAN (+) Ignition ON _____ Ignition OFF _____

    Interior CAN (−) Ignition ON _____ Ignition OFF _____

5. Disconnect the scan tool and allow the bus networks to enter sleep mode. How do you know that the bus networks entered sleep mode?

    _____

    _____

    _____

6. With the bus networks in sleep mode, perform each of the following while observing the voltmeters. Record which bus network becomes active after each event.

    Cycle the ignition key _____

    Open the passenger door _____

    Press the power door unlock button _____

Task 3

Resistance Measurements

This task uses the ohmmeter function of the DMM to measure the termination resistance of the CAN C and interior CAN buses.

1. Disconnect the battery.

2. Connect the J1962 BOB to the DLC.

3. Measure the resistance between CAN C (+) and CAN C (−): _____
    Are these readings within specifications?   ☐ Yes   ☐ No

4. Measure the resistance between interior CAN (+) and interior CAN (−) _____
    Are these readings within specifications?   ☐ Yes   ☐ No

Task 4

Testing CAN C for a Short to Ground

In this task you will diagnose a simulated communication network fault that was installed by your instructor.

1. Connect the J1962 BOB to the DLC and connect the scan tool to the BOB. Start a diagnostic session. Is communication with the vehicle possible? ☐ Yes ☐ No

2. List any communication network(s) not communicating. _____

_____

3. Were any active DTCs set? If yes, list them:

_____

_____

4. Using the service information, locate all communication network wiring schematics for the vehicle.

5. Using two voltmeters, measure the voltage levels on the CAN C bus with the ignition ON:

   CAN C (+) _____

   CAN C (–) _____

6. Using two voltmeters, measure the voltage levels on the interior CAN bus with the ignition ON:

   Interior CAN (+) _____

   Interior CAN (–) _____

7. Based on the voltage readings, what fault is present?

_____

_____

8. Refer to the communication network wiring schematics; where is the best place to disconnect and isolate the communication bus for diagnosis?

_____

_____

9. Disconnect the battery.

10. Connect the ohmmeter between the terminal at the J1962 BOB for the faulty circuit and chassis ground.

11. While observing the ohmmeter, disconnect each connector one at a time. Were you able to isolate the fault? ☐ Yes ☐ No

    If no, consult with your instructor.

    Where is the fault isolated to?

_____

_____

12. How would this fault be repaired?

_____

_____

_____

Instructor's Evaluation:

Grade	Rubric
0	No exposure due to: ☐ Absence ☐ Other:
1	Student unsuccessfully attempted the task
2	Student completed the task with significant instructor assistance.
3	Satisfactory performance of task. The student was able to perform elements of the task with minimal instructor assistance.
4	Competent/Proficient. Able to perform all elements of the task successfully and independently.

Instructor's Notes:

Name _____   Date _____

LIN BUS SYSTEM DIAGNOSIS

Upon completion of this job sheet, you will be able to determine normal LIN bus operation. You will also observe faulty bus system operation.

ASE Education Foundation Correlation

This job sheet addresses the following **MLR** tasks:

A.1.	Research vehicle service information such as fluid type, vehicle service history, service precautions, technical service bulletins, and recalls including vehicles with advanced driver assistance systems (ADAS). **(P-1)**
A.2.	Identify electrical/electronic system components and configuration. **(P-1)**
A.5.	Demonstrate proper use of a digital multimeter (DMM) when measuring voltage source, voltage drop (including grounds), current flow, and resistance. **(P-1)**
A.8.	Use fused jumper wires to check operation of electrical circuits per service information. **(P-2)**
A.9.	Use wiring diagrams to trace electrical/electronic circuits. **(P-1)**

This job sheet addresses the following **AST** tasks:

A.1.	Research vehicle service information such as fluid type, vehicle service history, service precautions, technical service bulletins, and recalls including vehicles with advanced driver assistance systems (ADAS). **(P-1)**
A.2.	Identify electrical/electronic system components and configuration. **(P-1)**
A.5.	Demonstrate proper use of a digital multimeter (DMM) when measuring voltage source, voltage drop (including grounds), current flow, and resistance. **(P-1)**
A.8.	Use fused jumper wires to check operation of electrical circuits per service information. **(P-1)**
A.9.	Use wiring diagrams during the diagnosis (troubleshooting) of electrical/electronic circuit problems. **(P-1)**
A.13.	Test and measure circuit using an oscilloscope and/or graphing multimeter (GMM); interpret results; determine needed action. **(P-2)**
G.8.	Diagnose body electronic systems circuits using a scan tool; check for module communication errors (data communication bus systems); determine needed action. **(P-2)**

This job sheet addresses the following **MAST** tasks:

A.1.	Research vehicle service information such as fluid type, vehicle service history, service precautions, technical service bulletins, and recalls including vehicles with advanced driver assistance systems (ADAS). **(P-1)**
A.2.	Identify electrical/electronic system components and configuration. **(P-1)**
A.5.	Demonstrate proper use of a digital multimeter (DMM) when measuring voltage source, voltage drop (including grounds), current flow, and resistance. **(P-1)**
A.8.	Use fused jumper wires to check operation of electrical circuits per service information. **(P-1)**
A.9.	Use wiring diagrams during the diagnosis (troubleshooting) of electrical/electronic circuit problems. **(P-1)**
A.13.	Test and measure circuit using an oscilloscope and/or graphing multimeter (GMM); interpret results; determine needed action. **(P-2)**
G.8.	Diagnose body electronic systems circuits using a scan tool; check for module communication errors (data communication bus systems); determine needed action. **(P-1)**

We Support

ASE | Education Foundation

Tools and Materials

- Scan tool
- DMM
- Lab scope
- Fused jumper wires
- Back probing tool
- Service information
- Vehicle with LIN bus system

Procedure

1. Use the service information to identify the function(s) of the LIN bus(es) on the assigned vehicle.

   _____

   _____

   _____

2. Identify the LIN bus cavities of the master module.

   _____

   _____

   _____

3. Identify the LIN bus cavities of the slave module.

   _____

   _____

   _____

4. Does the slave module have any inputs to it from other sources? ☐ Yes ☐ No

   If yes, list them.

   _____

   _____

5. With the ignition switch in the RUN position, connect the scan tool and navigate to the master module. Are any items available in the data display to assist in diagnosing a system failure? ☐ Yes ☐ No

   If yes, list them.

   _____

   _____

6. Press and release any switch inputs that were identified while observing the scan tool. Do the switch states change on the scan tool? ☐ Yes ☐ No

7. Press and hold one of the switch buttons for a minimum of 60 seconds and then release. Check and record any DTCs.

   _____

   _____

8. Is the DTC active or stored? _____

    a. If active, press and release the same button again. Did it change to stored now?

    ☐ Yes   ☐ No

9. Disconnect the electrical connector to a switch input to the slave module. Check and record any DTCs.

_____

_____

10. If no DTCs were set, why not?

_____

11. Press and release another switch input to the slave module while observing the scan tool. Does the scan tool indicate the switch presses?   ☐ Yes ☐ No

12. Explain your results.

_____

_____

13. With the slave module still disconnected, use the voltmeter function of the DMM to measure the voltage on the LIN bus. _____

14. Reconnect the slave module and record the voltage on the LIN bus. _____

15. Are the voltages the same?   ☐ Yes ☐ No

    If no, explain.

_____

16. Use a lab scope to observe the trace of the LIN bus voltages. Describe what is observed.

_____

17. Short the LIN bus circuit to ground by back probing the slave module connector and using a fused jumper wire connected to a chassis ground. Check and record any DTCs.

_____

_____

18. Caution: Verify circuit identification before performing this step, or vehicle damage could occur.

Short the LIN bus circuit to power by back probing the right steering wheel switch connector; use a fused jumper wire connected to vehicle power. Check and record any DTCs.

Instructor's Evaluation:

Grade	Rubric
0	No exposure due to: ☐ Absence ☐ Other:
1	Student unsuccessfully attempted the task
2	Student completed the task with significant instructor assistance.
3	Satisfactory performance of task. The student was able to perform elements of the task with minimal instructor assistance.
4	Competent/Proficient. Able to perform all elements of the task successfully and independently.

Instructor's Notes:

Name _____   Date _____

DATA BUS CIRCUIT WIRE REPAIR

Upon completion of this job sheet, you will be able to properly repair the wires of a data bus network circuit.

ASE Education Foundation Correlation

This job sheet addresses the following **MLR** tasks:

A.1.   Research vehicle service information such as fluid type, vehicle service history, service precautions, technical service bulletins, and recalls including vehicles with advanced driver assistance systems (ADAS). **(P-1)**

A.12.   Repair and/or replace connectors, terminal ends, and wiring of electrical/ electronic systems (including solder repair). **(P-2)**

This job sheet addresses the following **AST/MAST** tasks:

A.1.   Research vehicle service information such as fluid type, vehicle service history, service precautions, technical service bulletins, and recalls including vehicles with advanced driver assistance systems (ADAS). **(P-1)**

A.12.   Repair and/or replace connectors, terminal ends, and wiring of electrical/electronic systems (including solder repair); determine needed action. **(P-1)**

We Support
ASE | **Education Foundation**

Tools and Materials

- 100-watt soldering iron
- 60/40 rosin core solder
- Crimping tool
- Splice clip
- Heat shrink tube
- Heating gun
- Service information
- Vehicle with a data bus network system

Safety Considerations and PPE

- Safety glasses

Procedure

1. Refer to the service information and record the specification for the number of twists per meter or inch.

   _____

2. Cut out the portion of the damaged wiring being sure that the wire slices will be staggered.

3. What is the length of each wire that was removed?

   (+) _____

   (-) _____

4. Why is this measurement important?

   _____

   _____

   _____

5. Using the correct size stripper, remove about 1/2 inch (12 mm) of the insulation from the ends of all wires. Be careful not to nick or cut any of the wires.

6. Select the proper size splice clip to hold the splice.

7. Place the correct length and size of heat shrink tube over the wire. Slide the tube far enough away so the wires are not exposed to the heat of the soldering iron.

8. Overlap the two splice ends and hold in place your thumb and forefinger. While the wire ends are overlapped, center the splice clip around the wires and crimp into place.

9. Heat the splice clip with the soldering iron while applying solder to the opening in the back of the clip. Apply only enough solder to make a good connection. The solder should travel through the wire.

10. After the solder cools, slide the heat shrink tube over the splice and heat the tube with the hot air gun until it shrinks around the splice. Do not overheat the tube.

11. Repeat the soldering process for all wires.

12. Use the service information to identify DLC cavity assigned to the ISO-K circuit to the PCM.

    a. K-line

       Cavity number _____

       Circuit identification _____

       Color code _____

13. Connect the J1962 BOB to the DLC. Use a lab scope to trace the voltage on the K-line with the ignition switch in the OFF position. If available, use the print function to print the trace for the K-line circuit. If not available, describe what was observed.

    _____

14. Connect the scan tool, and with the ignition switch in the OFF position, use a lab scope to trace the voltage on the K-line. If available, use the print function to print the trace for the K-line circuit. If not available, describe what was observed.

    _____

15. With the ignition switch still in the OFF position, attempt to communicate with the PCM. Use a lab scope to trace the voltage on the K-line. If available, use the print function to print the trace for the K-line circuit. If not available, describe what was observed.

    _____

16. Turn the ignition switch to the RUN position. Establish communication between the scan tool and the PCM while observing the lab scope trace. If available, use the print function to print the trace for the K-line circuit. If not available, describe what was observed.

    _____

17. What is indicated by each of the voltage values seen on the trace?

    _____

    _____

    _____

Instructor's Evaluation:

Grade	Rubric
0	No exposure due to: ☐ Absence ☐ Other:
1	Student unsuccessfully attempted the task
2	Student completed the task with significant instructor assistance.
3	Satisfactory performance of task. The student was able to perform elements of the task with minimal instructor assistance.
4	Competent/Proficient. Able to perform all elements of the task successfully and independently.

Instructor's Notes:

Name _____   Date _____

SCAN TOOL AND TECHNICIAN OEM AUTHENTICATION

Upon completion of this job sheet, you will be able to identify the steps for authenticating the scan tool.

Tools and Materials

• Scan tool
• Computer with internet access

Procedure

1. Instructor assigned OEM. _____

2. Scan tool manufacturer and model. _____

   _____

   a. Do an internet search for authenticating a scan tool with the OEM assigned by your instructor. Does this OEM require scan tool authentication? Yes _____ No _____

3. Is your scan tool listed as an approved tool by the OEM ?   Yes _____   No _____

4. What If there are various options for the type of services provided, which would be the most fitting for you? _____

5. Is there a cost associated with the option chosen?   Yes _____ No _____

6. Briefly describe the process of getting an account. _____

   _____

Instructor's Evaluation:

Grade	Rubric
0	No exposure due to: ☐ Absence ☐ Other:
1	Student unsuccessfully attempted the task
2	Student completed the task with significant instructor assistance.
3	Satisfactory performance of task. The student was able to perform elements of the task with minimal instructor assistance.
4	Competent/Proficient. Able to perform all elements of the task successfully and independently.

Instructor's Notes:

DIAGNOSTIC CHART 10-1

PROBLEM AREA:	Vehicle communication systems.
SYMPTOMS:	Scan tool cannot communicate with a module.
POSSIBLE CAUSES:	**1.** Faulty battery feed circuit to module.
	2. Faulty ignition feed circuit to module.
	3. Poor module ground.
	4. Open data bus circuit to module.

DIAGNOSTIC CHART 10-2

PROBLEM AREA:	Vehicle communication systems.
SYMPTOMS:	Scan tool cannot communicate with any module.
POSSIBLE CAUSES:	**1.** Open bus circuit at DLC.
	2. Bus circuit shorted to ground.
	3. Bus circuit shorted to voltage.
	4. Internal failure in a module causing short to power.
	5. Internal failure in a module causing short to ground.
	6. Open termination resistors.
	7. Faulty scan tool.

DIAGNOSTIC CHART 10-3

PROBLEM AREA:	Vehicle communication systems.
SYMPTOMS:	Total bus failure indicated by default conditions and possible no-start.
POSSIBLE CAUSES:	**1.** Open bus circuit at DLC.
	2. Bus circuit shorted to ground.
	3. Bus circuit shorted to voltage.
	4. Internal failure in a module causing short to power.
	5. Internal failure in a module causing short to ground.
	6. Open termination resistors.

DIAGNOSTIC CHART 10-4

PROBLEM AREA:	Vehicle communication systems.
SYMPTOMS:	System fails to operate and/or warning lamp illumination.
POSSIBLE CAUSES:	**1.** Bus circuit shorted to ground.
	2. Bus circuit shorted to voltage.
	3. Open in bus circuit.
	4. Internal failure in a module causing short to power.
	5. Internal failure in a module causing short to ground.
	6. Open termination resistors.

CHAPTER 11

LIGHTING CIRCUITS DIAGNOSTICS AND REPAIR

Upon completion and review of this chapter, you should be able to:

- Correctly replace sealed-beam and composite headlights.
- Correctly service the high-intensity discharge (HID) lamp and ballast.
- Diagnose the computer-controlled headlight system.
- Diagnose the automatic headlight system.
- Perform a functional test of the automatic headlight system.
- Test and replace the automatic headlight system's photocell.
- Test the automatic headlight system's amplifier.
- Perform SmartBeam camera calibration and adjustment.
- Diagnose the automatic headlight leveling system.
- Diagnose the adaptive headlight system.
- Correctly aim sealed-beam and composite headlights.
- Diagnose the cause of brighter-than-normal lights.

- Diagnose the cause of dimmer-than-normal lights.
- Remove and replace dash-mounted and steering column–mounted headlight switches.
- Replace multifunction switches.
- Test and determine needed repairs of the dimmer switch and related circuits.
- Replace the dimmer switch.
- Diagnose incorrect taillight assembly operation.
- Diagnose the turn signal system for improper operation.
- Replace the turn signal switch.
- Diagnose the interior lights, including courtesy, instrument, and panel lights.
- Diagnose control module–operated illuminated entry systems.
- Diagnose BCM-controlled illuminated entry systems.
- Diagnose fiber-optic systems.

Terms to Know

Adaptive headlight system (AHS)	Dimmer switch	Optics test
	Feedback	Photocell resistance assembly
Curb height	Illuminated entry actuator	
Demonstration mode	Multifunction switch	

Introduction

Vehicle lighting systems are becoming very complex. Any failure of the lighting system requires a systematic approach to diagnose, locate, and correct the fault in the minimum amount of time.

Classroom Manual
Chapter 11, page 298

The importance of a proper operating lighting system cannot be overemphasized. Checking the lighting system should be done as a courtesy to your customer whenever they bring their vehicle to your shop for repairs. Often a customer may not be aware of a light failure. If a lighting circuit is not operating properly, there is a potential danger to the driver and other people. When today's technician performs repairs on the lighting systems, the repairs must assure vehicle safety and meet all applicable laws. Be sure to use the correct lamp type and size for the application (**Figure 11-1**).

Before performing any lighting system tests, check the battery's state of charge. Also, be sure all battery cable connections are clean and tight. Visually check wires for damaged insulation, loose connections, and improper routing.

When troubleshooting the lighting system, if only one bulb does not operate, it's usually faster to replace it with a known good unit first. Check the connector or socket for signs of corrosion. When testing the circuit with a voltmeter, ohmmeter, or test light, check the most easily reached components first. A troubleshooting chart for the lighting system is at the end of this chapter to assist in locating the fault.

AUTHOR'S NOTE Some headlight bulbs are very expensive. Before replacing these bulbs, be sure to perform complete circuit diagnostics.

Although federal laws largely regulate lighting circuits, and the systems are similar between the various manufacturers, there are variations in operation. Before attempting any repairs on an unfamiliar circuit, the technician should always refer to the manufacturer's service information.

Computer-controlled lighting systems aim to make system diagnosis as easy as possible. The controller may provide trouble codes to assist in diagnosis. Most manufacturers provide a detailed diagnostic chart for the most common symptoms. The most important thing to remember when diagnosing these systems is to follow the diagnostic procedures in order. Don't attempt to get ahead of the chart by assuming the outcome of a test. This will lead to the replacement of good parts and lost time.

Common automotive bulbs:
A, B – Miniature bayonet for indicator and instrument lights
C – Single-contact bayonet for license and courtesy lights
D – Double-contact bayonet for trunk and underhood lights
E – Double-contact bayonet with staggered indexing lugs for stop, turn, and brake lights
F – Cartridge type for dome lights
G – Wedge base for instrument lights
H – Blade double-contact for stop, turn, and brake lights

Figure 11-1 Correct selection of the lamp is important for proper operation of the system.

In this chapter, you will perform selected service samples on the computer-controlled concealed headlight, automatic headlight, and illuminated entry systems. It's out of the scope of this manual to provide service procedures for the different manufacturers that use these systems. Technicians must follow the procedure in the service information for the vehicle they are diagnosing.

Headlight Replacement

One of the most common lighting system repairs is replacing the headlight. Over time, the filament may burn through, or the lens may break. However, before replacing the headlight, confirm that voltage is present with a voltmeter or test light. Next, check the ground for proper connections. Also, confirm that the connector or socket has no corrosion and is in good condition. If these test good, the headlight is probably faulty and needs to be replaced. If there is no voltage t the connector, work back toward the switch and battery to locate the fault.

The procedure for replacing the headlight differs depending on the type of bulb used. Most conventional sealed-beam and halogen sealed-beam headlights have similar replacement procedures. Composite headlights require different procedures. Always refer to the service information for the vehicle you are working on.

Sealed-Beam Headlights

A broken lens of a conventional sealed-beam headlight will cause the filament to burn. Since the halogen lamp houses the filament in its own bulb envelope, cracking or breaking of the lens does not prevent headlight operation. The filament will continue to operate if the envelope does not break. However, a broken lens affects light quality, and the headlight should be replaced for the customer's safety.

Sealed-beam headlight replacement usually requires the removal of the bezel (**Figure 11-2**). Some vehicles may require removing the turn signal light assembly before the headlight is accessible. When removing the bezel, do not turn the two headlight-aiming adjustment screws. The headlight may have a retaining ring surrounding it that will require removal.

With the headlight removed and the plug disconnected, check the terminals of the plug for corrosion and other foreign material. Clean as necessary. Coat the connector terminals and the prongs of the new headlight with dielectric grease to prevent corrosion (**Figure 11-3**).

When positioning the headlight into the housing, be sure that the embossed number is at the top. Many headlights are marked "TOP." Install the retainer trim and the headlight bezel.

Special Tools
Fender covers
Safety glasses
Torx drivers

⚠ Caution
Because of the construction and placement of the prisms in the lens, the headlight must be installed in its proper position. The lens is usually marked "TOP" to indicate the proper installed position.

Classroom Manual
Chapter 11, page 300

Figure 11-2 Headlight bezel removal.

Figure 11-3 Apply dielectric grease to the terminals to help prevent corrosion.

Classroom Manual
Chapter 11, page 302

Because the composite headlight and HID systems use vented housings, condensation may develop inside the lens assembly. This condensation is not harmful to the bulb and does not affect headlight operation.

Check the operation of the headlight. After confirming proper headlight operation, check headlight aiming as described later in this chapter.

Composite Headlights

Photo Sequence 20 illustrates a typical procedure for replacing a composite headlight. Whenever replacing a halogen bulb, take care not to touch the envelope with your fingers. Staining the bulb with normal skin oil can substantially shorten the life of the bulb. Handle the lamp only by its base. Also, dispose of the old lamp properly. Wear safety glasses when handling halogen bulbs. In addition, be careful not to get any dielectric grease onto the bulb. This too will shorten the bulb's life. After replacing the bulb, check the headlight aiming and correct it if needed.

PHOTO SEQUENCE 20
Replacing a Composite Headlight Bulb

P20-1 Place fender covers around the work area.

P20-2 On some vehicles it may be necessary to remove the lamp unit.

P20-3 Disconnect the wiring harness from the socket.

P20-4 Unlock the socket by rotating the retaining ring 1/8 of a turn.

P20-5 Gently pull the bulb straight back out of the lamp unit. To prevent breaking the bulb and locating tabs, do not rotate it while pulling.

P20-6 Check the connector terminals for corrosion or other foreign material. Clean as needed.

P20-7 Coat the connector terminals and the prongs of the new headlight with dielectric grease to prevent corrosion.

P20-8 Align the socket with the keyed opening of the lamp unit. When properly aligned, the socket will seat completely onto the headlight unit.

P20-9 Insert the retaining ring into the lamp unit and rotate it to lock it in place.

P20-10 Reconnect the wiring harness to the socket.

P20-11 If removed, install the lamp unit.

P20-12 Check headlight operation.

HID Lamp and Ballast Service

⚡ **WARNING** **Before testing the HID system, remove the headlight assembly from the vehicle. The HID system operates on high voltage and current. Accidentally shorting the system can cause personal injury or death.**

The high-intensity discharge (HID) lamps require battery voltage and ground to operate, as any other lamp. The ballast has internal circuit protection to prevent damage resulting from an open or shorted circuit. In addition, the circuit protection will come into effect if an overcharge or undercharge condition exists. Any of these faults can cause the headlights not to turn on for that key cycle, or until the fault is corrected. If the customer states that both headlights don't operate, recycle the ignition or the headlight switch to see if the condition recurs. If the lights still don't come on, check the condition of the battery and the charging system. It's unlikely that both lamps failed at the same time, so check the switch input, HID relay, and common power feeds and grounds.

A burned-out lamp will appear black or smoky. If a visual inspection does not indicate a faulty lamp, check for battery voltage to the ballast. If battery voltage is within specifications, check the ground circuit for the ballast. If neither of these circuits has an open, short, or excessive resistance, substitute a known good ballast. If the headlight on the opposite side of the vehicle is operating properly, substitute its ballast with the side that does not operate. If the headlight illuminates, replace the ballast. If the light still does not come on, replace the bulb/ignitor element.

Depending on design, it may be necessary to remove the headlight assembly prior to replacing the bulb and ignitor. Some vehicles provide access that allows servicing the bulb and ignitor without removing the headlight assembly. Regardless of how the bulb element is accessed, replacement procedures are very similar.

Prior to servicing the headlight element, disconnect and isolate the battery negative cable or remove the HID relay. This will prevent the accidental operation of the headlights.

⚡ **WARNING** **The HID system produces a high voltage and current. Do not attempt to operate the headlights without all connections properly made.**

⚡ **WARNING** **Never attempt to open the ballast or ignitor assembly. The ignitor is a component of the bulb and cannot be separated.**

Classroom Manual
Chapter 11, page 302

Figure 11-4 A rubber boot on the back of the head-light assembly protests the HID element.

Figure 11-5 A cable connects the ignitor to the ballast.

To gain access to the bulb assembly, remove the rubber boot that protects the element (**Figure 11-4**). At the ignitor, disconnect the cable coming from the ballast (**Figure 11-5**). A wire clip may retain the element to the headlight assembly. Release the clip and remove the element by grabbing the ignitor and pulling the unit rearward. It may be necessary to rock the ignitor slightly while pulling it out.

To install the new element, push the bulb element assembly into the headlight unit. With the bulb element assembly properly seated, secure it with the retainer clip. Connect the cable from the ballast to the ignitor's connector. Install the protective rubber cover, making sure it is properly seated. Water may enter the element and damage the unit if the cover is not seated. If necessary, install the headlight assembly. Reconnect the battery or connect the relay and test for proper operation.

The ballast can be attached to the bottom of the headlight assembly, to the vehicle's frame, or to the inner fender well. A cable connects the ballast to the ignitor. In addition, another connector attaches the ballast to the vehicle's wiring system (**Figure 11-6**). It may be necessary to remove the headlight assembly to gain access to the ballast. Prior to removing the ballast, disconnect and isolate the battery negative cable or remove the HID relay.

Remove the screws that retain the ballast to the headlight assembly or frame. Next, disconnect the main electrical connector and the cable to the ignitor from the ballast. The cable may be secured by a wire lever that must be rotated down. Installation of the new ballast is in reverse order of the removal procedure. Be sure to properly secure the wire retainer around the cable connector.

Figure 11-6 HID ballast connectors. Notice the warning sticker on the ballast.

Testing Computer-Controlled Headlight Systems

A basic computer-controlled headlight system will use the headlight switch as an input to the BCM, which then activates the required relays to perform the requested function. The system may have separate relays for the park lights, fog lights, and headlights. Low-side drivers usually perform the circuit operation for each of these relays to complete the path to ground for the relay coil. The following will focus on testing the headlight system, although diagnostics for the park and fog light systems are similar.

Using **Figure 11-7** as a schematic of the system, we will go through the steps to diagnose a customer's concern that the headlights don't come on. Always refer to the proper wiring diagram for the vehicle you are working on. As with any electrical system problem, begin by confirming that the system is malfunctioning. Also, note any other systems that are not working properly. In the system shown, use the multifunction switch to activate the flash-to-pass feature. This will test the operation of the high-beam lights and relay.

Check any associated fuses and repair as needed. Next, conduct a visual inspection and repair any faults found. If this doesn't locate the cause of the problem, substitute the low-beam relay with a known good relay of the same type and retest operation. If the system works now, the relay was at fault and replacing the relay will complete the repair. If the system still does not work, begin testing the rest of the system.

Use a scan tool to check if there are any diagnostic trouble codes (DTCs). If so, use the diagnostic chart for the DTC to locate the problem. If there are no DTCs (not all systems will set codes), use the scan tool to perform an activation test of the head lights. If the headlights work during the activation test, the problem is in the input side of the system. Use the scan tool to monitor the headlight switch while moving it through all switch positions. Record the voltages for each position and compare them to specifications. In this example, if all positions display 5 volts, there is an open in the headlight switch MUX circuit. The open could be in the wire between the BCM and the switch, in the switch itself, or in the ground circuit. Test each of these components with an ohmmeter. The switch is a resistive multiplex switch that should have a different resistance value for each switch position. With the switch disconnected, attach the ohmmeter across terminals 1 and 2 of the switch. Place the switch in each position and record the results. Check the results with the service information. If the values are out of range, replace the switch.

Classroom Manual
Chapter 11, page 310

 Special Tools
Scan tool
DMM
Fused jumper wires
Test light

Figure 11-7 Schematic of a computer-controlled headlight system.

When testing the input signal with the scan tool, a voltage of 0 volt in all switch positions indicates either a short to ground in the circuit or a faulty BCM. Test the circuit for a short. If there are no shorts, check the voltage input and ground connections of the BCM. If these are good, replace the BCM.

If the headlights don't come on when the scan tool activates the system, the problem is in the relay control circuit, the BCM, the high-current circuit from the battery through the relay to the headlights, or in the headlight ground. To determine which circuit is at fault, remove the low-beam headlight relay. Use a voltmeter or test light to confirm battery voltage is present at pins 30 and 86 of the junction block terminals. If voltage is not present, there is an open or a short between the junction block and the battery. If voltage is present at both pins, use a fused jumper wire to connect pin 30 to pin 87. If the lights come on now, the problem is in the control side circuit of the relay (the relay was tested earlier when it was substituted).

SERVICE TIP If a scan tool is not available, the relay's input and the control sides can be tested by connecting a test light across terminals 85 and 86. If the test light comes on when the switch is in the headlamp position, the BCM receives the input and carries out the command.

To test the control side circuit, connect a test light between pins 86 and 85 of the junction block terminals (relay removed). Place the headlight switch into the headlamp position and observe the test light. Since the BCM should now complete the path to ground, the test light should illuminate. If the test light fails to turn on, go to pin 7 of the BCM and back probe a fused jumper wire into the connector. Touch the other end of the fused jumper wire to a good ground. If the test light comes on, check the ground connections of the BCM. If the grounds are good, replace the BCM. If the test light still does not come on, there is an open in the circuit between pin 85 of the junction block and the BCM. In this case, the BCM attaches to the back of the junction block, and the circuit wire is internal to the junction block. If this circuit is bad, the junction block will need to be replaced.

If the lights don't come on when the fused jumper wire is across pins 30 and 87 of the junction block, the problem is between the relay and the headlights. Connect a fused jumper wire from the battery positive post to the feed into one of the headlight connectors. If the headlight illuminates, check the circuit from the junction block to the headlight. If the circuit is good, the problem is probably at the splice in the junction block and the junction block will need to be replaced. If the headlight still does not turn on, check the ground connections. If the grounds are good, the bulb is burned out. It's unlikely that both headlight bulbs will burn out at the same time unless there are other problems. A charging system outputting too much voltage will likely cause both bulbs to burn out. Also, use a lab scope connected to the battery positive post and ground to look for voltage spikes above the charging system's target voltage value. Voltage spikes may indicate that the battery connections are not clean (remember, the battery acts as a buffer), the battery is bad, or there is resistance on the charging system's sense circuit.

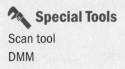

Special Tools
Scan tool
DMM

HSD-Controlled Headlights

Systems that use high-side drivers (HSDs) to operate the headlights typically use pulse width modulation (PWM) to keep the voltage to the bulbs at a consistent value. Based on the configuration, the HSD may provide short and open circuit diagnostics. The HSD can also act as the fuse and turn off if a short is determined. Some HSDs can detect circuit faults even with the headlights turned off. In HSD-controlled lighting systems, the driver uses a diagnostic voltage that is typically between 8 and 12 volts. When the driver is off, the diagnostic voltage is sent through the bulb to ground. The current in the circuit is too low to illuminate the bulb. An intact circuit will pull the voltage low. If there is an open, the voltage will be high, and the module will set a fault and (in some systems) send a message to the vehicle information module or Cabin Compartment Node (CCN) to turn on the lamp outage warning. This process of fault detection can be applied to all exterior lighting systems.

Diagnostics of the HSD headlight system is DTC driven. Detection of a fault results in the module recording a DTC. Once the scan tool retrieves the DTC, follow the troubleshooting tree to isolate the fault. Circuit testing is the same as with any other circuit. Confirm that the power and ground circuits are intact using voltage drop and resistance testing methods.

Classroom Manual
Chapter 11, page 313

Classroom Manual
Chapter 11, pages 348, 351

⚙️ **SERVICE TIP** Keep in mind that many HSD-controlled systems will disable the HSD when a fault is detected. The HSD may remain off for the remainder of the key cycle. Unplugging the connector to measure voltage will result in the system detecting the open and turning off the HSD. This can lead to misdiagnosis if you are not aware of this function. It may be necessary to apply an electrical load to the circuit to get accurate test results.

Automatic Headlight System Diagnosis

Classroom Manual
Chapter 11, page 314

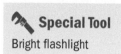

Special Tool

Bright flashlight

The automatic on-off with time delay has two functions: to turn on the headlights automatically when ambient light decreases to a predetermined level and to allow the head-lights to remain on for a certain amount of time after the vehicle has been turned off (providing light for the passengers exiting the vehicle). This system can be combined with the automatic dimming system.

The common components of the automatic on-off with time delay include the following:

1. Photocell and amplifier.
2. Power relay.
3. Timer control. Control of time delay can be a potentiometer incorporated into the headlight switch. It can also be a programmable feature through the CCN or overhead console. The timer control unit controls the automatic operation of the system and the length of time the headlights stay on after turning off the ignition.

If the headlights don't turn on, check the regular headlight system first before condemning the automatic headlight system. If the headlights don't illuminate with the automatic system turned off, the problem is in the basic circuit.

To perform a functional test of the automatic headlight system, turn the headlight switch into the appropriate position to activate the automatic headlight function. Cover the photocell and start the engine. Some systems activate with the ignition in the RUN position; however, many systems delay turning on the headlights until an engine run signal is received. The headlights should turn on within 30 seconds. Remove the cover you placed over the photocell and shine a bright light onto it. The headlights should turn off after 10 seconds but within 60 seconds. Cover the photocell again. When the head-lights turn on, wait for 15 seconds, then turn off the ignition switch. The headlights should turn off after the selected amount of time delay.

If indications point to the fault being within the automatic headlight system, make a few quick checks of the system. Inspect the photocell lens for obstructions. Check all connections from the headlight switch, as well as all fuses used in the system.

The following are some of the most common concerns that result from problems in the automatic headlight:

1. Lights turning on and off at the wrong ambient light levels.
2. Lights that don't turn on in darkness.
3. Lights not turning off in bright light.
4. Lights not staying on for an adjustable time after turning the ignition off.
5. Lights that don't turn off after turning off the ignition.

These problems can result from faults in the headlight switch, ignition switch, amplifier, or photocell. To locate the problem, perform the following test series. As a service sample of automatic headlight system diagnosis, refer to **Figure 11-8**. The following test

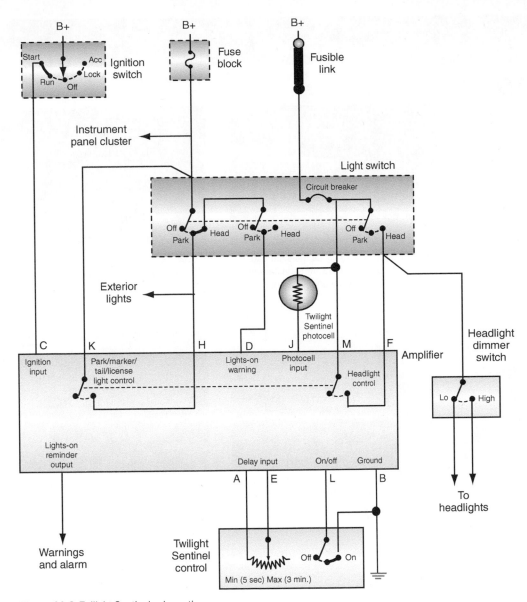

Figure 11-8 Twilight Sentinel schematic.

connections will refer to those this system uses. Use the correct service information for the system you are diagnosing.

Photocell Test

To test the photocell, use a scan tool to monitor photocell voltages or a DMM to measure the voltage of the photocell signal circuit to the BCM. With the engine running, observe the voltage readings. First, cover the photocell's lens with a piece of cardboard. The voltage should read 0 volt. Next, shine a flashlight onto the photocell's lens. The voltage should go up to 5 volts. If the voltages don't change, check the circuit, and if there are no problems, replace the photocell.

Another method of testing the photocell is to unplug the photocell connector and start the engine. Place the headlight switch in the correct position to turn on the automatic headlight feature. If the headlights turn on within 60 seconds, replace the photocell. If the lights don't turn on, test the amplifier (if equipped).

Photocell Replacement. Photo Sequence 21 illustrates the procedure for replacing the photocell.

Classroom Manual
Chapter 11, page 314,
and Chapter 9,
page 263

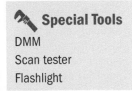 **Special Tools**
DMM
Scan tester
Flashlight

PHOTO SEQUENCE 21
Typical Procedure for Replacing a Photocell Assembly

P21-1 Place the fender covers over the vehicle's fenders and disconnect the battery negative terminal.

P21-2 The photocell may be installed into a defroster or speaker grill. Starting in a corner, use a trim stick to gently pry the grill up.

P21-3 Carefully work your way down the length of the grill, pulling it free from its seat.

P21-4 Gently work the grill up and out to access the photocell socket.

P21-5 Twist the socket to free it from the lock tabs and gently work the photocell free from the grill.

P21-6 Unplug the photocell connector.

Classroom Manual
Chapter 11, page 315

 Special Tools

Ohmmeter
Fused jumper wire
Test light
Fender covers
Safety glasses

Use a 20-amp (A) fuse in the jumper.

⚡ **Caution**

Skipping any of the tests will result in the replacement of the amplifier, even if it's good.

Amplifier Test

Refer to Figure 11-8 as an example of how to test the amplifier.

1. Disconnect the wire connector to the amplifier.
2. Turn the headlight switch to the OFF position.
3. Disconnect the negative battery terminal.
4. Turn the control switch to the ON position.
5. Use an ohmmeter to measure the resistance between the wire connector terminal L and ground. There should be 0 ohm of resistance. If there is more than 0 ohm of resistance, check the wire from amplifier terminal L to control switch terminal C for opens. Also, check the circuit from amplifier terminal B for opens. If the ohmmeter indicated 0 ohm of resistance, continue testing.
6. Turn the control switch to the OFF position.
7. The ohmmeter should read infinite when connected between the L terminal and ground. If there is low resistance, check the circuit for a short. If the ohmmeter indicates infinite resistance, continue testing.
8. Place the headlight switch in the PARK position and connect the ohmmeter between terminals H and D. There should be 0 ohm of resistance. If not, check the two circuits for an open. If the ohmmeter indicates 0 ohm of resistance, continue testing.
9. Place the control switch in the middle of MIN and MAX settings.

10. Connect the ohmmeter between terminals A and E. There should be between 500 and 250,000 ohms of resistance. If the resistance value is not within this range, check the circuits between the amplifier and the control switch for an open.
    - If the circuits are good, observe the ohmmeter while moving the control switch from the MIN position to the MAX position. The resistance value should change smoothly and consistently from one position to the next. If the ohmmeter indicates a resistance value out of limits, or the reading is erratic, replace the control switch.

11. If all of the resistance values tested are correct, continue testing.

12. Turn off all switches and reconnect the amplifier wire connector. Reconnect the battery negative terminal.

13. Turn the ignition switch to the RUN position. Turn the headlight switch to the OFF position, and turn the control switch to ON.

14. Connect a fused jumper wire between amplifier terminals J and M. The headlights should go off within 60 seconds. If the headlights go off within 60 seconds, but don't go off normally in bright light, check the circuit for opens. If the circuit is good, replace the photocell.

15. Turn off all switches.

16. Disconnect the wire connector to the amplifier.

17. Turn the ignition switch to RUN and turn the headlight switch to OFF.

18. Connect a test light between terminal M and ground. The test light should light. If not, there is a short or an open in the circuit between the battery and the amplifier or in the circuit between the amplifier and the photocell. If the test light illuminates, continue testing.

19. Connect the test light between terminals M and B. The test light should light. If not, check the ground circuit for an open. If the test light illuminates, continue testing.

20. Connect the test light between terminal K and ground. The test light should light. If not, check the fuse and the circuit for an open. If the test light illuminates, continue testing.

21. Connect the test light between terminal C and ground. The test light should light. If not, check the circuit for an open. If this circuit is good, check the ignition switch. If the test light illuminates, continue testing.

22. Turn the ignition switch to OFF.

23. With the test light connected as in step 21, the test light should turn off. If not, replace the ignition switch. If the test light turns off, continue testing.

24. Connect the test light between terminal H and ground. If the test light illuminates, replace the light switch. If the test light is off, continue testing.

25. Connect the test light between terminal F and ground. If the test light illuminates, replace the light switch. If the test light remains off, continue testing.

26. With the test light connected as in step 25, place the headlight switch in the HEAD position. If the test light does not turn on, check the circuit and the headlight switch for opens. If the test light illuminates, and all other tests have the correct results, replace the amplifier.

Resistance Assembly Test

Perform the resistance assembly test if the lights turn on and off at the wrong ambient light levels. Use **Figure 11-9** to construct a **photocell resistance assembly**, which replaces the photocell to produce predictable results.

Replace the photocell with the resistance assembly. Turn the resistance assembly switch to the OFF (open) position and start the engine. Replace the amplifier if the lights don't come on within 60 seconds.

Special Tool

Photocell resistance assembly

Figure 11-9 Photocell resistance assembly.

If the lights turn on, wait 30 seconds and turn the resistance assembly switch on. If the lights turn off within 60 seconds, replace the photocell. If the lights don't turn off within 60 seconds, replace the amplifier.

SmartBeam Diagnostics

Classroom Manual
Chapter 11, page 320

Special Tools

Scan tool
Calibration target
Grease pencil
Tape measure

SmartBeam is one of the systems that a manufacturer may use to control automatic high-beam operation. This system is presented as an example of the diagnostic procedures used to determine and repair faults.

Customer concerns related to the operation of SmartBeam will usually be in one of two categories: either the system is completely inoperative, or it's not performing properly. When diagnosing a concern that the SmartBeam system is inoperative, perform the following checks:

- Verify that the headlights work properly on both high and low beams when operated manually.
- Verify the power and ground circuits of the Automatic High-Beam Module (AHBM).
- Verify that the system status indicator light-emitting diode (LED) in the mirror (**Figure 11-10**) is on steady. A flashing LED indicates the detection of a fault.
- Verify the automatic high-beam/low-beam function has been enabled (**Figure 11-11**).
- Verify that the headlight switch is in the AUTO position.
- Verify that the headlight beam select switch is in the low-beam position.

Concerns related to poor system performance are usually associated with the sensitivity of the system. If the system is oversensitive, the customer may state that the high beams come on too late and go off too early. If the system is under-sensitive, then the

Figure 11-10 The LED can indicate auto high-beam faults.

Figure 11-11 For the auto high-beam function to operate, it must be activated.

high beams will come on too early and go off too late. These types of sensitivity problems can be caused by:

- Improperly aimed camera.
- Loose camera mounting or improperly positioned mirror button.
- Obstruction in front of the camera.
- Improper headlight aiming.
- Vehicle overloading causing the rear of the vehicle to sit lower than the front.

> **SERVICE TIP** Because the SmartBeam requires the function of several vehicle modules, diagnosis of the system may require verifying the correct operation of modules other than the AHBM itself.

Use a scan tool to verify system operation on stationary vehicles by manually activating the system.

> **CUSTOMER CARE** Inform the customer that the camera must have an unobstructed view from the front of the windshield for the system to perform properly. Hanging items from the mirror or placing toll road transponders in front of the camera will result in poor performance of the system.

> **SERVICE TIP** Remember that the SmartBeam system controls the high-beam portion of the headlamp system. The SmartBeam system varies the high-beam headlamp illumination intensity from low-beam level headlamp illumination intensity to full high-beam headlamp illumination intensity, and any level of headlamp illumination intensity in between.

The SmartBeam automatic high-beam system may have a **demonstration mode** that can assist in diagnostics. This mode demonstrates the function of the automatic high beams and the high-beam indicator while the vehicle is stationary and under any ambient lighting conditions. To initiate the demonstration mode function:

1. Begin with the ignition switch in the OFF position.
2. Depress and hold the AUTO button on the inside mirror (refer to Figure 11-10).
3. While continuing to depress the AUTO button, turn the ignition switch to the RUN position.
4. Continue to depress the AUTO button until the demonstration mode begins, as indicated by the high beams ramping up in intensity.
5. Release the button.

The system will complete three cycles of ramping up the headlamp high beams to full intensity and then ramping them down. During this time, the high-beam indicator should also come on and go off with the high beams. The high beams will cycle three times; then, the system will return to normal operation.

The LED in the rearview mirror can assist in diagnostics. Usually, this LED is on steady to indicate that the auto dimming function of the electrochromic mirror is on. A flashing LED indicates the system has detected a problem. The LED can flash at different frequencies and sequences to indicate a problem. For example, if the LED continuously flashes at a rate of 1 flash per second, the camera needs calibrating. To correct this condition, perform the camera calibration procedure.

If the LED flashes continuously at a rate of 2 flashes per second, this indicates that the system failed its last attempt to calibrate. To correct this condition, perform the camera calibration procedure.

The last possible LED indication is a series of flashes when the ignition is first placed in the RUN position, followed by the LED staying on steady. The flashing LED can indicate a hardware failure that may require AHBM replacement.

Camera Calibration Test and Adjustment

Classroom Manual
Chapter 11, page 322

System performance is seriously degraded if the camera's field of view is no longer within specifications. For proper operation, the camera's field of view must be maintained within 2° of the vehicle centerline and within 10° horizontally. If the alignment of the camera is within these specifications, the AHBM can adjust and fine-tune the alignment based on sensed lighting inputs while driving.

Perform the camera calibration procedure after replacing the inside rearview mirror, the rearview mirror mounting button, or the windshield. The calibration procedure ensures that the field of vision for the camera is aimed at the proper path ahead of the vehicle. If installing a new camera, it's shipped with the calibration mode initiated. After installing and connecting the new camera, the LED will flash once every second while the ignition is in the RUN position. Perform the following before attempting to calibrate the camera:

- Clean the windshield glass in front of the camera lens.
- Check for proper mounting of the mirror assembly and that the torque for the set screw that secures the assembly to the button is 15 inch pounds (1.7 Nm).
- Repair or replace any faulty, worn, or damaged suspension components.
- Verify proper tire inflation pressures.
- Verify that there is no load in the vehicle, except for the driver.
- The fuel tank should be full. Add 6.5 pounds (2.94 kg) of weight over the fuel tank for each gallon of missing fuel.

To calibrate the camera, the vehicle's centerline must be established and marked. A special alignment target is also required. The target consists of a black, square field containing three red LEDs positioned in a diagonal pattern (**Figure 11-12**). The target is placed a specified distance in front of the vehicle. With the proper height of the target established to match the height of the camera lens, position the target within 1/2° of the centerline of the vehicle. When in calibration mode, AHBM programming looks for

Figure 11-12 Calibration target.

the red LED light pattern of the target. The AHBM will identify the target, and the center of the target will become the new center of the field of view. The flashing green LED changes to a steady ON state when calibration is successful. **Photo Sequence 22** illustrates the typical procedure for setting up and calibrating the camera.

If the camera cannot locate the target's pattern, the calibration fails. The LED will blink at a 2-hertz rate, indicating that the calibration procedure has failed and that a fault has set. Incorrect positioning of the target or improper camera alignment may cause the calibration to fail.

PHOTO SEQUENCE 22
Automatic High-Beam Camera Calibration

P22-1 Locate the "ASI" mark on the windshield.

P22-2 Using the "ASI" marks on both sides of the windshield, measure across the windshield using the shaft of the tint band arrow as the reference points.

P22-3 Divide the measurement in half and mark that dimension near the lower edge of the tint band using a grease pencil.

PHOTO SEQUENCE 22 (CONTINUED)

P22-4 Measure across the upper edge of the blackout area at the bottom of the windshield using the inside corner of the intersection between the side blackout area and the lower blackout area as the reference points.

P22-5 Divide the lower measurement in half and mark that dimension near the upper edge of the blackout area using a grease pencil.

P22-6 To locate the centerline of the camera lens, measure and mark the glass 7/8 inch (21 mm) toward the passenger side of the windshield from the upper and lower glass centerline marks and draw a vertical line connecting these two marks.

P22-7 Remove the grease pencil mark in front of the lens.

P22-8 To locate the camera centerline on the rear glass, measure across the upper portion of the rear glass using the vertical edges of the body opening as the reference points.

P22-9 Divide the upper measurement by 2 and mark that dimension on the glass using a grease pencil.

P22-10 Measure across the rear glass at the bottom using the vertical edges of the body opening as the reference points.

P22-11 Divide the lower measurement in half and mark that dimension on the glass using a grease pencil.

P22-12 Measure and mark 7/8 inch (21 mm) toward the passenger side of the rear glass on both the upper and lower references and draw a vertical line connecting these two marks.

PHOTO SEQUENCE 22 (CONTINUED)

P22-13 Locate the calibration target 50 inches (127 cm) in front of the vehicle measuring from the foremost center of the front fascia.

P22-14 While sighting through the V-notch in the upper edge of the calibration target, move the target left or right to align the target to the camera centerline marks on both the windshield and the rear glass.

P22-15 Adjust the tripod so the center LED is on the target 57 inches (145 cm) from the floor.

P22-16 Use the scan tool to enter the SmartBeam unit into calibration mode. When in calibration mode, the LED in the mirror assembly will flash once per second.

P22-17 Turn on the LEDs in the calibration target.

P22-18 The LED in the mirror assembly should continue to flash for 5 to 10 seconds and then stop flashing to indicate that it has completed calibration.

If the camera fails to calibrate, a DTC will set. The DTC will indicate that the performance is off in any of the four directions. To correct this condition, you must verify that the camera mounting and the target's position are correct. If they are correct, then adjustment of the camera will be necessary.

⚙ SERVICE TIP If there is more than one high-beam camera alignment performance fault DTC, then one screw and one shim will be common to both faults. Make the correction at this location.

Each DTC for alignment describes the direction to move the camera. For example, if the DTC is "High-Beam Camera Alignment Performance Bottom," the camera locates the target but it's at the bottom of the field of view. This is corrected by tilting the camera down.

Figure 11-13 An adjustment shim for aligning the camera.

Figure 11-14 The adjustment shim.

A spacer between the camera mount and the camera housing aligns the camera (**Figure 11-13**). The adjustment spacer consists of four stepped shims connected by an integral plastic tree (**Figure 11-14**).

Determine the shim movements required to correct the alignment based on the DTCs retrieved (**Figure 11-15**). To adjust the camera alignment, move the inside rear view mirror head downward to access the two screws securing the rear cover. Remove the two screws and rotate the mirror head up toward the headliner to its uppermost position. Unsnap and pull the upper edge of the rear cover away from the housing far enough to disengage the tabs at the lower edge of the cover; then remove the rear cover.

With the mounting bracket now accessible, carefully cut the plastic tree for the shims at the appropriate location(s) to allow movement of the shims determined earlier. With the shims separated from the tree, loosen the attaching screws one-half turn. The shim's slotted screw holes allow the shims to slide. Slide the shim beneath each of the loosened screws to its most outboard position. With the shims in the desired position, torque the attaching screws to 7 inch pounds (0.8 Nm). Tightening the screws pulls the camera into the new alignment position. The total amount of correction is 2° to 2.5°. Only two shims can be moved for correction. The shims moved must be either of the following:

- Both top shims to tilt the camera upward.
- Both bottom shims to tilt the camera downward.
- Both left shims, to tilt the camera to the left.
- Both right shims, to tilt the camera to the right.

Reinstall the mirror assembly rear cover. Use the scan tool to erase any DTCs, and perform the calibration procedure again.

Camera Optics Test

If the automatic high-beam system is performing poorly, and the LED does not flash, the camera's optics may be dirty or obstructed. The **optics test** will confirm that the camera can recognize ambient light through the lens and the windshield. To perform the optics test, first initiate the automatic high-beam demonstration mode. While observing the LED, obstruct the view of the imager by placing a piece of cardboard between the camera lens and the windshield. The LED should turn off each time the cardboard obstructs the optics. The LED should illuminate again with the cardboard removed.

Failure of the LED to respond to these inputs indicates an obstruction of the imager optics. Clean the lens and the windshield glass and/or remove any obstructions from the windshield in front of the imager. To clean the imager lens, spray a small amount of glass cleaner onto a soft cloth and gently wipe the lens clean. After cleaning the lens and the windshield, repeat the test to confirm proper automatic high-beam system operation.

 Caution

Do not spray glass cleaners directly onto the imager lens. Doing so may damage the imager optics and electronics.

Figure 11-15 Adjustment chart and locations.

Camera adjustment table		
Fault	Tree cut location	Shim + screw location
Right	1	A + D
Left	1	B + C
Top	3 + 4	A + B
Bottom	2 + 5	C + D

Headlight Leveling Systems

The headlight leveling system adjusts the beam projection of the headlights based on vehicle load and other considerations. There are two common types of systems: driver-initiated level by use of a switch or thumb wheel, and automatic leveling.

The first requires the driver to adjust the headlights using a switch or thumb wheel to select the level. The switch (located in the instrument panel bezel) generally has four positions that will lower the headlight beam as the vehicle load increases. The switch is a MUX switch that sends a voltage signal to the controller board and logic circuitry of the

Classroom Manual
Chapter 11, page 323

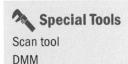

Special Tools

Scan tool
DMM

Figure 11-16 Headlight leveling motor and linkage.

headlamp leveling motor. The motor is installed in the headlight housing (**Figure 11-16**). When energized, the motor will extend or retract the pushrod through the integral screw-drive transmission. The other end of the pushrod snaps into a socket on the back of the reflector and moves the reflector as the pushrod extends or retracts. The reflector position changes the projected angle of the light from the headlamp.

The leveling motors and switch have a direct connection to chassis ground. Although there are differences between manufacturers concerning how power is supplied to the switch and motors, a common method is to supply power only with the headlights turned on. This can be accomplished by supplying voltage to the switch and motors through the fused park lamp relay output circuit.

The headlamp leveling switch, as well as the hard-wired inputs and outputs of the switch, may be diagnosed using a DMM. Use the appropriate wiring information when testing the circuits. The following is an example of testing the switch function using a DMM. **Figure 11-17** identifies the switch terminals.

Using the ohmmeter function of the DMM, perform the resistance tests at the terminal pins in the switch connector receptacle as shown in **Table 11-1**. If the switch fails any of the resistance tests, replace the switch.

Because of active electronic elements located in the leveling motor, conventional diagnostic tools and procedures do not apply. If you believe the headlamp leveling motor is defective, test the hard-wired headlamp leveling motor circuits and the leveling switch before considering motor replacement.

The second system automatically levels the headlamps, regardless of cargo, passenger loads, or road conditions. Inputs from the front and rear height sensors indicate changes

⚠ Caution

To avoid serious or fatal injury, disable the supplemental restraint system before attempting any component diagnosis or service. Disconnect and isolate the battery negative cable and wait at least 2 minutes to allow the system capacitor to discharge before performing further diagnosis or service.

Figure 11-17 Leveling switch terminal callouts.

Table 11-1   Table of Resistance Test for Leveling Switch

	Headlamp Leveling Switch Tests	
Switch Position	Resistance (ohms) ± 1% between Pins 6 and 9	Resistance (ohms) ± 1% between Pins 6 and 8
0	1,518	1,750
1	1,971	2,203
2	3,661	3,893
3	9,851	10,183

in the vehicle's load. The system makes headlight leveling adjustments until the feedback from the sensors is approximately 27.5° from its zero or neutral position.

One sensor is mounted on the spring link on the right side of the rear suspension (**Figure 11-18**). The other sensor is mounted on the upper control arm on the right side, in the front of the vehicle (**Figure 11-19**).

This system will use a headlamp stepper motor that adjusts the headlamp reflector up or down for proper headlamp aiming. While the vehicle is being driven, the headlamp level is adjusted (if required) based on sensor inputs. A headlight leveling module uses the input from the sensors to control the operation of the stepper motor. The module is also capable of monitoring the system for faults. If a system fault occurs, the module suspends system operation and stores a DTC.

Usually, the sensor linkage is not adjustable. However, anytime the sensors are removed or replaced, the system needs recalibration using a scan tool. This is also true after replacing the headlamp leveling module.

When performing the calibration function, the vehicle must be level. Place the ignition in the RUN position and confirm that the vehicle doors, trunk or liftgate, and hood are all closed. Make sure no one in the vehicle and that the vehicle is not disrupted (bounced or bumped) until the calibration procedure is complete. Also, the headlamps must be ON to power the stepper motors. Calibration will take approximately 12 to 15 seconds to complete. The scan tool will display a "status of calibration" message. The message will be either "still in progress," "passed," or "failed."

The headlight aiming module is also referred to as the high-intensity discharge translator module.

Figure 11-18 Rear sensor for automatic headlight leveling system.

Figure 11-19 Front sensor for automatic headlight leveling system.

If the calibration fails, use the wiring diagram to confirm that all circuits are functioning properly. If the circuits have no problems, the module is probably the fault. Confirm power and grounds to the module before replacing it.

Adaptive Headlight System Service

Classroom Manual
Chapter 11, page 324

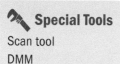

Special Tools
Scan tool
DMM

The **adaptive headlight system (AHS)** enhances night time safety by providing drivers additional time to act or react to approaching road hazards by predicting upcoming corners or turns in the road. If the system detects a fault, it reverts to a conventional system, sets a DTC, and notifies the driver by illuminating a warning light or displaying a message.

Like most computer-driven systems, adaptive headlight system diagnostics rely heavily on DTCs. Remember that DTCs don't always identify the component that has failed. Instead, they indicate a failed circuit, an invalid input received, no expected change after activation, or a rational test failure.

The scan tool can control the actuator motors using bi-directional controls. Most systems will not activate the adaptive headlight system unless vehicle speed is greater than a programmed threshold. Thus, the scan tool will be necessary to activate the motors while the vehicle is stationary. If the motor fails to operate when commanded, back probe the motor connector's power and ground circuits and connect a voltmeter across the terminals. Activate the motor with the scan tool and record the voltage. If the voltage is equal to battery voltage, the motor has failed. If there is no voltage, move the negative test lead to a chassis ground location and activate the motor again. Voltage at this time indicates an open ground circuit; no voltage indicates an open power circuit.

If the motors operate when using scan tool bidirectional controls, check the input sensors. Typical sensors required for operation include the vehicle speed sensor, steering angle sensor, and the yaw sensor. If all of the inputs are correct, the adaptive headlight module is faulty.

Replacement of any parts of the system may require initialization or calibration. Typically, these procedures will require the use of a scan tool.

Daytime Running Lamps

Classroom Manual
Chapter 11, page 327

Special Tools
Test light
DMM
Fused jumper wire

Some manufacturers use a daytime running lamp (DRL) module that receives inputs from either the headlight switch or the BCM to determine switch position. If the switch is in the OFF or AUTO position (and ambient light is high enough that the headlights don't need to be on), the DRL module will send a PWM signal to the headlight beam using HSDs. This turns the headlight on, but at reduced illumination levels.

Some vehicles use parking lamps for DRL function. In this case, the headlight position determines if the headlights are on. If they are not, the BCM or DRL module will turn on the park lamp relay.

As with any system, determine proper power and ground circuit operation. Diagnostics of either system type is performed by using a scan tool or DMM. Also, remember that the parking brake switch is an input. If the switch should fail and indicates to the DRL module that the parking brakes are applied, the DRL function is disabled.

Headlight Aiming

Classroom Manual
Chapter 11, page 300

Check headlight aiming anytime the lamps or lenses have been removed or replaced. Proper aiming is important for good light projection onto the road and to prevent discomfort and dangerous conditions for oncoming drivers.

CUSTOMER CARE The Department of Transportation reports that about 50% of all vehicles on the road have at least one improperly aimed headlight. This is a very dangerous condition. Educate your customers on the importance of proper headlight aiming. Suggest they have their headlight aiming checked at least once a year.

Correct headlight beam position is so critical that government regulations control limits for headlight aiming. For example, a headlight that is misaimed by 1° downward will reduce the vision distance by 156 feet (47.5 meters). The following are maximum allowable limits established by all states:

1. Low beam: In the horizontal plane, the left edge of the headlight high-intensity area should be within 4 inches (102 mm) to the right or left of the vertical centerline of the lamp. In the vertical plane, the top edges of the headlight high-intensity area should be within 4 inches (102 mm) above or below the horizontal centerline of the lamp (**Figure 11-20**).
2. High beam: In the horizontal plane, the center of the headlight high-intensity area should be within 4 inches (102 mm) to the right or left of the vertical centerline of the lamp. In the vertical plane, the center of the headlight high-intensity area should be within 4 inches (102 mm) above or below the horizontal centerline of the lamp (**Figure 11-21**).

Before aiming the headlights, check the vehicle for proper **curb height**. Curb height is the height of the vehicle when it has no passengers or loads and has normal fluid levels and tire pressure. This includes checking the suspension, tire inflation, removing any additional load in the vehicle, a half-filled fuel tank, and removing dirt, ice, snow, and so on from the vehicle.

Special Tools
Headlight-aiming unit
Torx driver
Fender covers
Safety glasses

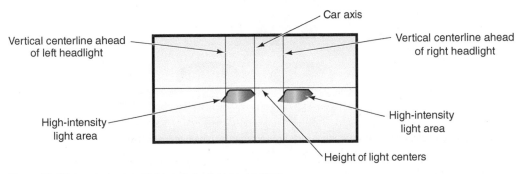

Figure 11-20 Low-beam headlight aiming adjustment pattern.

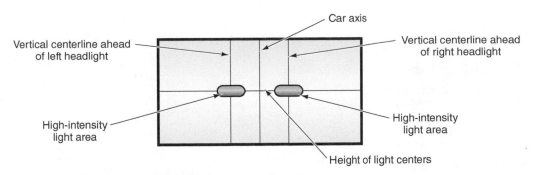

Figure 11-21 High-beam headlight aiming adjustment pattern.

Classroom Manual
Chapter 11, page 300

Sealed-Beam

Portable mechanical aiming units are capable of aiming most sealed-beam-type headlights (**Figure 11-22**). Suction cups secure the aiming units to the headlight lens (**Figure 11-23**). The aiming unit should have a variety of adapters to attach to the various styles of headlights. Before using the aiming equipment, be sure to follow the manufacturer's procedure for calibration. Park the vehicle on a level floor and place fender covers around the work area. It may be necessary to remove the trim and bezel from around the headlight. Using the correct adapter, connect the calibrated aimer units to the headlights. Be sure the adapters fit the headlight-aiming pads on the lens (**Figure 11-24**). Zero the horizontal adjustment dial. Confirm that the split-image target lines are visible in the view port (**Figure 11-25**). If the target lines are not seen, rotate the aimer unit. Turn the headlight horizontal adjusting screw until the split-image target lines align. Repeat for the headlight on the other side.

Figure 11-22 Typical portable mechanical headlight-aiming equipment and adapters.

Figure 11-23 The aiming units attach to the headlight lens with suction cups.

Figure 11-24 Headlight-aiming pads.

Split image
aligned

Split image
not aligned

Figure 11-25 Split image target.

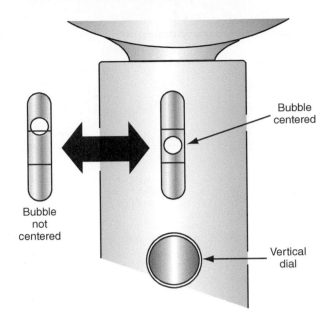

Bubble
centered

Bubble
not
centered

Vertical
dial

Vertical
dial

Horizontal
dial

Figure 11-26 Center the spirit level by turning the vertical aiming screw.

To set the vertical aim of the headlight, turn the vertical adjustment dial on the aiming unit to zero. Turn the vertical adjustment screw until the spirit-level bubble is centered (**Figure 11-26**). Recheck the horizontal aiming on each headlight. The vertical adjustment may have altered the original adjustments.

If the vehicle is equipped with a four-headlight system, repeat the procedures for the other pair of lamps.

Composite and HID

Some composite and HID headlight designs require special adapters (**Figure 11-27**). Also, the lens must have headlight-aiming pads to use a mechanical aiming unit. The headlight assembly will have a number molded on it. The adjustment rod setting must be set to that number and locked in place. The aiming unit attaches to the headlight lens in the same manner as with sealed-beam headlights (**Figure 11-28**).

Classroom Manual
Chapter 11, page 302

Figure 11-27 Special adapter for aiming composite headlights.

Figure 11-28 Connect the aiming equipment to the headlight lens. The lens must have aiming pads.

The adjustment procedure of composite and HID headlights is identical to that of the sealed-beam headlights. **Figure 11-29** shows the typical location of the headlight adjusting screws.

Many composite and HID headlight designs don't have alignment pads on the lens. These systems usually adjust the beam location by moving the reflector position. Since the lens does not move, conventional headlight aimers cannot be used. One method of aiming these systems is by locating the vehicle 25 feet (7.6 meters) away from a blank wall with the vehicle on a level surface. Using the centerline of the vehicle as a reference, mark the wall. Adjust the headlight beam until its location on the wall meets manufacturer's specifications. Some manufacturers may require that the headlights be adjusted with the high beams.

With complex lens designs and multiple bulb configurations, many service facilities use an optical aiming system (**Figure 11-30**). Also, states with mandatory safety

Figure 11-29 Composite headlight aiming screw locations.

Figure 11-30 Typical optical headlight aimer.

Figure 11-31 Optical aimer screen.

inspections that include checking the headlight aiming may require using optical headlight aimers. These units have an optical-grade lens that reproduces the headlight beam image. The image is transmitted to a screen to show the technician where the high intensity of the light is being projected (**Figure 11-31**).

Accurate setup of the optical sensor requires compensation for any slope in the bay floor. Typically, this procedure uses a laser light beam projected from the optical aimer unit to a measuring device at the rear wheel of the vehicle. An adjustment wheel is used to set the optical aimer unit so front to rear vehicle measurements are the same.

Once the initial set-up is complete, the center of the headlight bulb is determined and the aimer is aligned to it. Depending on the equipment, this may involve locating the grill's exact center and then the machine will automatically detect the beam center. Some equipment requires the beam center to be manually located and then the machine is aligned to the vehicle by using two symmetrical points on the vehicle (such as radiator support bolts, top of the headlight lens, or strut bolts). With the headlights on and set to low beam, the screen displays the high-intensity pattern. The low beams are aimed so the top edge of the high-intensity pattern is along the horizontal axis, and the left side of the zone pattern aligns with the vertical centerline (**Figure 11-32**). Typically, the headlights are aimed using only the low beams. The high beam pattern can be checked. The high-intensity zone should be center on the screen (**Figure 11-33**). Some aimers use a digital display to indicate if the beam is aligned. An "X" in the vertical and horizontal positions indicates the proper alignment of the beam. If the beam is out of alignment, the screen will indicate the direction of the misalignment.

> **SERVICE TIP** If the headlight lens is clear, the center of the bulb can be determined visually. Some lens will have a dot to indicate the location of the bulb's center. Another method to locate the center of the bulb is to cover the lens with a cloth shop rag and turn the headlights on. Since the center of the bulb will be dark, the light through the rag will identify the center of the bulb (**Figure 11-34**).

Shaded area indicates
high-intensity zone

Figure 11-32 Correct low-beam high-intensity location.

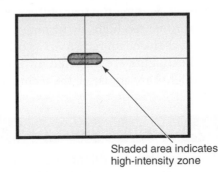

Shaded area indicates
high-intensity zone

Figure 11-33 Correct high-beam high-intensity location.

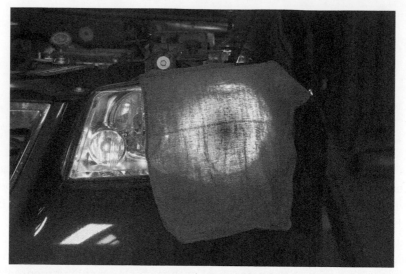

Figure 11-34 Using a shop rag to locate the center of the headlight bulb.

Figure 11-35 Some headlight assemblies are equipped with spirit levels to provide easy adjustments to offset vehicle loads.

Some late-model vehicles have spirit levels built into the headlamp assembly (**Figure 11-35**). These are not always used for initial headlamp adjustment, rather they are to assist the driver in adjusting their headlights based on the load in the vehicle. For example, loading the rear of the vehicle causes the front of the vehicle to lift and the light beam to be too high. By turning the adjuster wheels until the bubble is in the middle of the level, the beam returns to its original position. After removing the load from the trunk, the headlights are adjusted until the bubble is returned to the middle. Whenever the headlights are adjusted, the technician should adjust the spirit level also.

Diagnosing Dimmer- or Brighter-Than-Normal Lights

Classroom Manual
Chapter 11, page 309

The complete headlight circuit consists of the headlight switch, dimmer switch, high-beam indicator, and the headlights (**Figure 11-36**). Excessive resistance in these units, or at their connections, can result in lower illumination levels of the headlights.

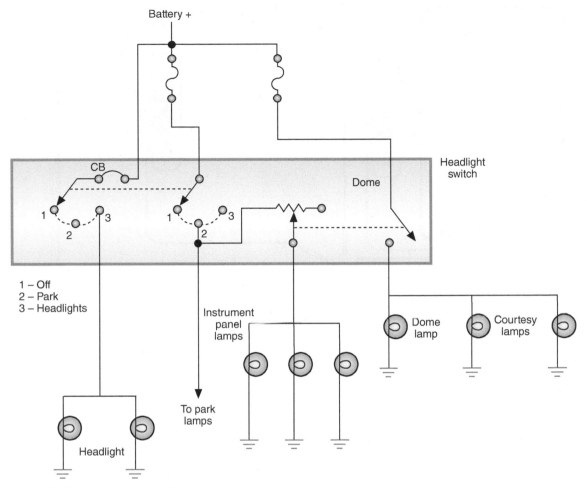

Figure 11-36 Headlight circuit components.

The extra resistance can be on the insulated side or the ground side of the circuit. To locate the excessive resistance, perform a voltage drop test (**Figure 11-37**). Consult the wiring diagram to determine the number of connectors and switches. This will provide you with the specification for maximum voltage drop. Start at the light and work toward the battery.

Most headlight systems are wired in parallel. If both headlights are dim, then the excessive resistance is in the common portions of the circuit. Dim headlights can also be the result of low-charging system output.

Other causes of dim lights can be the use of the wrong lamps, improper circuit routing, addition of extra electrical loads to the circuit, and wrong-size conductors.

Brighter-than-normal lights can be the result of higher-than-specified charging system output or improper lamp application. It's also possible that the dimmer switch contacts are stuck in the high-beam position.

> **SERVICE TIP** Although headlights can get dimmer with age, they do so at a slow rate and may not be noticeable. One of the headlight being noticeably dimmer than the other is caused by excessive voltage drop in the circuit of the dim light. During the process of installing a new headlight, the breaking and making of the socket connection may clean the terminals enough to make a good contact. Once installed the new headlight may operate properly. Don't be fooled. It was not the headlight that was at fault. It was the connection.

Special Tools
Voltmeter
Safety glasses

Although this procedure is being shown for the headlights, it's identical for all lighting systems. The only difference in the test results will be if the circuit uses insulated bulbs or grounded bulbs. If testing the turn signal circuit, bypass the flasher with a fused jumper wire.

Figure 11-37 Voltage drop testing the headlight system.

Headlight Switch Testing and Replacement

Classroom Manual
Chapter 11, page 309

The headlight switch controls most of the vehicle's lighting systems. The headlight switch will generally receive direct battery voltage to two of its terminals. Disconnect the battery before removing the headlight switch.

The headlight switch circuit incorporates a rheostat to control the instrument cluster illumination lamp brightness. Most dash-mounted headlight switches incorporate the rheostat into the switch assembly. Steering column–mounted switches may have the rheostat located on the dash.

Many customer concerns associated with the lighting systems can be the result of a faulty headlight switch. For example, dim or no instrument panel lights, dim or no headlights, dim or no parking lights, and improperly operating dome lights can all be caused by a faulty headlight switch.

 Special Tools

12-volt test light
Ohmmeter or self-
   powered test light
Fused jumper wire
Safety glasses
Battery terminal pliers
Terminal pullers

Dash-Mounted Switches

⚡**WARNING** **If the vehicle is equipped with air bags, disable the supplemental restraint system before attempting any component diagnosis or service. Failure to take proper precautions could result in accidental air bag deployment.**

Many methods are used to retain the headlight switch to the dash. Consult the service information of the vehicle you are working on. The following is a common method of removing the headlight switch:

1. Place fender covers on the fenders.
2. Install a memory keeper and disconnect and isolate the battery negative cable.
3. Remove the lower cluster bezel from the instrument panel.
4. Disconnect the wire harness connector from the back of the headlamp switch.
5. Remove the fasteners that secure the headlamp switch to the back of the cluster bezel.
6. Remove the headlamp switch from the cluster bezel.

> ⚙ **SERVICE TIP** Headlights that flash on and off as the vehicle goes over road irregularities indicates a loose connection. Headlights that flash on and off at a constant rate indicates the circuit breaker is being tripped and there is an electrical overload in the circuit.

With the switch removed, it can be tested for continuity and the connector plug will serve as a test point for the lighting circuits. First, test at the connector.

The following is a typical procedure for testing the headlight switch connector on the harness side. A test light and fused jumper wires are used to test the circuits. This procedure would be very similar for any non-computer-controlled headlight system—just use the service information to determine the function of each terminal. In this procedure, the terminals are identified as follows (**Figure 11-38**):

- Terminal B—battery.
- Terminal A—fuse.
- Terminal H—headlights.
- Terminal R—rear park and side marker lights.
- Terminal I—instrument panel lights.
- Terminal D1—dome light feed.
- Terminal D2—dome light.

Consult the service information for terminal identification for the vehicle you are working on. If this is not listed in a separate chart, you should be able to identify the circuits from the wiring diagrams.

Some headlight switches are retained by spring clips. Remove them by compressing the springs.

1. Connect the 12-volt (V) test light across terminal B and ground. The test light should light. If not, there is an open in the circuit back to the battery.

Switch terminals	Switch positions		
	Off	Park	Headlamp
B to H	No continuity	No continuity	Continuity
B to R	No continuity	No continuity	No continuity
B to A	No continuity	No continuity	No continuity
R to H	No continuity	No continuity	No continuity
R to A	No continuity	Continuity	Continuity
H to A	No continuity	No continuity	No continuity
D1 to D2	Continuity only when rheostat fully counterclockwise		
I to R	Continuity only when rheostat fully counterclockwise		
	Then slowly rotate clockwise and test lamp should dim		

Figure 11-38 Continuity test chart for the headlight switch.

2. Connect the test light across terminal A and ground. The test light should come on. If not, repair the circuit back to the fuse panel.

3. Connect a fused jumper wire between terminals B and H. The headlights should come on. If the headlights fail to turn on, trace the H circuit to the headlights. Also, check the ground circuit side from the headlights.

4. Connect a fused jumper wire between terminals A and R. The rear lamps should illuminate. If not, trace the circuit to the rear lights. Also, check the ground return path.

5. Connect a fused jumper wire between terminals A and I. The instrument panel lights should come on. If not, trace the circuit to the panel lights.

If all the tests performed at the switch connector pass, then the problem is in the headlight switch. Figure 11-38 indicates the test results that should be obtained when testing the headlight switch for continuity for the system just discussed. Use an ohmmeter or a self-powered test light to test the switch. Most service information will provide a table similar to that in **Figure 11-39**. If a chart is not available, use the wiring diagram to determine which terminals should or should not have continuity in the different switch positions.

Steering Column–Mounted Switches

On some vehicles, it's possible to test the steering column–mounted switch without removing it. Conduct the test at the connector at the base of the column (Figure 11-39). However, on some models, it's necessary to remove the column cover and/or the steering wheel to gain access.

A common procedure for removing and testing the **multifunction switch** is shown in **Photo Sequence 23**. The switch is called a multifunction switch because it can have a combination of any of the following switches in a single unit: headlights, turn signal, hazard, dimmer switch, horn, and flash-to-pass (**Figure 11-40**).

Dimmer Switch Testing and Replacement

Classroom Manual
Chapter 11, page 311

The **dimmer switch** is wired in series within the headlight circuit and controls the current path for high and low headlight beams. Most dimmer switches are located on the steering column or incorporated within the multifunction switch. Test the switch using a set of fused jumper wires to bypass the switch. If the headlights operate with the switch bypassed, the switch is faulty.

The steering column–mounted dimmer switch may use an actuator control rod from the lever to a remotely mounted switch (**Figure 11-41**). Another style incorporates the dimmer switch into the multifunction switch (**Figure 11-42**).

To remove the remote switch, first place fender covers on the fenders of the vehicle and disconnect the battery negative cable. Disconnect the wire connector at the switch. Remove the two switch mounting screws and disengage the switch from the actuator rod.

Special Tools

Fused jumper wires
Safety glasses
Battery terminal pliers
Terminal pullers

Use a memory keeper before disconnecting the battery.

When installing the switch, make sure the actuator rod firmly seats into the switch. During the installation, adjust the position of the switch so that all actuator rod slack is removed. If the switch has alignment holes, compress the switch until two appropriately sized dowels can be inserted into the alignment holes (**Figure 11-43**). While applying a slight rearward pressure, install and tighten the mounting bolts.

When the switch is adjusted properly, it will click when the lever is lifted and again when it's returned to its downward position. The second click should occur just before the stop.

If the dimmer switch is a part of the multifunction switch, follow the general procedure shown in Photo Sequence 24. To reinstall the multifunction switch, reverse the procedure. Torque the steering column–attaching nuts to the amount specified in the service information.

Figure 11-39 Using the connector to test the multifunction switch.

Figure 11-40 Multifunction switch.

Figure 11-41 Steering column–mounted dimmer switch.

Figure 11-42 Dimmer switch incorporated into the multifunction switch.

Figure 11-43 Insert a dowel into the alignment holes to adjust the dimmer switch.

PHOTO SEQUENCE 23
Removal and Testing of the Multifunction Switch

P23-1 Tools required to remove and test the multifunction switch: fender covers, battery terminal pliers, terminal pullers, assorted combination wrenches, torx drivers, and ohmmeter.

P23-2 Place the fender covers around the battery work area.

P23-3 Loosen the negative battery clamp bolt and remove the clamp using terminal pullers. Place the battery cable where it cannot contact the battery.

P23-4 Remove the shroud-retaining screws and remove the lower shroud from the column.

P23-5 If needed to gain access to the upper shroud, loosen the steering column–attaching nuts. Do not remove the nuts.

P23-6 Lower the steering column enough to remove the upper shroud.

P23-7 Remove the turn signal switch lever by slightly rotating the outer end of the lever and then pulling straight out on the lever. Some levers may be attached with fasteners.

P23-8 Peel back the foam shield from the turn signal switch.

P23-9 Disconnect the turn signal switch electrical connectors.

P23-10 Remove the screws attaching the turn signal switch to the lock cylinder assembly.

P23-11 Disengage the switch from the lock assembly.

P23-12 Use the ohmmeter to test the switch. Check for continuity from terminal 15 to 13 when the dimmer switch is in the low-beam position.

P23-13 With the switch in the low-beam position, the circuit should be open between terminals 15 and 12.

P23-14 The circuits between terminals 196 and 13 and 196 and 12 should be open in the low-beam position.

P23-15 With the switch in the high-beam position, continuity should be between terminals 15 and 12. Circuits 15 to 13, 196 to 13, and 196 to 12 should be open.

P23-16 When the dimmer switch is in the flash-to-pass position, there should be a closed circuit between terminals 196 and 12 and an open circuit between 196 and 13.

Taillight Assemblies

In a three-bulb taillight system, the brake light switch directly controls the brake lights (**Figure 11-44**). The brake lights on both sides of the vehicle are wired in parallel. Most brake light systems use dual-filament bulbs that perform multiple functions. In this type of circuit, the brake lights are wired through the turn signal and hazard switches (**Figure 11-45**).

If all of the taillights don't operate, check the condition of the fuse. If it's good, use a voltmeter to test the circuit. With the headlight switch in the PARK position (first detent), check for voltage at the last common connection between the switch and the lamps. If battery voltage is present, then the problem is in the individual circuits from that connector to the lamps. If no battery voltage is present, test for voltage from the switch terminal. If no voltage is present at this terminal yet the headlights operate when in the ON position, replace the switch. If battery voltage is present, the problem is between the switch and the last common connection. If there was no voltage present at the switch terminal, check for battery voltage into the switch.

Most taillight bulbs can be replaced without removal of the lens assembly. The bulb and socket are removed by twisting the socket slightly and pulling it out of the lens assembly (**Figure 11-46**). To remove the bulb from the socket, push in on the bulb slightly while turning it. When the lugs align with the channels of the socket, pull the bulb from the socket (**Figure 11-47**).

If the bulb is a blade base-type bulb, pull it straight out of the socket without twisting. On all bulb types, it's a good practice to use an oil-free rag or wear nylon gloves to grasp the bulb. This will keep oil off the bulb but also will protect you if the bulb should break.

Figure 11-48 shows how the lens assembly is fastened to the vehicle body. Remove the attachment nuts from the back of the assembly to remove it.

Turn Signals and Brake Lights

To test the turn signal switches, use a 12-volt test light to probe for voltage into and out of the switch. The ignition switch must be in the RUN position for the circuit to operate. If voltage is present on the input side of the switch but not on the output side, the switch is faulty.

Classroom Manual
Chapter 11, page 329

 Caution
Using the wrong type of lamp for the socket and application can result in "crazy lights." This is a result of feedback caused by the incorrect bulb (**Figure 11-49**).

Classroom Manual
Chapter 11, pages 329, 334

Figure 11-44 Three-bulb taillight circuit has individual control for each bulb.

Figure 11-45 Turn signal switch used in a two-bulb taillight circuit.

Figure 11-46 Bulb and socket removal from the taillight lens assembly.

Figure 11-47 Removing the bulb from the socket.

Figure 11-48 Taillight lens assembly.

Single-filament bulb   Double-filament bulb

Figure 11-49 Mistakenly installing a single-filament bulb into a socket designed for dual filament may cause the single contact of the bulb to short across the two contacts of the socket. This will result in lighting circuits operating when they are not supposed to.

Check brake light operation through the turn signal switch in the same manner. Also, check the brake light switch for proper adjustment and operations (**Figure 11-50**).

Many early vehicles use a turn signal switch that is separate from the multifunction switch. The steering wheel will have to be removed to gain access to the turn signal switch on these vehicles. The following procedure is a common method of turn signal switch replacement:

1. Place the fender covers on the fenders.
2. Install a memory keeper and disconnect the battery negative cable.

Special Tools

12-volt test light
Steering wheel puller
Lock ring compressor
Safety glasses

Figure 11-50 Typical brake light switch operation. Check for continuity in both positions. Should be open when at rest and closed when the pedal is depressed.

Figure 11-51 Steering wheel attachment.

⚡ **WARNING** If the vehicle is equipped with an air-bag system, wait for the recommended amount of time before removing any other components. Fifteen minutes or more may be required to discharge the capacitors that are used to fire the air bags.

3. Remove the steering column trim.
4. Remove the horn pad from the steering wheel and make alignment marks on the steering wheel and shaft (**Figure 11-51**).
5. If equipped, remove the steering shaft nut and horn collar.
6. Use a suitable puller to remove the steering wheel.
7. If needed, use a suitable compressor to compress the preload spring to the lock plate (**Figure 11-52**). Compress the spring only enough to remove the snap ring.

Figure 11-52 To remove the snap ring, use the compressing tool to relieve the pressure against the snap ring.

8. Use a pick and a small flat-blade screwdriver to remove the snap ring.
9. Remove the lock plate, horn contact carrier, and spring.
10. Remove the bolts at the upper steering column support and the upper mounting bracket from the column.
11. Disconnect the turn signal wiring connector.
12. Wrap tape around the wire and connector.
13. Remove the hazard warning knob from the column.
14. Remove the switch-retaining screws and remove the switch.

Most turn signal switches receive their voltage from the ignition switch when it's in the RUN position only. This prevents the turn signals from operating while the ignition switch is in the OFF position.

> ⚙️ **SERVICE TIP** Two-bulb system switches also control some brake light functions through a complex system of contacts. Many brake light problems are caused by worn contacts in the turn signal switch.

Flashers. The flasher uses a bimetallic strip and a heating coil to flash the turn signals. A common customer concern that is attributable to the flasher is related to the flashing speed. If the flasher is of the wrong type and rating, the amount of time required to heat the coil will differ from what the manufacturer designed into the circuit. Also, newer flashers that use electronic circuits will flash at an increased speed if one of the turn signal bulbs is burned out or the circuit has fault. If the flasher rating is higher than required, the flashing rate slows because it takes longer for the current to heat the coil.

Check the size and type of light bulbs in the circuit. Use only the lamp size recommended by the manufacturer. If these checks don't correct the problem, test the charging system output. Voltage output that is higher or lower than specified may cause the flasher rate to be incorrect. If the charging system output is within specifications, check for excessive resistance in the turn signal circuit. Check both sides of the circuit.

If none of the turn signals operate, check the fuse. Next, check the flasher. Remove the flasher from the fuse box (**Figure 11-53**). Connect a fused jumper wire across the fuse box terminals (**Figure 11-54**). If the turn signal lamps come on with the lever in either indicator position, the flasher is faulty. If the lights still don't illuminate, test the turn signal switch.

Classroom Manual
Chapter 11, page 335

🔧 **Special Tools**
Fused jumper wires
Safety glasses

Not all flashers are located in the fuse box. Use the component locator to find where the flasher is installed.

Figure 11-53 Many manufacturers locate the flasher unit in the fuse panel.

Figure 11-54 Connecting a fused jumper wire across the terminals to bypass the flasher unit.

SERVICE TIP If the turn signals operate properly in one direction but don't flash in the other, the problem is not in the flasher unit. A burned-out lamp filament will not cause enough current to flow to heat the bimetallic strip sufficiently to cause it to open. Thus, the lights don't flash. Locate the faulty bulb and replace it.

Interior Lights

Interior lighting includes courtesy lights, map lights, and instrument panel lights. These lights may also be incorporated into the illuminated entry feature.

Courtesy Lights

Classroom Manual
Chapter 11, page 342

Courtesy lights operate from the headlight and door switches. They receive their power source directly from a fused battery connection. The switches can control either the ground circuit (**Figure 11-55**) or the insulated circuit (**Figure 11-56**). The courtesy lights may also be activated from the headlight switch by turning the switch knob to the extreme counterclockwise position. The contacts in the switch close and complete the circuit.

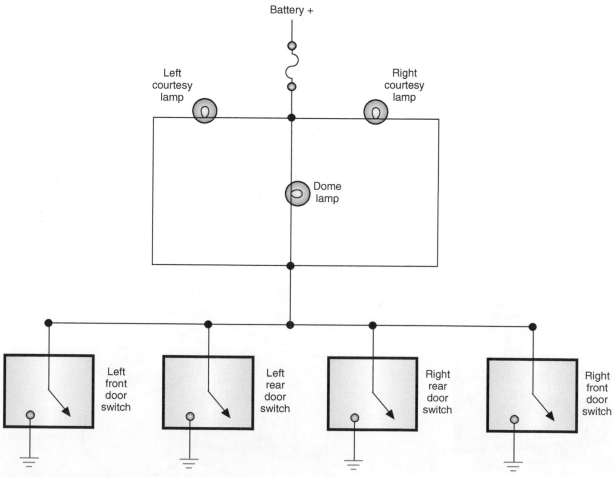

Figure 11-55 Courtesy lights using ground side switches.

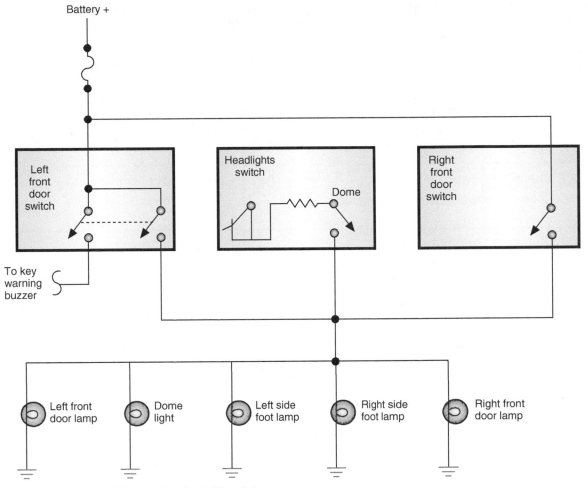

Figure 11-56 Courtesy lights using insulated side switches.

Figure 11-57 provides a systematic approach to troubleshooting courtesy lights. Follow the steps in proper order to locate the fault.

If all the lights of the circuit don't light, begin by checking the fuses. If the fuse is good, then use a voltmeter to check for battery voltage at the last common connection.

Any time a blown fuse is found, the cause of the circuit overload must be traced.

COURTESY LAMPS DO NOT TURN ON WHEN ONE DOOR IS OPENED
OK WHEN OTHER DOORS ARE OPENED

Test step	Result	Action to take
A0 **Verify condition**		Go to A1
A1 **Check power** Check for power at door switch	(ⓧK)	Check the power circuit back to fuse
	(OK)	Go to A2
A2 **Check the door switch** Check the door switch for proper operation	(ⓧK)	Replace the switch
	(OK)	Check circuit from switch to lamp

Figure 11-57 A diagnostic chart for the courtesy light system (continued on next page).

COURTESY LAMPS DO NOT COME ON WHEN HEADLAMP SWITCH IS TURNED COUNTERCLOCKWISE TO STOP

Test step	Result	Action to take
B0 Verify condition		Go to B1
B1 Check operation of door switches Check to see if courtesy lamps operate from door switches	(not OK)	Go to chart C
	(OK)	Go to B2
B2 Check for power Check for power at headlamp switch	(not OK)	Check circuit back to fuse
	(OK)	Go to B3
B3 Check for continuity Check continuity of headlamp switch	(not OK)	Replace headlamp switch
	(OK)	Check circuit from switch to lamp

COURTESY LAMP DOES NOT COME ON WHEN ALL DOORS ARE OPENED

Test step	Result	Action to take
C0 Verify condition	Vehicle with only one courtesy lamp	Go to C1
	Vehicle with more than one courtesy lamp	Go to C2
C1 Check operation of fuse circuit Check operation of other circuits that share the same fuse	(not OK)	Go to C4
	(OK)	Go to C2
C2 Check for power Check for power at the bulb	(not OK)	Replace bulb
	(OK)	Go to C3
C3 Check for continuity Check continuity of bulb	(not OK)	Replace bulb
	(OK)	Check bulb ground
C4 Check fuse Check continuity of fuse	(not OK)	Replace fuse
	(OK)	Go to C5
C2 Check for power Check for power through the fuse	(not OK)	Check power feed circuit
	(OK)	Check for open circuit between fuse and common point in lamp circuit

Figure 11-57 Continued

If voltage is present at the fuse box but not at the common connection, the problem is between these two points. If battery voltage is present at the common connection, trace the individual circuits until the cause(s) for the open is located.

If the courtesy lights don't come on when only one of the doors is opened but do come on when any of the other doors are opened, the problem is in the affected door's switch or circuit. In order to check the switch and circuit, bypass the switch with a fused jumper wire. If the lights come on with the switch bypassed, it's a faulty switch. If the lights don't come on with the switch bypassed, there is a problem in the circuit.

Illuminated Entry System Diagnostics

The diagnostic procedures for testing the illuminated entry system depend upon the manufacturer. Always refer to the service information for the vehicle you are working on. Always perform a visual inspection before performing any tests. First check the fuse. Then check to make sure all connections are tight and clean. Inspect the ground wires for good connections. Check all visible wires for fraying or damaged insulation, especially where they go through body parts. Make sure all doors are closed properly and the headlight switch is in the detent position.

The following are typical procedures for diagnosing the illuminated entry system. Systems activate when the outside-door handle is lifting use a contact switch that momentarily completes the ground circuit of the **illuminated entry actuator** module. The module activates the interior lights for 25 seconds, or until the ignition is placed in the RUN or ACC position.

A logic circuit in the module prevents battery drain if the door handle is held up for longer than 25 seconds. The system will operate as normal until the 25 seconds has elapsed, and then the module will turn off the lights. The lights will remain off and cannot be reactivated until the handle is returned to the released position.

This type of system has four main components: fuse, ignition switch, actuator module, and door handle switch assemblies (**Figure 11-58**). The door lock cylinder uses an LED to provide the illumination of the cylinder. The lens of the LED is built into the cylinder.

To test the system, disconnect the actuator harness from the actuator and connect the test light between terminal 8 and ground. The test light should illuminate. If the test light fails to come on, trace the circuit back to the battery to locate the problem.

Connect the test light between terminal 7 and ground. The test light should not glow with the ignition in the OFF position. When the ignition is in the RUN or ACC position, the test light should come on. If the test light does not turn on and off as described, trace the circuit to the fuse box and the ignition switch to locate the problem.

Connect a fused jumper wire between connector terminals 6 and 8. Make sure all doors are closed. The courtesy lights and door lock cylinders should be illuminated. If the lights don't operate, trace the circuit from terminal 6 to the LEDs and lights to locate the problem.

Connect an ohmmeter between connector terminal 2 and chassis ground. The ohmmeter should indicate an infinite reading. However, a minimum of 10,000 ohms is acceptable. Lift up on each of the outside-door handles to close the latch switch. Hold the handle up while observing the ohmmeter. The ohmmeter should indicate a resistance value of 50 ohms maximum. If either of the ohmmeter readings is out of specifications, trace the circuit to the latch switches. Also, test the latch switches for correct operation.

With the test light connected between connector terminals 1 and 8, the light should be on. If the test light fails to come on, trace the circuit from terminal 1 to ground.

If the preceding tests don't indicate any problems, the actuator module assembly is faulty. The module must be replaced.

Classroom Manual
Chapter 11, page 342

 Special Tools

DMM
Test light
Fused jumper wire
Safety glasses

Courtesy lights and illuminated entry lights usually use the same bulbs. Courtesy lights are lights that come on when the door is open or the switch is turned on. *Illuminated entry* is a term used for turning on lamps prior to entering the vehicle, such as when the handle is lifted, or remote keyless entry is used.

Figure 11-58 Illuminated entry system schematic.

Special Tools

Scan tool
Fused jumper wire
DMM
Test light

⚠ **Caution**

The normal operating voltage for the LED is 3 volts. The circuit uses a dropping resistor to protect the LED. Do not apply 12 volts to the LED circuit ahead of the resistor. If the resistor is bypassed, the LED will be destroyed. When applying voltage to test the LED, apply it only to the connector terminals.

Follow these steps to diagnose the system illustrated in **Figure 11-59**:

1. Move the dimmer control to the center position.
2. Open the driver-side door to activate the courtesy lights. If none of the courtesy lights turn on, continue testing. If only one bulb is inoperative, check its circuit and the bulb.
3. Lower the driver-side window and close all doors. Manually lock the driver's door. Wait 30 seconds with the ignition switch off.
4. Activate the illuminated system by lifting the driver's door handle. If the lights come on, repeat the test for the right-side door. If the system does not operate when the right-side door handle switch is closed, refer to the service information for the circuits to be tested. The procedure will be the same as when the left door is inoperative. However, the circuit designations are different.
5. If the lights don't turn on when the door handles are lifted, connect the scan tool to the diagnostic connector. Maneuver through the menu screens to locate the door handle switch state.
6. With the ignition switch on, observe the scan tool display while lifting the door handle. The display should indicate the switch closed when the handle is lifted. If the switch operates correctly, go to step 7. If the display does not indicate proper

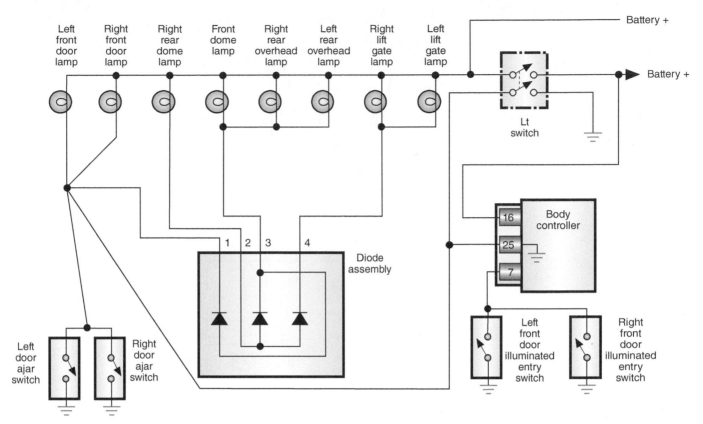

Figure 11-59 Schematic of an illuminated entry system.

switch operation, connect a fused jumper wire from the controller terminal identified in the service information to ground. This would be terminal 7 at the BCM in Figure 11-59. Observe the scan tool. If it indicates that the circuit is closed, follow the service information procedure for testing the door switches. If the display indicates that the circuit is open, replace the BCM.

7. Open the driver-side door. If the courtesy lights don't turn on, go to step 8. Close all doors and jumper the terminal of the BCM identified in the service information as lamp control (terminal 25 in Figure 11-59) to ground. If the lights don't turn on, there is an open circuit between the BCM and the lamps. If the lights turn on when the BCM is jumped, replace the BCM.

8. If the lights don't come on in step 7 when the door is opened, gain access to the driver-side door ajar switch harness and disconnect it. Use a fused jumper wire to jump across the two terminals of the connector on the harness side. This should complete the circuit to ground and the lights come on. If the lights don't turn on, check both circuits for an open. If the lights come on when the connector is jumped, replace the switch. Test the door ajar switch for the passenger side in the same manner. Use a fused jumper wire to jump across the two switch connector terminals on the harness.

Some BCM-controlled illuminated entry systems don't use a door handle switch. The system is activated when the doors are opened, or the remote keyless entry system is used to unlock the doors. The system uses the same lamp driver as the courtesy lamps. Check for operation of the system by opening each door, one at a time. Usually, the three-passenger door ajar switches are connected in parallel with the driver's door on its own circuit (**Figure 11-60**). If the system does not activate when the doors are

On some systems, terminal 4 is used. On these systems, jumping between terminals 6 and 8 will illuminate the courtesy lights. Jumping between terminals 4 and 8 will light only the door lock cylinders. If they don't, trace circuit 464.

⚠ **Caution**

The terminal to jump from the controller is different between years and models. Be sure to refer to the service information for the correct terminals.

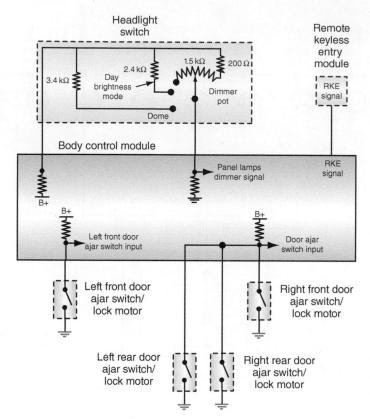

Figure 11-60 The inputs used for the illuminated entry system. Note that the passenger door ajar switches are in parallel.

opened, use the remote keyless entry fob and press the UNLOCK button. If the system works now, the problem is in the door ajar input circuits or the BCM. If using the fob does not activate the system, the fault is probably in the BCM or the control circuits to the lamps. However, it's possible that all inputs are missing. To confirm this, use the scan tool to activate the system. If the lamps illuminate now, each input will need to be tested.

To test the inputs to the BCM, use a scan tool to monitor the door ajar switches. The switches should change state as the doors are opened and closed. The scan tool will display the state of the switches, not the door. The switches are electrically open when the door is closed, and the switches are electrically closed when the door is open. If these is no change of state, evaluate the door ajar signal circuit by disconnecting the door ajar switch. Connect a fused jumper wire across the connector on the BCM side. If the illuminated entry system activates now (or the scan tool displays the switch as closed), there is a faulty switch. If this still does not activate the system, test for battery voltage to the switch connector. If battery voltage is not present, test the circuit back to the BCM by back probing for voltage at the BCM connector. If voltage is present now, there is an open in the wire between the BCM and the switch. If voltage is not present, check for good battery feed to the BCM and proper ground connections. If these are good, replace the BCM. If battery voltage is present at the switch connector when tested above, the fault is in the ground circuit for the switch.

> ⚙ **SERVICE TIP** You can also use the headlight switch dimmer function to test the lamps. Turn the switch to the DOME position. If the light comes on now, the output circuit from the BCM is working fine and the problem is in the inputs.

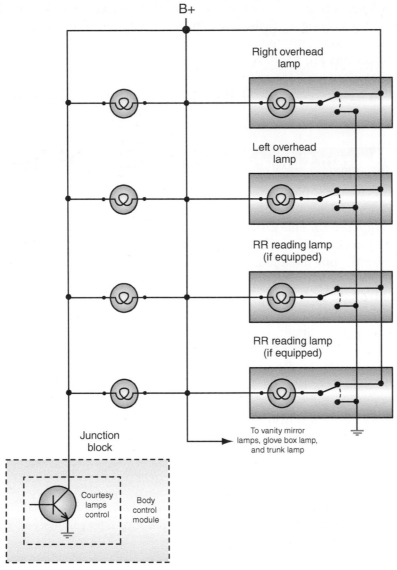

Figure 11-61 Output control of the illuminated entry system.

If the scan tool indicates proper input from the door ajar switches and the activation test fails to illuminate the lamps, then the output side of the system must be tested (**Figure 11-61**). Locate the proper wire into the BCM connector for the courtesy lamp control. Using a back-probing tool, jump this terminal to a good ground. If the lamps turn on now, check for proper battery feed and grounds to the BCM. If these are good, replace the BCM. If the lamps don't turn on, use the manual switches to turn on each light. If they don't turn on, trace the circuit from the battery to the common connection or splice.

Instrument Cluster and Panel Lights

In a conventional instrument cluster system, the headlight switch provides the power source for the instrument panel lights. The contacts are closed when the headlight switch is in the PARK or HEADLIGHT position. The current must flow through a variable resistor (rheostat) that is either a part of the headlight switch or a separate dial on the dash. The resistance of the rheostat is varied by turning the knob. By varying the resistance, changes in the current flow to the lamps control the brightness of the lights.

Classroom Manual
Chapter 11, page 345

 Special Tools
DMM
Safety glasses

Test for voltage output from the headlight switch to determine if the switch is operating properly. While varying the resistance in the rheostat, the voltage reading should vary. If voltage is present to the printed circuit, check the ground.

If all connections are good, remove the dash and test the printed circuit board. Use an ohmmeter to check for opens and shorts in the printed circuit board from the connector plug to the lamp sockets. If the printed circuit is bad, it must be repaired or replaced.

Lighting System Complexity

Today's vehicles have a sophisticated lighting system and electrical interconnections. It's possible to have problems with lights and accessories that cause them to operate when they are not supposed to. This is through a condition called **feedback**. Feedback occurs when electricity seeks a path of lower resistance. This alternate path operates another component than the one intended. If there is an open in the circuit, electricity will seek another path to follow. This may cause any lights or accessories in that path to turn on.

Figures 11-62 through **11-69** are examples of feedback and illustrate how feedback may cause undesired operation. **Figure 11-62** shows a system that has the dome light, taillight, and brake light circuits on one fuse; the cigar lighter and courtesy light circuits share a second fuse. The two fuses are in the main fuse block and share a common power bus bar.

If the dome light fuse blew and the headlight switch was in the PARK or HEADLIGHT position, the courtesy lights, dome light, taillight, parking lights, and instrument lights would all be very dimly lit (**Figure 11-63**). Current would flow through the cigar lighter fuse to the courtesy light and on to the door light switch. Current will then continue through the dome light to the headlight switch. Because the headlight switch is now closed, the instrument panel lights are also in the circuit. The lights are dim because all the bulbs are now connected in series.

If the dome light fuse is blown and the headlight switch is in the OFF position, all lights will turn off. However, if the door is opened, the courtesy lights will come on but the dome light will not (**Figure 11-64**).

Figure 11-62 A normally operating light circuit.

Figure 11-63 An open (blown) dome light fuse can cause feedback into other circuits when the headlights are turned on.

Figure 11-64 Circuit operation with a blown dome light fuse and the door switch closed.

With the same blown fuse and the brake light switch closed, the dome light, courtesy light, and brake light will all illuminate dimly because the loads are in series (**Figure 11-65**).

In this example, if the dome light and courtesy lights come on dimly when the cigar lighter is pushed in, the problem can be caused by a blown cigar lighter fuse (**Figure 11-66**). With the cigar lighter pushed in, a path to ground is completed. The lights and cigar lighter are now in series; thus, the lights are dim and there is not enough current to heat and release the cigar lighter. If the cigar lighter was left in this position, the battery would eventually drain down.

Figure 11-65 Feedback when brake light switch is closed.

Figure 11-66 Feedback as a result of the cigar lighter fuse being blown.

A blown cigar lighter fuse will also cause the dome light to get brighter when the doors are open, and the courtesy lights will go out (**Figure 11-67**). Also, if the lighter is pushed in and the brake light switch is closed, the dome and courtesy lights will go out (**Figure 11-68**).

Feedback can also be the result of a conductive corrosion that is developed at a connection. If the corrosion allows for current flow from one conductor to an adjacent

Figure 11-67 Opening the door will make the courtesy light go out and the dome light get brighter.

Figure 11-68 Dome and courtesy lights go out when the brakes are applied.

conductor in the connection, the other circuit will also be activated. **Figure 11-69** shows how corrosion in a common connector can cause the dome light to illuminate when the wiper motor is turned on. Because the wiper motor has a greater resistance than the light bulb, more voltage will flow through the bulb than through the motor. The bulb will light brightly, but the motor will turn very slowly or not at all. The same effect will result if the courtesy light switch is turned on with the motor switch off.

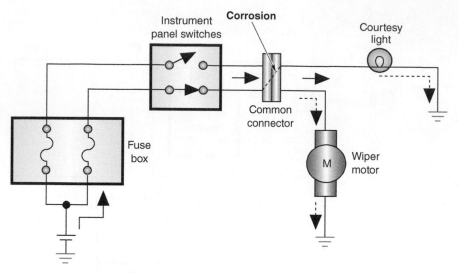

Figure 11-69 A corroded common connection can cause feedback.

AUTHOR'S NOTE Dual-filament light bulb sockets are subject to corrosion that can cause feedback. Also, dual-filament bulbs can have a filament burn out and attach to the other filament and result in feedback. The most common indicator of this problem is parking lights that illuminate when the brake pedal is applied.

Fiber Optics Diagnosis

Classroom Manual
Chapter 11, page 346

The fiber-optic system uses plastic strands to transmit light from the source to the object to be illuminated. The strands of plastic are sheathed by a polymer that insulates the light rays as they travel within the strands. The light rays travel through the strands by means of internal reflections. Fiber optics can be used to provide light in areas where bulbs would be inaccessible for service.

If the fiber optics don't illuminate, most likely the light source has failed. Check whether the bulb illuminates. If the bulb turns on, check that the fiber-optic lead is connected to the light source and to the lens. If these are good, the only other cause is that the cable is cut. It will need to be replaced.

CASE STUDY

A customer brings their vehicle into the shop because of an intermittent problem with the headlight delay feature. This person has taken the vehicle to other shops and spent several dollars in repair bills, but the problem has not been corrected. Using a systematic diagnostic approach and following the tests outlined in the service information lead the technician to test the potentiometer in the control switch. The resistance value is within specifications. However, while moving the potentiometer from the MIN to the MAX position, the ohmmeter reading is erratic in one portion. This area is the usual setting selected by the driver. The technician calls the customer and receives approval to replace the control switch. The new switch fixes the issue. The technician opens the old switch and finds that carbon from electrical arcing has built up in the problem area of the potentiometer.

ASE-Style Review Questions

1. The results of a functional test on the automatic headlight system are being discussed.
   Technician A says when the photocell is covered and the engine is running, the headlights should turn on within 30 seconds.
   Technician B says when a bright light is shone onto the photocell, the headlights should turn off within the specified period of time.
   Who is correct?
   A. A only
   B. B only
   C. Both A and B
   D. Neither A nor B

2. The automatic high-beam function does not operate properly. The LED in the mirror assembly is flashing continuously at a 1-hertz frequency. This can indicate:
   A. The mirror assembly needs to be calibrated.
   B. The system failed its last attempt to calibrate.
   C. A hardware failure has occurred.
   D. None of these choices.

3. An indicator that uses fiber optics is not functioning. This can be caused by:
   A. A bent fiber-optics cable.
   B. A faulty light source.
   C. Electromagnetic interference.
   D. None of these choices.

4. The most likely cause of the automatic headlights failing to activate in low ambient light conditions is:
   A. A faulty ignition switch.
   B. Camera angle alignment out of specifications.
   C. Burned-out headlight elements.
   D. Faulty headlight switch.

5. *Technician A* says problems with the automatic headlight system can be the fault of the headlight switch.
   Technician B says a bad ignition switch may cause the lights to not come on.
   Who is correct?
   A. A only
   B. B only
   C. Both A and B
   D. Neither A nor B

6. The headlights work in the manual position but don't turn on in the AUTO position. What is the most likely cause?
   A. A bad headlight ground.
   B. A faulty photocell assembly.
   C. A faulty headlight relay.
   D. A bad headlight relay ground connection.

7. *Technician A* says the photocell signal voltage should be 5 volts, with the lens covered.
   Technician B says the signal voltage should be 0 volt, with the flashlight shining on the lens.
   Who is correct?
   A. A only
   B. B only
   C. Both A and B
   D. Neither A nor B

8. When the headlights are switched from low beam to high beam, all headlights go out. All of the following can cause this **EXCEPT**:
   A. Faulty dimmer switch.
   B. Defective high-beam relay.
   C. Defective ignition switch.
   D. Open in high-beam circuit.

9. None of the turn signals operate. The **LEAST LIKELY** cause of this is:
   A. Burned-out bulbs.
   B. Faulty turn signal switch.
   C. Faulty turn signal flasher.
   D. Open circuit from ignition feed to turn signal.

10. A customer states that the headlights are brighter than normal and that they have to replace the lamps regularly.
    Technician A says this can be caused by too high charging system output.
    Technician B says this can be caused by excessive voltage drop in the circuit.
    Who is correct?
    A. A only
    B. B only
    C. Both A and B
    D. Neither A nor B

ASE Challenge Questions

1. The high-beam headlamps of a vehicle equipped with a Twilight Sentinel automatic headlamp system are inoperative; the low beams are working correctly.

   A voltmeter connected across the high-beam contacts of the dimmer switch indicates 0.15 volt when the high beams are "on."

   Technician A says the circuit from the Sentinel amplifier to the dimmer switch has excessive resistance.

   Technician B says the dimmer switch is faulty.
   Who is correct?
   A. A only
   B. B only
   C. Both A and B
   D. Neither A nor B

2. The headlights work normally, except they don't turn on in the automatic mode. What is the **LEAST LIKELY** cause?
   A. Faulty photocell.
   B. Faulty amplifier.
   C. Faulty headlight switch.
   D. Faulty headlight relay.

3. The brake lights of a vehicle equipped with dual-function turn signal/brake lamps are inoperative. The turn signals are functioning properly. However, there is no power at the brake light terminals when the brake pedal is depressed.

   Technician A says the brake light switch may be faulty.

   Technician B says the turn signal switch may be faulty.
   Who is correct?
   A. A only
   B. B only
   C. Both A and B
   D. Neither A nor B

4. The left turn signals of a vehicle are flashing very slowly; the right turn signals are operating correctly.

   Technician A says this may be caused by excessive circuit resistance on the left turn signal circuit.

   Technician B says this may be occurring because someone may have installed bulbs with higher-than-normal wattage ratings on the left side of the vehicle.
   Who is correct?
   A. A only
   B. B only
   C. Both A and B
   D. Neither A nor B

5. All of the following could cause premature failure of a composite bulb **EXCEPT**:
   A. High charging system voltage.
   B. Excessive bulb circuit ground resistance.
   C. Improper bulb handling.
   D. Cracked lamp housing.

Name _____   Date _____

DIAGNOSING LIGHTING SYSTEMS

Upon completion of this job sheet, you should be able to diagnose the cause of a no light operation, brighter-than-normal, and dimmer-than-normal operation; and determine needed repairs.

ASE Education Foundation Correlation

This job sheet addresses the following **MLR** tasks:

A.2.   Identify electrical/electronic system components and configuration. **(P-1)**

A.5.   Demonstrate proper use of a digital multimeter (DMM) when measuring source voltage, voltage drop (including grounds), current flow, and resistance. **(P-1)**

A.6.   Demonstrate knowledge of the causes and effects from shorts, grounds, opens, and resistance problems in electrical/electronic circuits. **(P-1)**

A.7.   Describe types of test lights; use appropriate test light to check operation of electrical circuits per service information. **(P-2)**

A.9.   Use wiring diagrams to trace electrical/electronic circuits. **(P-1)**

E.1.   Inspect interior and exterior lamps and sockets including headlights and auxiliary lights (fog lights/driving lights); replace as needed. **(P-1)**

This job sheet addresses the following **AST/MAST** tasks:

A.2.   Identify electrical/electronic system components and configuration. **(P-1)**

A.5.   Demonstrate proper use of a digital multimeter (DMM) when measuring source voltage, voltage drop (including grounds), current flow, and resistance. **(P-1)**

A.6.   Demonstrate knowledge of the causes and effects from shorts, grounds, opens, and resistance problems in electrical/electronic circuits. **(P-1)**

A.7.   Describe types of test lights; use appropriate test light to check operation of electrical circuits per service information. **(P-1)**

A.9.   Use wiring diagrams during the diagnosis (troubleshooting) of electrical/electronic circuit problems. **(P-1)**

E.1.   Inspect interior and exterior lamps and sockets including headlights and auxiliary lights (fog lights/driving lights); determine needed action. **(P-1)**

E.3.   Diagnose (troubleshoot) the causes of brighter-than-normal, intermittent, dim, or no light operation; determine needed action. **(P-1)**

We Support
ASE | Education Foundation

Tools and Materials

- Vehicle
- Wiring diagram for the vehicle
- Digital multimeter (DMM)
- Test light

Procedure

1. Record the concern the customer would have with the lighting system.

_____

_____

_____

2. Confirm the customer's concern by performing a check of the system and record your results.

_____

_____

_____

3. Determine if there are any other symptoms that may be related to the customer's concern by testing the operation of other systems. Record your results.

_____

_____

_____

_____

4. Analyze the symptoms and reference the wiring diagram to determine the most likely location of the fault. Record the component(s) and its (their) location(s).

_____

_____

_____

5. If the vehicle is equipped with HID headlights, record all safety concerns and warnings associated with the system.

_____

_____

_____

_____

_____

6. Locate a component from step 4 and return to the vehicle to test it. Describe the test procedure used and record the results.

7. Based on your diagnostic checks, was the cause of the problem located?

   ☐ Yes   ☐ No

   If yes, record your findings.

_____

_____

_____

   If no, continue to test other components identified in step 4.

8. If the fault is determined, make the repair and verify proper operation.

   Was the repair successful?   ☐ Yes   ☐ No

   If no, continue testing the system.

Instructor's Evaluation:

Grade	Rubric
0	No exposure due to: ☐ Absence ☐ Other:
1	Student unsuccessfully attempted the task
2	Student completed the task with significant instructor assistance.
3	Satisfactory performance of task. The student was able to perform elements of the task with minimal instructor assistance.
4	Competent/Proficient. Able to perform all elements of the task successfully and independently.

Instructor's Notes:

Name _____   Date _____

CHECKING A HEADLIGHT SWITCH

Upon completion of this job sheet, you should be able to check the operation of a headlight switch with an ohmmeter.

ASE Education Foundation Correlation

This job sheet addresses the following **MLR** tasks:

A.2.	Identify electrical/electronic system components and configuration. **(P-1)**
A.5.	Demonstrate proper use of a digital multimeter (DMM) when measuring source voltage, voltage drop (including grounds), current flow, and resistance. **(P-1)**
A.6.	Demonstrate knowledge of the causes and effects from shorts, grounds, opens, and resistance problems in electrical/electronic circuits. **(P-1)**
A.9.	Use wiring diagrams to trace electrical/electronic circuits. **(P-1)**
E.1.	Inspect interior and exterior lamps and sockets including headlights and auxiliary lights (fog lights/driving lights); replace as needed. **(P-1)**

This job sheet addresses the following **AST/MAST** tasks:

A.2.	Identify electrical/electronic system components and configuration. **(P-1)**
A.5.	Demonstrate proper use of a digital multimeter (DMM) when measuring source voltage, voltage drop (including grounds), current flow, and resistance. **(P-1)**
A.6.	Demonstrate knowledge of the causes and effects from shorts, grounds, opens, and resistance problems in electrical/electronic circuits. **(P-1)**
A.9.	Use wiring diagrams during the diagnosis (troubleshooting) of electrical/electronic circuit problems. **(P-1)**
E.1.	Inspect interior and exterior lamps and sockets including headlights and auxiliary lights (fog lights/driving lights); determine needed action. **(P-1)**
E.3.	Diagnose (troubleshoot) the causes of brighter-than-normal, intermittent, dim, or no light operation; determine needed action. **(P-1)**

We Support

ASE | Education Foundation

Tools and Materials

- A vehicle
- A wiring diagram for the vehicle
- A DMM

Procedure

1. Put the headlight switch in all possible positions and observe which lights are controlled by each position. List each position and the controlled lights here.

_____

_____

_____

_____

_____

_____

2. Locate the headlight switch in the wiring diagram and print out or draw the switch with each possible connection and possible position. Label the lights controlled by each position of the switch. Highlight the path of each position from the power source to ground. Highlight different circuitry with different colors.

3. Remove the circuit protection device to the headlights or disconnect the battery's negative cable. Remove the headlight switch according to the procedures outlined in the service information. Describe the procedure to your instructor before removing it.

   Instructor's OK to move to the next step. _____

4. Identify the various terminals of the switch and list the different terminals that should have continuity in the various switch positions.

_____

_____

_____

5. Connect the ohmmeter across these terminals, one switch position at a time, and record your readings.

_____

_____

_____

6. Based on the test, what are your conclusions about the switch?

_____

_____

_____

_____

7. Reinstall the switch and connect the negative battery cable or reinstall the fuse. Then check the operation of the headlights.

8. Instructor's verification that the vehicle is properly reassembled and the lighting system is operating properly.

Instructor's Evaluation:

Grade	Rubric
0	No exposure due to: ☐ Absence ☐ Other:
1	Student unsuccessfully attempted the task
2	Student completed the task with significant instructor assistance.
3	Satisfactory performance of task. The student was able to perform elements of the task with minimal instructor assistance.
4	Competent/Proficient. Able to perform all elements of the task successfully and independently.

Instructor's Notes:

Name _____   Date _____

HEADLIGHT AIMING

Upon completion of this job sheet, you should be able to adjust the aim of headlights using portable headlight-aiming equipment.

ASE Education Foundation Correlation

This job sheet addresses the following **MLR/AST/MAST** task:

E.2.   Aim headlights. **(P-2)**

We Support
ASE | Education Foundation

Tools and Materials

- A vehicle with adjustable headlights
- Portable headlight-aiming kit
- Hand tools

Procedure

1. Describe the type of headlights used on the vehicle.

   _____

2. Park the vehicle on a level floor.

   Install the calibrated aiming units to the headlights. (Make sure that the adapters fit the headlight-aiming pads on the lens.)

3. Zero the horizontal adjustment dial. Are the split-image target lines visible in the view port? _____

   If the lines cannot be seen, what should you do?

   _____

4. Turn the headlight horizontal adjusting screw until the split-image target lines are aligned. Then repeat this for the other headlight. List any problems you may have had doing this.

   _____

   _____

   _____

5. Turn the vertical adjustment dial on the aiming unit to zero. Turn the vertical adjustment screw until the spirit-level bubble is centered. Recheck your horizontal setting after adjusting the vertical.

6. List any problems you had making the vertical adjustment.

   _____

   _____

   _____

7. If the headlight assembly has four-lamp assemblies, repeat steps 4 and 5 on the other two lamps. List any problems you may have had doing this.

   _____

   _____

   _____

Instructor's Evaluation:

Grade	Rubric
0	No exposure due to: ☐ Absence ☐ Other:
1	Student unsuccessfully attempted the task
2	Student completed the task with significant instructor assistance.
3	Satisfactory performance of task. The student was able to perform elements of the task with minimal instructor assistance.
4	Competent/Proficient. Able to perform all elements of the task successfully and independently.

Instructor's Notes:

Name _____   Date _____

AIMING COMPOSITE OR HID HEADLIGHTS USING AN OPTICAL AIMER

Upon completion of this job sheet, you should be able to adjust the aim of headlights using portable headlight-aiming equipment.

ASE Education Foundation Correlation ——————————————————————

This job sheet addresses the following **MLR/AST/MAST** task:

E.2.   Aim headlights. **(P-2)**

We Support
ASE | Education Foundation

Tools and Materials

- A vehicle with adjustable headlights
- Optical headlight aimer
- Hand tools

Procedure

1. Describe the type of headlights used on the vehicle.

   _____

   _____

2. Park the vehicle on a level floor. Perform the process of floor slope compensation used by your assigned equipment. Record the dial settings.

   _____

   _____

3. What is the recommended distance from the headlight should the aimer be located?

   _____

   _____

4. Does your assigned equipment require locating the exact center of the grill?

   ☐ Yes   ☐ No

   If yes, perform the procedure; then go to step 5.

   If no, go to step 5.

5. Describe the process for locating the center of the beam.

   _____

   _____

   _____

   _____

6. Set the aligner in the proper position for the center of the beam and have your instructor confirm. Instructor's initials _____

7. If needed, use the alignment lens and two symmetrical locations on the vehicle to complete set up procedure.

8. Turn on the headlights and follow the equipment manufacturer's instructions for determining the current headlight beam aim. Record your results.

_____

_____

_____

9. If needed, adjust headlights.

Instructor's Evaluation:

Grade	Rubric
0	No exposure due to: ☐ Absence ☐ Other:
1	Student unsuccessfully attempted the task
2	Student completed the task with significant instructor assistance.
3	Satisfactory performance of task. The student was able to perform elements of the task with minimal instructor assistance.
4	Competent/Proficient. Able to perform all elements of the task successfully and independently.

Instructor's Notes:

Name _____   Date _____

TESTING AN AUTOMATIC HEADLIGHT SYSTEM

Upon completion of this job sheet, you should be able to diagnose an automatic headlight system and test the individual components of the system.

ASE Education Foundation Correlation —————————————————————

This job sheet addresses the following **MLR** tasks:

A.2.	Identify electrical/electronic system components and configuration. **(P-1)**
A.5.	Demonstrate proper use of a digital multimeter (DMM) when measuring source voltage, voltage drop (including grounds), current flow, and resistance. **(P-1)**
A.6.	Demonstrate knowledge of the causes and effects from shorts, grounds, opens, and resistance problems in electrical/electronic circuits. **(P-1)**
A.9.	Use wiring diagrams to trace electrical/electronic circuits. **(P-1)**
E.1.	Inspect interior and exterior lamps and sockets including headlights and auxiliary lights (fog lights/driving lights); replace as needed. **(P-1)**
G.1.	Demonstrate understanding of vehicle comfort, convenience, access, safety, and related systems operation. **(P-3)**

This job sheet addresses the following **AST** tasks:

A.2.	Identify electrical/electronic system components and configuration. **(P-1)**
A.5.	Demonstrate proper use of a digital multimeter (DMM) when measuring source voltage, voltage drop (including grounds), current flow, and resistance. **(P-1)**
A.6.	Demonstrate knowledge of the causes and effects from shorts, grounds, opens, and resistance problems in electrical/electronic circuits. **(P-1)**
A.9.	Use wiring diagrams to trace electrical/electronic circuits. **(P-1)**
E.1.	Inspect interior and exterior lamps and sockets including headlights and auxiliary lights (fog lights/driving lights); determine needed action. **(P-1)**
G.1.	Demonstrate understanding of vehicle comfort, convenience, access, safety, and related systems operation. **(P-2)**

This job sheet addresses the following **MAST** tasks:

A.2.	Identify electrical/electronic system components and configuration. **(P-1)**
A.5.	Demonstrate proper use of a digital multimeter (DMM) when measuring source voltage, voltage drop (including grounds), current flow, and resistance. **(P-1)**
A.6.	Demonstrate knowledge of the causes and effects from shorts, grounds, opens, and resistance problems in electrical/electronic circuits. **(P-1)**
A.9.	Use wiring diagrams to trace electrical/electronic circuits. **(P-1)**
E.1.	Inspect interior and exterior lamps and sockets including headlights and auxiliary lights (fog lights/driving lights); determine needed action. **(P-1)**
G.1.	Demonstrate understanding of vehicle comfort, convenience, access, safety, and related systems operation. **(P-2)**

We Support
ASE | Education Foundation

Tools and Materials
- A vehicle with an automatic headlight system
- Wiring diagram for the chosen vehicle
- Component locator for the chosen vehicle
- Service information for the chosen vehicle
- A DMM

Procedure

1. Locate the photocell. Then disconnect the connector to the photocell. Turn the ignition switch and the automatic headlamp control to ON.

   Turn the headlight switch off.

   If the lights come on within 60 seconds and the automatic headlights don't work as before, the photocell must be bad. If the lights still don't come on, test the amplifier unit.

   Describe what happened.

   _____

   _____

   _____

   _____

   _____

   _____

2. Turn the ignition off. Disconnect the negative cable from the battery. Locate the amplifier assembly and disconnect the electrical connector to it. Turn the headlight switch off.

   Turn the automatic headlight control to its ON position.

   Locate the resistance checks of the amplifier circuit in the service information. Briefly outline those procedures.

   _____

   _____

   _____

   _____

3. Follow the previously described procedures and list the results.

   _____

   _____

   _____

   _____

4. Turn the automatic headlight control to its ON position.

   Locate the resistance checks of the amplifier circuit in the service information. Briefly outline those procedures.

   _____

   _____

   _____

   _____

5. Follow the procedures in step 4 and list the results.

   _____

   _____

   _____

   _____

6. List any additional diagnostic steps that the manufacturer recommends in the case where the previously described tests did not identify the problem.

   _____

   _____

   _____

   _____

   _____

   _____

   _____

   _____

Instructor's Evaluation:

Grade	Rubric
0	No exposure due to: ☐ Absence ☐ Other:
1	Student unsuccessfully attempted the task
2	Student completed the task with significant instructor assistance.
3	Satisfactory performance of task. The student was able to perform elements of the task with minimal instructor assistance.
4	Competent/Proficient. Able to perform all elements of the task successfully and independently.

Instructor's Notes:

Name _____   Date _____

TESTING THE BCM-CONTROLLED HEADLIGHT SYSTEM

Upon completion of this job sheet, you should be able to test the computer-controlled headlight system and determine needed repairs.

ASE Education Foundation Correlation ──────────────────

This job sheet addresses the following **MLR** tasks:

A.2.	Identify electrical/electronic system components and configuration. **(P-1)**
A.5.	Demonstrate proper use of a digital multimeter (DMM) when measuring source voltage, voltage drop (including grounds), current flow, and resistance. **(P-1)**
A.6.	Demonstrate knowledge of the causes and effects from shorts, grounds, opens, and resistance problems in electrical/electronic circuits. **(P-1)**
A.7.	Describe types of test lights; use appropriate test light to check operation of electrical circuits per service information. **(P-2)**
A.8.	Use fused jumper wires to check operation of electrical circuits per service information. **(P-2)**
A.9.	Use wiring diagrams to trace electrical/electronic circuits. **(P-1)**
E.1.	Inspect interior and exterior lamps and sockets including headlights and auxiliary lights (fog lights/driving lights); replace as needed. **(P-1)**
G.1.	Demonstrate understanding of vehicle comfort, convenience, access, safety, and related systems operation. **(P-3)**

This job sheet addresses the following **AST** tasks:

A.2.	Identify electrical/electronic system components and configuration. **(P-1)**
A.5.	Demonstrate proper use of a digital multimeter (DMM) when measuring source voltage, voltage drop (including grounds), current flow, and resistance. **(P-1)**
A.6.	Demonstrate knowledge of the causes and effects from shorts, grounds, opens, and resistance problems in electrical/electronic circuits. **(P-1)**
A.7.	Describe types of test lights; use appropriate test light to check operation of electrical circuits per service information. **(P-1)**
A.8.	Use fused jumper wires to check operation of electrical circuits per service information. **(P-1)**
A.9.	Use wiring diagrams during the diagnosis (troubleshooting) of electrical/electronic circuit problems. **(P-1)**
E.1.	Inspect interior and exterior lamps and sockets including headlights and auxiliary lights (fog lights/driving lights); determine needed action. **(P-1)**
E.3.	Diagnose the causes of brighter-than-normal, intermittent, dim, or no light operation; determine needed action. **(P-1)**
G.1.	Diagnose vehicle comfort, convenience, access, safety, and related systems operation; determine needed action. **(P-2)**
G.8.	Diagnose body electronic system circuits using a scan tool; check for module communication errors (data communication bus systems); determine needed action. **(P-2)**

(Continued)

This job sheet addresses the following **MAST** tasks:

A.2.   Identify electrical/electronic system components and configuration. **(P-1)**

A.5.   Demonstrate proper use of a digital multimeter (DMM) when measuring source voltage, voltage drop (including grounds), current flow, and resistance. **(P-1)**

A.6.   Demonstrate knowledge of the causes and effects from shorts, grounds, opens, and resistance problems in electrical/electronic circuits. **(P-1)**

A.7.   Describe types of test lights; use appropriate test light to check operation of electrical circuits per service information. **(P-1)**

A.8.   Use fused jumper wires to check operation of electrical circuits per service information. **(P-1)**

A.9.   Use wiring diagrams during the diagnosis (troubleshooting) of electrical/electronic circuit problems. **(P-1)**

E.1.   Inspect interior and exterior lamps and sockets including headlights and auxiliary lights (fog lights/driving lights); determine needed action. **(P-1)**

E.3.   Diagnose the causes of brighter-than-normal, intermittent, dim, or no light operation; determine needed action. **(P-1)**

G.1.   Diagnose vehicle comfort, convenience, access, safety, and related systems operation; determine needed action. **(P-2)**

G.8.   Diagnose body electronic system circuits using a scan tool; check for module communication errors (data communication bus systems); determine needed action. **(P-1)**

We Support
ASE | Education Foundation

Tools and Materials

- A vehicle equipped with BCM-controlled headlights
- Scan tool
- DMM
- Test light
- Fused jumper wires
- Wiring diagram for the vehicle

Procedure

1. Test the operation of the headlights. Describe the symptoms.

_____

_____

2. Are there any other related symptoms? ☐ Yes   ☐ No

   If yes, describe the symptom(s).

_____

_____

3. Describe the basic operation of the system you are working on.

_____

_____

_____

4. Check any associated fuses and conduct a visual inspection of the system. List any problems found.

_____

_____

 Repair any faults found and test system operation. Did this fix the problem? ☐ Yes ☐ No

5. If used, substitute the headlamp relay with a known good relay of the same type and test operation. Test the system operation. Did this fix the problem? ☐ Yes ☐ No

6. Connect the scan tool and record any DTCs.

_____

_____

7. Use the scan tool to perform an activation test of the headlights. Did the headlamps come on during the activation? ☐ Yes ☐ No

 What do you know about the problem so far?

_____

_____

8. Use the scan tool to monitor the headlight switch as it's placed in all switch positions.

 Record the voltages for each position and compare to specifications.

_____

_____

9. What do you know about the system so far?

_____

_____

10. Remove the headlight relay and use a voltmeter or test light to confirm battery voltage is present at pins 30 and 86 of the junction block terminals. Is voltage present at these terminals? ☐ Yes ☐ No

 If *no*, what would you test next?

_____

_____

11. If voltage is present at both terminals in step 10, use a fused jumper wire to connect pin 30 to pin 87. Did the lights come on? ☐ Yes ☐ No

 What do you know about the circuit now?

_____

_____

12. Connect a test light between pins 86 and 85 of the junction block terminals (relay removed). Place the headlight switch into the headlamp position and observe the test light. What does this test check?

_____

_____

13. Based on your results for step 12, what do you know about the problem?

_____

_____

14. What other checks need to be performed to pinpoint the fault?

_____

_____

15. Perform the tests you listed and record your findings and recommendations.

_____

_____

Instructor's Evaluation:

Grade	Rubric
0	No exposure due to: ☐ Absence ☐ Other:
1	Student unsuccessfully attempted the task
2	Student completed the task with significant instructor assistance.
3	Satisfactory performance of task. The student was able to perform elements of the task with minimal instructor assistance.
4	Competent/Proficient. Able to perform all elements of the task successfully and independently.

Instructor's Notes:

_____

Name _____   Date _____

DIAGNOSING HSD-CONTROLLED LIGHTING SYSTEMS

Upon completion of this job sheet, you should be able to diagnose the cause or causes of brighter-than-normal, intermittent, dim, or no light operation and determine necessary action for HSD-controlled lighting systems.

ASE Education Foundation Correlation

This job sheet addresses the following **MLR** tasks:

A.2.	Identify electrical/electronic system components and configuration. **(P-1)**
A.5.	Demonstrate proper use of a digital multimeter (DMM) when measuring source voltage, voltage drop (including grounds), current flow, and resistance. **(P-1)**
A.6.	Demonstrate knowledge of the causes and effects from shorts, grounds, opens, and resistance problems in electrical/electronic circuits. **(P-1)**
A.9.	Use wiring diagrams to trace electrical/electronic circuits. **(P-1)**
E.1.	Inspect interior and exterior lamps and sockets including headlights and auxiliary lights (fog lights/driving lights); replace as needed. **(P-1)**
G.1.	Demonstrate understanding of vehicle comfort, convenience, access, safety, and related systems operation. **(P-3)**

This job sheet addresses the following **AST** tasks:

A.2.	Identify electrical/electronic system components and configuration. **(P-1)**
A.5.	Demonstrate proper use of a digital multimeter (DMM) when measuring source voltage, voltage drop (including grounds), current flow, and resistance. **(P-1)**
A.6.	Demonstrate knowledge of the causes and effects from shorts, grounds, opens, and resistance problems in electrical/electronic circuits. **(P-1)**
A.9.	Use wiring diagrams during the diagnosis (troubleshooting) of electrical/electronic circuit problems. **(P-1)**
E.1.	Inspect interior and exterior lamps and sockets including headlights and auxiliary lights (fog lights/driving lights); determine needed action. **(P-1)**
E.3.	Diagnose (troubleshoot) the causes of brighter-than-normal, intermittent, dim, or no light operation; determine needed action. **(P-1)**
G.1.	Diagnose vehicle comfort, convenience, access, safety, and related systems operation; determine needed action. **(P-2)**
G.8.	Diagnose body electronic system circuits using a scan tool; check for module communication errors (data communication bus systems); determine needed action. **(P-2)**

This job sheet addresses the following **MAST** tasks:

A.2.	Identify electrical/electronic system components and configuration. **(P-1)**
A.5.	Demonstrate proper use of a digital multimeter (DMM) when measuring source voltage, voltage drop (including grounds), current flow, and resistance. **(P-1)**
A.6.	Demonstrate knowledge of the causes and effects from shorts, grounds, opens, and resistance problems in electrical/electronic circuits. **(P-1)**
A.7.	Describe types of test lights; use appropriate test light to check operation of electrical circuits per service information. **(P-1)**

(Continued)

A.8.   Use fused jumper wires to check operation of electrical circuits per service information. **(P-1)**

A.9.   Use wiring diagrams during the diagnosis (troubleshooting) of electrical/ electronic circuit problems. **(P-1)**

E.1.   Inspect interior and exterior lamps and sockets including headlights and auxiliary lights (fog lights/driving lights); determine needed action. **(P-1)**

E.3.   Diagnose (troubleshoot) the causes of brighter-than-normal, intermittent, dim, or no light operation; determine needed action. **(P-1)**

G.1.   Diagnose vehicle comfort, convenience, access, safety, and related systems operation; determine needed action. **(P-2)**

G.8.   Diagnose body electronic system circuits using a scan tool; check for module communication errors (data communication bus systems); determine needed action. **(P-1)**

We Support
ASE | Education Foundation

Tools and Materials

- Vehicle with HSD-controlled headlights
- Scan tool
- DMM
- Back-probing tools
- Service information

Procedure

For this task you will identify the circuits of the front lighting system.

Task One

1. Use the service information to identify the right headlight and park/turn assembly connector and the circuit information. Record the information in the following table:

Connector	CAVITY	Circuit ID	Wire Color
Headlamp low beam			
Low-beam ground			
Headlamp high beam			
High-beam ground			
Park lamp			
Turn signal			
Park/turn ground			

Task Two

For this task, you will identify the operational characteristics of the HSD-controlled lighting system.

1. Connect the scan tool and navigate to the module responsible for headlight control. Which module is responsible? _____

2. Clear any DTCs.

3. List any actuator tests that are available for the exterior lighting and perform the tests. Record your results.

_____

_____

_____

_____

_____

4. With the headlamps turned off, disconnect the right headlamp connector from the low-beam bulb socket. Are any DTCs set?   ☐ Yes   ☐ No

If so, record them. _____

5. Are there any observable symptoms?

_____

_____

6. Turn on the headlamps. Are there any DTCs?   ☐ Yes   ☐ No

If so, record them.

_____

_____

_____

7. Is there a lamp-out indicator illuminated?   ☐ Yes   ☐ No

8. Turn off the headlights and reconnect the headlamp low-beam connector.

9. With the headlamps off, disconnect the right high-beam connector. Are there any DTCs?   ☐ Yes   ☐ No

If so, record them.

_____

_____

_____

10. Turn on the high beams. Are there any DTCs?   ☐ Yes   ☐ No

If so, record them.

_____

_____

_____

11. Is there a lamp-out indicator illuminated?   ☐ Yes   ☐ No

12. Turn off the headlights and connect the headlamp high beam. Are there any DTCs?

☐ Yes   ☐ No

If so, record them.

_____

_____

_____

13. Turn on the park lamps and right turn signal. Are there any DTCs?   ☐ Yes   ☐ No

If so, record them.

_____

_____

_____

14. Is there a lamp-out indicator illuminated?   ☐ Yes   ☐ No

15. Is there any indication that a lamp is out?   ☐ Yes   ☐ No

If so, what is it?

_____

_____

Task Three

For this task you will determine the voltages applied to the HSD-controlled lighting system.

1. Refer to the table in Task One. Use a DMM to measure the voltages at the headlight assembly connector(s). Measure with the connector disconnected and then with it reconnected. Measure voltage on the headlamps; turn lamp and park lamp circuits. Record the results in the following table:

	Lamp-Connected Volts	**Lamp-Disconnected Volts**
Low beam		
High beam		
Turn		
Park		

2. Explain your results.

_____

_____

_____

_____

_____

3. When are the HSD circuits monitored?

_____

_____

_____

4. When do the DTCs set?

_____

_____

_____

5. Clear all fault codes.

_____

_____

_____

Instructor's Evaluation:

Grade	Rubric
0	No exposure due to: ☐ Absence ☐ Other:
1	Student unsuccessfully attempted the task
2	Student completed the task with significant instructor assistance.
3	Satisfactory performance of task. The student was able to perform elements of the task with minimal instructor assistance.
4	Competent/Proficient. Able to perform all elements of the task successfully and independently.

Instructor's Notes:

Name _____   Date _____

DIAGNOSING AUTOMATIC HIGH-BEAM SYSTEMS

Upon completion of this job sheet, you should be able to diagnose the cause of poor automatic high-beam system operation and determine needed repairs.

ASE Education Foundation Correlation

This job sheet addresses the following **MLR** tasks:

A.2.   Identify electrical/electronic system components and configuration. **(P-1)**

G.1.   Demonstrate understanding of vehicle comfort, convenience, access, safety, and related systems operation. **(P-3)**

This job sheet addresses the following **AST** tasks:

A.2.   Identify electrical/electronic system components and configuration. **(P-1)**

G.1.   Diagnose vehicle comfort, convenience, access, safety, and related systems operation; determine needed action. **(P-2)**

G.8.   Diagnose body electronic system circuits using a scan tool; check for module communication errors (data communication bus systems); determine needed action. **(P-2)**

This job sheet addresses the following **MAST** tasks:

A.2.   Identify electrical/electronic system components and configuration. **(P-1)**

G.1.   Diagnose vehicle comfort, convenience, access, safety, and related systems operation; determine needed action. **(P-2)**

G.8.   Diagnose body electronic system circuits using a scan tool; check for module communication errors (data communication bus systems); determine needed action. **(P-1)**

We Support
ASE | Education Foundation

Tools and Materials

- Vehicle with camera sensing Automatic Headlight Dimming (SmartBeam)
- Calibration target
- Grease pencil
- Scan tool
- Tape measure
- Piece of cardboard
- Service information

Procedure

For this job sheet task, you will identify the operational characteristics of the SmartBeam automatic high-beam system and calibrate the camera.

1. Turn on the ignition. Observe the SmartBeam camera status LED and determine if the camera is calibrated. Is the SmartBeam camera calibrated?   ☐ Yes   ☐ No

   NOTE: If the camera is not calibrated, it must be calibrated prior to entering demonstration mode.

2. Turn off the ignition. Press and hold the AUTO button on the rearview mirror while turning the ignition switch to the RUN position. Continue to depress the AUTO button until the LED blinks and the high beams turn on; then release the button. Describe what happens.

_____

_____

_____

3. Repeat the steps to enter the demonstration mode again. While the lamps are illuminated during the demonstration mode, slide a piece of cardboard between the camera lens and the windshield. Describe the results.

_____

_____

4. Use the service information and determine the alignment specification of the camera. Record the specifications.

_____

_____

5. Connect the scan tool and navigate to the "automatic high-beam module (AHBM)." Locate the camera aim test function and activate the test. Describe the status of the LED.

_____

6. What does this indicate?

_____

7. Turn off the ignition, and use a tape measure and service information procedures to determine the centerline of the camera lens; mark the windshield and rear glass with a grease pencil.

8. Using a tape measure, determine a location in front of the vehicle which is the specified distance from the vehicle. What is the specified distance? _____

9. Locate the calibration target in front of the vehicle, and align it with the marks on the windshield.

10. Using a tape measure, determine the distance the target's center LED should be from the floor. Record this distance. _____

11. Adjust the calibration target to the required height.

12. Turn on the power supply to the calibration target, and then place the ignition switch in the RUN position. Describe the results.

_____

_____

13. Use the scan tool to determine and record the calibration status.

14. Check for DTCs and record.

_____

_____

_____

15. Clear any DTCs.

Instructor's Evaluation:

Grade	Rubric
0	No exposure due to: ☐ Absence ☐ Other:
1	Student unsuccessfully attempted the task
2	Student completed the task with significant instructor assistance.
3	Satisfactory performance of task. The student was able to perform elements of the task with minimal instructor assistance.
4	Competent/Proficient. Able to perform all elements of the task successfully and independently.

Instructor's Notes:

DIAGNOSTIC CHART 11-1

PROBLEM AREA:	Headlights—park/taillights.
SYMPTOMS:	Bright illumination, early bulb failure.
POSSIBLE CAUSES:	**1.** Alternator output too high. **2.** Defective dimmer switch.

DIAGNOSTIC CHART 11-2

PROBLEM AREA:	Headlights—park/taillights.
SYMPTOMS:	Intermittent headlight or park/taillights operation, headlights flicker.
POSSIBLE CAUSES:	**1.** Defective circuit breaker. **2.** Overload in circuit. **3.** Improper connection. **4.** Defective switch. **5.** Poor ground. **6.** Excessive resistance.

DIAGNOSTIC CHART 11-3

PROBLEM AREA:	Headlights.
SYMPTOMS:	Dim headlight illumination.
POSSIBLE CAUSES:	**1.** Poor ground connection. **2.** Corroded headlight socket. **3.** Poor battery cable connections. **4.** Low generator output. **5.** Loose or broken generator drive belt.

DIAGNOSTIC CHART 11-4

PROBLEM AREA:	Headlights—park/taillights.
SYMPTOMS:	No or improper headlight or park/taillights operation.
POSSIBLE CAUSES:	**1.** Burned-out headlights. **2.** Defective headlight switch. **3.** Open circuit. **4.** Defective circuit breaker. **5.** Overload in circuit. **6.** Improper or poor connection. **7.** Poor ground. **8.** Excessive resistance. **9.** Defective relay. **10.** Blown fuse. **11.** Faulty dimmer switch. **12.** Short in insulated circuit. **13.** Improper bulb application. **14.** Improper headlight aiming.

DIAGNOSTIC CHART 11-5

PROBLEM AREA:	Active Headlights.
SYMPTOMS:	Headlights fail to flow into turn.
POSSIBLE CAUSES:	**1.** Headlight swivel motor or circuits. **2.** Yaw sensor input missing or incorrect. **3.** Steering angle sensor input missing or incorrect. **4.** Steering angle sensor requires calibration. **5.** Active headlight ECU.

DIAGNOSTIC CHART 11-6

PROBLEM AREA:	Instrument cluster lighting.
SYMPTOMS:	Intermittent brightness control of instrument cluster light circuits. Dash lights flicker.
POSSIBLE CAUSES:	**1.** Improper connection. **2.** Defective headlight switch rheostat. **3.** Poor ground. **4.** Excessive resistance. **5.** Faulty printed circuit.

DIAGNOSTIC CHART 11-7

PROBLEM AREA:	Instrument cluster lighting.
SYMPTOMS:	Low-level light intensity from panel illumination lights.
POSSIBLE CAUSES:	**1.** Burned-out bulbs. **2.** Defective headlight switch rheostat. **3.** Improper bulb application. **4.** Improper connection. **5.** Poor ground. **6.** Excessive resistance. **7.** Defective or faulty printed circuit.

DIAGNOSTIC CHART 11-8

PROBLEM AREA:	Instrument cluster lighting.
SYMPTOMS:	No bulb illumination.
POSSIBLE CAUSES:	**1.** Blown circuit protection device. **2.** Burned-out bulbs. **3.** Defective headlight switch rheostat. **4.** Open circuit. **5.** Improper connection. **6.** Poor ground. **7.** Excessive resistance. **8.** Improper bulb application. **9.** Defective printed circuit.

DIAGNOSTIC CHART 11-9

PROBLEM AREA:	Instrument cluster lighting.
SYMPTOMS:	No dash light brightness control.
POSSIBLE CAUSES:	**1.** Defective headlight switch rheostat.

DIAGNOSTIC CHART 11-10

PROBLEM AREA:	Courtesy lights.
SYMPTOMS:	Intermittent courtesy light operation.
POSSIBLE CAUSES:	**1.** Improper connection. **2.** Defective headlight switch. **3.** Defective door jamb switch. **4.** Defective or sticking door switch. **5.** Poor ground. **6.** Excessive resistance.

DIAGNOSTIC CHART 11-11

PROBLEM AREA:	Courtesy lights.
SYMPTOMS:	Dimmer-than-normal courtesy lights. Battery condition good.
POSSIBLE CAUSES:	**1.** Improper bulb application. **2.** Improper connection. **3.** Poor ground. **4.** Excessive resistance.

DIAGNOSTIC CHART 11-12

PROBLEM AREA:	Courtesy light operation.
SYMPTOMS:	No courtesy light illumination.
POSSIBLE CAUSES:	**1.** Blown circuit protection device. **2.** Burned-out bulbs. **3.** Defective headlight switch. **4.** Defective door switches. **5.** Open circuit. **6.** Improper connection. **7.** Poor ground. **8.** Excessive resistance. **9.** Improper bulb application.

DIAGNOSTIC CHART 11-13

PROBLEM AREA:	Courtesy lights.
SYMPTOMS:	Courtesy lights stay on all of the time.
POSSIBLE CAUSES:	**1.** Defective door jamb switch. **2.** Defective headlight switch. **3.** Shorted circuit.

DIAGNOSTIC CHART 11-14

PROBLEM AREA:	Stop (brake) lamp operation.
SYMPTOMS:	Intermittent stop lamp operation.
POSSIBLE CAUSES:	**1.** Misadjusted brake light switch. **2.** Poor ground connection. **3.** Excessive resistance. **4.** Faulty sockets. **5.** Poor connections. **6.** Faulty turn signal switch contacts. **7.** Defective brake light switch.

DIAGNOSTIC CHART 11-15

PROBLEM AREA:	Stop (brake) lamp operation.
SYMPTOMS:	Dimmer-than-normal stop lights.
POSSIBLE CAUSES:	**1.** Excessive circuit resistance. **2.** Poor ground connection. **3.** Improper bulb application. **4.** Improper connections. **5.** Faulty turn signal switch contacts. **6.** Improper bulb application.

DIAGNOSTIC CHART 11-16

PROBLEM AREA:	Stop (brake) lamp operation.
SYMPTOMS:	No stop lamps illuminate. Stop lights fail to illuminate when the brakes are applied.
POSSIBLE CAUSES:	**1.** Faulty brake light switch. **2.** Open in the circuit. **3.** Improper bulb application. **4.** Faulty turn signal switch. **5.** Improper common ground connection. **6.** Burned-out light bulbs.

DIAGNOSTIC CHART 11-17

PROBLEM AREA:	Turn signal operation.
SYMPTOMS:	Turn signals don't operate in either direction.
POSSIBLE CAUSES:	**1.** Blown fuse. **2.** Defective or worn flasher unit. **3.** Defective or faulty turn signal switch. **4.** Open circuit.

DIAGNOSTIC CHART 11-18

PROBLEM AREA:	Turn signals.
SYMPTOMS:	Turn signal lamp does not illuminate.
	Turn signal indicator illuminates but does not flash.
POSSIBLE CAUSES:	**1.** Improper bulb. **2.** Burned-out bulb. **3.** Open circuit. **4.** Failed flasher unit.

DIAGNOSTIC CHART 11-19

PROBLEM AREA:	Hazard light operation.
SYMPTOMS:	Hazard lights fail to operate when activated.
POSSIBLE CAUSES:	**1.** Blown fuse. **2.** Defective or worn flasher unit. **3.** Defective or faulty hazard light switch. **4.** Open circuit. **5.** Defective turn signal switch.

DIAGNOSTIC CHART 11-20

PROBLEM AREA:	Back-up light operation.
SYMPTOMS:	Back-up lights fail to operate some of the time.
POSSIBLE CAUSES:	**1.** Misadjusted back-up light switch. **2.** Poor ground connection. **3.** Excessive resistance. **4.** Faulty sockets. **5.** Poor connections.

DIAGNOSTIC CHART 11-21

PROBLEM AREA:	Back-up light operation.
SYMPTOMS:	Dimmer-than-normal back-up lights
POSSIBLE CAUSES:	**1.** Excessive circuit resistance. **2.** Poor ground connection. **3.** Improper bulb application. **4.** Improper connections. **5.** Faulty back-up switch contacts.

DIAGNOSTIC CHART 11-22

PROBLEM AREA:	Back-up light operation.
SYMPTOMS:	Back-up lights fail to illuminate when the transmission is in reverse.
POSSIBLE CAUSES:	**1.** Faulty back-up light switch. **2.** Misadjusted back-up light switch. **3.** Blown fuse. **4.** Open in the circuit. **5.** Improper bulb application. **6.** Improper common ground connection. **7.** Burned-out light bulbs.

DIAGNOSTIC CHART 11-23

PROBLEM AREA:	Back-up light does not operate.
SYMPTOMS:	One back-up light fails to illuminate when the transmission is in reverse.
POSSIBLE CAUSES:	**1.** Burned-out lamp. **2.** Loose connection. **3.** Open circuit to lamp.

DIAGNOSTIC CHART 11-24

PROBLEM AREA:	Concealed headlight system.
SYMPTOMS:	Headlight doors fail to open or close.
POSSIBLE CAUSES:	**1.** Faulty headlight switch. **2.** Power feed to relay or relay module. **3.** Faulty relay or relay module. **4.** Faulty door motor. **5.** Open, short, or high resistance in motor control circuits. **6.** Controller power and ground circuits. **7.** Faulty controller.

DIAGNOSTIC CHART 11-25

PROBLEM AREA:	Computer-controlled headlight system.
SYMPTOMS:	Headlights fail to turn on.
POSSIBLE CAUSES:	**1.** Faulty headlight switch. **2.** Power feed to relays. **3.** Faulty relay(s). **4.** Open, short, or high resistance in relay control circuit. **5.** Headlamp circuit failure. **6.** Burned-out headlamp elements. **7.** Controller power and ground circuits. **8.** Faulty controller.

DIAGNOSTIC CHART 11-26

PROBLEM AREA:	Automatic headlight system.
SYMPTOMS:	Headlights fail to turn on in automatic mode; headlights work in manual mode.
POSSIBLE CAUSES:	1. Open input circuit from switch to control module. 2. Faulty module. 3. Faulty switch.

DIAGNOSTIC CHART 11-27

PROBLEM AREA:	Automatic headlight system.
SYMPTOMS:	Headlights turn on in daytime when switch is in AUTO mode.
POSSIBLE CAUSES:	1. Faulty photocell. 2. Open photocell circuit. 3. Shorted photocell circuit. 4. Faulty control module. 5. Bus communications error. 6. Immobilizer system inoperative.

DIAGNOSTIC CHART 11-28

PROBLEM AREA:	Automatic high/low beam.
SYMPTOMS:	Headlights fail to automatically switch to high beam.
POSSIBLE CAUSES:	1. Obstruction in front of camera. 2. Improper headlight aiming. 3. System not initialized. 4. Camera alignment. 5. System voltage to module. 6. Module ground circuit. 7. Bus network failure. 8. Faulty controller.

DIAGNOSTIC CHART 11-29

PROBLEM AREA:	Headlight leveling system.
SYMPTOMS:	Headlights fail to level properly.
POSSIBLE CAUSES:	1. Improper headlight aiming. 2. Faulty switch input or circuit. 3. Faulty headlight level sensor or sensor circuits. 4. Disconnected link to headlight level sensor(s). 5. Circuits to motor(s). 6. Defective motor(s). 7. System voltage to module. 8. Module ground circuit. 9. Bus network failure. 10. System not calibrated. 11. Faulty controller.

DIAGNOSTIC CHART 11-30

PROBLEM AREA:	Daytime running lights.
SYMPTOMS:	DRL fail to turn on.
POSSIBLE CAUSES:	**1.** Faulty headlight switch input or circuit. **2.** Power circuit to DRL relay or relay module. **3.** Defective relay or relay module. **4.** Parking brake switch or circuit. **5.** System voltage to module. **6.** Module ground circuit. **7.** Bus network failure. **8.** Faulty controller.

DIAGNOSTIC CHART 11-31

PROBLEM AREA:	Illuminate entry system.
SYMPTOMS:	Courtesy lights fail to turn on.
POSSIBLE CAUSES:	**1.** Faulty headlight switch input or circuit. **2.** Faulty door ajar switches or circuits. **3.** Dead battery in remote FOB. **4.** Defective FOB. **5.** Open, short, or high resistance in the lamp driver circuit. **6.** Power circuit to the module. **7.** Defective relay or relay module. **8.** Module ground circuit. **9.** Bus network failure. **10.** Burned-out lamp elements. **11.** Faulty controller.

DIAGNOSTIC CHART 11-32

PROBLEM AREA:	Illuminate entry system.
SYMPTOMS:	Courtesy lights fail to turn off.
POSSIBLE CAUSES:	**1.** Faulty headlight switch input or circuit. **2.** Faulty door switches or circuits. **3.** Short to ground in the lamp driver circuit. **4.** Power circuit to the module. **5.** Defective relay or relay module. **6.** Module ground circuit. **7.** Bus network failure. **8.** Faulty controller.

CHAPTER 12

INSTRUMENTATION AND WARNING LAMP SYSTEM DIAGNOSIS AND REPAIR

Upon completion and review of this chapter, you should be able to:

- Remove and replace the instrument cluster.
- Remove and replace the printed circuit.
- Diagnose and repair causes for erratic and inaccurate speedometer readings.
- Diagnose and repair causes for tachometer malfunctions.
- Diagnose and repair faulty gauge circuits.
- Diagnose and repair the cause of multiple gauge failure.
- Diagnose sender units, including thermistors, piezoresistive, and mechanical variable styles.
- Diagnose and repair warning light circuits.
- Diagnose and repair the cause of multiple warning light failures.

- Enter diagnostic mode and retrieve trouble codes from BCM-controlled electronic instrument clusters.
- Perform self-diagnostic tests on the electronic instrumentation system.
- Determine faults as indicated by the self-test.
- Diagnose computer-driven speedometer and odometer instrumentation malfunctions.
- Test magnetic pickup speed sensors.
- Test optical-type speed sensors.
- Determine the cause of constant low gauge readings in computer-controlled instrumentation systems.
- Diagnose and locate the cause for constant high gauge readings in computer-controlled instrumentation systems.
- Diagnose head-up display operation.

Terms To Know

Electromagnetic gauges

Odometer

Prove-out circuit

Quartz swing needle

Resistive shorts

Tachometer

Introduction

The instrument panel gauges and warning lamps monitor the various vehicle operating systems and provide information to the driver about their operation. An open circuit is a common cause of gauge or warning lamp issues. The open can occur in the wiring, in the sensor, or in the printed circuit board. Other issues include improper gauge calibration, loose connections, excessive circuit resistance, and defective bulbs, gauges, or sending units.

In this chapter you will learn how to diagnose the gauge, lamp, and sending unit of the various styles of conventional instrument panels. You will learn how to remove the instrument panel to replace the printed circuit, and how to repair the speedometer cable core.

This chapter will also introduce you to common service procedures used to diagnose and repair computer-driven instrumentation systems. These systems include the speedometer, odometer, fuel, and engine instrumentation. The computer-driven instrument cluster uses a microprocessor (µP) to process information from various sensors and to control the gauge display. Depending on the manufacturer, the microprocessor can be a separate computer that receives direct information from the sensors and makes the calculations, or the body control module (BCM) is used to perform all functions. The computer may control a digital or a quartz swing needle instrument cluster.

It's out of the scope of any textbook to cover the service procedures for every type of instrument cluster. To illustrate, in one model year alone, Chrysler offered two different electronic clusters (Huntsville and Motorola). The Huntsville cluster was available in four different variations, one with a message center, one with a trip computer, and two options that offered tachometers. In addition, the system could also have a 24-voice alert function. In the same year, Ford offered five different electronic clusters, and each division of General Motors (GM) offered its own electronic instrument cluster. Add to this the many different import vehicles that use their own systems. This chapter will familiarize you with general procedures; however, it's important to remember that each system uses its own diagnostic procedures.

Usually, your responsibility is to isolate the faulty component and replace it. Most instrument panel components are not repaired or serviced in the shop but sent back to the manufacturer or to a specialty shop for rebuilding.

Instrument Panel and Printed Circuit Removal

It may be necessary to remove the instrument panel to replace defective gauges, lamps, or printed circuits. Before removing the instrument panel, always disconnect the negative battery cable. Consult the service information for the procedure of the vehicle you are working on. The following is a common method of removing the instrument cluster and printed circuit:

Special Tools
Fender covers
Safety glasses
Battery terminal
Pliers

1. Place fender covers on the fenders and disconnect the battery negative cable.
2. Remove the retaining screws to the steering column cover and remove the cover.
3. Remove the finish panel retaining screws and remove the finish panel.
4. Remove the retaining bolts that hold the cluster to the dash.
5. Gently pull the cluster away from the dash, reach behind the instrument panel, and disconnect the speedometer cable (if applicable) and the cluster feed plug(s) from the printed circuit receptacle. Be careful not to damage the printed circuit.
6. Remove the instrument voltage regulator (IVR) and all illumination and indicator lamp sockets (**Figure 12-1**).
7. Remove the charging system warning lamp resistor if applicable.
8. Remove all printed circuit attaching nuts and remove the printed circuit.

The IVR provides a constant voltage to the gauge regardless of the voltage output of the charging system.

Many instrument clusters must be replaced as a unit and don't allow for servicing individual gauges, lamps, or the circuit board. Since the cluster may also be a module, it's important to check the service procedures before replacing the cluster. It may be necessary to retrieve data from the original cluster and transfer it to the new cluster. This data may include mileage, vehicle identification number (VIN), theft deterrent codes, vehicle build configurations, and so forth. If the data transfer procedure isn't properly done, the vehicle may not start.

Figure 12-1 Instrument panel printed circuit board.

> **AUTHOR'S NOTE** A common issue with Dodge trucks is the tachometer behaving erratically. Flexing of the circuit board over time stresses the solder connections of the driver chips on the back of the board. Removing the printed circuit isn't recommended, but heating the pins of the driver chips with the tip of a small soldering iron to reflow the solder renews the connections. It only takes about a second or two at each pin to get the necessary flow. Don't leave the tip in place too long. This renews the instrument cluster without having to perform odometer and vehicle build configuration procedures. Perform this repair only If authorized by your manager.

BCM Diagnostics

The BCM may be capable of running diagnostic checks of the electronic instrument cluster to determine if a fault is present. Values received from monitored functions outside of programmed parameters set a diagnostic trouble code (DTC). Retrieval of this code can aid in troubleshooting.

Using the scan tool to force gauge movement to selected positions separates the gauge from the input side of the system. If the gauge works using this scan tool function, the input to the BCM needs further testing.

Self-Diagnostics

Classroom Manual
Chapter 12, page 367

Special Tool
Scan tool

Most instrument panel display modules have a diagnostic mode within their programming. The diagnostic mode allows the module to isolate faults within the instrument panel cluster. In most systems, if the module cannot complete its self-diagnostic test, the fault is within the module and must be replaced. Successful completion of the self-diagnostic test indicates the problem isn't in the module. The following are examples of self-diagnostic procedures.

Diagnosis of a Typical Electronic Instrument Cluster

All electronic instrument clusters (EICs) are sensitive to static electricity damage, and EIC cartons usually have a static electricity warning label. When servicing EICs:

1. Don't open the EIC carton until ready to install the component.
2. Ground the carton to a known good ground before opening the package.
3. Always touch a known good ground before handling the component.
4. Don't touch EIC terminals with your fingers.
5. Follow all service precautions and procedures in the vehicle manufacturer's service information.

Prove-Out Display. Most EICs have a prove-out display each time the ignition is turned on. During this display, all the EIC segments are illuminated and then turned off momentarily (**Figure 12-2**). The EIC returns to normal display after the prove-out. If the EIC doesn't illuminate during the prove-out display, check the EIC's power supply and ground circuits. If these are good, replace the EIC.

If some of the segments don't illuminate during the prove-out display, the EIC is defective and needs replacement. The turn signal and high-beam indicators may not illuminate during the prove-out mode. Other indicator lights remain on when the EIC display is turned off momentarily in the prove-out mode. After completion of the prove-out mode, the indicator lights go out shortly, and the EIC returns to normal display.

Function Diagnostic Mode. Some EICs have a function diagnostic mode that displays diagnostic information in the display readings if certain defects occur in the system. For example, if the coolant temperature sender has a shorted circuit, the two top and bottom bars are illuminated in the temperature gauge and the ISO symbol is extinguished (**Figure 12-3**). If the engine coolant never reaches normal operating temperature, or the coolant temperature sender circuit has an open circuit, the bottom bar in the temperature gauge and the ISO symbol illuminates.

If the fuel gauge sender develops a short or open circuit, the two top and bottom bars in the fuel gauge illuminate; however, the ISO symbol doesn't illuminate. A shorted fuel gauge sender causes CS to be displayed in the fuel remaining or distance to empty displays. An open fuel gauge sender circuit will illuminate "CO" in the fuel remaining

Caution
Even the slightest physical shock to the VFD display can damage it. When handling EICs, don't drop or jar them.

Classroom Manual
Chapter 12, page 369

Caution
When servicing EICs, follow all service precautions related to static discharge in the vehicle manufacturer's service information to avoid EIC damage.

The diagnostic procedure for EICs varies depending on the vehicle make and model year. Always follow the diagnostic procedures in the vehicle manufacturer's service information.

Figure 12-2 All EIC segments are illuminated during the prove-out display.

Engine temperature sensor input
short circuited lights two top and
bottom bars and extinguishes
temperature ISO ( ) symbol and legend

Figure 12-3 EIC function diagnostic mode.

and the distance to empty displays. When the function diagnostic mode indicates short or open circuits in the inputs, use a voltmeter and ohmmeter to test the circuit with the indicated problem. The vehicle manufacturer's service information will provide instructions for performing these voltmeter and ohmmeter tests.

When the word *ERROR* appears in the odometer display, the EIC computer cannot read valid odometer information from the nonvolatile memory chip.

Special Test Mode. Many EICs have a special test mode to determine if the display is working properly. Typically this mode is entered by pressing buttons on the EIC. For example, some systems enter this mode by pressing the E/M and SELECT buttons simultaneously while placing the ignition from the OFF to the RUN position. When this action is complete, a number appears in the speedometer display and two numbers illuminate in the odometer display. The gauges and message center displays are not illuminated. If any of the numbers are flashing in the speedometer or odometer displays, the EIC is defective and requires replacement.

⚠ WARNING **If the odometer has been repaired or replaced and it cannot indicate the same mileage as before its removal, most municipalities require an odometer mileage label to be attached to the left front door frame. Failure to comply with this procedure could lead to court action.**

Electromechanical Cluster Self-Test

Classroom Manual
Chapter 12, page 357

Like the EIC test, button pushes and cycling of the ignition switch initiates the electromechanical instrument cluster (MIC) self-test. **Photo Sequence 24** illustrates a procedure for entering the MIC self-test and the resulting displays as the test progresses. During the actuator test, the circuitry positions each gauge needle at various calibration points, illuminates each of the segments in the VFD units, and turns all of the indicators ON and OFF again.

Confirmation of proper cluster operation indicates that there may be problems elsewhere in the system. Use the service information to determine where each message or hard wire input originates. Problems with incorrect or erratic gauge and indicator operation may be caused by faults in the data bus, the originating control module, or the inputs to one of the control modules. Use a diagnostic scan tool to diagnose these components.

PHOTO SEQUENCE 24
Self-Test of a MIC

P24-1 With the ignition in the OFF position, press and hold the trip odometer reset button.

P24-2 While continuing to hold the reset button, place the ignition in the RUN position.

P24-3 Release the odometer/trip odometer switch button. The instrument cluster will perform a bulb check of each LED indicator, and each gauge needle sweeps to its high calibration point.

P24-4 The VFD segments and LED indicators remain illuminated as the gauges move to their next calibration points.

P24-5 As the gauges move to additional calibration points, the VF displays perform a segment test.

P24-6 The gauges will continue to move to each calibration point.

P24-7 The gauges then move to their low calibration points.

P24-8 The VF display continues to perform its segment test.

P24-9 At the completion of the test, the cluster will automatically exit the self-diagnostic mode and return to normal operation.

Chrysler Electromechanical Cluster Self-Test

Many Chrysler MICs provide a self-test function of the cluster. This test does not check any of the inputs, only the cluster.

To enter self-diagnostics, the ignition must be in the LOCK position. Push and hold the TRIP and RESET buttons on the cluster at the same time. While continuing to hold these buttons, place the ignition in the RUN position. Note that the cluster will illuminate in the UNLOCK position but will not activate self-diagnosis. Continue to hold the two buttons until CODE is displayed in the odometer. Release the buttons. If there are any fault codes, they are displayed in the odometer. A code 999 means there are no faults. If fault codes are present, use the correct diagnostic manual to diagnose the system.

After the codes are displayed, the cluster goes through a series of tests as follows:

Check 0. Tests all of the vacuum fluorescent (VF) display segments in the odometer and PRND3L. All segments should illuminate.
Check 1. Tests the operation of all gauges. The gauge swing needles will move to programmed values.
Check 2. Illuminates each odometer VF segment individually.
Check 3. Tests the PRND3L display.
Check 4. Illuminates all of the warning lamps controlled by the MIC.

Observe the operation of the MIC during each test. Replace or repair the MIC if any of these tests fail proper operation.

Special Tool

Scan tool

An additional feature of this cluster is the ability to calibrate the gauges. Calibration requires the use of a factory level scan tool. All gauges are calibrated in the same manner. The following steps provide a guide to recalibrating the speedometer.

1. Plug the scan tool cable into the DLC. The ignition switch does not need to be in the RUN position. However, the bus must be active. Opening a door will awake the BCM.
2. Once the scan tool powers up the MAIN MENU is displayed.
3. Select BODY.
4. Select ELECTRO/MECH CLUSTER from the BODY menu.
5. Select MISCELLANEOUS.
6. Select CALIBRATE GAUGES.
7. The scan tool will ask if the cluster has a tachometer. Answer with the YES or NO key.
8. The scan tool will ask if the vehicle is a diesel. Answer with the YES or NO key.
9. The scan tool will ask if the cluster units are in MPH. Answer with the YES or NO key.
10. Place the ignition switch into the UNLOCK position.

This places the scan tool in the mode to run gauge calibration. The first gauge to be calibrated will be the speedometer. The scan tool screen displays that it's sending a signal to the MIC to set the mph at 0. If the needle isn't aligned with the 0, then use the up or down arrow keys to move the needle until it's aligned. With the gauge calibrated to 0, press the enter key, and the scan tool will move to the next calibration unit. The next unit is 20 mph. Follow the same procedure to align the needle with the 20-mph mark on the cluster. After each calibration, press the enter key to move to the next unit. The other calibration units are 55 and 75 mph.

After calibrating all the gauges, the scan tool instructs the MIC to write the new values to memory.

Ford Electronic Cluster Self-Diagnostic Test

The electronic cluster is capable of indicating a fault and providing an explanation of the cause. Use **Figure 12-4** as a guide to the function of the gauges when in diagnostic mode. Use the gauge display to determine the nature of the fault. Then refer to the service information for diagnostic charts to locate the problem.

Oil gauge	H ⬒ N o r m L ⬒	Oil pressure sensor input short circuited lights two top and bottom bars and extinguishes oil can ISO symbol	H N o r m L	Low oil pressure warning or oil pressure sensor input open circuited lights bottom bar and flashes ISO symbol
Temp gauge	H ⬒ N o r m C ⬒	Engine temperature sensor input short circuited lights two top and bottom bars and extinguishes ISO symbol	H N o r m C	Cold engine temperature indication or engine temperature sensor input open circuited lights bottom bar and ISO symbol
Fuel gauge	H ⬒ N o r m L ⬒	Fuel level sender input short circuited or open lights two top and bottom bars and extinguishes ISO symbol	CO CS	Fuel level sender input short or open circuited displays "CS" (short) or "CO" (open) in driver information center
Odometer	55 ERROR			Odometer malfunction displays ERROR in odometer display
Fuel computer	FFS			Fuel flow signal short or open circuited displays FFS in driver information center

Figure 12-4 Gauge readout indicates the nature of the fault when the system is in the diagnostic mode.

Speedometer Diagnostics and Repair

Speedometer complaints range from chattering noises when cold, to inaccurate readings, to not operating at all. Diagnosis and repair of the speedometer depends on the type: conventional or electronic.

Conventional Speedometer

Issues of noise, erratic, or inaccurate readings in instrument clusters using conventional speedometers can often be corrected by lubricating the cable with an approved lubricant. If the cable is dry, it will bind as it attempts to rotate in the housing. However, just applying lubricant may only stop the noise temporarily. It's a good practice to remove the cable core to clean and inspect it before adding lubricant. If the cable is well lubricated and the problem is still present, check the condition of the speedometer drive gear. If the cable and the gears are not the problem, the speedometer head may be faulty and need to be replaced.

To remove the cable core, disconnect the speedometer cable assembly from the back of the speedometer head. For most vehicles, this is done by reaching behind the instrument panel and pressing down on the flat surface of the plastic quick connect. Some vehicle makes will require the removal of the instrument panel to gain access to the speedometer cable.

Classroom Manual
Chapter 12, page 361

Special Tools
Fender covers
Safety glasses
Battery terminal pliers
Terminal pliers

Caution

It's possible to short out wires while reaching behind the instrument panel. Disconnect the battery negative cable before removing the speedometer cable assembly.

⚡ **Caution**

Changing tire size and differential gear ratios from original equipment specifications results in speedometer inaccuracy. In some States it's illegal to calibrate the speedometer unless performed by a shop that is qualified to perform this task.

⚡ **Caution**

Federal and state laws prohibit tampering with the correct mileage as indicated on the odometer. If the odometer needs replacing, it must display the reading of the original odometer, or a door sticker must be installed indicating the reading of the odometer when it was replaced (**Figure 12-6**).

Classroom Manual
Chapter 12, page 361

Figure 12-5 Speedometer connection at the transaxle.

CUSTOMER CARE The speedometer cable should be lubricated every 10,000 miles. This practice reduces speedometer cable problems that cause noisy, erratic, or inaccurate readings.

It may be possible to remove the core by pulling it out of the housing. If core removal cannot be done this way, disconnect the speedometer retainer from the transaxle (**Figure 12-5**). Pull the core out of the speed sensor. With the cable core removed from the housing, clean it with solvent and wipe it dry. Place the core on a flat surface and stretch it out straight. Roll it back and forth while looking for signs of kinks or other damage. Replace the core if it's damaged.

Figure 12-6 An odometer repair label.

Tachometer   Speedometer

Voltmeter

Engine temperature

Warning lights

Fuel level

Oil pressure

Figure 12-7 Quartz swing needle speedometer used with conventional gauges.

If the speedometer assembly needs to be replaced, usually a new **odometer** is included with the assembly. The odometer is a mechanical counter in the speedometer unit that indicates the total miles driven by the vehicle. Be sure to follow the manufacturer's procedures for setting the odometer to the correct reading.

Electronic Speedometers and Odometers

Diagnosis of the sending units and input circuits to the gauges of electromechanical instrument clusters typically follow normal testing procedures. The following is an example of diagnosing a **quartz swing needle** electronic speedometer and odometer gauges of an electromechanical cluster (**Figure 12-7**). Computer-driven quartz swing needle displays are similar in design to the air-core electromagnetic gauges used in conventional analog instrument panels.

Classroom Manual
Chapter 12,
pages 361, 367, 369

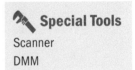 **Special Tools**
Scanner
DMM

> ⚙ SERVICE TIP If the electronic speedometer uses a cable-driven sensor, service the cable as described in the previous section.

Most electronic speedometer and odometer systems receive their input from the vehicle speed sensor (VSS). If there is a fault with the speed sensor, other systems (such as cruise control) will also be affected. While test driving the vehicle, attempt to activate the cruise control system to determine if it's operating properly. If the cruise control system fails, the vehicle speed sensor, its circuit, BCM, the ECM, or the data bus network can be the problem.

Some systems may provide for the replacement of the stepper motor or odometer chip separate from the cluster. Always refer to the service information for the vehicle you are diagnosing.

Generally, if the BCM and/or cluster module pass their self-diagnostic tests, the fault will be in the speed sensor circuit. Common test procedures for the speed sensor are presented later in this chapter.

If the speedometer isn't operating, but the odometer works properly, the fault is in the instrument cluster. Likewise, if the speedometer operates but the odometer fails, the fault is in the cluster. In either case, replace the cluster. If the speedometer and/or odometer are inaccurate, or both don't operate, check the following items:

1. If the system uses an optical vehicle speed sensor, check the speedometer cable for kinks, twists, or other defects that will cause an inaccurate reading.
2. See if there are shorts in the wiring circuit of the speed sensor.
3. Make sure of proper gear ratio and tire size. Both of these items will affect correct speedometer and odometer operation. Changing tire size from that intended by the manufacturer has the same effect as changing gear ratios in the differential.

If the speedometer and/or odometer display illuminates but remains at 0 (or any other digit), or operates erratically or intermittently, check the connector at the speed sensor for proper installation and corrosion. Next, check the wiring circuit for any shorts or opens. If these checks don't isolate the problem, test the speed sensor. Testing of the speed sensor will depend on the type of sensor used.

Classroom Manual
Chapter 12, pages
362, 368

Special Tools

Fused jumper wires
Ohmmeter
Lab scope

Magnetic Pickup Speed Sensor Testing. Disconnect the wire connector at the vehicle speed sensor. With the ignition placed in the RUN position, use a fused jumper wire to make and break the connection between the two wires (**Figure 12-8**). Doing this should cause the speedometer display to change. Changing the rate of speed at which you make and break the connection should show different speeds in the display. The faster you make and break the connection, the higher the speedometer reading.

If there is no change in the speedometer display, check for opens and shorts in the sensor circuit. If the cluster passes its self-test (and you did not skip any steps), this is the only area in which the fault can be located.

If the speedometer changes speed, the problem is in the speed sensor. To test the speed sensor, remove it from the transaxle. Connect an ohmmeter to the connector terminals of the sensor and select the lowest scale (**Figure 12-9**). Rotate the sensor gear while observing the ohmmeter. The ohmmeter should detect distinct pulses. Compare the number of pulses per revolution with specifications. Also, compare the resistance value with specifications. The sensor is good if the number of pulses and resistance values are within specifications.

Figure 12-8 Testing the magnetic pickup speed sensor circuit. Making and breaking the connection should produce a reading in the speedometer window.

Caution:
Do not use a test light to check the distance sensor. Damage to the sensor may result.

Figure 12-9 Using an ohmmeter to test the speed sensor.

Figure 12-10 Optical speed sensor circuit.

Optical Speed Sensor Testing. Disconnect the speedometer cable at the transaxle and rotate the cable in its housing as fast as possible. If the speedometer display operates properly, check the speedometer pinion and drive gear for damage.

Classroom Manual
Chapter 12, page 367

If there is no speedometer operation, check for a broken speedometer cable. This can usually be determined by feeling for resistance while turning the cable by hand. Little resistance indicates a broken cable. Excessive resistance indicates a damaged cable or sensor head.

If the cable is good, the problem is in the sensor or in the wiring between the sensor and the speedometer. Follow the manufacturer's procedure for removing the instrument cluster. Connect a digital multimeter (DMM) to read the pulsed speed signal from the sensor (**Figure 12-10**). Rotate the speedometer cable while observing the voltmeter. Compare the pulses per revolution and pulse output values with specifications. Replace the sensor if the values are not within specifications.

Hall-Effect Sensors. Diagnostics of the Hall-effect sensor were covered In Chapter 9. As discussed, using a lab scope provides a fast and accurate test of the switch. Review Chapter 9 for the test procedures of this type of speed sensor.

Classroom Manual
Chapter 12, page 369

🔧 SERVICE TIP A reversible, variable speed drill can be used to rotate the speedometer cable if you are sure there are no twists or kinks in the cable.

CUSTOMER CARE Often, inaccurate speedometers are caused by improper tire sizes installed on the vehicle. The larger the tire diameter over original equipment, the slower the speedometer will read. Not all manufacturers provide for entering different tire sizes into the computer, and aftermarket "black boxes" may be illegal to install (depending on state and local ordinances). In most cases, it's in the customer's best interest to keep original equipment size tires on the vehicle.

Tachometer

Classroom Manual
Chapter 12, page 371

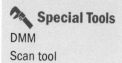

Special Tools

DMM
Scan tool

The **tachometer** is a gauge instrument used to display the engine's speed in revolutions per minute (rpm). Most electrically operated tachometers receive their reference pulses from the ignition system. If the tachometer is faulty, it requires replacement since there is no servicing of the meter itself. **Figure 12-11** illustrates a troubleshooting chart used to diagnose the electrically operated tachometer.

Computer-driven tachometers receive their signals from the crankshaft position sensor (CKP). If the tachometer isn't functioning, but the engine starts, check the circuit

TACHOMETER INOPERATIVE, ERRATIC, WRONG OPERATION

Test step	Result	Action to take
1. **Check operation** Check tachometer operation.	⊗	Go to 2.
	OK	Test complete. Check for intermittent operation.
2. **Check fuse** Check tachometer fuse.	⊗	Replace fuse and determine cause.
	OK	Go to 3.
3. **Check wiring** Check connections and wiring in engine compartment and at instrument cluster.	⊗	Repair connections or wiring.
	OK	Go to 4.
4. **Check resistance and voltage** Disconnect battery. Remove instrument cluster and make resistance and voltage checks at lower wire harness connector as follows: (1) Check pin 5 resistance to ground. Should read 1 Ω or less. (2) Check pin 17 resistance to negative terminal of ignition coil. Should read 15 Ω or less. (3) Connect battery. Turn ignition switch to RUN position. Check for +12 V at pin 14. Turn ignition off and disconnect battery.	⊗	Repair wiring for open or high resistance.
	OK	Go to 5.
5. **Check fasteners** Check for loose fasteners on rear of instrument cluster, or damaged printed circuit.	⊗	Tighten fasteners/Replace printed circuit.
	OK	Replace tachometer.

Figure 12-11 Troubleshooting chart for testing the tachometer.

from the sensor to the instrument cluster. If the instrument cluster receives the signal from the data bus, use a scan tool to determine if the signal is being sent correctly. Usually, the powertrain control module (PCM) monitors the CKP sensor and sends the rpm signal over the data bus. If the message is sent first to the BCM, then to the instrument cluster, check the input message to both modules. If the bus message is being sent properly to these two modules, the gauge or the instrument cluster is the problem. Perform the self-diagnostic test, or scan tool activation test, for the tachometer. If the tachometer fails to operate during this test, replace the gauge or cluster. Some manufacturers don't allow for individual gauge replacement and require the replacement of the entire instrument cluster as a unit.

Conventional Instrument Cluster Gauge Diagnosis

Electromagnetic gauges produce needle movement by magnetic forces. With the exception of the voltmeter and ammeter, all **electromagnetic gauges** use a variable resistance sending unit. The types of tests performed depend on the nature of the problem and if the system uses an IVR.

> **SERVICE TIP** It's common for the fuel gauge sender unit to get corrosion on the ground wire connection. Before replacing the sending unit, clean the ground connections and test for proper operation.

Single Gauge Failure

If the gauge system does not use an IVR, check the gauge for proper operation as follows:

1. Check the fuse panel for any blown fuses. The gauge that isn't operating may share a fuse with another circuit that is separate from the other gauges.
2. Disconnect the wire connector from the sending unit of the malfunctioning gauge.
3. Check the terminal connectors for signs of corrosion or damage.
4. Use a test light to confirm that voltage is present to the connector with the ignition switch in the RUN position. If the test light does not illuminate, check the circuit back to the gauge and battery.
5. Connect a 10-ohm resistor in series with the lead from the gauge to the sending unit. Connect the lead to ground (**Figure 12-12**).
6. With the ignition switch in the RUN position, watch the gauge. Depending on the gauge design, the needle should indicate either high or low on the scale. Check the service information for correct results.
7. Remove the 10-ohm resistor and replace it with a 73-ohm resistor between the sensor lead and ground. Repeat step 6.
8. If the test results are in the acceptable range, the sending unit is faulty.
9. If the gauge did not operate properly in steps 5 and 7, check the wiring to the gauge. If the wiring is good, replace the gauge.

If the gauge circuits use an IVR, follow steps 1 through 4 as described. The test light should flicker on and off. If it does not illuminate, reconnect the sending unit lead and check for voltage at the sender unit side of the gauge (**Figure 12-13**). If there is voltage at this point, repair the circuit between the gauge and the sending unit. If there is no voltage at this location, test for voltage at the battery side of the gauge. If voltage is present

Classroom Manual
Chapter 12, page 357

Caution

These gauges are called analog because they use needle movement to indicate current levels. However, many modern instrument panels use computer-driven analog gauges that operate under different principles. It's important that you follow the manufacturer's procedures for testing the gauges or gauge damage will result.

Special Tools
DMM
12-volt test light
Fender covers
Safety glasses
Shop oil pressure
Gauge

Figure 12-12 Testing gauge operation by putting the gauge lead to ground through a resistor. The resistor protects the circuit.

Figure 12-13 Checking for regulated voltage on the sender unit side of the gauge.

at this point, the gauge is defective. If no voltage is present at this terminal, continue to check the circuit between the battery and the gauge.

If the IVR is working properly and voltage is present at the sender unit, follow steps 5 through 9.

> **SERVICE TIP** It's unlikely that all of the gauges would fail at the same time. If the diagnostic tests indicate defective gauges, bench test them before replacing them. Use an ohmmeter to check the resistance. Most electromagnetic gauges should read between 10 and 14 ohms. On systems that don't use an IVR and all of the gauges are defective, check the charging system for excessive output.

Figure 12-14 Check the last common connection in the circuit for voltage.

Multiple Gauge Failure

If all gauges fail to operate properly, begin by checking the circuit fuse. Test for voltage to the fuse. If voltage isn't present at this point, the fault is between the fuse and the battery. Remember, most systems supply battery voltage to the instrument panel gauges through the ignition switch. If voltage is present at the fuse, continue through the circuit by testing for voltage at the last common circuit point (**Figure 12-14**). If voltage isn't present at this point, work toward the fuse to find the fault. Keep in mind that this common connection point may be on the printed circuit board. If this is the case, test for voltage at the instrument panel connector first.

If the system uses an IVR, use a voltmeter to test for regulated voltage at a common point to the gauges (**Figure 12-15**). If the voltage is out of specifications, check the ground circuit of the IVR. If that is good, replace the IVR. If there is no voltage present

Special Tools

12-volt test light
10-ohm resistor
73-ohm resistor
Fender covers
Safety glasses

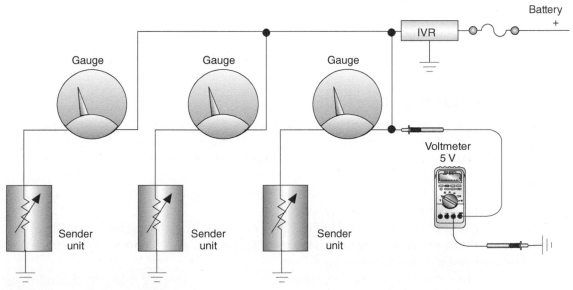

Figure 12-15 Testing for correct IVR operation.

at the common point, check for voltage on the battery side of the IVR. If voltage is present at this point, then replace the IVR. If battery voltage isn't present on the battery side of the IVR, the problem is in the circuit between the fuse and the IVR.

If regulated voltage is within specifications, test the printed circuit from the IVR to the gauges. If there is an open in the printed circuit, replace the board.

Electronic Gauge Diagnosis

Classroom Manual
Chapter 12,
pages 361, 366

Special Tools
Fused jumper wires
DMM

The following are guidelines for testing the individual gauges for proper operation. Do these tests if the self-diagnostic test indicates no problems with the cluster or the module. Test procedures will vary between manufacturers. To properly troubleshoot the gauges, you will need the manufacturer's diagnostic procedure, specifications, and circuit diagram.

> **AUTHOR'S NOTE** Electronic instrument clusters rely on data communication with other models. This applies to both gauge and warning light operations. Using the scan tool to perform a diagnostic module scan to determine if "Loss of Communication" fault codes are recorded may quickly identify the cause of faulty cluster operation. Diagnosing and repair of the data bus networks were covered in Chapter 10.

Gauge Reads Low Constantly

A gauge that constantly remains at its lowest reading and remains there when the ignition is in the RUN position indicates an open in the gauge circuit. To locate the open, follow these steps:

1. Disconnect the wire harness from the sending unit.
2. Connect a fused jumper wire between the wire circuit from the gauge and ground.
3. Place the ignition in the RUN position. The gauge should indicate maximum.

If the gauge reads high, check the sending unit ground connection. If the ground is good, the sending unit is faulty.

If the gauge continues to read low, follow the circuit diagram for the vehicle you are servicing to test for opens in the circuit from the sending unit. If the circuit is good, test the control module following recommended diagnostic procedures.

If the gauge reads lower than it should at all times, this indicates excessive resistance in the circuit. In addition, the gauge itself may be faulty. Performing a gauge self-test or a scan tool bi-directional control test of the gauge will confirm if the gauge is functioning properly.

Gauge Reads High Constantly

A gauge that reads at the maximum limit when the ignition is in the RUN position indicates a short to ground in the circuit, or a faulty gauge. To test the gauge, use the self-test feature or the scan tool's bi-directional control feature.

Test the circuit by disconnecting the wire harness at the gauge sending unit. Place the ignition switch in the RUN position while observing the gauge. If the gauge reads low, the sending unit is faulty and needs replacement.

If the gauge continues to read high, use the circuit diagram to test for shorts to ground in the circuit from the sending unit. If the circuit is good, test the control module following recommended diagnostic procedures.

A resistive short to ground can cause the gauge to read higher than actual conditions would indicate. Check the wiring harness for indications of worn, cracked, or frayed insulation. Also, checks connectors for corrosion that may cause a short to an adjacent circuit.

> 🛠 **SERVICE TIP** Although some service information procedures don't require it, connecting a 10-ohm resistor into the jumper wire is a good practice when performing these tests. This prevents a nonresistive short to ground, yet does not noticeably affect gauge operation.

Inaccurate Gauge Readings

Faulty sending units usually cause inaccurate gauge readings. To test the operation of the gauge, you will need the manufacturer's specifications concerning resistance values as they relate to gauge readings. Gauge testers are available to test the units as different resistance values are changed.

Other reasons for inaccurate gauge readings include poor connections, **resistive shorts**, and poor grounds. Resistive shorts are shorts to ground that pass through a form of resistance first. Also, look for damage around the sending unit. For example, a damaged fuel tank can result in inaccurate gauge readings.

Gauge Sending Units

Three types of sending units are associated with electromechanical gauges: a thermistor, a piezoresistive sensor, and a mechanical variable resistor. Most of these can be tested before replacement to confirm the fault.

Testing the fuel level sending unit does not always require removing it from the tank. If testing it in the tank, add and remove fuel to change the level. If the sending unit can be removed easily from the tank, bench testing the unit is the better option. **Photo Sequence 25** illustrates how to bench test a mechanical variable resistor sending unit.

To test the coolant temperature sensing unit, use an ohmmeter to measure the resistance between the terminal lead and the ground terminal (**Figure 12-16**). The resistance value of the variable resistor should change in proportion to the coolant temperature. Check the test results with the manufacturer's specifications.

To test a piezoresistive sensor sending unit used for oil pressure gauges, connect the ohmmeter to the sending unit terminal and ground (**Figure 12-17**). Check the resistance

Classroom Manual
Chapter 12,
pages 364, 372

🔧 **Special Tools**

12-volt test light
10-ohm resistor
73-ohm resistor
Fender covers
Safety glasses
DMM

Figure 12-16 Testing a thermistor with an ohmmeter.

Figure 12-17 Using an ohmmeter to test a piezoresistive sensor.

with the engine off and compare to specifications. Start the engine and allow it to idle. Check the resistance value and compare to specifications. Before replacing the sending unit, connect a shop oil pressure gauge to the engine to confirm that the lubrication system produces adequate oil pressure.

Warning Lamps

Classroom Manual
Chapter 12, page 377

A warning light may indicate low oil pressure, high coolant temperature, a defective charging system, or a brake failure. Unlike gauge sending units, the sending units for a warning light are nothing more than simple switches. The style of the switch can be either normally open or normally closed, depending on the monitored system.

PHOTO SEQUENCE 25
Bench Testing the Fuel Level Sender Unit

P25-1 Tools required to perform this task: DMM, jumper wires, and service information.

P25-2 Select the ohmmeter function of the DMM.

P25-3 Connect the negative test lead of DMM to the ground terminal of the sender unit.

P25-4 Connect the positive test lead to the variable resistor terminal.

P25-5 Holding the sender unit in its normal position, place the float rod against the empty stop.

P25-6 Read the ohmmeter and check the results with specifications. UNF p 581-7

P25-7 Slowly move the float toward the full stop while observing the ohmmeter. The resistance change should be smooth and consistent.

P25-8 Check the resistance value while holding the float against the full stop. Check the results with specifications.

P25-9 Check that the float isn't filled with fuel, distorted, or loose.

It isn't likely that all of the warning lights would fail at the same time. Check the fuse if all of the lights are not operating properly. Next, check for voltage at the last common connection. If voltage isn't present, trace the circuit back toward the battery. If voltage is present at the common connection, test each circuit branch in the same manner as described here for individual lamps.

To test a faulty warning lamp on a system with a normally open switch (sending unit), place the ignition in the START position. The **prove-out circuit** should light the warning lamp. A prove-out circuit completes the warning light circuit to ground through the ignition switch when it's in the START position. The warning light is on during engine cranking to indicate to the driver that the bulb is working properly. If the light does not come on during the prove-out, disconnect the sender switch lead (**Figure 12-18**). Use a fused jumper wire to connect the sender switch lead to ground. With the ignition in the RUN position, the warning lamp should light. If the lamp illuminates, test the prove-out circuit for an open. If the light does not come on, either the bulb is burned out or the wiring is damaged. Use a test light to confirm that there is voltage present at the sensor terminal connector. If there is voltage, the bulb is probably bad. If voltage isn't present to the sending unit, the bulb may be burned out. At this point, remove the instrument cluster and check for battery voltage to the panel connector. If voltage is present, substitute a known good bulb and test again.

If the customer states that the oil pressure light does not go out after starting the engine, use a shop oil pressure gauge to confirm adequate oil pressure.

If the system uses a normally closed switch, test in the same manner. However, there will not be a separate prove-out circuit.

Classroom Manual
Chapter 12,
pages 377, 379

 Special Tools

Fused jumper wires
Scan tool
DMM

Figure 12-18 Coolant temperature sensor switch and lead.

If the customer states that the warning light stays on, test in the following manner: disconnect the lead to the sender switch. The light should go out with the ignition switch in the RUN position. If it does not, there is a short to ground in the wiring from the sender switch to the lamp. If the light goes out, replace the sender switch.

Classroom Manual
Chapter 12, page 380

Diagnose computer-driven instrument cluster warning lamps using a scan tool. If the customer states that the lamp does not operate, use the scan tool's bi-directional controls to command the lamp's activation. If the lamp does light when commanded, the fault is in the signal to the instrument cluster. If the lamp does not come on when commanded, the problem is a faulty lamp, circuit board, or instrument cluster module.

Since the sending unit is used as an input and does not directly turn on the warning light, use the wiring diagram to determine if it is a normally open or normally closed switch. The type of switch will determine how the control module (and the scan tool) interprets the condition of the monitored system. For example, if a normally open switch is used to monitor oil pressure, the control module will receive a high sensed voltage when the engine is off, or if oil pressure drops too low. When the module receives a sensed voltage that has pulled low, the oil pressure has increased high enough to close the switch.

Computer-driven clusters may not have dedicated warning light prove-out circuits. The control module commands the light on when the ignition is in the RUN position. After starting the engine, the control module commands the warning lamp off if the monitored system is operating properly.

The sending unit switch circuit can be tested by creating an open and a short condition. Using the normally open switch system as an example, with the engine running, disconnecting the switch should cause the warning light to come on. To diagnose a concern that the light stays on all the time, first confirm that the engine's lubrication system is developing pressure. Next, disconnect the wire from the switch and short the wire to ground. The light should turn off. If it does, the switch is faulty. If the warning light is still on, there is an open between the switch and the control module.

If the customer states that the lamp is on all of the time, this indicates a problem in the monitored system. Use the scan tool to check for DTCs and follow the service information to diagnose the fault associated with the DTC. If there are no problems in the monitored circuit, you will need to diagnose the module that receives the lamp's sensor input. If no problem is found in this module, diagnose the instrument cluster module. Follow the service information procedures for all diagnostic tests.

Trip Computers

Classroom Manual
Chapter 12, page 376

Trip computer systems may have several different names. Traveler, Vehicle Information Center, and Driver Information Center are the most common.

Simple trip computers use inputs from the speed sensor and the fuel gauge to perform their functions. Like most electronic instrument clusters, trip computers usually have a self-diagnostic test procedure. Follow the manufacturer's procedure for initiating this test. If the trip computer passes the self-test, check the fuel gauge and speed sensor inputs.

Complex trip computers may receive inputs from several different areas of the vehicle (**Figure 12-19**). These systems require using specific manufacturer diagnostic charts, specifications, and procedures. Using the skills you have acquired, you should be able to perform these tests competently.

Maintenance Minders

Most manufacturers will display a warning lamp or message when an oil change or other services are due (**Figure 12-20**). The determination of when the service is due can be

Figure 12-19 Complex information centers use many inputs to monitor the vehicle's subsystems.

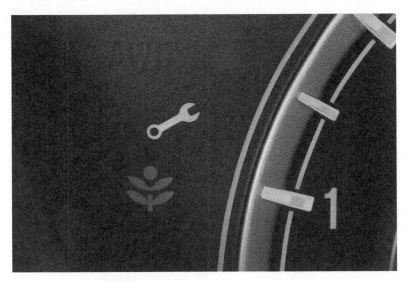

Figure 12-20 Maintenance minder indicator.

calculated strictly off of mileage driven; however, many systems consider many factors. Some systems will calculate the maintenance service requirement based on vehicle mileage, engine run hours, engine load, climatic conditions, and drive cycles.

Resetting the maintenance indicator may require the use of a scan tool. However, most manufacturers will provide an alternate method. Use the service information, or owner's manual, to determine the process for resetting the minder. Various methods in use range from a simple button press to complex activation of several inputs in a specific order. For example, one manufacturer requires placing the ignition in the RUN position and pressing the accelerator pedal three times within 10 seconds, followed by turning off the ignition.

Some maintenance minder systems may allow the vehicle operator to change the mileage intervals between services. Usually, this will not alter calculations that use factors other than just vehicle mileage. For example, even with the mileage interval set for 6,000 miles (9,656 Km), the minder may still come on at just 3,000 miles (4,828 Km) based on heavy engine loads in a cold climate.

Classroom Manual
Chapter 12, page 375

Special Tools

Scan tool

DMM

Back probe tools

Test light

Head-Up Display Diagnostics

We will look at the head-up display (HUD) used by General Motors as an example of typical diagnostic procedures for this system (**Figure 12-21**). The head-up display is a slave device to the instrument cluster and communicates directly with the instrument cluster via the serial data bus. The image information is received from the instrument cluster through a video cable. Since the HUD module is a slave, the system has limited self-diagnosis capacity. Head-up display DTCs are reported through the instrument cluster.

Improper multiplexed switch inputs will set DTC B361B. Use the scan tool to monitor the switch data. With the ignition ON, hold each of the available switch positions. While holding the switch position, the scan tool should report that switch as being activated.

If the scan tool parameters don't equal the expected result with the switch in the appropriate state, test the switch circuits. With the ignition ON back probe the switch harness connector terminal 3. Use a test light with the clamp attached to battery positive and locate the probe onto the backprobe tool. The test light should illuminate. Use a DMM to measure the voltage at the tip of the test light probe. The circuit has excessive resistance if the reading is greater than 0.02 volt. Move the backprobe tool to terminal 8 of the HUD module and retest. If the voltage is still high, test the HUD ground circuit. If the ground circuit is good, replace the HUD module.

If the switch return circuit is good, turn the ignition OFF and test for less than 1 volt between the switch signal circuit terminal 4 and ground. If the voltage exceeds 1 volt, disconnect the HUD module connector and measure the voltage at terminal 4 of the harness connector. If voltage is present, the circuit is shorted to voltage.

If voltage was not present at terminal 4 of the switch connector, turn the ignition to the RUN position and measure for battery voltage. If less than battery voltage, test the circuit between the HUD module and the switch for a short to ground and/or an open.

Figure 12-21 Head-up display schematic.

If battery voltage is present at switch connector terminal 4, remove the switch and test the resistance values between terminals 3 and 4 while each switch position is activated. If the resistance values don't match specifications, replace the switch. Diagnostics of the HUD module is within the instrument cluster. Other than testing for good voltage feed and ground circuits, there are no external diagnostics.

CASE STUDY

A customer brings their vehicle into the shop because the fuel level gauge remains on empty all the time, regardless of how full the tank is. The technician checks the fuse box for any blown fuses and finds that all are good. The technician disconnects the lead wire to the fuel tank sending unit and probes for voltage. The test light comes on when the ignition is in the RUN position. After connecting a 10-ohm resistor between the sending unit lead and ground, the gauge needle moved to the FULL position. Upon checking the service information, the technician determines this is normal operation. Before draining and removing the fuel tank, the technician uses a fused jumper wire to jump from the ground terminal of the sending unit to a known good ground on the frame. When the lead wire is reconnected and the ignition placed in the RUN position, the gauge works properly. The technician cleans the ground connections for the sending unit and returns the vehicle to the customer.

ASE-Style Review Questions

1. In a conventional instrument cluster, all of the gauges are inoperable. What is the **LEAST LIKELY** cause of this?

   A. Faulty IVR.

   B. Blown fuse.

   C. Shorted sending unit.

   D. Faulty printed circuit.

2. All of the following statements concerning gauge sending units are true **EXECPT**:

   A. A gauge can use a switch as a sensor.

   B. A gauge can use a thermistor as a sensor.

   C. A gauge can use a piezoresistive sensor.

   D. A gauge can use a variable resistor as a sensor.

3. The oil pressure warning light circuit with a normally open sending unit will not turn off with the engine running. A shop gauge confirms good oil pressure. What is the most likely cause?

   A. Faulty IVR.

   B. Damaged printed circuit board.

   C. Open in the circuit to the oil pressure switch.

   D. Short to ground in the wire to the oil pressure switch.

4. The tachometer of a computer-driven instrument cluster is inoperative.

   Technician A says if the gauge displays the commanded values while using scan tool bi-directional controls this indicates the gauge is operating properly.

   Technician B says if scan tool bi-directional controls fail to change gauge reading, the CKP sensor may be faulty.

   Who is correct?

   A. A only   C. Both A and B

   B. B only   D. Neither A nor B

5. *Technician A* says to use a voltmeter to test a thermistor sensor unit.

   Technician B says the thermistor sensor measures pressure.

   Who is correct?

   A. A only   C. Both A and B

   B. B only   D. Neither A nor B

6. *Technician A* says when the ignition switch is in the START position, the proving circuit should light the warning lamps.

   Technician B says the warning lamp should come on if the sender switch lead is grounded and the ignition switch is in the RUN position.

   Who is correct?

   A. A only   C. Both A and B

   B. B only   D. Neither A nor B

7. A digital speedometer constantly reads 0 mph.

   Technician A says the speed sensor may be faulty.

   Technician B says the throttle position sensor may have an open.

   Who is correct?

   A. A only   C. Both A and B

   B. B only   D. Neither A nor B

8. All gauges read low in a computer-controlled instrument cluster.

   Technician A says the connector to the cluster may be loose.

   Technician B says the cluster module may be at fault.

   Who is correct?

   A. A only   C. Both A and B

   B. B only   D. Neither A nor B

9. An electronically controlled instrument cluster is being diagnosed for no speedometer operation. Which of the following statements is most correct?

   A. If the BCM and/or cluster module passes its self-test, the fault is probably in the vehicle speed sensor circuit.

   B. If the speedometer does not work, but the odometer does, then the problem is probably in the vehicle speed sensor circuit.

   C. If the speedometer fails to move during the self-test, the problem is probably in the vehicle speed sensor circuit.

   D. All of these choices.

10. Ford diagnostic mode displays are being discussed.

    Technician A says a CO displayed in the fuel gauge window indicates that the fuel level is near empty.

    Technician B says a CS indicates an open in the sender circuit.

    Who is correct?

    A. A only   C. Both A and B

    B. B only   D. Neither A nor B

ASE Challenge Questions

1. The odometer of a vehicle equipped with a cable-driven speedometer is inoperative; however, the speedometer is working correctly.

   Technician A says that the speedometer cable may be faulty.

   Technician B says that the speedometer drive gear at the transmission may be stripped.

   Who is correct?

   A. A only

   B. B only

   C. Both A and B

   D. Neither A nor B

2. An electronic fuel gauge is reading higher than normal.

   Technician A says the wire leading to the gauge sender unit may be open.

   Technician B says that the resistance of the sender unit may be higher than normal.

   Who is correct?

   A. A only

   B. B only

   C. Both A and B

   D. Neither A nor B

3. The coolant temperature warning light is on whenever the ignition is in the RUN position; when the engine is started, the light remains illuminated.

   Technician A says that the sending unit wire may be shorted to ground.

   Technician B says that the coolant temperature switch may be electrically open.

   Who is correct?

   A. A only
   B. B only
   C. Both A and B
   D. Neither A nor B

4. The diagnosis of a digital instrument cluster missing two segments in its display is being discussed.

   Technician A says that the cluster has excessive ground circuit resistance.

   Technician B says that the cluster should be replaced.

   Who is correct?

   A. A only
   B. B only
   C. Both A and B
   D. Neither A nor B

5. A vehicle equipped with an electronic instrument cluster has the malfunction indicator lamp (MIL) illuminated; a speed sensor code (indicating the PCM does not "see" vehicle speed) is stored in the PCM. The speedometer works correctly; however, a scan tool connected to the PCM indicates 0 mph whenever the vehicle is in motion.

   Technician A says that the vehicle speed sensor may be faulty.

   Technician B says that the PCM may be faulty.

   Who is correct?

   A. A only
   B. B only
   C. Both A and B
   D. Neither A nor B

Name _____   Date _____

CHECKING A FUEL GAUGE

Upon completion of this job sheet, you should be able to diagnose an inaccurate or inoperative fuel gauge.

ASE Education Foundation Correlation ————————————————

This job sheet addresses the following **MLR** tasks:

A.1.   Research vehicle service information such as fluid type, vehicle service history, service precautions, technical service bulletins, and recalls including vehicles with advanced driver assistance systems (ADAS). **(P-1)**

A.2.   Identify electrical/electronic system components and configuration. **(P-1)**

A.5.   Demonstrate proper use of a digital multimeter (DMM) when measuring source voltage, voltage drop (including grounds), current flow, and resistance. **(P-1)**

A.6.   Demonstrate knowledge of the causes and effects from shorts, grounds, opens, and resistance problems in electrical/electronic circuits. **(P-1)**

A.9.   Use wiring diagrams to trace electrical/electronic circuits. **(P-1)**

F.1.   Verify operation of instrument panel gauges and warning/indicator lights; reset maintenance indicators as required. **(P-1)**

This job sheet addresses the following **AST** tasks:

A.1.   Research vehicle service information such as fluid type, vehicle service history, service precautions, technical service bulletins, and recalls including vehicles with advanced driver assistance systems (ADAS). **(P-1)**

A.2.   Identify electrical/electronic system components and configuration. **(P-1)**

A.5.   Demonstrate proper use of a digital multimeter (DMM) when measuring source voltage, voltage drop (including grounds), current flow, and resistance. **(P-1)**

A.6.   Demonstrate knowledge of the causes and effects from shorts, grounds, opens, and resistance problems in electrical/electronic circuits. **(P-1)**

A.9.   Use wiring diagrams during the diagnosis (troubleshooting) of electrical/electronic circuit problems. **(P-1)**

F.1.   Verify operation of instrument panel gauges and warning/indicator lights; reset maintenance indicators as required. **(P-1)**

F.2.   Inspect and test gauges and gauge sending units for causes of abnormal gauge readings; determine needed action. **(P-2)**

This job sheet addresses the following **MAST** tasks:

A.1.   Research vehicle service information such as fluid type, vehicle service history, service precautions, technical service bulletins, and recalls including vehicles with advanced driver assistance systems (ADAS). **(P-1)**

A.2.   Identify electrical/electronic system components and configuration. **(P-1)**

A.5.   Demonstrate proper use of a digital multimeter (DMM) when measuring source voltage, voltage drop (including grounds), current flow, and resistance. **(P-1)**

A.6.   Demonstrate knowledge of the causes and effects from shorts, grounds, opens, and resistance problems in electrical/electronic circuits. **(P-1)**

A.9.   Use wiring diagrams during the diagnosis (troubleshooting) of electrical/electronic circuit problems. **(P-1)**

F.1.   Verify operation of instrument panel gauges and warning/indicator lights; reset maintenance indicators as required. **(P-1)**

F.2.   Inspect and test gauges and gauge sending units for causes of abnormal gauge readings; determine needed action. **(P-1)**

We Support

Education Foundation

Tools and Materials

- A vehicle
- DMM
- Service information for the chosen vehicle
- Wiring diagram for the chosen vehicle
- Vehicle hoist

Procedure

1. From the procedures listed in the service information, describe the procedures for testing the fuel gauge and fuel gauge sending unit on this vehicle.

   _____

   _____

   _____

   _____

   _____

2. Locate the fuel gauge circuit wiring diagram and print out the schematic. Map the circuit from power to ground.

3. Follow the procedure to simulate a full fuel tank. Describe what happened when you followed this procedure; include your conclusions from this test in your answer.

   _____

   _____

   _____

   _____

4. Follow the procedure to simulate an empty fuel tank. Describe what happened when you followed this procedure; include your conclusions from this test in your answer.

   _____

   _____

   _____

   _____

Instructor's Evaluation:

Grade	Rubric
0	No exposure due to: ☐ Absence ☐ Other:
1	Student unsuccessfully attempted the task
2	Student completed the task with significant instructor assistance.
3	Satisfactory performance of task. The student was able to perform elements of the task with minimal instructor assistance.
4	Competent/Proficient. Able to perform all elements of the task successfully and independently.

Instructor's Notes:

Name _____   Date _____

TESTING AN OIL PRESSURE WARNING LIGHT CIRCUIT

Upon completion of this job sheet, you should be able to test the oil pressure warning circuit and determine needed repairs.

ASE Education Foundation Correlation

This job sheet addresses the following **MLR** tasks:

A.1. Research vehicle service information such as fluid type, vehicle service history, service precautions, technical service bulletins, and recalls including vehicles with advanced driver assistance systems (ADAS). **(P-1)**

A.2. Identify electrical/electronic system components and configuration. **(P-1)**

A.5. Demonstrate proper use of a digital multimeter (DMM) when measuring source voltage, voltage drop (including grounds), current flow, and resistance. **(P-1)**

A.6. Demonstrate knowledge of the causes and effects from shorts, grounds, opens, and resistance problems in electrical/electronic circuits. **(P-1)**

A.9. Use wiring diagrams to trace electrical/electronic circuits. **(P-1)**

F.1. Verify operation of instrument panel gauges and warning/indicator lights; reset maintenance indicators as required. **(P-1)**

This job sheet addresses the following **AST** tasks:

A.1. Research vehicle service information such as fluid type, vehicle service history, service precautions, technical service bulletins, and recalls including vehicles with advanced driver assistance systems (ADAS). **(P-1)**

A.2. Identify electrical/electronic system components and configuration. **(P-1)**

A.5. Demonstrate proper use of a digital multimeter (DMM) when measuring source voltage, voltage drop (including grounds), current flow, and resistance. **(P-1)**

A.6. Demonstrate knowledge of the causes and effects from shorts, grounds, opens, and resistance problems in electrical/electronic circuits. **(P-1)**

A.9. Use wiring diagrams during the diagnosis (troubleshooting) of electrical/electronic circuit problems. **(P-1)**

F.1. Verify operation of instrument panel gauges and warning/indicator lights; reset maintenance indicators as required. **(P-1)**

F.2. Inspect and test gauges and gauge sending units for causes of abnormal gauge readings; determine needed action. **(P-2)**

This job sheet addresses the following **MAST** tasks:

A.1. Research vehicle service information such as fluid type, vehicle service history, service precautions, technical service bulletins, and recalls including vehicles with advanced driver assistance systems (ADAS). **(P-1)**

A.2. Identify electrical/electronic system components and configuration. **(P-1)**

A.5. Demonstrate proper use of a digital multimeter (DMM) when measuring source voltage, voltage drop (including grounds), current flow, and resistance. **(P-1)**

A.6. Demonstrate knowledge of the causes and effects from shorts, grounds, opens, and resistance problems in electrical/electronic circuits. **(P-1)**

A.9. Use wiring diagrams during the diagnosis (troubleshooting) of electrical/electronic circuit problems. **(P-1)**

F.1. Verify operation of instrument panel gauges and warning/indicator lights; reset maintenance indicators as required. **(P-1)**

F.2. Inspect and test gauges and gauge sending units for causes of abnormal gauge readings; determine needed action. **(P-1)**

Tools and Materials
- Vehicle with conventional analog instrument cluster with oil pressure warning lamp
- Wiring Diagram
- DMM

Procedure

1. Describe the normal conditions in which the oil pressure warning lamp should be on.

   _____

   _____

   _____

2. At what pressure should the oil pressure warning lamp turn off?

   _____

3. Referring to the service information and wiring diagrams, is the oil pressure warning lamp sending unit a normally open or normally closed switch?

   _____

4. Locate the oil pressure warning lamp circuit and print it out. Map the circuit from power to ground, identifying components of the circuit. Attach the printed copy to this job sheet.

5. List all possible causes for an oil pressure warning lamp that does not turn on.

   _____

   _____

   _____

   _____

   _____

   _____

6. Unplug the sending unit connector and use the DMM to test for voltage with the ignition key in the RUN position with the engine off. Is voltage present?

   _____

7. If voltage isn't present, what will be your next step?

   _____

   _____

   _____

8. If voltage isn't present in step 6, test the circuit. What are your results and recommendations?

   _____

   _____

   _____

9. If voltage was present in step 6, describe the method to test the sending unit.

_____

_____

_____

_____

_____

10. Perform the test described in step 9 and record your results and recommendations.

_____

_____

_____

_____

_____

Instructor's Evaluation:

Grade	Rubric
0	No exposure due to: ☐ Absence ☐ Other:
1	Student unsuccessfully attempted the task
2	Student completed the task with significant instructor assistance.
3	Satisfactory performance of task. The student was able to perform elements of the task with minimal instructor assistance.
4	Competent/Proficient. Able to perform all elements of the task successfully and independently.

Instructor's Notes:

Name _____   Date _____

TESTING THE ELECTRONIC INSTRUMENT CLUSTER

Upon completion of this job sheet, you should be able to test the electronic instrument cluster using a scan tool or stand-alone diagnostic routines and determine needed repairs.

ASE Education Foundation Correlation

This job sheet addresses the following **MLR/AST/MAST** tasks:

A.1.   Research vehicle service information such as fluid type, vehicle service history, service precautions, technical service bulletins, and recalls including vehicles with advanced driver assistance systems (ADAS). **(P-1)**

A.2.   Identify electrical/electronic system components and configuration. **(P-1)**

F.1.   Verify operation of instrument panel gauges and warning/indicator lights; reset maintenance indicators as required. **(P-1)**

This job sheet addresses the following **AST** tasks:

F.2.   Inspect and test gauges and gauge sending units for causes of abnormal gauge readings; determine needed action. **(P-2)**

G.8.   Diagnose body electronic system circuits using a scan tool; check for module communication errors (data communication bus systems); determine needed action. **(P-2)**

This job sheet addresses the following **MAST** tasks:

F.2.   Inspect and test gauges and gauge sending units for causes of abnormal gauge readings; determine needed action. **(P-1)**

G.8.   Diagnose body electronic system circuits using a scan tool; check for module communication errors (data communication bus systems); determine needed action. **(P-1)**

We Support
ASE | Education Foundation

Tools and Materials

- A vehicle equipped with electronic instrument cluster
- Scan tool
- Service information

Procedure

1. Test the operation of the gauges. Describe the symptoms.

   _____

   _____

2. Are there any other related symptoms? ☐ Yes   ☐ No

   If yes, describe the symptom.

   _____

   _____

3. Refer to the proper service information to determine if a self-diagnostic routine can be performed. If so, describe how to enter the diagnostics.

   _____

   _____

   _____

4. Is the self-diagnostic routine capable of displaying fault codes?

☐ Yes   ☐ No

5. Perform the procedure listed in step 3 and record the results and any DTCs.

_____

_____

6. Use a scan tool and perform a vehicle module scan and save the report.

Does the scan tool indicate that DTCs are present?

☐ Yes   ☐ No

If yes, record the DTCs.

_____

_____

7. Based on the tests so far, which circuit(s) cause the fault?

☐ CLUSTER   ☐ SENSOR   ☐ DATA BUS

8. Based on the results so far, what tests need to be performed to find the cause of the fault?

_____

_____

9. Perform the tests listed in step 8. What is your determination and recommendation?

_____

_____

_____

Instructor's Evaluation:

Grade	Rubric
0	No exposure due to: ☐ Absence ☐ Other:
1	Student unsuccessfully attempted the task
2	Student completed the task with significant instructor assistance.
3	Satisfactory performance of task. The student was able to perform elements of the task with minimal instructor assistance.
4	Competent/Proficient. Able to perform all elements of the task successfully and independently.

Instructor's Notes:

_____

Name _____   Date _____

DIAGNOSING THE HEAD-UP DISPLAY

Upon completion of this job sheet, you should be able to diagnose the HUD system.

ASE Education Foundation Correlation

This job sheet addresses the following **MLR** tasks:

A.1.   Research vehicle service information such as fluid type, vehicle service history, service precautions, technical service bulletins, and recalls including vehicles with advanced driver assistance systems (ADAS). **(P-1)**

A.2.   Identify electrical/electronic system components and configuration. **(P-1)**

A.5.   Demonstrate proper use of a digital multimeter (DMM) when measuring source voltage, voltage drop (including grounds), current flow, and resistance. **(P-1)**

A.6.   Demonstrate knowledge of the causes and effects from shorts, grounds, opens, and resistance problems in electrical/electronic circuits. **(P-1)**

A.7.   Describe types of test lights; use appropriate test light to check operation of electrical circuits per service information. **(P-2)**

A.9.   Use wiring diagrams to trace electrical/electronic circuits. **(P-1)**

G.1.   Demonstrate understanding of vehicle comfort, convenience, access, safety, and related systems operation. **(P-3)**

This job sheet addresses the following **AST** tasks:

A.1.   Research vehicle service information such as fluid type, vehicle service history, service precautions, technical service bulletins, and recalls including vehicles with advanced driver assistance systems (ADAS). **(P-1)**

A.2.   Identify electrical/electronic system components and configuration. **(P-1)**

A.5.   Demonstrate proper use of a digital multimeter (DMM) when measuring source voltage, voltage drop (including grounds), current flow, and resistance. **(P-1)**

A.6.   Demonstrate knowledge of the causes and effects from shorts, grounds, opens, and resistance problems in electrical/electronic circuits. **(P-1)**

A.7.   Describe types of test lights; use appropriate test light to check operation of electrical circuits per service information. **(P-1)**

A.9.   Use wiring diagrams during the diagnosis (troubleshooting) of electrical/ electronic circuit problems. **(P-1)**

F.3.   Diagnose the causes of incorrect operation of warning devices and other driver information systems; determine needed action. **(P-2)**

G.1.   Diagnose vehicle comfort, convenience, access, safety, and related systems operation; determine needed action. **(P-2)**

G.7.   Diagnose operation of safety systems and related circuits (such as horn, airbags, seat belt pretensioners, occupancy classification, wipers, washers, speed control/collision avoidance, heads-up display, park assist, and back-up camera); determine needed repairs. **(P-2)**

G.8.   Diagnose body electronic system circuits using a scan tool; check for module communication errors (data communication bus systems); determine needed action. **(P-2)**

(Continued)

This job sheet addresses the following **MAST** tasks:

A.1.   Research vehicle service information such as fluid type, vehicle service history, service precautions, technical service bulletins, and recalls including vehicles with advanced driver assistance systems (ADAS). **(P-1)**

A.2.   Identify electrical/electronic system components and configuration. **(P-1)**

A.5.   Demonstrate proper use of a digital multimeter (DMM) when measuring source voltage, voltage drop (including grounds), current flow, and resistance. **(P-1)**

A.6.   Demonstrate knowledge of the causes and effects from shorts, grounds, opens, and resistance problems in electrical/electronic circuits. **(P-1)**

A.7.   Describe types of test lights; use appropriate test light to check operation of electrical circuits per service information. **(P-1)**

A.9.   Use wiring diagrams during the diagnosis (troubleshooting) of electrical/electronic circuit problems. **(P-1)**

F.3.   Diagnose the causes of incorrect operation of warning devices and other driver information systems; determine needed action. **(P-1)**

G.1.   Diagnose vehicle comfort, convenience, access, safety, and related systems operation; determine needed action. **(P-2)**

G.7.   Diagnose operation of safety systems and related circuits (such as horn, airbags, seat belt pretensioners, occupancy classification, wipers, washers, speed control/collision avoidance, heads-up display, park assist, and back-up camera); determine needed repairs. **(P-1)**

G.8.   Diagnose body electronic system circuits using a scan tool; check for module communication errors (data communication bus systems); determine needed action. **(P-1)**

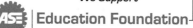
We Support
Education Foundation

Tools and Materials

- A vehicle equipped with HUD
- Scan tool
- DMM
- Test light
- Service information for the chosen vehicle
- Wiring diagram for the chosen vehicle

Procedure

1. Use the scan tool to monitor the switch data while pressing each switch. While holding each switch listed below, verify the scan tool HUD switch state parameter is as follows:

   DIM + switch: _____

   DIM – switch: _____

   UP switch: _____

   DOWN switch: _____

   PAGE switch: _____

2. Are all switch parameters the expected value?

Yes ☐   No ☐

3. Identify the switch terminal for the switch return circuit.

_____

4. Identify the switch terminal for the switch sense circuit.

_____

Ignition OFF/Vehicle OFF and all vehicle systems OFF, disconnect the harness connector at the S27 Head-up display switch. It may take up to 2 minutes for all vehicle systems to power down.

5. Use a test light and DMM to test the switch return circuit. Record your results.

_____

6. If the voltage reading in step 5 was out of specifications, what would be your next step?

_____

_____

_____

7. With the ignition in the OFF position, measure the voltage on the switch sense circuit. Record your results.

_____

8. What do the test results in step 7 indicate?

_____

9. Turn the ignition OFF and remove the HUD switch assembly.

10. Measure the resistance between the sense circuit terminal and the return terminal in all switch positions:

No switch pressed: _____

With the DIM + switch pressed _____

With the DIM − switch pressed _____

With the UP switch pressed _____

With the DOWN switch pressed _____

With the PAGE switch pressed _____

11. Based on the test results, what are your conclusions concerning the HUD switch function?

_____

Instructor's Evaluation:

Grade	Rubric
0	No exposure due to: ☐ Absence ☐ Other:
1	Student unsuccessfully attempted the task
2	Student completed the task with significant instructor assistance.
3	Satisfactory performance of task. The student was able to perform elements of the task with minimal instructor assistance.
4	Competent/Proficient. Able to perform all elements of the task successfully and independently.

Instructor's Notes:

Name _____   Date _____

RESETTING MAINTENANCE INDICATORS

Upon completion of this job sheet, you should be able to properly reset the maintenance indicator.

ASE Education Foundation Correlation _____

This job sheet addresses the following **MLR/AST/MAST** tasks:

A.1.   Research vehicle service information such as fluid type, vehicle service history, service precautions, technical service bulletins, and recalls including vehicles with advanced driver assistance systems (ADAS). **(P-1)**

F.1.   Verify operation of instrument panel gauges and warning/indicator lights; reset maintenance indicators as required. **(P-1)**

We Support
ASE | Education Foundation

Tools and Materials

- A vehicle equipped with maintenance indicator
- Scan tool
- Service information

Procedure

1. Reference the service information to determine the process for resetting the maintenance indicator. Record the procedure below:

   _____

   _____

   _____

   _____

2. Can the reset procedure be performed with a scan tool?

   ☐ Yes   ☐ No

   If yes, describe the procedure.

   _____

   _____

3. Can the service interval mileage be changed?

   ☐ Yes   ☐ No

   If yes, describe the procedure.

   _____

   _____

4. Perform the maintenance reset procedure. How do you know the procedure was successful?

   _____

   _____

   _____

   _____

Instructor's Evaluation:

Grade	Rubric
0	No exposure due to: ☐ Absence ☐ Other:
1	Student unsuccessfully attempted the task
2	Student completed the task with significant instructor assistance.
3	Satisfactory performance of task. The student was able to perform elements of the task with minimal instructor assistance.
4	Competent/Proficient. Able to perform all elements of the task successfully and independently.

Instructor's Notes:

DIAGNOSTIC CHART 12-1

PROBLEM AREA:	Instrument cluster gauges.
SYMPTOMS:	One or all gauges fluctuate from low or high to normal readings.
POSSIBLE CAUSES:	**1.** Poor ground connection. **2.** Excessive resistance. **3.** Poor connections. **4.** Faulty sending unit. **5.** Defective printed circuit.

DIAGNOSTIC CHART 12-2

PROBLEM AREA:	Instrument cluster gauges.
SYMPTOMS:	One or all gauges read high.
POSSIBLE CAUSES:	**1.** Faulty instrument voltage regulator. **2.** Shorted printed circuit. **3.** Faulty sending unit. **4.** Short to ground in sending unit circuit. **5.** Faulty gauge. **6.** Poor sending unit ground connection.

DIAGNOSTIC CHART 12-3

PROBLEM AREA:	Instrument cluster gauges.
SYMPTOMS:	One or all gauges read low.
POSSIBLE CAUSES:	**1.** Faulty instrument voltage regulator. **2.** Poor sending unit ground connection. **3.** Improper bulb application. **4.** Improper connections..

DIAGNOSTIC CHART 12-4

PROBLEM AREA:	Instrument cluster gauges.
SYMPTOMS:	One or all gauges fail to read.
POSSIBLE CAUSES:	**1.** Blown fuse. **2.** Open in the printed circuit. **3.** Faulty gauge. **4.** Poor common ground connection. **5.** Open in the sending unit circuit. **6.** Faulty sending unit. **7.** Faulty instrument voltage regulator. **8.** Poor electrical connection to the cluster. **9.** Shorted sending unit circuit.

DIAGNOSTIC CHART 12-5

PROBLEM AREA: SYMPTOMS:	Warning light operation. Warning light remains on all the time.
POSSIBLE CAUSES:	**1.** Sending unit circuit grounded. **2.** Faulty sending unit switch.

DIAGNOSTIC CHART 12-6

PROBLEM AREA:	Warning light operation.
SYMPTOMS:	Warning light fails to operate on an intermittent basis.
POSSIBLE CAUSES:	**1.** Loose sending unit circuit connections. **2.** Faulty sending unit.

DIAGNOSTIC CHART 12-7

PROBLEM AREA:	Warning light operation.
SYMPTOMS:	One or all warning lights fail to operate.
POSSIBLE CAUSES:	**1.** Blown fuse. **2.** Burned-out bulb. **3.** Open in the circuit. **4.** Defective sending unit switches.

DIAGNOSTIC CHART 12-8

PROBLEM AREA:	Electronic instrument panel.
SYMPTOMS:	Digital display does not light.
POSSIBLE CAUSES:	**1.** Blown fuse. **2.** Inoperative power and ground circuit. **3.** Faulty instrument panel.

DIAGNOSTIC CHART 12-9

PROBLEM AREA:	Electronic instrument panel.
SYMPTOMS:	Speedometer reads the wrong speed.
POSSIBLE CAUSES:	**1.** Faulty speedometer. **2.** Wrong gear on vehicle speed sensor (VSS). **3.** Wrong tire size.

DIAGNOSTIC CHART 12-10

PROBLEM AREA:	Electronic instrument panel.
SYMPTOMS:	Fuel gauge displays two top and bottom bars.
POSSIBLE CAUSES:	**1.** Open or short in the circuit.

DIAGNOSTIC CHART 12-11

PROBLEM AREA:	Electronic instrument panel.
SYMPTOMS:	Fuel computer displays CS or CO.
POSSIBLE CAUSES:	**1.** Open or short in fuel gauge sender. **2.** Inoperative instrument panel.

DIAGNOSTIC CHART 12-12

PROBLEM AREA:	Electronic instrument panel.
SYMPTOMS:	Odometer displays error.
POSSIBLE CAUSES:	**1.** Inoperative odometer memory module in the instrument panel.

DIAGNOSTIC CHART 12-13

PROBLEM AREA:	Electronic instrument panel.
SYMPTOMS:	Fuel gauge display is erratic.
POSSIBLE CAUSES:	**1.** Sticky or inoperative fuel gauge sender. **2.** Fault in the circuit. **3.** Inoperative fuel gauge.

DIAGNOSTIC CHART 12-14

PROBLEM AREA:	Electronic instrument panel.
SYMPTOMS:	Fuel gauge will not display FULL or EMPTY.
POSSIBLE CAUSES:	**1.** Sticky or inoperative fuel gauge sender.

DIAGNOSTIC CHART 12-15

PROBLEM AREA:	Electronic instrument panel.
SYMPTOMS:	Fuel economy function of the message center is erratic or inoperative.
POSSIBLE CAUSES:	**1.** Inoperative fuel flow signal. **2.** Faulty wiring. **3.** Inoperative instrument panel.

DIAGNOSTIC CHART 12-16

PROBLEM AREA:	Electronic instrument panel.
SYMPTOMS:	Extra or missing display segments.
POSSIBLE CAUSES:	**1.** Inoperative instrument panel.

DIAGNOSTIC CHART 12-17

PROBLEM AREA:	Electronic instrument panel.
SYMPTOMS:	Speedometer always reads zero.
POSSIBLE CAUSES:	**1.** Faulty wiring. **2.** Inoperative instrument panel.

DIAGNOSTIC CHART 12-18

PROBLEM AREA:	Electronic instrument panel.
SYMPTOMS:	Temperature gauge displays two top and bottom bars.
POSSIBLE CAUSES:	**1.** Short in the circuit. **2.** Inoperative coolant temperature sender. **3.** Inoperative instrument panel.

CHAPTER 13

ELECTRICAL ACCESSORIES DIAGNOSIS AND REPAIR

Upon completion and review of this chapter, you should be able to:

- Identify the causes of no operation, intermittent operation, or constant horn operation.
- Diagnose the cause of poor sound quality from the horn system.
- Perform diagnosis and repair of no–windshield wiper operation.
- Identify causes for slower-than-normal wiper operation.
- Identify the causes of continuous wiper operation.
- Diagnose faulty intermittent wiper system operation.
- Remove and install wiper motors and wiper switches.
- Diagnose computer-controlled wiper systems.
- Determine the causes for improper operation of the windshield washer system and replace the pump if required.
- Perform diagnosis of problems associated with blower motor circuits.

- Diagnose and repair electric rear window defoggers.
- Diagnose the power window system.
- Diagnose common problems associated with the power seat system.
- Diagnose the memory seat function.
- Diagnose climate-controlled seat systems.
- Perform diagnostics of the power door lock and automatic door lock systems.
- Diagnose vehicle alarm systems.
- Use self-diagnostic tests on alarm systems that provide this feature.
- Perform self-diagnostic procedures on electronic cruise control systems.
- Diagnose the adaptive cruise control system.
- Determine the circuit fault causing a malfunction in sunroof operation.

Terms to Know

Adaptive cruise control (ACC)	KOEO	Resistor block
	KOER	Servo
Dump valve	Park switch	Tin
False wipes	Pinpoint test	

Introduction

The electrical accessories included in this chapter represent the most often performed electrical repairs. Most of the systems discussed don't provide for the rebuilding of components. A technician must be capable of diagnosing faults quickly and accurately and then replacing the defective part. As with any electrical system, always use a systematic diagnostic approach to finding the cause. Refer to the service information to obtain information concerning correct system operation. The fault is easier to locate once you understand how the system is supposed to operate.

This chapter includes diagnostic and repair procedures for horn systems, windshield wipers and washer systems, blower motors, electric defogger systems, power seat and window systems, and power door locks.

Also, you will learn how to service electronic accessories designed to increase passenger comfort, provide ease of operation, and increase passenger safety. These systems include electronic cruise control, memory seats, sunroof, automatic door locks, keyless entry, and antitheft systems.

Most of these systems are additions to existing systems. For example, the memory seat feature is an addition to the conventional power seat system. As vehicles and accessories become more sophisticated, these luxury features will become more commonplace.

Although the procedures presented here are typical, always refer to the proper service information for the vehicle you are servicing.

Horn Diagnosis

Customer complaints associated with the horn system can include no operation, intermittent operation, continuous operation, or poor sound quality. Testing of the horn system varies between systems that do and don't use a relay.

No Horn Operation

Classroom Manual
Chapter 13,
pages 385, 386, 387

Systems with Relay. Begin diagnostics of no horn operation by confirming the complaint by depressing the horn button. If the switch is mounted in the steering wheel, rotate the steering wheel from stop to stop while depressing the horn button. If the horn sounds intermittently while rotating the steering wheel, the problem is probably a faulty clockspring. Check the airbag system for diagnostic trouble codes (DTCs) associated with the clockspring circuit. If the horn does not sound during this test, continue to check the system as follows:

Special Tools

Fused jumper wires
DMM
Test light
Fender covers

1. Check the fuse or fusible link. If defective, replace as needed. After replacing the fuse, operate all other circuits it protects. It's possible that another circuit is faulty, but the customer has not noticed. Isolate the cause of the blown fuse or fusible link.
2. Connect a fused jumper wire from the battery positive terminal to the horn terminal. If the horn sounds, continue testing; if the horn does not sound, check the ground connection. Replace the horn if the ground is good.
3. Remove the relay from its connector and check for voltage at terminals 30 and 86 (**Figure 13-1**). If there is no voltage at these points, trace the wiring from the relay to the battery to locate the problem. Continue testing if voltage is present at the power feed terminals.
4. With the relay removed, connect a test light across terminals 85 and 86 of the relay connector (**Figure 13-2**). Press the horn switch and observe the test light. The test light should illuminate. If it does not, the horn switch or its circuit is faulty—go to step 7. If the light illuminates, continue testing.

Figure 13-1 Testing for voltage to the relay connector.

Figure 13-2 Using a test light to test the horn switch circuit.

5. Use a fused jumper wire and jump terminals 30 and 87 (**Figure 13-3**). If the horns sound, replace the relay. If the horns don't sound, continue to test.

6. In a multiple-horn system, test for voltage at the last common connection between the horn relay and the horns (**Figure 13-4**). On a single-horn system, test for voltage at the horn terminal. Voltage should only be present at this connection when the horn button is depressed. If there is no voltage at this connection, repair the open between the relay and the common connection. If voltage is present, check the individual circuits from the connection to the horns; repair as needed.

Figure 13-3 Using a fused jumper wire to test the horn circuit. If the horn sounds, the relay is faulty.

Figure 13-4 Testing for voltage at the last common connection.

7. Check for voltage on the battery side of the horn switch. If there is no voltage at this loca-tion, the fault is between the relay and the switch. Continue testing if voltage is present.
8. Check for continuity through the switch. If good, check the ground connection for the switch. Then recheck operation. Replace the horn switch if there is no continuity when the button is depressed.

Systems without Relay. Follow the steps described under "Systems with Relay" to confirm the complaint. If this step confirms that the horn is not operational, perform the following tests to locate the fault:

1. Check the fuse or fusible link. If defective, replace as needed. After replacing the fuse, operate all other circuits it protects. It's possible that another circuit is faulty, but the customer has not noticed. Isolate the cause of the blown fuse or fusible link.

2. Connect a fused jumper wire from the battery positive terminal to the horn terminal. If the horn sounds, continue testing; if the horn does not sound, check the ground connection. Replace the horn if the ground is good.

3. In a multiple-horn system, test for voltage at the last common connection between the horn switch and the horns. On a single-horn system, test for voltage at the horn terminal. Voltage should only be present at this connection when the horn button is depressed. Continue testing if there is no voltage at this connection. If voltage is present, check the individual circuits from the connection to the horns; repair as needed.

4. Check for voltage at the horn side of the switch when the button is depressed. If voltage is present, the problem is in the circuit from the switch to the horn(s). Continue testing if there is no voltage at this connection.

5. Check for voltage at the battery side of the switch. If voltage is present, the switch is faulty and requires replacement. If there is no voltage at this terminal, the circuit from the battery to the switch is the problem.

Poor Sound Quality

Poor sound quality can be the result of several factors. In a multiple-horn system, if one of the horns does not operate, the horn sound quality may suffer. Other reasons include a damaged diaphragm, excessive circuit resistance, poor ground connections, or improperly adjusted horns.

Classroom Manual
Chapter 13, page 387

If one horn of a multiple-horn system does not operate, use a fused jumper wire from the battery positive terminal to the horn terminal to determine whether the fault is in the horn or the circuit. If the horn sounds, the problem is in the circuit between the last common connection and the affected horn.

Special Tools

DMM
Fused jumper wires
Fender covers

If one or all horns produce poor-quality sound, use a voltmeter to measure the voltage at the horn terminal when the switch is closed. The voltage should be within 0.5 volt (V) of battery voltage. If the voltage measured is less than the battery voltage, there is excessive voltage drop in the circuit. Work back through the circuit, measuring the voltage drop across connectors, relays, and switches to find the source of the high resistance.

If the voltage to the horn is of proper value, test the ground circuit. Use the voltage drop method while the horn switch is closed. If the voltage drop is excessive, check the connections to ground.

If the circuits test good, connect a fused jumper wire between the battery positive terminal and the horn terminal to activate the horn. With the horn sounding, turn the adjusting screw counterclockwise one-quarter to three-eighths of a turn. Replace the horn if the sound quality cannot be improved.

Horn Sounds Continuously

A sticking horn switch or sticking contact points in the relay are common causes for a horn that does not shut off. To find the fault, disconnect the horn relay from the circuit. Use an ohmmeter to check for continuity from the relay's battery feed terminal to its horn circuit terminal (**Figure 13-5**). If there is continuity, the relay is defective. If there is no continuity through the relay, test the switch.

Classroom Manual
Chapter 13, page 386

The easiest way to test the switch circuit is to measure for continuity between the relay connector terminal 85 and ground. With the horn switch released, there should not be continuity to ground. Continuity indicates a short to ground in the circuit to the horn switch, or the switch contacts are stuck. Use an ohmmeter to test the switch in the normal manner.

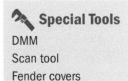

Special Tools

DMM
Scan tool
Fender covers

Figure 13-5 Ohmmeter tests to find the cause of continuous horn operation.

Computer-Controlled Horn Systems

Classroom Manual
Chapter 13, page 387

> ⚙️ **SERVICE TIP** Most modules that use high-side drivers can detect an open circuit. Often, when a fault in the circuit is determined, the HSD is turned off for the remainder of that ignition cycle. If you disconnect the circuit from the module to the horn, the HSD may shut down, and you will need to cycle the ignition switch between each test.

Systems that control the horn function by computers have the microprocessor control the relay coil ground circuit, or use high-side drivers (HSDs) to operate the horns. Use the bidirectional activation feature of the scan tool to actuate the horns. If the horns sound when activated with the scan tool, the problem is in the inputs to the computer. If the system does not operate when actuated with the scan tool, the problem may be the driver circuits, faulty horn, faulty horn ground circuit, or defective module. Testing the circuits from the module to the relays (if used) and the horns is the same as just described.

Wiper System Service

Classroom Manual
Chapter 13,
pages 389, 390, 392

Customer complaints concerning windshield wiper operation can include no operation, intermittent operation, continuous operation, and wipers that will not park. Other complaints involve blade adjustment (such as blades slapping the molding or one blade that parks lower than the other).

When a customer brings the vehicle into the shop because of faulty windshield wiper operation, you need to determine if it's an electrical or a mechanical problem. To do this, disconnect the arms to the wiper blades from the motor (**Figure 13-6**). Turn on the wiper system. Observe the operation of the motor. The problem is mechanical if the motor operates properly.

1. Lift locking tab

2. Slide clip

Figure 13-6 To remove the clip, lift the locking tab and pull the clip.

No Operation in One Speed Only

Problems that cause the system to not operate in only one switch position are usually electrical. Use the wiring schematic to determine proper operation. For example, use the three-speed wiper system schematic illustrated in **Figure 13-7** to determine the cause of a motor that does not operate only in the MEDIUM-speed position. The problem is that the 7-ohm (Ω) resistor is open. The problem could not be the shunt field in the motor because LOW and HIGH speeds operate, nor could the problem be in the wiring to the motor since all speeds use this circuit.

Verify an opened resistor by using a voltmeter to measure the voltage at the terminal leading to the shunt field. If the voltage drops to 0 volt (V) in the MEDIUM position, replace the switch. By properly using the wiring schematic and understanding the correct operation of the system, you can diagnose this problem quickly.

Classroom Manual
Chapter 13, page 393

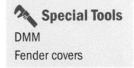 **Special Tools**
DMM
Fender covers

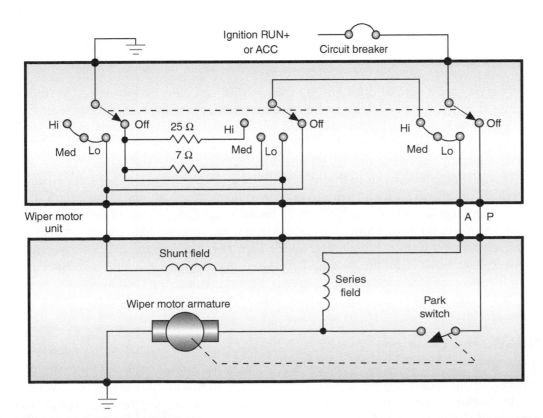

Figure 13-7 Three-speed wiper system schematic.

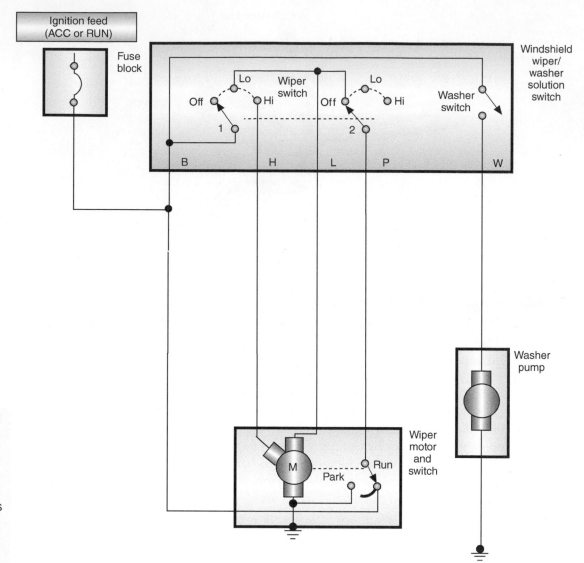

Figure 13-8 Two-speed wiper circuit.

Classroom Manual
Chapter 13, pages 390, 392

There can be several different fault areas that can cause a two-speed system to operate at only one speed. Diagnosing this issue requires using wiring schematics and being able to identify what the expected voltage at different locations in the circuit should be. Use **Figure 13-8** to step through a common test sequence to locate why the motor does not operate in the HIGH position.

Turn the ignition switch to the ACC or RUN position and place the wiper switch in the HIGH position. Use a voltmeter to test for voltage at the high-speed connector of the motor. If voltage is present, the high-speed brush is worn, or the wire from the terminal to the brush is open. Most shops don't rebuild the wiper motor; replacement is usually the preferred service. If there is no voltage at the high-speed connector, check for voltage at connector terminal H for the switch. If voltage is present here, the fault is in the circuit from the switch to the motor. If no voltage is present, replace the switch.

To test for no LOW-speed operation only, use the same procedure to test the low-speed circuit.

No Wiper Operation

Classroom Manual
Chapter 13, page 389

If the wiper motor does not operate in any speed position, check the fuse or circuit breaker. If the fuse is blown, replace it and test the operation of the motor. The cause for the blown fuse or tripped circuit breaker must be identified. Check the mechanical portion of the system for binding that can cause excessive current flow.

If the fuse is good, check the motor ground using a fused jumper wire from the motor body to a good chassis ground. If the motor operates when the ignition switch is in the RUN position and the wiper switch is placed in all speed positions, repair the ground connection. Continue testing if the motor does not operate.

Use a voltmeter to check for voltage at the low-speed terminal of the motor with the ignition in the RUN or ACC position and the wiper switch in the LOW position. If there is no voltage at this point, test for voltage on the low-speed terminal of the wiper switch. If the voltmeter indicates battery voltage at this terminal, the fault is in the circuit between the switch and the motor. Look for indications of burned insulation or other damage that would affect both the high- and low-speed circuits.

No voltage at the low-speed terminal of the wiper switch indicates that the fault may be in the switch or the power feed circuit. Test for battery voltage at the battery supply terminal of the switch. If there is voltage at this point, the switch is faulty. If no voltage is at the supply terminal, trace the circuit back to the battery to locate the fault.

If battery voltage is present at the low-speed terminal of the motor, check for voltage at the high-speed terminal. Voltage at both of these terminals indicates that the motor is faulty. If there is no voltage at the high-speed terminal, use the procedure just described to trace the high-speed circuit.

> ⚙ **SERVICE TIP** No voltage at either terminal means the problem is probably at a shared location. In most systems, this would be the power supply portion of the wiper switch and the switch itself. If the switch is good, the problem is in the wiring loom.

Slower-than-Normal Wiper Speeds

Electrical or mechanical faults can cause slower-than-normal wiper speeds. Use an ammeter to determine the current draw of the motor. Measure amperage with and without the mechanical portion of the system connected. If the draw changes substantially with the mechanical portion removed, the fault is in the arms and/or wiper blades.

Most electrical circuit faults that cause slow wiper operation result from excessive resistance. If the complaint is that all speeds are slow, use the voltage drop test procedure to check for resistance in the power feed supply circuit to the wiper switch. If the power supply circuit is good, check the switch for excessive resistance.

If the insulated side voltage tests fail to locate the problem, check the voltage drop on the ground side of the wiper motor. Connect the voltmeter's positive lead to the ground terminal of the motor (or motor body) and the negative lead to the vehicle chassis. The voltage drop should be no more than 0.1 volt. If excessive, repair the ground circuit connections. If voltage drop on both the insulated and the ground sides of the motor is within specifications, the fault is in the motor.

Wipers Will Not Park

The most common complaint associated with a faulty **park switch** is that the wipers stop immediately when the switch is turned off and don't return to the parked position. This may not be the direct fault of the **park switch**, however. The park switch supplies current to the motor after placing the wiper control switch into the OFF position. This allows the motor to continue operating until the wipers have reached their park position.

Removing the motor cover may permit observing the operation of the park switch (**Figure 13-9**). Operate the wipers through three or four cycles while observing the latch arm. With the wiper switch in the OFF position, the park switch latch must be in position to catch the drive pawl. Check to make sure the drive pawl is not bent. If good, replace the park switch.

A faulty wiper switch can also cause the park feature not to operate. Using **Figure 13-10**, if wiper 2 is bent or broken so that it does not make an electrical connection with the contacts,

Special Tools

DMM
Fender covers
Fused jumper wire

Classroom Manual
Chapter 13,
pages 390, 392

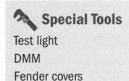

Special Tools

Test light
DMM
Fender covers

Drive pawl

Rotation

Latch arm (must be
out of the way when
gear is removed)

Relay slot

Figure 13-9 Checking the operation of the park switch while the motor is operating.

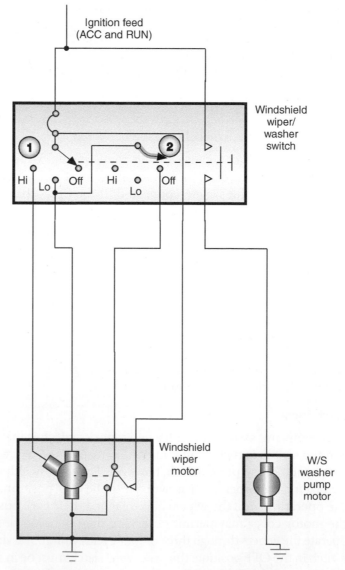

Ignition feed
(ACC and RUN)

Windshield
wiper/
washer
switch

1

2

Hi   Lo   Off   Hi   Off
                  Lo

Windshield
wiper
motor

W/S
washer
pump
motor

Figure 13-10 A faulty contact wiper in the switch can cause the park
feature not to operate.

the wipers will not park even with the park switch in the PARK position, as shown. To test the switch, check for voltage at the low-speed circuit while moving the switch from LOW to OFF. If the switch operates properly, voltage should be present for a few seconds after the switch is in the OFF position. No voltage indicates that the problem may be in the switch.

Checking the operation of the park switch can also be done by using a test light to probe for voltage on the park switch circuit when the wiper switch is turned off. Probing for voltage at this circuit should produce a pulsating light when the motor is running.

If the wiper blades continue to operate with the wiper switch in the OFF position, the most probable cause is welded contacts in the park switch (**Figure 13-11**). If the park switch does not open, current continues to flow to the wiper motor. The only way to turn off the wipers is to turn off the ignition switch, physically remove the wires to the motor, or pull the fuse.

Some motors provide for the replacement of the park switch. However, most shops replace the motor assembly.

Figure 13-11 Sticking contacts in the park switch can cause the wipers to operate even after the switch is turned off.

Classroom Manual
Chapter 13, page 394

Intermittent Wiper System Diagnosis

The illustration shown in **Figure 13-12** is a schematic of one of Ford's intermittent wiper systems. If the intermittent function is the only portion of the system that fails to operate properly, begin by checking the ground connection for the timer module. If the ground is good, perform a continuity test of the switch using an ohmmeter (**Figure 13-13**). Replace the module if the switch is good and all wires and connections are good.

Figure 13-12 Intermittent wiper system schematic.

Figure 13-13 Wiper switch continuity chart.

Wiper Motor Removal and Installation

Removal procedures differ among manufacturers. Some motors are tucked in areas that may require removing several engine compartment components. Always refer to the correct service information for the recommended procedure. **Photo Sequence 27** provides a common method of removing a wiper motor.

Special Tools
Fender covers
Battery terminal puller
Battery terminal pliers

PHOTO SEQUENCE 26
Wiper Motor Removal

P26-1 Place fender covers over the vehicle's fenders.

P26-2 Disconnect the battery negative cable.

P26-3 Disconnect the wiper arms from the linkage.

P26-4 To gain access to the motor, remove the shield cover from the cowl.

P26-5 Disconnect the wire connector at the motor.

P26-6 Remove the linkage from the motor.

P26-7 Remove the attaching bolts from the motor assembly.

P26-8 Remove the motor.

Caution

The internal permanent magnets of the motor are a ceramic material. Use care while handling the motor to avoid damaging the magnets.

On some vehicles, the linkage is removed by lifting the locking tab and pulling the clip away from the pin (refer to Figure 13-6). Installation is basically the reverse of the removal procedure. Be sure to attach the ground wire to one of the mounting bolts during installation.

Wiper Switch Removal and Installation

The wiper switch removal procedure differs among manufacturers and depends on switch location. The procedures presented here are common. However, always refer to the manufacturer's service information for correct procedures. Protect the customer's investment by using fender covers while disconnecting the battery negative cable.

Dash-Mounted Switches. Depending on the location of the switch control, it may be necessary to remove the finish panel. Usually, the finish panel is held in place by a combination of fasteners and clips.

Remove the switch housing retaining screws. Then remove the housing. Pull off the wiper switch knob. Disconnect the wire connectors from the switch. Remove the switch from the dash.

Reverse the procedure to install the switch.

Steering Column–Mounted Switches. Remove the upper and lower steering column shrouds to expose the switch. Disconnect the wire connectors to the switch. It may be necessary to peel back the foam to gain access to the retaining screws. Remove the screws and the switch.

Diagnosing Computer-Controlled Wiper Systems

Classroom Manual
Chapter 13, page 398

Most computer-controlled wiper systems can perform a diagnostic routine and set DTCs upon identifying a fault. The following are some common DTCs associated with the wiper system and common diagnostic procedures for locating the fault. Use **Figure 13-14** as an example of the system.

Wiper Park Switch Input Circuit Low

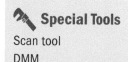 **Special Tools**

Scan tool
DMM

The BCM can perform the functions of the FCM described here.

Whenever the ignition is in the RUN position and the wipers are requested ON, the front control module (FCM) monitors the park switch circuit. As the wipers sweep across the windshield and return to their parked position, the switch should cycle from open to close. If the FCM detects a short to ground condition, the "Wiper park switch input circuit low" fault sets. This DTC can indicate a short to ground in the front wiper park switch sense circuit, faulty wiper motor (internal park switch failure), or a fault in the FCM. The customer would probably state that the wipers turn off immediately after the wiper switch is set to the OFF position, regardless of the position of the wipers on the windshield.

Begin by inspecting the related wiring for any chafed, pierced, pinched, or partially broken wires. If there are no wiring faults, erase the fault codes and disconnect the front wiper motor harness connector. Turn the ignition to the RUN position, activate the wipers for about 30 seconds, and then check for a "Wiper park switch input circuit high" DTC. If this fault code sets, replace the wiper motor.

If the circuit low fault returns, measure the resistance between ground and the wiper park switch sense circuit. Circuit resistance less than 5 ohms (Ω) indicates a short to ground. If the resistance is more than 5 ohms, replace the FCM.

Wiper Park Switch Input Circuit High

The "Wiper park switch input circuit high" fault sets if the FCM detects an open circuit. This DTC can also set if the park switch sense circuit shorts to voltage. This DTC can

Figure 13-14 Computer-controlled wiper system.

indicate a short to ground in the front wiper park switch sense circuit, faulty wiper motor (internal park switch failure), or a fault in the FCM. The customer would probably state that the wipers turn off immediately after the switch is set to the OFF position, regardless of the position of the wipers on the windshield.

Begin by inspecting the related wiring for any chafed, pierced, pinched, or partially broken wires. If there are no wiring faults, erase the fault codes and disconnect the front wiper motor harness connector and the FCM connector. Measure for voltage on the park switch sense circuit. Since both connectors are disconnected, there should be 0 volt on the circuit. The park switch sense circuit is shorted to a voltage supply circuit if there is measured voltage.

With the connectors still disconnected, measure the resistance in the park switch circuit. The resistance should be close to 0 ohm. Repair the sense circuit if the ohmmeter indicates high resistance or an open circuit. If the problem still has not been located, test the park switch ground circuit.

If all tests do not isolate the problem, the fault is in the FCM. Be sure to test all connections, power, and ground feeds before condemning the FCM.

Wiper On–Off Control Circuit Low

The "Wiper On–Off control circuit low" DTC sets if the FCM detects a short to ground or low-voltage condition in the ON relay's control circuit. This could be caused by an open in the fused battery feed circuit to the relay, a faulty relay coil, an open in the control circuit between the relay and the FCM, or a short to ground in the control circuit. In addition, the FCM may be the faulty component. The customer would probably state that the wipers don't work in any switch position.

Begin by checking the condition of the fuse that protects the voltage supply circuit to the relay. A blown fuse indicates a short to ground in the circuit between the fuse and the relay or a shorted relay coil. If the fuse is good, replace the relay with a known good unit and test operation. If the wipers still don't operate, continue testing by removing the relay and turning the ignition to the RUN position. Measure for voltage at cavity 86 of the relay. If the voltage is not the same as battery voltage, there is an open or a high resistance in the circuit. It's always a good practice to load the circuit when testing. Removing the relay means that the test is an open-circuit test. Load the circuit with a test light and measure available voltage.

If the fault is still not located, reinstall the relay and backprobe the relay control circuit at the FCM. Measure the voltage at this location. If the voltmeter reads 0 volt, there is an open between the relay and the FCM. If the voltage reads battery voltage, use a fused jumper wire to jump the backprobed connector to ground. If the wipers come on during this test, the FCM has a faulty driver.

> **AUTHOR'S NOTE** The input from the multifunction switch is not being tested because of the fault code retrieved. In this instance, the DTC indicates the problem area as being the relay control circuit. If the wipers did not operate and there were no DTCs, the input side would be tested.

Wiper On–Off Control Circuit High

The "Wiper On–Off control circuit high" DTC sets if the FCM detects a short to voltage condition in the ON relay's control circuit. This could result from a shorted relay coil or an internal fault in the FCM. The customer would probably state that the wipers don't work in any switch position.

To test for this type, begin by erasing the fault codes and replacing the relay with a known good unit. Operate the system for about 30 seconds and then turn it off. Use the scan tool to see if the DTC returns. If it does not return, the relay is the problem. If the DTC returns, the FCM is faulty. Be sure to test all connections, powers, and grounds before faulting the FCM.

Wiper HI/LOW Control Circuit Low

The "Wiper HI/LOW control circuit low" DTC sets when the FCM detects a short to ground or low-voltage condition on the HI/LOW relay control circuit. When the FCM's driver turns off, the FCM should sense 12 volts on the control circuit. Turning the driver on should increase the voltage to about 700 mV. If the voltage is always low on this circuit, the FCM will set the code. The customer will probably state that their wipers work on low and intermittent speeds, but not on high speed. Causes for setting this fault include an open battery feed circuit, open relay coil, short to ground in the control circuit, and an internal problem within the FCM.

> **AUTHOR'S NOTE** If the control circuit is shorted to ground between the relay and FCM, the customer will probably state that the wipers appear to be working only on high speed.

Begin by testing all fuses associated with the relay control circuit. Referring to Figure 13-14, if the wipers work on low speed but not on high speed, the problem is between the common splice in the 12-volt feed circuit and the FCM.

Remove the HI/LOW relay and measure the voltage at cavity 86 with the ignition in the RUN position. It would be best to load the circuit with a test light when measuring this voltage. If the voltage is about equal to battery voltage, the circuit to this point is good. If the voltage is 0 volt, there is an open in the circuit toward the common splice.

Replace the relay with a known good unit and clear all DTCs. Attempt to operate the wipers on high speed; then turn the wipers off again. Use the scan tool to read DTCs; if the fault code does not come back, the relay is the problem; if the fault comes back, then check the circuit between the relay and the FCM for an open or a short to ground.

Backprobe the FCM connector for the HI/LOW relay control circuit. With the wipers operating at low speed, jump the backprobed terminal to ground. If the wipers go to high speed, the fault is in the FCM. If the wipers still don't operate at high speed, check for an open in the relay control circuit between the relay and the FCM.

Wiper HI/LOW Control Circuit High

The "Wiper HI/LOW control circuit high" DTC sets when the FCM detects a short to voltage or high-voltage condition on the HI/LOW relay control circuit. With the FCM's driver turned off, the FCM should sense 12 volts on the control circuit. Turning on the driver should increase the voltage to about 700 mV. If the voltage is always high on this circuit, the FCM will set the code. The customer will probably state that the wipers work on low and intermittent speeds, but not on high speed. Causes for setting this fault include a short to battery voltage on the circuit between the relay and the FCM, a faulty relay, or an internal problem within the FCM.

Begin erasing all DTCs and replacing the relay with a known good unit. Operate the wipers for about 30 seconds, and then turn them off. Check to see if the DTC returns; if not, the relay is the problem. If the DTC returns, test for voltage on the circuit between the relay and the FCM by disconnecting the connector on both ends and measuring for voltage. If there is a measured voltage, isolate the shorted circuit and repair it. If the circuit is not shorted to voltage, replace the FCM.

Automatic Wiper System Diagnosis

A vehicle equipped with the automatic wiper system uses the same type of computer-controlled wipers, but adds a rain sensor. Also, when automatic wipers are enabled, the intermittent position on the multifunction switch control knob is used to set the desired system sensitivity. Using the schematic of an automatic wiper system (**Figure 13-15**), you can see that the right multifunction switch is a MUX switch to the left multifunction switch, which is a LIN slave module to the cabin compartment node (CCN). The electronic wiper switch sensitivity message is sent to the CCN over the LIN data bus, and then the CCN relays the message to the rain sensor module (RSM) over the CAN B data bus.

Classroom Manual
Chapter 13, page 399

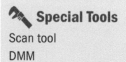
Special Tools
Scan tool
DMM

Figure 13-15 Automatic wiper system.

The RSM monitors an area within the wipe pattern of the windshield glass for the accumulation of moisture. Based upon internal programming and the selected sensitivity level, when sufficient moisture has accumulated, the RSM sends the appropriate electronic wipe command messages to the FCM over the CAN B data bus. The FCM operates the front wiper system accordingly.

AUTHOR'S NOTE The CCN is another term used to describe the electronic instrument cluster when it's a dominant module on the CAN bus network.

If the wipers don't operate properly in either the manual mode or the automatic mode, the problem is in the basic wiper system. If the wipers operate properly in the manual mode but not in the automatic mode, then the problem is in the automatic function components. This includes the RSM and the data bus network to it. It's also possible that the system is disabled.

⚠ WARNING **To avoid serious or fatal injury, disable the supplemental restraint system (SRS) before attempting to diagnose or remove parts on the steering wheel, steering column, or instrument panel. Disconnect and isolate the battery negative (ground) cable; then wait 2 minutes for the system capacitor to discharge before performing further diagnosis or service. Failure to follow this warning could result in accidental airbag deployment.**

Diagnosing the hardwired wiper system circuits and components uses conventional diagnostic tools and procedures. However, conventional diagnostic methods will not prove conclusive in diagnosing the front wiper and washer system, electronic controls, or communication between other modules and devices that provide the automatic wiper feature. The most reliable, efficient, and accurate means to diagnose the front wiper and washer system or the electronic controls and communication related to front wiper and washer system operation require a scan tool.

A problem the customer may experience with automatic wiper systems is **false wipes**. False wipes are unnecessary wipes that occur when the automatic wiper system is enabled. When diagnosing this type of complaint, keep in mind that the system design is to operate whenever it detects moisture. Any road spray, bug splatters, or mist from passing vehicles may cause the wipers to cycle. These are normal characteristics of this system and are not false wipes.

False wipes are usually the result of foreign material on the windshield. In some instances, false wipes can result from flaws in the windshield that interfere with the system optics. The optics for this system include the lenses of the RSM, the lenses of the RSM bracket, the adhesive pad layer that bonds the bracket to the inside of the windshield, and the windshield glass. Anything that distorts the intensity of the IR light beams or affects the ability of the photo diodes to measure the returning beams accurately can result in the RSM logic misinterpreting the input data as moisture on the windshield.

Usually, the system can detect faults in the electronics or bus communications that may cause the system not to function properly. Use a scan tool to check for DTCs associated with the front wiper system, the multifunction switch, the CCN, and the RSM. If there are DTCs in the components that operate within the system, use the appropriate diagnostic procedures to isolate the problem. If no DTCs are present, perform the routine illustrated in **Photo Sequence 28**. If faulty, the RSM mounting bracket is usually serviced as a unit with the windshield glass. If either the bracket or the windshield glass is the cause of the system problem, replace the entire RSM bracket and windshield glass unit.

Windshield Washer System Service

Many windshield washer problems are due to restrictions in the delivery system. To check for restrictions, remove the hose from the pump and operate the system. If the pump ejects a stream of fluid, then the fault is in the delivery system. If the pump does not deliver a spray of fluid, continue testing using the following procedure:

1. Make checks of obvious conditions such as low fluid level, blown fuses, or disconnected wires.
2. Activate the washer switch while observing the motor. If the motor operates but does not squirt fluid, check for blockage at the pump. Remove any foreign material. If there is no blockage, replace the motor.

Classroom Manual
Chapter 13, page 400

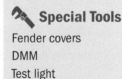

Special Tools
Fender covers
DMM
Test light
Safety glasses

3. If the motor does not operate, use a voltmeter or test light to check for voltage at the washer pump motor with the switch closed. If there is voltage, then check the ground circuit with an ohmmeter. If the ground connection is good, then replace the pump motor.

4. If there is no voltage to the pump motor in step 3, trace the circuit back to the switch. Test the switch for proper operation. If there is power into the switch but not out of it to the motor, replace the switch.

If the motor requires replacement, follow this procedure for pumps installed in the reservoir (**Figure 13-16**). Disconnect the wire connector and hoses from the pump. Remove the reservoir assembly from the vehicle. Use a small blade screwdriver to pry out the retainer ring.

PHOTO SEQUENCE 27
Inspection of Rain Sensor Module

P27-1 Carefully inspect the outer surface of the windshield glass for physical damage, including scratches, cracks, or chips in the vicinity of the RSM mounting bracket lenses. If damage is present, replace the windshield.

P27-2 From the outside of the windshield glass, carefully inspect the adhesive layer between the windshield glass and the RSM bracket for any voids greater than 0.040 inch (1 mm). If an adhesive void is detected, replace the flawed RSM mounting bracket and windshield unit.

P27-3 Remove the inside rearview mirror.

P27-4 Lightly pull the RSM away from the windshield bracket to confirm that both module sliding cam locks have fully engaged all four pins of the mounting bracket on the inside of the windshield glass. If needed, reinstall the module onto the bracket.

P27-5 Remove the RSM from the mounting bracket.

P27-6 Inspect the RSM lenses and the mounting bracket lenses for contamination.

PHOTO SEQUENCE 27 (CONTINUED)

P27-7 Clean any foreign material from each lens using isopropyl alcohol and a lint-free cloth.

P27-8 Carefully inspect the RSM for any physical damage, including scratches on the RSM lenses.

P27-9 Carefully inspect the RSM mounting bracket for any physical damage, including scratches on the RSM bracket lenses. If there is any damage, replace the bracket and windshield as a unit.

Figure 13-16 Reservoir-mounted washer pump and motor.

⚡ **WARNING** Wear safety glasses to prevent the ring from striking your eyes. Also, be careful to position the palm of your hand so that if the screwdriver slips it will not puncture your skin.

Use a pair of pliers to grip one of the walls surrounding the terminals. Pull out the motor, seal, and impeller.

Before installing the pump assembly, lubricate the seal with a dry lubricant to prevent the seal from sticking to the reservoir's wall. Align the small projection on the motor with the slot in the reservoir and assemble. Make sure the seal seats against the bottom of the motor cavity. Use a 12-point, 1-inch socket to hand press the retaining ring into place.

Replace the reservoir assembly in the vehicle. Reconnect the hose and wires. When refilling the reservoir, do so slowly to prevent air from being trapped in the reservoir. Check system operation while checking for leaks.

⚠ **Caution**

Do not operate the washer pump without fluid. Doing so may damage the new pump motor.

Classroom Manual
Chapter 13, page 403

Special Tools

Fused jumper wires

Test light

DMM

Fender covers

Blower Motor Service

Fan speed control of a conventional blower system sends current through a **resistor block**. The resistor block is a series of resistors with different values. There is usually one less resistor than fan speed positions because the high-speed circuit bypasses the resistors. The higher the resistance value, the slower the fan speed. The position of the switch determines the total resistance of the circuit. Circuits can use either ground side switches or insulated side switches.

If the customer complaint is that the fan operates in only a couple of speed positions, the most likely cause is an open resistor in the resistor block. Using the illustration in **Figure 13-17**, if resistor 1 is open, the motor will not operate in any position except high speed. If resistor 2 is open, the motor will operate only in high and M2 speeds. If resistor 3 is open, the motor will operate in all speeds except low.

If the motor operates in any one of the speed select positions, the fault is not in the motor. If the motor fails to operate at all, begin by inspecting the fuse. If the fuse is good, use the correct wiring schematic to determine if the system uses ground or insulated side switches. The diagnostic procedure used depends on the circuit design.

Figure 13-17 Typical wiring for a four-speed fan motor circuit.

Inoperable Motor with Insulated Switches

Use a fused jumper wire to bypass the switch and resistor block to check motor operation. Connect the fused jumper wire from a battery positive supply to the motor terminal. If the motor does not operate, connect a second fused jumper wire from the motor body to a good ground. Replace the motor if it still does not operate.

If the motor operates with the switch and the resistor block bypassed, trace up the circuit toward the switch. Use a voltmeter or test light to check for voltage in and out of the blower speed control switch. The switch is faulty if voltage is at the input terminal but not at any of the output terminals. No voltage at the input terminal indicates an open in the circuit between the battery and the switch.

Inoperable Motor with Ground Side Switches

Using **Figure 13-18** as an example of a negative side switch blower motor circuit to test the motor, connect the fused jumper wire from the motor negative terminal to a good ground. This bypasses the switch and resistor block. If the motor does not operate, use a voltmeter or test light to check for voltage at the battery terminal of the motor. If voltage is present, then the motor is defective.

If there is no voltage at the battery terminal of the motor, the problem is in the circuit from the battery to the motor. Be sure to check the circuit breaker.

Figure 13-18 Blower motor circuit using negative side switch.

⚙️ **SERVICE TIP** Because the high-speed circuit bypasses the resistor block, it's doubtful that no motor operation would be the fault of open resistors. However, an open in the wire from the block to the motor can cause the problem. Most likely the switch is bad and in need of replacement. Always confirm your diagnosis by doing a continuity test on the switch.

Operation of the motor with the jumper wire connected to the ground terminal indicates the problem is in the switch side of the circuit. Use an ohmmeter to check the ground connection for the switch. A fused jumper wire or test light can also be used to test this connection.

If the ground connection is good, use a voltmeter or test light to probe for voltage at any of the circuits from the resistor block to the switch. Replace the switch if there is power to this point. Replace the resistor block if there is no power at these points.

Constantly Operating Blower Fans

General Motors has designed blower fan systems that constantly run on low speed, even with the switch in the off position. Don't confuse this normal operation with a circuit defect.

Constantly running blower motors are more common in ground side switch systems. A short to ground at any point on the ground side of the circuit will cause the motor to run. Other areas to check include the switch and the circuit between the switch and the resistor block.

In insulated side switch circuits, check for copper-to-copper shorts in the power side of the system. If the motor receives power from another circuit, due to a copper-to-copper short, the motor will continue to run whenever current flows through that circuit. Some systems may incorporate a relay, and the motor will continue to operate if the contact points fuse together.

Computer-Controlled Fans

Classroom Manual
Chapter 13, page 405

Most computer-controlled motor circuits control motor speed by duty cycling. Typically, a power module (**Figure 13-19**) controls the motor based on drive signals from the

Figure 13-19 Power module.

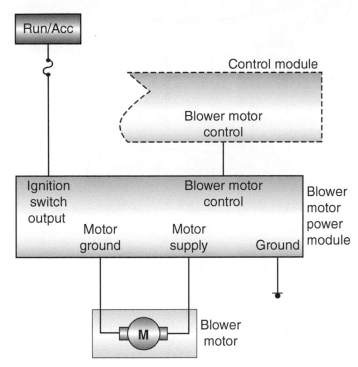

Figure 13-20 Example of a motor circuit using a power module.

control module. The control module sends a pulse-width modulated (PWM) signal to the power module, and the power module amplifies the signals to provide the variable motor speed.

Use the scan tool's bidirectional controls to activate the motor at different speeds. If the blower motor does not operate properly by scan tool activation, use a DMM or lab scope to isolate the root cause. Consider the system illustrated by the wiring diagram in **Figure 13-20**. Backprobe the power module at the terminals for both the control and motor ground circuits. Using two DMMs set to read voltage, connect one between the control circuit terminal and chassis ground, and the second between the motor ground terminal and chassis ground. A dual-channel lab scope can be set up in this manner also.

For this example system, low-side switching controls the blower motor. The expected input voltage and the voltage on the ground terminal with the blower motor off should be 12 volts. The control circuit voltage reading should drop as the blower motor switches from LOW speed to HIGH speed. Using a lab scope will display a duty cycle. The duty cycle controls the voltage level across the blower motor. The voltage on the motor ground circuit will also decrease as the blower motor speed increases.

 Special Tools
Scan tool
DMM (2)
Back-probing tools
Fused jumper wires

Electric Defogger Diagnosis and Service

The rear window defogger grids are a series of horizontal, ceramic, silver-compounded lines baked into the surface of the window. If the rear window defogger fails to operate when the switch is activated, use a test light to test the grids. Under normal conditions, the test light should be bright on one side of the grid and off on the other side. If the test light has full brilliance on both sides of the grid, then the ground connection for the grid is broken.

If the test light does not illuminate at any position on the grid, use normal test procedures to check the switch and relay circuits. There may be several fuses involved in the system. Use the correct wiring diagram to determine the fuse identification.

Classroom Manual
Chapter 13, page 406

Some manufacturers refer to the rear window defogger system as electric backlight.

Figure 13-21 Zones of test light brilliance while probing a rear window defogger grid.

Figure 13-22 Test light brilliance when passed over a break.

Special Tools

Test light
Grease pencil
Heat gun or blow-dryer
Defogger repair kit
Test light
Masking tape
Clean cloth
Wooden stick
Isopropyl alcohol
Steel wool

 Caution

Be careful not to tear the grid with the test light. Only a light touch on the grid is required.

Special Tools

Soldering iron
Rosin flux paste
3% silver solder
Steel wool

 To **tin** the tip, apply solder to the tip of the soldering iron. This provides better heating control.

Most rear window defogger complaints are associated with broken grids. The complaint will be only a portion of the window is cleared while the rest remains foggy. Some grid wire breaks are easily detected by visual inspection. However, many are too small to see. To test the grid lines, start the engine and activate the system (remember that the system is controlled by a timer). Use a test light to check each grid wire to locate the breaks. Test each grid in at least two places—one on each side of the centerline. **Figure 13-21** illustrates the expected test results for each grid.

If the test light does not indicate normal operation on a specific grid line, place the test light probe on the grid at the left bus bar and work toward the right until the light goes out. The point where the light goes out is the location of a break (**Figure 13-22**). Mark the location of the break with a grease pencil on the *outside* of the glass.

The rear window defogger should turn off about 10 minutes after activation. If the circuit fails to turn off, check the ground for the control module (**Figure 13-23**). If the ground is good, replace the module.

Grid Wire Repair

It's possible to repair a broken grid wire with a special repair kit. Follow the procedures in **Photo Sequence 29** to repair the grid.

Bus Bar Lead Repair

The bus bar lead wire can be resoldered using a solder containing 3% silver and a rosin flux paste. Clean the repair area using a steel wool pad. Apply a small quantity of rosin flux paste to the wire lead and bus bar. **Tin** the solder iron tip with the solder. Finish the repair by soldering the wires to the bus bar. Be careful not to overheat the wire.

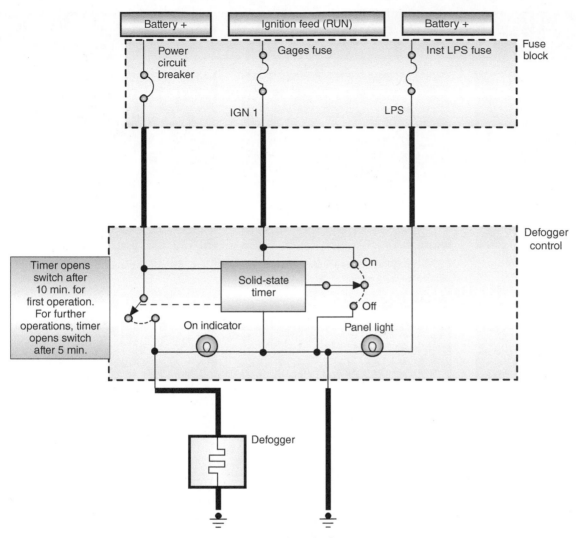

Figure 13-23 The defogger control incorporates a solid-state timer circuit.

PHOTO SEQUENCE 28
Typical Procedure for Grid Wire Repair

P28-1 Tools required to perform this task include masking tape, repair kit, 500°F heat gun, test light, steel wool, Isopropyl alcohol, and a clean cloth.

P28-2 Clean the grid line area to be repaired by buffing it with steel wool and wiping it clean with a cloth dampened with Isopropyl alcohol. Clean an area about 1/4 inch (6 mm) on each side of the break.

P28-3 Position a piece of tape above and below the grid to control the width of the repair.

PHOTO SEQUENCE 28 (CONTINUED)

P28-4 Mix the hardener and silver plastic thoroughly. If the hardener has crystallized, immerse the packet in hot water.

P28-5 Apply the grid repair material to the repair area using a small stick.

P28-6 Remove the tape.

P28-7 Apply heat to the repair area for 2 minutes. Hold heat gun 1 inch (25 mm) from the repair.

P28-8 Inspect the repair. If it's discolored, apply a coat of tincture of iodine to the repair. Allow to dry for 30 seconds, and then wipe off the excess with a cloth.

P28-9 Test the repair with a test light. Note: it takes 24 hours for the repair to fully cure.

Power Window Diagnosis

Classroom Manual
Chapter 13, page 410

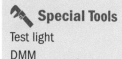 **Special Tools**

Test light
DMM
Fused jumper wires

Usually, removal of the door panel is necessary to gain access to the window motor and regulator. There are several methods used to attach the trim panel to the door frame. Common methods include screws, push pins, clips, and "L" brackets.

Figure 13-24 illustrates a typical door panel removal procedure. Begin by locating and removing any screws that fasten the panel to the frame. This panel uses "L" hooks molded into the panel. This requires the panel to be lifted off the attachment hooks and the lock rod at the same time in order to remove the panel. Once the panel is no longer attached to the door frame, disconnect any electrical connectors and actuating rods to free the panel. Lay the panel on a clean work bench and protect it from damage.

If push pins attach the trim panel, use a trim stick to gently pry the panel from the frame. Be sure to have the push pins properly aligned with their receiver holes prior to pushing the panel back into position on the frame.

Use the illustration in **Figure 13-25** as a guide to diagnosing the power window circuit. If the window does not operate, begin by testing the circuit breaker. Use a test

Figure 13-24 Door panel attachment.

Figure 13-25 Simplified power window circuit.

light or voltmeter to test for voltage on both sides of the circuit breaker. If voltage is present on both sides, the circuit breaker is good. If there is voltage into the circuit breaker but not out of it, it is faulty or tripped due to a short. If there is no voltage into the circuit breaker, then there is an open in the feed from the battery.

If the circuit breaker is good, use fused jumper wires to test the motor. The motor is a reversible motor, so connections to the motor terminals are not polarity sensitive. Disconnect the wire connectors to the motor. Connect battery positive to one of the terminals and ground the other. If the motor does not operate, reverse the fused jumper wire connections. The motor should reverse directions when the polarity is reversed. If the motor does not operate in one or both directions, it's defective and needs replacing.

⚠ **Caution**

Study the service information before attempting to remove the panel. Sometimes there will be hidden screws. Failure to remove these screws before attempting to pull the panel will damage the panel.

⚠ **WARNING** Do not place your hands into the window's operating area. Make the final test connections outside of the door where there is no danger of getting caught in the window track.

If the motor operates with the switches bypassed, the problem is in the control circuit. To test the master switch, connect the test light between terminals 1 and 2 (**Figure 13-26**). The test light should illuminate when the master switch is in the rest (OFF) position. If the light does not glow, there is an open in the circuit to the window switch or from the window switch to ground at terminal 4. Check the ground at terminal 4 for good connections. If the connections are good, continue testing.

If the test light illuminates when connected across terminals 1 and 2, place the switch in the UP position. The test light will go out if the switch is good. Repeat the test between terminals 1 and 3. Place the switch in the DOWN position.

If the master switch is good, test the window switch. Battery voltage should be present at terminal 6. If not, check to see if the lockout switch is closed. Check the circuit from terminal 6 to the circuit breaker. Connect the test light across terminals 5 and 6 (**Figure 13-27**). The light should come on. Move your test light to connect between terminals 6 and 7. Again, the test light should come on. Next, connect the test light between terminals 8 and 9 of the window switch. The test light should be off with the switch in the at-rest position. Placing the window switch in either the UP or the DOWN position should illuminate the test light. If the light does not come on, you need to isolate the problem. It may be the switch, the circuits between the switch and the motor, or the motor itself. Use common test methods to determine the fault.

Slower-than-normal operating speeds indicate excessive resistance or binding in the mechanical linkage. Use the voltage drop test method to locate the cause of excessive resistance. Excessive resistance can be in the switch circuits, the ground circuit, or the motor. If the problem is mechanical, lubricate the track and check for binding or bent linkage.

The circuit used in this example is typical. However, use the service information for the vehicle you are working on to get the correct wiring schematic.

Figure 13-26 Using a test light to check the operation of a power window master switch.

Figure 13-27 Test light connections for testing the window switch.

⚡ **WARNING** Follow the manufacturer's recommended procedure when removing the power window motor. The springs used in window regulators can cause serious injury if removed improperly.

> **CUSTOMER CARE** Testing and repairing the power window system usually requires the removal of the door panels. Always refer to proper service information to determine the correct methods of removal and installation of the panel to prevent damage. Also, use new clips (if applicable) to assure a tight connection and eliminate noise from the panel. Most doors will have a sound dampening material behind the panel. Therefore, you must remember to reinstall this material.

Computer-Controlled Window Systems

Computer-controlled window systems have the same type of motor and regulator as the standard power window system. Consequently, diagnosing these components uses the same methods already discussed. The main differences between the two systems are that the switches on computer-controlled windows are inputs, and motor control is by commands from the module. Like most computer-controlled systems, system faults may result in setting DTCs. Also, the scan tool can perform activation tests of the system.

Figure 13-28 illustrates a system that uses door modules to control the operation of windows and other accessories. Begin by accessing the door module with the scan tool and seeing if DTCs have set. If there are DTCs, then follow the diagnostic procedure for that code. If there are no DTCs, then access the actuation mode of the scan tool and attempt to command the window up and down. If the window operates properly, the problem is likely within the input side of the system. Use the scan tool to see the data stream from the input switches. If the module is not seeing the input, use a DMM to confirm proper voltage to and from the switch.

If the activation fails to operate the window motor, then use a DMM to monitor voltages while performing the activation. Be sure to test both sides of the circuit since the module typically uses H-gate circuits. Connect a DMM test lead to each terminal at the motor. When actuating the window motor, the DMM should show battery voltage. When the motor operates in the opposite direction, the DMM should read a reversed polarity with the same voltage value. If proper voltages are present, replace the motor. If the voltage is not correct, trace the circuit between the motor and the module for opens or shorts. If the circuits are good, replace the module.

⚡ **WARNING** Be careful when using the scan tool to command window operation. Do not place your hands, arms, or head where they can become pinched or trapped when the window moves.

Power Seat Diagnosis

The power seat system is usually very simple to troubleshoot. Test for voltage to the input of the switch control. If voltage is available to the switch, remove it from the seat or arm rest. Using a continuity chart from the service information, test the switch for proper operation. If the switch operates properly, it may be necessary to remove the seat to test the motors and circuits to the motors.

Testing the power seat motors is the same as the power window motor. Be sure to test each armature of the trimotor. If any of the armatures fail to operate, replace the trimotor as a unit.

Classroom Manual
Chapter 13, page 413

Special Tools
Scan tool
DMM
Back-probing tools

Classroom Manual
Chapter 13, page 416

Special Tools
Fused jumper wires
DMM
Test light

Figure 13-28 Computer-controlled power window system.

⚠️ **WARNING** Be careful when making the fused jumper wire connection to test the motor. Do not place your hands where they can become pinched or trapped when the seat moves.

⚠️**WARNING** When replacing the trimotor, follow the manufacturer's service procedures closely. Improper removal of the springs may result in personal injury.

Noisy operation of the seat can generate from the motor, transmission, or cable. If the motor or transmission is the cause of the noise, replace it. A noisy cable can usually be remedied with dry lubricant, provided the cable is not damaged.

Memory Seat Diagnosis

If the seat motors fail to operate under any condition, test the motors and switches as outlined earlier. This section relates only to that portion of the system that operates the memory function.

Classroom Manual
Chapter 13, page 418

 Special Tools
Test light
DMM
Scan tool

Although not all systems store fault codes, you can usually use the scan tool to check inputs and perform actuator tests. If the seat operates properly when using the power seat switches but not the memory feature, the problem is either the inputs or the memory seat control module. Use the scan tool to check the inputs from all switches. Also, test the input from the keyless entry system if it's tied into the memory seats. If the scan tool indicates that the inputs are operating properly, check the connector from the control module to the motors. Also, check the powers and ground for the module. If these are good, replace the module.

If the memory seat system does not support scan tool diagnostics, the problem must be isolated using basic electrical diagnostics. You need to reference the proper service information for the vehicle to determine voltage values and which circuits to test. The following is a typical example of testing the system.

Using the illustration in **Figure 13-29** of a memory seat circuit, all tests are performed at memory seat module connectors C1 and C3. The connectors are disconnected from the module to perform the tests. Place the ignition in the RUN position and the gear selector in the PARK position.

The lamp should illuminate with the test light between C1 connector terminal B and ground. If the light does not come on, check for a circuit fault in the battery feed circuit. Connect the test light between terminals A and B at the C1 connector. If the test light fails to illuminate, check the ground circuit for an open. Move the test light between circuit 39 and ground. The light should turn on. If not, there is an open in circuit 39.

Connect the test light between terminal D of connector C3 and ground. If the light does not turn on, there is a problem in the neutral safety switch circuit. Check the adjustment of the neutral safety switch and circuits 75 and 275 for opens. With the test light connected between terminal B of the C3 connector and ground, the test light should remain off. An illuminated light indicates a faulty left-seat switch assembly.

Leave the test light connected between terminal B of the C3 connector and ground. Place the memory select switch in position 1. If the test light does not illuminate, check circuit 615 for an open. If the wire is good, use an ohmmeter to test the memory select switch for an open. Release the memory select switch and press the EXIT button. The test light should light. If the light fails to illuminate, there is a fault in the left-seat switch assembly. It, therefore, must be replaced.

Move the probe of the test light to terminal A of the C3 connector and press the exit button. If the light fails to illuminate, the problem is in circuit 616 or in the exit switch.

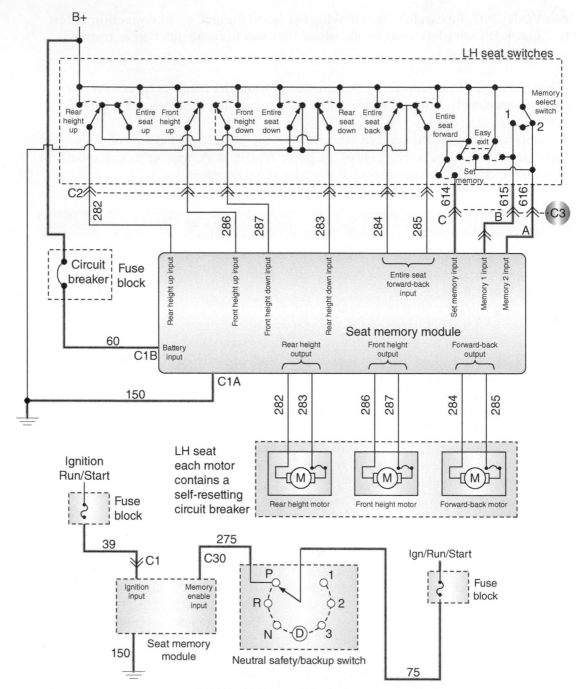

Figure 13-29 A circuit schematic is required to diagnose the memory seat feature.

With the test light still connected to the A terminal of the C3 connector, release the EXIT button. The test light should turn off. If the test light remains illuminated, replace the left-seat switch assembly.

Continue to leave the test light connected to terminal A. Place the memory select switch in the number 2 position. The test light should light. If not, replace the left-seat switch assembly.

With the test light connected between C3 connector terminal C and ground and the memory select switch released, the test light should be off. If it remains on, the seat switch assembly is defective. Press the set memory switch. The test light should illuminate. If not, check the set memory switch and circuit 614 for an open.

The fault is in the control module if all the test results are correct.

Climate-Controlled Seat Service

Changing the duty cycle to an output changes the average voltage applied to a device. This is the premise used to control the temperature of heated seats. In the example given here, we will discuss the diagnostic procedure for the positive temperature coefficient-type heater element.

The intent of the seat heater grid is to get the seat cushions up to operating temperature as fast as possible and then to maintain the heat level. Many heated seats use an HSD-controlled heating circuit. Use a scan tool to diagnose most of these systems since it records DTCs for out-of-range voltage values that may indicate electrical shorts or opens. The system may also diagnose the input switches. Most climate-controlled seat systems rely on several modules to perform their functions. The data bus network transfers the information between these modules. Be sure to include testing of the data bus when diagnosing the system. The following are typical diagnostic procedures for heated seats with a grid in the cushion and seat back (**Figure 13-30**). Note that this system uses a cushion and seat back grid wired in parallel. If the seat cushion warms but the seat back does not (or vice versa), the module and ground are not at fault. The fault would be in the shunt circuit to the inoperative grid or the affected grid itself. Some systems use a series connection between the seat cushion and seat back grids.

Classroom Manual
Chapter 13, page 420

⚒ Special Tools

Scan tool
DMM
Lab scope
Back-probing tools
Fused jumper wire

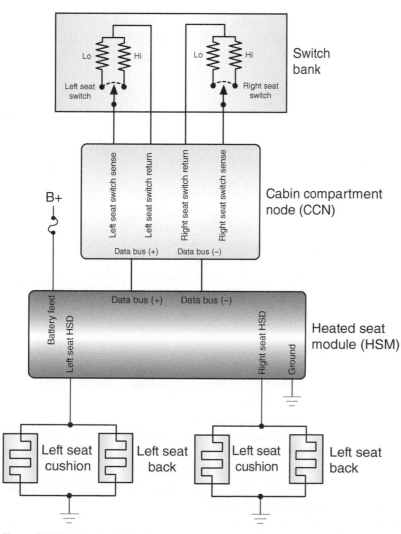

Figure 13-30 Heated seat circuit.

> ⚙️ **SERVICE TIP** Some manufacturers use a low voltage cut-off feature that turns off the heated seat system if battery voltage drops below a specified value. Be certain to check the vehicle's electrical system for proper voltage anytime the power seat system appears inoperative, especially for intermittent operation complaints.

> ⚙️ **SERVICE TIP** When measuring the resistance of the heating element, stress the grid by putting pressure against the cushion. This will help determine if an intermittent open or short in the element is present.

Since this system uses HSDs, they may deactivate the system if a fault is detected. For example, if an open is detected in the circuit before any of the elements or on the common ground so no current flows, HSDs will turn off the system. If you were to check for voltage from the heated seat module (HSM) at this time, the reading would be 0.0 volt. This may fool you into thinking the HSM is defective when in actuality, it's not. For this reason, it's best to test for continuity through the heating element circuits using an ohmmeter. To test the grids, access the wire harness connectors for the heating elements. Usually, this connection is under the seat. With the seat element connector disconnected, measure the resistance between the circuits leading into and out of the inoperative heated seat element. Compare the reading to specifications. The grid wires of each element are wired in parallel, so the measured resistance must fall within the specification range listed. A higher-than-specified reading indicates one or more grid wires are open, and a lower-than-specified reading indicates the grid wires are shorted.

> ⚙️ **SERVICE TIP** On some vehicles, the HSDs are supplied voltage from a relay. If all other circuits that receive voltage from this relay function properly, the relay is not at fault.

To test the heater circuit with a DMM or lab scope, the harness connector needs to be backprobed. If the circuit is disconnected, the HSD fault detection will determine that there is an open circuit and shut off the HSD. Be sure to allow the grid to cool completely. With the DMM or lab scope connected between the heater voltage supply wire and ground, with the ignition in the RUN position, and the heated seats turned to the HIGH position, the circuit should have the specified voltage. If the voltage is less than specification, repair the supply circuit. Some systems will supply a constant 12 volts, while others use PWM or duty cycle.

If circuit control is by PWM or duty cycle, the measured voltage on the supply circuit should begin to drop as the heater grid warms. This drop in voltage is due to the decrease in the duty cycle since the seat is heating. If the voltage does not drop, the heater element or ground circuit is suspect. Use a DMM to perform a voltage drop test of the ground circuit and repair any excessive resistance faults. Retest the heater.

A scan tool may also have the capabilities of monitoring the duty cycle of the heater circuit. The duty cycle should decrease as the heater grid warms. Some systems also display the temperature of the grid on the scan tool. Both indicate whether the heater element and circuit are operating properly.

If the heating element is faulty, it can usually be replaced without replacing the entire cushion. Some systems require removing the faulty grid from underneath the seat trim cover, and a new unit be attached. Other systems leave the old grid in place and apply the new element directly on top of the old element. In this case, the seat trim cover is removed, and the wires that lead from the inoperative heating element are cut off flush with the edge of the old heating element. The new element is installed by peeling off the adhesive backing and sticking the element directly on top of the original heating element. It's important to take care not to fold or crease the element assembly while attaching it. If this occurs, premature failure of the new element may result.

If one or both seats fail to operate, diagnose the input switches for proper operation and transmission of data over the vehicle data bus network. Many systems don't wire the switches directly to the HSM. Instead, the switches are hardwired to a close module and bussed. Study the wiring diagram and service information for the system to determine what modules are involved. Perform switch diagnostics through the module that receives the hardwired input with a scan tool. Use a DMM to check the integrity of the circuit between the switch and the module. Also, an ohmmeter can be used to test the switches.

> ⚙ **SERVICE TIP** Not all HSD heaters are duty-cycle controlled. Be sure to check with proper service information.

If switch inputs are being received by the HSM and the elements are not faulty, the module is suspect. Most systems use one control module for both front seats. If one seat operates but the other does not, then the power and ground circuits to the HSM are not at fault. However, if both seats fail to operate, check powers and grounds before deciding that the module is faulty.

Seats can be heated or cooled using a Peltier element. When used with a climate-controlled seat, the Peltier element is integral to the climate controller (**Figure 13-31**). The controller cools or warms the airflow from the climate control fan motor based on the climate control ECU activation. A temperature sensor monitors the surface temperature of the seat back and cushion. A fan motor provides airflow to the seat cushion and seat back (**Figure 13-32**).

Classroom Manual
Chapter 13, page 421

Figure 13-31 Climate-controlled seat using a Peltier element and a fan.

Figure 13-32 Airflow is directed through the seat cushion by the fan.

⚙ **SERVICE TIP** Some systems don't turn on the heated seat function unless the seat is occupied. A faulty sensor or switch used to determine seat occupancy may prevent the heated seat system from operating. Also, during diagnostics, weight may need to be added to the seat for the heaters to turn on.

Diagnostics of the Peltier element–controlled device is DTC driven. If the circuits are good, the Peltier element is suspect. The module/element is usually serviced as an assembly. Like any other diagnosis routine, this is a process of elimination. By eliminating all other possibilities, such as switch inputs and circuits, the faulty component is the only thing left.

The climate-controlled seats may also use a fan. Most systems use the fan only for cooling, but some also use it for heating. Consider **Figure 13-33** of a seat fan system as an example for diagnosing the circuit. In this example, the HSM controls fan operation. The customer's concern is that the right front seat cooling operation does not function properly.

Use a scan tool to record any DTCs recorded by the HSM. If there are active codes, follow the diagnostic routine in the service information for the retrieved code(s). If the codes are stored, this may indicate that an intermittent problem exists. In this case, erase the DTCs with the scan tool, cycle the ignition to OFF, and then start the engine. Turn the seat cooling ON, and wait for at least 12 seconds before proceeding. Use the scan tool to see if any DTCs reset. If the DTCs don't reset, the conditions that originally caused the codes to set are not present. Verify that the seat harness is routed correctly to the HSM. Check for any harness damage or pinching possibilities along the seat track.

Figure 13-33 Seat fan circuit.

Confirm that the connector terminals are not damaged or pulled out slightly. Also, look for any corroded or contaminated terminals.

> ⚙ **SERVICE TIP** The battery must be fully charged and proper operation of the charging system verified before diagnosing the fan circuit.

> ⚙ **SERVICE TIP** Adjust the seat into several different positions to see if the DTC resets. This would indicate that the harness may be pulled.

If an active DTC returns, turn the ignition switch to the OFF position and disconnect the HSM C5 connector. Inspect the connector and repair it if necessary. If there is no obvious problem, connect an ammeter between the return circuit (P922) and chassis ground at the HSM C5 connector. Connect a fused jumper wire from battery voltage to the motor feed circuit (P422) terminal at the HSM C5 connector. Once the jumper wire is connected, both the seat cushion and seat back fans should operate. In this system, the normal amperage reading should be between 720 and 880 mA. If 0 amp (A) is indicated, one of the circuits is open. Lower-than-normal amperage indicates high resistance in the circuit. Higher-than-normal amperage indicates that a motor may have a shorted winding. If the amperage is per specifications, the HSM is suspect. Confirm proper powers and grounds to the HSM before condemning the module.

If the amperage is not within specifications, isolate whether the problem is with the seat cushion fan or the seat back fan circuit. Use the fused jumper wires and the ammeter to measure the amperage of the seat back motor by itself at connector C351. The reading should be between 320 and 480 mA. If the reading is not within the specifications, the problem is in the circuit or fan of the seat back. If the reading is within specifications, the problem is in the circuit or fan of the seat cushion. Test the circuits using either voltage drop or an ohmmeter. If the circuit is good, replace the fan.

Power Door Lock Diagnosis

To test the door lock motor, apply 12 volts directly to the motor terminals. The actuator rod should complete its travel in less than 1 second. Reverse polarity to test operation in both directions.

Use an ohmmeter to check the switch for continuity. There should be no continuity between any terminals when the switch is in its neutral position. Use the circuit schematic to determine when there should be continuity between terminals.

If the system uses a relay, use the schematic to determine relay circuit operation. In this example, battery voltage should be present at terminal 4 of the connector. Using an ohmmeter, check the ground connections of terminals 1 and 5 of the connector. To test the relay, connect a test light across terminal 3 and ground. Ground terminal 1 and apply power to terminals 2 and 4. The test light will light if the relay is good.

Automatic Door Lock System Troubleshooting

Some systems offer self-diagnostics through the body computer. The service information will provide the steps required to enter diagnostics on these vehicles. The following is an example of locating the fault in vehicles that don't provide this feature when the door locks work but they don't lock or unlock automatically. As with any electrical diagnosis,

Classroom Manual
Chapter 13, page 422

Special Tools
Fused jumper wires
DMM
Test light

Classroom Manual
Chapter 13, page 425

you will need the circuit diagram for the system you are working on. The following steps relate to the system shown (**Figure 13-34**):

Special Tools

DMM
Scan tool
Test light

1. Locate the controller and backprobe for voltage at the power input terminal D with the ignition switch in the RUN position. If there is no voltage, there is an open in circuit 39.
2. Backprobe for voltage between terminals A and D. If there is no voltage, check for an open in circuit 150.
3. Make sure the courtesy lights are off, and all doors are closed. With the gear selector in PARK, turn the ignition to the RUN position.
4. Connect a test light between controller terminal B and a good ground. If the neutral safety switch circuit operates properly, the test light illuminates.
5. With the test light connected as in step 4, move the gear selector to any other position. The test light should go out. If the light does not go out, check the neutral safety switch. It may be out of adjustment or faulty.
6. Leave the gear selector as in step 5 and connect the test light between terminals C and D. The test light should not illuminate. If it does, check circuit 156, the light switch, and the doorjamb switches.
7. Return the gear selector to the PARK position. Connect the test light between terminal H and ground.
8. Observe the test light while the gear selector is moved from PARK to REVERSE. The test light should flash once. If not, replace the controller.

If the circuits pass all tests, check circuits 207 and 195, and the left switch assembly for opens.

Antitheft System Troubleshooting

Classroom Manual
Chapter 13, page 427

This system is also referred to as vehicle security system and vehicle theft security system.

As with many electrical systems, manufacturers take many approaches in designing their antitheft system. Most relay, switch, and circuit tests require only basic electrical troubleshooting capabilities. Use the troubleshooting chart as a guide in locating the fault. Refer to the service information for the correct procedure for arming the system you are diagnosing.

Self-Diagnostic Systems

Some antitheft systems offer self-diagnostic capabilities. The following is a typical example of entering diagnostics. Follow the service procedures for the proper method of entering diagnostics for the vehicle you are working on.

Some vehicle theft security systems enter the diagnostic mode when the ignition is cycled three times from the OFF to the ACC position. When the vehicle theft security system is in diagnostic mode, the horn should sound twice, and the park lamps and tail lamps should flash. If the horn does not sound or the lights don't flash, voltmeter and ohmmeter tests are required to locate the cause of the problem.

Diagnosing many vehicle theft alarm systems requires a scan tool. Follow the scan tool manufacturer's recommended procedure to enter the vehicle theft alarm system diagnostic mode. Entering the diagnostic mode may be indicated by the horn honking twice. This confirms that the trunk lock cylinder is in the proper position. Putting the key in the ignition switch should cause the park lamps and tail lamps to flash.

The following procedures should cause the horn to sound once if the system is operating normally:

1. Activating the power door locks to the LOCKED and UNLOCKED positions.
2. Using the key to lock and unlock each front door.
3. Turning on the ignition to exit the diagnostic mode.

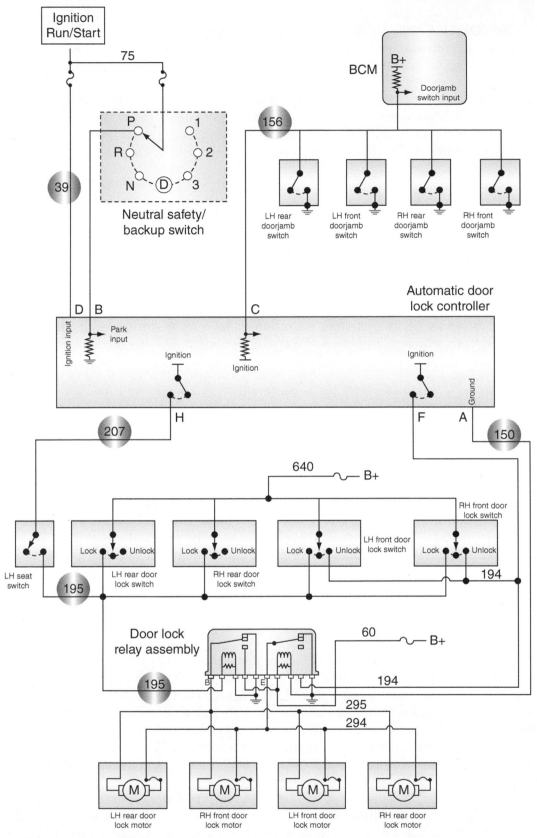

Figure 13-34 Automatic door lock schematic.

CUSTOMER CARE Always check the indicator lights in a customer's vehicle. These lights may signal a dangerous situation, but the customer may not have noticed them. For example, the vehicle theft security system light may not flash after performing the normal system arming procedure. This indicates an inoperative security system. If someone broke into the car, the alarms might not trigger. The customer has paid a considerable amount of money to have this system on the car; therefore, it should be working. Bringing this defect to the customer's attention will probably result in you repairing the system and appreciation from your customer.

Alarm Sounds for No Apparent Reason

Mechanical and corrosion factors on the cylinder tamper switches can cause the system to activate for no apparent reason. If the customer complains of this condition, check the lock cylinder for looseness. Any looseness of the cylinder can cause the switch to activate the alarm.

Other causes of alarm system activation include loose, corroded, or improperly adjusted jamb switches. Proper adjustment of the switches ensures they remain in the OFF position when the doors are fully closed. Switch adjustment is by a nut located at the base of the switch.

⚠ Caution

Failure to disconnect both harnesses will lead to false test indications.

Controller Test

To test the controller used in **Figure 13-35**, disconnect the harness from the controller and the harness to the relay. Connect the test light between the N terminal and ground. The test light should illuminate to indicate voltage to the horns and controller.

Move the test light probe to terminal M. The light should turn on only when the electrical door lock switch is moved to the UNLOCK position. Next, connect the test light between terminal B and a 12-volt source. The light should illuminate only if the doors are locked. The light should go out if any doors are unlocked.

Probe terminal K for voltage. The test light should illuminate only when the ignition is in RUN. Check for a blown fuse or an open circuit if the test light does not illuminate.

To test whether the cylinders are operating properly and have not been tampered with, connect the test light between terminal J and a 12-volt source. The test light should illuminate only if a door is open. If it glows with the doors closed, inspect the lock cylinders for damage.

With the test light connected between terminal H and a 12-volt source, the test light should be on only when the outside door key is in the unlock position. Move the test light between terminal G and a good ground. The light should illuminate while operating the electric door lock switch. This indicates that there is electrical power to the switch. The test light should come on in the LOCK position and go out in the UNLOCK position.

Reconnect the relay harness. With the test light connected between the F terminal of the controller connector and ground the horn should sound, and the lights should turn on. This indicates that the relay coil is functioning.

Next, with the test light connected between terminal E of the controller connector and ground, place the ignition in the RUN position. Use a voltmeter to measure the voltage to the starter. There should be 0 volt. This indicates the starter interrupt relay is opening to prevent the engine from starting.

Connect a fused jumper wire between terminal D and a good ground while observing the security warning light. The warning light should be on. Connect a test light between terminal A of the controller connector and a 12-volt power supply. The test light should illuminate. If not, there is a problem in the ground circuit.

Figure 13-35 Circuit schematic of an antitheft system.

⚙ SERVICE TIP Sometimes it may be easier to attempt starting the engine than to check for voltage at the starter. If the relay works properly, the engine will not start.

Diagnosis and Service of Electronic Cruise Control Systems

Classroom Manual
Chapter 13, page 432

Cruise control systems are also called speed control.

The cruise control system is one of the most popular electronic accessories installed on today's vehicles. During open-road driving, it will maintain a constant vehicle speed without the continued effort of the driver. This reduces driver fatigue and increases fuel economy.

Problems with the system can vary from no operation, to intermittent operation, to not disengaging. To diagnose these system complaints, today's technicians must be able to rely on their knowledge and diagnostic capabilities. Most of the system is tested using familiar diagnostic procedures. Build on this knowledge and ability to diagnose cruise control problems. Use system schematics, troubleshooting diagnostics, and switch continuity charts to assist in isolating the cause of the fault.

⚡ WARNING **When servicing and testing the cruise control system, you will be working close to the airbag and the antilock brake systems. The service manual will instruct you when to disarm and/or depressurize these systems. Failure to follow these procedures can result in injury and additional costly repairs to the vehicle.**

Self-Diagnostics

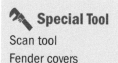
Special Tool
Scan tool

Most cruise control systems have self-diagnostics capabilities. This allows for retrieving trouble codes to assist you in locating system faults. In addition, the electronic cruise control system may support bidirectional control for performing diagnostic tests.

⚠ **Caution**

Do not attempt to place the transmission back into PARK at any time during the test without first stopping the drive wheels with the brakes. Doing so may result in damage to the transmission.

On any vehicle, perform a visual inspection of the system. On electromechanical systems, check the vacuum hoses for disconnects, pinches, loose connections, and so forth. Check and adjust linkage cables or chains, if needed. On all systems, inspect the wiring for tight, clean connections. Also, look for good insulation and proper wire routing. Check the fuses for opens and replace them as needed. Some manufacturers require additional preliminary checks before entering diagnostics. In addition, perform a road test (or simulated road test) in compliance with the service information to confirm the complaint.

Ford IVSC System Diagnostics. Ford's integrated vehicle speed control (IVSC) system has self-test capabilities within the **KOEO** and **KOER** routine of the electronic control assembly (ECA) (**Figure 13-36**). KOEO stands for Key On, Engine Off. It's a static test of the IVSC inputs and outputs. KOER stands for Key On, Engine Running. It's a dynamic check of the engine in operation. Testing of the IVSC system is broken down into two divisions: quick tests and pinpoint tests.

Special Tool
Scan tool
Fender covers

Perform the quick test first. The quick test will check the operation and function of all system components except the vehicle speed sensor. The processor stores the self-test program within its memory. When this test is activated, the processor initiates a function test of the IVSC system to verify that the sensors and actuators are connected and operating properly. It is possible to perform the quick test with a scan tool.

The quick test detects faults that are present at the time of the test. It will not store history codes. Follow the troubleshooting procedures in the service information for any DTCs retrieved. Also, perform the KOEO, KOER, and intermittent (wiggle) test procedures.

After performing service to the system, do the quick test to verify proper operation.

If any failure codes are displayed, perform the **pinpoint test**. This is a specific component test service. If there is a complaint with the cruise control system, and the quick test does not indicate any faults, test the speed sensor.

Diagnosing Systems without Trouble Codes

Systems that don't provide for trouble code diagnostics require you to perform a series of diagnostic tests. The test performed will depend on the symptom. The following sections discuss areas of generic troubleshooting procedures for all types of systems.

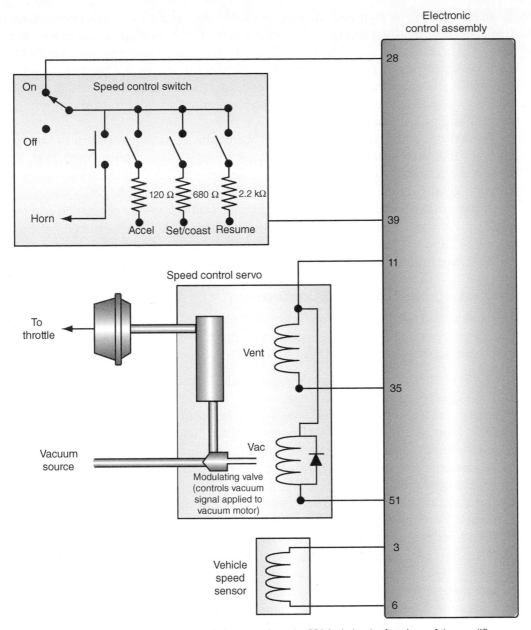

Figure 13-36 The IVSC system is an integrated system where the ECA includes the functions of the amplifier.

Simulated Road Test. The simulated road test is a method of performing a road test without leaving the shop. Before performing this test, connect the shop's ventilation system to the vehicle's exhaust pipe. Lift the drive wheels from the floor and place jack stands under the vehicle. If the vehicle has CV joint shafts, place the jack stands under the lower control arms so the shafts are in their normal drive position. If the vehicle is rear-wheel drive with a solid axle, place the jack stands under the axle.

⚡ **WARNING** Block the wheels that are to remain on the ground. The wheels must remain blocked throughout the test.

Start the engine and place the transmission into drive. You may need to turn off the traction control system so engine torque management doesn't prevent the engine speed from increasing. Turn the speed control switch into the ON position. Accelerate and hold the speed at 35 mph (56 km/h). Press and release the SET ACCEL button. Maintain a slight foot pressure on the accelerator. The speed should be maintained at 35 mph (56 km/h) for a short period of time and then gradually start to surge. The engine surge is caused by operating the system while there is no load on the engine and is normal.

⚠ **Caution**

During this test, the engine may overspeed. If the system should appear to go out of control, you must be ready to turn it off.

⚠ **Caution**

Do not exceed 50 mph (80.5 Km/h), or damage to the differential assembly may result.

Press the OFF button and the engine should decelerate to an idle speed. Stop the drive wheels by lightly applying the brakes. Press the ON button and accelerate to 35 mph (56 km/h). Press and hold the SET/ACCEL button and gradually remove your foot from the accelerator pedal. The engine rpm should begin to increase. Continue to hold the SET ACCEL button until the indicated speed reaches 50 mph (80.4 km/h); then release the button. Vehicle speed should remain at 50 mph (80.4 km/h) for a short period and then the engine will start to surge.

Press the COAST button and hold it. The engine rpm should return to idle speed. Allow the indicated speed to slow to 35 mph (56 km/h) without applying the brakes. When the speed is returned to 35 mph (56 km/h), release the COAST button. Hold the speed at 35 mph (56 km/h) for a short period of time; then the engine will begin to surge.

Tap the brake pedal, which will cause the speed control system to shut off and engine speed to return to idle. Set the indicated speed to 50 mph (80.4 km/h); then use the brakes to slow to 35 mph (56 km/h). Maintain 35 mph (56 km/h) using the accelerator. Depress the RESUME button and the speed should climb to 50 mph (80.4 km/h).

Special Tool

DMM

Vacuum pump

Diagnosing No-Operation Concerns

The first step in a verified no-operation concern is to check all fuses. Next, visually inspect the system for any obvious problems. If the visual inspection does not pinpoint the problem, perform the following steps:

1. Apply the brake pedal to observe proper brake light operation. If the brake lights don't operate, check the switch and circuit. Some brake switches have multiple internal switches; for example, one contact for the brake lights, another contact used as an input to the PCM or TCM, and a third for the cruise control servo circuit. On these systems, just because the brake lights come on does not mean the brake switch is good. Use a scan tool to monitor the different brake switch inputs to confirm proper operation.
2. If the vehicle has a manual transmission, check to ensure that the clutch deactivator switch is operating properly. Use an ohmmeter or voltmeter to test its operation.
3. Check for proper operation of the actuator lever and throttle linkage.
4. Disconnect the vacuum hose between the check valve and the servo (on the servo side of the check valve). Apply 18 inches of vacuum to the open end of the hose to test the check valve. It should hold the vacuum. If not, replace the check valve.
5. Check the vacuum dump valve for proper operation.
6. Test control switches and circuits following the normal electrical diagnostic procedure already learned. Use the circuit diagram and switch continuity charts to aid in testing.
7. Test servo operation.
8. Test speed sensor operation.
9. If all tests indicate proper operation, yet the system is not operational, replace the controller.

Diagnosing Continuously Changing Speeds

If the vehicle speed changes up and down while the cruise control is on, use the following steps to locate the problem:

1. Check the actuator linkage for smooth operation.
2. Test the servo.
3. Check the speed sensor.
4. Check the operation of the vacuum dump valve.
5. Check all electrical connections.
6. If none of these tests locate the fault, replace the amplifier controller.

Diagnosing Intermittent Operation

Intermittent operation is usually caused by loose electrical or vacuum connections. If a visual inspection fails to locate the fault, test drive the vehicle and identify when the intermittent problem occurs. If the problem occurs during normal cruising, begin at step 1. If the problem occurs when operating the control buttons or when rotating the steering wheel, begin with step 3.

1. Connect the vacuum gauge to the hose entering the servo. With the engine running, there should be at least 2.5 inches Hg (8.5 kPa) of vacuum.
2. Test the servo assembly.
3. Use the service information's switch continuity chart and system schematic to test switch operation. Turn the steering wheel through its full range while testing the switches. For example, using the Ford system shown in **Figure 13-37**, this test would be conducted by disconnecting the connector at the amplifier and connecting an ohm-meter between the terminal for circuit 151 and ground (with the ignition switch off). While rotating the steering wheel throughout its full range, make the following checks:

   - Depress the OFF button; the ohmmeter reading should read between 0 and 1 ohm.
   - Depress the SET/ACCEL button and check for a reading between 646 and 714 ohms.
   - Depress the COAST button and the ohmmeter should read between 126 and 114 ohms.
   - When the RESUME button is depressed, the reading should be between 2,310 and 2,090 ohms.

If the resistance values fluctuate while turning the steering wheel, the most likely cause is contamination on the slip rings or faulty clockspring. If the resistance values are above specifications, check the switches and ground circuit.

If the preceding tests (or the road test) fail to identify the fault, conduct a simulated road test while wiggling the electrical and vacuum connections. Monitor the brake switch input with a scan tool. A misadjusted or faulty brake switch can cause the system to shut off if it receives a brake applied input.

Component Testing

Use the normal testing procedures you have already learned to test the safety switches and circuits. Testing of the servo assembly, dump valve, and speed sensor are included to familiarize you with these procedures.

Classroom Manual
Chapter 13, page 433

Servo Assembly. Actuator tests vary depending on design. Some manufacturers use vacuum servos, and others use stepper motors. Be sure to follow the service procedures for the vehicle you are diagnosing. The **servo** controls the position of the throttle by receiving a controlled amount of vacuum. The following servo assembly test is a common test for Ford's cruise control system. Use the schematic in Figure 13-37 to perform the following test.

Disconnect the eight-pin connector to the amplifier. Connect an ohmmeter between circuits 144 and 145. The resistance value should be between 40 and 125 ohms. Move the lead from circuit 145 to circuit 146. The resistance value should read between 60 and 190 ohms. If the resistance levels are out of specifications, check and repair the wiring between the amplifier and the servo. If the wiring is good, replace the servo.

If the resistance values are within specifications, leave the amplifier disconnected and start the engine. Jump 12 volts to circuit 144 and jump circuit 146 to ground. Momentarily jump circuit 145 to ground. The servo actuator arm should pull in, and the engine speed should increase.

Caution
Be ready to abort the test by turning off the ignition if engine rpm should rise to a level where internal damage may result.

WARNING Be sure to have the transmission in PARK or NEUTRAL. Block the wheels and set the parking brake before performing the servo test.

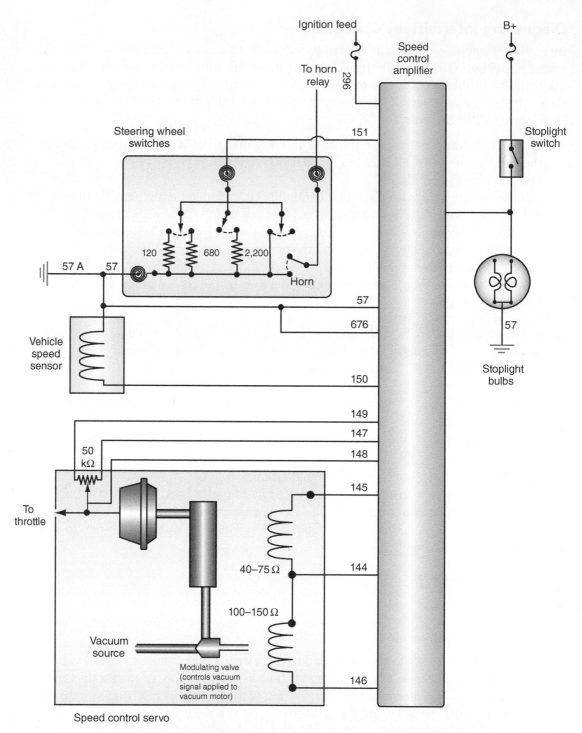

Figure 13-37 Speed control electrical schematic.

Remove the jumper to ground on circuit 146. The servo should release, and engine peed should return to idle. Replace the servo if it does not operate as described.

Dump Valve. The **dump valve** is a safety switch that releases vacuum to the servo when the brake pedal is pressed. A dump valve that is stuck open or leaks will cause a no-operation or an erratic-operation complaint. Failure of the dump valve to release vacuum may not, by itself, be noticed by the driver. It's part of a fail-safe system. If the dump valve does not release, the electrical switch signal also disengages the cruise control system when the brakes are applied. It's good practice to test the dump valve.

Special Tool

Hand vacuum pump

To test the dump valve, disconnect the vacuum hose from the servo assembly to the dump valve. Connect a hand vacuum pump to the hose and apply vacuum to the dump valve. If vacuum cannot be applied, either the hose or the dump valve is defective.

If the valve holds vacuum, press the brake pedal. The vacuum should be released. If not, adjust the dump valve according to the service procedures. Replace the dump valve if it fails to release the vacuum, and it's properly adjusted.

Speed Sensor. Disconnect the connector from the amplifier (Figure 13-37) and connect an ohmmeter between circuits 150 and 57A. The resistance should be approximately 200 ohms. If the resistance value is less than 200 ohms, check for a short in the circuits between the amplifier and the speed sensor. If there is no wiring problem, there is a short in the sensor's coil. If the resistance value is infinite, there is an open in the wires or in the sensor coil.

To test the sensor separate from the wiring harness, disconnect the wire connector from the sensor and connect the ohmmeter between the two terminals. Do this test after testing at the amplifier connector to determine if there is a fault in the entire circuit.

Special Tool

DMM

Adaptive Cruise Control System Service

The **adaptive cruise control (ACC)** system maintains both the vehicle speed and a set distance between vehicles. The ACC system uses laser radar or infrared transceiver sensors to determine vehicle-to-vehicle distances and relational speeds.

The laser radar or infrared sensor is mounted behind the front grill or fascia (**Figure 13-38**). The infrared system uses an integrated sensor and module assembly, while the laser radar sensor is usually remotely mounted from the distance module.

Hardwire circuits are diagnosed with the DMM using conventional methods. Since the system uses data from several different modules and relies on several modules to activate separate functions of the system, diagnostics will require the use of a scan tool. You will need to access all modules involved with the ACC system and retrieve any DTCs they contain. Use the vehicle module scan feature of the scan tool for this. Remedy the cause of any DTCs. Also, communication errors will result in the ACC system not operating.

Classroom Manual
Chapter 13, page 434

Special Tools

Scan tool
DMM
Test light
Alignment screen
Masking tape
Grease pencil
Tape measure

ACC sensor

Figure 13-38 Usually, the ACC sensor is located behind the front grill.

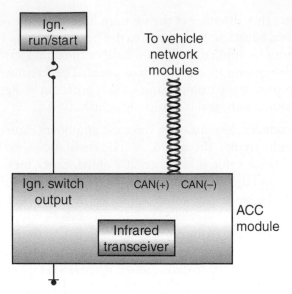

Figure 13-39 ACC sensor circuit.

The following discusses the diagnosis of different faults associated with an infrared system interrogated with the ACC module (**Figure 13-39**). The diagnostics will be similar to systems that use laser radar sensors.

If the voltage input to the ACC module drops below 9 volts while the engine is running, a "voltage low" DTC sets. The module will disable the ACC system as long as the fault is active. This condition can result in intermittent ACC operation due to loose or corroded terminals and connectors. Also, this code can set due to a charging system with intermittent problems. For intermittent conditions, use the wiring information to guide you through a wiggle test and inspection of all connectors and power and ground circuits.

> **CUSTOMER CARE** When servicing the ACC system, talk with the customer to find out the type of driving environment they were experiencing when the reported problem occurred. Heavy rain, snow, ice, and fog can cause the ACC system to be deactivated. System shutdown for these reasons is considered normal and should be properly explained to the customer.

The voltage low fault can result from a faulty ignition feed circuit or a faulty ground circuit. Test both of these circuits while they are under load.

To test the ground circuit, disconnect the ACC module harness connector and use a test light connected to battery voltage. Probe the ground circuit with the test light. The test light should illuminate brightly. Use a voltmeter to read the voltage on the test light probe. If the circuit is good, the voltmeter should read close to 0 volt (**Figure 13-40**). If the test light does not illuminate and the voltmeter shows 12 volts, the ground circuit is open. If the test light did illuminate and the voltmeter reads above 300 mV, there is resistance in the ground circuit.

The ignition feed circuit is tested by attaching the test light to a good ground and moving the probe to the ignition feed circuit terminal in the ACC module harness connector. The test light should illuminate brightly with the ignition placed in the RUN position.

1 Ground
2 Empty
3 CAN (−)
4 Empty
5 Ignition switch output
6 Empty
7 CAN (+)
8 Empty

Figure 13-40 Load testing the ground circuit to the ACC module.

1 Ground
2 Empty
3 CAN (−)
4 Empty
5 Ignition switch output
6 Empty
7 CAN (+)
8 Empty

Figure 13-41 Load testing the power supply circuit to the ACC module.

A voltage reading at the test light probe should read battery voltage (**Figure 13-41**). If the light does not illuminate and the voltmeter reads 0 volt, there is an open in the circuit. If the test light is dim, and the voltmeter reads less than battery voltage, there is high resistance in the circuit.

If the ground and the ignition feed circuits test good (and the voltage low code is still active), the module is faulty and requires repair or replacement.

If the voltage to the module exceeds 16 volts, the module will disable the ACC system and set a "Voltage High" DTC. This fault can be caused by a charging system problem. Use the scan tool and check for any charging system–related DTCs stored in the PCM.

If the charging system operates properly, test the ignition feed and ground circuits as described. If there is no problem in these circuits, the module is faulty.

If the ACC module detects that the sensor lens is damaged or missing, it will disable the ACC system and set a DTC. This DTC can be the result of dirt or debris on the lens, suspension modifications, installation of an aftermarket grill, or the installation of accessories in front of the sensor.

Begin by performing a thorough visual and physical inspection of the ACC sensor. Look for a missing or damaged sensor lens. If the lens is damaged or missing, replace the sensor/ module.

If the lens is not damaged, inspect the ACC module harness connector terminals for corrosion, damage, or terminal pushout. Also, use the wiring information as a guide to inspecting the wiring harness and connectors. Carefully inspect all power and ground circuits. If all tests don't isolate a root cause for the fault, replace the module.

Removing and reinstalling, or replacing the sensor/module assembly requires adjusting the sensor. If the module detects that the sensor is out of adjustment, it will disable the system and set a DTC. This procedure may require a special alignment screen (**Figure 13-42**). In this example, the screen is black with specifically located reflective tape strips. The tape is applied, so there is a single vertical stripe and two inverted "T" shapes.

In preparation for performing the alignment procedure, check and correct the following (if necessary):

1. There is no damage to the suspension, fascia, or grill.
2. The vehicle has a proper wheel alignment.
3. Tire inflation pressures are correct.
4. The vehicle is level.
5. The ACC lens is clean and free from obstructions.
6. There is no load in the vehicle.
7. Sensor assembly mounting screws are properly torqued.

 Caution

Ensure that there are no reflective surfaces or objects within 2 meters (6.56 feet) behind the aiming board. These objects may cause an error during the aiming process and be interpreted as a board misalignment.

Figure 13-42 ACC alignment screen.

Figure 13-43 Vehicle and alignment screen set-up.

Figure 13-44 Sensor height measurement location.

8. The fuel tank is FULL. Add 6.5 pounds (2.94 kg) of weight over the fuel tank for each gallon of missing fuel.
9. Close and latch the hood.

Locate the vehicle's centerline and mark the location on the windshield and rear window. The alignment screen is set 157.5 inches (4 meters) from the sensor (**Figure 13-43**). Use the notch on the top of the screen to align the screen with the center of the vehicle. Adjust the height of the screen to specifications. The height specification depends on the measured distance from the ground to the center of the sensor's vertical adjustment ball screw (**Figure 13-44**). Use the measured distance to determine the height to adjust the screen (**Table 13-1**). The screen height must be equal at both upper corners.

Table 13-1   Specifications for Alignment Screen Height Based on the Height of the Sensor

Sensor Height	Board Height
Height Specifications	Measured in Millimeters
651	853
652	854
653, 654	855
655, 656	856
657, 658	857
659	858
660, 661	859
662, 663	860
664	861
665, 666	862
667, 668	863
669	864
670, 671	865
672, 673	866
674	867
675, 676	868
677, 678	869

Table 13-1   Continued

679, 680	870
681	871
682, 683	872
684, 685	873
686	874
687, 688	875
689, 690	876
691	877
692, 693	878
694, 695	879
696, 697	880
698	881
699, 700	882
701, 702	883
703	884
704, 705	885
706, 707	886
708	887
709, 710	888
711, 712	889
713	890
714, 715	891
716, 717	892
718, 719	893
720	894
721, 722	895
723, 724	896
725	897
726, 727	898
728, 729	899
730, 731	900
732, 733	901
734	902
735, 736	903
737, 738	904
739, 740	905
741	906
742	907

The first step in the procedure is doing a rough alignment. Access the "Align ACC sensor" function in the scan tool. The vertical alignment cover is placed on the alignment screen to cover both of the alignment "T's." This forces the sensor to aim at only the vertical stripe. Use the adjusting screws to adjust the sensor (**Figure 13-45**) until the scan tool indicates the completion of the rough alignment.

Next, remove the vertical alignment cover and perform the fine alignment procedure. This forces the sensor to locate all of the reflections from the tape. Upon completion of this step, the scan tool will clear the internal compensation and inform you that the alignment procedure was successful.

Figure 13-45 ACC sensor/module adjustment locations.

Electronic Sunroof Diagnosis

Troubleshooting the causes of slow, intermittent, or no sunroof operation is a relatively simple procedure. Although there are exceptions, most sunroof systems are stand-alone and not integrated with other systems. Because the system stands alone, diagnostics is like testing any other motor-driven accessory. The following is the diagnostic procedure for troubleshooting the electronic sunroof system shown (**Figure 13-46**). In addition, refer to the diagnostic charts at the end of this chapter to determine the causes of other system malfunctions.

Slow sunroof operation results from excessive resistance in the circuit or motor. Resistance inside the motor results from brush wear, bushing wear, and corroded connections.

Classroom Manual
Chapter 13, page 438

 Special Tools
Test light
DMM
Ammeter

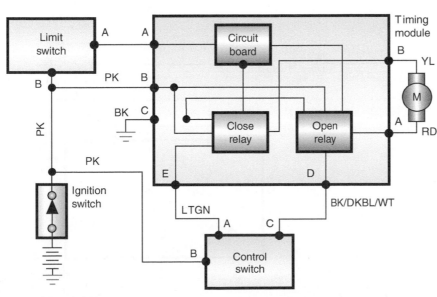

Figure 13-46 Electronic sunroof schematic.

Shorted armature or field coils can also result in slow motor operation. Use an ammeter to test the current draw of the circuit to determine motor condition.

Voltage drop testing isolates the location of excessive circuit resistance. Obtain the correct schematic for the vehicle and follow through the circuit to locate the cause of the excessive resistance. If testing the motor and its circuits does not locate the cause of slow operation, the problem is mechanical. Check the drive cable and tracks for any signs of wear or damage.

Intermittent problems may be the result of loose or corroded connections. To locate the cause of an intermittent fault, operate the system while wiggling the wires. This will assist in isolating the location of the poor connection. Some systems use a circuit breaker to protect the motor. The circuit breaker may be overheating and tripping prematurely, or there may be resistance to glass movement in the rails. The circuit breaker will trip and then cool down and reset. Check that the glass can move easily in the rails. In addition, many intermittent problems result from a faulty control switch. Operate the control switch several times while performing the circuit test. Replace the switch if it fails at any time during the test.

Follow the procedures listed in the service information to locate an intermittent or no-operation fault within the circuit. Perform the usual visual inspections of the circuit before continuing. Be sure to check for proper system grounds.

CASE STUDY

The customer states that although the fan blower motor operates well in low- and medium-speed positions, it runs slow in the high-speed position. The circuit's wiring schematic (**Figure 13-47**) indicates that the resistor block does not control high-speed operation. Because the motor operates properly in low and medium speeds, the fault is not in the motor or the power supply feed. By examining the schematic, the only place that there could be resistance in the high-speed circuit is at the switch wiper or in the circuit between the switch and the resistor block. A voltage drop test confirms that the location of the resistance is the connector to the switch.

Figure 13-47 Three-speed blower motor schematic.

CASE STUDY

The owner of a 2016 Charger says the automatic door lock feature fails to operate. The technician connects a scan tool to the diagnostic link connector (DLC) and initiates the actuator output test. The door locks operate properly while in this test mode. The technician then initiates the switch test. The scan tool indicates that the driver's side door ajar switch is constantly reading open. The service information instructs the technician to disconnect the door ajar switch connector. When disconnected, the scan tool indicates that the switch is open. Next, the technician connects a fused jumper wire across the harness side connector terminals. The scan tool still indicates that the switch is open. According to the service information, the input state should indicate closed when the jumper wire is connected.

The technician tests for voltage from the cabin compartment node (CCN) to the switch connector. The voltage should be 12 volts, but 0 volt is measured. Inspection of the wire between the CCN and the connector is good. According to the service information, the CCN must be replaced since it's not sending out the 12-volt reference signal to the left-front door switch.

The resistance of the terminal lead from the CCN to the switch is measured. Per the service manual specifications, the resistance value is too high, and the CCN requires replacement.

ASE-Style Review Questions

1. All of the following can cause the horn to sound continuously **EXCEPT:**
   A. A short to power in the circuit between the relay and horn.
   B. A short to ground between relay and horn.
   C. The horn switch contacts are stuck closed.
   D. A short to ground between the relay coil and the horn switch.

2. The two-speed windshield wiper operates in HIGH position only.
   Technician A says the low-speed brush may be worn.
   Technician B says the motor has a faulty ground connection.
   Who is correct?
   A. A only
   B. B only
   C. Both A and B
   D. Neither A nor B

3. What is the **LEAST LIKELY** cause of slow windshield wiper operation?
   A. Binding mechanical linkage.
   B. Excessive resistance in the motor's ground circuit.
   C. Short to power at the low-speed brush.
   D. Worn common brush.

4. Aligning the ACC sensor is being discussed.
   Technician A says the sensor's height determines the screen's height.
   Technician B says the screen is positioned directly in front of the sensor.
   Who is correct?
   A. A only
   B. B only
   C. Both A and B
   D. Neither A nor B

5. A customer states that the vehicle alarm will trip when there is no apparent attempt of entry.
   Technician A says the fault may be a loose lock cylinder.
   Technician B says a misadjusted jamb switch may be the cause.
   Who is correct?
   A. A only
   B. B only
   C. Both A and B
   D. Neither A nor B

6. The heater fan motor does not operate in high-speed position.
   Technician A says the cause is a faulty resistor block.
   Technician B says the motor is defective.
   Who is correct?
   A. A only
   B. B only
   C. Both A and B
   D. Neither A nor B

7. The grid of a rear window defogger removes only some areas of fog from the window.
Technician A says the timer circuit is faulty.
Technician B says the grid is damaged.
Who is correct?
   A. A only
   B. B only
   C. Both A and B
   D. Neither A nor B

8. The passenger-side power window does not operate in either direction, despite which switch is used (master or window).
Technician A says the problem is a faulty master switch.
Technician B says the problem is a worn motor.
Who is correct?
   A. A only
   B. B only
   C. Both A and B
   D. Neither A nor B

9. A sunroof is opening slower than normal and appears to be jerky in its operation. The most likely cause of this problem is:
   A. A shorted armature or field coil in the sunroof motor.
   B. A stuck open limit switch.
   C. A stuck closed limit switch.
   D. A stuck OPEN switch.

10. The power door locks will lock the door, but they don't unlock it.
Technician A says the motor is faulty.
Technician B says the unlock relay is faulty.
Who is correct?
   A. A only
   B. B only
   C. Both A and B
   D. Neither A nor B

ASE Challenge Questions

1. The driver-side heated seat shown in **Figure 13-48** fails to warm at any level request. The harness connector for the left heated seat is disconnected, the system activated, and voltage measured. The voltmeter reads 0 volt. What is the **MOST LIKELY** cause of this reading?
   A. The circuit between the HSM and the connector is open.
   B. The circuit between the HSM and the connector is shorted to ground.
   C. The HSM is faulty.
   D. Improper diagnostics of an HSD system.

2. A vehicle's cruise control system is inoperative. Which of the following will not cause this problem?
   A. Stuck closed brake pedal vacuum switch.
   B. Open stop lamp switch.
   C. Inoperative speedometer.
   D. Open cruise engagement switch.

Figure 13-48

3. The computer-controlled power window for the driver's side does not operate in either direction. Measuring the voltage on the switch MUX input circuit shows 12 volts regardless of the position of the switch. Which of the following would be the **MOST LIKELY** cause?

   A. This is normal operation.

   B. Open in the MUX return circuit.

   C. A shorted reference voltage on the MUX circuit.

   D. A short to ground on the MUX return circuit.

4. A completely inoperative GM automatic door lock system is being discussed; the door locks will not open or close manually or automatically.

   Technician A says there may be an open in the power feed circuit to the automatic door lock *controller*.

   Technician B says that one or more doorjamb switches may be faulty.

   Who is correct?

   A. A only   C. Both A and B

   B. B only   D. Neither A nor B

5. The blower motor in **Figure 13-49** operates at a high speed only. The **LEAST LIKELY** cause of this problem would be:

   A. An open in the motor ground circuit.

   B. An open in the control circuit.

   C. A short to ground in the control circuit.

   D. A short to ground in the motor ground circuit.

6. The horn of a vehicle equipped with a horn relay sounds continuously. Which of the following could be the cause of this problem?

   A. Shorted horn relay coil.

   B. Open horn switch.

   C. Grounded wire on the switched side of the horn relay.

   D. Open horn relay coil.

7. The high-speed position of a two-speed windshield wiper system is inoperative; the low-speed position is fine.

   Which of the following could cause this problem?

   A. High resistance on the ground side of the motor.

   B. Open motor resistor.

   C. Faulty park switch.

   D. Worn motor brush.

8. An A/C blower motor turns slowly. A voltmeter connected across the power and ground terminals of the motor indicates 13.6 volts when the blower speed switch is in the HIGH position, and the engine is running; an ammeter indicates that the motor is drawing about 3 amps.

   Technician A says that the blower motor ground circuit may have excessive resistance.

   Technician B says that the blower motor armature may be binding.

   Who is correct?

   A. A only   C. Both A and B

   B. B only   D. Neither A nor B

9. The blower motor of a vehicle equipped with a ground-side-controlled motor circuit runs at high speed whenever the engine runs; the blower switch has no control of the motor speed.

   Technician A says that the ground side of the motor may be shorted to ground.

   Technician B says that the blower switch contacts may have excessive resistance.

   Who is correct?

   A. A only   C. Both A and B

   B. B only   D. Neither A nor B

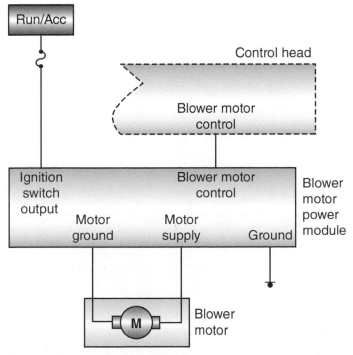

Figure 13-49

10. A power window motor is inoperative. A voltmeter that is connected across the power and ground terminals of the motor indicates 0 volt when the window switch is moved to the DOWN position; when the ground lead of the voltmeter is connected to a chassis ground (and the switch is in the DOWN position), the voltmeter then indicates 12.6 volts.

Technician A says that the window motor has an open internal circuit.

Technician B says that the window switch may have an open contact.

Who is correct?

A. A only

B. B only

C. Both A and B

D. Neither A nor B

Name _____   Date _____

DIAGNOSING A WINDSHIELD WIPER CIRCUIT

Upon completion of this job sheet, you should be able to diagnose an inoperative or poorly working windshield wiper system.

ASE Education Foundation Correlation

This job sheet addresses the following **MLR** tasks:

A.2.	Identify electrical/electronic system components and configuration. **(P-1)**
A.5.	Demonstrate proper use of a digital multimeter (DMM) when measuring source voltage, voltage drop (including grounds), current flow, and resistance. **(P-1)**
A.6.	Demonstrate knowledge of the causes and effects from shorts, grounds, opens, and resistance problems in electrical/electronic circuits. **(P-1)**
A.9.	Use wiring diagrams to trace electrical/electronic circuits. **(P-1)**
G.1.	Demonstrate understanding of vehicle comfort, convenience, access, safety, and related systems operation. **(P-3)**
G.5.	Verify windshield wiper and washer operation; replace wiper blades. **(P-1)**

This job sheet addresses the following **AST** tasks:

A.2.	Identify electrical/electronic system components and configuration. **(P-1)**
A.5.	Demonstrate proper use of a digital multimeter (DMM) when measuring source voltage, voltage drop (including grounds), current flow, and resistance. **(P-1)**
A.6.	Demonstrate knowledge of the causes and effects from shorts, grounds, opens, and resistance problems in electrical/electronic circuits. **(P-1)**
A.9.	Use wiring diagrams during the diagnosis (troubleshooting) of electrical/electronic circuit problems. **(P-1)**
G.1.	Diagnose vehicle comfort, convenience, access, safety, and related systems operation; determine needed action. **(P-2)**
G.5.	Verify windshield wiper and washer operation; replace wiper blades. **(P-1)**
G.7.	Diagnose operation of safety systems and related circuits (such as horn, airbags, seat belt pretensioners, occupancy classification, wipers, washers, speed control/collision avoidance, heads-up display, park assist, and back-up camera); determine needed repairs. **(P-2)**

This job sheet addresses the following **MAST** tasks:

A.2.	Identify electrical/electronic system components and configuration. **(P-1)**
A.5.	Demonstrate proper use of a digital multimeter (DMM) when measuring source voltage, voltage drop (including grounds), current flow, and resistance. **(P-1)**
A.6.	Demonstrate knowledge of the causes and effects from shorts, grounds, opens, and resistance problems in electrical/electronic circuits. **(P-1)**
A.9.	Use wiring diagrams during the diagnosis (troubleshooting) of electrical/electronic circuit problems. **(P-1)**
G.1.	Diagnose vehicle comfort, convenience, access, safety, and related systems operation; determine needed action. **(P-2)**
G.5.	Verify windshield wiper and washer operation; replace wiper blades. **(P-1)**
G.7.	Diagnose operation of safety systems and related circuits (such as horn, airbags, seat belt pretensioners, occupancy classification, wipers, washers, speed control/collision avoidance, heads-up display, park assist, and back-up camera); determine needed repairs. **(P-1)**

We Support

ASE | Education Foundation

Tools and Materials

- A vehicle
- Wiring diagram for the chosen vehicle
- Service information for the chosen vehicle
- DMM

Procedure

1. Describe the general operation of the windshield wipers. Check the operation in all speeds and modes.

   _____

   _____

   _____

   _____

2. Check the mechanical linkages for evidence of binding or breakage. Record your findings.

   _____

   _____

   _____

   _____

3. Locate the wiring diagram for the windshield wiper circuit. Print out the diagram and highlight the power, grounds, and the circuit controls. Attach the printout to this job sheet.

4. Describe how the motor is controlled to operate at different speeds.

   _____

   _____

   _____

   _____

5. Connect the voltmeter across the ground circuit; energize the motor. What was your reading on the meter? What does this indicate?

   _____

   _____

   _____

   _____

6. Probe the power feed to the motor in the various switch positions. Observe your voltmeter readings. What were they? What is indicated by these readings?

   _____

   _____

   _____

   _____

7. Describe the general operation of the windshield wiper motor.

_____

_____

_____

_____

Instructor's Evaluation:

Grade	Rubric
0	No exposure due to: ☐ Absence ☐ Other:
1	Student unsuccessfully attempted the task
2	Student completed the task with significant instructor assistance.
3	Satisfactory performance of task. The student was able to perform elements of the task with minimal instructor assistance.
4	Competent/Proficient. Able to perform all elements of the task successfully and independently.

Instructor's Notes:

Name _____   Date _____

TESTING THE WINDSHIELD WASHER CIRCUIT

Upon completion of this job sheet, you should be able to diagnose problems in the windshield washer circuit.

ASE Education Foundation Correlation

This job sheet addresses the following **MLR** tasks:

A.2. Identify electrical/electronic system components and configuration. **(P-1)**

A.5. Demonstrate proper use of a digital multimeter (DMM) when measuring source voltage, voltage drop (including grounds), current flow, and resistance. **(P-1)**

A.6. Demonstrate knowledge of the causes and effects from shorts, grounds, opens, and resistance problems in electrical/electronic circuits. **(P-1)**

A.8. Use fused jumper wires to check operation of electrical circuits per service information. **(P-2)**

A.9. Use wiring diagrams to trace electrical/electronic circuits. **(P-1)**

G.1. Demonstrate understanding of vehicle comfort, convenience, access, safety, and related systems operation. **(P-3)**

G.5. Verify windshield wiper and washer operation; replace wiper blades. **(P-1)**

This job sheet addresses the following **AST** tasks:

A.2. Identify electrical/electronic system components and configuration. **(P-1)**

A.5. Demonstrate proper use of a digital multimeter (DMM) when measuring source voltage, voltage drop (including grounds), current flow, and resistance. **(P-1)**

A.6. Demonstrate knowledge of the causes and effects from shorts, grounds, opens, and resistance problems in electrical/electronic circuits. **(P-1)**

A.8. Use fused jumper wires to check operation of electrical circuits per service information. **(P-1)**

A.9. Use wiring diagrams during the diagnosis (troubleshooting) of electrical/electronic circuit problems. **(P-1)**

G.1. Diagnose vehicle comfort, convenience, access, safety, and related systems operation; determine needed action. **(P-2)**

G.5. Verify windshield wiper and washer operation; replace wiper blades. **(P-1)**

G.7. Diagnose operation of safety systems and related circuits (such as horn, airbags, seat belt pretensioners, occupancy classification, wipers, washers, speed control/collision avoidance, heads-up display, park assist, and back-up camera); determine needed repairs. **(P-2)**

This job sheet addresses the following **MAST** tasks:

A.2. Identify electrical/electronic system components and configuration. **(P-1)**

A.5. Demonstrate proper use of a digital multimeter (DMM) when measuring source voltage, voltage drop (including grounds), current flow, and resistance. **(P-1)**

A.6. Demonstrate knowledge of the causes and effects from shorts, grounds, opens, and resistance problems in electrical/electronic circuits. **(P-1)**

A.8. Use fused jumper wires to check operation of electrical circuits per service information. **(P-1)**

(Continued)

A.9.   Use wiring diagrams during the diagnosis (troubleshooting) of electrical/electronic circuit problems. **(P-1)**

G.1.   Diagnose vehicle comfort, convenience, access, safety, and related systems operation; determine needed action. **(P-2)**

G.5.   Verify windshield wiper and washer operation; replace wiper blades. **(P-1)**

G.7.   Diagnose operation of safety systems and related circuits (such as horn, airbags, seat belt pretensioners, occupancy classification, wipers, washers, speed control/collision avoidance, heads-up display, park assist, and back-up camera); determine needed repairs. **(P-1)**

We Support
| Education Foundation

Tools and Materials

- A vehicle
- Wiring diagram for the chosen vehicle
- DMM
- Hand tools
- Fused jumper wire

Procedure

1. Describe the general operation of the windshield wipers.

    _____

    _____

    _____

    _____

2. Check the fluid level in the washer fluid reservoir.

    Replenish the level, if necessary.

3. Remove the fluid output line from the washer pump. Activate the washer pump. Does fluid come out of the pump? _____

    What are your conclusions from this?

    _____

    _____

    _____

4. If fluid comes out of the pump but none sprays on the windshield, disconnect the fluid lines from the washer fluid nozzles. Activate the pump. Does fluid come out of the pump? _____

    What are your conclusions from this?

    _____

    _____

    _____

    _____

5. If the pump does not operate when switched on, run a fused jumper wire from the battery to the pump. What happened? What are your conclusions from this test?

_____

_____

_____

_____

6. Connect the voltmeter across the ground of the pump. Activate the pump. Describe what happened. What are your conclusions from this?

_____

_____

_____

_____

7. What are your recommendations about the windshield washer system?

_____

_____

_____

_____

Instructor's Evaluation:

Grade	Rubric
0	No exposure due to: ☐ Absence ☐ Other:
1	Student unsuccessfully attempted the task
2	Student completed the task with significant instructor assistance.
3	Satisfactory performance of task. The student was able to perform elements of the task with minimal instructor assistance.
4	Competent/Proficient. Able to perform all elements of the task successfully and independently.

Instructor's Notes:

Name _____   Date _____

TESTING A REAR WINDOW DEFOGGER

Upon completion of this job sheet, you should be able to diagnose inoperative and poorly working rear window defogger units.

ASE Education Foundation Correlation ————————————————————

This job sheet addresses the following **MLR** tasks:

A.2.   Identify electrical/electronic system components and configuration. **(P-1)**

A.5.   Demonstrate proper use of a digital multimeter (DMM) when measuring source voltage, voltage drop (including grounds), current flow, and resistance. **(P-1)**

A.6.   Demonstrate knowledge of the causes and effects from shorts, grounds, opens, and resistance problems in electrical/electronic circuits. **(P-1)**

A.9.   Use wiring diagrams to trace electrical/electronic circuits. **(P-1)**

G.1.   Demonstrate understanding of vehicle comfort, convenience, access, safety, and related systems operation. **(P-3)**

This job sheet addresses the following **AST** tasks:

A.2.   Identify electrical/electronic system components and configuration. **(P-1)**

A.5.   Demonstrate proper use of a digital multimeter (DMM) when measuring source voltage, voltage drop (including grounds), current flow, and resistance. **(P-1)**

A.6.   Demonstrate knowledge of the causes and effects from shorts, grounds, opens, and resistance problems in electrical/electronic circuits. **(P-1)**

A.9.   Use wiring diagrams during the diagnosis (troubleshooting) of electrical/electronic circuit problems. **(P-1)**

G.1.   Diagnose vehicle comfort, convenience, access, safety, and related systems operation; determine needed action. **(P-2)**

G.7.   Diagnose operation of safety systems and related circuits (such as horn, airbags, seat belt pretensioners, occupancy classification, wipers, washers, speed control/collision avoidance, heads-up display, park assist, and back-up camera); determine needed repairs. **(P-2)**

This job sheet addresses the following **MAST** tasks:

A.2.   Identify electrical/electronic system components and configuration. **(P-1)**

A.5.   Demonstrate proper use of a digital multimeter (DMM) when measuring source voltage, voltage drop (including grounds), current flow, and resistance. **(P-1)**

A.6.   Demonstrate knowledge of the causes and effects from shorts, grounds, opens, and resistance problems in electrical/electronic circuits. **(P-1)**

A.9.   Use wiring diagrams during the diagnosis (troubleshooting) of electrical/electronic circuit problems. **(P-1)**

G.1.   Diagnose vehicle comfort, convenience, access, safety, and related systems operation; determine needed action. **(P-2)**

G.7.   Diagnose operation of safety systems and related circuits (such as horn, airbags, seat belt pretensioners, occupancy classification, wipers, washers, speed control/collision avoidance, heads-up display, park assist, and back-up camera); determine needed repairs. **(P-1)**

Tools and Materials

- A vehicle with a rear window defogger
- Wiring diagram for the chosen vehicle
- DMM

Procedure

1. From the wiring diagram, identify the power feed to the rear window defogger. With the circuit activated, check for voltage at the power terminal to the defogger unit. Is battery voltage present? _____

   If no voltage is present, continue your testing at the control switch.

   If much less than battery voltage is present at the terminal, check for excessive resistance in the power circuit.

2. Identify the ground circuit for the defogger; measure the voltage drop across this part of the circuit. Your reading is _____. If it's more than 0.1 volt, check the wire, connectors, and ground connection for the cause of higher-than-normal resistance.

3. If all of the grids in the defogger circuit are good, applied voltage should be dropped across each grid. To test the condition of each grid, measure the voltage drop across each grid, starting from the top grid and moving down. List your findings.

   _____

   _____

   _____

   _____

   _____

4. If one or more of the grids don't drop applied voltage, continue your diagnosis by moving the negative meter lead toward the positive side of the grid and watch the meter. When battery voltage is measured, this indicates that there is an open after that point. The grid should be repaired.

5. On a working grid, connect the positive lead of the meter to the power feed terminal and move the negative lead to the following positions; record your measured voltage.

   3/4 of the way across the grid: _____

   1/2 of the way across the grid: _____

   1/4 of the way across the grid: _____

   Describe what is happening.

   _____

   _____

   _____

   _____

6. Turn off the circuit, measure the resistance across the following points of the grid, and record your resistance readings.

3/4 of the way across the grid: _____

1/2 of the way across the grid: _____

1/4 of the way across the grid: _____

Describe why you had these readings.

_____

_____

_____

_____

Instructor's Evaluation:

Grade	Rubric
0	No exposure due to: ☐ Absence ☐ Other:
1	Student unsuccessfully attempted the task
2	Student completed the task with significant instructor assistance.
3	Satisfactory performance of task. The student was able to perform elements of the task with minimal instructor assistance.
4	Competent/Proficient. Able to perform all elements of the task successfully and independently.

Instructor's Notes:

Name _____   **Date** _____

PWM HIGH-SIDE-DRIVER-ACTUATED HEATED SEAT CIRCUIT TESTING

Upon completion of this job sheet, you will be able to determine proper operation of the heated seat circuit.

ASE Education Foundation Correlation

This job sheet addresses the following **MLR** tasks:

A.2.   Identify electrical/electronic system components and configuration. **(P-1)**

A.5.   Demonstrate proper use of a digital multimeter (DMM) when measuring source voltage, voltage drop (including grounds), current flow, and resistance. **(P-1)**

A.6.   Demonstrate knowledge of the causes and effects from shorts, grounds, opens, and resistance problems in electrical/electronic circuits. **(P-1)**

A.9.   Use wiring diagrams to trace electrical/electronic circuits. **(P-1)**

G.1.   Demonstrate understanding of vehicle comfort, convenience, access, safety, and related systems operation. **(P-3)**

This job sheet addresses the following **AST** tasks:

A.2.   Identify electrical/electronic system components and configuration. **(P-1)**

A.5.   Demonstrate proper use of a digital multimeter (DMM) when measuring source voltage, voltage drop (including grounds), current flow, and resistance. **(P-1)**

A.6.   Demonstrate knowledge of the causes and effects from shorts, grounds, opens, and resistance problems in electrical/electronic circuits. **(P-1)**

A.9.   Use wiring diagrams during the diagnosis (troubleshooting) of electrical/electronic circuit problems. **(P-1)**

A.13.   Test and measure circuit using an oscilloscope and/or graphing multimeter (GMM); interpret results; determine needed action. **(P-2)**

G.1.   Diagnose vehicle comfort, convenience, access, safety, and related systems operation; determine needed action. **(P-2)**

G.7.   Diagnose operation of safety systems and related circuits (such as horn, airbags, seat belt pretensioners, occupancy classification, wipers, washers, speed control/collision avoidance, heads-up display, park assist, and back-up camera); determine needed repairs. **(P-2)**

G.8.   Diagnose body electronic system circuits using a scan tool; check for module communication errors (data communication bus systems); determine needed action. **(P-2)**

This job sheet addresses the following **MAST** tasks:

A.2.   Identify electrical/electronic system components and configuration. **(P-1)**

A.5.   Demonstrate proper use of a digital multimeter (DMM) when measuring source voltage, voltage drop (including grounds), current flow, and resistance. **(P-1)**

(Continued)

A.6. Demonstrate knowledge of the causes and effects from shorts, grounds, opens, and resistance problems in electrical/electronic circuits. **(P-1)**

A.9. Use wiring diagrams during the diagnosis (troubleshooting) of electrical/electronic circuit problems. **(P-1)**

A.13. Test and measure circuit using an oscilloscope and/or graphing multimeter (GMM); interpret results; determine needed action. **(P-1)**

G.1. Diagnose vehicle comfort, convenience, access, safety, and related systems operation; determine needed action. **(P-2)**

G.7. Diagnose operation of safety systems and related circuits (such as horn, airbags, seat belt pretensioners, occupancy classification, wipers, washers, speed control/collision avoidance, heads-up display, park assist, and back-up camera); determine needed repairs. **(P-1)**

G.8. Diagnose body electronic system circuits using a scan tool; check for module communication errors (data communication bus systems); determine needed action. **(P-1)**

We Support
ASE | Education Foundation

Tools and Materials

- Scan tool
- Lab scope
- DMM (2)
- Service information
- Appropriate vehicle

Procedure

1. Using appropriate service information, identify the driver-side heated seat circuits.

   _____

   _____

2. Backprobe the control circuit terminal of the heater seat circuit at either the control module or the grid connectors. Connect two DMMs to the back-probing tools. Set one DMM to read voltage and the second DMM on percent duty cycle (+ trigger).

3. Connect a scan tool to the vehicle, and access the heated seat data.

4. With the ignition switch in the RUN position, record the voltage and heater duty cycle as shown on the scan tool.

   _____

   _____

5. Use the scan tool to start the actuation for the seat heater. Observe the voltage and percent duty-cycle reading on the voltmeters. Record these values when the actuator is started.

   _____

   _____

   _____

6. Observe the DMM volts and percent duty cycle, and record these values after the actuator test has been run for about 30 seconds.

   _____

   _____

7. What does this test prove?

_____

_____

_____

_____

8. Stop the activation and start the engine. Use the heated seat switch to turn the system onto the HI setting. Record the voltage and the percent duty cycle when the heater is first turned on.

_____

_____

_____

9. Record the voltage and the percent duty cycle after the voltages have stabilized.

_____

_____

_____

10. How is this duty cycle different from the duty cycle after the heater has been on for a short time?

_____

_____

_____

_____

Instructor's Evaluation:

Grade	Rubric
0	No exposure due to: ☐ Absence ☐ Other:
1	Student unsuccessfully attempted the task
2	Student completed the task with significant instructor assistance.
3	Satisfactory performance of task. The student was able to perform elements of the task with minimal instructor assistance.
4	Competent/Proficient. Able to perform all elements of the task successfully and independently.

Instructor's Notes:

Name _____   Date _____

LAB SCOPE TESTING OF A BLOWER MOTOR POWER MODULE

Upon completion of this job sheet, you will be able to determine proper operation of the blower motor power module.

ASE Education Foundation Correlation

This job sheet addresses the following **MLR** tasks:

A.2.   Identify electrical/electronic system components and configuration. **(P-1)**

A.5.   Demonstrate proper use of a digital multimeter (DMM) when measuring source voltage, voltage drop (including grounds), current flow, and resistance. **(P-1)**

A.6.   Demonstrate knowledge of the causes and effects from shorts, grounds, opens, and resistance problems in electrical/electronic circuits. **(P-1)**

A.9.   Use wiring diagrams to trace electrical/electronic circuits. **(P-1)**

G.1.   Demonstrate understanding of vehicle comfort, convenience, access, safety, and related systems operation. **(P-3)**

This job sheet addresses the following **AST** tasks:

A.2.   Identify electrical/electronic system components and configuration. **(P-1)**

A.5.   Demonstrate proper use of a digital multimeter (DMM) when measuring source voltage, voltage drop (including grounds), current flow, and resistance. **(P-1)**

A.6.   Demonstrate knowledge of the causes and effects from shorts, grounds, opens, and resistance problems in electrical/electronic circuits. **(P-1)**

A.9.   Use wiring diagrams during the diagnosis (troubleshooting) of electrical/electronic circuit problems. **(P-1)**

A.13.   Test and measure circuit using an oscilloscope and/or graphing multimeter (GMM); interpret results; determine needed action. **(P-2)**

G.1.   Diagnose vehicle comfort, convenience, access, safety, and related systems operation; determine needed action. **(P-2)**

G.8.   Diagnose body electronic system circuits using a scan tool; check for module communication errors (data communication bus systems); determine needed action. **(P-2)**

This job sheet addresses the following **MAST** tasks:

A.2.   Identify electrical/electronic system components and configuration. **(P-1)**

A.5.   Demonstrate proper use of a digital multimeter (DMM) when measuring source voltage, voltage drop (including grounds), current flow, and resistance. **(P-1)**

A.6.   Demonstrate knowledge of the causes and effects from shorts, grounds, opens, and resistance problems in electrical/electronic circuits. **(P-1)**

A.9.   Use wiring diagrams during the diagnosis (troubleshooting) of electrical/electronic circuit problems. **(P-1)**

A.13.   Test and measure circuit using an oscilloscope and/or graphing multimeter (GMM); interpret results; determine needed action. **(P-1)**

G.1.   Diagnose vehicle comfort, convenience, access, safety, and related systems operation; determine needed action. **(P-2)**

G.8.   Diagnose body electronic system circuits using a scan tool; check for module communication errors (data communication bus systems); determine needed action. **(P-1)**

Tools and Materials
- Scan tool
- Lab scope
- DMM
- Service information
- Appropriate vehicle

Procedure

1. Use the appropriate service information and identify the input circuit from the control module to the power module and the output motor circuits.

   _____

   _____

   _____

2. Is the motor speed controlled by PWM of the power side or ground side of the motor?

   _____

3. Set up the lab scope for two channels. Both channels should be set to read 12-volt square-wave patterns.

4. Using back-probing tools, connect channel 1 red lead to the power module terminal for the input circuit. Connect channel 2 red lead to the blower motor output control circuit identified in step 2.

5. With the blower motor turned OFF, record the voltage on.

   Channel 1:_____

   Channel 2:_____

6. Slowly turn the blower switch from LOW to HI while observing the scope pattern. Describe your results.

   _____

   _____

   _____

7. Based on your observations, if the input circuit to the power module is shorted to ground, the blower speed is:

   ☐ HIGH   ☐ LOW   ☐ OFF

8. If the input circuit has an open, the blower operates only at what speed:

   ☐ HIGH   ☐ LOW   ☐ OFF

Instructor's Evaluation:

Grade	Rubric
0	No exposure due to: ☐ Absence ☐ Other:
1	Student unsuccessfully attempted the task
2	Student completed the task with significant instructor assistance.
3	Satisfactory performance of task. The student was able to perform elements of the task with minimal instructor assistance.
4	Competent/Proficient. Able to perform all elements of the task successfully and independently.

Instructor's Notes:

Name _____   Date _____

TESTING THE ELECTRONIC CRUISE CONTROL SYSTEM

Upon completion of this job sheet, you should be able to test the electronic cruise control system using a scan tool or stand-alone diagnostic routines and determine needed repairs.

ASE Education Foundation Correlation

This job sheet addresses the following **MLR** tasks:

A.2.	Identify electrical/electronic system components and configuration. **(P-1)**
A.5.	Demonstrate proper use of a digital multimeter (DMM) when measuring source voltage, voltage drop (including grounds), current flow, and resistance. **(P-1)**
A.6.	Demonstrate knowledge of the causes and effects from shorts, grounds, opens, and resistance problems in electrical/electronic circuits. **(P-1)**
A.9.	Use wiring diagrams to trace electrical/electronic circuits. **(P-1)**
G.1.	Demonstrate understanding of vehicle comfort, convenience, access, safety, and related systems operation. **(P-3)**

This job sheet addresses the following **AST** tasks:

A.2.	Identify electrical/electronic system components and configuration. **(P-1)**
A.5.	Demonstrate proper use of a digital multimeter (DMM) when measuring source voltage, voltage drop (including grounds), current flow, and resistance. **(P-1)**
A.6.	Demonstrate knowledge of the causes and effects from shorts, grounds, opens, and resistance problems in electrical/electronic circuits. **(P-1)**
A.9.	Use wiring diagrams during the diagnosis (troubleshooting) of electrical/electronic circuit problems. **(P-1)**
G.1.	Diagnose vehicle comfort, convenience, access, safety, and related systems operation; determine needed action. **(P-2)**
G.7.	Diagnose operation of safety systems and related circuits (such as horn, airbags, seat belt pretensioners, occupancy classification, wipers, washers, speed control/collision avoidance, heads-up display, park assist, and back-up camera); determine needed repairs. **(P-2)**
G.8.	Diagnose body electronic system circuits using a scan tool; check for module communication errors (data communication bus systems); determine needed action. **(P-2)**

This job sheet addresses the following **MAST** tasks:

A.2.	Identify electrical/electronic system components and configuration. **(P-1)**
A.5.	Demonstrate proper use of a digital multimeter (DMM) when measuring source voltage, voltage drop (including grounds), current flow, and resistance. **(P-1)**
A.6.	Demonstrate knowledge of the causes and effects from shorts, grounds, opens, and resistance problems in electrical/electronic circuits. **(P-1)**
A.9.	Use wiring diagrams during the diagnosis (troubleshooting) of electrical/electronic circuit problems. **(P-1)**
G.1.	Diagnose vehicle comfort, convenience, access, safety, and related systems operation; determine needed action. **(P-2)**
G.7.	Diagnose operation of safety systems and related circuits (such as horn, airbags, seat belt pretensioners, occupancy classification, wipers, washers, speed control/collision avoidance, heads-up display, park assist, and back-up camera); determine needed repairs. **(P-1)**
G.8.	Diagnose body electronic system circuits using a scan tool; check for module communication errors (data communication bus systems); determine needed action. **(P-1)**

We Support

Education Foundation

Tools and Materials
- A vehicle equipped with electronic cruise control
- Scan tool
- Service information

Procedure

1. Does the system have a stand-alone computer, or is the system a part of the PCM functions? ☐ STAND-ALONE   ☐ PCM

2. If possible, perform the simulated road test and record your results.

   _____

   _____

3. Are there any other related symptoms? ☐ Yes   ☐ No

   If yes, describe the symptom.

   _____

   _____

4. Refer to proper service information to determine if a self-diagnostic routine can be performed. If so, describe how to enter diagnostics.

   _____

   _____

   _____

   _____

5. Is the self-diagnostic routine capable of displaying fault codes? ☐ Yes   ☐ No

6. Perform the procedure listed in step 4 and record the results and any DTCs.

   _____

   _____

7. Use a scan tool and access the electronic cruise control system. Does the scan tool indicate that DTCs are present? ☐ Yes   ☐ No

   If yes, record the DTCs.

   _____

   _____

8. Based on the results so far, what tests need to be performed to find the cause of the fault?

   _____

   _____

9. Perform the tests listed in step 8. What is your determination and recommendation?

   _____

   _____

Instructor's Evaluation:

Grade	Rubric
0	No exposure due to: ☐ Absence ☐ Other:
1	Student unsuccessfully attempted the task
2	Student completed the task with significant instructor assistance.
3	Satisfactory performance of task. The student was able to perform elements of the task with minimal instructor assistance.
4	Competent/Proficient. Able to perform all elements of the task successfully and independently.

Instructor's Notes:

Name _____   Date _____

ADAPTIVE CRUISE CONTROL SERVICE

Upon completion of this job sheet, you will verify operation of the laser radar sensor used in adaptive cruise control systems.

ASE Education Foundation Correlation ————————————————

This job sheet addresses the following **MLR** tasks:

A.2.	Identify electrical/electronic system components and configuration. **(P-1)**
A.5.	Demonstrate proper use of a digital multimeter (DMM) when measuring source voltage, voltage drop (including grounds), current flow, and resistance. **(P-1)**
A.6.	Demonstrate knowledge of the causes and effects from shorts, grounds, opens, and resistance problems in electrical/electronic circuits. **(P-1)**
A.9.	Use wiring diagrams to trace electrical/electronic circuits. **(P-1)**
G.1.	Demonstrate understanding of vehicle comfort, convenience, access, safety, and related systems operation. **(P-3)**

This job sheet addresses the following **AST** tasks:

A.2.	Identify electrical/electronic system components and configuration. **(P-1)**
A.5.	Demonstrate proper use of a digital multimeter (DMM) when measuring source voltage, voltage drop (including grounds), current flow, and resistance. **(P-1)**
A.6.	Demonstrate knowledge of the causes and effects from shorts, grounds, opens, and resistance problems in electrical/electronic circuits. **(P-1)**
A.9.	Use wiring diagrams during the diagnosis (troubleshooting) of electrical/ electronic circuit problems. **(P-1)**
G.1.	Diagnose vehicle comfort, convenience, access, safety, and related systems operation; determine needed action. **(P-2)**
G.7.	Diagnose operation of safety systems and related circuits (such as horn, airbags, seat belt pretensioners, occupancy classification, wipers, washers, speed control/collision avoidance, heads-up display, park assist, and back-up camera); determine needed repairs. **(P-2)**
G.8.	Diagnose body electronic system circuits using a scan tool; check for module communication errors (data communication bus systems); determine needed action. **(P-2)**

This job sheet addresses the following **MAST** tasks:

A.2.	Identify electrical/electronic system components and configuration. **(P-1)**
A.5.	Demonstrate proper use of a digital multimeter (DMM) when measuring source voltage, voltage drop (including grounds), current flow, and resistance. **(P-1)**
A.6.	Demonstrate knowledge of the causes and effects from shorts, grounds, opens, and resistance problems in electrical/electronic circuits. **(P-1)**
A.9.	Use wiring diagrams during the diagnosis (troubleshooting) of electrical/ electronic circuit problems. **(P-1)**
G.1.	Diagnose vehicle comfort, convenience, access, safety, and related systems operation; determine needed action. **(P-2)**
G.7.	Diagnose operation of safety systems and related circuits (such as horn, airbags, seat belt pretensioners, occupancy classification, wipers, washers, speed control/collision avoidance, heads-up display, park assist, and back-up camera); determine needed repairs. **(P-1)**
G.8.	Diagnose body electronic system circuits using a scan tool; check for module communication errors (data communication bus systems); determine needed action. **(P-1)**

Tools and Materials

- Scan tool
- DMM
- Test light
- Alignment screen
- Tape
- Grease pencil
- Tape measure
- Wiring diagram
- Vehicle-equipped adaptive cruise control

Procedure

Task 1

1. Where is the sensor located?

   _____

   _____

   _____

2. Is the sensor integrated with the module? ☐ Yes   ☐ No

   If no, where is the module? located?

   _____

   _____

   _____

3. What other modules are used by the ACC system for operation or inputs?

   _____

   _____

   _____

4. Identify the circuits to the sensors.

   _____

   _____

   _____

   _____

5. What supplies the voltage to the sensor? _____

6. Where is the ground for the sensor located? _____

7. Disconnect the sensor harness connector, and use a DMM to measure the voltage on the supply circuit with the ignition switch in the RUN position (engine off). What is your reading? _____

8. Using a test light, load the supply circuit and use the voltmeter to read the voltage at the test light probe. _____

9. If the readings between steps 7 and 8 are different, explain why.

_____

_____

_____

10. Describe how you will test the ground circuit.

_____

_____

_____

11. Test the ground circuit using the method described in step 9, and record your results.

_____

_____

_____

12. With the sensor harness disconnected, use the scan tool to retrieve fault codes.

_____

_____

_____

13. What are the observable symptoms of the ACC system with the sensor disconnected?

_____

_____

_____

14. Reconnect all connectors and clear any DTCs.

Task 2

1. When is the sensor calibration procedure to be performed?

_____

_____

2. What is the required height of the alignment screen?

_____

3. How far from the sensor is the screen to be located?

_____

4. Determine the centerline of the vehicle, and locate the screen in the proper position.

5. What other preparations must be made prior to aligning the sensor?

_____

_____

_____

6. Describe how the alignment procedure is to be performed.

_____

_____

_____

_____

7. Perform the alignment procedure. Was the procedure successful? ☐ Yes   ☐ No

If no, what must be done to correct the problem?

_____

_____

Instructor's Evaluation:

Grade	Rubric
0	No exposure due to: ☐ Absence ☐ Other:
1	Student unsuccessfully attempted the task
2	Student completed the task with significant instructor assistance.
3	Satisfactory performance of task. The student was able to perform elements of the task with minimal instructor assistance.
4	Competent/Proficient. Able to perform all elements of the task successfully and independently.

Instructor's Notes:

_____

Name _____   Date _____

TESTING THE ANTITHEFT SYSTEM OPERATION

Upon completion of this job sheet, you should be able to test the antitheft system operation through the self-test diagnostic routine and determine needed repairs.

ASE Education Foundation Correlation ──────────────────────────────

This job sheet addresses the following **MLR** tasks:

A.2.	Identify electrical/electronic system components and configuration. **(P-1)**
A.5.	Demonstrate proper use of a digital multimeter (DMM) when measuring source voltage, voltage drop (including grounds), current flow, and resistance. **(P-1)**
A.6.	Demonstrate knowledge of the causes and effects from shorts, grounds, opens, and resistance problems in electrical/electronic circuits. **(P-1)**
A.9.	Use wiring diagrams to trace electrical/electronic circuits. **(P-1)**
G.1.	Demonstrate understanding of vehicle comfort, convenience, access, safety, and related systems operation. **(P-3)**

This job sheet addresses the following **AST** tasks:

A.2.	Identify electrical/electronic system components and configuration. **(P-1)**
A.5.	Demonstrate proper use of a digital multimeter (DMM) when measuring source voltage, voltage drop (including grounds), current flow, and resistance. **(P-1)**
A.6.	Demonstrate knowledge of the causes and effects from shorts, grounds, opens, and resistance problems in electrical/electronic circuits. **(P-1)**
A.9.	Use wiring diagrams during the diagnosis (troubleshooting) of electrical/electronic circuit problems. **(P-1)**
G.1.	Diagnose vehicle comfort, convenience, access, safety, and related systems operation; determine needed action. **(P-2)**
G.3.	Diagnose operation of security/antitheft systems and related circuits (such as: theft deterrent, door locks, remote keyless entry, remote start, and starter/fuel disable); determine needed action. **(P-2)**
G.8.	Diagnose body electronic system circuits using a scan tool; check for module communication errors (data communication bus systems); determine needed action. **(P-2)**

This job sheet addresses the following **MAST** tasks:

A.2.	Identify electrical/electronic system components and configuration. **(P-1)**
A.5.	Demonstrate proper use of a digital multimeter (DMM) when measuring source voltage, voltage drop (including grounds), current flow, and resistance. **(P-1)**
A.6.	Demonstrate knowledge of the causes and effects from shorts, grounds, opens, and resistance problems in electrical/electronic circuits. **(P-1)**
A.9.	Use wiring diagrams during the diagnosis (troubleshooting) of electrical/electronic circuit problems. **(P-1)**
G.1.	Diagnose vehicle comfort, convenience, access, safety, and related systems operation; determine needed action. **(P-2)**
G.3.	Diagnose operation of security/antitheft systems and related circuits (such as: theft deterrent, door locks, remote keyless entry, remote start, and starter/fuel disable); determine needed action. **(P-1)**
G.8.	Diagnose body electronic system circuits using a scan tool; check for module communication errors (data communication bus systems); determine needed action. **(P-1)**

Tools and Materials

- A vehicle equipped with an antitheft system
- Scan tool
- Service information

Procedure

1. Can the system self-diagnostic test be performed stand-alone or by using a scan tool? ☐ STAND-ALONE ☐ SCAN TOOL

2. If it can be performed stand-alone, describe how the process is performed.

   _____

   _____

   _____

3. If a scan tool must be used, list the screen menu selections required to perform the self-test.

   _____

   _____

   _____

   _____

4. Enter the diagnostic self-test and record your results.

   _____

   _____

   _____

5. Based on the results so far, what tests need to be performed to find the cause of the fault?

   _____

   _____

6. Perform the tests listed in step 5. What is your determination and recommendation?

   _____

   _____

Instructor's Evaluation:

Grade	Rubric
0	No exposure due to: ☐ Absence ☐ Other:
1	Student unsuccessfully attempted the task
2	Student completed the task with significant instructor assistance.
3	Satisfactory performance of task. The student was able to perform elements of the task with minimal instructor assistance.
4	Competent/Proficient. Able to perform all elements of the task successfully and independently.

Instructor's Notes:

_____

DIAGNOSTIC CHART 13-1

PROBLEM AREA:	Horn system.
SYMPTOMS:	Horn sounds even when switch is not activated.
POSSIBLE CAUSES:	**1.** Faulty horn switch. **2.** Horn control circuit shorted to ground. **3.** Faulty horn relay.

DIAGNOSTIC CHART 13-2

PROBLEM AREA:	Horn system.
SYMPTOMS:	Intermittent horn operation.
POSSIBLE CAUSES:	**1.** Faulty horn switch. **2.** Poor clockspring contacts. **3.** Faulty horn relay. **4.** Poor ground connection at horn. **5.** Poor ground connection at switch.

DIAGNOSTIC CHART 13-3

PROBLEM AREA:	Horn system.
SYMPTOMS:	Horn fails to sound when the horn switch is activated.
POSSIBLE CAUSES:	**1.** Blown fuse. **2.** Faulty horn switch. **3.** Poor clockspring contacts. **4.** Faulty horn relay. **5.** Poor ground connection at horn. **6.** Poor ground connection at switch.

DIAGNOSTIC CHART 13-4

PROBLEM AREA:	Wiper system operation.
SYMPTOMS:	Wipers operate any time the ignition switch is in the RUN position.
POSSIBLE CAUSES:	**1.** Faulty wiper switch. **2.** Defective park switch or activation arm. **3.** Shorted control circuit.

DIAGNOSTIC CHART 13-5

PROBLEM AREA:	Wiper system operation.
SYMPTOMS:	Wipers operate some of the time.
POSSIBLE CAUSES:	**1.** Poor ground connection. **2.** Poor control circuit connection. **3.** Faulty switch. **4.** Worn motor contacts or brushes.

DIAGNOSTIC CHART 13-6

PROBLEM AREA:	Wiper system operation.
SYMPTOMS:	Wipers fail to function when switch is activated.
POSSIBLE CAUSES:	**1.** Blown fuse. **2.** Open in the control circuit. **3.** Short in the control circuit. **4.** Faulty wiper motor. **5.** Poor ground connection. **6.** Mechanical linkage binding. **7.** Faulty wiper switch.

DIAGNOSTIC CHART 13-7

PROBLEM AREA:	Wiper system operation.
SYMPTOMS:	**1.** Wipers operate only at low speed. **2.** Wipers operate only at high speed.
POSSIBLE CAUSES:	**1.** Faulty wiper switch. **2.** Worn brushes. **3.** Poor control circuit connections. **4.** Open in control circuit.

DIAGNOSTIC CHART 13-8

PROBLEM AREA:	Wiper system operation.
SYMPTOMS:	**1.** Wipers stop on the windshield when switch is turned off. **2.** Wipers remain on when wiper switch is turned off.
POSSIBLE CAUSES:	**1.** Faulty park switch. **2.** Activation arm broken or out of adjustment.

DIAGNOSTIC CHART 13-9

PROBLEM AREA:	Wiper system operation.
SYMPTOMS:	Wipers will not shut off.
POSSIBLE CAUSES:	**1.** Faulty switch. **2.** Faulty park switch or circuit.

DIAGNOSTIC CHART 13-10

PROBLEM AREA:	Intermittent wiper system operation.
SYMPTOMS:	No intermittent wiper operation, low and high speeds operate normally.
POSSIBLE CAUSES:	Faulty switch.

DIAGNOSTIC CHART 13-11

PROBLEM AREA:	Wiper motor operation.
SYMPTOMS:	Slow or no wiper operation.
POSSIBLE CAUSES:	**1.** Faulty wiper motor. **2.** Binding wiper linkage.

DIAGNOSTIC CHART 13-12

PROBLEM AREA:	Washer system operation.
SYMPTOMS:	Washer operates without switch activation.
POSSIBLE CAUSES:	**1.** Faulty switch. **2.** Short in control circuit.

DIAGNOSTIC CHART 13-13

PROBLEM AREA:	Washer system operation.
SYMPTOMS:	Intermittent washers operation.
POSSIBLE CAUSES:	**1.** Faulty switch. **2.** Faulty pump. **3.** Poor control circuit connections. **4.** Poor ground connection.

DIAGNOSTIC CHART 13-14

PROBLEM AREA:	Washer system operation.
SYMPTOMS:	Windshield washer fails to operate.
POSSIBLE CAUSES:	**1.** Blown fuse. **2.** Faulty switch. **3.** Faulty motor. **4.** Poor ground connection. **5.** Open or short in control circuit. **6.** Restriction in hoses.

DIAGNOSTIC CHART 13-15

PROBLEM AREA:	Power side window operation.
SYMPTOMS:	Power windows operate slower than normal.
POSSIBLE CAUSES:	**1.** Control circuit resistance. **2.** Faulty motor. **3.** Poor ground connection. **4.** Improperly adjusted regulators. **5.** Binding linkages.

DIAGNOSTIC CHART 13-16

PROBLEM AREA:	Power side window operation.
SYMPTOMS:	Intermittent power window operation.
POSSIBLE CAUSES:	**1.** Faulty switch. **2.** Faulty motor. **3.** Poor ground connections. **4.** Poor control circuit connections. **5.** Binding linkage. **6.** Faulty circuit breaker.

DIAGNOSTIC CHART 13-17

PROBLEM AREA:	Power side window operation.
SYMPTOMS:	Power windows fail to operate when switch is activated.
POSSIBLE CAUSES:	**1.** Faulty switch. **2.** Faulty motor. **3.** Poor ground connections. **4.** Poor control circuit connections. **5.** Binding linkage. **6.** Faulty circuit breaker.

DIAGNOSTIC CHART 13-18

PROBLEM AREA:	Power seat operation.
SYMPTOMS:	Power seats operate slower than normal.
POSSIBLE CAUSES:	**1.** Control circuit resistance. **2.** Faulty motor. **3.** Poor ground connection. **4.** Binding linkages. **5.** Faulty motor transmission.

DIAGNOSTIC CHART 13-19

PROBLEM AREA:	Power seat operation.
SYMPTOMS:	Intermittent power seat operation.
POSSIBLE CAUSES:	**1.** Faulty switch. **2.** Faulty motor. **3.** Poor ground connections. **4.** Poor control circuit connections. **5.** Binding linkage. **6.** Faulty circuit breaker. **7.** Faulty transmission.

DIAGNOSTIC CHART 13-20

PROBLEM AREA:	Power seat operation.
SYMPTOMS:	Power windows fail to operate when switch is activated.
POSSIBLE CAUSES:	**1.** Faulty switch. **2.** Faulty motor. **3.** Poor ground connections. **4.** Poor control circuit connections. **5.** Binding linkage. **6.** Faulty circuit breaker. **7.** Faulty transmission.

DIAGNOSTIC CHART 13-21

PROBLEM AREA:	Memory seat system operation.
SYMPTOMS:	Seat does not move to preset positions; power seat works normally.
POSSIBLE CAUSES:	**1.** Faulty motor position sensor. **2.** Faulty switch. **3.** Faulty control module. **4.** Open control module power feed circuit. **5.** Poor control module ground circuit. **6.** Improper park/neutral switch input. **7.** Bus communications error.

DIAGNOSTIC CHART 13-22

PROBLEM AREA:	Heated seat.
SYMPTOMS:	Implausible voltage on control circuit.
POSSIBLE CAUSES:	**1.** Control circuit shorted to voltage. **2.** Control circuit shorted to ground. **3.** Control circuit open or high resistance. **4.** Heater element. **5.** Control module.

DIAGNOSTIC CHART 13-23

PROBLEM AREA:	Heated seat.
SYMPTOMS:	Voltage on control circuit too low.
POSSIBLE CAUSES:	**1.** Control circuit shorted to ground. **2.** Control circuit open or high resistance. **3.** Heater element. **4.** Control module.

DIAGNOSTIC CHART 13-24

PROBLEM AREA:	Heated seat.
SYMPTOMS:	Voltage on control circuit high.
POSSIBLE CAUSES:	**1.** Control circuit shorted to voltage. **2.** Control circuit open or high resistance. **3.** Heater element. **4.** Control module.

DIAGNOSTIC CHART 13-25

PROBLEM AREA:	Seat fan.
SYMPTOMS:	Seat fails to cool properly.
POSSIBLE CAUSES:	**1.** Control circuit shorted to voltage. **2.** Control circuit open. **3.** Control circuit high resistance. **4.** Fan return circuit open. **5.** Faulty fan motor. **6.** Control module.

DIAGNOSTIC CHART 13-26

PROBLEM AREA:	Rear window defogger operation.
SYMPTOMS:	Rear window defogger fails to clear the window.
POSSIBLE CAUSES:	**1.** Open in the grid. **2.** Excessive circuit resistance. **3.** Faulty switch. **4.** Blown fuse. **5.** Open in control circuit. **6.** Grid connection loose. **7.** Poor ground connection.

DIAGNOSTIC CHART 13-27

PROBLEM AREA:	Rear window defogger operation.
SYMPTOMS:	Intermittent rear window defogger operation.
POSSIBLE CAUSES:	**1.** Faulty switch. **2.** Poor circuit control connection. **3.** Poor ground connection. **4.** Loose connection to grid.

DIAGNOSTIC CHART 13-28

PROBLEM AREA:	Power door lock system operation.
SYMPTOMS:	Power door locks fail to operate some of the time.
POSSIBLE CAUSES:	**1.** Faulty circuit breaker. **2.** Faulty switch. **3.** Poor control circuit connections. **4.** Poor ground connections. **5.** Faulty motor or solenoid. **6.** Faulty relay.

DIAGNOSTIC CHART 13-29

PROBLEM AREA:	Power door lock system operation.
SYMPTOMS:	All door locks don't operate in either direction.
POSSIBLE CAUSES:	**1.** Faulty circuit breaker. **2.** Faulty master switch. **3.** Open in control circuit. **4.** Short in control circuit. **5.** Poor ground connection. **6.** Faulty relay. **7.** Poor relay control circuit connections.

DIAGNOSTIC CHART 13-30

PROBLEM AREA:	Power door lock system operation.
SYMPTOMS:	One door lock does not operate in either direction.
POSSIBLE CAUSES:	**1.** Obstruction or binding of linkage. **2.** Open in control circuit. **3.** Short in control circuit. **4.** Poor ground connection. **5.** Faulty motor or solenoid.

DIAGNOSTIC CHART 13-31

PROBLEM AREA:	Power door lock system operation.
SYMPTOMS:	All locks work from one switch only.
POSSIBLE CAUSES:	**1.** Faulty switch. **2.** Open in control circuit. **3.** Short in control circuit.

DIAGNOSTIC CHART 13-32

PROBLEM AREA:	Cruise control operation.
SYMPTOMS:	Cruise control speed changes over or below set requests.
POSSIBLE CAUSES:	1. Faulty servo. 2. Defective controller. 3. Faulty speed sensor. 4. Throttle linkage adjustment. 5. Faulty amplifier. 6. Faulty dump valve.

DIAGNOSTIC CHART 13-33

PROBLEM AREA:	Cruise control operation.
SYMPTOMS:	Cruise control fails to set.
POSSIBLE CAUSES:	1. Faulty switch or circuit. 2. Faulty servo or circuit. 3. Defective controller. 4. Poor ground connection. 5. Poor control circuit connections. 6. Faulty relay or circuit. 7. Faulty speed sensor. 8. Faulty speed sensor circuit. 9. Throttle linkage adjustment. 10. Faulty or misadjusted brake switch.

DIAGNOSTIC CHART 13-34

PROBLEM AREA:	Adaptive cruise control.
SYMPTOMS:	Disabling of system.
POSSIBLE CAUSES:	1. Wiring harness, terminal, connector damage. 2. Damaged sensor lens. 3. Obstruction in front of sensor. 4. Improper sensor alignment. 5. High resistance in sensor voltage supply circuit. 6. Open in sensor voltage supply circuit. 7. High resistance in sensor ground circuit. 8. Open in sensor ground circuit. 9. Sensor supply circuit shorted to sensor ground circuit. 10. Sensor supply circuit shorted to ground. 11. Faulty sensor. 12. Internal controller fault.

DIAGNOSTIC CHART 13-35

PROBLEM AREA:	Sunroof operation.
SYMPTOMS:	Sunroof opens or closes slower than normal.
POSSIBLE CAUSES:	**1.** Binding track. **2.** Excessive circuit resistance. **3.** Worn motor. **4.** Misadjusted linkage. **5.** Trim panel mispositioned. **6.** Cable guides mispositioned. **7.** Slipping motor clutch.

DIAGNOSTIC CHART 13-36

PROBLEM AREA:	Sunroof operation.
SYMPTOMS:	Intermittent sunroof operation.
POSSIBLE CAUSES:	**1.** Worn or defective motor. **2.** Faulty controller. **3.** Loose ground connection. **4.** Poor control circuit connection. **5.** Binding linkage and/or tracks.

DIAGNOSTIC CHART 13-37

PROBLEM AREA:	Sunroof operation.
SYMPTOMS:	Sunroof fails to move in either direction.
POSSIBLE CAUSES:	**1.** Worn or defective motor. **2.** Faulty controller. **3.** Loose ground connection. **4.** Poor control circuit connection. **5.** Binding linkage and/or tracks. **6.** Faulty circuit breaker or fuse. **7.** Trim panel mispositioned. **8.** Cable guides mispositioned. **9.** Slipping motor clutch.

DIAGNOSTIC CHART 13-38

PROBLEM AREA:	Antitheft system operation.
SYMPTOMS:	System fails to arm or to operate some of the time.
POSSIBLE CAUSES:	**1.** Faulty controller. **2.** Defective switches. **3.** Poor wire connections. **4.** Poor ground connections.

DIAGNOSTIC CHART 13-39

PROBLEM AREA:	Antitheft system operation.
SYMPTOMS:	System won't disarm.
POSSIBLE CAUSES:	**1.** Lock cylinder disarm switch loose. **2.** Open in the lock cylinder disarm switch circuit. **3.** Defective lock cylinder disarm switch. **4.** Defective lock cylinder. **5.** Faulty control module power and ground. **6.** Faulty control module.

DIAGNOSTIC CHART 13-40

PROBLEM AREA:	Antitheft system operation.
SYMPTOMS:	System trips and sounds alarm by itself.
POSSIBLE CAUSES:	**1.** Doorjamb, hood, or trunk switch loose. **2.** Loose connection in doorjamb switch, hood switch, or trunk switch circuit. **3.** Defective doorjamb, hood, or trunk switch. **4.** Faulty control module.

DIAGNOSTIC CHART 13-41

PROBLEM AREA:	Antitheft system operation.
SYMPTOMS:	System won't trip when door is opened.
POSSIBLE CAUSES:	**1.** Doorjamb switch circuit open. **2.** Faulty doorjamb switch. **3.** Faulty control module.

CHAPTER 14

SERVICING RADIO FREQUENCY AND INFOTAINMENT SYSTEMS

Upon completion and review of this chapter, you should be able to:

- Diagnose remote keyless entry systems.
- Program transmitters into the RKE system.
- Diagnose the immobilizer system.
- Program keys to the immobilizer system.
- Diagnose the tire pressure monitoring system.
- Diagnose the vehicle's audio and video entertainment system for no operation.

- Diagnose the vehicle's audio and video entertainment system for causes of radio noise.
- Diagnose the audio system for noise concerns.
- Service the speakers.
- Diagnose navigational system faults.

🔧 **Basic Tool**

Basic mechanic's tool set

Service information

Terms to Know

Conducted interference	Radio choke	Speakers
Immobilizer system	Radio interference	Tire pressure monitoring (TPM) system
PIN code	Radio noise	
Radiated noise	Remote keyless entry (RKE)	

Introduction

This chapter discusses the diagnosis and servicing of the vehicle infotainment system as well as other radio frequency–operated systems. Usually, the mention of radio frequency (RF) conjures thoughts of audio systems. Although this is one use of RF, today's vehicles can have several systems that operate with RF. These include the remote keyless entry (RKE), remote start, immobilizer, and tire pressure monitoring (TPM) systems.

Remote Keyless Entry and Remote Start

Many modern vehicles utilize a **remote keyless entry (RKE)** system to lock and unlock the doors, turn on the interior lights, and release the trunk latch before approaching the vehicle. A handheld key FOB is the transmitter assembly (**Figure 14-1**) and sends data messages via RF to a small receiver installed inside the vehicle. Pressing a button on the FOB will allow the operation of the RKE system, which is typically a function of the body control module (BCM) or cabin compartment node (CNN). The electronics of the RKE system are responsible for translating coded input from the key FOB into commands for module driver outputs. If the code is accepted, the control module generates a bus message command that is sent to other modules, or performs a direct output to the drivers to execute the requested function. **Figure 14-2** is a simple schematic of an RKE system.

Classroom Manual
Chapter 14, page 450

🔧 **Special Tools**

Scan tool

RF signal meter

DMM

Figure 14-1 RKE FOB transmitter.

Figure 14-2 RKE system schematic.

To diagnose an inoperative system, begin by confirming that the door locks, trunk latch, and interior lamps work normally when manually activated. If the base systems operate as intended, a detailed remote system diagnosis is necessary. Follow the specific steps in the diagnostic procedures for the problem experienced by the customer.

The transponder and receiver system operate at a fixed radio frequency. If the unit does not work from a normal distance, check for two conditions: weak batteries in the remote transmitter or a stronger radio transmitter close by (radio station, airport transmitter, etc.). Use an RF signal meter (**Figure 14-3**) to test the operation of the transponder. Turn on the RF signal meter and operate the transponder. The light-emitting diodes (LEDs) display the signal strength. If the transponder fails to indicate a signal, test the transponder batteries. Also, battery installation is polarity sensitive. If the transponders test good, confirm they are properly programmed to the system.

Confirm that any supporting modules receive the bus communication messages from the control module. These will usually set "Loss of communication" fault codes. The cause can be bus circuit issues or faulty control modules. Always confirm proper power and ground circuit to the control module before condemning them.

Figure 14-3 RF detector used to determine FOB transmission signal strength.

Some perceived problems with the remote start feature may actually be normal operation. For example, most systems will only allow the engine to run for 5 to 10 minutes, and then the engine is shut off. The customer may think the engine did not start if they approach the vehicle after the expiration of the timeout period. In addition, the engine typically can only be started a couple of times without the ignition physically being placed in the RUN position. Starting the engine remotely too many times results in the denial of subsequent requests. The horn will double chirp to indicate the remote start function has been disabled. Reactivating the remote start function requires physically placing the ignition into the RUN position. The customer may mistake this operation of the system as an intermittent fault.

Other inputs to the remote start system include door ajar and hood switches. The remote start function is disabled if a door or the hood isn't properly closed. If the remote start feature isn't functional, confirm the operation of all inputs that can result in disabling the system.

Diagnosing Systems with Aftermarket Kits

Diagnosing problems with aftermarket installations of remote keyless entry and start systems can present some challenges. First, it's completely up to the installer to determine where to locate the components in the vehicle. The control module can be located in the trunk, in one of the side kick panels, behind the dash, under a seat, and so forth. The same is true for the location of relays, diodes, program switches, power feeds, and grounds. The next challenge is determining the function of each circuit to and from the control module. A search on the Web will often provide installation guides for the installed unit. In addition, copies of the owner's manual may provide instruction on how to use the system and program remote transmitters. Also, the installed system may have more than one relay in a single circuit. Finally, one of the most challenging aspects of add-on kits is that the system is designed to work on several different applications instead of being vehicle specific. This means there are different configurations of positive and ground circuits, relay output circuits, alarm disarm circuits, and switched inputs (**Figure 14-4**).

Some systems will have diagnostic capabilities. If a fault is detected, they may flash an LED indicating a problem. By using the program switch to enter diagnostic mode, the LED will flash a trouble code. The process varies based on the manufacturer.

Many of the control circuits operate bypass relays. These relays allow the remote keyless entry/remote start electronic control unit (ECU) to operate the systems by bypassing the factory systems. When diagnosing the systems, first determine if the system works at all. For example, if the customer's issue is that the dome lights don't come on when pressing the FOB's UNLOCK button, you need to confirm the operation of the base system. If opening the doors, or turning on the dome light switch, illuminates the dome lights, then the problem probably lies in the add-on system.

If the base system does not operate at all, disconnect the aftermarket ECU and see if the problem goes away. For example, a problem with the add-on RKE system can cause no, intermittent, or irrational power door lock operation. If unplugging the module (or removing power to it) results in proper power door lock operation, the problem is within the add-on system.

If the aftermarket system is the cause of the issue, use normal voltage and ground testing techniques to isolate the problem. If an installation guide with the manufacturer's wiring diagram isn't available, you should try to trace the circuits to and from the ECM and create your own schematic.

If you determine the problem is within a FOB, refer to the owner's guide to attempt reprogramming the FOBs. This may require the use of a Bitwrite program on some systems. The system may be equipped with a program button that, when activated according to instructions, will erase the FOBs and then allow reprogramming.

Classroom Manual
Chapter 14, page 451

Classroom Manual
Chapter 14, page 452

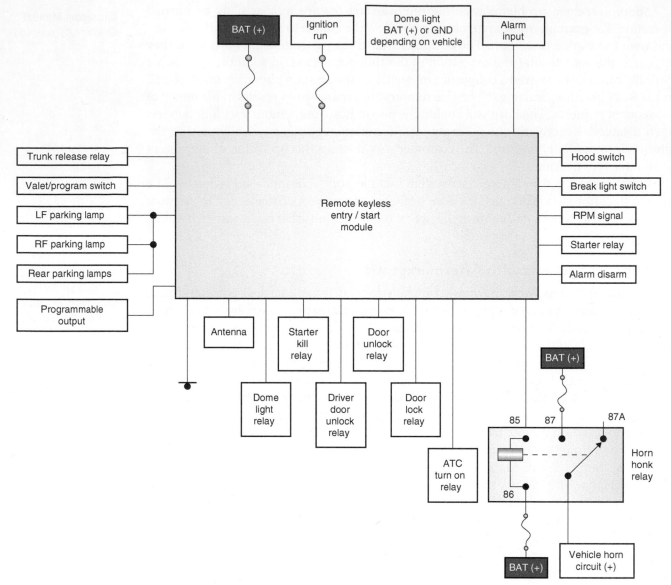

Figure 14-4 Aftermarket remote keyless entry and start systems can present some diagnostic challenges.

Immobilizer System Diagnosis and Service

Classroom Manual
Chapter 14, page 458

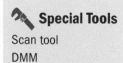 **Special Tools**

Scan tool
DMM

The **immobilizer system** protects the vehicle from unauthorized use by disabling the engine if the use of an invalid key, or an attempt to hot-wire the ignition system, is made. Most systems consist of an immobilizer module, ignition key with transponder chip, the powertrain control module (PCM), and an indicator lamp. The special keys have a transponder chip under the covering (**Figure 14-5**). The immobilizer module radiates an RF signal through its antenna to the key. The transponder signals back to the antenna its identification number. If the message identifies the key as valid, the immobilizer module sends a message over the data bus to the PCM authorizing engine start.

AUTHOR'S NOTE The following service procedures are examples. Different manufacturers will have varying procedures and operation of their systems.

Figure 14-5 Internal transponder located under the cover of the key.

If the response received from the key transponder is missing or identifies the key as invalid, the immobilizer module sends an "invalid key" message to the PCM over the data bus. This prevents the engine from starting. In addition, the indicator lamp provides system status messages. Placing the ignition into RUN initiates a bulb check that normally illuminates the indicator lamp for 3 seconds. After the bulb check is complete, the lamp should go out. A lamp that remains on after the bulb check indicates the immobilizer module has detected a system malfunction, or that the system has become inoperative. A lamp that flashes after completion of the bulb check indicates the detection of an invalid key or that a key-related fault is present.

A common service to the immobilizer system is the addition of keys. Usually, the system can record several different key codes. However, the manufacturer may only provide two keys at the time of vehicle purchase. The purchase of additional keys will require programming them into the system prior to use.

The components of the immobilizer system must become members of a team. At the factory, an End Of Line (EOL) procedure registers all components to the system.

A unique transponder identification code is permanently programmed into each key. During the EOL procedure, the immobilizer module is programmed with a secret key code, Vehicle Identification Number (VIN), and a Personal Identification Number **(PIN) code**. The PIN is a security code necessary to perform most immobilizer system services. You can obtain the PIN from the vehicle owner, the original vehicle invoice, or the vehicle manufacturer.

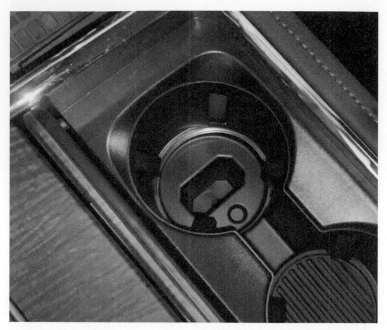

Figure 14-6 The backup start location allows a FOB with a dead battery to start the engine.

The EOL process programs the transponder identification code for each key to the immobilizer's memory. In addition, the key learns the secret key code from the immobilizer module and programs it into its transponder memory. Also, the Secret Key code is programmed into another module on the vehicle as a backup. The backup data may be programmed into a replacement immobilizer module.

Adding or replacing a key to the immobilizer system involves using many of these EOL procedures. The process is automatic during the customer programming routine. When using a scan tool, the process is selected.

Most immobilizer system transponders don't use batteries. If the immobilizer transponder shares the same FOB as the RKE system's transponder, the batteries only support the RKE functions.

Systems that don't require batteries use a piezoelectric crystal that is "excited" when hit by the radio frequencies from the control module's antenna. The movement of the crystal generates a voltage to power the transponder.

In systems that share the batteries between the immobilizer and RKE systems, weak batteries can result in the engine not starting. These systems typically monitor the battery strength and display battery replacement warning messages in the instrument panel or vehicle information center. If the FOB batteries are discharged, the vehicle may have a dedicated backup start location to allow the engine to start (**Figure 14-6**). Typical backup locations include the halo around the Start/Stop button and at the bottom of the center console cupholder. The backup location has an antenna in close proximity that can excite the FOB's transponder chip, allowing bi-directional communications between the FOB and the immobilizer so the engine can start.

Customer Routine for Adding Keys

AUTHOR'S NOTE Customer programming procedures differ between manufacturers. The examples provided here are a couple of methods. Always refer to the proper service information for the exact procedure of the vehicle you are servicing.

Most manufacturers provide a method the customer can follow to program new keys. This option isn't available on all systems. The following is one method for customer programming of the keys.

Performing the customer programming routine requires two valid keys. The immobilizer module needs to see two unique key IDs before it allows programming. Using the same key twice will not work. Also, some systems require that the key being programmed must be a blank key. A key already programmed to a vehicle might not be able to be programmed to any other vehicle. Not following the programming sequence results in the immobilizer aborting the programming process. The following procedure is one method for performing customer programming of a new key:

1. Insert one of the two valid keys into the ignition switch, and place the ignition in the RUN position. Leave the ignition in the RUN position for at least 3 seconds but not more than 15 seconds.
2. Cycle the ignition to the OFF position and remove the key.
3. Within 15 seconds, insert the second valid key into the ignition switch and place the ignition in the RUN position. Leave the ignition in the RUN position while observing the immobilizer warning lamp.
4. In about 10 seconds, the security indicator in the instrument cluster starts to flash, and a single chime sounds. This indicates that the immobilizer system is in programming mode.
5. Within 60 seconds, turn the ignition switch to the OFF position and remove the key.
6. Insert the blank key into the ignition switch, and turn the ignition switch to the RUN position.
7. In about 10 seconds, an audible chime sounds and the security indicator stays illuminated solid for 3 seconds and then turns off. This confirms the successful programming of the new key.
8. Cycle the ignition switch to the OFF position before attempting to start the engine.

Programming FOBs to Start/Stop button systems typically uses the backup start location antenna. If the backup location is the Start/Stop button halo, using the end of the FOB to push the button initiates the backup antenna (**Figure 14-7**). Not all FOBs require the backup location. Some systems only need the FOB to be in the vehicle. Be sure that only one FOB is in the vehicle during the process.

Figure 14-7 Backup start using the top of the key to press the Start/Stop button.

The customer programming of Ford's Intelligent Access PATS begins by placing a programmed FOB into the backup start location and placing the ignition into the ON position for 5 seconds. Turn the ignition OFF and remove the FOB. Within 10 seconds, place the second programmed FOB into the backup starting location and turn the ignition ON again. Wait 5 seconds and turn the ignition OFF and remove the FOB. Wait 5 seconds, place the unprogrammed FOB into the backup location, and turn the ignition ON again. When completed, the information cluster displays a message that confirms successful programming of the key, the hazard lights flash, and the horn sounds.

Secure Service Procedures

Service involving more than adding new transponders usually requires a scan or a specialized immobilizer programming tool. Performing most functions with these tools requires inputting the PIN access code. The scan or programming tool may be able to retrieve the code. If not, you may need to request the code through the automaker's website or Customer Care Center. You may need a Locksmith Identification (LSID) card to obtain the code.

The National Automotive Service Task Force (NASTF) offers the Vehicle Security Professional (VSP) Registry that supplies access to the manufacturer's security information, including key codes, PINs, and immobilizer reset information. The repair facility can specify a dedicated individual who can access the system.

Read all instructions, warnings, and precautions carefully and follow them to the tee (**Figure 14-8**). For example, a warning screen may notify you that the engine will not start unless at least two transponders are registered to the system. In this case, you do not want to erase existing transponders unless you have as least two to program back to the system.

Programming the Ford PATS requires the programming tool to set up a secure session. This requires a reliable high-speed internet connection. Some PATS versions use an INCODE that is generated from an OUTCODE. Others only require a 10-minute "Gaining Security Access" timer. After the timer expires, you have a limited time to perform one of the available options.

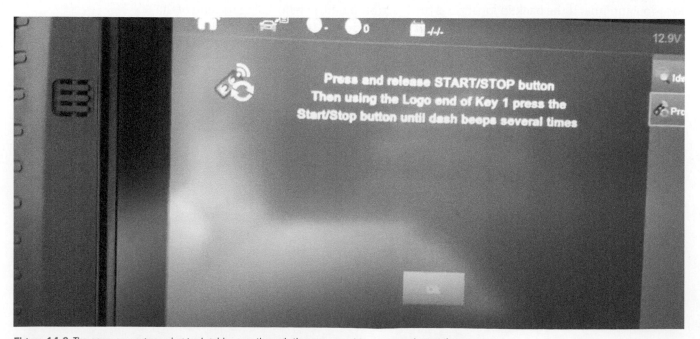

Figure 14-8 The scan or programming tool guides you through the process with on-screen instructions.

Figure 14-9 Special programming tool for accessing the EEPROM.

To program an additional transponder, select "Program Additional Ignition Key." If programming a blade key, insert it into the ignition switch and placed the ignition switch into the RUN position. If a programming a FOB, place it in the backup start location.

The process of programming each additional key completes within a few seconds. The door locks and hazard flashers may cycle to indicate the key is successfully programmed.

If the engine does not start after trying to program a transponder, perform a Parameter Reset or Module Initialization for the BCM, PCM, or ABS module as necessary.

Most immobilizer systems use OBD II DLC for transponder or module programming. However, some European vehicles may require direct programming of the Electrically Erasable Programmable Read-Only Memory (EEPROM) chip) **(Figure 14-9)**. Some EEPROM chips must be removed from the board, programmed, and then resoldered to the board.

If the vehicle owner loses a key, or needs to replace a damaged key, the new transponder IDs will need to programming. Unless two additional keys are still programmed, the technician will need to use the secure method of programming the new key. **Photo Sequence 29** shows a common procedure for programming additional keys.

> Programming additional keys may require repeating the entire procedure for each key. The system automatically exits programming mode after the programming of a blank key.

⚙ **SERVICE TIP** It is best practice to have all previously programmed keys available when programming new keys. It may be necessary to reprogram the original keys also.

Problems with the immobilizer system can result in an engine no-start condition. Since the system uses many hardwired components and circuits, these can be diagnosed and tested using standard diagnostic tools and procedures. However, to reliably diagnose the electronic message inputs used to provide the electronic features of the immobilizer system requires a diagnostic scan tool.

P29-1 Connect the scan tool to access the immobilizer module, and locate the function for programming additional keys.

P29-2 Pressing the ENTER key on the scan tool will start the programming function.

P29-3 Enter the PIN when prompted.

P29-4 Confirm that the PIN entered is correct.

P29-5 Insert the blank key into the ignition switch and turn to the RUN position.

P29-6 Pressing the NEXT key on the scan tool will start the procedure for programming the key.

P29-7 Upon receiving a successful key programming confirmation, you are asked if you wish to program another key.

P29-8 To program another key by repeating the same steps.

P29-9 When all keys are programmed, turn the ignition to the OFF position before attempting to start the engine.

If the diagnostics lead to replacing any module involved with immobilizer operation, there may be specific module replacement and programming procedures. Usually, the replacement module requires initialization. This will require using a scan tool and access to the PIN. This process transfers the required data between modules, so the system is operative, and the existing keys can still be used. Reprogramming the keys to the new immobilizer module may be necessary after the completion of the initialization procedure. However, reusing the same keys is possible since the secret code data matches.

CUSTOMER CARE If the customer has an issue with intermittent no-starts, one problem may be that they have more than one transponder on their key ring. If the transponders are close enough, they may send messages simultaneously. The immobilizer control module will receive these corrupted messages and broadcast an invalid key message, and set a DTC.

Tire Pressure Monitoring Systems

The **tire pressure monitoring (TPM) system** is a safety system that notifies the driver if a tire is underinflated or overinflated. Although the tire pressure sensor (TPS) is usually attached to the valve stem (**Figure 14-10**), some are attached to the wheel with an adjustable strap. The TPS's internal transmitter broadcasts tire pressure and temperature information to a dedicated ECU.

When diagnosing a concern with the TPM system, you need to determine when the system is in operation. Some systems will only transmit tire inflation information when the vehicle is in motion, while others continue to monitor tire pressure even with the vehicle stationary.

The receiver module can set a diagnostic trouble code (DTC) when one of the transmitters fails to produce a signal. The ID number may need programming into the TPM ECU after installing a replacement transmitter in the tire. However, some systems self-learn the new transmitter when the vehicle is driven. Self-learning begins when the ECU receives an unrecognized signal at the same transmission intervals as the recognized transmitters. When this occurs, the ECU stores the ID and monitors that transmitter.

AUTHOR'S NOTE It may be necessary to drive the vehicle up to 10 miles (16 kilometers) to learn the replacement transmitter ID.

When a sensor requires training to the system depends on the type of TPMS. Some indirect TPMSs require recalibration whenever the tires are inflated, rotated, or replaced. How to initiate the training procedure also varies. For example, the indirect TPMS used on the Honda Insight is recalibrated by pressing the TPMS button and holding it down until the indicator light blinks twice. The vehicle is then driven at speeds between 30 and 60 mph (48 and 96 km/h) for approximately a half hour to complete the process.

Classroom Manual
Chapter 14, page 455

 Special Tools
Scan tool
DMM
TPM analyzer
Training magnet

Tire pressure
sensor/transmitter

Figure 14-10 The TPS is attached to the rim as part of the valve stem.

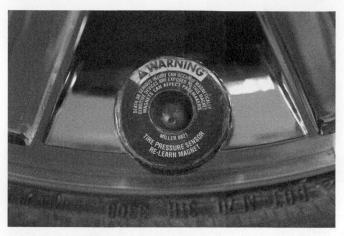

Figure 14-11 Placing the training magnet around the valve stem causes the sensor to transmit.

Figure 14-12 A transponder inside the inner fender well identifies sensor location.

Some systems require training the sensor's location on the vehicle into the receiving module. To learn the sensor's location, some systems require using a scan tool to enter the location learning training mode and then placing a magnet around the tire's valve stem (**Figure 14-11**) in a specified order. Placing the magnet around the valve stem pulls a reed switch closed and causes the TPS to send its data. After learning the location, the ID is locked to that position. Rotating the tires to a different location on the vehicle requires repeating this procedure. Systems that use transponders (**Figure 14-12**) relearn tire position each time the vehicle is in motion.

AUTHOR'S NOTE Tire pressure sensor ID programming is different than location programming.

When received, a new direct TPMS sensor/transmitter may be in "sleep mode" to prevent battery drain. Before the sensor sends data, it needs to be awakened. The process of doing this depends on the manufacturer. Some will require a scan tool, some may require a special frequency generator, while others only require the vehicle to be driven at a specified speed and distance.

Premium systems with frequency transponders may require the use of a tire TPMS sensor/transmitter relearn and analyzer tool (**Figure 14-13**) or a scan tool. The TPMS tool will excite the sensor/transmitter to allow the control module to learn its transponder ID and location.

A special TPS analyzer evaluates a tire pressure sensor's ability to send data. The analyzer interacts with the tire pressure sensor without contact through wireless communication. The analyzer excites the TPS and captures its data for display (**Figure 14-14**). The display includes the transponder ID, the tire pressure reading, and the status of the sensor.

If there is no response from the TPS, the analyzer illuminates the FAIL light. This means either the TPS is faulty or the installed sensor is incompatible with the system.

Some analyzers will communicate with the scan tool to allow for transferring the tire pressure sensor identities and vehicle wheel position to the scan tool for programming into the TPM ECU.

Figure 14-13 TPM analyzer.

Figure 14-14 The TPM analyzer displays the sensor data.

If the system uses tire pressure transponders to excite the sensors, use a scan tool to test transponder operation. Consider **Figure 14-15** of a transponder system. Only three transponders are used since the ECU infers the fourth position. If a problem with the transponder is suspected, use the scan tool to initiate the TPM trigger module test (**Figure 14-16**). This test will verify the communications between the transponder and the TPM ECU. After completing the test, the scan tool displays the results (**Figure 14-17**). Notice that the right front module did not acknowledge the ECU. Referring back to Figure 14-15, there is no transponder in this location.

⚠ **Caution**
Don't use tire sealants that enter by way of the valve stem. The sealant will destroy the sensor.

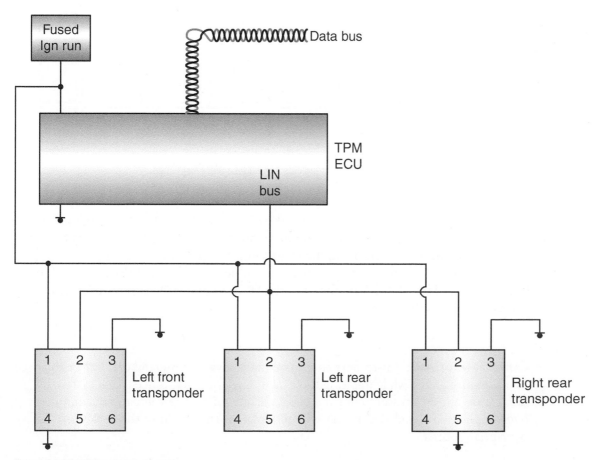

Figure 14-15 TPM system schematic.

Figure 14-16 Trigger module test used to verify communications between transponder and ECU.

Figure 14-17 Results of trigger module tests indicate if the transponder responded.

If a transponder does not acknowledge the ECU, test the ignition run and the ground circuits. If these test good, test the local interconnect network (LIN) bus circuit. If all of these tests don't indicate a problem, the transponder is suspect.

If none of the transponders respond, the problem is likely a short to ground or to voltage in the LIN bus. If this isn't the case, the ECU is faulty. Check powers and grounds to the ECU before condemning it.

CUSTOMER CARE Customers should be made aware that using the TPM display as a pressure gauge while inflating their tires may not work. Since many systems don't transmit air pressure data at a constant rate unless the vehicle is driving over a specified speed, the display will not change. This could result in excessively overinflated tires.

Infotainment System Service

Today's audio/video systems integrate several different functions. They may include AM/FM radio, satellite audio and video, CD player, DVD player, MP3 player, iPod connectivity, hands-free cell phone, navigational systems, cellular services, and Wi-Fi connectivity. A central telematics module may act as a gateway for all the individual systems that encompass the telematics system as a whole. Since the system sets fault codes when a failure is determined, diagnosing the system requires a scan tool.

Vehicle Audio System Service

Warranty and shop policies dictate what repairs are allowed for fixing internal radio faults. It may be possible to sublet repairs of damaged or faulty radio and navigational monitors to a specialty shop. The replacement of backlighting bulbs and LEDs is often an in-house activity. Just because you may not repair internal problems does not excuse you from doing proper diagnostics. Often the tendency is to remove the unit when the customer has described having a particular problem before performing a thorough prediagnosis routine. In many cases, the units are returned to the dealership or shop with "NO TROUBLE FOUND." The problem could have been solved in many cases without removing the unit.

> **SERVICE TIP** Don't blame the radio unless you eliminate all other possible problem sources.

Before removing the radio/component, consider the following:

- Test the vehicle's radio system outdoors and away from tall buildings.
- Most radio noise can be located on weak amplitude modulation (AM) stations at the low-frequency end of the tuning band.
- When using a test antenna, properly ground the base to the vehicle's body. Don't hold the antenna mast in your hands.
- Most noises heard in the radio system enter through the antenna.

External problems such as faulty speakers, broken wires, antenna issues, and so forth can be diagnosed and repaired by the automobile technician. The technician must determine the exact nature of the problem when performing a diagnosis. Determining whether the problem is intermittent or constant, or whether the problem occurs when the vehicle is moving or stationary, will help pinpoint the nature of the problem.

The diagnostic charts at the end of this chapter provide a guide to diagnosing poor radio performance complaints. Remember that the infotainment and telematics systems rely on data bus communications between individual modules. Be sure to confirm the proper operation of the data bus communication system.

If the infotainment or telematics component does not appear to power on, the first step is to check for available voltage and good grounds. Battery voltage must be available to the radio, navigational module, hands-free module, and so forth on the power terminal. Test for voltage on the circuit while the circuit is loaded; don't trust open circuit testing. Most radios require two voltage feeds: direct battery and ignition run (**Figure 14-18**). Also, perform a voltage drop test of the ground circuit. If these circuits are good, an internal problem exists in the radio tuner.

Classroom Manual
Chapter 14, page 462

Special Tools
Scan tool
DMM

Figure 14-18 Audio system circuits.

SERVICE TIP Just because the clock works does not mean there is power to the radio's internal circuitry. Always check for a blown fuse or disconnected wire when diagnosing an inoperative radio.

Wiring diagrams for component systems are more complex. In addition to power and audio signal wires, note that some systems have a serial data wire for microprocessor communication between components to control unit functions.

Most vehicles equipped with an amplifier provide on-board diagnostics using a scan tool. On these systems, diagnosing speaker-related problems is done through the amplifier. For vehicles without an amplifier, speaker-related problems are diagnosed through the radio. Typical fault conditions that the system can detect are speaker circuits shorted to ground, shorted to voltage, shorted together, or open.

Like the radio, the CD/DVD system components generate DTCs for faults that can be retrieved using the scan tool. Use the diagnostic charts at the end of this chapter to assist in identifying common customer complaints for poor CD system performance.

The satellite radio system will also generate DTCs as it performs diagnostic system checks for audio output, presence of the antenna, antenna signal, and current subscription status. If there is no satellite radio audio output from the speakers, first check for an active satellite radio subscription. Also, confirm that there are obstructions to the satellite antenna's field of view. It's best to have the vehicle in an area that provides an unobstructed line of sight to the entire horizon when diagnosing a no-reception complaint. If you determine the subscription is valid and the antenna is unobstructed, the fault is likely a hardware failure. This will require repair or replacement of the satellite radio module, the satellite radio antenna, the vehicle's radio, the amplifier, or the wire harness as directed in the service manual diagnostic procedures.

Another concern may be the display of "Updating Channels" on the radio screen. This may indicate a hardware communication failure. Confirm that the satellite radio module is communicating on the data bus.

If there are no DTCs, begin diagnostics by inspecting the antenna connection. Disconnect the antenna from the telematics module and inspect for damage, corrosion, and proper fit. If this inspection fails to identify an issue, use a test antenna (**Figure 14-19**) connected directly to the telematics module. Attempt to tune in several satellite stations. Replace the telematics module if the reception is still unavailable or of poor quality.

Classroom Manual
Chapter 14, page 465

Classroom Manual
Chapter 14, page 468

Classroom Manual
Chapter 14, page 467

> ⚙ SERVICE TIP The antenna test kit includes a test antenna and different cable connection adapters. Properly ground the test antenna's base to the vehicle. If the test antenna improves reception, the antenna or cable is faulty.

If the system operates properly with the test antenna, work from the telematics module toward the antenna by connecting the test antenna at all antenna connections. This will isolate the location of the issue and determine if it's a cable or antenna fault.

Figure 14-19 Test antenna kit.

Replacing Radios Equipped with HDDs

Radios equipped with internal hard drives can hold over two thousand songs and other data. When installing a replacement radio, you need to attempt to backup the hard drive. After installing the new radio, transfer the backup data to the new radio. Depending on the system, the backup procedure may be available only through the use of a scan tool. Hard drive backup is by using an external hard drive attached to the radio's universal serial bus (USB) port. The following are the typical steps to backing up the HDD:

1. Connect a battery charger to the vehicle's battery and set it to maintain 13 volts. The process of backing up the HDD can take several hours.
2. Turn the ignition to the RUN position with the engine OFF.
3. Power on the external HDD.
4. Connect the external HDD to the radio using the USB cable. The radio display will change to the "Manage My Files" screen.
5. Connect the scan tool and perform the steps to place the radio into "Dealership Mode."
6. Follow the on-screen instructions on the radio to back up the radio HDD.

Antenna Service

Classroom Manual
Chapter 14,
pages 447, 448

Most AM/FM radio complaints are associated with poor reception. An improper ground connection of the radio's antenna is a leading cause of poor reception. The base of the antenna provides a path-to-chassis ground. Resistance in any portion of the vehicle's ground path can affect the overall performance of the audio system. Check the following conditions whenever diagnosing a poor radio reception concern:

- Loose or corroded battery cable terminals.
- High-resistance body and engine grounds.
- Loss of antenna and audio system grounds.

The radio is attached to the antenna lead by a coaxial cable that carries the signal through the center conductor. The cable's outer conductor acts as a shield to protect the radio signal from interference. A secure ground at the body connection is necessary for the cable to be effective as a shield. A poor ground causes excessive ignition noise in AM reception, or erratic sound. To check the antenna ground with a fixed antenna, do the following:

1. Tune the radio for a weak AM station or signal.
2. Unscrew and remove the antenna mast.
3. Connect a short jumper wire between the mast mounting stud and the cable lead.

If the radio no longer receives the radio station, the antenna ground is good. If the radio station is still received, the ground is poor or open.

For vehicles that have a power antenna, check the ground using the following procedures:

1. Fully lower the antenna and then separate the motor harness connector. You may need to unplug the motor while it's running to keep the antenna about an inch from its base.
2. Connect an alligator clip near the top of the antenna. This will serve as the antenna.
3. Tune the radio to a weak AM station.
4. Remove the alligator clip and connect a short jumper between the mounting bracket and the top of the mast.

Like the test for the fixed mast antenna, the antenna ground is good if the radio no longer receives the radio station. If the radio station is still received, the ground is poor or open.

Additional tests of the antenna include verification that there is continuity between the tip of the antenna mast to the tip of the radio connector. Also, verify there is no continuity between the antenna mast and chassis ground.

A test antenna kit is available for determining problems with the antenna or its cable (refer to Figure 14-19). The kit includes a test antenna and different cable connection adapters. The test antenna must be properly grounded to the vehicle. If antenna reception improves while using the test, the antenna or cable is faulty.

Radio Noise Diagnostics

Radio noise is undesired interference or static (popping, clicking, or crackling) obstructing the normal sound of the radio station. Reception to the radio receiver from the antenna can be interfered with by noise resulting from radio frequency interference (RFI). This is especially true of the AM band. **Radio interference** generally refers to all undesirable signals that enter the radio and are heard as static or noise.

Radiated noise results from an accessory that produces a signal that the antenna detects. The noise is picked up by the radio receiver and amplified through the audio circuits. The FM band is susceptible to electromagnetic interference (EMI), but usually isn't as noticeable compared to AM. Control of RFI and EMI noise is done by proper ground connections at the radio antenna base, radio receiver, engine-to-body, and the heater core. In addition, using resistor-type spark plugs and radio suppression secondary ignition wiring reduces radio noise. The radio itself also has internal suppression devices, such as capacitors that shunt AC noise to ground and slows sudden changes of voltage in a circuit.

Common sources of radiated interference include the following:

- Defective ignition components.
- Heater blower, wiper, fan, power seat, and power window motors.
- Electric fuel pumps.
- A loose ground strap or loose antenna mount.
- Noise-producing RFI interference resulting from the operation of an electrical circuit (such as turn signals) may cause a popping noise.

Some systems use a **radio choke**. The choke is a winding of wire. In a direct current (DC) circuit the choke acts like a short, but in an AC circuit it acts as a high resistance. The choke blocks the noisy AC but allows the DC to pass normally. The radio choke blocks **conducted interference**. Conductive interference is audible at very low volume levels and results from problems with the power circuit. The radio choke is installed in the ignition switch output circuit to prevent unwanted and random signals on the power circuit from reaching the radio. Sources of conducted interference can come from the generator, courtesy lamps, or control modules. In addition, nearby high-power transmission lines may produce radio static.

To inspect for the source of radio noise resulting from RFI or EMI, attempt to identify the component, which is the source. For example, an ignition system problem can cause radio noise that is engine speed related. Once the source is identified, check the ground

Classroom Manual
Chapter 14, page 466

path and connections to that component. Test any capacitors in the circuit. Before replacing any components, verify all ground connections. Some of the grounds and connections to inspect include the following:

- Radio antenna base ground.
- Radio receiver chassis ground.
- Generator.
- Engine-to-body ground straps.
- Electric fuel pump.
- Ignition module.
- Heater core ground strap.
- Wiper motors.
- Blower motor.
- Exhaust system ground straps.

In addition, check spark plugs and spark plug wire routing and condition.

> ⚙ **SERVICE TIP** If you suspect a noise suppression capacitor is the cause of radio interference, a clip-on capacitor can be used to test the noise suppressor. Connect the test capacitor in place of the suppressor. If this reduces the radio static, a new suppressing capacitor is required.

A "diagnostic RF sniffer" tool can be made from an old piece of antenna lead-in from a mast or power antenna (**Figure 14-20**). This sniffer, along with the radio, locates "hot spots" that generate RFI noise. The noise can be in wiring harnesses, in the upper part of the dash, or even between the hood and the windshield. When checking for noise on a wire, it's best to hold the sniffer parallel and close to the wire.

Antenna
lead in

50 mm (2 inches)
of coax shield removed

White plastic
insulator

Center wire

Figure 14-20 Making an RF sniffer from an old antenna cable.

Speaker Service

Speakers turn the electrical energy from the radio receiver amplifier into acoustical energy. A speaker moves the air using a permanent magnet and an electromagnet (**Figure 14-21**). The electromagnet is energized when the amplifier delivers current to the voice coil at the speaker.

A faulty speaker distorts the sound of the radio or may be totally inoperative. A failure with the speaker requires its replacement since it's not serviceable. The exception would be to solder a terminal connection that became loose or broken.

> ⚙ SERVICE TIP Speakers that are "blown" may result from not being matched with the radio/amplifier. The wattage rating of the speakers must be the same or higher than the radio/amplifier. Turning up the volume on speakers with lower ratings cause them to blow. Manufacturers match their factory-installed systems; however, aftermarket-installed radios may not be.

To test an inoperative speaker, disconnect the harness connector and momentarily connect an AA battery across the speaker terminals. Make and break the connection quickly. The speaker should make a popping noise as the battery is connected and disconnected. If the speaker makes the popping noise, it passes, and the problem is in the speaker's wiring. If there is no noise, the speaker is defective and requires replacement.

Disconnect the harness connector to the speaker, and measure the speaker's resistance. The resistance of the electromagnetic voice coil is measured using an ohmmeter. The ohmmeter should read the low resistance of the speaker voice coil (about 8 ohms [Ω]). Excessive, or infinite, readings indicate the need for speaker replacement.

Telematics Connectivity Issues

Connectivity issues associated with cell phones and audio/video devices (iPod®, iPad®, and MP3 players) can be very difficult to diagnose. The problem lies in determining if the problem is within the vehicle's system or the customer's equipment. First, confirm that the device to be paired to the telematics system is compatible. Manufacturers publish a list of compatible devices that will work with their system on their websites (**Figure 14-22**). Also, confirm the total number of devices currently paired to the telematics system. The telematics system limits the number of paired components.

Classroom Manual
Chapter 14, page 462

Special Tools
AA batteries
DMM

Classroom Manual
Chapter 14, page 469

Special Tools
Multimedia test kit
Scan tool

Figure 14-21 Speaker components.

Figure 14-22 Manufacturer websites provide information concerning device compatibility.

Figure 14-23 Multimedia test kit isolates if the fault is in the vehicle system or the device.

Special multimedia test kits are available to help diagnose customer concerns with features of the telematics systems (**Figure 14-23**). The purpose of this tool is to simulate a known good cell phone or a known good media source. The test kit can determine if the fault is related to the vehicle's system or the device.

Use the test kit to diagnose the following:

- The AUX port.
- The USB port (both power and audio input).
- Bluetooth phone pairing issues.
- Bluetooth hands free calling.
- Bluetooth streaming audio.
- Bluetooth SMS messaging (send and receive).

The Bluetooth pairing test validates the ability of the telematics system to pair with a device by simulating the use of a known good compatible phone to test the system operation. **Photo Sequence 30** illustrates how to perform this process.

If the telematics system pairs with the multimedia test tool, the connectivity function is working properly, and the customer's device is the problem. If pairing does not occur, confirm that the telematics software is up to date. Also, refer to the appropriate service information and Technical Service Bulletins (TSBs) for relevant diagnostic information.

If the customer states that the telematics system was working with their device but stopped working, determine if a recent update was applied to the device. Updates to device software or firmware may cause it to become out of sync or incompatible with the telematics system.

PHOTO SEQUENCE 30
Performing a Bluetooth Pairing Test

P30-1 Connect the power outlet power adapter into the 12-volt receptacle.

P30-2 Connect the USB cable into the power outlet power adapter to utilize a Bluetooth connection.

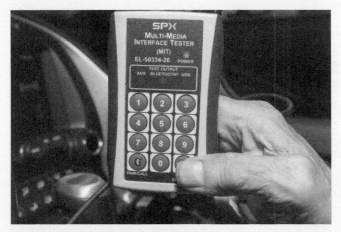

P30-3 Place the tool into Bluetooth mode.

P30-4 Use the vehicle controls to place the system into pairing mode.

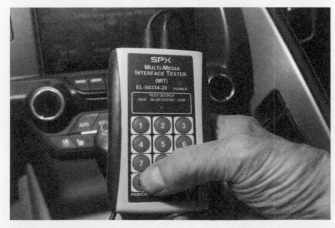

P30-5 Press the PAIR button on the test tool. The tool may identify the telematics system.

P30-6 Enter the pairing PIN. The test tool should pair with the telematics system.

Next, clear the device pairing from the telematics system and reboot the customer's device. Attempt to pair the device to the telematics system again and check the operation. In addition, check to see if there are any software updates to the telematics system.

Wi-Fi Connectivity Diagnostics

Classroom Manual
Chapter 14, page 472

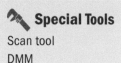

Special Tools

Scan tool
DMM

Like a home Wi-Fi system, the telematics Wi-Fi uses a wireless router to connect to an Internet service. The Hotspot feature of the Wi-Fi connection is limited to a radius of about 100 feet (30.48 meters) from the router. Test the hardwired circuits between components related to the Wi-Fi router and the Wi-Fi system using a DMM and conventional procedures. Reference the appropriate wiring diagrams to determine test locations and expected test values. However, conclusive diagnosis of the telematics module and the bus communication between modules of the Wi-Fi system will require a scan tool.

Usually, faulty components of the Wi-Fi system are not serviceable. For example, the antenna may be internal to the telematics module and, if faulty, will require the module to be replaced. If the system uses a separate antenna, it may be possible to replace it by itself. Also, the Wi-Fi antenna may double as the cellular antenna. Determining any related symptoms between different features of the telematics system can help isolate the root cause of the customer's concern.

Navigational System Service

Classroom Manual
Chapter 14, page 470

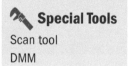

Special Tools

Scan tool
DMM

The navigational system receives global positioning system (GPS) signals from up to eight satellites to display the position and direction of the vehicle. A gyro-sensor, along with the vehicle speed and steering angle sensors, enables the system to display the present vehicle position even in locations where GPS signals are blocked.

Diagnosis of the navigational system is scan tool and DTC driven. Using the DTC, refer to the appropriate service information for the diagnostic routine. All electrical components of the system must have confirmed power and ground circuits before replacement.

Components of the system are generally replaced and not serviced. A typical procedure for replacing the navigational system GPS antenna (**Figure 14-24**) is:

1. Disconnect and isolate the battery negative cable.
2. Remove the instrument panel top cover.
3. Remove the radio.

Figure 14-24 Navigational system GPS antenna.

4. Remove the antenna mounting fasteners.
5. In some cases, each end of the cable is cut, and the remaining portion of the cable is left in the instrument panel.
6. Position the new antenna and feed the wire harness through the opening by carefully pulling up on the instrument panel.
7. Install the antenna mounting fasteners.
8. Secure the antenna wire harness into place.
9. Install the radio.
10. Install the instrument panel top cover.
11. Reconnect the battery negative cable.

CASE STUDY

A customer brought their vehicle into the service facility because of poor radio reception. The technician noticed interference noises coming from all of the speakers, and that the noise appeared to be engine speed related.

While inspecting for the cause of the interference, the technician noticed that the coil-on-plug ignition system used on this engine had capacitors for each bank. The capacitor on the left bank was plugged in, but the capacitor for the right bank was not connected. After plugging the capacitor in, the radio noise ceased.

CASE STUDY

A customer brought their vehicle into the service facility because they cannot pair their cell phone to the telematics system. Following the published procedure, the technician attempted to pair the phone. All attempts to pair the phone failed. Next, the technician confirmed that the cell phone was compatible with the telematics system. Finally, the technician checked for any software updates to the telematics system, but there were none.

The technician connected the scan tool to determine the number of paired devices. The scan tool reported that only three devices were paired to the system. The service information indicated that the system could hold 12 paired devices. In addition, the scan tool did not report any DTCs recorded in the telematics system module.

To determine if the pairing issue is a fault in the telematics system of the customer's device, the technician connected the multimedia test kit. After performing the set-up steps and entering the PIN, the tool could not pair to the system. This isolated the fault as being in the telematics system and not the device. Following the diagnostic routine in the service information led to determining that the telematics module required replacement. After replacing the module, the technician confirmed proper system operation and returned the vehicle to the customer.

ASE-Style Review Questions

1. The doors fail to lock or unlock when using the RKE system; however, they lock and unlock when the door switch is activated. Which is the **LEAST LIKELY** cause?

   A. FOB battery voltage is too low.

   B. RKE antenna faulty.

   C. Faulty lock/unlock relay.

   D. Data bus communication circuit fault.

2. *Technician A* says most amplified audio systems provide diagnostics of speaker-related problems with a scan tool through the amplifier.

   Technician B says if "Updating Channels" on the radio screen is constantly displayed, the subscription for satellite radio has expired.

   Who is correct?

   A. A only   C. Both A and B

   B. B only   D. Neither A nor B

3. *Technician A* says replacing the immobilizer module requires replacing the keys also.

   Technician B says procedures that copy data to the new immobilizer module may need to be performed for the vehicle to start.

   Who is correct?

   A. A only   C. Both A and B

   B. B only   D. Neither A nor B

4. A new tire pressure sensor has been installed onto a wheel.

   Technician A says the ID number may need to be programmed into the TPM ECU.

   Technician B says the system may learn the new transmitter ID while driving the vehicle.

   Who is correct?

   A. A only   C. Both A and B

   B. B only   D. Neither A nor B

5. TPM systems that don't use a transponder are being discussed.

   Technician A says TPM sensors may require location training by use of a magnet.

   Technician B says the training magnet procedure needs to be performed whenever tire pressure is corrected.

   Who is correct?

   A. A only   C. Both A and B

   B. B only   D. Neither A nor B

6. A customer states that the engine starts and then dies almost immediately.

   Technician A says the immobilizer key may be faulty.

   Technician B says a data bus circuit may be open.

   Who is correct?

   A. A only   C. Both A and B

   B. B only   D. Neither A nor B

7. *Technician A* says the resistance between the antenna tip and chassis ground should be less than 1 ohm.

   Technician B says the resistance between the terminals of a speaker should be very high.

   Who is correct?

   A. A only   C. Both A and B

   B. B only   D. Neither A nor B

8. TPM systems that use transponders are being discussed.

   Technician A says a fault code indicating no response from a transponder can be due to a faulty tire pressure sensor.

   Technician B says the scan tool may support a transponder trigger test to isolate the problem.

   Who is correct?

   A. A only   C. Both A and B

   B. B only   D. Neither A nor B

9. To program ignition keys into the immobilizer system, all of the following may be required **EXCEPT:**

   A. PIN.

   B. Scan tool.

   C. DMM.

   D. All previously programmed keys.

10. Using a TPM analyzer results in the illumination of the FAIL light when attempting to retrieve data from the sensor. Which of the following is the **MOST LIKELY** cause?

    A. Too low of tire pressure.

    B. Transponder power supply circuit open.

    C. Data circuit between transponder and TPM ECU defective.

    D. Improper tire pressure sensor installed.

ASE Challenge Questions

1. The RKE system fails to operate. While pressing a FOB button, the RF sniffer tool does not indicate the reception of a signal. Which of the following is the **LEAST LIKELY** cause?

   A. FOB not programmed.

   B. Batteries too low.

   C. Damaged FOB.

   D. Batteries installed wrong.

2. On a vehicle equipped with an immobilizer system, the engine starts for a few seconds and then dies. What is the **LEAST LIKELY** cause of this?

   A. Using an invalid key.

   B. The immobilizer module has not been initialized.

   C. Lost communication between the immobilizer module and PCM.

   D. A faulty starter relay.

3. *Technician A* says weak and fading AM signals could result from a bad or ungrounded antenna. *Technician B* says most radio noise enters by way of the radio.
   Who is correct?

   A. A only

   B. B only

   C. Both A and B

   D. Neither A nor B

4. *Technician A* says most complaints associated with poor reception from the AM/FM radio, satellite radio, navigational system, or cellular phone are usually the result of an improper ground connection of the antenna.
   Technician B says resistance in any portion of the vehicle's ground path can affect the overall performance of the telematics system.
   Who is correct?

   A. A only

   B. B only

   C. Both A and B

   D. Neither A nor B

5. A customer states that the engine starts and then dies. This seems to be intermittent. The customer makes arrangements to leave the vehicle for three days to be diagnosed and repaired. During these three days, the vehicle does not have any problems. The immobilizer module does have a stored "Invalid Key." What is the **MOST LIKELY** cause of this problem?

   A. The battery in the key is low.

   B. The customer has more than one transponder key on the ring.

   C. The cut of the key blade is worn.

   D. None of these choices

Name _____   Date _____

ANALYZING THE RKE SYSTEM

Upon completion of this job sheet, you should be able to determine proper operation of the RKE system and perform diagnostics.

ASE Education Foundation Correlation

This job sheet addresses the following **MLR** tasks:

A.1.   Research vehicle service information such as fluid type, vehicle service history, service precautions, technical service bulletins, and recalls including vehicles with advanced driver assistance systems (ADAS). **(P-1)**

A.2.   Identify electrical/electronic system components and configuration. **(P-1)**

A.5.   Demonstrate proper use of a digital multimeter (DMM) when measuring source voltage, voltage drop (including grounds), current flow, and resistance. **(P-1)**

A.6.   Demonstrate knowledge of the causes and effects from shorts, grounds, opens, and resistance problems in electrical/electronic circuits. **(P-1)**

A.9.   Use wiring diagrams to trace electrical/electronic circuits. **(P-1)**

G.1.   Demonstrate understanding of vehicle comfort, convenience, access, safety, and related systems operation. **(P-3)**

G.3.   Describe operation of keyless entry/remote start systems. **(P-3)**

This job sheet addresses the following **AST** tasks:

A.1.   Research vehicle service information such as fluid type, vehicle service history, service precautions, technical service bulletins, and recalls including vehicles with advanced driver assistance systems (ADAS). **(P-1)**

A.2.   Identify electrical/electronic system components and configuration. **(P-1)**

A.5.   Demonstrate proper use of a digital multimeter (DMM) when measuring source voltage, voltage drop (including grounds), current flow, and resistance. **(P-1)**

A.6.   Demonstrate knowledge of the causes and effects from shorts, grounds, opens, and resistance problems in electrical/electronic circuits. **(P-1)**

A.9.   Use wiring diagrams during the diagnosis (troubleshooting) of electrical/electronic circuit problems. **(P-1)**

A.13.   Test and measure circuit using an oscilloscope and/or graphing multimeter (GMM); interpret results; determine needed action. **(P-2)**

G.1.   Diagnose vehicle comfort, convenience, access, safety, and related systems operation; determine needed action. **(P-2)**

G.3.   Diagnose operation of security/antitheft systems and related circuits (such as: theft deterrent, door locks, remote keyless entry, remote start, and starter/fuel disable); determine needed action. **(P-2)**

G.8.   Diagnose body electronic system circuits using a scan tool; check for module communication errors (data communication bus systems); determine needed action. **(P-2)**

(Continued)

This job sheet addresses the following **MAST** tasks:

A.1.   Research vehicle service information such as fluid type, vehicle service history, service precautions, technical service bulletins, and recalls including vehicles with advanced driver assistance systems (ADAS). **(P-1)**

A.2.   Identify electrical/electronic system components and configuration. **(P-1)**

A.5.   Demonstrate proper use of a digital multimeter (DMM) when measuring source voltage, voltage drop (including grounds), current flow, and resistance. **(P-1)**

A.6.   Demonstrate knowledge of the causes and effects from shorts, grounds, opens, and resistance problems in electrical/electronic circuits. **(P-1)**

A.9.   Use wiring diagrams during the diagnosis (troubleshooting) of electrical/electronic circuit problems. **(P-1)**

A.13.   Test and measure circuit using an oscilloscope and/or graphing multimeter (GMM); interpret results; determine needed action. **(P-1)**

G.1.   Diagnose vehicle comfort, convenience, access, safety, and related systems operation; determine needed action. **(P-2)**

G.3.   Diagnose operation of security/antitheft systems and related circuits (such as: theft deterrent, door locks, remote keyless entry, remote start, and starter/fuel disable); determine needed action. **(P-1)**

G.8.   Diagnose body electronic system circuits using a scan tool; check for module communication errors (data communication bus systems); determine needed action. **(P-1)**

We Support
Education Foundation

Tools and Materials

- Scan tool
- Lab scope
- DMM
- Extra FOBs
- Wiring diagram

Procedure

Task 1

This task involves programming of the RKE FOBs to the system using a scan tool.

1. Using the service information, determine the FOB programming function using the scan tool and describe the procedure.

_____

_____

_____

2. Do all FOBs have to be programmed at the same time, or can individual FOBs be entered?

_____

3. Connect the scan tool to the vehicle, access the RKE ECU, and select the function to program RKE.

4. Follow the instructions given by the scan tool to an additional FOB. How can you tell that the RKE module has been programmed successfully?

_____

_____

5. How many FOBs can be programmed to the system? _____

6. How many FOBs can be programmed to operate with the memory seat function?

_____

7. When is it necessary to program the RKE module?

_____

_____

Task 2

The following task involves programming the RKE FOB without using the scan tool.

1. Look up the procedure for programming the RKE system without the use of the scan tool (customer programming mode). Describe the procedure.

_____

_____

_____

2. Perform the procedure to program the additional FOB. How do you know if the RKE module has been programmed successfully?

_____

_____

Task 3

For this task, you will analyze the input and output signal transmission of the RKE module.

1. Determine all of the modules used for RKE operation and provide their function.

_____

_____

_____

2. How does the RKE module communicate with the other modules?

_____

_____

3. Identify the circuits that are attached to the RKE module.

_____

_____

_____

4. Set up the lab scope to view a 12-volt square wave at 0.4 s/Div, and connect the lead to the program circuit. Set the record trigger to falling edge.

_____

5. Enter customer programming mode and record the results of the lab scope reading.

_____

_____

6. Move the lab scope lead to the RKE interface circuit, and change the time display to 20 ms/Div. Cycle the ignition switch from OFF to RUN, and then press the Lock or Unlock button on the key FOB. Describe the lab scope sine wave.

_____

7. Where is the voltage coming from for the two circuits just tested?

_____

_____

8. What would be the result if the interface circuit was opened?

_____

_____

Instructor's Evaluation:

Grade	Rubric
0	No exposure due to: ☐ Absence ☐ Other:
1	Student unsuccessfully attempted the task
2	Student completed the task with significant instructor assistance.
3	Satisfactory performance of task. The student was able to perform elements of the task with minimal instructor assistance.
4	Competent/Proficient. Able to perform all elements of the task successfully and independently.

Instructor's Notes:

_____

Name _____   Date _____

PROGRAMMING AN IMMOBILIZER KEY

Upon completion of this job sheet, you should be able to determine operating characteristics of the immobilizer system and properly program keys.

ASE Education Foundation Correlation

This job sheet addresses the following **MLR** tasks:

A.1.   Research vehicle service information such as fluid type, vehicle service history, service precautions, technical service bulletins, and recalls including vehicles with advanced driver assistance systems (ADAS). **(P-1)**

G.1.   Demonstrate understanding of vehicle comfort, convenience, access, safety, and related systems operation. **(P-3)**

G.3.   Describe operation of keyless entry/remote start systems. **(P-3)**

This job sheet addresses the following **AST** tasks:

A.1.   Research vehicle service information such as fluid type, vehicle service history, service precautions, technical service bulletins, and recalls including vehicles with advanced driver assistance systems (ADAS). **(P-1)**

G.1.   Diagnose vehicle comfort, convenience, access, safety, and related systems operation; determine needed action. **(P-2)**

G.3.   Diagnose operation of security/antitheft systems and related circuits (such as: theft deterrent, door locks, remote keyless entry, remote start, and starter/fuel disable); determine needed action. **(P-2)**

G.8.   Diagnose body electronic system circuits using a scan tool; check for module communication errors (data communication bus systems); determine needed action. **(P-2)**

This job sheet addresses the following **MAST** tasks:

A.1.   Research vehicle service information such as fluid type, vehicle service history, service precautions, technical service bulletins, and recalls including vehicles with advanced driver assistance systems (ADAS). **(P-1)**

G.1.   Diagnose vehicle comfort, convenience, access, safety, and related systems operation; determine needed action. **(P-2)**

G.3.   Diagnose operation of security/antitheft systems and related circuits (such as: theft deterrent, door locks, remote keyless entry, remote start, and starter/fuel disable); determine needed action. **(P-1)**

G.8.   Diagnose body electronic system circuits using a scan tool; check for module communication errors (data communication bus systems); determine needed action. **(P-1)**

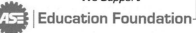

We Support
ASE | Education Foundation

Tools and Materials

- Vehicle equipped with an immobilizer system
- Extra keys
- Scan tool
- PIN (if required)
- Service information

Procedure

Task 1

1. According to the service information, does the immobilizer system of the assigned vehicle support customer programming of extra keys?   ☐ Yes   ☐ No

2. If yes, describe the procedure.

   _____

   _____

   _____

3. Obtain an extra key from your instructor, and attempt to start the vehicle with the new key. Describe the results.

   _____

   _____

4. Perform the customer programming method described in step 1 to program a new key.

5. How do you know the key was successfully programmed?

   _____

   _____

Task 2

1. Connect the scan tool to the DLC, and access the immobilizer system. _____

2. If possible, perform a Module ID and record the following items:

   Version: _____ Part Number: _____ Country Code: _____

3. Does the immobilizer system on the assigned vehicle require the use of a PIN?

   ☐ Yes   ☐ No

   If yes, obtain the PIN for the vehicle and record it: _____

4. How do you obtain the PIN?

   _____

   _____

5. If an incorrect PIN is entered or the correct procedure isn't followed, are there any actions that must be done to reenter the programming mode?   ☐ Yes   ☐ No

   If yes, describe what must be done.

   _____

   _____

6. Is there a special procedure that must be followed if the PCM is replaced on the assigned vehicle?   ☐ Yes   ☐ No

   If yes, describe the procedure.

   _____

   _____

   _____

7. Is there a special procedure that must be followed if the immobilizer module is replaced on the assigned vehicle?   ☐ Yes   ☐ No

If yes, describe the procedure.

_____

_____

8. Do the keys have to be reprogrammed after the immobilizer module is replaced?
☐ Yes   ☐ No

9. Use the scan tool to erase all current ignition keys.

10. Attempt to start the engine, and record the results.

_____

_____

11. Use the scan tool to read DTCs and record.

_____

_____

12. Using the scan tool, access the function to program ignition keys and follow the instructions on the scan tool or in the service manual.

13. Does the vehicle start?   ☐ Yes   ☐ No

14. Clear any DTCs that were set.

Task 3

1. Wrap one of the programmed keys with aluminum foil, and attempt to start the engine. Record your results.

_____

_____

2. Use the scan tool to read any DTCs in the system and record.

_____

_____

3. According to the service manual, what could cause the DTCs to set?

_____

_____

_____

Instructor's Evaluation:

Grade	Rubric
0	No exposure due to: ☐ Absence ☐ Other:
1	Student unsuccessfully attempted the task
2	Student completed the task with significant instructor assistance.
3	Satisfactory performance of task. The student was able to perform elements of the task with minimal instructor assistance.
4	Competent/Proficient. Able to perform all elements of the task successfully and independently.

Instructor's Notes:

Name _____   Date _____

TIRE PRESSURE MONITORING SYSTEM TESTING

Upon completion of this job sheet, you should be able to determine proper operation of the TPM system and determine the location of a faulty transmitter.

ASE Education Foundation Correlation —————————————————————————

This job sheet addresses the following **MLR** tasks:

A.1.   Research vehicle service information such as fluid type, vehicle service history, service precautions, technical service bulletins, and recalls including vehicles with advanced driver assistance systems (ADAS). **(P-1)**

G.1.   Demonstrate understanding of vehicle comfort, convenience, access, safety, and related systems operation. **(P-3)**

This job sheet addresses the following **AST** tasks:

A.1.   Research vehicle service information such as fluid type, vehicle service history, service precautions, technical service bulletins, and recalls including vehicles with advanced driver assistance systems (ADAS). **(P-1)**

G.1.   Diagnose vehicle comfort, convenience, access, safety, and related systems operation; determine needed action. **(P-2)**

G.8.   Diagnose body electronic system circuits using a scan tool; check for module communication errors (data communication bus systems); determine needed action. **(P-2)**

This job sheet addresses the following **MAST** tasks:

A.1.   Research vehicle service information such as fluid type, vehicle service history, service precautions, technical service bulletins, and recalls including vehicles with advanced driver assistance systems (ADAS). **(P-1)**

G.1.   Diagnose vehicle comfort, convenience, access, safety, and related systems

G.8.   Diagnose body electronic system circuits using a scan tool; check for module communication errors (data communication bus systems); determine needed action. **(P-1)**

We Support

ASE | Education Foundation

Tools and Materials

- Vehicle equipped with TPM system
- Scan tool

Procedure

1. Locate the pressure thresholds in the service information for the vehicle you are assigned.

   _____

   _____

2. Use a handheld tire pressure gauge to measure the actual tire pressures in the classroom vehicle and record your results.

   _____

   _____

3. Turn the ignition switch to the RUN position. Lower the windows in order to be able to hear the audible low tire pressure warning from outside the vehicle.

4. Quickly release air pressure from one of the tires until the low tire pressure warning sounds. Measure and record the tire pressure.

_____

5. Why did the low tire pressure warning sound if the tire pressure is still above the low pressure ON threshold?

_____

_____

6. If a vehicle is displaying a Low Pressure warning message and all tire pressures check within the low- and high-pressure thresholds, how could this feature be used to locate the faulty transmitter?

_____

_____

Instructor's Evaluation:

Grade	Rubric
0	No exposure due to: ☐ Absence ☐ Other:
1	Student unsuccessfully attempted the task
2	Student completed the task with significant instructor assistance.
3	Satisfactory performance of task. The student was able to perform elements of the task with minimal instructor assistance.
4	Competent/Proficient. Able to perform all elements of the task successfully and independently.

Instructor's Notes:

_____

Name _____   Date _____

USING THE TPM ANALYZER

Upon completion of this job sheet, you should be able to determine proper operation of the TPM system and determine the location of a faulty transmitter.

ASE Education Foundation Correlation

This job sheet addresses the following **MLR** task:

A.1.   Research vehicle service information such as fluid type, vehicle service history, service precautions, technical service bulletins, and recalls including vehicles with advanced driver assistance systems (ADAS). **(P-1)**

This job sheet addresses the following **AST** tasks:

A.1.   Research vehicle service information such as fluid type, vehicle service history, service precautions, technical service bulletins, and recalls including vehicles with advanced driver assistance systems (ADAS). **(P-1)**

G.7.   Diagnose operation of safety systems and related circuits (such as horn, air-bags, seat belt pretensioners, occupancy classification, wipers, washers, speed control/collision avoidance, heads-up display, park assist, and back-up camera); determine needed repairs. **(P-2)**

G.8.   Demonstrate testing of body electronic systems circuits using a scan tool; check for module communication errors (data communication bus systems); determine needed action. **(P-2)**

This job sheet addresses the following **MAST** tasks:

A.1.   Research vehicle service information such as fluid type, vehicle service history, service precautions, technical service bulletins, and recalls including vehicles with advanced driver assistance systems (ADAS). **(P-1)**

G.7.   Diagnose operation of safety systems and related circuits (such as horn, air-bags, seat belt pretensioners, occupancy classification, wipers, washers, speed control/collision avoidance, heads-up display, park assist, and back-up camera); determine needed repairs. **(P-1)**

G.8.   Diagnose body electronic system circuits using a scan tool; check for module communication errors (data communication bus systems); determine needed action. **(P-1)**

We Support

ASE | **Education Foundation**

Tools and Materials

• Vehicle equipped with TPM system
• Scan tool
• TPM analyzer

Procedure

1. With the analyzer's power turned on, select the model year and vehicle type from the selection screens.

2. Place the analyzer near the valve stem of one of the tires and start the procedure to retrieve the sensor information. Record the information displayed by the analyzer.

_____

_____

_____

3. Lock the senor location into the analyzer.

4. Repeat for all sensors.

_____

_____

_____

_____

_____

5. What could be the cause of the analyzer to display a fail condition when attempting to identify the sensor?

_____

_____

_____

6. Connect the analyzer to the scan tool using the USB port or wireless communications.

7. Select the function on the scan tool for programming the sensor location into the TPM ECU using the analyzer and follow the instructions on the screen. Describe the procedure.

_____

_____

_____

Instructor's Evaluation:

Grade	Rubric
0	No exposure due to: ☐ Absence ☐ Other:
1	Student unsuccessfully attempted the task
2	Student completed the task with significant instructor assistance.
3	Satisfactory performance of task. The student was able to perform elements of the task with minimal instructor assistance.
4	Competent/Proficient. Able to perform all elements of the task successfully and independently.

Instructor's Notes:

_____

Name _____   Date _____

TESTING THE AUDIO SYSTEM

Upon completion of this job sheet, you should be able to determine proper operation of the audio system.

ASE Education Foundation Correlation ——————————————————

This job sheet addresses the following **MLR** tasks:

A.1. Research vehicle service information such as fluid type, vehicle service history, service precautions, technical service bulletins, and recalls including vehicles with advanced driver assistance systems (ADAS). **(P-1)**

A.5. Demonstrate proper use of a digital multimeter (DMM) when measuring source voltage, voltage drop (including grounds), current flow, and resistance. **(P-1)**

A.6. Demonstrate knowledge of the causes and effects from shorts, grounds, opens, and resistance problems in electrical/electronic circuits. **(P-1)**

A.9. Use wiring diagrams to trace electrical/electronic circuits. **(P-1)**

G.2. Remove and reinstall door panel. **(P-2)**

This job sheet addresses the following **AST** tasks:

A.1. Research vehicle service information such as fluid type, vehicle service history, service precautions, technical service bulletins, and recalls including vehicles with advanced driver assistance systems (ADAS). **(P-1)**

A.5. Demonstrate proper use of a digital multimeter (DMM) when measuring source voltage, voltage drop (including grounds), current flow, and resistance. **(P-1)**

A.6. Demonstrate knowledge of the causes and effects from shorts, grounds, opens, and resistance problems in electrical/electronic circuits. **(P-1)**

A.9. Use wiring diagrams during the diagnosis (troubleshooting) of electrical/electronic circuit problems. **(P-1)**

G.2. Remove and reinstall door panel. **(P-1)**

G.6. Diagnose operation of entertainment and related circuits (such as: radio, DCD, remote CD changer, navigation, amplifiers, speakers, antennas, and voice activated accessories); determine needed action. **(P-2)**

G.8. Diagnose body electronic system circuits using a scan tool; check for module communication errors (data communication bus systems); determine needed action. **(P-2)**

This job sheet addresses the following **MAST** tasks:

A.1. Research vehicle service information such as fluid type, vehicle service history, service precautions, technical service bulletins, and recalls including vehicles with advanced driver assistance systems (ADAS). **(P-1)**

A.5. Demonstrate proper use of a digital multimeter (DMM) when measuring source voltage, voltage drop (including grounds), current flow, and resistance. **(P-1)**

A.6. Demonstrate knowledge of the causes and effects from shorts, grounds, opens, and resistance problems in electrical/electronic circuits. **(P-1)**

A.9. Use wiring diagrams during the diagnosis (troubleshooting) of electrical/electronic circuit problems. **(P-1)**

G.2. Remove and reinstall door panel. **(P-1)**

(Continued)

G.6.   Diagnose operation of entertainment and related circuits (such as: radio, DCD, remote CD changer, navigation, amplifiers, speakers, antennas, and voice activated accessories); determine needed action. **(P-2)**

G.8.   Diagnose body electronic system circuits using a scan tool; check for module communication errors (data communication bus systems); determine needed action. **(P-1)**

We Support

ASE | Education Foundation

Tools and Materials

- Scan tool
- DMM
- Test antenna kit
- Wiring diagram

Procedure

Task 1

For this task, you will identify the capabilities of the scan tool to diagnose and test audio system inputs and outputs.

1. Connect the scan tool to the assigned vehicle, and establish communication. Turn the radio on, perform a Module ID, and record the software version of the radio below:

   _____

   _____

2. Access the amplifier and connect a jumper wire between the power amplifier RR speaker input circuit from the radio and chassis ground.

3. Center the balance controls on the radio. What is the effect on the audio system?

   _____

   _____

4. Read radio DTCs. Are any detected?   ☐ Yes   ☐ No

5. Disconnect the short from the amplifier connector, and move it to the RR speaker output circuit and ground.

6. What is the effect on the audio system?

   _____

   _____

7. Cycle the ignition switch two times and read DTCs. Are any detected?   ☐ Yes   ☐ No

   If yes, list the DTCs that were set.

   _____

   _____

   Why did these DTCs set?

   _____

   _____

8. Explain your results.

   _____

   _____

   _____

9. Restore all equipment to its original condition.

10. Is the vehicle equipped with a DVD player?   ☐ Yes   ☐ No

    If yes, are there any tests available?   ☐ Yes   ☐ No

    If yes, list the available tests.

    _____

    _____

    Perform the tests and describe your results.

    _____

    _____

    _____

11. What components can be tested using the actuator test function in the scan tool?

    _____

    _____

    _____

12. What data information is available on the scan tool?

    _____

    _____

    _____

Task 2

For this task, you will identify the capabilities of the scan tool to diagnose and test satellite radio system inputs and outputs.

1. Connect the scan tool to the assigned vehicle and establish communication with the satellite radio system.

2. Are there any satellite radio system tests available?   ☐ Yes   ☐ No

    If yes, list the available tests, perform the tests, and record the results.

    _____

    _____

    _____

3. Clear any DTCs that are set.

4. At the satellite radio unit, disconnect one of the antenna cables.

5. Check for DTCs and list any that were set.

    _____

    _____

6. Reconnect the antenna.

7. Can the scan tool be used to verify that the signal is present?   ☐ Yes   ☐ No

8. Can the scan tool be used to verify that the subscription is present?   ☐ Yes   ☐ No

9. Clear any DTCs that were set.

10. Explain your results.

_____

_____

_____

Task 3

For this task, you will analyze radio reception performance, identify mast antenna faults, and perform antenna diagnostic tests.

1. Move the vehicle to a location that provides good reception.

2. Use a jumper wire to short the antenna mast to ground. When tuned to a strong AM station, what is the effect of the shorted antenna mast?

_____

_____

3. Switch to an FM station. What is the effect of the shorted antenna mast?

_____

_____

4. Remove the jumper wire and disconnect the antenna lead from the radio. What is the effect on FM and AM reception?

_____

_____

5. Locate the resistance specifications for the antenna in the service information.

_____

_____

6. Using a DMM, measure the resistance between the coax (outer conductor) connector and negative battery cable. What is the reading?

_____

7. If resistance is excessive, what could be the cause?

_____

_____

8. Using a DMM, measure the resistance between the tip of the radio antenna connector (inner conductor) and antenna mast. What is the reading? _____

9. If resistance is excessive, what could be the cause?

_____

_____

10. Test the resistance between the tip of the radio antenna connector (inner conductor) and ground. What is the reading?

_____

11. List some items that could cause radio frequency interference in the audio system.

_____

_____

_____

Task 4

For this task, you will analyze speaker performance and diagnostics.

1. Using the service information, identify the circuits to the LR speaker.

_____

2. Connect the scan tool to the DLC of the assigned vehicle and establish communication with the radio.

3. With the ignition switch in the OFF position, disconnect the LR speaker connector and connect a jumper between the LR Speaker (+) circuit and ground.

4. Turn the ignition switch to the RUN position and turn the radio on. What is the effect on the audio system?

_____

_____

Note: Cycle the ignition switch three times and leave the radio on. It may take some time for a code to set.

5. Record DTCs.

_____

_____

6. Remove the short and erase DTCs.

7. Move the short to the speaker (−) circuit to ground and cycle the ignition key three times with the radio on. What is the effect on the audio system?

_____

_____

8. Are there any DTCs set?   ☐ Yes   ☐ No

   If yes, list them.

_____

_____

   If no, why not?

_____

_____

9. Remove the short and reconnect the speaker.

Instructor's Evaluation:

Grade	Rubric
0	No exposure due to: ☐ Absence ☐ Other:
1	Student unsuccessfully attempted the task
2	Student completed the task with significant instructor assistance.
3	Satisfactory performance of task. The student was able to perform elements of the task with minimal instructor assistance.
4	Competent/Proficient. Able to perform all elements of the task successfully and independently.

Instructor's Notes:

Name _____   Date _____

USB INPUT/OUTPUT TEST

Upon completion of this job sheet, you should be able to validate the ability of the telematics system to play audio from a customer's device that has been connected to the USB port. The test will also determine if the USB port can provide power to a connected device.

ASE Education Foundation Correlation

This job sheet addresses the following **MLR** tasks:

A.1. Research vehicle service information such as fluid type, vehicle service history, service precautions, technical service bulletins, and recalls including vehicles with advanced driver assistance systems (ADAS). **(P-1)**

A.5. Demonstrate proper use of a digital multimeter (DMM) when measuring source voltage, voltage drop (including grounds), current flow, and resistance. **(P-1)**

A.6. Demonstrate knowledge of the causes and effects from shorts, grounds, opens, and resistance problems in electrical/electronic circuits. **(P-1)**

A.9. Use wiring diagrams to trace electrical/electronic circuits. **(P-1)**

A.12. Repair and/or replace connectors, terminal ends, and wiring of electrical/ electronic systems (including solder repair). **(P-2)**

This job sheet addresses the following **AST** tasks:

A.1. Research vehicle service information such as fluid type, vehicle service history, service precautions, technical service bulletins, and recalls including vehicles with advanced driver assistance systems (ADAS). **(P-1)**

A.5. Demonstrate proper use of a digital multimeter (DMM) when measuring source voltage, voltage drop (including grounds), current flow, and resistance. **(P-1)**

A.6. Demonstrate knowledge of the causes and effects from shorts, grounds, opens, and resistance problems in electrical/electronic circuits. **(P-1)**

A.9. Use wiring diagrams during the diagnosis (troubleshooting) of electrical/ electronic circuit problems. **(P-1)**

A.12. Repair and/or replace connectors, terminal ends, and wiring of electrical/ electronic systems (including solder repair). **(P-1)**

G.6. Diagnose operation of entertainment and related circuits (such as: radio, DCD, remote CD changer, navigation, amplifiers, speakers, antennas, and voice activated accessories); determine needed action. **(P-2)**

G.8. Diagnose body electronic system circuits using a scan tool; check for module communication errors (data communication bus systems); determine needed action. **(P-2)**

This job sheet addresses the following **MAST** tasks:

A.1. Research vehicle service information such as fluid type, vehicle service history, service precautions, technical service bulletins, and recalls including vehicles with advanced driver assistance systems (ADAS). **(P-1)**

A.5. Demonstrate proper use of a digital multimeter (DMM) when measuring source voltage, voltage drop (including grounds), current flow, and resistance. **(P-1)**

A.6. Demonstrate knowledge of the causes and effects from shorts, grounds, opens, and resistance problems in electrical/electronic circuits. **(P-1)**

A.9. Use wiring diagrams during the diagnosis (troubleshooting) of electrical/ electronic circuit problems. **(P-1)**

A.12. Repair and/or replace connectors, terminal ends, and wiring of electrical/ electronic systems (including solder repair). **(P-1)**

(Continued)

G.6.   Diagnose operation of entertainment and related circuits (such as: radio, DCD, remote CD changer, navigation, amplifiers, speakers, antennas, and voice activated accessories); determine needed action. **(P-2)**

G.8.   Diagnose body electronic system circuits using a scan tool; check for module communication errors (data communication bus systems); determine needed action. **(P-1)**

We Support

ASE | **Education Foundation**

Tools and Materials

- Vehicle equipped with a telematics system with USB port
- Multimedia test kit
- Scan tool
- Service information

Procedure

1. Use the service information and record the USB port's voltage and amperage output.

    _____

    _____

    _____

2. What is the voltage and amperage requirement of the customer's device for operation and charging?

    _____

    _____

    _____

3. Connect the multimedia tester's USB cable into the vehicle's USB port.

4. Does the tester power on?   ☐ Yes   ☐ No

5. Based on the results of step 4, what does this indicate?

    _____

    _____

    _____

6. Place the multimedia test tool into USB mode.

    NOTE: If the system does not automatically enter USB mode, use the vehicle controls to place the telematics system into USB mode.

7. Does the telematics system play audio from the multimedia tester?

    Yes _____, go to step 8.

    No _____, go to step 12.

8. What does the test confirm?

_____

_____

_____

_____

_____

9. Use the service information or vehicle manufacturer's website and confirm the device is compatible with the telematics system. Is the device compatible?

☐ Yes   ☐ No

10. Are there any updates for the customer device?   ☐ Yes   ☐ No

11. Check the following items and record the results.

Check the multimedia ports for damaged connection points or debris in the ports.

_____

_____

_____

Check wiring circuits and connections to the USB port for concerns.

_____

_____

_____

Check for related TSBs.

_____

_____

_____

Check for vehicle-related system updates.

_____

_____

_____

12. What is your conclusion concerning the operation of the USB port?

_____

_____

_____

_____

Instructor's Evaluation:

Grade	Rubric
0	No exposure due to: ☐ Absence ☐ Other:
1	Student unsuccessfully attempted the task
2	Student completed the task with significant instructor assistance.
3	Satisfactory performance of task. The student was able to perform elements of the task with minimal instructor assistance.
4	Competent/Proficient. Able to perform all elements of the task successfully and independently.

Instructor's Notes:

DIAGNOSTIC CHART 14-1

PROBLEM AREA:	Keyless entry/remote start system operation.
SYMPTOMS:	Keyless entry fails to operate intermittently.
POSSIBLE CAUSES:	**1.** Defective key pad. **2.** Open circuit. **3.** Shorted circuit. **4.** Defective controller.

DIAGNOSTIC CHART 14-2

PROBLEM AREA:	Keyless entry/remote start system operation.
SYMPTOMS:	Keyless entry fails to operate.
POSSIBLE CAUSES:	**1.** Defective key pad. **2.** Open circuit. **3.** Shorted circuit. **4.** Defective controller.

DIAGNOSTIC CHART 14-3

PROBLEM AREA:	Immobilizer system.
SYMPTOMS:	Indicator lamp does not illuminate.
POSSIBLE CAUSES:	**1.** Faulty LED. **2.** Blown fuse. **3.** Faulty ground. **4.** Faulty battery feed. **5.** Faulty ignition feed.

DIAGNOSTIC CHART 14-4

PROBLEM AREA:	Immobilizer system.
SYMPTOMS:	Indicator lamp flashes during bulb check. Vehicle starts and dies.
POSSIBLE CAUSES:	**1.** Invalid key. **2.** Faulty key. **3.** Faulty antenna. **4.** Faulty module.

DIAGNOSTIC CHART 14-5

PROBLEM AREA:	Immobilizer system.
SYMPTOMS:	Indicator lamp stays on after bulb check.
POSSIBLE CAUSES:	**1.** Immobilizer fault has been detected. **2.** Immobilizer system inoperative.

DIAGNOSTIC CHART 14-6

PROBLEM AREA:	Tire pressure monitoring system operation.
SYMPTOMS:	Tire pressure light is on or display shows a low tire.
POSSIBLE CAUSES:	**1.** Low tire pressure. **2.** Faulty tire sensor. **3.** Faulty control module. **4.** Control module power feed circuit open or high resistance. **5.** Control module ground circuit fault. **6.** Bus communications error. **7.** Transponder failure. **8.** Display module fault.

DIAGNOSTIC CHART 14-7

PROBLEM AREA:	Audio system.
SYMPTOMS:	Radio will not turn on.
POSSIBLE CAUSES:	**1.** Open power battery feed circuit. **2.** Open ignition feed circuit. **3.** Poor radio ground circuit. **4.** Loss of bus communications.

DIAGNOSTIC CHART 14-8

PROBLEM AREA:	Audio system.
SYMPTOMS:	Radio will not produce sound.
POSSIBLE CAUSES:	**1.** Open power battery feed circuit. **2.** Open ignition feed circuit. **3.** Poor ground connection. **4.** Defective radio. **5.** Open speak circuit. **6.** Defective amplifier. **7.** Open power feed to amplifier. **8.** Poor amplifier ground circuit. **9.** Bus communications error. **10.** Stuck MUTE button.

DIAGNOSTIC CHART 14-9

PROBLEM AREA:	Audio system.
SYMPTOMS:	No sound in AM or FM mode. CD audio operates normally.
POSSIBLE CAUSES:	**1.** Faulty antenna connection. **2.** Poor antenna ground. **3.** Faulty radio.

DIAGNOSTIC CHART 14-10

PROBLEM AREA:	Audio system.
SYMPTOMS:	Excessive noise heard in AM audio.
POSSIBLE CAUSES:	**1.** Faulty antenna connection. **2.** Poor antenna ground. **3.** Faulty engine to chassis ground.

DIAGNOSTIC CHART 14-11

PROBLEM AREA:	Audio system.
SYMPTOMS:	Poor radio reception.
POSSIBLE CAUSES:	**1.** Faulty antenna connection. **2.** Poor antenna ground. **3.** Faulty radio.

DIAGNOSTIC CHART 14-12

PROBLEM AREA:	Audio CD system.
SYMPTOMS:	Radio will not go into CD mode.
POSSIBLE CAUSES:	**1.** CD not inserted into player. **2.** Stuck MODE button. **3.** Faulty player.

DIAGNOSTIC CHART 14-13

PROBLEM AREA:	Audio CD system.
SYMPTOMS:	CD will not eject.
POSSIBLE CAUSES:	**1.** CD adhesive label came loose. **2.** Warped CD. **3.** Faulty player.

DIAGNOSTIC CHART 14-14

PROBLEM AREA:	Audio CD system.
SYMPTOMS:	CD will not play.
POSSIBLE CAUSES:	**1.** Faulty power feed circuit. **2.** Poor ground connection. **3.** Defective player.

DIAGNOSTIC CHART 14-15

PROBLEM AREA:	Audio CD system.
SYMPTOMS:	CD will not insert.
POSSIBLE CAUSES:	**1.** Damaged edges of CD. **2.** CD surface defects. **3.** Wrong disc size. **4.** Faulty player.

DIAGNOSTIC CHART 14-16

PROBLEM AREA:	Audio CD system.
SYMPTOMS:	CD skips or jumps tracks.
POSSIBLE CAUSES:	**1.** CD surface defects. **2.** Excessive vehicle vibration. **3.** Loose mounting bolts. **4.** Faulty player.

DIAGNOSTIC CHART 14-17

PROBLEM AREA:	Video display.
SYMPTOMS:	DVD screen does not power on.
POSSIBLE CAUSES:	**1.** Power and/or CAN circuit maybe open. **2.** Screen software issue.

DIAGNOSTIC CHART 14-18

PROBLEM AREA:	Video display.
SYMPTOMS:	Green DVD screen.
POSSIBLE CAUSES:	Defective satellite video receiver module.

DIAGNOSTIC CHART 14-19

PROBLEM AREA:	Video display.
SYMPTOMS:	Black DVD screen.
POSSIBLE CAUSES:	**1.** Defective satellite video receiver module. **2.** Video circuit damaged.

DIAGNOSTIC CHART 14-20

PROBLEM AREA:	Satellite radio.
SYMPTOMS:	Poor or no satellite audio reception. Radio will not go into CD mode.
POSSIBLE CAUSES:	**1.** Satellite antenna. **2.** Satellite antenna connection. **3.** Open or shorted satellite antenna cable. **4.** Telematics module.

DIAGNOSTIC CHART 14-21

PROBLEM AREA:	Satellite video.
SYMPTOMS:	Satellite video has no video.
POSSIBLE CAUSES:	**1.** Obstructed line of sight to the sky. **2.** Antenna or antenna cable inoperative.

DIAGNOSTIC CHART 14-22

PROBLEM AREA:	Satellite video display.
SYMPTOMS:	Screen does not power on.
POSSIBLE CAUSES:	**1.** Power and/or CAN circuit may be open. **2.** Screen software issue.

DIAGNOSTIC CHART 14-23

PROBLEM AREA:	Satellite video display.
SYMPTOMS:	Green screen.
POSSIBLE CAUSES:	Defective satellite video receiver module.

DIAGNOSTIC CHART 14-24

PROBLEM AREA:	Satellite video display.
SYMPTOMS:	Black screen.
POSSIBLE CAUSES:	**1.** Defective satellite video receiver module. **2.** Video circuit damaged.

DIAGNOSTIC CHART 14-25

PROBLEM AREA:	Satellite video.
SYMPTOMS:	Satellite video has no video.
POSSIBLE CAUSES:	**1.** Obstructed line of sight to the sky. **2.** Antenna or antenna cable inoperative.

DIAGNOSTIC CHART 14-26

PROBLEM AREA:	Telecommunication system.
SYMPTOMS:	Failed phone pairing attempts.
POSSIBLE CAUSES:	**1.** Phone does not support Hands-Free Profile. **2.** Phone not Bluetooth enabled. **3.** PIN numbers between phone and system not matching. **4.** Phone has reached maximum number of allowed devices paired.

DIAGNOSTIC CHART 14-27

PROBLEM AREA:	Telecommunication system.
SYMPTOMS:	Poor phone audio quality.
POSSIBLE CAUSES:	**1.** Microphone failure. **2.** Rearview mirror not properly fixed to mounting button.

DIAGNOSTIC CHART 14-28

PROBLEM AREA:	Cellular service.
SYMPTOMS:	Poor or no cellular reception.
POSSIBLE CAUSES:	**1.** Cellular antenna connection. **2.** Open or shorted cellular antenna cables. **3.** Open or shorted cellular antenna. **4.** Cellular antenna. **5.** Cellular module.

DIAGNOSTIC CHART 14-29

PROBLEM AREA:	Wi-Fi hotspot.
SYMPTOMS:	Wi-Fi hotspot inoperative.
POSSIBLE CAUSES:	**1.** Wi-Fi service registration incomplete. **2.** Wi-Fi subscription service is inactive, expired, or not installed. **3.** Wi-Fi capability isn't enabled in user's device. **4.** Selected Wireless Access Point (WAP) is incorrect. **5.** Wireless Access Point (WAP) password or security settings incorrect. **6.** Cellular antenna.

DIAGNOSTIC CHART 14-30

PROBLEM AREA:	Navigation.
SYMPTOMS:	Poor or no GPS reception.
POSSIBLE CAUSES:	**1.** GPS antenna connection. **2.** Open or shorted GPS antenna cables. **3.** Open or shorted GPS antenna. **4.** GPS antenna. **5.** GPS module.

CHAPTER 15
SERVICING PASSIVE RESTRAINT AND OCCUPANT SAFETY SYSTEMS

Upon completion and review of this chapter, you should be able to:

- Diagnose automatic seat belt systems.
- Understand the safety concerns associated with the air bag system and safely service the air bag system.
- Enter air bag system self-diagnostics.
- Use a scan tool to retrieve air bag system trouble codes properly.
- Properly use the air bag simulator load tool.
- Comprehend the complexities of diagnosing intermittent faults.
- Properly utilize all tools and tool features to isolate the cause of an intermittent fault.

- Replace the air bag module according to manufacturer's service information standards.
- Replace the clockspring assembly according to manufacturer's service information standards.
- Properly diagnose air bag system pressure sensors and their circuits.
- Properly diagnose air bag system accelerometer sensors.
- Perform squib circuit diagnostics.
- Properly diagnose the occupant classification system.
- Validate the occupant classification system.

Terms to Know

Air bag simulator load tool	Floating	OCS service kit
Automatic seat belt system	Live module	OCS validation test

Introduction

This chapter details the proper and safe servicing of automatic passive restraint systems. Passive restraints operate automatically with no action required on the part of the driver or occupant. The safety of the vehicle occupants depends on the ability of the technician to diagnose and repair these systems properly. In this chapter, we discuss air bag diagnostic procedures and the procedures for replacing the air bag module and the clockspring. In addition, there are typical procedures for validating the occupant classification system (OCS).

Automatic Seat Belt Service

The **automatic seat belt system** automatically puts the shoulder and/or lap belt around the driver or occupant. The automatic seat belt uses direct current (DC) motors to move the belts via carriers on tracks. Even though the components used in the

Classroom Manual
Chapter 15, page 480

automatic seat belt system vary according to the manufacturer, the basic principles of locating and repairing the cause of a problem are similar. Refer to the service information to obtain a circuit diagram of the system. The circuit schematics, troubleshooting charts, and diagnostic charts will assist you in finding the fault.

Special Tools
Scan tool
DMM

The automatic seat belt system troubleshooting chart guides you through the process of troubleshooting the circuits, switches, lamps, and motors. In addition, an operational logic chart provides helpful information when testing switches. The skills indicated in these charts require only those described in previous chapters.

Some systems are capable of storing diagnostic trouble codes (DTCs) that are retrieved with a scan tool. Use the diagnostic chart associated with the DTC to test the operation of the system and to determine faults.

Use the scan tool's bi-directional control feature to perform functional tests of the system. To perform this test, turn the ignition to the RUN position and connect the scan tool to the diagnostic connector. Select the system test and follow the scan tool instructions. If there is a failure in the system, the scan tool will display the code.

Drive Motor Assembly Replacement

A faulty drive motor can cause slow or no operation of the automatic seat belts. If you diagnose the component as faulty, replace the motor. The following describes a typical procedure for motor assembly replacement.

Begin by disconnecting the negative battery cable. If the vehicle has side air bag curtains, wait for the required amount of time specified by the manufacturer to allow the air bag system to discharge. Remove the B-pillar trim to gain access to the drive motor. Disconnect the electrical connector from the seat belt motor and locate the wiring harness out of the area. Remove the screws that attach the drive belt track to the motor. Next, remove the screw that attaches the vertical guide to the pillar. Slide the belt down far enough to disengage the sprockets from the motor drive gear teeth. Finally, remove the screws that attach the motor to the pillar.

Air Bag Safety and Service Warnings

Classroom Manual
Chapter 15, page 483

When working on or around the air bag system, it's important to follow some safety warnings. There are safety concerns with both deployed and live air bag modules. The air bag module comprises the air bag and inflator assembly packaged in a single module. Be aware of the following safety concerns:

Special Tool
Safety glasses

1. Wear safety glasses when servicing the air bag system.
2. Wear safety glasses when handling an air bag module.
3. Always disconnect the battery negative cable, isolate the cable end, and wait for the amount of time specified by the vehicle manufacturer before proceeding with the necessary diagnosis or service. The required amount of time may be as much as 15 minutes. Failure to observe this precaution may cause accidental air bag deployment and personal injury.
4. Handle all sensors with care. Do not strike or jar a sensor in such a manner that deployment may occur.
5. Replacement air bag system parts must be the same quality as the original. Do not use replacement parts of lesser or questionable quality. Improper or inferior components may result in improper air bag deployment and injury to the vehicle occupants.
6. Do not strike or jar a sensor or an air bag system occupant restraint controller (ORC). This may cause air bag deployment or the sensor to become inoperative.

7. Before an air bag system powers up, all sensors and mounting brackets must be properly mounted and torqued to ensure correct sensor operation. If sensor fasteners do not have the proper torque, improper air bag deployment may result in injury to the vehicle occupants.
8. When carrying a **live module** (undeployed module), face the trim and bag away from your body.
9. Do not carry the module by its wires or connector.
10. When placing a live module on a bench, face the trim and air bag up.
11. The deployment reaction produces sodium hydroxide that is converted to sodium carbonate when it comes into contact with atmospheric moisture. It's unlikely that sodium hydroxide will still be present. Some deployed air bags may have a powdery residue from the corn starch. Wear safety glasses and gloves when handling a deployed air bag. Wash hands after handling.
12. Dispose of deployed air bags in accordance with EPA and manufacturer procedures. Deploy live air bag modules before disposal. Because the deployment of an air bag is an explosive process, improper disposal may result in injury and in fines.
13. Do not use a battery- or AC-powered voltmeter, ohmmeter, or any other type of test equipment not specified in the service information. Never use a test light to probe for voltage.
14. Never reach across the steering wheel to turn the ignition switch.
15. Air bag circuits are usually located in a YELLOW conduit, use YELLOW wires, and/or have YELLOW connectors. Never work on or around these circuits without first disconnecting the battery and letting the system disarm.

Scan Tool Diagnostics

Use a DMM or lab scope to diagnose the hardwired circuits between modules and components related to the restraint systems following conventional procedures. The appropriate wiring information will prove to be a great asset in performing these diagnostic procedures.

Classroom Manual
Chapter 15, page 489

However, conventional diagnostic methods will not prove conclusive when diagnosing the electronic controls or communication networks between other modules and devices that provide features of the air bag system. For this reason, many diagnostic procedures will require the use of a diagnostic scan tool. The scan tool is the most reliable and accurate method to diagnose the electronic controls and communications related to system operation, as well as the retrieval or erasure of DTCs.

Special Tool

Scan tool

Retrieving Diagnostic Trouble Codes

⚡ WARNING **When turning the ignition on, do not reach across the steering wheel. Make sure that your body is away from any air bag modules that may deploy.**

The scan tool will display both history and current codes. Some manufacturers also display how long (in minutes) a code was active. Some systems store the DTCs in nonvolatile memory. These codes do not get erased due to battery disconnect or damaged. The only way to erase these codes is by use of the scan tool.

Some fault codes require replacing the ORC before performing any further diagnostics. For example, in some GM systems, a displayed code 52 indicates enough accident information is stored in the EEPROM to fill its memory. It is not possible to erase this code and requires the replacement of the ORC.

After retrieving all DTCs, use the diagnostic chart for the code to troubleshoot the system. After a repair or component replacement, the diagnostic chart will instruct you to verify the system. To perform this check, turn the ignition to the RUN position. If there are any additional trouble codes, the light will remain illuminated after completing the bulb check.

> **AUTHOR'S NOTE** Most of today's vehicles connect the scan tool to the air bag system through the J1962 DLC. However, many older vehicles will use a dedicated DLC.

Clearing Trouble Codes

In most systems, it's not possible to clear active codes; the condition that caused the code must be corrected first. After completing the repair, all current trouble codes become stored codes. After the codes have moved to the stored status, it is possible to erase them. In most cases, erasing DTCs requires a scan tool since DTCs are saved in nonvolatile memory and are retained even after disconnecting the battery.

Air Bag System Testing

Classroom Manual
Chapter 15, page 483

Special Tools
DMM
Test harness

Classroom Manual
Chapter 15, page 491

It's very important to follow *all* procedures listed in the diagnostic chart or service information. If the chart calls for the use of an ohmmeter to test a circuit, previous steps involving disconnecting harness connectors prepared the circuit for this test. Ignoring these steps and connecting the ohmmeter to the circuit may result in the air bag(s) deploying.

Since most systems use shorting bars at the connectors, you must remember to lift these up to properly diagnose the harness. Most manufacturers provide test connections or test harnesses that plug into the harness connectors (**Figure 15-1**) and lift the shorting bar. The connectors also provide test locations for connecting the digital multimeter (DMM). Before connecting the special wiring harness, remove and isolate the battery negative cable. Then wait for the time period specified by the vehicle manufacturer.

Figure 15-1 Test harness connected at the SRS terminals.

⚡ WARNING Use only the vehicle manufacturer's recommended tools and equipment for air bag system service and diagnosis. Failure to observe this precaution may result in unwarranted air bag deployment and personal injury.

⚡ WARNING Do not use battery- or AC-powered voltmeters or ohmmeters except those meters specified by the vehicle manufacturer. Failure to observe this precaution may result in unwarranted air bag deployment and personal injury.

⚡ WARNING Do not use nonpowered probe-type or self-powered test lights to diagnose the air bag system. Unwarranted air bag deployment and personal injury may result.

⚡ WARNING Follow the vehicle manufacturer's service and diagnostic procedures. Failure to observe these precautions may cause inaccurate diagnosis, unnecessary repairs, or unwarranted air bag deployment resulting in personal injury.

⚡ WARNING To avoid serious or fatal injury, disable the air bag system (and any other supplemental restraint systems) before attempting any steering wheel, steering column, air bag, seat belt tensioner, or impact sensor diagnosis or service. Also, disable the system before performing any electrical diagnosis or repair on or around the air bag system. Disable the system by disconnecting and isolating the battery negative (ground) cable and waiting the specified time for the system capacitor to discharge. This is the only sure way to disable the system. Failure to take the proper precautions could result in accidental air bag deployment.

Air Bag Simulator Load Tools

Most automotive manufacturers, and many after-market suppliers, developed an air bag diagnostics method using an **air bag simulator load tool** (**Figure 15-2**). The simulator connects in place of air bag components to represent a known good circuit and resistances. The simulator allows for a safe test sequence that should identify any circuit problems. If a circuit problem exists that would potentially cause the air bag to deploy, the simulator will indicate it and will remain in safe mode until repairs to fix the problem are completed.

If an active code goes stored after installing the simulator, the problem is with the disconnected component. For example, if there is an active DTC for "Driver's side squib circuit open," but removing the air bag module and installing the simulator in its place causes the DTC to become stored, then the open is within the air bag module.

🔧 **Special Tools**
Air bag load simulator
Scan Tool

Figure 15-2 Air bag simulator tool.

⚡ **WARNING** Never attempt to measure the resistance of an air bag module; deployment may occur.

⚡ **WARNING** Never attempt to repair an air bag module.

Squib Circuit Diagnosis

Classroom Manual
Chapter 15, page 484

🔧 **Special Tools**

Air bag load simulator
Scan Tool

When the ignition is in the RUN position, the ORC will send a 40 mA test current to the squibs of the various air bags and pretensioners on the system every 500 ms. The test current verifies the integrity of the squib and the circuit wiring. A DTC sets when the ORC detects a short to ground, short to voltage, squib circuits shorted together, or an open squib circuit for more than 2.5 seconds.

The squib circuits are maintained in a floating configuration at the module to prevent inadvertent deployment of the air bags, curtains, or pretensioners. Floating means that the circuit isn't connected to either power or ground. Using the air bag simulator load tool will provide a substitute for suspect squib with a known good component. The load tool adapters provide test points and a method for opening the shorting bar connections.

Study **Photo Sequence 31** as an example of using a load tool to diagnose a side curtain squib circuit shorted to ground fault. Keep in mind that this same routine, along with conventional DMM testing, is used to locate the cause of any air bag system squib circuit faults.

PHOTO SEQUENCE 31
Using the Airbag Simulator Load Tool

P31-1 Use the scan tool to retrieve fault codes and note if they are active or stored. Also, be sure to record all DTCs.

P31-2 After disconnecting the battery and waiting the required amount of time, disconnect the appropriate squib connector.

P31-3 Connect the load tool to the squib connector.

P31-4 With the load tool connected in place of the air bag squib, place the ignition in the RUN position and reconnect the battery.

P31-5 With the scan tool, read the active DTCs. If the code doesn't reappear, the fault is in the squib, and the air bag must be replaced. If the code reappears with the simulator connected, continue testing.

P31-6 With the ignition in the OFF position and the battery disconnected for the required amount of time, remove the load simulator tool from the squib connector and disconnect the ORC connector.

PHOTO SEQUENCE 31 (CONTINUED)

P31-7 Connect the load tool ORC 8443-28 adaptor to the ORC connectors.

P31-8 Measure the resistance of each squib circuit to ground. If the resistance is below specification, repair the circuit for a short-to-ground condition.

P31-9 If the circuit is good, replace the ORC after confirming good power and ground circuits.

Intermittent Faults

One of the most aggravating situations for the vehicle owner is to have their vehicle returned to them with "No Problem Found" stamped on the repair order. It's also very aggravating to the technician who cannot duplicate the problem the customer is experiencing. Intermittent problems challenge the technician to fully examine the system to determine possible causes for the fault. The answer isn't to throw parts at it; instead, the answer is to utilize a systematic diagnostic routine.

Intermittent faults with sensors, and most air bag system circuits, will usually set a DTC. However, the DTC will be stored instead of active when the technician retrieves them. This means the fault occurred at some point but isn't active right now, and that the conditions necessary to set this DTC are not currently present.

Begin by reviewing any technical service bulletins related to the problem the customer is experiencing. Perform any repair directed by the bulletin. If there is no bulletin, you must continue with the diagnostics. This includes a thorough visual inspection of the circuit components that are setting the DTC. For example, if the DTC is for a sensor fault, the sensor, sensor circuit wires, connectors, and terminals require inspection. Most intermittent conditions result from a poor connection somewhere in the circuit. It's a good practice to disconnect and clean all grounds involved in the system. If there are multiple stored DTCs, use the wiring information to determine common connections or grounds for all of the affected circuits. Further testing is necessary if the visual inspection fails to isolate the cause.

If the scan tool provides environmental data information associated with the stored DTC, this may help duplicate the problem. Environmental data is a snapshot of the conditions when the fault occurred. This information will assist you in determining the operating conditions that set the code so you can attempt to duplicate the conditions. It will also indicate how many ignition cycles ago the fault set and how long it was an active fault (**Figure 15-3**).

Next, use the scan tool to clear the DTC. With the ignition in the RUN position, monitor the scan tool for at least two minutes while wiggling the harness and connectors for the sensor circuits. Monitor the scan tool while looking for the sensor data to change or for the DTC to reset. If this doesn't cause a change in the sensor data or the DTC to return, cycle the ignition key between OFF and RUN several times, leaving the ignition in the RUN position for at least 10 seconds at a time. If the DTC doesn't return after this test, start the engine and allow it to reach normal operating temperature.

Special Tools
Scan Tool
Lab scope
DMM

EV Data - ORC			
Name	Value	Unit	Toggle Row Height
Lost Communication w/ Up-Front Right Satellite Acceleration Sensor	U0171		
Number of DTC	1		
DTC	C1 71		
DTC Readiness Flag	Complete		
DTC Storage State	Active		
Warning Indicator Request State	On		
Occurrence Flag	Fault		
Reserved	Undefined		
Original Odometer	11642	miles	
Most Recent Odometer	11642	miles	
Frequency Counter	1		
Operation Cycle Counter	0		
Accumulation Timer	0	minutes	

Figure 15-3 Scan tool showing environmental data associated with DTCs.

⚙ **SERVICE TIP** Do not erase the fault code until you have reviewed the environmental or freeze frame data. Clearing the DTC will also clear this data.

If the code appears during any of the tests, perform the diagnostic routines for an active code. If the code doesn't return, duplicate the conditions indicated by the environmental data. Using the data recording function of the scan tool may assist the technician in locating the fault (**Figure 15-4**). Also, a lab scope is a very useful tool to

Figure 15-4 Partial screen capture of scan tool display for a freeze frame.

help diagnose intermittent conditions since it will indicate any glitches, spikes, voltage dropouts, and so forth.

> **CUSTOMER CARE** If you cannot duplicate an intermittent fault, do not attempt the repair by guessing. Some scan tool and OEM manufacturers provide a data recording tool (separate from the scan tool) that you can leave with eh customer. The owner pushes a button to save the data recording when a problem occurs. Upon returning the vehicle to the shop, you can retrieve the information and pinpoint the root cause.

Inspection After an Accident

Any time the vehicle is involved in an accident, even if the air bag did not deploy, all air bag system components need inspecting. The wiring harness must be inspected for damage and repaired or replaced as needed. Replace any damaged or dented components. Don't attempt to repair any of the sensors or modules. Service of these components is by replacement only.

    If the air bags deployed, the service information will provide a list of components that must be replaced. The list of components will vary depending on the manufacturer.

Classroom Manual
Chapter 15, page 493

 Special Tools

Safety glasses
Particle dust mask
Protective clothing
   with long sleeves
Tape
Shop vacuum cleaner

Cleanup Procedure After Deployment

After an air bag deployment, remove the residue from inside the passenger compartment before entering the vehicle. Tape the air bag exhaust vents closed to prevent additional powder from escaping. Use the shop vacuum cleaner to remove all powder from the vehicle's interior. Work from the outside to the center of the vehicle. Vacuum the heater and air-conditioning (A/C) vents. Run the heater fan blower motor on low speed and vacuum any powder blown from the plenum.

Component Replacement

None of the sensors, air bag modules, or controllers used in the air bag system is repairable. In addition, seat belt pretensioners are not serviceable. If any of these components are defective, replacement of the component is the only fix.

    If wiring repair is required, first refer to the proper service information. Some manufacturers don't recommend wire harness repair, and others may have very specific procedures they require.

Air Bag Module Replacement

⚡ WARNING **Before replacing any component of the air bag system, follow the service information procedure for disarming the system—even if the air bag was deployed.**

⚡ WARNING **Wear safety glasses when working on the air bag system.**

Classroom Manual
Chapter 15, page 484

> ⚙ SERVICE TIP Anytime the steering column is removed from the vehicle, either remove the clockspring or lock the steering wheel so it cannot rotate. This will prevent accidentally extending the clockspring beyond its travel limits.

Follow **Photo Sequence 32** as a guide to replacing an air bag module. When removing the air bag module, maintain the clockspring's correct index position. Failure to do so can cause damage to the enclosure, wiring, or module. Any of these situations can cause the air bag system to default into a non-operative mode. Always follow the service information for the vehicle you are working on.

Torque the module fasteners to the specific value. Usually, a new air bag module includes new fasteners.

Reverse the procedure to reinstall the module. To rearm the system, connect the yellow two-way electrical connector at the base of the steering wheel and install the connector position assurance (CPA). Turn the ignition to the RUN position, replace the SRS fuse, and connect the battery. This procedure keeps you out of the vehicle as the system is powered up.

With the ignition in the RUN position, observe the SRS warning light. It should perform its bulb test and then shut off. Perform the SRS diagnostic system check to confirm proper operation.

PHOTO SEQUENCE 32
Typical Procedure for Removing the Air Bag Module

P32-1 Place the wheels in the straight ahead position and center the steering wheel.

P32-2 Disconnect the negative battery cable.

P32-3 Tape the cable terminal to prevent accidental connection with the battery post. Wait the specified time to allow the reserve energy to dissipate.

P32-4 Use a trim stick to remove the access cover.

P32-5 With the access cover removed, you should be able to determine how the air bag module is attached.

P32-6 Remove the fasteners that secure the module from the rear of the steering wheel.

PHOTO SEQUENCE 32 (CONTINUED)

P32-7 Disconnect the connector for the horn.

P32-8 Release the connector position assurance lock tabs.

P32-9 Disconnect the squib connectors and remove the module.

Clockspring Replacement

Inspect the clockspring (**Figure 15-5**) any time the vehicle has been involved in an accident, even if the air bag did not deploy. If the driver side air bag deployed, the heat generated might damage the clockspring. For this reason, replacing the clockspring after air bag deployment is a good practice. Exact procedures vary according to the manufacturer.

⚠ WARNING **Wear safety glasses when servicing the air bag system.**

In most cases (but not all), the front wheels are placed into the straight ahead position. After properly disarming the air bag system, remove the air bag module. Mark the steering shaft and steering wheel with index marks for installation. After removing the lock nut, a special puller may have to be used to pull off the steering wheel (**Figure 15-6**).

Classroom Manual
Chapter 15, page 488

Special Tools
Steering wheel
puller set

Figure 15-5 Clockspring.

Figure 15-6 Removing the steering wheel may require a puller.

Notch   Red mark

Figure 15-7 The index marks should align if the clockspring is properly centered.

With the upper and lower steering column shrouds removed, disconnect the lower clockspring connector. If reusing the clockspring, tape it to prevent rotation. Remove the retaining fasteners and the clockspring.

Installing the clockspring is in the reverse order. Some replacement clocksprings have a locking insert to prevent the rotation of the rotor. Don't remove this insert until the clockspring is secured onto the column by the retaining screws. If the clockspring has rotated for some reason, it must be re-centered. Usually, this is done by gently rotating the rotor in one direction. Don't use excessive force, or damage to the clockspring may occur. Once the clockspring is in this full stop position, rotate the rotor in the opposite direction while counting the turns until you reach the other stop. Next, rotate the rotor back half the number of turns counted so it's in the middle of its travel. At this time, some form of index mark (**Figure 15-7**) or tabs should align. Torque the new steering wheel attaching nut to specifications. After installing the air bag module, rearm the system and perform the verification test.

Classroom Manual
Chapter 15,
pages 489, 496

 Caution

All restraint system sensors and modules must be properly fastened. Most sensors require proper orientation. In addition, all fasteners must be properly torqued. Often the manufacturer will discourage the reuse of fasteners.

Diagnosing Air Bag System Pressure Sensors

All sensor diagnostics begin with a thorough inspection of the sensor and its related circuits. This includes looking for physical damage and proper electrical connection. Carefully examine the wiring harness and connectors between the sensor and the control module, including any dedicated ground circuits. Test all connectors and connector terminals for looseness, pullout, or damage. If the sensor has a hose connected, confirm the hose's condition.

This section discusses the diagnostics of different types of pressure sensors. Although the diagnostic routines discussed may pertain to a specific function, the diagnostics of the circuits of most pressure sensors will be compatible.

Diagnosing MEMS Pressure Sensors

MEMS pressure sensors determine and confirm side impacts (**Figure 15-8**). Seldom is this the only input used to determine an impact event. Other inputs include vehicle motion, yaw, lateral sensor, and accelerometer inputs. MEMS pressure sensors are not

Figure 15-8 MEMS pressure sensor mounted inside the door to measure sudden pressure changes.

diagnosed in the same manner as most strain or pressure sensors since they have their own internal processing units and will usually communicate data to the ORC differently than a typical pressure sensor.

The MEMS sensor contains an electronic communication chip that allows it to communicate sensor status and sensor fault information to the ORC. The impact sensors perform their own self-test using power supplied by the ORC and then communicate this status back to the ORC with periodic updates. The ORC microprocessor continuously monitors the electrical circuits to determine the system's readiness. If the ORC detects a monitored system fault, it sets a DTC and illuminates the air bag indicator lamp. The impact sensor receives battery current and ground through dedicated sensor plus and minus circuits from the ORC. The impact sensors and the ORC communicate by modulating the current in the sensor plus circuit (**Figure 15-9**).

Special Tools
Scan tool
DMM
Jumper wires
Backprobing tools
Service information

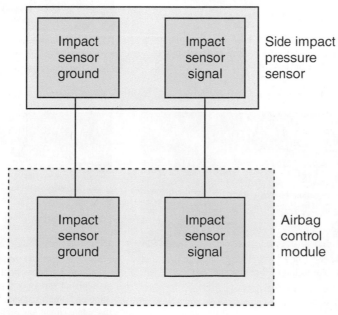

Figure 15-9 MEMS pressure sensor circuit.

Diagnose the hardwired circuits between the pressure-type impact sensors and the ORC using conventional diagnostic tools and procedures. However, a scan tool is the most common method for diagnosing the impact sensor, the electronic controls, or the communication network between the sensor and the ORC.

Use the scan tool to access DTCs stored in the ORC. The following are some diagnostic examples for different DTCs that you may encounter.

Lost Communication with Side-Impact Sensor. Repairing this fault condition involves verifying the integrity of the wiring between the ORC and the sensor. The ORC monitors communication with the impact sensor continuously and will set this type of DTC when it no longer receives notification of the status update from the impact sensor. Possible causes for the setting of this DTC include the following:

- Side-impact sensor signal circuit open
- Side-impact sensor ground circuit open
- Side-impact sensor signal and ground circuit shorted together
- Side-impact sensor signal circuit shorted to ground
- Side-impact sensor signal circuit shorted to voltage
- Side-impact sensor ground circuit shorted to voltage
- Faulty impact sensor
- Faulty ORC

Photo Sequence 33 details a typical procedure for diagnosing the MEMS pressure sensor for this fault. The diagnostics used for this fault include those used for several other faults as well.

PHOTO SEQUENCE 33
Diagnosing the MEMS Pressure Sensor

P33-1 When retrieving the DTCs, pay attention to if they are active or stored faults. Also, record all of the DTCs to help determine if a common fault is responsible.

P33-2 If the DTC is active, turn the ignition OFF, disconnect the battery, and wait for the specified amount of time before proceeding.

P33-3 Disconnect the side-impact sensor connector. Check connectors and terminals. Clean and repair as necessary.

P33-4 Disconnect the ORC connector and inspect the connector and terminals for damage. Clean and repair as necessary.

P33-5 Connect the air bag simulator load tool adaptor to the ORC connectors.

P33-6 Measure the resistance of the side-impact sensor signal circuit between the side-impact sensor connector and the load tool adaptor. If the resistance is more than 1.0 ohm, the signal circuit has excessive resistance or an open condition that requires repair.

PHOTO SEQUENCE 33 (CONTINUED)

P33-7 Measure the resistance of the impact sensor ground circuit between the sensor connector and the load tool adapter. If the resistance exceeds 1.0 ohm, the ground circuit has excessive resistance or an open condition that requires repair.

P33-8 Measure the resistance between the side-impact sensor signal circuit and the sensor ground circuit at the sensor connector. If the resistance is low, repair the condition causing the short between the two circuits.

P33-9 Move the test leads to measure the resistance between the sensor signal circuit and chassis ground. If the resistance reading is below specifications, repair the signal circuit for a short to ground.

P33-10 Turn the ignition to the RUN position and reconnect the battery. Measure the voltage on the signal circuit. If voltage is present, repair the signal circuit for a short-to-voltage condition.

P33-11 Move the test leads to measure the voltage on the sensor ground circuit between the load tool adaptor and chassis ground. If voltage is present, repair the sensor ground circuit for a short-to-voltage condition.

P33-12 If all tests pass, turn the ignition to the OFF position and disconnect the battery. After waiting the specified amount of time, replace the impact sensor.

P33-13 Remove any special tools used and reconnect all disconnected components (except the battery).

P33-14 Turn the ignition to the RUN position and reconnect the battery. Use the scan tool to clear all codes in the ORC.

P33-15 If needed, perform the module reset function with the scan tool.

P33-16 After waiting one minute, recheck for any active DTCs. If the same code for loss of communication returns, replace the ORC. If the code doesn't return, the repair is complete.

Side-Impact Sensor Configuration Mismatch. When the ORC is powered up, it sends a test current to the impact sensors. During the test, the ORC also confirms if the impact sensors that it discovered agree with the preprogrammed vehicle configuration of components assigned to the ORC. The DTC sets when the ORC finds an impact sensor, but the module configuration doesn't indicate that the device should be present. Diagnosing this type of fault is centered on confirming the use of the correct ORC and wiring harness. Usually, this fault sets after someone did some swapping of modules or attempted to install a used module. Also, installing the incorrect wiring harness will set this fault.

Diagnosing Air Bag System Accelerometer Sensors

Classroom Manual
Chapter 15,
pages 489, 491

Electro-mechanical accelerometers are tested by removing them from the system and measuring for open and shorted conditions. The circuits to the electro-mechanical sensors are tested for continuity, excessive resistance, and shorts like any other circuit. Be sure to follow all safety warnings and cautions associated with the test procedures.

MEMS-type accelerometers are diagnosed in much the same way as the MEMS pressure sensors just discussed. The MEMS accelerometer performs its own self-test routine upon receiving power from the ORC. The status of the diagnostic routine is communicated to the ORC. The diagnostic routine constantly runs when the ignition is on. A DTC sets, and the SRS warning lamp illuminates if the routine determines a fault at any time.

The diagnostic process of checking circuit integrity is identical to that explored previously in Photo Sequence 34 for the MEMS pressure sensor. This code requires that the entire circuit be tested and will encompass most of the diagnostics that may be associated with other DTCs. Loss of communication between the accelerometer and the ORC can result from any of the following conditions:

- Impact sensor signal circuit open
- Impact sensor ground circuit open
- Impact sensor signal circuit shorted to sensor ground circuit
- Impact sensor signal circuit shorted to ground
- Impact sensor signal circuit shorted to voltage
- Impact sensor ground circuit shorted to voltage
- Faulty impact sensor
- Faulty ORC

Occupant Classification System Service

Classroom Manual
Chapter 15, page 501

Special Tools

Scan tool
DMM

Classroom Manual
Chapter 15, page 502

Diagnosing the electrical functions of the occupant classification system (OCS) is performed using a scan tool. Detection of a fault will set and record DTCs. In addition, upon detection of a fault, the passenger air bag disable lamp (PADL) turns on, indicating to the driver that the air bags are suppressed.

The Wheatstone bridge is used in the OCS to determine if an occupant is in the passenger side front seat and to classify them according to their weight. There are two main types of systems used: the bladder and the strain gauge system.

The bladder system uses a silicone-filled bladder positioned between the seat foam and the seat support (**Figure 15-10**). A hose connects the pressure sensor to the bladder (**Figure 15-11**). Adding weight to the seat bottom causes the bladder to disperse the silicone, and the pressure sensor reads the increase in pressure. The pressure reading is input to the OCM (**Figure 15-12**).

Figure 15-10 Bladder-type occupant classification system.

Figure 15-11 The pressure sensor changes voltage signals as weight on the bladder changes.

Because seat occupation—and ultimately occupant classification—is determined by pressure, the system corrects for changes in atmospheric pressures. In addition, natural aging of the seat foam is also learned by monitoring gradual changes. When the seat is empty, the OCM compares the sensor voltage with the value stored in memory. Adding weight to the seat changes the voltage values measured by the sensor. Based on the change of voltage from the sensor, the OCM can determine the weight of the occupant.

The bladder system typically uses a strain gauge connected to the seat belt anchor (**Figure 15-13**). Because a securely strapped infant seat causes an increase in downward pressure on the bladder, the system includes a belt tension sensor (BTS) to determine that the seat belt is tight around an object (about 24–26 lbs [11–12 kg] of force). The reading will indicate that the belt is tighter than what it normally would be for a belt around a person. This would indicate that the belt is securing a child safety seat.

Belt Tension Sensor Diagnostic Test

The following is a typical procedure for testing the belt tension sensor (BTS) used on the Delphi bladder OCS. Since this sensor determines if an infant seat is on the front passenger-side seat, it's critical to follow proper diagnostics procedures. Always refer to the service information for the vehicle you are working on.

With the scan tool connected, monitor the output of the BTS. Tracking may use a method involving counts (**Figure 15-14**). With the front passenger-side seat unoccupied and the BTS in its static state, the counts should be between 39 and 69. If the counts are outside of this range, replace the seat belt retractor and the BTS assembly.

If the counts are within the specified range, use a pull scale (**Figure 15-15**) to apply a load to the BTS. Connect the hook of the scale through the BTS webbing loop. Grab the pull handle at the top of the spring scale and pull the spring scale straight up, keeping the horizontal bar level with the door sill and the spring scale in line with the BTS. Apply and hold 25 pounds (11.5 kg) of pressure on the sensor while monitoring the BTS counts on the scan tool. The counts should increase to between 153 and 204. If the counts are not within this range, replace the passenger seat belt retractor and BTS assembly.

Classroom Manual
Chapter 15, page 504

 Special Tools
Scan tool
Pull scale

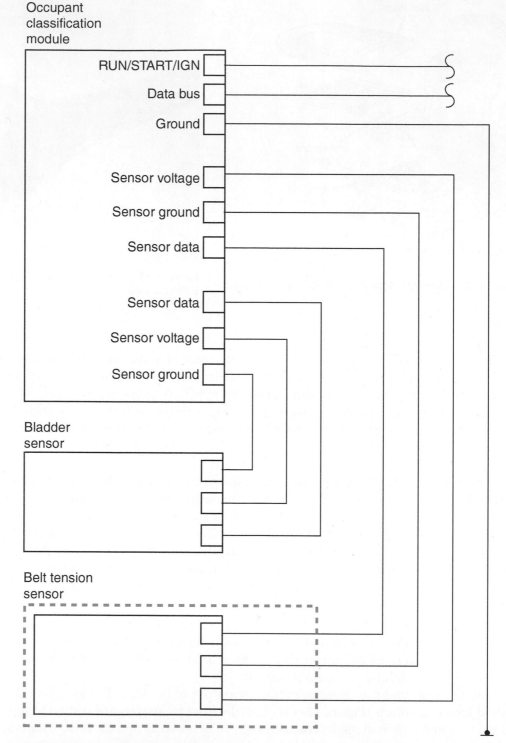

Figure 15-12 Electrical schematic of the bladder occupant classification system.

If the counts are within range, release the pressure applied with the pull scale and monitor the BTS counts again. With no load applied to the BTS, the output should return to between 39 and 69 counts within 20 seconds. If the counts are not within this range, replace the passenger seat belt retractor and BTS assembly.

If all tests discussed pass, then the BTS is operating as intended. Confirm all electrical connections before returning the vehicle to the customer.

Figure 15-13 Belt tension sensor incorporated within the webbing loop mount.

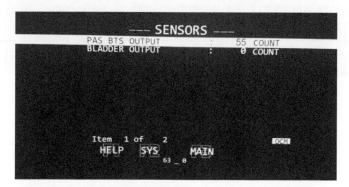

Figure 15-14 Scan tool will show the BTS counts.

Figure 15-15 Using the pull scale to obtain pressure against the BTS.

Circuit Performance Fault Code Diagnostics

Depending on the system, the ORC will usually sample the signal voltage from the sensor when the ignition is in the RUN position. If the voltage value indicates a pressure difference that is above or below an acceptable level, a DTC will be set and cause the air bag system warning lamp to illuminate. A circuit performance fault-type code sets if the circuit is over or under voltage for a specified time.

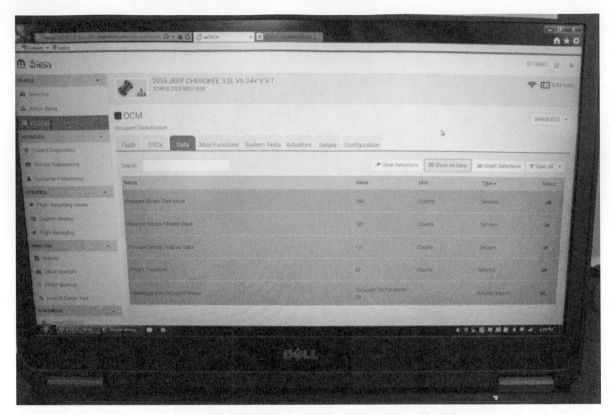

Figure 15-16 Strain gauge values as seen on a scan tool.

Although manufacturers that use a Wheatstone bridge–type sensor will have differing diagnoses of the sensors, if the sensor signal voltage is within the programmed value range of the ORC, a DTC may not set. For example, an open signal circuit will set a DTC, but resistance in the circuit that causes the voltage to be off by a few millivolts may not set a fault code, or it may trigger intermittent faults.

In some systems, testing the Wheatstone bridge sensor is a simple way of monitoring the sensor data output and voltage with a scan tool (**Figure 15-16**). With the ignition in the RUN position, record the pressure reading and voltage value. The pressure reading should represent static pressure (such as atmospheric pressure or no strain), and the voltage should confirm the reading. Compare this reading with the actual pressure, the two should be equal. A sensor that doesn't read correct static pressure cannot accurately read sudden pressure changes. If the reading isn't correct, verify the condition of all the circuits to the sensor, as discussed next. If the circuits are good, replace the sensor.

A "Sensor Circuit High" fault results from the signal voltage exceeding a specified value. The circuit high fault is the result of any of the following conditions:

- Signal circuit open or high resistance
- Return circuit open or high resistance
- Signal circuit shorted to voltage
- 5-volt supply circuit shorted to voltage
- Faulty sensor
- Internal ORC fault

To perform a quick test of the circuits, consider the schematic of a sensor circuit (**Figure 15-17**). If the scan tool indicates a voltage reading of 5 volts, any of the conditions listed above could be the cause. Unplug the sensor and use a jumper wire to jump the signal circuit to ground. The scan tool should now read 0 volt. If it still reads 5 volts, the circuit is open between the ORC and the connector. If the voltage is above 0 volt, there is excessive resistance in the signal circuit between the ORC and the connector.

Figure 15-17 Typical pressure sensor circuit.

If the reading was 0 volt, move the jumper wire to short the signal circuit to the return circuit. The scan tool should now read 0 volt in a properly operating circuit. If the scan tool displays 5 volts, the return circuit is open. If the reading is above 0 volt, there is excessive resistance in the return circuit. If all of the readings are normal, the sensor is faulty.

> ⚙️ **SERVICE TIP** If the sensor return circuit is faulty within the ORC, all sensors that use this circuit would have a high voltage reading. If the return circuit of the sensor indicates an open circuit, check the return circuit from another input component. If this circuit shows good continuity, the problem is between a splice or in-line connector in the return circuit and the sensor connector. Use the wiring diagrams to locate these.

A sensor circuit low fault results from the signal voltage being below a specified value. The circuit low fault is the result of any of the following conditions:

- 5-volt supply circuit open or high resistance
- 5-volt supply circuit shorted to ground
- 5-volt supply circuit shorted to the sensor return circuit
- Signal circuit shorted to ground
- Signal circuit shorted to the sensor return circuit
- Faulty sensor
- Internal ORC fault

If the scan tool reading with the ignition in the RUN position and sensor connected was 0 volt, any of the above conditions could be the cause. The sensor is faulty if the voltage changes to 5 volts when the sensor is unplugged. If the voltage remains at 0 volt when the sensor is unplugged, reconnect the sensor and backprobe the supply circuit. Use a voltmeter to check the supply circuit for 5 volts. It's important to test this voltage with the sensor connected to provide a load against the supply circuit. If the voltage is low, there is either high resistance, an open, or a fault in the ORC. Isolate these problems by using a voltmeter to perform a voltage drop test of the circuit.

A voltage exceeding 5 volts, indicates a short to voltage condition. Isolate the offending circuit by removing fuses and relays one at a time while monitoring the voltage on the supply circuit. If the voltage returns to normal after pulling a fuse, use the wiring diagram to determine if any connectors are common between the two circuits. Next, inspect the wiring harness for damage or any condition that can cause the short.

> ⚙ **SERVICE TIP** Anytime you use a scan tool to monitor a pressure reading, don't accept the value until you confirm the voltage. By correlating the signal voltage to the pressure reading, a fault in the circuit can be determined.

To test the sensor circuit shown in Figure 15-17 using a DMM, backprobe the 5-volt supply circuit wire (connector plugged into the sensor) with a voltmeter with the ignition in the RUN position. The voltage should be between 4.5 and 5.2 volts. If the voltage is too high, check for a short-to-voltage condition.

If the voltage is too low, backprobe the ORC connector to check the voltage on the circuit at the ORC. If the voltage is within specifications at the ORC but low at the sensor, repair the 5-volt supply circuit wire for high resistance. If this voltage is low at the computer, turn the ignition off, disconnect the battery, and wait for 2 minutes. Disconnect the ORC connector and use an ohmmeter to test for a short to ground. If the resistance is below specifications, this indicates that the circuit is shorted to chassis ground. If the reading exceeds the minimum specifications, use the ohmmeter to measure the resistance between the 5-volt supply circuit and the sensor return circuit. If the resistance is below specifications, locate the short between the two circuits in the wiring harness. If no problem is found, inspect the wiring and connectors between the sensor and the ORC while looking for any chafed, pierced, pinched, or partially broken wires. Inspect for broken, bent, pushed out, or corroded terminals. Be sure to review any technical service bulletins (TSBs) that may apply. If all of these fail to isolate the cause, replace the ORC.

If the voltage supply circuit is good, test the sensor return (ground) circuit. With the ignition in the RUN position, backprobe the ground circuit terminal at the connector (with the connector plugged into the sensor). Connect the voltmeter from the sensor ground wire to the battery ground. If the voltage drop across this circuit exceeds specifications, test the voltage drop on the ORC's ground circuits. If this is good, repair the ground wire from the sensor to the computer.

If the first test of the 5-volt supply circuit indicates a voltage reading between 4.5 and 5.2 volts, test the signal circuit for an open or high resistance. To test the sensor signal circuit, backprobe the sensor signal wire with the sensor connected and turn the ignition to the RUN position (engine off). Connect the DMM from the signal wire to ground. The voltage reading indicates the barometric pressure signal from the sensor to the ORC. If the signal circuit voltage, as compared to actual barometric pressure, confirms that the sensor's voltage supply and return circuits are good, replace the sensor.

The barometric pressure voltage signal varies depending on altitude and atmospheric conditions. Follow this calculation to obtain an accurate barometric pressure reading:

1. Determine present barometric pressure; for example, 29.85 inches.
2. Multiply your altitude by 0.001; for example, 600 feet times 0.001 = 0.6.
3. Subtract the altitude correction from the present barometric pressure; for example, 29.85 − 0.6 = 29.79.
4. Check the vehicle manufacturer's specifications to obtain the proper barometric pressure voltage signal in relation to the present barometric pressure.

AUTHOR'S NOTE Some sensor signal circuits may have 0 volt when the connector is separated. Always confirm the expected voltage values with the proper service information.

If the 5-volt supply, signal, and return circuits test good, reconnect the ORC harness connector and the battery. Use a jumper wire at the sensor harness connector to connect the signal circuit and the return circuit. With the ignition in the RUN position, view the sensor voltage displayed on the scan tool. It should read 0 volt and set a DTC for a circuit low code. If this test passes, replace the sensor. If this test doesn't pass, perform all of the inspections discussed earlier and then replace the ORC.

Sensor Performance Fault Diagnostics

A sensor performance fault code sets when the voltage of the signal circuit is within acceptable parameters; however, the reading isn't acceptable for the conditions. In addition, the ORC may determine blockage at the ports of the sensor if a reading is different than expected. For example, if the pressure sensors in the two front doors don't indicate the same amount of pressure, the sensor that shows a change from the reading when the ignition was turned off as compared to when it was turned on again may set a performance code.

SERVICE TIP Be sure to properly identify the fault code. There is a difference between diagnostic routines for a *sensor* performance fault code and a *circuit* performance fault code.

The pressure sensor will usually set a performance fault if the sensor provides abnormal readings over a specified amount of time (**Figure 15-18**). To diagnose this type of fault, start with a visual inspection of the sensor, paying close attention to the sample port opening. If the visual inspection fails to isolate the cause of the performance fault code, use the following steps to determine the root cause. Refer to the proper service information to determine the conditions that cause the DTC to set. Also, refer to the wiring diagram for the system. For this example, consider the wiring diagram in **Figure 15-19**.

With the ignition in the OFF position and the battery disconnected for two minutes, disconnect the pressure sensor connector and the ORC connector. Measure the resistance of each sensor circuit between the two connectors. If the resistance in any of the circuits

Caution
Use any special tools, such as breakout boxes, which the manufacturer calls for in the service information. Failure to follow these guidelines may result in damaged terminals.

Absolute BARO reading	Lowest allowable voltage at −40°F	Lowest allowable voltage at 257°F	Lowest allowable voltage at 77°F	Designed output voltage	Highest allowable voltage at 77°F	Highest allowable voltage at 257°F	Highest allowable voltage at −40°F
31.0"	4.548 V	4.632 V	4.716 V	4.800 V	4.884 V	4.968 V	5.052 V
30.9"	4.531 V	4.615 V	4.699 V	4.783 V	4.867 V	4.951 V	5.035 V
30.8"	4.514 V	4.598 V	4.682 V	4.766 V	4.850 V	4.934 V	5.018 V
30.7"	4.497 V	4.581 V	4.665 V	4.749 V	4.833 V	4.917 V	5.001 V
30.6"	4.480 V	4.564 V	4.648 V	4.732 V	4.816 V	4.900 V	4.984 V
30.5"	4.463 V	4.547 V	4.631 V	4.715 V	4.799 V	4.883 V	4.967 V
30.4"	4.446 V	4.530 V	4.614 V	4.698 V	4.782 V	4.866 V	4.950 V
30.3"	4.430 V	4.514 V	4.598 V	4.682 V	4.766 V	4.850 V	4.934 V
30.2"	4.413 V	4,497 V	4.581 V	4.665 V	4.749 V	4.833 V	4.917 V
30.1"	4.396 V	4.480 V	4.564 V	4.648 V	4.732 V	4.816 V	4.900 V
30.0"	4.379 V	4.463 V	4.547 V	4.631 V	4.715 V	4.799 V	4.883 V

Figure 15-18 Barometric pressure voltage values at various pressures.

Figure 15-19 Differential pressure sensor (DPS) circuit.

is greater than that specified by the manufacturer, repair that circuit. Usually, the maximum amount of resistance allowed will be about 1 ohm.

If the circuits all test within specifications, replace the sensor. Follow the service information requirements to perform a verification test of the system. If the performance fault code returns, the service information will generally lead you to replacing the ORC. Be sure that all of your testing up to this point has been complete and thorough. Do not replace the ORC before determining it to be the last possible cause of the problem.

Bladder-type systems produced by Delphi have some unique service requirements. If diagnostics indicate that the occupant classification module (OCM) requires replacement, this is possible only if the original seat is in the vehicle. Since the ORC stores the aging calibrations of the seat, the installation of a new OCM will be recognized by the ORC, and it will continue to use the old seat wear calibrations.

If diagnostics of a bladder-type system direct you to replace any of the following items, install an **OCS service kit**:

- Bladder
- Sensor
- Seat foam
- Cloth seat cover

The OCS service kit consists of the seat foam, the bladder, the pressure sensor, the OCM, and the wiring. The service kit is calibrated as an assembly. The service kit OCM has a special identification data bit that is transmitted once it's connected. After installation of the service kit, the ORC will identify the service kit OCM and clear all calibration data related to the system components.

The wiring of the service kit is hot glued to the module and the sensor to prevent the separation of the components. Also, the wiring harness may have a tag attached to it to identify it as a service kit.

The service kit wiring uses a single connector that mates to the existing vehicle wiring harness. The connection on the vehicle harness that originally connected to the OCM now connects to the service kit harness. The connector on the vehicle harness that originally connected it to the pressure sensor is not used with the service kit. Tie this wire to the harness to prevent it from getting caught in the seat tracks.

Classroom Manual
Chapter 15, page 502

⚡ **Caution**

Since the ORC stores the seat-aging calibrations, swapping seats isn't a good practice. The ORC will not have the correct calibrations for the replacement seat, and the system will not work as intended. This may lead to air bag deployment when an infant is in the seat.

Strain gauge systems by TRW don't have these service requirements. However, replacing components on either system requires validating the system before returning the vehicle to the owner.

Classroom Manual
Chapter 15, page 504

OCS Validation

Many OCSs require validating the OCS virtually any time the passenger-side front seat has any type of service performed to it. Loosening the passenger-side front-seat retaining bolts, removing the seat, replacing the seat, replacing the seat trim, replacing a sensor, or replacing the OCM requires performing an **OCS validation test**. The OCS validation test confirms that the system can properly classify the occupant. This task usually requires the use of special weights (**Figure 15-20**).

The special weight set has three parts. The tool approximates federal standards for the weight of occupants:

1. The base weight of the tool weighs 37 pounds (17 kg) to validate for the classification of a rear-facing infant seat (RFIS) weight.
2. The addition of the 10-pound (4.5-kg) weight to the base validates for the weight classification of a child.
3. The addition of the 52-pound (24-kg) weight to the assembly validates for the weight classification of the fifth percentile female.

The total weight of the tool is 99 pounds (45 kg). Weights are added to the base in the proper order and placed in the correct position with the dowel pins. Differences between the bladder-type and strain gauge—type systems affect which weights to use. It's imperative to follow the correct procedure.

Photo Sequence 34 illustrates a typical procedure for validating the OCS.

Special Tools
Validation weight set
Scan tool

⚠ **Caution**
If the seat requires removal from the vehicle, the strain gauges are susceptible to damage. Do not drop the seat or allow anyone to sit in it while it's out of the vehicle.

Figure 15-20 Special OCS weights used to validate the system.

PHOTO SEQUENCE 34
Occupant Classification Validation

P34-1 Tools required to perform this task include a scan tool and special verification weights.

P34-2 Connect the scan tool and check for active battery voltage, OCM internal failure, or communication DTCs. Correct any of these conditions before continuing.

P34-3 Make sure the seat is empty and is in its full rearward position, with the back rest in a normal upright position.

P34-4 Use the scan tool to access the "OCM Verification" screen and follow the instructions to begin the test.

P34-5 Once the test has started, verify that the PADL is illuminated.

P34-6 When instructed, add the correct weight amount to the seat. Follow the scan tool prompts to complete this phase of the process. The scan tool should confirm the successful completion of this phase.

P34-7 When instructed, add the additional weight and follow the prompts to complete this phase. The scan tool should confirm the successful completion of this phase.

P34-8 Confirm that the PADL is now off.

P34-9 Use the scan tool to access DTCs and check for any active codes. If active codes are present, repair this condition and repeat the system verification procedure. Clear any stored codes.

Classroom Manual
Chapter 15, page 506

 Special Tool

Scan tool

Diagnosing Rollover Sensors

Usually, the rollover sensor used for the rollover protection system is internal to the ORC. However, it may be a part of a sensor group that is also used for rollover mitigation and other ABS functions. This section will focus on the ORC internal sensor. Chapter 16 covers cluster sensor diagnosis.

A fault code for "Roll-Over Internal" is set when the ORC detects corruption in the data and circuitry related to the internal rollover sensor. On many systems, the detection of a failure of the rollover sensor in two consecutive ignition cycles sets an indefinite DTC. If the code is active when retrieving codes with the scan tool, the repair is to replace the ORC. If the code is stored, the failure occurred but not twice in a row. Perform the diagnostic routine described earlier under the heading of "Intermittent Faults."

CASE STUDY

A customer brings their vehicle to the service shop because they have observed that the air bag warning light is always illuminated. The technician confirms the warning lamp operation by performing the diagnostic system check. Next, the technician uses the scan tool to retrieve fault codes. The scan tool displays a current code for an open driver-side air bag squib circuit. Following the service information procedure, the technician removes the air bag module and tests the clockspring circuit. Since the clockspring circuit proves to be in good condition, the technician replaces the air bag module. After installing the new module, the technician runs the diagnostic system test again to confirm that the repair was successful.

Determining the need to replace the air bag module resulted from using the process of elimination. Since the clockspring circuit proved to be good, the air bag module had to be faulty. The resistance of the air bag module's squib was not tested because this can cause the air bag to deploy, and the service information provided several warnings against it.

CASE STUDY

A customer brought their Dodge Durango to the shop because the air bag warning light was always on. Upon receipt of the repair order, the technician verifies that the lamp is indeed on. Next, the technician connects the scan tool and retrieves DTCs. There are three DTCs, all of which are active codes for the loss of communications with the three left side-impact sensors for the left side air bag curtain.

Upon investigation of the wiring diagrams, the technician notes that the three sensors for this vehicle are wired in series with each other. This is an unusual method for wiring the side-impact sensors since most are wired directly to the ORC on dedicated circuits.

Realizing this important fact concerning the operation of the sensors, the technician was then able to discern that when the ORC performed its diagnostic routine, a fault in the sensor wired first to the ORC would also result in the other sensors failing.

The technician follows the diagnostic routines and safety precautions to diagnose the circuits between the ORC and the first sensor. Using the appropriate test equipment, there was no problem found in the wiring, and the sensor was replaced. After reconnecting all of the connectors and removing all test equipment, the technician erases the DTCs and then rechecks to see if any return. The absence of new DTCs confirmed that the sensor was faulty and the new sensor is operating properly.

By using the wiring information correctly, the technician could determine how the ORC would set loss of communication fault codes for all three sensors. This allowed the technician to locate the problem area without having to remove trim panels to access sensors and wires that could not cause the problem.

ASE-Style Review Questions

1. The occupant classification system must be validated whenever:
   A. The seat has been removed from the vehicle.
   B. The OCM has been replaced.
   C. A new sensor has been installed.
   D. All of these choices.

2. A low voltage on the pressure sensor's supply circuit could be caused by any of the following EXCEPT:
   A. An open in the supply circuit wire.
   B. A short to ground in the supply circuit wire.
   C. High internal resistance in the sensor.
   D. Excessive voltage drop on the supply circuit.

3. The PADL is off when the seat is empty. This may indicate:
   A. A faulty OCM.
   B. Normal operation.
   C. The BTS is indicating the seat belt is cinched.
   D. All of these choices.

4. Replacement of an air bag module is being discussed.
   Technician A says to follow the service information procedure for disarming the system.
   Technician B says to wear safety glasses.
   Who is correct?
   A. A only   C. Both A and B
   B. B only   D. Neither A nor B

5. *Technician A* says the clockspring should be inspected whenever the vehicle has been involved in an accident, even if the air bag was not deployed.
   Technician B says the heat generated when an air bag deploys may damage the clockspring.
   Who is correct?
   A. A only   C. Both A and B
   B. B only   D. Neither A nor B

6. *Technician A* says air bag residue should be swept from the vehicle's interior using a whisk broom.
   Technician B says the air bag control module must be replaced whenever a vehicle is involved in an accident.
   Who is correct?
   A. A only   C. Both A and B
   B. B only   D. Neither A nor B

7. Air bag system service is being discussed.
   Technician A says before an air bag system component is replaced, the negative battery cable should be disconnected and the technician should wait the specified time advised in the service information.
   Technician B says this waiting period is necessary to dissipate the reserve energy in the air bag system computer.
   Who is correct?
   A. A only   C. Both A and B
   B. B only   D. Neither A nor B

8. Air bag sensor service is being discussed.
   Technician A says incorrect torque on air bag sensor fasteners may cause improper air bag deployment.
   Technician B says the arrow on an air bag sensor must face toward the driver's side of the vehicle.
   Who is correct?
   A. A only   C. Both A and B
   B. B only   D. Neither A nor B

9. *Technician A* says to test the MEMS sensor, use a voltmeter connected to the communication line to interpret the transmission code of the sensor.
   Technician B says to use a scan tool to test communications and electronic controls of the MEMS sensor. Who is correct?
   A. A only
   B. B only
   C. Both A and B
   D. Neither A nor B

10. The purpose of the air bag load tool is being discussed.
   Technician A says the load tool is used to load the air bag system circuits to prevent accidental deployment.
   Technician B says the load tool is used to simulate known good resistances of the circuit.
   Who is correct?
   A. A only
   B. B only
   C. Both A and B
   D. Neither A nor B

ASE Challenge Questions

1. A customer states that the "Pass Air Bag Off" light comes on whenever they lay their briefcase on the front passenger seat.

   Technician A says the OCS is too sensitive and needs to be validated.

   Technician B says this is normal since the weight of the briefcase is matching the weight of a small child.

   Who is correct?

   A. A only   C. Both A and B

   B. B only   D. Neither A nor B

2. The customer is concerned about the "Pass Air Bag Off" light not illuminating when their eight-year-old child sits on the seat. The vehicle is equipped with a bladder-type system.

   Technician A says this is normal operation.

   Technician B says a DTC will set for this condition.

   Who is correct?

   A. A only   C. Both A and B

   B. B only   D. Neither A nor B

3. A customer states that while driving the vehicle, the air bag warning lamp illuminates intermittently.

   Technician A says this can be caused by a loose connection to one of the system's sensors.

   Technician B says this may indicate a defect that will set a trouble code.

   Who is correct?

   A. A only   C. Both A and B

   B. B only   D. Neither A nor B

4. The strain gauges of the OCS on the passenger-side seat all indicate about 2.5 volts with the seat empty. Technician A says this indicates a normal static reading. Technician B says this indicates the common supply circuit or return circuit has excessive resistance. Who is correct?

   A. A only

   B. B only

   C. Both A and B

   D. Neither A nor B

5. Side-impact air bags are being discussed.

   Technician A says most systems have a control module or sensor located in the B-pillar.

   Technician B says the side air bags only deploy when the front air bags deploy.

   Who is correct?

   A. A only

   B. B only

   C. Both A and B

   D. Neither A nor B

Name _____   Date _____

WORKING SAFELY AROUND AIR BAGS

Upon completion of this job sheet, you should be able to work safely around and with air bag systems.

ASE Education Foundation Correlation ————————————————

This job sheet addresses the following **MLR** tasks:

A.1.   Research vehicle service information such as fluid type, vehicle service history, service precautions, technical service bulletins, and recalls including vehicles with advanced driver assistance systems (ADAS). **(P-1)**

A.2.   Identify electrical/electronic system components and configuration. **(P-1)**

G.1.   Demonstrate understanding of vehicle comfort, convenience, access, safety, and related systems operation. **(P-3)**

This job sheet addresses the following **AST** tasks:

A.1.   Research vehicle service information such as fluid type, vehicle service history, service precautions, technical service bulletins, and recalls including vehicles with advanced driver assistance systems (ADAS). **(P-1)**

A.2.   Identify electrical/electronic system components and configuration. **(P-1)**

G.1.   Diagnose vehicle comfort, convenience, access, safety, and related systems operation; determine needed action. **(P-2)**

G.7.   Diagnose operation of safety systems and related circuits (such as horn, airbags, seat belt pretensioners, occupancy classification, wipers, washers, speed control/collision avoidance, heads-up display, park assist, and back-up camera); determine needed repairs. **(P-2)**

This job sheet addresses the following **MAST** tasks:

A.1.   Research vehicle service information such as fluid type, vehicle service history, service precautions, technical service bulletins, and recalls including vehicles with advanced driver assistance systems (ADAS). **(P-1)**

A.2.   Identify electrical/electronic system components and configuration. **(P-1)**

G.1.   Diagnose vehicle comfort, convenience, access, safety, and related systems operation; determine needed action. **(P-2)**

G.7.   Diagnose operation of safety systems and related circuits (such as horn, airbags, seat belt pretensioners, occupancy classification, wipers, washers, speed control/collision avoidance, heads-up display, park assist, and back-up camera); determine needed repairs. **(P-1)**

We Support
ASE | **Education Foundation**

Tools and Materials

- A vehicle with air bags
- Service information for the chosen vehicle
- Component locator for the chosen vehicle

Safety Considerations and PPE

- Safety glasses

Procedure

1. Locate the information about the air bag system in the service information. How are the critical parts of the system identified in the vehicle?

   _____

   _____

   _____

   _____

   _____

   _____

2. List the main components of the air bag system and describe their locations.

   _____

   _____

   _____

   _____

   _____

   _____

3. There are some very important guidelines to follow when working with and around air bag systems. These are listed here with some key words left out. Read through these and fill in the blanks with the correct words.

   a. Wear _____ _____ when servicing an air bag system and when handling an air bag module.

   b. Wait at least _____ minutes after disconnecting the battery before beginning any service. The reserve _____ module is capable of storing enough energy to deploy the air bag for up to _____ minutes after battery voltage is lost.

   c. Always handle all _____ and other components with extreme care. Never strike or jar a sensor, especially when the battery is connected; this can cause deployment of the air bag.

   d. Never carry an air bag module by its _____ or _____, and, when carrying it, always face the trim and air bag _____ from your body. When placing a module on a bench, always face the trim and air bag.

   e. _____ air bags may have a powdery residue on them. _____ is produced by the deployment reaction and is converted to_____when it comes in contact with the moisture in the atmosphere.

      Although it's unlikely that harmful chemicals will still be on the bag, it's wise to wear _____ _____ and _____ when handling a deployed air bag. Wash your hands immediately after handling a deployed air bag.

f. A live air bag must be _____ before it's disposed of. A deployed air bag should be disposed of in a manner consistent with the _____ and the manufacturer's recommended procedures.

g. Never use a battery- or A/C-powered _____, _____, or any other type of test equipment in the system unless the manufacturer specifically says to. Never probe with a _____ _____ for voltage.

Instructor's Evaluation:

Grade	Rubric
0	No exposure due to: ☐ Absence ☐ Other:
1	Student unsuccessfully attempted the task
2	Student completed the task with significant instructor assistance.
3	Satisfactory performance of task. The student was able to perform elements of the task with minimal instructor assistance.
4	Competent/Proficient. Able to perform all elements of the task successfully and independently.

Instructor's Notes:

Name _____   Date _____

DIAGNOSING "DRIVER SQUIB CIRCUIT OPEN" FAULT

Upon completion of this job sheet, you should be able to diagnose the cause of a "driver squib circuit open" fault in an air bag system and determine needed repairs.

ASE Education Foundation Correlation

This job sheet addresses the following **MLR** tasks:

A.1.	Research vehicle service information such as fluid type, vehicle service history, service precautions, technical service bulletins, and recalls including vehicles with advanced driver assistance systems (ADAS). **(P-1)**
A.2.	Identify electrical/electronic system components and configuration. **(P-1)**
A.6.	Demonstrate knowledge of the causes and effects from shorts, grounds, opens, and resistance problems in electrical/electronic circuits. **(P-1)**
A.8.	Use fused jumper wires to check operation of electrical circuits per service information. **(P-2)**
A.9.	Use wiring diagrams to trace electrical/electronic circuits. **(P-1)**
A.12.	Repair and/or replace connectors, terminal ends, and wiring of electrical/electronic systems (including solder repair). **(P-2)**
G.1.	Demonstrate understanding of vehicle comfort, convenience, access, safety, and related systems operation. **(P-3)**
G.4.	Describe disabling and enabling procedures for supplemental restraint system (SRS); verify indicator lamp operation. **(P-2)**

This job sheet addresses the following **AST** tasks:

A.1.	Research vehicle service information such as fluid type, vehicle service history, service precautions, technical service bulletins, and recalls including vehicles with advanced driver assistance systems (ADAS). **(P-1)**
A.2.	Identify electrical/electronic system components and configuration. **(P-1)**
A.6.	Demonstrate knowledge of the causes and effects from shorts, grounds, opens, and resistance problems in electrical/electronic circuits. **(P-1)**
A.8.	Use fused jumper wires to check operation of electrical circuits per service information. **(P-1)**
A.9.	Use wiring diagrams during the diagnosis (troubleshooting) of electrical/electronic circuit problems. **(P-1)**
A.12.	Repair and/or replace connectors, terminal ends, and wiring of electrical/electronic systems (including solder repair). **(P-1)**
G.1.	Diagnose vehicle comfort, convenience, access, safety, and related systems operation; determine needed action. **(P-2)**
G.4.	Describe disabling and enabling procedures for supplemental restraint system (SRS); verify indicator lamp operation. **(P-1)**
G.7.	Diagnose operation of safety systems and related circuits (such as horn, airbags, seat belt pretensioners, occupancy classification, wipers, washers, speed control/collision avoidance, heads-up display, park assist, and back-up camera); determine needed repairs. **(P-2)**
G.8.	Diagnose body electronic system circuits using a scan tool; check for module communication errors (data communication bus systems); determine needed action. **(P-2)**

(Continued)

This job sheet addresses the following **MAST** tasks:

A.1. Research vehicle service information such as fluid type, vehicle service history, service precautions, technical service bulletins, and recalls including vehicles with advanced driver assistance systems (ADAS). **(P-1)**

A.2. Identify electrical/electronic system components and configuration. **(P-1)**

A.6. Demonstrate knowledge of the causes and effects from shorts, grounds, opens, and resistance problems in electrical/electronic circuits. **(P-1)**

A.8. Use fused jumper wires to check operation of electrical circuits per service information. **(P-1)**

A.9. Use wiring diagrams during the diagnosis (troubleshooting) of electrical/electronic circuit problems. **(P-1)**

A.12. Repair and/or replace connectors, terminal ends, and wiring of electrical/electronic systems (including solder repair). **(P-1)**

G.1. Diagnose vehicle comfort, convenience, access, safety, and related systems operation; determine needed action. **(P-2)**

G.4. Describe disabling and enabling procedures for supplemental restraint system (SRS); verify indicator lamp operation. **(P-1)**

G.7. Diagnose operation of safety systems and related circuits (such as horn, airbags, seat belt pretensioners, occupancy classification, wipers, washers, speed control/collision avoidance, heads-up display, park assist, and back-up camera); determine needed repairs. **(P-1)**

G.8. Diagnose body electronic system circuits using a scan tool; check for module communication errors (data communication bus systems); determine needed action. **(P-1)**

We Support
ASE | Education Foundation

Tools and Materials

A vehicle equipped with a driver-side air bag (bugged by instructor prior to performing this task)

• Service information for chosen vehicle
• Battery terminal puller
• Fused jumper wires
• Scan tool

Safety Considerations and PPE

• Safety glasses

Procedure

1. Describe the normal conditions in which the air bag warning lamp should operate.

   _____

   _____

   _____

2. Perform the diagnostic system check. How did the air bag warning light respond?

   _____

   _____

3. Follow all safety warnings and cautions listed in the service information and connect the scan tool. Record all fault codes the scan tool displays.

Active: _____

_____

Stored: _____

_____

4. Is the code for "Driver-Side Squib Circuit" or equivalent active?   ☐ Yes   ☐ No
If no, consult your instructor.

5. List all possible causes that can set this fault code.

_____

_____

_____

_____

6. Disconnect the negative battery cable and isolate it.

7. How long does the service information say you must wait before proceeding?

_____

_____

8. Follow the service information instructions to remove the driver-side air bag module from the steering wheel.

9. Connect a jumper wire across the upper connector of the clockspring.

10. With the ignition in the RUN position, connect the battery ground cable.

11. Use the scan tool and record the active fault code.

_____

_____

12. Does the fault code indicate the circuit is shorted?   ☐ Yes   ☐ No

13. If you answered yes to question 12, what is the faulty component?

_____

_____

14. If you answered no to question 12, locate the lower clockspring connector. Describe this connector's location.

_____

_____

_____

15. Disconnect the negative battery cable and isolate it. Wait the recommended amount of time before proceeding.

16. Disconnect the lower clockspring connector and connect a jumper wire across the control module side of the harness connector.

17. With the ignition in the RUN position, reconnect the battery ground cable.

18. Use the scan tool to retrieve active fault codes and record them.

_____

19. Does the fault code indicate the circuit is shorted?   ☐ Yes   ☐ No

20. If you answered yes to question 19, what is the faulty component?

_____

21. If you answered no to question 19, locate the connector to the air bag control module. Describe this connector's location.

_____

_____

_____

22. Disconnect the negative battery cable and isolate it. Wait for the recommended amount of time before proceeding.

23. Disconnect the air bag control module connector and locate the wires of the driver-side squib circuit. What are the color codes of the wires?

_____

_____

_____

24. Use an ohmmeter to test the wire harness between the air bag control module connector and the lower clockspring connector. Record your results.

_____

_____

25. Based on your results, what have you determined to be the location of the fault?

_____

_____

Instructor's Evaluation:

Grade	Rubric
0	No exposure due to: ☐ Absence ☐ Other:
1	Student unsuccessfully attempted the task
2	Student completed the task with significant instructor assistance.
3	Satisfactory performance of task. The student was able to perform elements of the task with minimal instructor assistance.
4	Competent/Proficient. Able to perform all elements of the task successfully and independently.

Instructor's Notes:

_____

Name _____   Date _____

DIAGNOSING AIR BAG PRESSURE SENSOR CIRCUIT PERFORMANCE FAULT

Upon completion of this job sheet, you should be able to diagnose the cause of a pressure sensor circuit fault code.

ASE Education Foundation Correlation

This job sheet addresses the following **MLR** tasks:

A.1.   Research vehicle service information such as fluid type, vehicle service history, service precautions, technical service bulletins, and recalls including vehicles with advanced driver assistance systems (ADAS). **(P-1)**

A.2.   Identify electrical/electronic system components and configuration. **(P-1)**

A.5.   Demonstrate proper use of a digital multimeter (DMM) when measuring source voltage, voltage drop (including grounds), current flow, and resistance. **(P-1)**

A.6.   Demonstrate knowledge of the causes and effects from shorts, grounds, opens, and resistance problems in electrical/electronic circuits. **(P-1)**

A.8.   Use fused jumper wires to check operation of electrical circuits per service information. **(P-2)**

A.9.   Use wiring diagrams to trace electrical/electronic circuits. **(P-1)**

G.1.   Demonstrate understanding of vehicle comfort, convenience, access, safety, and related systems operation. **(P-3)**

G.2   Remove and reinstall door panel. **(P-2)**

This job sheet addresses the following **AST** tasks:

A.1.   Research vehicle service information such as fluid type, vehicle service history, service precautions, technical service bulletins, and recalls including vehicles with advanced driver assistance systems (ADAS). **(P-1)**

A.2.   Identify electrical/electronic system components and configuration. **(P-1)**

A.5.   Demonstrate proper use of a digital multimeter (DMM) when measuring source voltage, voltage drop (including grounds), current flow, and resistance. **(P-1)**

A.6.   Demonstrate knowledge of the causes and effects from shorts, grounds, opens, and resistance problems in electrical/electronic circuits. **(P-1)**

A.8.   Use fused jumper wires to check operation of electrical circuits per service information. **(P-1)**

A.9.   Use wiring diagrams during the diagnosis (troubleshooting) of electrical/electronic circuit problems. **(P-1)**

G.1.   Diagnose vehicle comfort, convenience, access, safety, and related systems operation; determine needed action. **(P-2)**

G.2   Remove and reinstall door panel. **(P-1)**

G.7.   Diagnose operation of safety systems and related circuits (such as horn, airbags, seat belt pretensioners, occupancy classification, wipers, washers, speed control/collision avoidance, heads-up display, park assist, and back-up camera); determine needed repairs. **(P-2)**

G.8.   Diagnose body electronic system circuits using a scan tool; check for module communication errors (data communication bus systems); determine needed action. **(P-2)**

(Continued)

This job sheet addresses the following **MAST** tasks:

A.1.   Research vehicle service information such as fluid type, vehicle service history, service precautions, technical service bulletins, and recalls including vehicles with advanced driver assistance systems (ADAS). **(P-1)**

A.2.   Identify electrical/electronic system components and configuration. **(P-1)**

A.5.   Demonstrate proper use of a digital multimeter (DMM) when measuring source voltage, voltage drop (including grounds), current flow, and resistance. **(P-1)**

A.6.   Demonstrate knowledge of the causes and effects from shorts, grounds, opens, and resistance problems in electrical/electronic circuits. **(P-1)**

A.8.   Use fused jumper wires to check operation of electrical circuits per service information. **(P-1)**

A.9.   Use wiring diagrams during the diagnosis (troubleshooting) of electrical/ electronic circuit problems. **(P-1)**

G.1.   Diagnose vehicle comfort, convenience, access, safety, and related systems operation; determine needed action. **(P-2)**

G.2.   Remove and reinstall door panel. **(P-1)**

G.7.   Diagnose operation of safety systems and related circuits (such as horn, airbags, seat belt pretensioners, occupancy classification, wipers, washers, speed control/collision avoidance, heads-up display, park assist, and back-up camera); determine needed repairs. **(P-1)**

G.8.   Diagnose body electronic system circuits using a scan tool; check for module communication errors (data communication bus systems); determine needed action. **(P-1)**

Tools and Materials

Vehicle equipped with air bag system pressure sensor, such as the strain gauges used for OCS (bugged by instructor prior to performing the task). Note: Although it would be ideal to use a vehicle with pressure sensors associated with the air bag system, this task can be performed on most transducer-type pressure sensor circuits if needed.

- Service information
- Fused jumper wire
- Backprobing tools
- Scan tool

Safety Considerations and PPE

- Safety glasses

Procedure

1. Read and record any fault codes in the air bag system.

   Active: _____

   Stored: _____

2. Monitoring the sensor data output and voltage with a scan tool with the ignition in the RUN position, record the pressure reading and voltage value.

_____

3. What do these values represent?

_____

_____

_____

4. Do the readings in step 2 indicate actual conditions or are they substitute values?

_____

5. What can cause the "Sensor Circuit High" fault to set?

_____

_____

_____

6. Unplug the sensor and use a fused jumper wire to jump the signal circuit to ground. What does the scan tool indicate for the voltage value?

_____

_____

_____

7. Based on this reading, can you determine the location of the fault?

_____

8. If the voltage is above 0 volt, what would this indicate?

_____

_____

_____

9. If the reading in step 6 is 0 volt, what would be your next diagnostic step?

_____

_____

_____

10. Based on your readings, what is the fault?

_____

_____

_____

Instructor's Evaluation:

Grade	Rubric
0	No exposure due to: ☐ Absence ☐ Other:
1	Student unsuccessfully attempted the task
2	Student completed the task with significant instructor assistance.
3	Satisfactory performance of task. The student was able to perform elements of the task with minimal instructor assistance.
4	Competent/Proficient. Able to perform all elements of the task successfully and independently.

Instructor's Notes:

Name _____   Date _____

TESTING THE MEMS SENSOR

Upon completion of this job sheet, you should be able to diagnose the cause of a MEMS sensor loss of communication fault code.

ASE Education Foundation Correlation —————————————

This job sheet addresses the following **MLR** tasks:

A.1. Research vehicle service information such as fluid type, vehicle service history, service precautions, technical service bulletins, and recalls including vehicles with advanced driver assistance systems (ADAS). **(P-1)**

A.2. Identify electrical/electronic system components and configuration. **(P-1)**

A.5. Demonstrate proper use of a digital multimeter (DMM) when measuring source voltage, voltage drop (including grounds), current flow, and resistance. **(P-1)**

A.6. Demonstrate knowledge of the causes and effects from shorts, grounds, opens, and resistance problems in electrical/electronic circuits. **(P-1)**

A.9. Use wiring diagrams to trace electrical/electronic circuits. **(P-1)**

A.12. Repair and/or replace connectors, terminal ends, and wiring of electrical/electronic systems (including solder repair). **(P-2)**

G.1. Demonstrate understanding of vehicle comfort, convenience, access, safety, and related systems operation. **(P-3)**

G.2. Remove and reinstall door panel. **(P-2)**

This job sheet addresses the following **AST** tasks:

A.1. Research vehicle service information such as fluid type, vehicle service history, service precautions, technical service bulletins, and recalls including vehicles with advanced driver assistance systems (ADAS). **(P-1)**

A.2. Identify electrical/electronic system components and configuration. **(P-1)**

A.5. Demonstrate proper use of a digital multimeter (DMM) when measuring source voltage, voltage drop (including grounds), current flow, and resistance. **(P-1)**

A.6. Demonstrate knowledge of the causes and effects from shorts, grounds, opens, and resistance problems in electrical/electronic circuits. **(P-1)**

A.9. Use wiring diagrams during the diagnosis (troubleshooting) of electrical/electronic circuit problems. **(P-1)**

A.12. Repair and/or replace connectors, terminal ends, and wiring of electrical/electronic systems (including solder repair). **(P-1)**

G.1. Diagnose vehicle comfort, convenience, access, safety, and related systems operation; determine needed action. **(P-2)**

G.2. Remove and reinstall door panel. **(P-1)**

G.7. Diagnose operation of safety systems and related circuits (such as horn, airbags, seat belt pretensioners, occupancy classification, wipers, washers, speed control/collision avoidance, heads-up display, park assist, and back-up camera); determine needed repairs. **(P-2)**

G.8. Diagnose body electronic system circuits using a scan tool; check for module communication errors (data communication bus systems); determine needed action. **(P-2)**

This job sheet addresses the following **MAST** tasks:

A.1. Research vehicle service information such as fluid type, vehicle service history, service precautions, technical service bulletins, and recalls including vehicles with advanced driver assistance systems (ADAS). **(P-1)**

(Continued)

A.2. Identify electrical/electronic system components and configuration. **(P-1)**

A.5. Demonstrate proper use of a digital multimeter (DMM) when measuring source voltage, voltage drop (including grounds), current flow, and resistance. **(P-1)**

A.6. Demonstrate knowledge of the causes and effects from shorts, grounds, opens, and resistance problems in electrical/electronic circuits. **(P-1)**

A.9. Use wiring diagrams during the diagnosis (troubleshooting) of electrical/electronic circuit problems. **(P-1)**

A.12. Repair and/or replace connectors, terminal ends, and wiring of electrical/electronic systems (including solder repair). **(P-1)**

G.1. Diagnose vehicle comfort, convenience, access, safety, and related systems operation; determine needed action. **(P-2)**

G.2. Remove and reinstall door panel. **(P-1)**

G.7. Diagnose operation of safety systems and related circuits (such as horn, airbags, seat belt pretensioners, occupancy classification, wipers, washers, speed control/collision avoidance, heads-up display, park assist, and back-up camera); determine needed repairs. **(P-1)**

G.8. Diagnose body electronic system circuits using a scan tool; check for module communication errors (data communication bus systems); determine needed action. **(P-1)**

We Support
ASE | Education Foundation

Tools and Materials

Vehicle equipped with a MEMS pressure or accelerometer sensor (bugged by instructor prior to performing task)

• Service information
• Air bag simulator load tool and adapters
• Scan tool
• DMM
• Battery terminal puller

Safety Considerations and PPE

• Safety glasses

Procedure

1. Record DTCs

   Active: _____

   Stored: _____

2. Disconnect the battery and wait for the specified amount of time before proceeding.

3. Follow the procedure in the service information to remove the door panel (if necessary).

4. Disconnect the connector of the sensor identified by the DTC. Check the condition of the connectors and terminals.

5. Disconnect the ORC connector and inspect the connector and terminals for damage.

6. Connect the air bag simulator load tool adaptor to the ORC connectors, and measure the resistance of the sensor signal circuit between the sensor connector and the load tool adaptor.

7. If the resistance is too high, what would this indicate?

_____

_____

_____

8. Measure the resistance of the sensor ground circuit between the sensor connector and the load tool adapter.

_____

9. If the resistance is too high, what would this indicate?

_____

_____

_____

10. Measure the resistance between the sensor signal circuit and the sensor ground circuit at the sensor connector.

_____

11. What would a low reading in step 9 indicate?

_____

_____

_____

12. Move the test leads to measure the resistance between the sensor signal circuit and chassis ground.

_____

13. If the resistance reading is below specifications, what would this indicate?

_____

_____

_____

14. Turn the ignition to the RUN position, and then reconnect the battery. Measure the voltage on the signal circuit.

_____

15. If voltage is present, what is the next step?

_____

_____

_____

16. Move the test leads to measure the voltage on the sensor ground circuit between the load tool adaptor and chassis ground. Record your results.

_____

17. If voltage is present, what is the next step?

_____

_____

_____

18. If all tests pass, what is the likely problem?

_____

_____

_____

_____

Instructor's Evaluation:

Grade	Rubric
0	No exposure due to: ☐ Absence ☐ Other:
1	Student unsuccessfully attempted the task
2	Student completed the task with significant instructor assistance.
3	Satisfactory performance of task. The student was able to perform elements of the task with minimal instructor assistance.
4	Competent/Proficient. Able to perform all elements of the task successfully and independently.

Instructor's Notes:

Name _____   Date _____

SIDE-IMPACT SENSOR REPLACEMENT

Upon completion of this job sheet, you should be able to properly replace the side-impact sensor.

ASE Education Foundation Correlation

This job sheet addresses the following **MLR** tasks:

A.1.   Research vehicle service information such as fluid type, vehicle service history, service precautions, technical service bulletins, and recalls including vehicles with advanced driver assistance systems (ADAS). **(P-1)**

A.2.   Identify electrical/electronic system components and configuration. **(P-1)**

G.1.   Demonstrate understanding of vehicle comfort, convenience, access, safety, and related systems operation. **(P-3)**

G.2.   Remove and reinstall door panel. **(P-2)**

This job sheet addresses the following **AST** tasks:

A.1.   Research vehicle service information such as fluid type, vehicle service history, service precautions, technical service bulletins, and recalls including vehicles with advanced driver assistance systems (ADAS). **(P-1)**

G.1.   Diagnose vehicle comfort, convenience, access, safety, and related systems operation; determine needed action. **(P-2)**

G.2.   Remove and reinstall door panel. **(P-1)**

This job sheet addresses the following **MAST** tasks:

A.1.   Research vehicle service information such as fluid type, vehicle service history, service precautions, technical service bulletins, and recalls including vehicles with advanced driver assistance systems (ADAS). **(P-1)**

G.1.   Diagnose vehicle comfort, convenience, access, safety, and related systems operation; determine needed action. **(P-2)**

G.2.   Remove and reinstall door panel. **(P-1)**

We Support
ASE | **Education Foundation**

Tools and Materials
- Vehicle equipped with side-impact sensors
- Service information
- Scan tool
- Torque wrench
- Safety glasses
- Battery terminal puller

Safety Considerations and PPE
- Safety glasses

Procedure

1. Locate the fastener torque specifications in the service information and record.

_____

2. Are there any safety warnings or cautions listed in the service information concerning the installation procedures of the sensor?   ☐ Yes   ☐ No

3. If yes, describe.

_____

_____

_____

4. Follow the service information procedures to remove the door panel.

5. Is there a requirement for proper orientation of the sensor?   ☐ Yes   ☐ No

If yes, describe how to determine proper orientation.

_____

_____

_____

6. Is a gasket used between the sensor and the mounting surface area?   ☐ Yes   ☐ No

7. Is the gasket in good condition?   ☐ Yes   ☐ No

8. Install the sensor.

Instructor's Evaluation:

Grade	Rubric
0	No exposure due to: ☐ Absence ☐ Other:
1	Student unsuccessfully attempted the task
2	Student completed the task with significant instructor assistance.
3	Satisfactory performance of task. The student was able to perform elements of the task with minimal instructor assistance.
4	Competent/Proficient. Able to perform all elements of the task successfully and independently.

Instructor's Notes:

_____

Name _____   Date _____

CLOCKSPRING REPLACEMENT AND CENTERING

Upon completion of this job sheet, you should be able to remove, replace, and properly center the clockspring used in an air bag system.

ASE Education Foundation Correlation

This job sheet addresses the following **MLR** tasks:

A.1.   Research vehicle service information such as fluid type, vehicle service history, service precautions, technical service bulletins, and recalls including vehicles with advanced driver assistance systems (ADAS). **(P-1)**

A.2.   Identify electrical/electronic system components and configuration. **(P-1)**

G.1.   Demonstrate understanding of vehicle comfort, convenience, access, safety, and related systems operation. **(P-3)**

G.4.   Describe disabling and enabling procedures for supplemental restraint system (SRS); verify indicator lamp operation. **(P-2)**

This job sheet addresses the following **AST** tasks:

A.1.   Research vehicle service information such as fluid type, vehicle service history, service precautions, technical service bulletins, and recalls including vehicles with advanced driver assistance systems (ADAS). **(P-1)**

A.2.   Identify electrical/electronic system components and configuration. **(P-1)**

G.1.   Diagnose vehicle comfort, convenience, access, safety, and related systems operation; determine needed action. **(P-2)**

G.4.   Describe disabling and enabling procedures for supplemental restraint system (SRS); verify indicator lamp operation. **(P-1)**

This job sheet addresses the following **MAST** tasks:

A.1.   Research vehicle service information such as fluid type, vehicle service history, service precautions, technical service bulletins, and recalls including vehicles with advanced driver assistance systems (ADAS). **(P-1)**

A.2.   Identify electrical/electronic system components and configuration. **(P-1)**

G.1.   Diagnose vehicle comfort, convenience, access, safety, and related systems operation; determine needed action. **(P-2)**

G.4.   Describe disabling and enabling procedures for supplemental restraint system (SRS); verify indicator lamp operation. **(P-1)**

We Support
ASE | **Education Foundation**

Tools and Materials

- A vehicle equipped with a driver-side air bag
- Service information for chosen vehicle
- Battery terminal puller
- Steering wheel puller

Safety Considerations and PPE

- Safety glasses

Procedure

1. According to the service information, in what position must the front wheels be before beginning?

_____

_____

2. Place the front wheels in the position described in step 1.

3. Disconnect the battery negative cable and isolate it.

4. According to the service information, how long must you wait before proceeding?

_____

_____

5. Remove the air bag module from the steering wheel.

6. Remove the steering wheel attaching bolt or nut. Can this fastener be reused?
☐ Yes   ☐ No

7. Mark the shaft and steering wheel with index marks for reinstallation.

8. Use a steering wheel puller to remove the steering wheel from the shaft.

9. Remove the upper and lower steering column shrouds.

10. Disconnect the clockspring connector from the steering column harness.

11. Remove the retaining screws (or release the locking tabs) and the clockspring.

12. If the same clockspring is to be reinstalled, describe the procedure for centering the clockspring.

_____

_____

_____

13. In what position must the front wheel and steering column be to install the clockspring?

_____

14. Install all components. What is the torque specification for the clockspring fasteners (if used)?

_____

What is the torque specification for the steering wheel fastener?

_____

What is the torque specification for the air bag module fasteners?

_____

Instructor's Evaluation:

Grade	Rubric
0	No exposure due to: ☐ Absence ☐ Other:
1	Student unsuccessfully attempted the task
2	Student completed the task with significant instructor assistance.
3	Satisfactory performance of task. The student was able to perform elements of the task with minimal instructor assistance.
4	Competent/Proficient. Able to perform all elements of the task successfully and independently.

Instructor's Notes:

_____

Name _____   Date _____

OCCUPANT CLASSIFICATION SYSTEM VALIDATION PROCEDURE

Upon completion of this job sheet, you should be able to properly perform the OCS validation procedure.

ASE Education Foundation Correlation

This job sheet addresses the following **MLR** tasks:

A.1. Research vehicle service information such as fluid type, vehicle service history, service precautions, technical service bulletins, and recalls including vehicles with advanced driver assistance systems (ADAS). **(P-1)**

G.1. Demonstrate understanding of vehicle comfort, convenience, access, safety, and related systems operation. **(P-3)**

This job sheet addresses the following **AST** tasks:

A.1. Research vehicle service information such as fluid type, vehicle service history, service precautions, technical service bulletins, and recalls including vehicles with advanced driver assistance systems (ADAS). **(P-1)**

G.1. Diagnose vehicle comfort, convenience, access, safety, and related systems operation; determine needed action. **(P-2)**

G.7. Diagnose operation of safety systems and related circuits (such as horn, airbags, seat belt pretensioners, occupancy classification, wipers, washers, speed control/collision avoidance, heads-up display, park assist, and back-up camera); determine needed repairs. **(P-2)**

G.8. Diagnose body electronic system circuits using a scan tool; check for module communication errors (data communication bus systems); determine needed action. **(P-2)**

This job sheet addresses the following **MAST** tasks:

A.1. Research vehicle service information such as fluid type, vehicle service history, service precautions, technical service bulletins, and recalls including vehicles with advanced driver assistance systems (ADAS). **(P-1)**

G.1. Diagnose vehicle comfort, convenience, access, safety, and related systems operation; determine needed action. **(P-2)**

G.7. Diagnose operation of safety systems and related circuits (such as horn, airbags, seat belt pretensioners, occupancy classification, wipers, washers, speed control/collision avoidance, heads-up display, park assist, and back-up camera); determine needed repairs. **(P-1)**

G.8. Diagnose body electronic system circuits using a scan tool; check for module communication errors (data communication bus systems); determine needed action. **(P-1)**

We Support
ASE | Education Foundation

Tools and Materials
- Vehicle with OCS
- Validation weight set
- Battery charger

Safety Considerations and PPE
- Safety glasses

Procedure

1. Connect a battery charger to the vehicle's battery and set so 13 volts is read across the terminals.

2. Is the system on the assigned vehicle a strain gauge or bladder system?

   ☐ Strain gauge   ☐ Bladder system

3. Connect the scan tool to the DLC and access the OCM module. Provide the following information (if available):

   OCM part number: _____

   Software version: _____

4. Make sure the passenger front seat is empty. What is the state of the PADL?

   _____

5. Sit in the seat and describe the state of the PADL.

   _____

6. Use the scan tool to access the sensor display function. Record the values while remaining in the seat.

   _____

   _____

7. Move out of the seat and record the values.

   _____

   _____

8. Assure that the seat is empty and navigate the scan tool to the validation function. What instructions are provided on the scan tool screen when the test is started (if any)?

   _____

   _____

   _____

9. According to the scan tool, what is the first step that needs to be performed?

   _____

   _____

10. What is the first weight amount to be placed on the seat?

   _____

11. During the validation procedure, what is the state of the PADL?

   _____

   _____

12. Complete the validation procedure while listing the required weight amounts.

   _____

   _____

   _____

13. How do you know that the procedure was completed?

_____

14. What is the state of the PADL once the procedure is completed?

_____

15. Clear any DTCs that may have been set.

16. Disconnect a sensor and record any DTCs.

_____

_____

17. Reconnect the sensor. Did the fault go from active to stored?

_____

18. What needs to be done to clear the DTC?

_____

19. Perform the required task list in step 18.

Instructor's Evaluation:

Grade	Rubric
0	No exposure due to: ☐ Absence ☐ Other:
1	Student unsuccessfully attempted the task
2	Student completed the task with significant instructor assistance.
3	Satisfactory performance of task. The student was able to perform elements of the task with minimal instructor assistance.
4	Competent/Proficient. Able to perform all elements of the task successfully and independently.

Instructor's Notes:

DIAGNOSTIC CHART 15-1

PROBLEM AREA:	Air bag system operation.
SYMPTOMS:	Air bag warning lamp illuminated.
POSSIBLE CAUSES:	1. Squib circuit shorted to ground. 2. Defective clockspring. 3. Squib circuit open. 4. Sensor circuit open. 5. Sensor circuit shorted. 6. Faulty module. 7. Poor battery feed circuit to control module. 8. Poor ignition feed circuit to control module. 9. Poor control module ground circuit. 10. Loss of bus communications. 11. Faulty instrument cluster lamp circuit.

DIAGNOSTIC CHART 15-2

PROBLEM AREA:	Loss of communication between MEMS sensor and the ORC.
SYMPTOMS:	Air bag lamp illuminated.
POSSIBLE CAUSES:	1. Sensor signal circuit open. 2. Sensor ground circuit open. 3. Sensor signal circuit shorted to sensor ground circuit. 4. Sensor signal circuit shorted to ground. 5. Sensor signal circuit shorted to voltage. 6. Sensor ground circuit shorted to voltage. 7. Faulty impact sensor. 8. Faulty ORC.

DIAGNOSTIC CHART 15-3

PROBLEM AREA:	Side curtain squib circuit open.
SYMPTOMS:	Air bag lamp illuminated.
POSSIBLE CAUSES:	1. Squib line 1 circuit open. 2. Squib line 2 circuit open. 3. Failure of side curtain air bag squib. 4. Faulty control module.

DIAGNOSTIC CHART 15-4

PROBLEM AREA:	Squib circuit shorted to voltage.
SYMPTOMS:	Air bag lamp illuminated.
POSSIBLE CAUSES:	1. Squib line 1 circuit shorted to voltage. 2. Squib line 2 circuit shorted to voltage. 3. Failure of side curtain air bag squib. 4. Faulty control module.

DIAGNOSTIC CHART 15-5

PROBLEM AREA:	Squib circuit shorted to ground.
SYMPTOMS:	Air bag lamp illuminated.
POSSIBLE CAUSES:	**1.** Squib line 1 circuit shorted to ground. **2.** Squib line 2 circuit shorted to ground. **3.** Failure of side curtain air bag squib. **4.** Faulty control module.

DIAGNOSTIC CHART 15-6

PROBLEM AREA:	Squib circuits shorted together.
SYMPTOMS:	Air bag lamp illuminated.
POSSIBLE CAUSES:	**1.** Squib line 1 circuit shorted to squib line 2. **2.** Failure of side curtain air bag squib. **3.** Faulty control module.

CHAPTER 16
SERVICING ADAS

Upon completion and review of this chapter, you should be able to:

- Identify ADAS components used in vehicle pre-repair scan.
- Identify damage to ADAS components.
- Demonstrate the use of special equipment necessary to properly service ADASs.
- Diagnose the Steering Angle Sensor (SAS) used by various ADASs.
- Properly diagnose the park assist system and demonstrate the proper use of

- DMMs and scan tools to diagnose the ultrasonic sensor.
- Properly diagnose the back-up camera system.
- Diagnose and service the self-park system.
- Use a scan tool and DMM to diagnose the common sensors used by Electronic Stability Program (ESP) and Electronic Roll Mitigation (ERM) systems.

Terms to Know

Park Assist System (PAS)

Sensor cluster

Steering Angle Sensor (SAS)

Introduction

This chapter discusses diagnostics of vehicle safety and Advanced Driver Assistance Systems (ADASs), such as lane change warning, park assist, self-park and obstruction detection. In addition, you will become familiar with the specialty equipment that may be necessary to service ADASs. There is no attempt to discuss every ADAS individually. Since the different ADASs use the same types of components, we will use some system examples. The knowledge gained from the process of diagnosing one system transfers to other ADASs.

Identify Information, Tools and Equipment Requirements

When a vehicle arrives for service with a customer's concern associated with an ADAS, you must be able to identify what the job requirements are. The repair order may only indicate that the customer has observed the warning light or message. It's also important to remember that many of the vehicle's powertrain, safety, body, suspension, and accessory systems are interconnected, so diagnostics and repairs to seemingly unrelated systems may actually involve the vehicle's ADAS.

Diagnostics and repair of an ADAS begin with a productive customer interview. You will need a clear complaint description. Also, it would help if you inquired about any recently performed services such as windshield replacement, collision repairs, mirror replacement, brake work, or suspension work.

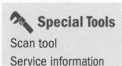

Perform a visual inspection of the vehicle, looking for body damage of any type that can affect an ADAS sensor's operation. Sensors or cameras mount to (or behind) body panels, grills, mirrors, and windshields. Damage to these components may interfere with the proper operation of the system.

When inspecting the vehicle, make a note of any vehicle modifications. Aftermarket grills, shaded windshields, and so forth can inhibit sensor performance. Also, changing tire size from original equipment may change the vehicle's stance enough to cause issues.

A fault in an ADAS will usually result in the disabling of the system, turning on a warning light or message, and setting a DTC. Dirt accumulation on the lens, physical damage, poor electrical connections, or circuit damage can prevent a sensor from operating properly.

To evaluate system operation, you must obtain the correct information. As simple as this sounds, you can save time by confirming what ADASs the vehicle you are servicing has. Use the proper service information or the National Highway Traffic and Safety Administration (NHTSA) VIN Decoder app to acquire this information. Obtaining the correct service information is more assured by entering the Vehicle Identification Number (VIN). If having to enter the vehicle information manually, be sure to accurately determine the vehicle's year, make, and model. You may also need to input the trim package on the vehicle.

A test drive may be necessary to confirm the customer's concern. This gives you the opportunity to experience what the customer has. Pay close attention to how the ADAS is functioning. For example, note how equally the Lane Departure Warning (LDW) detects lane markings on each side of the road. If the LDW system is more sensitive to one side than the other, this may indicate a sensor is out of calibration, but it can also indicate a wheel alignment problem. Note if the steering wheel is straight, how it returns to center when coming out of a turn, if the vehicle pulls or drifts to one side, etc.

In many cases, you will find that checking the alignment will be one of the first steps you perform. Look at tire wear patterns that indicate improper alignment. Proper thrust angle is essential for many ADAS sensors to work. A thrust angle that is off by 1 degree translates into 1 foot (305 mm) deviation in 60 feet (18 meters). A vehicle traveling at 60 mph (96.5 kph) will cover 441 feet (134.4 meters) in 5 seconds. This means the vehicle can migrate over 7 feet (2 meters) from center within the 5 seconds.

With the customer's concern verified, reference any applicable Technical Service Bulletins (TSBs). The TSB may note updates to specifications, special tool needs, and service procedures. In addition, the TSB may identify updated computer software and parts. The TSB will indicate the problem's symptoms and the vehicles the bulletin applies to. These elements must match what you are working on.

The diagnostic scan may quickly identify if a component is out of calibration. Use the scan tool to observe the live data PIDs. Look for data that "does not make sense." For example, the steering wheel is straight, and the wheels are pointing straight ahead, but the steering angle sensor reads 60 degrees. While observing the live data of the proximity sensors, move a pole in front of each sensor and confirm that the data is correct.

Some scan tools provide graphing of the data PIDs. This helps focus your attention on any input or output that drops out. Another helpful feature is to use the record function. This will allow you to playback the recording so you can analyze the data.

DTCs may not directly identify the faulty component. However, if the fault identifies that a specific input sensor or actuator is not functioning correctly, locate the component on the vehicle and perform a visual inspection. Inputs like cameras, radar, and laser sensors can stop functioning due to dirt accumulation on the lens. Also, damage to the grill, windshield, or other parts in front of the sensor may interfere with its ability to perform. In addition, inspect the component's wiring connector. Look for physical damage, corrosion, and loose terminals.

For the ADAS component to perform correctly requires it to be properly mounted. Make sure there is no damage to braces or mounting fixtures. If the vehicle was involved in an accident, the body repairs might not have included sensor or actuator mounting brackets.

It should be obvious that safety is the primary concern. Your knowledge and competency level determine how much diagnostics and service of an ADAS you perform. Your workplace's policies are a significant factor in determining the depth of diagnostics and service you can do. You must be familiar with and follow these policies. Lack of the correct tools, equipment, and/or training can lead to unsafe vehicle operation.

ADAS Sensor Alignment Screens and Special Tools

Special Tools

Scan tool
Calibration frame with
    screens and targets
Service information

Vehicle repairs and services not directly related to ADAS may require ADAS component calibration and/or initialization. For example, the Steering Angle Sensor may need recalibration after performing a wheel alignment otherwise the Lane Departure Warning, Lane Centering, Self Parking, and other systems may not operate.

Replacing the windshield may require special considerations, special calibration procedures, and equipment. If the vehicle has a Forward-Facing Camera (FFC) mounted to the inside of the windshield, the correct windshield is a must. There may be a slight difference in the size of the black mask surrounding the perimeter of the windshield between the windshield equipped with FFC and those without. In some instances, the difference in mask placement between the two windshields equates to only 1 mm. This is just enough that installing a non-FFC windshield may interfere with the FFC.

After replacing the windshield and installing the FFC, verification of the FFC may be required. Many cameras will also require calibration using special target screens. Often these screens are vehicle specific.

Most service information systems will have a "Special Tool and Equipment" section. Reference this area to determine if you have the required tools and equipment. There may also be a separate "Specifications" section that you should reference so you become aware of any necessary measurements. In addition, the "Component Locator" section of the service information will assist you in finding the location of the electrical components, connectors, and harnesses.

A sensor may require adjustment anytime it's removed or replaced. It's impossible to describe all the different procedures here. You must follow the guidelines as outlined in the service information. Sensor adjustment may only require a tape measure and an angle gauge. However, some may require specialized targets, adjustment shims, and special scan tool functionality. The inability to adjust a sensor indicates improper sensor mounting, body damage, or suspension component wear.

Since many ADAS components require calibration and/or initialization, it may be necessary to use special calibration target screens (**Figure 16-1**) and other special tools. Shops not equipped with these specialty items must not attempt to perform any service requiring them.

Figure 16-1 Special calibration screens and other equipment may be needed to service an ADAS.

> ⚙️ **SERVICE TIP** Do NOT try to perform ADAS calibration procedures without an approved battery maintainer/power supply attached to the vehicle. Dropping battery voltage will cause calibration issues.

Diagnosing Photoelectric SAS Sensors

ESP and ERM-equipped vehicles use the optical **Steering Angle Sensor (SAS)** to determine steering wheel position and degrees of rotation (**Figure 16-2**). The SAS is usually a direct input to the Antilock Brake Module (ABM), which broadcasts the data over the CAN bus. This input helps determine the driver's intended path of travel.

> ⚡ **WARNING** **To avoid personal injury or death on vehicles equipped with airbags, you must disable the system before attempting any steering wheel, steering column, or airbag component diagnosis or service. Disconnect and isolate the battery negative cable, and wait two minutes for the system capacitor to discharge before performing further diagnosis or service. Failure to take the proper precautions could result in accidental airbag deployment.**

Whenever there is a suspected problem with the SAS output signal, there are some basic tests and checks to perform. Keep in mind that several factors can cause the output signal to be erroneous. For example, if the vehicle was involved in an accident that causes it not to track properly, this will influence the position of the SAS while in operation. Inspect for any damage to the steering column and the steering gear mounting. Proper wheel alignment is necessary for the SAS to provide accurate data. Also, proper mounting of the SAS is critical and requires verification. In addition, inspect all related wiring for pinched, chafed, pierced, and partially broken wires. Inspect all connectors for broken, bent, pushed-out, and corroded terminals. Finally, perform a voltage drop test on the supply and ground circuit.

> **AUTHOR'S NOTE** After performing a wheel alignment on a vehicle using a SAS, you need to refer to the proper service information to determine if the sensor needs calibrating. Not performing this step may lead to the customer returning to the shop with warning lights and/or messages indicating the deactivation of ADASs.

Classroom Manual
Chapter 16, page 522

 Special Tools
Scan tool
DMM
Service information

Figure 16-2 Optical Steering Angle Sensor.

If the ABM detects that the calculated steering wheel angle exceeds what is physically possible, it will set a SAS performance or intermittent fault. **Photo Sequence 35** illustrates how to use a scan tool to test the SAS. If the sensor does not pass, it's faulty. If all of these tests fail to identify the problem, the ABM is faulty.

PHOTO SEQUENCE 35
Testing the Steering Angle Sensor

P35-1 You will need to use a scan tool to diagnose the SAS.

P35-2 Place the wheels in the straight-ahead position.

P35-3 Use the scan tool to read the SAS position. If the steering wheel is straight ahead, the reading should be +/- 15° of 0°.

P35-4 Slowly rotate the steering wheel from left to right while observing the scan tool display.

P35-5 The position displayed on the scan tool should decrease when rotating the steering wheel to the right.

P35-6 The position display on the scan tool should increase when rotating the steering wheel to the left. Monitor for a smooth change in SAS position while performing these tests.

P35-7 Rotate the steering wheel all the way to the right stop. Note the SAS position on the scan tool.

P35-8 Rotate the steering wheel all the way to the left stop, and note the SAS position as displayed on the scan tool.

P35-9 The scan tool reading should indicate a travel to 720° while rotating the steering wheel from lock to lock.

Diagnosing Park Assist Systems

The **Park Assist System (PAS)** provides an example for diagnosing systems that use ultrasonic sensors (**Figure 16-3**). Upon the detection of an object, the system uses an LED display and warning chimes to provide the driver with visual and audible warnings of the object's presence.

Ultrasonic transceiver sensors in the bumpers locate and identify the proximity of obstacles in the path of the vehicle. Each sensor receives battery voltage and ground from the park assist module. Each sensor has a dedicated serial bus communication circuit to the module (**Figure 16-4**).

If a customer states that their park assist system is not functioning properly, begin your testing by confirming that dirt or other objects are covering the sensors. Also, confirm that the parking brake is not applied and the parking brake switch is operating properly.

The microprocessor in the park assist module contains the park assist system logic circuits and uses on-board diagnostics to continuously monitor the entire park assist system electrical circuits and components. If the park assist module determines a problem in any park assist system circuits or components, it will store a DTC.

Diagnose the hardwired circuits connecting components related to the park assist display using a DMM, and conventional diagnostic procedures using the proper wiring information. However, DMM diagnostic methods will not diagnose problems associated with the park assist display, the electronic controls, or communication between modules and other devices that provide some features of the park assist system. Problems in these areas are more easily identified and repaired using a scan tool.

The park assist module controls the park assist display. In addition, the display receives voltage and ground from the module (**Figure 16-5**). A dedicated serial bus provides bidirectional communication between the display and the module.

When diagnosing problems associated with the display, keep in mind that the microprocessor in the park assist module has complete control over the LED display. The module continually monitors the display status and will store a DTC upon detection of a fault.

Classroom Manual
Chapter 16, pages 514, 525

The **park assist system (PAS)** is a parking aid that alerts the driver to obstacles located in the path immediately behind the vehicle.

 Special Tools

Scan tool
Service information

Ultrasonic sensor locations

Figure 16-3 Ultrasonic sensors used for back-up obstacle detection are usually located in the rear fascia/bumper.

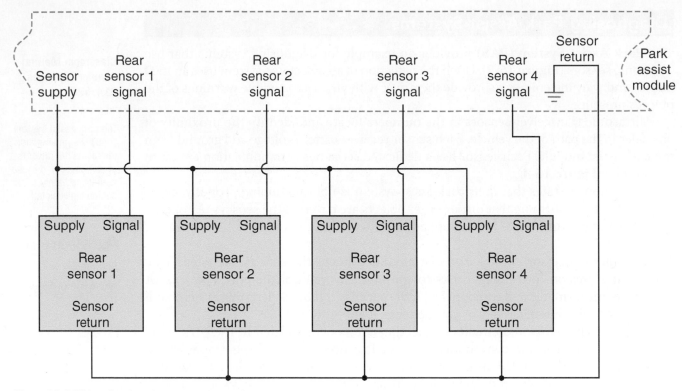

Figure 16-4 Wiring of park assist system transponders.

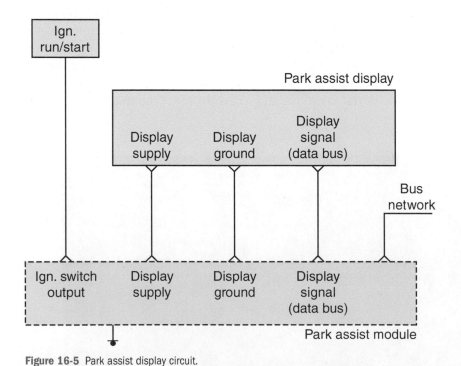

Figure 16-5 Park assist display circuit.

Classroom Manual
Chapter 16, pages
520, 527

Special Tools

Scan tool DMM
Service information

Back-Up Camera Systems

The back-up camera assists the driver in providing additional viewing of the area behind their vehicle. Putting the vehicle's transmission into reverse changes the radio modes and uses the rear camera (**Figure 16-6**). The radio monitor displays the rear

Figure 16-6 Rearview camera.

camera video (**Figure 16-7**). Activation of the camera is by a feed from the back-up lamp circuit (**Figure 16-8**). The camera sends a standard NTSC (TV) video signal to the radio.

Most problems with the back-up camera system will result in the setting of a fault code that can be accessed by a scan tool. Also, the scan tool may provide for activation functions to help isolate the problem area. **Table 16-1** is a diagnostic chart that directs the technician to possible causes for the condition.

Figure 16-7 Rearview camera display on the radio screen.

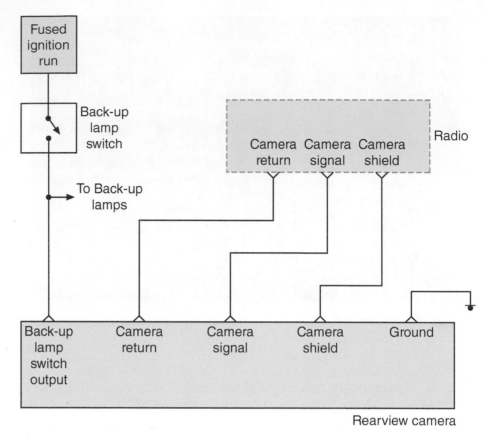

Figure 16-8 Rearview camera circuit.

Table 16-1 Diagnostic Chart for the Rear View Camera System

Condition	Possible Causes	Correction
Camera Inoperative	1. Fuse inoperative. 2. Wiring damaged. 3. Camera ground damaged. 4. Video wiring damaged. 5. Camera inoperative.	1. Check the back-up fuse. Replace the fuse, if required, and repair the cause of the blown fuse. 2. Check for voltage at the camera connector for back-up lamp switch circuit. Repair wiring if required. 3. Check for continuity between the ground wire at the camera connector and ground. Repair wiring if required. 4. Check for continuity between the radio connector and the camera connector. There should be continuity. 5. Replace the camera. Refer to the appropriate Service information.
No Video Display	1. Fuse inoperative. 2. Radio connector damaged. 3. Wiring damaged. 4. Radio ground damaged. 5. Video wiring damaged. 6. Radio inoperative.	1. Check radio fuses. Replace fuses, if required, and repair the cause of the blown fuse. 2. Check for loose or corroded radio connectors. Repair, as necessary. 3. Check for battery voltage at the radio connector. Repair wiring if required. 4. Check for continuity between radio chassis and a known good ground. There should be continuity. Repair ground if required. 5. Check for continuity between the radio connector and the camera connector. There should be continuity. 6. Refer to appropriate diagnostic service information for diagnosing the radio system.

Diagnosing Self-Park Systems

The self-park system encompasses many different systems and inputs for proper operation. The ABS, electric steering, cameras, monitors, and radar sensors all make up the system. A fault in any of these systems or components can result in the self-parking system shutting down.

The steering wheel angle sensor monitors the rotation of the steering wheel during the parking maneuvers. Disconnecting the battery may necessitate the initialization of the steering angle sensor before the self-parking system will operate. Steering sensor initialization may require the use of a scan tool. However, many systems are initialized in the following manner:

1. Turn the ignition to the RUN position.
2. Move the shift lever to the "R" position to confirm that the self-park system is inoperative. "System initializing" may appear in the rearview monitor.
3. Move the shift lever to the "P" position and turn the steering wheel lock to lock.
4. Return the shift lever to the "R" position.

If the system is initialized, the displaying of a "Check Surroundings for Safety" or a similar message indicates that the system is ready.

The self-park system runs continuous diagnostics of the module logic, data bus, actuators, and inputs. Faults will set DTCs; thus most diagnostics are DTC driven. Use the scan tool to monitor data PIDs and perform bidirectional diagnostics. However, before condemning any modules or components, confirm that the power and ground circuits are functioning properly.

To test the detection range of the sensors, place the ignition in the RUN position, and switch the self-park system ON. Different gear positions will be selected during the test, so be sure to block the tires. For example, when measuring the front corner sensors, the gear selection can be in any gear except PARK. The front center sensor is tested in any gear position other than PARK or REVERSE. The rear corner and rear center sensors require the gear selection to be in REVERSE.

To test the range detection of the corner sensors, move a 2.36-inch (60-mm) pole outward from the center of the sensor. Move the pole in the direction the sensor is pointing. The pole can be held either parallel or perpendicular to the ground, but not at a diagonal. The sensor should detect the pole when within the specified range. Move the pole left and right of the sensor's centerline to determine the span of the detectable range. The scan tool data PID will indicate the presence of the pole and the approximate range it's from the sensor. Alternately, the system display and warning buzzer should indicate that the sensors have detected the pole.

To test the detection range of the front and rear center sensors, use a wall or barrier (**Figure 16-9**). Be sure to test for proper sensor pattern width. Typically, the sensor pattern will extend further than the sides of the vehicle.

The ultrasonic sensors require adjustment anytime they are removed or replaced. The actual procedure will vary depending on the manufacturer of the self-park system and the sensor used. The following procedure for the Toyota Prius is an example of the process.

First, confirm the work area is level, and the vehicle is at the correct ride height. Next, measure the distance from the floor to each of the sensors and compare with specifications (**Figure 16-10**). If the measurement is not correct, confirm proper mounting. Some sensors can be adjusted slightly, but check for body damage and suspension component wear if the measurement and specification difference is sufficient.

With proper sensor height confirmed, attach the digital angle gauge to the sensor (**Figure 16-11**). The angle gauge must read within the acceptable range as indicated in the service information. Check for proper sensor mounting and body damage if the angle is not within tolerance.

Classroom Manual
Chapter 16, pages
515, 528, 531

 Special Tools

Scan tool
60-mm diameter pole
Digital angle gauge
Tape measure
Alignment screens

Classroom Manual
Chapter 16, page 514

Figure 16-9 Range of sensor detection.

Wall ──
Approx.
1,750 mm
(68.9 in.)
Approx. 1,130 mm
(44.5 in.)
Approx. 1,500 mm
(59.1 in.)

Wall ──
Approx.
1,430 mm
(56.3 in.)
Approx. 810 mm
(31.9 in.)
Approx. 1,000 mm
(39.4 in.)

Figure 16-10 Measure the distance from the floor to the sensor.

Height

Figure 16-11 The digital angle gauge will display the orientation of the sensor.

Digital angle gauge

ESP and ERM System Diagnostics

Classroom Manual
Chapter 16, page 531

Special Tools

Scan tool
DMM
Backprobing tools
Jumper wires
Test light
Service information

The first indication to the vehicle operator that a problem is present in the ESP or the electronic rollover mitigation (ERM) system is the illumination of the warning light (**Figure 16-12**). This light may illuminate along with the antilock brake warning light, the red brake warning light, and the brake assist system (BAS) warning light. Illumination of the red brake warning lamp means there is a problem with the base brake system. In this case, the other associated amber warning lights may also illuminate since the brake problem affects the entire system. The ABM, and other modules used within the systems, continuously monitor the circuits for continuity faults along with short activation tests of solenoids, pumps, and other actuators. The detection of a fault illuminates the amber warning lamps, indicating disabling of the system for the rest of that ignition cycle. When the vehicle is restarted, and the system's self-test determines there are no faults, the warning lamps turn off after completing the bulb check, and the system will be operational. If a fault is detected (or appears anytime during the drive cycle), the warning lamps illuminate again, and the system becomes nonoperational. Anytime the system computer detects a fault, a DTC is recorded.

The **sensor cluster** on ESP- and ERM-equipped vehicles contains both the lateral (side-to-side) and yaw (vehicle rotational) sensors. The ABM uses the data supplied by the sensor cluster to determine how to control engine torque and which brake(s) to apply to maintain vehicle control during an ESP or rollover mitigation event. The sensor cluster receives its power and ground from the ABM. It also has a dedicated data circuit to transmit the information (**Figure 16-13**). The sensor cluster is centrally located in the vehicle, typically under the center console (**Figure 16-14**).

The sensor cluster is also referred to as a dynamic sensor.

Figure 16-12 The detection of a fault in the ESP system illuminates the warning lamps.

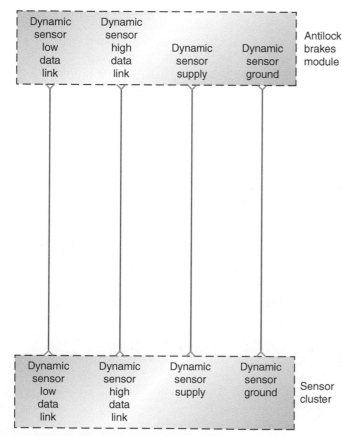

Figure 16-13 ESP sensor cluster circuit.

Figure 16-14 The sensor cluster is mounted on the centerline of the vehicle, usually under the console.

Both the lateral and yaw sensors operate on an internal 5-volt reference. The neutral position has an average of 2.5 volts. **Figure 16-15** illustrates the typical voltage of a lateral acceleration sensor signal, and **Figure 16-16** illustrates that of a yaw sensor signal. Test the sensor cluster by removing its fasteners, observing the scan tool display while moving the sensor back and forth (**Figure 16-17**), and turning it around its center (**Figure 16-18**).

A dedicated CAN C bus transmits instantaneous data to the ABM. Because of this data transmission procedure, it may not be possible to use a DMM or lab scope to monitor the voltage outputs. Diagnosis of the sensor requires a scan tool.

The sensor is initialized when it's first installed. It's possible that it will go to an uninitialized state if the supply voltage drops below an acceptable level. If this occurs, the warning lamp will illuminate, and system operation may be inhibited. This can be caused by damage to the wires, terminals, or connectors. It can also result from high resistance in the supply and ground circuits. If there are no circuit failures, the sensor or the ABM is faulty.

Figure 16-15 Voltage output of the lateral sensor.

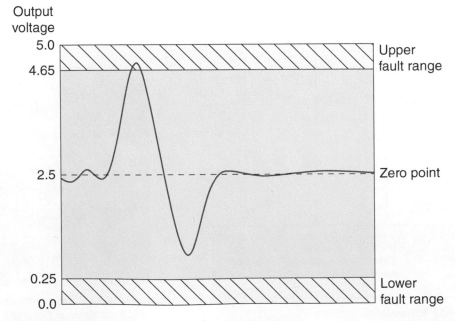

Figure 16-16 Voltage output of the yaw sensor.

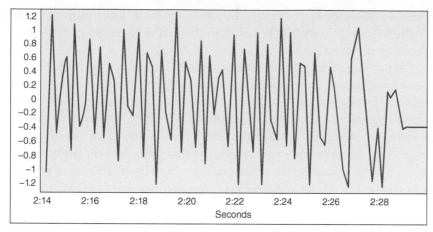

Figure 16-17 Scan tool display of the sensor cluster output while moving the sensor back and forth.

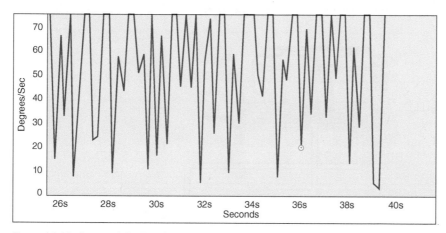

Figure 16-18 Scan tool display of the sensor cluster output while rotating the sensor around its center.

To test the supply and ground circuits, disconnect the harness connector at the sensor and connect a voltmeter across the supply and ground circuit terminals. With the ignition switch in the RUN position, the voltage should be higher than 10.5 volts. If the voltage is correct, the sensor is faulty.

If the voltage reading was lower than 10.5 volts, disconnect the ABM harness connector (leave the sensor connector unplugged) and use an ohmmeter to check the supply circuit for high resistance. If the supply circuit has no problem, use the ohmmeter to measure the resistance of the ground circuit. If these don't indicate a circuit problem, the supply circuit may be shorted to another circuit in the harness. Use the ohmmeter to measure the resistance between the supply circuit and the other circuits in the harness. Any reading below 10K ohms indicates that the circuits are shorted together.

If the cause of the low-voltage reading is not isolated, the problem is a faulty ABM. Be sure to test power and ground circuits before performing repairs or replacing the ABM.

A DTC will be set for supply voltage high if the ABM detects voltage on the sensor cluster supply circuit after turning off power. A short to voltage in the wiring harness or connector damage can cause this condition. If the circuit does not have a defect, the ABM is faulty since it's not properly powering down the circuit.

Systems that use G-sensors will also set fault codes if the measured acceleration signal is higher than specifications for too long. This type of fault can result from wiring harness, terminal, or connector damage, requiring a thorough inspection. Also, high resistance in the supply circuit or the ground circuit can cause this fault. Keep in mind that improper torquing of the sensor fasteners may result in the ABM setting a code. Finally, a faulty sensor or ABM may be the cause.

Because the rollover mitigation system ties to the ESP system, which uses the ABS hydraulic control unit (HCU) and its solenoids, along with the ABS pump motor or active brake booster to initiate the application of individual brake channels (**Figure 16-19**), diagnostics of the actuators are based on ABS diagnostic routines. The ERM system uses these same ABS components to apply an outside brake caliper during a sideways slide to work to prevent the vehicle from rolling. In some slide conditions, the solenoids prevent pressurized hydraulic fluid from entering the caliper, so the wheels will continue to roll instead of grabbing.

Failure of a solenoid circuit will set a DTC. Remember that the DTC directs the technician to the circuit that has the fault, not necessarily to the actual component. For example, a fault code that indicates the circuit is open may result from a broken wire or a faulty solenoid. It's the responsibility of the technician to use the information from the DTC and the appropriate service information to diagnose and repair the condition properly.

A lab scope is excellent for testing the HCU solenoids. **Figure 16-20** illustrates a pattern obtained from a normally functioning solenoid valve. Using the solenoid activation

Figure 16-19 Hydraulic control solenoids used to control ABS and ESP functions.

Figure 16-20 Lab scope trace pattern as ABS solenoid is activated during wheel lockup.

test function in the scan tool and a lab scope, observe the voltage levels and inductive kicks. The inductive kicks indicate that the solenoid is working properly.

> **AUTHOR'S NOTE** Obtaining a lab scope pattern of the solenoid activation is only possible if there is a harness between the ABM and the HCU. If the ABM and HCU are integrated, you will not be able to access the solenoid circuit.

A DTC sets when the ABM determines that the active brake booster solenoid circuit has failed the diagnostic test. This can be the result of any of the following conditions:

- Wiring harness, terminal, or connector damage
- Active brake booster solenoid (1) circuit and active brake booster solenoid (2) circuit shorted together
- Active brake booster solenoid (1) circuit shorted to ground, voltage, or open
- Active brake booster solenoid (2) circuit shorted to ground, voltage, or open
- Faulty active brake booster solenoid
- Faulty ABM

Consider **Figure 16-21** as an example of testing the active booster solenoid. Using the scan tool, read and record all DTCs and any freeze frame or snapshot data. If the code is active, inspect the connectors of the ABM and the active booster solenoid (**Figure 16-22**) for broken, bent, pushed-out, or corroded terminals and repair any problems found.

To test for solenoid activation, disconnect the harness connector at the ABM (be sure the solenoid harness is connected). With the ignition in the RUN position, connect a

Figure 16-21 Wiring schematic of the command circuit between the ABM and the active booster solenoid.

Module-antilock brakes

Booster-active brake (ESP)

Figure 16-22 Example of connector terminal locations for active brake booster system.

jumper wire between the solenoid (2) and chassis ground at the ABM harness connector. Connect a fused jumper wire between the solenoid (1) circuit and a 12-volt supply at the ABM harness connector for several seconds. Brake application should occur when making these connections. If there is, the ABM is defective and needs replacing. If there is no activation, diagnose the circuits and the solenoid.

Test the resistance of the solenoid with an ohmmeter. Usually, these solenoids have a low resistance value that is within the range of 1 to 2 ohms. Always confirm the specified resistance value with the proper service information. If the resistance value is out of specifications, replace the solenoid.

With the solenoid connector disconnected, connect a voltmeter between the voltage supply circuit and chassis ground. Turn the ignition switch to the RUN position and observe the reading. If the voltage is 0 volt, there is an open, a short to ground, or the two circuits are shorted together. Use normal ohmmeter test procedures to determine the type of fault and to locate the cause. In addition, a test light connected to 12 volts can determine if the circuit is shorted to ground.

To check for an open circuit, disconnect the solenoid and the ABM harness connectors. Connect a jumper wire between the solenoid (1) circuit and ground. Using a 12-volt test light connected to 12 volts, check the solenoid (1) circuit. If the test light illuminates, replace the ABM. If it does not, the circuit has an open.

If the initial voltage test records a voltage reading, remember that this was an open circuit test and is not conclusive. Load the supply circuit with a test light, and measure the voltage again. If the voltage is less than battery volts, the circuit has high resistance. If the supply circuit checks good, test the ground circuit for shorts and opens. With both the ABM and solenoid harness connectors disconnected, use a test light connected to 12 volts to check the solenoid (2) circuit for a short to ground. If the test light illuminates, repair the short-to-ground condition.

If the test light does not illuminate, test the circuit for a short to voltage by connecting the test light to chassis ground and probing the solenoid (2) circuit. If the light comes on, repair the short-to-voltage condition.

To test for an open, disconnect the ABM and the solenoid harness connectors. Connect a jumper wire between the solenoid (2) circuit and chassis ground. Using a test light connected to 12 volts, check the solenoid (2) circuit. If the test light does not illuminate, there is an open in the circuit.

Faults within the ABS pump motor circuit can be the result of any of the following conditions:

- ABS pump motor fuse open
- Wiring harness, terminal, or connector damage
- Fused Battery circuit shorted to ground or open
- Ground circuit open
- Faulty ABM

Use the scan tool to read and record all DTCs. Reference any freeze frame or snapshot data that are available. Erase the DTCs. Next, drive the vehicle above 25 mph (40 Km/h) and then park it and cycle the ignition. Use the scan tool again to read and record any DTCs and other associated information. If DTCs return, follow the service information for the diagnostic routine for the specific DTC. The following is a typical procedure for diagnosing the ABS pump circuits.

Begin by using the scan tool to perform an actuation test of the pump motor. If the pump motor actuates, but you just recorded DTCs for the circuit, the problem is intermittent. Carefully examine all connectors, terminals, and wiring for damage.

If the pump motor fails to actuate, check the voltage supply circuit (**Figure 16-23**). First, confirm that the fuse is not open. If the fuse is blown, locate and repair the cause of the excessive amperage. Since the pump motor is usually an integral component of the HCU, there are internal circuits that may not be easily tested. If the pump motor is a stand-alone component, be sure to test all of the circuits from the ABM to the motor.

> **AUTHOR'S NOTE** See *Today's Technician: Automotive Suspension and Steering Systems* Shop Manual for descriptions of diagnosing and testing the various automatic suspension system actuators that may be used by ESP and ERM systems.

If the fuse is good, check for proper voltage applied to the fused B(1) circuit. Disconnect the ABM harness connector, and load the circuit with a test light. The voltage should be close to battery voltage. If it's less than battery voltage, there is resistance in the circuit. If the voltage reading was 0 volt, the circuit is open (a short to ground was eliminated by checking the fuse).

If there is proper voltage on the supply circuit, check the ground circuit next. With the ABM harness connector disconnected, use a test light attached to a 12-volt source and test the ground circuit(s). If the test light illuminates, measure the loaded voltage on the circuit (**Figure 16-24**). If the voltmeter reads more than 100 mV, there is excessive resistance on the ground circuit(s). If the test light does not illuminate, locate and repair the open in the ground circuit.

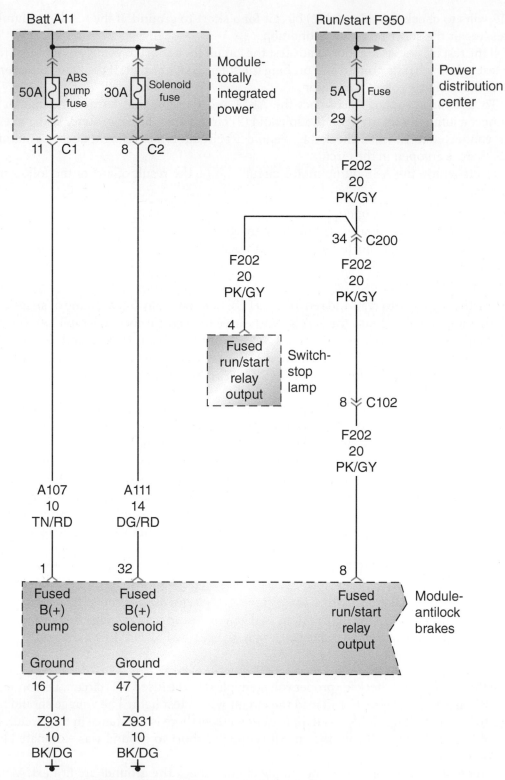

Figure 16-23 Wiring diagram of power and ground circuits for the ABM.

Figure 16-24 To accurately test the ground circuit, test it under load.

Emergency Braking System Service

Classroom Manual
Chapter 16,
pages 513, 530

The Automatic Emergency Braking (AEB) system is interlinked to other safety systems on the car. A vehicle equipped with AEB will also have Forward Collision Warning (FCW) and Adaptive Cruise Control (ACC). These three technologies can utilize the same components and communicate with the ABS and other modules over a high-speed data bus network.

It's unlikely that you will test the operation of the AEB system under "real world" conditions. If the AEB system detects a defect in a sensor or actuator, it will set a DTC and illuminate a warning light. Typically, the setting of a DTC will prohibit the operation of the AEB system. Use normal troubleshooting techniques to identify and correct the defect. The diagnostics may lead to checking sensor positioning and calibration.

Proper brake system service is critical for the AEB system to function as designed. In addition, camera and/or sensor faults must be identified, diagnosed, and corrected for proper system operation.

Most diagnostics of the AEB system will be DTC based. As with any other electrical and electronic system, circuit integrity, powers, and grounds must all be confirmed prior to replacing a sensor or control module. After retrieving the DTC, use the appropriate service information to determine the set conditions for the fault and the diagnostic steps necessary to isolate the root cause of the failure.

 Special Tools
Scan tool
DMM
Service information

Since this system is an extension of ACC, sensor diagnostics were covered in Chapter 13. Also, testing the ABS actuators used by this system was discussed in the previous section.

ADAS Calibration Procedures

Sensors may require calibration or initialization after the vehicle has been involved in a collision, been repainted, and if any of the following components are replacement:

- Door, hoods, etc.
- Bumpers
- Windshields
- Mirrors

 Special Tools
Scan tool
Service information
Alignment screens

After a battery disconnect, sensors may require initialization before some ADASs will operate. In addition, components may require initialization or recalibration after suspension system repairs or a wheel alignment. Also, removing the camera or sensor may require calibration.

If you determine that sensor or camera calibration is the cause of the customer's concern, it's best to check and correct the wheel alignment first. As mentioned, a vehicle that "dog tracks" due to an incorrect thrust angle can cause sensors to aim into the wrong detection zone. It may not be possible to calibrate the sensor if the thrust angle is incorrect.

There are 2 different calibration/recalibration modes: dynamic and static. The type used depends on the vehicle manufacturer and the ADAS.

The dynamic calibration procedure involves driving the vehicle at a specified speed and direction within an allotted time. This procedure may require a scan tool to initiate the process and to indicate when the calibration is complete.

To illustrate a static calibration procedure, consider a 2018 Jeep Grand Cherokee that was involved in a front-end collision. The scan tool reports an active DTC C2212-00 for a fault in the front radar sensor or circuit. The front sensor on this vehicle is in the front grille.

The service information provides the step-by-step procedure for diagnosing and repairing this issue. The front sensor of this vehicle requires an initial vertical alignment. With the vehicle parked on a level surface, a vertical alignment tool is attached to the sensor, and the vertical adjustment is made (**Figure 16-25**).

With the initial adjustment made, drive the vehicle to initiate the auto-alignment and self-calibrate the sensor. Monitor the progress of the auto-alignment with a scan tool. To complete the auto-alignment procedure, the sensor must see the correct roadside objects. The procedure also requires driving the vehicle at different speeds.

The scan tool will indicate each stage's completion and what events prevent the calibration procedure from continuing. The scan tool will report the completion of the alignment routine. It's possible that additional adjustments may need to be performed. For example, the scan tool may report the need to adjust the horizontal alignment. If additional adjustments are necessary, repeat the auto-alignment procedure.

Static calibration is performed in the service bay. Usually, special targets or calibration equipment is necessary to perform the procedure. Also, a scan tool or dedicated calibration tool may be required.

AUTHOR'S NOTE The area used to perform static calibrations must be properly lighted. There should not be any glass or reflective materials in the area. Block ambient light from open overhead doors and windows. Things like posters, calendars, metallic objects, and so forth can cause a reflection that interferes with the calibration procedure and could cause the system to calibrate improperly.

Figure 16-25 Additional specialty tools may be necessary. Here a vertical alignment tool is used to set the initial adjustment of the front radar sensor.

The actual procedure for sensor or camera calibration varies depending on the manufacturer of the ADAS and the sensor used. However, all static ADAS calibration procedures require the vehicle to be on a level surface and at the correct ride height. Some manufacturers may require the vehicle be loaded during this process.

The same sensor may have different calibration procedures. The scan tool may require you to input the reason for the calibration, which drives the specific calibration procedure. In addition, this becomes part of the post scan report, so the customer knows the reason for doing the procedure.

High level scan tools or special ADAS tablets will display the required equipment to perform the selected aiming procedure (**Figure 16-26**). The target and calibration screen manufacturer may have their own scan tools with the required software for using this equipment. Some may provide a subscription service to use a tablet.

Follow the instructions to position the vehicle in front of the calibration frame properly. This may involve measuring distances between the vehicle and the frame. Some calibration equipment does not require making physical measurements; instead, they use a camera to determine the appropriate frame to vehicle position. If your scan tool does not provide these instructional screens, you will need to reference the proper service information.

The calibration frame is adjusted for the sensor being calibrated. Some will use laser beams aimed at rulers on the crossbar. Additional adjustments bring the laser to reflect back to the rear wheel scale boards. After completing the setup process, the final preparation step is the installation of the target boards.

During the calibration process, the scan tool may display a progress bar. The calibration procedure may take a long time to complete, so be patient. The scan tool will indicate the completion of the routine.

Although the scan tool reports the sensor or camera aiming and calibration procedures as "Complete," does that mean it's correct? It's possible to set up the equipment incorrectly and still achieve a "Complete" result. Remember that ADAS is all about safety. This means any service procedure performed must be accurate. Proper preparation and setup are essential to assuring precise calibration results. Take the time to set the equipment up correctly; this is not the time to hurry the job. As with any repair, you must verify that the calibration is correct and complete.

After completing the calibration procedure, cycle the ignition and perform a module scan. Next, take the vehicle on a test drive to verify the system is operating, and there are no warning messages or lights. After the test drive, perform another module scan to make sure there are no DTCs.

Figure 16-26 A scan tool displaying the preparations necessary to perform the calibration procedure and the instructions for setting up the calibration frame.

Case Study

A customer brought their 2021 Jeep Grand Cherokee into the shop because the "ESP" light was always on. The technician verified the condition and used their scan tool to retrieve DTCs. A code C2114 was set for "dynamic sensor supply voltage low." After consulting the service information, the technician accessed the sensor cluster and the ABM. While taking out the center console to gain access to the sensor cluster, it was discovered that some additional electrical accessories were installed. These included a power outlet for a DVD player and a hands-free module. Following the diagnostic procedures, the technician tested the circuits and found resistance in the sensor ground circuit. A closer inspection of the circuit wire revealed a damaged section of wire. Apparently, this wire was damaged and not repaired during the installation of the extra accessories. After cutting out the damaged section and soldering in a new wire, the system fault was no longer active. After test driving the vehicle to confirm the repair, the vehicle was returned to its owner.

Case Study

A customer has brought their vehicle into the shop because the self-park system does not operate. When the system is activated, a warning buzzer sounds and the message center displays, "Park System Not Available." The technician confirms the customer's concern and connects their scan tool to check for DTCs. DTC C1694 for "Steering Angle Initialization Incomplete" was reported. The technician performed the calibration procedure by turning the steering wheel from full left to full right. The DTC was cleared, and the system confirmation routine was performed. However, the DTC reappeared, and the system would not initiate. Following the service information system procedures, the technician performed the "Steering Center Memorize" function of the scan tool. Next, they performed the "Max Steering Angle Memorize" function. The system verification was performed and was successful. When the customer returned to pick up the vehicle, they informed the technician that another shop replaced the battery the day before the self-park system stopped working. With today's vehicles, the technician must reference all applicable service information when performing even the most routine of service procedures, including battery replacement.

ASE-Style Review Questions

1. A customer with an ESP-equipped vehicle is experiencing the activation of the vehicle's ESP system when accelerating from a stop. The vehicle has no active or stored faults. What could cause this condition?

   A. Improperly mounted steering angle sensor.

   B. Damaged steering column.

   C. Improper suspension alignment.

   D. All of these choices.

2. The self-park system fails to activate.

   Technician A says this can be due to a steering angle sensor that is not initialized.

   Technician B says this can be caused by a sensor that is covered by debris.

   Who is correct?

   A. A only   C. Both A and B

   B. B only   D. Neither A nor B

3. A "Roll-Over Internal" fault code usually indicates which type of repair?

   A. Replacement of the control module.

   B. Replenishing the mercury vapors into the sensor.

   C. Removing the sensor from the control module and rotating it 180 degrees.

   D. All of these choices.

4. *Technician A* says the sensor cluster output of an ESP system can be tested with a lab scope. *Technician B* says the sensor cluster can be diagnosed only with a scan tool. Who is correct??

   A. A only   C. Both A and B

   B. B only   D. Neither A nor B

5. Diagnostics of a park assist system that uses ultrasonic sensors is being discussed.

   Technician A says the sensors are diagnosed using a DMM.

   Technician B says the module controls the operation of the sensors.

   Who is correct?

   A. A only   C. Both A and B

   B. B only   D. Neither A nor B

6. The self-parking system does not perform a parallel parking maneuverer with a displayed message that the parking space is not of sufficient size. However, the selected parking spaces are large enough to park the vehicle.

   Technician A says this can be caused by a misaligned ultrasonic corner sensor.

   Technician B says this can be caused by a disconnected rear center sensor.

   Who is correct?

   A. A only

   B. B only

   C. Both A and B

   D. Neither A nor B

7. *Technician A* says that replacement windshields may have specific applications between those equipped with forward facing cameras and those that don't.

   Technician B says windshield tinting will not affect forward facing cameras since they use infrared.

   Who is correct?

   A. A only

   B. B only

   C. Both A and B

   D. Neither A nor B

8. Setting up the calibration frame is being discussed.

   Technician A says to set the frame up in an area that is flooded with ambient light.

   Technician B the scan tool or tablet with the proper software will guide the technician through the setup process.

   Who is correct?

   A. A only

   B. B only

   C. Both A and B

   D. Neither A nor B

9. *Technician A* says it's best practice to check and correct wheel alignment before calibrating sensors or cameras.

Technician B says some sensors may need to be calibrated after a wheel alignment.

Who is correct?

A. A only
B. B only
C. Both A and B
D. Neither A nor B

10. The steering angle sensor has set a fault code for not being initialized. Technician A says this could be caused by a loss of battery voltage to the sensor. Technician B says this could be caused by excessive resistance in the sensor ground circuit. Who is correct?

A. A only
B. B only
C. Both A and B
D. Neither A nor B

ASE Challenge Questions

1. All of the LEDs of the park assist display are illuminated along with a constant audio tone emitted whenever the vehicle is put into reverse. Technician A says the voltage supply circuit to the display is shorted to ground. Technician B says the sensors may be dirty. Who is correct?

A. A only
B. B only
C. Both A and B
D. Neither A nor B

2. The Lane Keep System (LKS) warning light comes on intermittently.

Technician A says the mounting for the sensor(s) may be loose.

Technician B says a fault in the electric steering gear may be the cause.

Who is correct?

A. A only
B. B only
C. Both A and B
D. Neither A nor B

3. *Technician A* says The Automatic Emergency Braking (AEB) system is interlinked with Forward Collision Warning (FCW) and Adaptive Cruise Control (ACC). These three technologies can utilize the same components and will communicate with the ABS and other modules over a high-speed data bus network.

Technician B says modules that control the AEB function use a dedicated LIN bus for communications.

Who is correct?

A. A only
B. B only
C. Both A and B
D. Neither A nor B

4. *Technician A* says windshield replacement may require recalibration of the Forward Facing Camera.

Technician B says all recalibrating procedures should be performed outside in bright light.

Who is correct?

A. A only
B. B only
C. Both A and B
D. Neither A nor B

5. ADAS calibration procedures are being discussed.

Technician A says all static ADAS calibration procedures require the vehicle to be on a level surface and at the correct ride height.

Technician B says sensors may have different calibration procedures based on the reason for the calibration.

Who is correct?

A. A only
B. B only
C. Both A and B
D. Neither A nor B

Name _____   Date _____

ANALYZING THE STEERING ANGLE SENSOR

Upon completion of this job sheet, you will have become familiar with steering angle sensor (SAS) operation and interaction with the ESP system.

ASE Correlation ───

This job sheet addresses the following **MLR** tasks:

A.1.   Research vehicle service information such as fluid type, vehicle service history, service precautions, technical service bulletins, and recalls including vehicles with advanced driver assistance systems (ADAS). **(P-1)**

G.1.   Demonstrate understanding of vehicle comfort, convenience, access, safety, and related systems operation. **(P-3)**

This job sheet addresses the following **AST** tasks:

A.1.   Research vehicle service information such as fluid type, vehicle service history, service precautions, technical service bulletins, and recalls including vehicles with advanced driver assistance systems (ADAS). **(P-1)**

G.1.   Diagnose vehicle comfort, convenience, access, safety, and related systems operation; determine needed action. **(P-2)**

G.7.   Diagnose operation of safety systems and related circuits (such as horn, airbags, seat belt pretensioners, occupancy classification, wipers, washers, speed control/collision avoidance, heads-up display, park assist, and back-up camera); determine needed repairs. **(P-2)**

G.8.   Diagnose body electronic system circuits using a scan tool; check for module communication errors (data communication bus systems); determine needed action. **(P-2)**

This job sheet addresses the following **MAST** tasks:

A.1.   Research vehicle service information such as fluid type, vehicle service history, service precautions, technical service bulletins, and recalls including vehicles with advanced driver assistance systems (ADAS). **(P-1)**

G.1.   Diagnose vehicle comfort, convenience, access, safety, and related systems operation; determine needed action. **(P-2)**

G.7.   Diagnose operation of safety systems and related circuits (such as horn, airbags, seat belt pretensioners, occupancy classification, wipers, washers, speed control/collision avoidance, heads-up display, park assist, and back-up camera); determine needed repairs. **(P-1)**

G.8.   Diagnose body electronic system circuits using a scan tool; check for module communication errors (data communication bus systems); determine needed action. **(P-1)**

We Support
ASE | Education Foundation

Tools and Materials

- Scan tool
- Service information

Procedure

1. Connect the scan tool and navigate to the steering angle position sensor data.

2. Record the approximate steering angle position sensor degree reading with the front wheels positioned straight ahead.

_____

3. What happens to the steering angle sensor data when the steering wheel is rotated to the right?

_____

_____

_____

_____

4. What happens to the steering angle sensor data when the steering wheel is rotated to the left?

_____

_____

_____

_____

5. Turn the steering wheel to the full left position and record the data reading.

_____

6. Turn the steering wheel to the full right position and record the data reading.

_____

7. What was the full amount of rotation displayed by the scan tool?

_____

Specification: _____

8. If the degrees of rotation are not within specifications, what could be the possible causes?

_____

_____

_____

_____

Instructor's Evaluation:

Grade	Rubric
0	No exposure due to: ☐ Absence ☐ Other:
1	Student unsuccessfully attempted the task
2	Student completed the task with significant instructor assistance.
3	Satisfactory performance of task. The student was able to perform elements of the task with minimal instructor assistance.
4	Competent/Proficient. Able to perform all elements of the task successfully and independently.

Instructor's Notes:

Name _____   Date _____

PERFORMING STEERING ANGLE SETTING PROCEDURE (PARK GUIDANCE SYSTEM)

Upon completion of this job sheet, you should be able to properly perform the steering angle setting procedure on a vehicle equipped with the park assist guidance system.

ASE Education Foundation Correlation

This job sheet addresses the following **MLR** tasks:

A.1. Research vehicle service information such as fluid type, vehicle service history, service precautions, technical service bulletins, and recalls including vehicles with advanced driver assistance systems (ADAS). **(P-1)**

G.1. Demonstrate understanding of vehicle comfort, convenience, access, safety, and related systems operation. **(P-3)**

This job sheet addresses the following **AST** tasks:

A.1. Research vehicle service information such as fluid type, vehicle service history, service precautions, technical service bulletins, and recalls including vehicles with advanced driver assistance systems (ADAS). **(P-1)**

G.1. Diagnose vehicle comfort, convenience, access, safety, and related systems operation; determine needed action. **(P-2)**

G.7. Diagnose operation of safety systems and related circuits (such as horn, airbags, seat belt pretensioners, occupancy classification, wipers, washers, speed control/collision avoidance, heads-up display, park assist, and back-up camera); determine needed repairs. **(P-2)**

G.8. Diagnose body electronic system circuits using a scan tool; check for module communication errors (data communication bus systems); determine needed action. **(P-2)**

This job sheet addresses the following **MAST** tasks:

A.1. Research vehicle service information such as fluid type, vehicle service history, service precautions, technical service bulletins, and recalls including vehicles with advanced driver assistance systems (ADAS). **(P-1)**

G.1. Diagnose vehicle comfort, convenience, access, safety, and related systems operation; determine needed action. **(P-2)**

G.7. Diagnose operation of safety systems and related circuits (such as horn, airbags, seat belt pretensioners, occupancy classification, wipers, washers, speed control/collision avoidance, heads-up display, park assist, and back-up camera); determine needed repairs. **(P-1)**

G.8. Diagnose body electronic system circuits using a scan tool; check for module communication errors (data communication bus systems); determine needed action. **(P-1)**

We Support
ASE | Education Foundation

Tools and Materials

• Vehicle equipped with park assist guidance
• Scan tool
• Service information

Procedure

For this job sheet task, you will perform the steering angle sensor initialization, center memorization, and maximum angle memorization routines.

1. Using the service information system, locate the steering angle sensor initialization procedure. List the steps below:

_____

_____

_____

2. How do you confirm that the steering angle sensor has been properly initialized?

_____

_____

_____

3. Use the scan tool to access the "Steering angle setting" function.

4. Center the steering wheel. What is the maximum tolerance the steering wheel can be off-center?

_____

5. Select "Steering center memorized."

6. Turn the steering wheel to the left lock position and then to the right lock position and select "Max steering memorize."

7. How is proper steering angle adjustment confirmed?

_____

_____

_____

Instructor's Evaluation:

Grade	Rubric
0	No exposure due to: ☐ Absence ☐ Other:
1	Student unsuccessfully attempted the task
2	Student completed the task with significant instructor assistance.
3	Satisfactory performance of task. The student was able to perform elements of the task with minimal instructor assistance.
4	Competent/Proficient. Able to perform all elements of the task successfully and independently.

Instructor's Notes:

_____

Name _____   Date _____

ANALYZING THE SENSOR CLUSTER (DYNAMIC SENSOR)

Upon completion of this job sheet, you will verify operation of the dynamic sensor, diagnose faults, and explain how they affect ESP system operation.

ASE Correlation

This job sheet addresses the following **MLR** tasks:

A.1. Research vehicle service information such as fluid type, vehicle service history, service precautions, technical service bulletins, and recalls including vehicles with advanced driver assistance systems (ADAS). **(P-1)**

A.2. Identify electrical/electronic system components and configuration. **(P-1)**

A.5. Demonstrate proper use of a digital multimeter (DMM) when measuring source voltage, voltage drop (including grounds), current flow, and resistance. **(P-1)**

A.6. Demonstrate knowledge of the causes and effects from shorts, grounds, opens, and resistance problems in electrical/electronic circuits. **(P-1)**

A.7. Describe types of test lights; use appropriate test light to check operation of electrical circuits per service information. **(P-2)**

A.8. Use fused jumper wires to check operation of electrical circuits per service information. **(P-2)**

A.9. Use wiring diagrams to trace electrical/electronic circuits. **(P-1)**

A.12. Repair and/or replace connectors, terminal ends, and wiring of electrical/electronic systems (including solder repair). **(P-2)**

G.1. Demonstrate understanding of vehicle comfort, convenience, access, safety, and related systems operation. **(P-3)**

This job sheet addresses the following **AST** tasks:

A.1. Research vehicle service information such as fluid type, vehicle service history, service precautions, technical service bulletins, and recalls including vehicles with advanced driver assistance systems (ADAS). **(P-1)**

A.2. Identify electrical/electronic system components and configuration. **(P-1)**

A.5. Demonstrate proper use of a digital multimeter (DMM) when measuring source voltage, voltage drop (including grounds), current flow, and resistance. **(P-1)**

A.6. Demonstrate knowledge of the causes and effects from shorts, grounds, opens, and resistance problems in electrical/electronic circuits. **(P-1)**

A.7. Describe types of test lights; use appropriate test light to check operation of electrical circuits per service information. **(P-1)**

A.8. Use fused jumper wires to check operation of electrical circuits per service information. **(P-1)**

A.9. Use wiring diagrams during the diagnosis (troubleshooting) of electrical/electronic circuit problems. **(P-1)**

A.12. Repair and/or replace connectors, terminal ends, and wiring of electrical/electronic systems (including solder repair). **(P-1)**

A.13. Test and measure circuit using an oscilloscope and/or graphing multimeter (GMM); interpret results; determine needed action. **(P-2)**

G.1. Diagnose vehicle comfort, convenience, access, safety, and related systems operation; determine needed action. **(P-2)**

(Continued)

G.7. Diagnose operation of safety systems and related circuits (such as horn, airbags, seat belt pretensioners, occupancy classification, wipers, washers, speed control/collision avoidance, heads-up display, park assist, and back-up camera); determine needed repairs. **(P-2)**

G.8. Diagnose body electronic system circuits using a scan tool; check for module communication errors (data communication bus systems); determine needed action. **(P-2)**

This job sheet addresses the following **MAST** tasks:

A.1. Research vehicle service information such as fluid type, vehicle service history, service precautions, technical service bulletins, and recalls including vehicles with advanced driver assistance systems (ADAS). **(P-1)**

A.2. Identify electrical/electronic system components and configuration. **(P-1)**

A.5. Demonstrate proper use of a digital multimeter (DMM) when measuring source voltage, voltage drop (including grounds), current flow, and resistance. **(P-1)**

A.6. Demonstrate knowledge of the causes and effects from shorts, grounds, opens, and resistance problems in electrical/electronic circuits. **(P-1)**

A.7. Describe types of test lights; use appropriate test light to check operation of electrical circuits per service information. **(P-1)**

A.8. Use fused jumper wires to check operation of electrical circuits per service information. **(P-1)**

A.9. Use wiring diagrams during the diagnosis (troubleshooting) of electrical/ electronic circuit problems. **(P-1)**

A.12. Repair and/or replace connectors, terminal ends, and wiring of electrical/ electronic systems (including solder repair). **(P-1)**

A.13. Test and measure circuit using an oscilloscope and/or graphing multimeter (GMM); interpret results; determine needed action. **(P-1)**

G.1. Diagnose vehicle comfort, convenience, access, safety, and related systems operation; determine needed action. **(P-2)**

G.7. Diagnose operation of safety systems and related circuits (such as horn, airbags, seat belt pretensioners, occupancy classification, wipers, washers, speed control/collision avoidance, heads-up display, park assist, and back-up camera); determine needed repairs. **(P-1)**

G.8. Diagnose body electronic system circuits using a scan tool; check for module communication errors (data communication bus systems); determine needed action. **(P-1)**

Tools and Materials
- Scan tool
- DMM
- Lab scope
- 12-volt test light
- Fused jumper wires
- Wiring diagram

Procedure

1. Using the wiring diagram, identify the circuits to the sensor cluster.

   _____

   _____

   _____

   _____

2. Establish communication with the ABM using the scan tool and check for DTCs. Are any DTCs present?   ☐ Yes ☐ No

   If yes, record them.

   _____

   List the possible causes for the DTC(s).

   _____

   _____

   _____

   _____

3. Visually inspect the sensor wiring. Check for any chafed, broken, or pierced wires. Also, visually inspect related harness connectors for broken, bent, spread, or pushed-out terminals. Record your results.

   _____

   _____

   _____

   _____

4. With the ignition in the OFF position, disconnect the ABM harness connector and the sensor cluster harness connector.

5. With the ignition in the RUN position, use a test light connected to ground and probe the sensor cluster high data link circuit.

   Did the test light illuminate?   ☐ Yes ☐ No

   What does this indicate?

   _____

   _____

   _____

   _____

6. Connect the test light to battery positive and probe the sensor cluster high data link circuit. Did the test light illuminate?   ☐ Yes ☐ No

What does this indicate?

_____

_____

_____

_____

7. Connect a fused jumper wire between the sensor cluster high data link circuit at the sensor harness connector and ground. Using a 12-volt test light connected to B(+), probe the dynamic sensor high data link circuit at the ABM harness connector. Did the test light illuminate?   ☐ Yes ☐ No

What does this indicate?

_____

_____

_____

_____

8. Connect the test light to B(+) and probe the sensor cluster low data link circuit. Did the test light illuminate?   ☐ Yes ☐ No

What does this indicate?

_____

_____

_____

_____

9. Connect a fused jumper wire between the sensor low data link circuit at the sensor harness connector and ground. Using a test light connected to B(+), probe the dynamic sensor low data link circuit at the ABM harness connector. Did the test light illuminate?   ☐ Yes ☐ No

What does this indicate?

_____

_____

_____

_____

10. Reconnect the ABM harness connectors.

11. At the sensor cluster harness connector, use a DMM or lab scope connected to chassis ground and measure the voltage on the sensor supply circuit. What was the measured voltage?

_____

12. Measure the voltage across the sensor supply circuit and the sensor ground circuit. What was the measured voltage?

_____

13. What conclusion can be drawn from the measured results?

_____

_____

_____

_____

14. Reconnect the scan tool and clear all DTCs. Do any DTCs return?   ☐ Yes ☐ No

Instructor's Evaluation:

Grade	Rubric
0	No exposure due to: ☐ Absence ☐ Other:
1	Student unsuccessfully attempted the task
2	Student completed the task with significant instructor assistance.
3	Satisfactory performance of task. The student was able to perform elements of the task with minimal instructor assistance.
4	Competent/Proficient. Able to perform all elements of the task successfully and independently.

Instructor's Notes:

Name _____   Date _____

ANALYZING THE PARK ASSIST SYSTEM

Upon completion of this job sheet, you will be able to verify operation of the ultrasonic sensor.

ASE Education Foundation Correlation

This job sheet addresses the following **MLR** tasks:

A.1.   Research vehicle service information such as fluid type, vehicle service history, service precautions, technical service bulletins, and recalls including vehicles with advanced driver assistance systems (ADAS). **(P-1)**

A.2.   Identify electrical/electronic system components and configuration. **(P-1)**

A.5.   Demonstrate proper use of a digital multimeter (DMM) when measuring source voltage, voltage drop (including grounds), current flow, and resistance. **(P-1)**

A.6.   Demonstrate knowledge of the causes and effects from shorts, grounds, opens, and resistance problems in electrical/electronic circuits. **(P-1)**

A.9.   Use wiring diagrams to trace electrical/electronic circuits. **(P-1)**

G.1.   Demonstrate understanding of vehicle comfort, convenience, access, safety, and related systems operation. **(P-3)**

This job sheet addresses the following **AST** tasks:

A.1.   Research vehicle service information such as fluid type, vehicle service history, service precautions, technical service bulletins, and recalls including vehicles with advanced driver assistance systems (ADAS). **(P-1)**

A.2.   Identify electrical/electronic system components and configuration. **(P-1)**

A.5.   Demonstrate proper use of a digital multimeter (DMM) when measuring source voltage, voltage drop (including grounds), current flow, and resistance. **(P-1)**

A.6.   Demonstrate knowledge of the causes and effects from shorts, grounds, opens, and resistance problems in electrical/electronic circuits. **(P-1)**

A.9.   Use wiring diagrams during the diagnosis (troubleshooting) of electrical/electronic circuit problems. **(P-1)**

A.13. Test and measure circuit using an oscilloscope and/or graphing multimeter (GMM); interpret results; determine needed action. **(P-2)**

G.1.   Diagnose vehicle comfort, convenience, access, safety, and related systems operation; determine needed action. **(P-2)**

G.7.   Diagnose operation of safety systems and related circuits (such as horn, airbags, seat belt pretensioners, occupancy classification, wipers, washers, speed control/collision avoidance, heads-up display, park assist, and back-up camera); determine needed repairs. **(P-2)**

G.8.   Diagnose body electronic system circuits using a scan tool; check for module communication errors (data communication bus systems); determine needed action. **(P-2)**

This job sheet addresses the following **MAST** tasks:

A.1.   Research vehicle service information such as fluid type, vehicle service history, service precautions, technical service bulletins, and recalls including vehicles with advanced driver assistance systems (ADAS). **(P-1)**

A.2.   Identify electrical/electronic system components and configuration. **(P-1)**

(Continued)

A.5.   Demonstrate proper use of a digital multimeter (DMM) when measuring source voltage, voltage drop (including grounds), current flow, and resistance. **(P-1)**

A.6.   Demonstrate knowledge of the causes and effects from shorts, grounds, opens, and resistance problems in electrical/electronic circuits. **(P-1)**

A.9.   Use wiring diagrams during the diagnosis (troubleshooting) of electrical/electronic circuit problems. **(P-1)**

A.13.   Test and measure circuit using an oscilloscope and/or graphing multimeter (GMM); interpret results; determine needed action. **(P-1)**

G.1.   Diagnose vehicle comfort, convenience, access, safety, and related systems operation; determine needed action. **(P-2)**

G.7.   Diagnose operation of safety systems and related circuits (such as horn, airbags, seat belt pretensioners, occupancy classification, wipers, washers, speed control/collision avoidance, heads-up display, park assist, and back-up camera); determine needed repairs. **(P-1)**

G.8.   Diagnose body electronic system circuits using a scan tool; check for module communication errors (data communication bus systems); determine needed action. **(P-1)**

We Support
ASE | Education Foundation

Tools and Materials

- Scan tool
- DMM
- Lab scope
- Backprobing tools
- Wiring diagram
- Vehicle equipped with a park assist system

Procedure

1. Where are the ultrasonic transceiver sensors located on the assigned vehicle?

   _____

   _____

   _____

2. Identify the circuits to the sensors.

   _____

   _____

   _____

3. What supplies the voltage to the sensor?

   _____

4. Where is the ground for the sensor located?

   _____

5. Block the wheels and set the parking brake.

6. Disconnect a sensor harness connector, and use a DMM or lab scope to measure the voltage on the supply circuit with the ignition in the RUN position (engine off) and the transmission in reverse. What is your reading?

_____

7. Reconnect the sensor, backprobe the connector for the voltage supply circuit, and measure voltage with the ignition in the RUN position (engine off) and the transmission in reverse. What is your reading?

_____

8. If the readings between steps 6 and 7 are different, explain why.

_____

_____

_____

9. Describe how you will test the ground circuit.

_____

_____

_____

10. Test the ground circuit using the method described in step 9, and record your results.

_____

11. With the sensor harness disconnected, use the scan tool to retrieve fault codes.

12. Was the module capable of determining the sensor circuit open?   ☐ Yes ☐ No

If so, explain how this can be done.

_____

_____

_____

13. What are the observable symptoms with the park assist system with the sensor disconnected?

_____

_____

_____

Instructor's Evaluation:

Grade	Rubric
0	No exposure due to: ☐ Absence ☐ Other:
1	Student unsuccessfully attempted the task
2	Student completed the task with significant instructor assistance.
3	Satisfactory performance of task. The student was able to perform elements of the task with minimal instructor assistance.
4	Competent/Proficient. Able to perform all elements of the task successfully and independently.

Instructor's Notes:

Name _____   Date _____

PARK ASSIST GUIDANCE SYSTEM OPERATION CHECK

Upon completion of this job sheet, you should be able to properly perform an operation check of the park assist guidance system and recognize fault detection characteristics.

ASE Education Foundation Correlation ─────────────────────────

This job sheet addresses the following **MLR** tasks:

A.1.   Research vehicle service information such as fluid type, vehicle service history, service precautions, technical service bulletins, and recalls including vehicles with advanced driver assistance systems (ADAS). **(P-1)**

G.1.   Demonstrate understanding of vehicle comfort, convenience, access, safety, and related systems operation. **(P-3)**

This job sheet addresses the following **AST** tasks:

A.1.   Research vehicle service information such as fluid type, vehicle service history, service precautions, technical service bulletins, and recalls including vehicles with advanced driver assistance systems (ADAS). **(P-1)**

G.1.   Diagnose vehicle comfort, convenience, access, safety, and related systems operation; determine needed action. **(P-2)**

G.7.   Diagnose operation of safety systems and related circuits (such as horn, airbags, seat belt pretensioners, occupancy classification, wipers, washers, speed control/collision avoidance, heads-up display, park assist, and back-up camera); determine needed repairs. **(P-1)**

G.8.   Diagnose body electronic system circuits using a scan tool; check for module communication errors (data communication bus systems); determine needed action. **(P-2)**

This job sheet addresses the following **MAST** tasks:

A.1.   Research vehicle service information such as fluid type, vehicle service history, service precautions, technical service bulletins, and recalls including vehicles with advanced driver assistance systems (ADAS). **(P-1)**

G.1.   Diagnose vehicle comfort, convenience, access, safety, and related systems operation; determine needed action. **(P-2)**

G.7.   Diagnose operation of safety systems and related circuits (such as horn, airbags, seat belt pretensioners, occupancy classification, wipers, washers, speed control/collision avoidance, heads-up display, park assist, and back-up camera); determine needed repairs. **(P-1)**

G.8.   Diagnose body electronic system circuits using a scan tool; check for module communication errors (data communication bus systems); determine needed action. **(P-1)**

We Support

Tools and Materials

- Vehicle equipped with park assist guidance.
- Scan tool
- 60-mm diameter pole
- Tape measure
- Service information

Procedure

For this job sheet task, you will perform an operational check of the park assist guidance system and observe system fault detection.

1. Unplug the right front corner sensor.

2. Attempt to activate the system to perform a parking maneuver. What methods are used to indicate a fault in the system?

   _____

   _____

   _____

3. Record any DTCs that were set.

   _____

   _____

   _____

4. Reconnect the sensor and clear any DTCs. Confirm the system is operating properly.

5. Cover a rear corner sensor.

6. Attempt to activate the system to perform a parking maneuver. What methods are used to indicate a fault in the system?

   _____

   _____

   _____

7. Record any DTCs that were set.

   _____

   _____

   _____

8. Uncover the sensor and clear any DTCs. Confirm that the system is operating properly.

9. Place the ignition in the RUN position and make sure that the guidance system is activated.

10. Record the gear selection required to measure the detection distance for each sensor:

    Left front: _____

    Right front: _____

    Center front: _____

    Left rear: _____

    Right rear: _____

    Center rear: _____

11. Move a 2.36-inch (60-mm) diameter pole near each corner sensor to measure its detection range. Record your results.

    Left front: _____

    Right front: _____

    Left rear: _____

    Right rear: _____

12. Move the pole to determine the sensor span for each of the corner sensors. Record your results.

    Left front: _____

    Right front: _____

    Left rear: _____

    Right rear: _____

13. Are the corner sensor detection ranges within specifications?   ☐ Yes ☐ No

Instructor's Evaluation:

Grade	Rubric
0	No exposure due to: ☐ Absence ☐ Other:
1	Student unsuccessfully attempted the task
2	Student completed the task with significant instructor assistance.
3	Satisfactory performance of task. The student was able to perform elements of the task with minimal instructor assistance.
4	Competent/Proficient. Able to perform all elements of the task successfully and independently.

Instructor's Notes:

DIAGNOSTIC CHART 16-1

PROBLEM AREA:	Ultrasonic module.
SYMPTOMS:	Disabling of system.
POSSIBLE CAUSES:	1. Wiring harness, terminal, connector damage. 2. High resistance in module voltage supply circuit. 3. High resistance in module ground circuit. 4. Charging system failure. 5. Internal controller fault.

DIAGNOSTIC CHART 16-2

PROBLEM AREA:	Infrared sensor circuit failure.
SYMPTOMS:	Disabling of system.
POSSIBLE CAUSES:	1. Wiring harness, terminal, connector damage. 2. Damaged lens. 3. Obstruction in front of sensor. 4. Improper sensor alignment. 5. High resistance in sensor voltage supply circuit. 6. Open in sensor voltage supply circuit. 7. High resistance in sensor ground circuit. 8. Open in sensor ground circuit. 9. Sensor supply circuit shorted to sensor ground circuit. 10. Sensor supply circuit shorted to ground. 11. Faulty sensor. 12. Internal controller fault.

DIAGNOSTIC CHART 16-3

PROBLEM AREA:	Camera system.
SYMPTOMS:	Disabling of system. Lack of image display.
POSSIBLE CAUSES:	1. Wiring harness, terminal, connector damage. 2. Damaged lens. 3. Obstruction in front of camera. 4. Improper camera calibration/alignment. 5. High resistance in voltage supply circuit. 6. Open in voltage supply circuit. 7. High resistance in ground circuit. 8. Open in ground circuit. 9. Damaged camera video wiring. 10. Faulty camera.

DIAGNOSTIC CHART 16-4

PROBLEM AREA:	Park assist.
SYMPTOMS:	When reverse (R) has been selected, an image of the area behind the vehicle is not displayed.
POSSIBLE CAUSES:	**1.** Parking assist ECU. **2.** Power source circuit. **3.** Camera feed circuit. **4.** Reverse signal circuit.

DIAGNOSTIC CHART 16-5

PROBLEM AREA:	Park assist.
SYMPTOMS:	When the target parking position is set, the initial display position of the target parking space is very far from the actual parking space.
POSSIBLE CAUSES	**1.** Ultrasonic sensor. **2.** Front bumper cover. **3.** Parking assist ECU.

DIAGNOSTIC CHART 16-6

PROBLEM AREA:	Park assist.
SYMPTOMS:	The vehicle width extension lines don't overlap with the predicted path lines when the steering wheel is centered.
POSSIBLE CAUSES:	**1.** Steering angle setting. **2.** Improper mounting of the rear camera assembly. **3.** Height control sensor/vehicle height difference setting. **4.** Rear camera optical axis adjustment. **5.** Parking assist ECU.

DIAGNOSTIC CHART 16-7

PROBLEM AREA:	Park assist.
SYMPTOMS:	"System initializing" is displayed.
POSSIBLE CAUSES:	**1.** Steering angle setting. **2.** Spiral cable with sensor subassembly. **3.** Parking assist ECU.

DIAGNOSTIC CHART 16-8

PROBLEM AREA:	Park assist.
SYMPTOMS:	Assist operation stops frequently.
POSSIBLE CAUSES:	**1.** Inspect the tires (tire size, tire pressure, and tire wear). **2.** Rear television camera optical axis adjustment. **3.** Parking assist ECU.

DIAGNOSTIC CHART 16-9

PROBLEM AREA:	Sensor cluster input missing
SYMPTOMS:	Supply voltage low Dynamics sensor status changes from initialized to not initialized
POSSIBLE CAUSES:	**1.** Wiring harness, terminal, connector damage **2.** High resistance in sensor voltage supply circuit **3.** High resistance in sensor ground circuit **4.** Faulty sensor **5.** Internal controller fault

DIAGNOSTIC CHART 16-10

PROBLEM AREA:	Sensor cluster input missing
SYMPTOMS:	Supply voltage high Voltage detected on the dynamics sensor supply circuit when the circuit's power is turned off.
POSSIBLE CAUSES:	**1.** Wiring harness, terminal, connector damage **2.** Sensor supply circuit shorted to voltage **3.** Internal controller fault

DIAGNOSTIC CHART 16-11

PROBLEM AREA:	Sensor cluster circuit performance
SYMPTOMS:	Measured acceleration signal is higher than specifications for longer than the specified detection time
POSSIBLE CAUSES:	**1.** Wiring harness, terminal, connector damage **2.** High resistance in sensor voltage supply circuit **3.** High resistance in sensor ground circuit **4.** Improper sensor mounting **5.** Faulty sensor **6.** Internal controller fault

DIAGNOSTIC CHART 16-12

PROBLEM AREA:	Active brake booster solenoid and circuit
SYMPTOMS:	ABM detects a circuit fault and illuminates warning lamps
POSSIBLE CAUSES:	**1.** Wiring Harness, terminal, connector damage **2.** Active brake booster solenoid (+) circuit and active brake booster solenoid (−) circuit shorted together **3.** Active brake booster solenoid (+) circuit shorted to ground, voltage, or open **4.** Active brake booster solenoid (−) circuit shorted to ground, voltage, or open **5.** Faulty active brake booster solenoid **6.** Faulty antilock brake module

DIAGNOSTIC CHART 16-13

PROBLEM AREA:	ABS pump motor control circuit
SYMPTOMS:	ABM detects a circuit fault and illuminates warning lamps
POSSIBLE CAUSES:	1. Wiring harness, terminal, connector damage 2. ABS pump motor fuse open 3. Fused B(+) circuit shorted to ground 4. Fused B(+) circuit open 5. Ground circuit open 6. Faulty antilock brake module

CHAPTER 17

HYBRID AND HIGH-VOLTAGE SYSTEM SERVICE

Basic Tools

Basic mechanic's tool set

Safety glasses

Service information

Upon completion and review of this chapter, you should be able to:

■ Demonstrate proper safety precautions associated with servicing the hybrid electric vehicle and the electric vehicle.

■ Properly remove and install the high-voltage service plug.

■ Access and interpret DTCs, information codes, freeze frame data, and history data.

■ Use scan tool data to determine HV battery cell condition.

■ Use a DMM to determine HV battery cell block voltages.

■ Determine the cause of HV battery failures.

■ Charge and balance the HV battery.

■ Properly remove and install the HV battery.

■ Perform HV battery refurbishing procedures.

■ Determine failures of the inverter/converter assembly.

■ Use a lab scope to monitor inverter voltage and amperage waveforms.

■ Determine the cause of system main relay failures.

■ Replace the system main relays.

■ Properly perform isolation tests on HV system circuits.

Terms to Know

Battery ECU

Converter

Freeze frame data

High-voltage electronic control unit (HV ECU)

High-voltage service plug

History data

Information codes

Inverter

System main relay (SMR)

Introduction

Classroom Manual
Chapter 17, page 537

The ability to acquire parts to repair some of the HV components has begun to improve and many shops are now offering repair services to the HV system. This chapter discusses some of the diagnostic and service procedures for the high-voltage systems of the HEV, PHEV, and EV. A main focus of this chapter will be on safety concerns associated with the high-voltage system.

Safety Precautions

Since the hybrid system can use voltages in excess of 500 volts (both DC and AC), it's vital that the service technician be familiar with, and follow, all safety precautions. Failure to perform the correct procedures can result in electrical shock, battery leakage, or an explosion. The following are the safety procedures for servicing the HEV's high-voltage systems:

■ Remove the key from the ignition.

■ Disconnect the negative (−) terminal of the auxiliary (12-volt) battery. Refer to the proper service information since you may need to disconnect the auxiliary battery prior to removing the high-voltage service plug.

■ Remove the high-voltage service plug and put it where it cannot accidentally be reinstalled by someone else.

■ Cover the high-voltage service plug receptacle with insulation tape.

■ Do not attempt to test or service the system for 5 minutes after removing the high-voltage service plug. It takes at least 5 minutes for the high-voltage capacitors inside the inverter to discharge.

■ Test the integrity of the insulating gloves prior to use.

■ Wear high-voltage insulating gloves when disconnecting the service plug.

■ Follow the "One-Hand Rule" until HV power down has been confirmed.

■ Never cut the orange or blue high-voltage power cables. Orange or blue wire harnesses, terminals, and connectors identify the high-voltage system. In addition, high-voltage components may have a "High Voltage" caution label attached to them.

■ Cover the terminals of a disconnected connector with insulation tape.

■ Use a CAT III, 1,000-volt DMM to confirm that high-voltage circuits have 0 volts before performing any service procedure. The DMM must be 1,000-volt CAT III to protect against an eight-thousand-volt transient (voltage spike).

■ If using an oscilloscope, differential probes are mandatory to connect to high voltage, non-vehicle grounded circuits safely.

■ Use insulated tools when available.

■ Do not wear metallic objects that may cause electrical shorts.

■ Follow the service information diagnostic procedures.

■ Wear protective safety goggles when inspecting the HV battery.

■ Before touching any high-voltage system wires or components, wear insulating gloves, ensure the power switch is off, and disconnect the service disconnect plug.

■ Disconnect the service plug prior to performing a resistance check of any HV circuit.

■ Turn the power switch to the OFF position prior to disconnecting or reconnecting any connectors or components.

■ Isolate the ends of any disconnected high-voltage wires with insulation tape.

■ Properly torque the high-voltage terminals.

> Review Chapter 1 of this manual concerning the safety procedures associated with working on or around the HV system.

High-Voltage Service Plug

HEVs, PHEVs, and EVs have a **high-voltage service plug** that disconnects the HV battery from the system. Usually, this plug is located at the battery (**Figure 17-1**). Confirm you turned the vehicle off before disconnecting the high-voltage service plug. You may also need to disconnect the negative terminal of the auxiliary battery. Removing the high-voltage service plug splits the high-voltage circuit at the intermediate position of the HV battery.

The high-voltage service plug assembly contains a safety interlock reed switch. Lifting the clip on the high-voltage service plug opens the reed switch and turns off power to the System Main Relay (SMR) or contactors. The main fuse for the high-voltage circuit is inside the high-voltage service plug assembly.

However, never assume that the high-voltage circuits are off. Removing the high-voltage service plug doesn't disable the individual HV batteries. Use a CAT III DMM to verify that the system has powered down to 0 volts before beginning service. Set the voltmeter to the 400 VDC scale when testing the circuit for voltage.

After removing the high-voltage service plug, wait a minimum of 5 minutes before beginning service on the system to allow the high voltage to discharge from the capacitors in the inverter circuit.

Special Tools

CAT III DMM capable of reading 400 VDC
Insulating gloves
Insulating tape

Caution

Do not operate the power switch with the high-voltage service plug removed. Doing so may damage the HV ECU.

Figure 17-1 The high-voltage service plug is usually located at the HV battery.

Figure 17-2 To install the service plug, make sure the lever is down and then fully locked once installed.

To install the high-voltage service plug, first lock the lever in the DOWN position (**Figure 17-2**). Next, slide the plug into the receptacle, and lock it in place by lifting the lever upward. Once it's locked in place, it closes the reed switch.

> **SERVICE TIP** Disconnecting the batteries may erase the DTCs. Before disconnecting the system, be sure to check and record DTCs.

Self-Diagnosis Capabilities

Classroom Manual
Chapter 17, pages 538, 552

Special Tool
Scan tool

The **high-voltage electronic control unit (HV ECU)** controls the two motor generators (MG1 and MG2) and the ICE based on torque demand, regenerative brake control, and the HV battery's state of charge (SOC) (**Figure 17-3**). It's the responsibility of the HV ECU to provide a reliable circuit shutdown in the event of a malfunction. The HV ECU uses three relays housed in the SMR assembly to connect and disconnect the high-voltage circuit. Upon detecting a malfunction, the HV ECU uses the relays to control the system based on programmed instructions stored in its memory. If the system is malfunctioning, the HV ECU will illuminate the warning lamp in the instrument cluster. In addition, it may illuminate the HV system warning, the HV battery warning, or the discharge warning lamps. Diagnostic trouble codes (DTCs) set when the fault occurs.

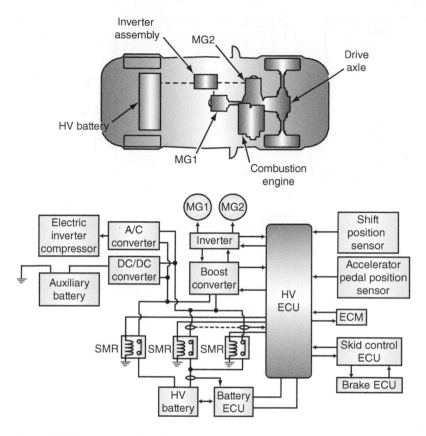

Figure 17-3 Components of an HEV system.

Information codes are additional codes that provide more information and freeze frame data concerning the DTCs. Information codes, along with the DTC, indicate more precisely the location of the fault. Access these codes using the scan tool while in the HV ECU system screen (**Figure 17-4**).

The **freeze frame data** records the driving condition when the malfunction occurred. This is useful for determining how the vehicle was operating at the time and for locating any input or output values that are out of range.

History data information can be useful for determining if a customer's concern is actually a problem with the system. It provides a means of determining if the vehicle owner's driving habits may cause the problem. The data will display information such as gear shift lever movement before the vehicle was ready, the transmission being shifted into park while the vehicle was moving, if the accelerator pedal was depressed while in the NEUTRAL position, and so on. Use the correct service information for details on using this data.

In addition, the scan tool may provide activation tests of the HV ECU. Different tests are available. Using the service information, determine the required test for the specific test routine.

The following are typical HEV/PHEV mode functions:

- Mode 1. This mode runs the ICE continuously in the PARK position to check HV ECU operation.
- Mode 2. Cancels the traction control.
- Inverter stop. Keeps the inverter power transistor on to determine if there is an internal leak in the inverter or the HV ECU.
- Cranking request. Activates the motor generator continuously to crank the ICE in order to measure the compression.

The DTCs associated with the high-voltage system may include both SAE and manufacturer codes. SAE codes must set as prescribed by the SAE, while manufacturers have the freedom to determine their own criteria for setting manufacturer codes.

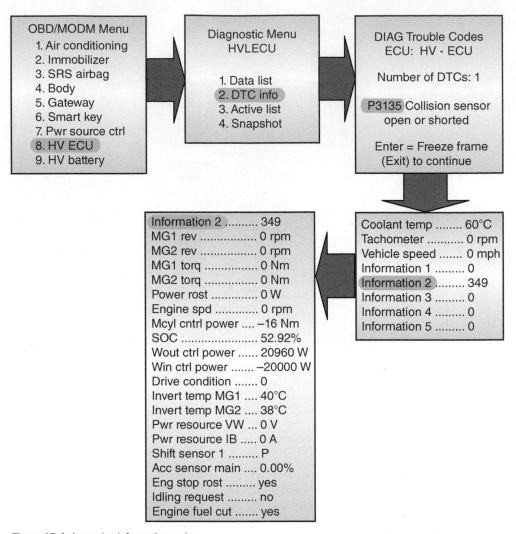

Figure 17-4 Accessing information codes.

High-Voltage Battery Service

Classroom Manual
Chapter 17, pages
537, 552

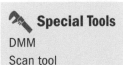

Special Tools

DMM
Scan tool

Often, the indications to a driver of an HEV or PHEV of an improperly charged HV battery (or that it is failing) is a reduction in fuel economy due to the ICE running more often, lack of acceleration, shortened EV mode, charge level indicators in the instrument panel dropping quickly. An ECU may set DTCs if the SOC drops too low or block voltages differ greater than the threshold.

The service life of an HV battery starts on its first day after production. This means that it does have an expiration date or a life span. Cell imbalance shortens the battery's life expectancy. Voltage depression (crystal formation) will inevitably occur within the cells; however, the process accelerates if the cells are imbalanced. Over time, cell capacity diminishes, and the cell's voltage will drift out of range. Drift greater than the threshold will set DTCs.

A **battery ECU** monitors the condition of the HV battery assembly (**Figure 17-5**). The battery ECU determines the SOC of the HV battery by monitoring voltage, current, and temperature. The battery ECU collects data and transmits it to the HV ECU, which determines proper charge and discharge control. The battery ECU also controls the operation of the battery blower motor to maintain proper HV battery temperature.

Classroom Manual
Chapter 17, pages
538, 552

Driving the vehicle subjects the HV battery to repetitive charge and discharge cycles. The electric motor(s) discharge the HV battery during acceleration mode and then recharges the HV battery by regenerative braking. An amperage sensor (**Figure 17-6**) provides charge and discharge data to the battery ECU, which transmits requests to the

Figure 17-6 The amperage sensor.

Figure 17-5 The battery ECU.

HV ECU to maintain the SOC of the HV battery. The battery ECU attempts to keep the SOS between 40% and 80%. The battery ECU also monitors delta SOC to determine if it can maintain acceptable levels of charge. The normal, low-to-high SOC delta is 20%.

If the battery ECU sends abnormal messages to the HV ECU, the HV ECU illuminates the warning light and enters fail-safe control. Fail-safe control can result in the battery ECU restricting or stopping the charging and discharging of the HV battery. DTCs and informational codes may set along with freeze frame data.

If there is a leak in the high-voltage system insulation, the system will enter fail-safe control and set DTCs. This will occur if the battery ECU determines the insulation resistance of the power cable to be 100 kΩ or less.

> ⚙ **SERVICE TIP** Some battery blocks may read 0 volts when the service plug is removed. Be sure to confirm proper voltages with the service information.

Whenever an HV battery malfunction occurs, use the scan tool to view the "HV Battery Data List." This provides all HV battery system information. Use the scan tool to view the HV battery's SOC and module voltage PIDs (**Figure 17-7**). Refer to the proper

Name	Value	Unit
State of Charge	45.9	%
Battery Voltage (cell sum)	303.5	Volts
Min Block Voltage	15.2	Volts
Max Block Voltage	15.2	Volts
Block 1 Voltage	15.2	Volts
Block 2 Voltage	15.2	Volts
Block 3 Voltage	15.2	Volts
Block 4 Voltage	15.2	Volts
Block 5 Voltage	15.2	Volts
Block 6 Voltage	15.2	Volts
Block 7 Voltage	15.2	Volts
Block 8 Voltage	15.2	Volts
Block 9 Voltage	15.2	Volts
Block 10 Voltage	15.2	Volts
Block 11 Voltage	15.2	Volts
Block 12 Voltage	15.2	Volts
Block 14 Voltage	15.2	Volts
Block 15 Voltage	15.2	Volts
Block 16 Voltage	15.2	Volts
Block 17 Voltage	15.2	Volts
Block 19 Voltage	15.2	Volts
Block 20 Voltage	15.2	Volts
High Voltage Sensor	303.0	Volts

Unable to read DTCs.

Figure 17-7 Scan tool reading of HV battery voltages.

service information to determine the target SOC. Typically the HV battery is never charged to 100%. The acceptable SOC also depends on the type of battery. Most manufacturers specify a SOC of between 40% and 80% range. However, Honda allows for a 20% to 80% range.

Compare each block voltage readings with all others. They should be within 0.3 volts of each other when the contactors are open. If any of the voltage readings are higher or lower than the others, that module or block may be damaged.

If the scan tool doesn't display the HV battery voltages, it will be necessary to test the battery with a DMM. This may require removing the HV battery pack from the vehicle and separating the battery's case.

Also, refer to the data PIDs for block internal resistance (**Figure 17-8**). HV batteries have low internal resistance, allowing very fast current flow in and out of the battery. Cell material is the main contributor that determines the battery's internal resistance. However, contaminated terminal connections, electrolyte leakage, or cell damage can cause an increase in resistance. The resistance of each block should be very close to each other.

High-Voltage Battery Charging And Balancing

If the SOC of the HEV's HV battery drops too low to allow the ICE to run, the HV battery will need recharging. On HEVs this will require using a special HV battery charger (**Figure 17-9**).

During the charging process, the HV battery will generate heat. During the battery charging and balancing procedures, the vehicle's battery ventilation system cools the battery. A special, vehicle-specific wiring harness is installed (**Figure 17-10**) so the fan operates during the charging and balancing procedures and prevents battery overheating.

⚠️ WARNING **Active ventilation of the battery requires using the manufacturer's designed airflow ductwork passages and fan since the airflow must pass between the modules and along the cell walls for proper cooling. Using a shop fan to blow air across the top of the battery is not sufficient. Using anything other than the vehicle's original battery fan and duct work may damage the battery.**

Classroom Manual
Chapter 17, pages
552, 556

Special Tools

Scan tool
High-voltage battery
    charger
Cooling fan harness

Refer to Chapter 5
concerning HV battery
service procedures.

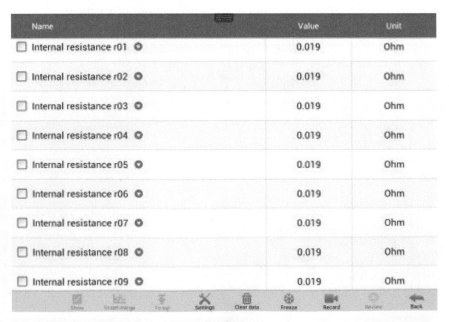

Name		Value	Unit
☐ Internal resistance r01 ⊙		0.019	Ohm
☐ Internal resistance r02 ⊙		0.019	Ohm
☐ Internal resistance r03 ⊙		0.019	Ohm
☐ Internal resistance r04 ⊙		0.019	Ohm
☐ Internal resistance r05 ⊙		0.019	Ohm
☐ Internal resistance r06 ⊙		0.019	Ohm
☐ Internal resistance r07 ⊙		0.019	Ohm
☐ Internal resistance r08 ⊙		0.019	Ohm
☐ Internal resistance r09 ⊙		0.019	Ohm

Figure 17-8 Scan tool readings of the block resistance.

Figure 17-9 ProlongPro® battery charger for HEV HV battery charging and reconditioning.

Figure 17-10 HV battery charger connection.

⚡ **WARNING** **Do not attempt to recharge the HV battery with a standard 12-volt battery charger. HV battery chargers are now available from aftermarket suppliers that will recharge most HV batteries.**

Turn the vehicle off while charging the HV battery. The first phase of charging is 'filling' the battery. The charging process will begin at the present SOC of the HV battery, and the charger's voltage display will slowly climb at a steady rate to a 100% charge level, then level off. This first charge fill can take between 3 and 16 hours (depending on the initial SOC).

CUSTOMER CARE It is recommended that battery balancing be repeated every 6 to 8 months to extend the battery's life and keep it operating efficiently.

Classroom Manual
Chapter 17, page 556

The balancing process starts when the initial fill phase completes. The charger may display a steady voltage reading or a very slow voltage oscillation between 1 and 3 volts for several minutes. The initial balancing phase can take between 4 and 12 hours. During this time, the SOC of each cell rises to a true 100%. When the peak voltage is reached, the cell warms. The energy conversion to heat causes the cell voltage to drop. Since the energy conversion stops, the cell cools and then charges to 100% SOC again. The cells with the higher voltage cells will undergo this thermal cycle several times as the cells with lower voltages continue to fill in an attempt to match them. Completion of the charging and balancing process is when there is no longer any voltage fluctuation (or the set time elapsed). The entire process of charging and balancing takes about 24 hours.

If additional battery services are not going to be performed, disconnect the charger, and wait 30 minutes before driving the vehicle. This is unnecessary if the HEV can reset the hybrid system so the new charge level of the battery pack can be recognized and used immediately. For example, the hybrid system may reset by disconnecting the auxiliary battery for 60seconds, reconnecting it, and then starting the vehicle to allow the ICE to ideal. It is normal for the vehicle to force charge the battery for two to ten minutes as it recalibrates and recognizes the new charge.

Before returning the vehicle to the customer, it is a good practice to test drive the vehicle to confirm the HV system is operating properly. While driving the vehicle, provide several opportunities for regenerative braking. Monitor the SOC of the HV battery to ensure that the SOC is increasing after each regenerative braking event.

Classroom Manual
Chapter 17, page 554

PHEVs and EVs use charging stations to charge their HV batteries. However, this is the same method the vehicle owner would charge the battery. If they exclusively use a Level 1 station to charge the HV battery, it's possible that the charge time is not long enough to replenish the battery. It's also possible that they may be attempting to charge the battery during cold weather while the vehicle is outside. If these are not an issue, use a Level 3 station to charge the HV battery. The Level 3 station sends DC voltage directly to the HV battery. The problem may be the onboard charger if the battery can be charged using the Level 3 station.

HV Battery Removal

Removal of the HV battery may be necessary to replace components contained in the battery case or to DMM test the battery modules. Since the removal process requires disconnecting the HV cables, follow all safety procedures. The terminals connecting the HV battery to the HV system via the orange wires are usually located under a cover and are close to the service disconnect plug (**Figure 17-11**). Remove the service disconnect plug and test for voltages before touching the cable terminals. **Photo Sequence 36** is a typical example of removing the HV battery from a vehicle. In this case, the HV battery is under the rear seat. When removed from the vehicle, place the HV battery onto a clean, insulated work surface.

Figure 17-11 HV battery cable connections.

PHOTO SEQUENCE 36
Removing the HV Battery

P36-1 Gain access to the service disconnect and unplug it from the HV battery. Wait five minutes before continuing to work.

P36-2 Remove the rear seat.

P36-3 Remove the terminal block cover assembly.

P36-4 Remove the retainers from the wire terminals connecting the HV wires to the battery pack.

P36-5 Remove any battery pack cooling ducts and vent tubes.

P36-6 Disconnect any battery pack electrical connectors.

P36-7 Remove the battery pack mounting retainers.

P36-8 Attach the battery lifting fixture to the battery pack.

P36-9 Remove the battery pack from the vehicle.

When installing the battery, it's important to adhere to all torque specifications. This includes the fasteners that attach the battery to the vehicle and the terminal fasteners. After installing the battery, perform the isolation check before powering up the high-voltage system.

DMM Testing of the HV Battery

⚡ **WARNING** **Even with the service disconnect removed, high voltage is still present in the battery pack. When performing any diagnostic or service procedure, use extreme care with tools and other test equipment around insulated terminals and wires. Failure to take proper precautions could result in serious or fatal injury.**

If a scan tool is unavailable, or cannot display module or block voltages of the HV battery, a DMM can be used. After removing the HV battery from the vehicle, take off its case cover. Remove any protective covers that are over the terminals of the modules. Use the DMM to measure the voltage for each module or block (**Figure 17-12**). A module or block with higher or lower voltages than the others is suspect.

Battery Refurbishing

NiMh Hybrid electric vehicle battery reconditioning consists of the combination of two principal parts. The first is charging and cell balancing the battery pack, as discussed. The second is deep discharging the battery pack to progressively lower voltage levels. Doing these two activities multiple times may recondition the battery. By cycling between the charging/balancing and discharging processes, the cells' memory effect (voltage depression) breaks down, and the cell voltage levels equalize. When completed, the battery will have an increased usable capacity.

If you have completed the battery charge and balance procedure, disconnect the charger, and connect the discharger (**Figure 17-13**) to the wire harness. The following is the process

Figure 17-12 Using a DMM to measure block voltages of the HV battery.

Figure 17-13 HV battery discharger used to balance the battery cells.

suggested by the ProlongPro Battery System® to perform one battery reconditioning treatment after completing the initial charge and balance:

1. Discharge the battery to at least 0.8V per cell average. For example, if the battery pack has 168 cells, the Discharge Termination Voltage selected would be 168 × 0.8 = 134 Volts.
2. Charge the battery using the 'Charge & balance' setting. Select a Cell Balancing Time of at least 1 hour.
3. Discharge the battery to at least 0.5V per cell average. This would be 84 volts in our previous example of a 168 cell battery.
4. Charge the battery using the 'Charge & balance' setting. Select a Cell Balancing Time of at least 1 hour.
5. Discharge the battery to at least 0.3V per cell average. This would be 50 volts in our previous example of a 168 cell battery.
6. Charge the battery using the 'Charge & balance' setting. Select a Cell Balancing Time of at least 4 hours.
7. Disconnect the charger and allow the battery to rest for at least one hour before attempting to start the vehicle.

Batteries that need to be removed from the car and reconditioned (rebuilt) should be allowed to sit for a couple of days. This will help identify high self-discharging modules. Upon disassembly, measure each module's voltage. Replace any modules that are 1 volt lower than the pack mean.

After reassembling the battery, install it into the vehicle and connect the special wire harness cable to the battery cooling fan. Charge and discharge the battery as described. End with a charge of at least 4 hours. After charging, disconnect the charger and allow the battery to rest for at least one hour before attempting to start the vehicle.

HV Battery Capacity Test

The capacity test evaluates the energy potential of each cell at a given voltage/current. The first step is pre-charge the battery to full. Connect the capacity tester to the wiring harness and start the test. During the capacity test, view the module progress noting the lowest and the highest module voltage (**Figure 17-14**). The higher the Ah of capacity, the better the condition of the cells within a module. The lower the Ah of capacity, the worse the condition of the cells within a module.

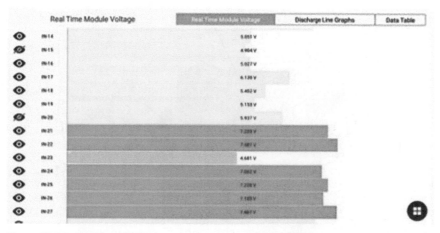

Figure 17-14 Bar graph showing cell voltages during the capacity test.

Inverter/Converter and Motor Control

The **inverter** (**Figure 17-15**) controls the current flow between the electric motor(s) and the HV battery. The inverter converts HV DC battery voltage into three-phase AC that is synchronized to the operating speed of the traction motor(s). During regenerative braking, the inverter converts the AC produced in the stator coils into DC for storage in the HV battery. The HV ECU controls the activation of the power transistors to perform these functions (**Figure 17-16**).

The inverter and motor control processors (MCPs) are slave modules to the HV ECU. Based on commands of the HV ECU, each MCP will control the speed, direction, and output torque of its respective motor. The MCPs control motor operation through the sequencing actuation of high-current switching transistors.

Use a scan tool for diagnostics of the inverter and MCPs. Since the inverter and MCPs are slave modules, it may not be possible to access these components directly. If not, all scan tool functions are performed through the HV ECU. The diagnostic tree for a recorded DTC generally leads the technician through a series of visual inspections, continuity tests, shorted circuit tests, and voltage tests.

The scan tool may be able to read the total direct current at the HV battery and the alternating current between each MCP and its respective motor. With this information, you can observe the electrical interaction of the HV battery, inverter, MCPs, and motors under different operating conditions.

The motors use resolver sensors to measure the rotor's position, speed, and direction. These sensors are necessary to control the operation of the motors. The resolver is an integrated component of each motor assembly and is not serviceable. Loss of any of the resolver signals results in the motors not operating.

As with most systems of the HEV, DTCs drive the diagnostic routine for the resolvers. The tests usually direct the technician toward resistance and voltage testing of the circuits. Measure the resistance of the resolvers across the pins for each resolver coil. Each coil should have very little resistance. If the resolver is damaged, the coil will likely fail open or short to the grounding shielding circuit.

The **converter** is a bidirectional DC/DC transformer (**Figure 17-17**). It converts the high-voltage DC to low-voltage DC to recharge the auxiliary battery and to power 12-volt electrical components. If the DC/DC converter malfunctions, the auxiliary battery voltage

Figure 17-15 The inverter assembly.

Figure 17-16 Inverter assembly internal electrical circuit.

Figure 17-17 The internal circuit of the DC/DC converter.

will drop until it's no longer possible to drive the vehicle. The HV ECU monitors the operation of the DC/DC converter, illuminates the warning lamp, and sets a DTC if a failure is determined. This means that most of the diagnostics will be based on the DTC retrieved and following the steps outlined by the manufacturer. Keep in mind you must confirm the condition of the HV cables, HV connections, and LV connections and circuits as in any system.

With the system properly powered down and confirmed, the inspection of the circuits can begin. This usually involves visually checking the condition of all connections in the system. Also, check that the torque of all terminal fasteners is at the proper specifications. Use an ohmmeter to test the condition of the conductors. Test for excessive resistance, shorts to ground, shorts to voltage, and circuits that are shorted together. Performing these procedures uses the same methods used on any electrical circuit. If the circuit does not have any faults, the problem is likely in the inverter/converter module. Often, these modules reside within the same housing as the HEV module. Confirm the ability to separate the modules or if they require replacement as a unit before attempting to remove any covers or fasteners.

The ultra-capacitors can also fail. In many cases these are serviceable items of the inverter/converter. Use the capacitor test function of the DMM to confirm their condition.

Classroom Manual
Chapter 17, page 538

Lab Scope Waveforms

You can use a lab scope to observe the voltage and amperage waveform of the inverter under different operating conditions. You want to determine if the waveforms are out of phase or if there is excessive variance in amplitudes.

Most lab scopes do not have a CAT III rating and require a differential (isolator) probe (**Figure 17-18**) to protect the lab scope and the user. To view voltage on the lab scope, connect the differential probe between the lab scope and the inverter terminals with the input of the isolator probe attached to two of the three-phase terminals at the inverter. Attach the output of the isolator probe to the lab scope. Select the desired attenuation ratio.

The voltage waveform is a simulated AC signal sent from the inverter to the three-phase motor. The voltage waveform can take different shapes depending on the desired torque output. The waveform starts as a rectangular waveform (**Figure 17-19**) and then switches to a stepped waveform (**Figure 17-20**).

The inverter uses PWM to control the amperage to the motor. Initially, the current increases in the windings as the voltage on-time increases. Reducing the on-time drops the amperage, simulating an AC waveform.

Special Tools

Lab scope
CAT III Differential probe
CAT III amp probes
Class '0' gloves (properly leak tested)

Figure 17-18 The differential (isolation) probe must be used to prevent a short to chassis ground through the lab scope.

Figure 17-19 This voltage waveform shows the PWM signal from the inverter to the motor and is a rectangular simulated AC signal. Since the attenuation ratio is set to 20:1, the voltage on the scale must be multiplied by 20.

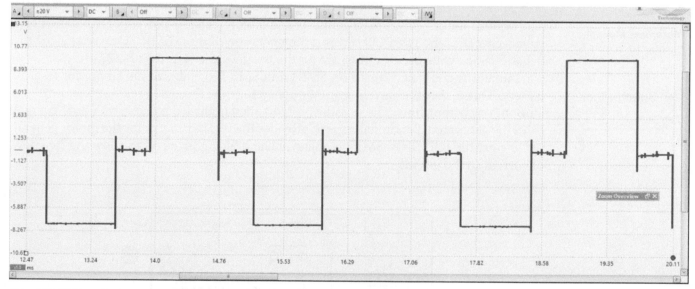

Figure 17-20 The voltage waveform changes to a stepped waveform.

Viewing the amperage waveforms may necessitate using CAT III amp probes attached around jumper wires. Disconnect the battery service plug and let the system power down. After confirming no voltage is in the HV system, disconnect the HV cables from the inverter. Attach the jumper wires between the inverter and the cable ends and place a CAT III amp probe around each of the jumper leads. Reinstall the service plug and power on the vehicle.

The amperage waveform will change in amplitude depending on torque demand (**Figure 17-21**). The higher the torque of the motor, the higher the amplitude. The frequency of the waveform depicts the speed of the motor.

The amplitude of all three phases should be even, and each waveform should have a 120° offset. Further testing will be necessary if one phase is consistently much lower in amplitude than the other two or if the waveforms are out of phase from one another. The problem may lie in the inverter, the harness, or the motor.

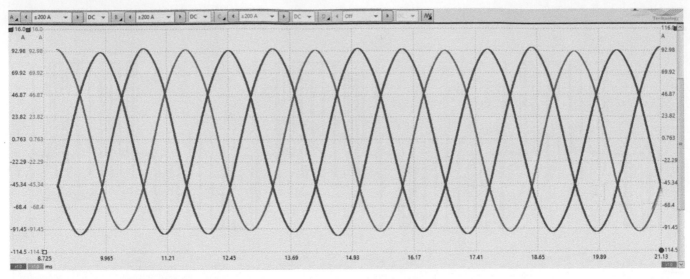

Figure 17-21 Inverter amperage waveform.

System Main Relay

Classroom Manual
Chapter 17, page 552

Special Tool

Scan tool

The **system main relay (SMR)** connects and disconnects the power source of the high-voltage system. The HV ECU controls the operation of three SMRs. One relay is for the negative side, while the other two are for the positive side (**Figure 17-22**).

Initially, the HV ECU turns on SMR1 and SMR3 to energize the circuit for the HV battery. This makes the current from the generator flow through a resistor to control the amount of current flow. The resistor protects the circuit from excessive initial current from the generator. This initial current provides the needed energy for the motor generator to start the ICE. Then SMR2 is energized, and SMR1 turns off to allow the free flow of current in the circuit.

Figure 17-22 Schematic of the SMR.

During shutdown, the HV ECU turns off SMR2 first, then SMR3. This provides a means for the HV ECU to verify that the relays have properly turned off. Detection of a fault will set a DTC and store information data about the failure. The following is a typical procedure if diagnostics lead to the replacement of an SMR:

1. Secure the proper service information, tested insulating gloves, and insulating tape.
2. Gain access to the auxiliary battery and disconnect the negative terminal.
3. Wearing the insulated gloves, remove the high-voltage service plug.
4. Use insulating tape to cover the terminals of the plug receptacle.
5. After gaining access to the battery carrier panel, remove the junction terminal. Be sure to wear your insulating gloves.
6. Remove the ground wire, SMR2 cover, and disconnect the relay connector to remove SMR2.
7. Remove SMR3 by disconnecting the main battery cable, disconnecting the connector, and removing the fasteners.
8. Disconnect the connector and the ground terminal for SMR1.
9. Remove the fasteners and SMR1.

Isolation Testing

The HV circuit is isolated from the vehicle's chassis ground. If there is a loss of isolation, a DTC sets, and the HV system may shut down and prevent a vehicle restart. The vehicle may still be drivable until it is shut off, then it may not restart if the problem is still present. In an HEV, the detection of an isolation fault sets DTC P0AA6. Some manufacturers will also set a location-specific code that identifies which section of the high voltage system the isolation breakdown occurred. If a location code is not available, use the isolation tester (**Figure 17-23**) to find the location or the portion of the circuit the fault lies. The insulation tester induces high voltage at very low amperage into the circuits of the HV system to get very accurate readings on all HV circuit isolation conditions (**Figure 17-24**).

Special Tools
Class '0' gloves (properly leak tested)
Insulation tester
CAT III DMM
Scan tool

The isolation test is also called the insulation test or the megaohm test.

SERVICE TIP When road testing the vehicle for a loss of isolation fault, be sure to turn on the air conditioning since an electric A/C compressor may be the source

Figure 17-23 Use the isolation tester to determine the location of the fault.

Figure 17-24 The meter on the left indicates good isolation. The meter on the right indicates a short to chassis ground.

After retrieving the DTCs and reviewing the service information, test and put on the isolation gloves, then remove the battery disconnect. Access the terminals where the service information has you test for voltage in the HV system. Use a CAT III DMM to measure the voltage. When the voltage drops to 0, you can begin testing the HV system.

Select the appropriate voltage range as detailed in the service information for the diagnostic procedure for the DTC. If the voltage range is too low, the meter may report the circuit as good when it is not. Selecting a too high of a voltage range can damage the inverter/converter or A/C compressor.

Place the red test lead of the isolation tester onto one of the HV cable terminals and the black lead to good chassis ground. Press the "Isolation Test" button and read the resistance and voltage values displayed by the meter. Compare the resistance reading to specifications. If the reading is lower than specifications, there is a short to ground. Any reading lower than specifications requires further diagnosing.

If there is no location code, you will need to isolate the HV circuits from each other. As you break down the circuits, test each one individually. For example, disconnect the HV cables from the inverter and test the cables again. If the test passes, test the terminals at the inverter. If both pass, continue testing the HV battery and the motor(s).

When testing HV cables that are isolated from any components, test for loss of isolation to chassis ground and loss of isolation between the cables. To test for loss of isolation between cables, connect the red test lead to one of the cables and the black to another cable (instead of chassis ground). Repeat the test for all possible combinations to check all of the cables.

⚡ WARNING **When testing the HV battery for isolation faults, wear Class 0 gloves.**

To test the battery, you will need to access the SMR or contactors. You may also need to disconnect the battery ECU. Connect the red test lead to the battery side of the SMR and the black lead to chassis ground. If the battery has been removed from the vehicle, connect the negative lead to the battery case. Test all SMRs or contactors. If the resistance is below specifications, inspect the battery cable terminals for corrosion (**Figure 17-25**). Also, look for traces of leakage from the battery cells onto the case (**Figure 17-26**).

Figure 17-25 Corrosion of the HV battery cable terminals results in loss of isolation faults.

Cell leakage

Figure 17-26 Cell leakage onto the battery case causes a low resistance path to chassis ground.

To test for loss of isolation within a motor, remove the terminal cover to gain access to the motor's cables. Disconnect the cables to isolate the motor. Although the legs are electrically connected, test each winding leg for a short to ground. Do not test for shorts between legs since this would not be any benefit.

Test the circuits to the A/C compressor in the same manner. A loss of isolation fault within the compressor can result from using the incorrect compressor oil. Changing the oil is not the fix since damage to the winding's insulation has already occurred.

Case Study

An owner of a 2015 Prius has brought their vehicle into the shop claiming that the warning light comes on every once in a while but goes out after a couple of starts. The technician retrieves the DTCs from the HV ECU and finds a "Shift before ready" code. Upon investigation of the information code, the technician also refers to history data. Here it's determined that the cause of the fault was that the customer was not waiting for the ready light to stop flashing before placing the transmission into drive. The warning light would go out after three starts if the same condition did not reoccur. The technician took the time to review the proper startup sequence with the vehicle owner to avoid future problems.

ASE-Style Review Questions

1. All of the following statements concerning high-voltage system safety are true **EXCEPT**:

   A. Disconnect the motor generators prior to turning the ignition off.

   B. You may need to disconnect the negative (−) terminal of the auxiliary battery before removing the service plug.

   C. Allow the capacitors the allotted time to discharge after removing the high-voltage service plug.

   D. Remove the service plug prior to performing a resistance check of any HV circuit.

2. *Technician A* says HEV batteries can provide over 270 volts.

   Technician B says the HEV high voltage from the MG1 and MG2 to the inverter/converter can be more than 500 volts.

   Who is correct?

   A. A only
   B. B only
   C. Both A and B
   D. Neither A nor B

3. When working on the high-voltage system, which of the following should be done?

   A. Always place the high-voltage service plug where someone will not accidentally reinstall it.

   B. Use a CAT III DMM set on 400 VDC to determine if the high-voltage system voltage is at 0 volts.

   C. Test the integrity of the insulating gloves prior to use.

   D. All of these choices.

4. Charging the HV battery of an HEV is being discussed.

   Technician A says a special harness is connected to the vehicle to operate the vehicle's battery ventilation system to cool the battery.

   Technician B says during the rebalancing phase, the cells with higher voltages thermal cycle several times as the cells with lower voltages continue to fill in an attempt to match them.

   Who is correct?

   A. A only
   B. B only
   C. Both A and B
   D. Neither A nor B

5. Which of the following can result in a loss of isolation fault?

   A. Electrolyte leakage from the cell to the battery case.

   B. Corrosion of the HV battery terminals.

   C. Insulation breakdown of a motor winding.

   D. All of these choices.

6. The isolation tester is being used to test the cables from the inverter to the HV battery. All statements are true **EXCEPT**:

   A. Disconnect the cables from the inverter/converter.

   B. Set the tester voltage to the value specified in the service information.

   C. A meter reading of 0.0 ohms indicates no loss of isolation.

   D. Test for loss of isolation between the cables.

7. *Technician A* says once the service plug is disconnected, there is no high voltage in the vehicle systems.

   Technician B says turn the vehicle off before disconnecting the high-voltage service plug.

   Who is correct?

   A. A only   C. Both A and B

   B. B only   D. Neither A nor B

8. Scan tool data PIDs concerning HV battery cell blocks are being discussed.

   Technician A says the block voltages should be within 3.0 volts of each other.

   Technician B says cells with high resistance data PIDs may indicate electrolyte leakage.

   Who is correct?

   A. A only   C. Both A and B

   B. B only   D. Neither A nor B

9. *Technician A* says the HV battery can be charged with a conventional flooded battery charger set at 3.5 amps.

   Technician B says the HV battery can only be charged if it's removed from the vehicle.

   Who is correct?

   A. A only   C. Both A and B

   B. B only   D. Neither A nor B

10. *Technician A* says information codes can provide more specific indications of the fault location.

    Technician B says the freeze frame data is a recording of the driving condition when the malfunction occurred.

    Who is correct?

    A. A only   C. Both A and B

    B. B only   D. Neither A nor B

ASE Challenge Questions

1. *Technician A* says if an intermittent fault code is being set in the high-voltage system, water may be entering a connector.

   Technician B says to locate the cause of current leakage through the high-voltage cable insulation, spray the cable with water.

   Who is correct?

   A. A only   C. Both A and B

   B. B only   D. Neither A nor B

2. *Technician A* says if the vehicle fails to start, the system main relay may have a fault.

   Technician B says if the vehicle fails to start, the HV battery SOC may be too low.

   Who is correct?

   A. A only   C. Both A and B

   B. B only   D. Neither A nor B

3. *Technician A* says a faulty reed switch in the service plug may cause the vehicle not to start.

   Technician B says a water cooler pump failure may result in an "HV battery too hot" fault code.

   Who is correct?

   A. A only   C. Both A and B

   B. B only   D. Neither A nor B

4. *Technician A* says the auxiliary battery starts the ICE if the HV battery SOC is below 15%.

   Technician B says the inverter may have malfunctioned if the ICE fails to start.

   Who is correct?

   A. A only

   B. B only

   C. Both A and B

   D. Neither A nor B

5. *Technician A* says if the vehicle operates slowly from a stop, the ICE may require a tune-up.

   Technician B says the HEV system doesn't provide a method of performing a compression test on the ICE.

   Who is correct?

   A. A only

   B. B only

   C. Both A and B

   D. Neither A nor B

Name _____   Date _____

HYBRID SAFETY

Upon completion of this job sheet, you should be familiar with the critical safety procedures involved in servicing a high-voltage hybrid system.

ASE Education Foundation Correlation ——————————————————————

This job sheet addresses the following foundational task:

Shop and Personal Safety

14. Identify vehicle systems which pose a safety hazard during service due to high voltage such as: hybrid/electric drivetrain, lighting systems, ignition systems, A/C systems, injection systems, etc.

This job sheet addresses the following **MLR/AST/MAST** task:

A.1. Research vehicle service information such as fluid type, vehicle service history, service precautions, technical service bulletins, and recalls including vehicles with advanced driver assistance systems (ADAS). **(P-1)**

We Support
ASE | **Education Foundation**

Tools and Materials

- HEV
- Service information

Safety Considerations and PPE

- Insulating gloves
- Eye protection

WARNING: You will be working around high voltage while doing this job sheet. Follow all warnings and cautions in the manufacturer's service procedures. Even though the service disconnect is removed, the HV battery is still "live."

Procedure

_____

_____

1. What must be done prior to disconnecting the service plug?

_____

_____

2. How long must you wait after the plug is disconnected before servicing the system?

_____

_____

3. Access the 12-volt auxiliary battery and remove the negative terminal.

4. Test the insulating gloves for leaks. Are the gloves safe to use?   ☐ Yes   ☐ No

   If No, inform your instructor.

5. Put on the insulating gloves and eye protection.

6. Remove the service plug. What device(s) are integrated into the service plug assembly?

_____

_____

7. Reinstall the service plug.

8. Review your observations with your instructor.

Instructor's Evaluation:

Grade	Rubric
0	No exposure due to: ☐ Absence ☐ Other:
1	Student unsuccessfully attempted the task
2	Student completed the task with significant instructor assistance.
3	Satisfactory performance of task. The student was able to perform elements of the task with minimal instructor assistance.
4	Competent/Proficient. Able to perform all elements of the task successfully and independently.

Instructor's Notes:

Name _____   Date _____

HYBRID SYSTEM DTCS

Upon completion of this job sheet, you should be able to diagnose hybrid faults by retrieving DTCs, information codes, and data values.

ASE Education Foundation Correlation ─────────────────────

This job sheet addresses the following **MLR** tasks:

A.1.   Research vehicle service information such as fluid type, vehicle service history, service precautions, technical service bulletins, and recalls including vehicles with advanced driver assistance systems (ADAS). **(P-1)**

A.3.   Retrieve and record DTCs, OBD monitor status, and freeze frame data; clear codes and data when directed. **(P-1)**

This job sheet addresses the following **AST** tasks:

A.1.   Research vehicle service information such as fluid type, vehicle service history, service precautions, technical service bulletins, and recalls including vehicles with advanced driver assistance systems (ADAS). **(P-1)**

A.3.   Retrieve and record DTCs, OBD monitor status, and freeze frame data; clear codes and data when directed. **(P-1)**

G.8.   Diagnose body electronic system circuits using a scan tool; check for module communication errors (data communication bus systems); determine needed action. **(P-2)**

This job sheet addresses the following **MAST** tasks:

A.1.   Research vehicle service information such as fluid type, vehicle service history, service precautions, technical service bulletins, and recalls including vehicles with advanced driver assistance systems (ADAS). **(P-1)**

A.3.   Retrieve and record DTCs, OBD monitor status, and freeze frame data; clear codes and data when directed. **(P-1)**

G.8.   Diagnose body electronic system circuits using a scan tool; check for module communication errors (data communication bus systems); determine needed action. **(P-1)**

We Support
ASE | Education Foundation

Tools and Materials
- HEV
- Scan tool
- Service information

Procedure

1. With the vehicle started, observe if any warning lamps are illuminated. If so, which ones?

   _____

   _____

2. Connect the scan tool and check for DTCs in the PCM and the HV ECU. Record any DTCs displayed.

   _____

   _____

3. Access and record any information codes that are displayed.

   _____

   _____

4. Use the service information to determine what system component is affected. Record your results.

_____

_____

5. Access the "HV ECU Data List" screen and record any values that are out of range.

_____

_____

6. Based on your results, what would be the next logical approach to take?

_____

_____

Instructor's Evaluation:

Grade	Rubric
0	No exposure due to: ☐ Absence ☐ Other:
1	Student unsuccessfully attempted the task
2	Student completed the task with significant instructor assistance.
3	Satisfactory performance of task. The student was able to perform elements of the task with minimal instructor assistance.
4	Competent/Proficient. Able to perform all elements of the task successfully and independently.

Instructor's Notes:

Name _____   Date _____

HV BATTERY REMOVAL AND SERVICE

Upon completion of this job sheet, you should be able to diagnose the condition of the HV battery.

ASE Education Foundation Correlation ———————————————

This job sheet addresses the following foundational and Workplace task:

Shop and Personal Safety

14.   Identify vehicle systems which pose a safety hazard during service due to high voltage such as: hybrid/electric drivetrain, lighting systems, ignition systems, A/C systems, injection systems, etc.

This job sheet addresses the following **MLR** tasks:

A.1.   Research vehicle service information such as fluid type, vehicle service history, service precautions, technical service bulletins, and recalls including vehicles with advanced driver assistance systems (ADAS). **(P-1)**

A.5.   Demonstrate proper use of a digital multimeter (DMM) when measuring source voltage, voltage drop (including grounds), current flow, and resistance. **(P-1)**

This job sheet addresses the following **AST** tasks:

A.1.   Research vehicle service information such as fluid type, vehicle service history, service precautions, technical service bulletins, and recalls including vehicles with advanced driver assistance systems (ADAS). **(P-1)**

A.5.   Demonstrate proper use of a digital multimeter (DMM) when measuring source voltage, voltage drop (including grounds), current flow, and resistance. **(P-1)**

G.8.   Diagnose body electronic system circuits using a scan tool; check for module communication errors (data communication bus systems); determine needed action. **(P-2)**

This job sheet addresses the following **MAST** tasks:

A.1.   Research vehicle service information such as fluid type, vehicle service history, service precautions, technical service bulletins, and recalls including vehicles with advanced driver assistance systems (ADAS). **(P-1)**

A.5.   Demonstrate proper use of a digital multimeter (DMM) when measuring source voltage, voltage drop (including grounds), current flow, and resistance. **(P-1)**

G.8.   Diagnose body electronic system circuits using a scan tool; check for module communication errors (data communication bus systems); determine needed action. **(P-1)**

We Support
ASE | Education Foundation

Tools and Materials

- CAT III DMM
- Scan tool
- Battery lifting fixture
- Service information

Safety Considerations and PPE

- Class "0" gloves
- Safety glasses

WARNING You will be working around high voltage while doing this job sheet. Follow all warnings and cautions in the manufacturer's service procedures. Even though the service disconnect is removed, the HV battery is still "live."

Task 1

HV Battery Removal

Procedure

1. Using the correct service information, locate the HV battery removal and installation procedure.

2. What essential tool(s) are required when removing the HV battery?

   _____

   _____

   _____

3. Secure a CAT III DMM and verify the operation of the DMM. How should you verify operation of the DMM?

   _____

   _____

   _____

4. Why is it important to verify DMM operation prior to use?

   _____

   _____

   _____

5. Test the class "0" gloves for leaks before continuing. Pass _____   Fail _____

6. Remove the service disconnect plug and wait five minutes.

7. Remove any trim or seats required to access the HV battery.

8. Gain access to the cable connection terminals. While wearing the gloves, use the CAT III DMM to measure the voltage at the high-voltage cable terminal connections to the battery. Record your results.

   _____

   _____

   _____

9. If safe, disconnect the HV cables and any other connectors attached to the HV battery.

10. Attach the lifting fixture (if needed), and remove the battery. Place the battery on a shop table that has electrically insulated padding on it.

Task 2

Identify High-Voltage Points in the HV Battery Case

WARNING You will be working around high voltage while doing this job sheet. Follow all warnings and cautions in the manufacturer's service procedures. Even though the service disconnect is removed, the HV battery is still "live."

Procedure

1. Secure a CAT III DMM and verify the operation of the DMM.

   Pass _____   Fail _____

2. While using the class "0" gloves, test for HV at the cable connections and record your readings.

_____

3. Remove the cover from the battery case.

4. Visually inspect the condition of the HV battery modules and interior components. Record any concerns that you have seen.

_____

_____

_____

5. Using the service information, identify and test each test point listed below with the DMM to determine where HV is present within the HV battery pack. Use the class "0" gloves and the one-hand rule to test for HV.

(+) to (−) contactor (system side) _____

(+) to (−) contactor (battery side) _____

(+) to (−) HV battery post _____

(+) HV battery post to (−) at service disconnect split _____

(−) HV battery post to (+) at service disconnect split _____

(+) to (−) at service disconnect split _____

6. Measure the observed voltage for all modules or blocks. Record your results in the table below.

Module	Voltage	Module	Voltage
1		11	
2		12	
3		13	
4		14	
5		15	
6		16	
7		17	
8		18	
9		19	
10		20	

7. Are any of the readings different than the others?   ☐ Yes ☐ No

If yes, why?

_____

_____

_____

8. What are your conclusions about the SOC of the HV battery?

_____

_____

_____

Instructor's Evaluation:

Grade	Rubric
0	No exposure due to: ☐ Absence ☐ Other:
1	Student unsuccessfully attempted the task
2	Student completed the task with significant instructor assistance.
3	Satisfactory performance of task. The student was able to perform elements of the task with minimal instructor assistance.
4	Competent/Proficient. Able to perform all elements of the task successfully and independently.

Instructor's Notes:

Name _____   Date _____

INVERTER AND MOTOR MODULE OPERATION

Upon completion of this job sheet, you should be able to observe the operation of the inverter and motor modules through the use of the scan tool data functions.

ASE Education Foundation Correlation

This job sheet addresses the following **MLR** task:

A.1.   Research vehicle service information such as fluid type, vehicle service history, service precautions, technical service bulletins, and recalls including vehicles with advanced driver assistance systems (ADAS). **(P-1)**

This job sheet addresses the following **AST** tasks:

A.1.   Research vehicle service information such as fluid type, vehicle service history, service precautions, technical service bulletins, and recalls including vehicles with advanced driver assistance systems (ADAS). **(P-1)**

G.8.   Diagnose body electronic system circuits using a scan tool; check for module communication errors (data communication bus systems); determine needed action. **(P-2)**

This job sheet addresses the following **MAST** tasks:

A.1.   Research vehicle service information such as fluid type, vehicle service history, service precautions, technical service bulletins, and recalls including vehicles with advanced driver assistance systems (ADAS). **(P-1)**

G.8.   Diagnose body electronic system circuits using a scan tool; check for module communication errors (data communication bus systems); determine needed action. **(P-1)**

We Support
ASE | Education Foundation

Tools and Materials

- Scan tool
- HEV
- Service information

Procedure

1. Locate the inverter and the motor control modules on the vehicle, and describe their location(s).

   _____

   _____

   _____

   _____

2. Are these modules contained within a single housing, or are they separate?

   _____

   Where is the HV ECU located?

3. How are the modules grounded?

   _____

4. Identify the low-voltage connections to the inverter module. How many pins are on each connector?

   _____

   _____

   _____

5. Identify the high-voltage DC connections to the inverter module. How many cables are in each set?

_____

6. Identify the high-voltage AC connections to the inverter module. How many cables are in each set?

_____

7. Connect the scan tool to the DLC; can the HV ECU and motor control modules be accessed by the scan tool?   ☐ Yes   ☐ No

   If no, talk to your instructor to get further instructions.

8. Look at the data points for the first motor. What data are provided that would indicate how the motor is being used while the vehicle is being driven?

_____

_____

_____

_____

9. How is the speed of the motors measured?

_____

_____

_____

_____

10. Run the e-motors with the scan tool. What does this test tell you about motor operation?

_____

_____

_____

_____

Instructor's Evaluation:

Grade	Rubric
0	No exposure due to: ☐ Absence ☐ Other:
1	Student unsuccessfully attempted the task
2	Student completed the task with significant instructor assistance.
3	Satisfactory performance of task. The student was able to perform elements of the task with minimal instructor assistance.
4	Competent/Proficient. Able to perform all elements of the task successfully and independently.

Instructor's Notes:

Name _____   Date _____

INVERTER TESTING PROCEDURES

Upon completion of this job sheet, you should be able to use a lab scope to observe the operation of the inverter and motor modules.

ASE Education Foundation Correlation ———————————————————————

This job sheet addresses the following **MLR** task:

A.1.   Research vehicle service information such as fluid type, vehicle service history, service precautions, technical service bulletins, and recalls including vehicles with advanced driver assistance systems (ADAS). **(P-1)**

This job sheet addresses the following **AST** tasks:

A.1.   Research vehicle service information such as fluid type, vehicle service history, service precautions, technical service bulletins, and recalls including vehicles with advanced driver assistance systems (ADAS). **(P-1)**

A.13.   Test and measure circuit using an oscilloscope and/or graphing multimeter (GMM); interpret results; determine needed action. **(P-2)**

G.8.   Diagnose body electronic system circuits using a scan tool; check for module communication errors (data communication bus systems); determine needed action. **(P-2)**

This job sheet addresses the following **MAST** tasks:

A.1.   Research vehicle service information such as fluid type, vehicle service history, service precautions, technical service bulletins, and recalls including vehicles with advanced driver assistance systems (ADAS). **(P-1)**

A.13.   Test and measure circuit using an oscilloscope and/or graphing multimeter (GMM); interpret results; determine needed action. **(P-1)**

G.8.   Diagnose body electronic system circuits using a scan tool; check for module communication errors (data communication bus systems); determine needed action. **(P-1)**

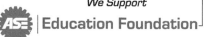

We Support
|Education Foundation

Tools and Materials

- Lab scope
- CAT III amp probes (3)
- Differential probe
- Set of 3 jumper wires
- HEV, PHEV, or EV
- Service information

Safety Considerations and PPE

- Safety glasses
- Class 0 insulation gloves (properly tested)

WARNING You will be working around high voltage while doing this job sheet. Follow all warnings and cautions in the manufacturer's service procedures. Even though the service disconnect is removed, the HV battery is still "live."

WARNING Do not use a lab scope without the proper differential (isolation) probe and the CAT III amp probes.

Obtaining Voltage Waveform

Procedure

1. Place the vehicle on a hoist so you can simulate driving the vehicle. May need to disable traction control.

2. Remove the service plug and ensure voltage is drained before making any connections. Confirmed _____

3. Connect the differential probe to two of the three motor phase cables.

4. Connect other end of differential probe to lab scope.

5. Put on the gloves and reinstall the service plug.

6. Select the proper attenuator ratio._____

7. Select the proper scale and time frame. Scale _____ Time _____

8. Obtain the voltage waveform while simulating a drive.

9. Record the highest voltage amplitude: _____ volts

10. Record the lowest voltage amplitude: _____ volts

11. Is the waveform a sine _____ rectangular _____ stepped _____

12. Does the voltage waveform appear normal? Yes_____ No _____

    If no, why not?

    _____

    _____

    _____

Amperage Test

1. With the service plug removed and confirmation that the HV is drained, disconnect the cables to the motor. This can be done at the inverter or at the motor.

2. Install three pre-made jumper wires between the cables and the terminals of the inverter or motor.

3. Install a CAT III amp probe around each of the jumper wires.

4. Put on the gloves and reinstall the service plug.

5. Obtain the amperage waveforms while simulating a drive.

6. Record the highest amperage amplitude: _____ amps

7. Record the lowest amperage amplitude: _____ amps

8. Does the amperage waveform appear normal? Yes_____ No _____

    If no, why not?

    _____

    _____

    _____

Instructor's Evaluation:

Grade	Rubric
0	No exposure due to: ☐ Absence ☐ Other:
1	Student unsuccessfully attempted the task
2	Student completed the task with significant instructor assistance.
3	Satisfactory performance of task. The student was able to perform elements of the task with minimal instructor assistance.
4	Competent/Proficient. Able to perform all elements of the task successfully and independently.

Instructor's Notes:

DIAGNOSTIC CHART 17-1

PROBLEM AREA:	High-voltage system.
SYMPTOMS:	ICE will not start. Warning lamp illumination.
POSSIBLE CAUSES:	**1.** Auxiliary battery low. **2.** HV battery low SOC. **3.** Faulty motor generator. **4.** Faulty HV ECU. **5.** Poor electrical connections in HV cable. **6.** Service plug not installed.

DIAGNOSTIC CHART 17-2

PROBLEM AREA:	HV battery.
SYMPTOMS:	Low SOC. Warning lamp illumination.
POSSIBLE CAUSES:	**1.** Poor electrical connections. **2.** Faulty motor generator or circuits. **3.** Faulty HV ECU or circuits. **4.** Faulty battery ECU or circuits. **5.** Faulty amperage sensor or circuits. **6.** Faulty HV battery.

DIAGNOSTIC CHART 17-3

PROBLEM AREA:	HEV inverter/converter over current
SYMPTOMS:	HV battery and LV battery fails to charge No HEV operation MIL or HEV warning lamps
POSSIBLE CAUSES:	**1.** 12-volt battery failure **2.** Low SOC of 12-volt battery **3.** High resistance in 12-volt battery cables or connections **4.** 12-volt battery cable shorted to ground **5.** Improper inverter/converter module ground connection **6.** Poor or improper inverter/converter module LV connections **7.** Poor or improper HV cable connections **8.** Excessive LV system loads **9.** Faulty inverter/converter module

DIAGNOSTIC CHART 17-4

PROBLEM AREA:	HEV 12-volt battery low SOC
SYMPTOMS:	LV battery fails to charge No HEV operation MIL or HEV warning lamps
POSSIBLE CAUSES:	**1.** 12-volt battery failure **2.** Low SOC of 12-volt battery **3.** High resistance in 12-volt battery cables or connections **4.** Improper inverter/converter module ground connection **5.** Poor or improper inverter/converter module LV connections **6.** Excessive LV system loads **7.** Faulty inverter/converter module

DIAGNOSTIC CHART 17-5

PROBLEM AREA:	HEV 12-volt battery high SOC
SYMPTOMS:	LV battery overcharged No HEV operation MIL or HEV warning lamps
POSSIBLE CAUSES:	**1.** 12-volt battery failure **2.** Improper jump-start procedures **3.** Overcharging with a battery charger **4.** High resistance in 12-volt battery cables or connections **5.** Improper inverter/converter module ground connection **6.** Poor or improper inverter/converter module LV connections **7.** Faulty inverter/converter module

DIAGNOSTIC CHART 17-6

PROBLEM AREA:	HEV drive motor inverter performance
SYMPTOMS:	No HEV operation MIL or HEV warning lamps
POSSIBLE CAUSES:	**1.** High resistance in the three-phase cable assembly **2.** Short in the three-phase cable assembly **3.** Improper terminal connections at the three-phase cable assembly **4.** Faulty inverter/converter module **5.** Faulty electric motor

1. The current draw of a window motor is being measured.

   Technician A says connect the ammeter onto the power supply side of the motor.

   Technician B says connect the ammeter onto the ground side of the motor.

   Who is correct?

   A. A only   C. Both A and B

   B. B only   D. Neither A nor B

2. The digital readout of an auto-ranging DVOM set to the "volts" position displays "13.7." Which of the following statements is true about this measurement?

   A. The actual value is 13.7 volts.

   B. The actual value is 13.7 mV.

   C. The actual value is 137 mV.

   D. More information is needed to determine the actual value.

3. A voltmeter that is connected across the input and output terminals of an instrument cluster illumination lamp rheostat indicates 12.6 volts, with the switch in the minimum brightness position and the engine off. Which of the following statements is true?

   A The voltage available at the lamps will be about 12.6 volts.

   B. The voltage available at the lamps will be 0.0 volt.

   C. The rheostat is operating correctly.

   D. More information is needed in order to determine whether the lamps are operating correctly.

4. A voltmeter indicates 0.45 volt. Which of the following represents this measurement?

   A. 0.450 mV.   C. 4.5 mV.

   B. 450 mV.   D. 0.045 volt.

5. The following information about a solenoid control signal has been gathered using a lab scope: Frequency is 10 hertz and pulse width is 5 ms.

   Technician A says this means that the solenoid is being turned on and off 10 times per second and that the length of time the injector is open during each "on" pulse is 5 ms.

   Technician B says this means that the solenoid is being turned on and off 10 times per second and that the length of time the solenoid is closed during each "off" cycle is 95 ms.

   Who is correct?

   A. A only   C. Both A and B

   B. B only   D. Neither A nor B

6. Lab scope testing is being discussed.

   Technician A says the preferred method to use when observing the output of a potentiometer is to select an external trigger.

   Technician B says select a high trigger level when analyzing the output of a low-voltage computer input sensor.

   Who is correct?

   A. A only   C. Both A and B

   B. B only   D. Neither A nor B

7. The left-rear and right-rear taillights and the left-rear brake light of a vehicle turn on dimly whenever the brake pedal is depressed; however, the right-rear brake light operates at the correct brightness.

   Technician A says the left-rear taillight and brake light may have a poor ground connection.

   Technician B says the brake light switch may have excessive resistance.

   Who is correct?

   A. A only   C. Both A and B

   B. B only   D. Neither A nor B

8. The horn of a vehicle equipped with a horn relay sounds weak and distorted whenever it is applied. Which of the following is the **LEAST LIKELY** cause of this problem?

   A. High resistance in the relay load circuit.

   B. High resistance in the horn ground circuit.

   C. Excessive voltage drop between the relay load contact and the horn.

   D. A 12-volt drop across the relay coil winding.

9. The circuit breaker that protects an electric window circuit opens whenever attempting to lower the window.

   Technician A says the internal resistance of the motor is too high.

   Technician B says the window regulator may be sticking.

   Who is correct?

   A. A only   C. Both A and B
   B. B only   D. Neither A nor B

10. The relay coil resistance of an inoperative electric antenna circuit is 0.5 kΩ.

    Technician A says this could prevent the proper operation of the antenna motor.

    Technician B says this could result in the antenna rising but not retracting.

    Who is correct?

    A. A only   C. Both A and B
    B. B only   D. Neither A nor B

    NOTE: Question 11 refers to the horn circuit on page 656.

11. The horn circuit wiring diagram is being discussed.

    Technician A says the wire that provides the ground path for the high-note horn is an 18-gauge black wire that is part of circuit Z1.

    Technician B says the wire that provides power to the horn relay coil is a violet wire connected to terminal 87 of the relay.

    Who is correct?

    A. A only   C. Both A and B
    B. B only   D. Neither A nor B

    NOTE: Question 12 refers to the rear wiper/washer on page 657.

12. The wiring diagram is being discussed.

    Technician A says the wire that activates the rear wiper motor module originates at terminal G of the rear window/wiper switch.

    Technician B says only three wires are connected to this vehicle's accessory switch panel.

    Who is correct?

    A. A only   C. Both A and B
    B. B only   D. Neither A nor B

13. Domestic and import wire sizes are being discussed.

    Technician A says a 14-gauge wire is larger than a 16-gauge wire.

Technician B says a 0.8 mm^2 wire is smaller than a 1.0 mm^2 wire.

Who is correct?

A. A only   C. Both A and B
B. B only   D. Neither A nor B

14. What is the expected total current flow through a 12-volt circuit with two 12-ohm resistors connected in parallel?

    A. 0.5 amp   C. 2 amps
    B. 1 amp   D. 32 amps

15. A hydrometer that is being used to measure the specific gravity of a battery indicates 1.240.

    Technician A says if the ambient temperature is 70°F, the corrected specific gravity reading will be 1.236.

    Technician B says if the battery temperature is 60°F, the corrected specific gravity reading will be 1.232.

    Who is correct?

    A. A only   C. Both A and B
    B. B only   D. Neither A nor B

16. The specific gravity of each of the cells of a battery is as follows: 1.220, 1.245, 1.190, 1.205, 1.210, and 1.215. Which of the following procedures should be performed?

    A. Charge the battery and then retest the specific gravity.

    B. Perform a battery capacity test.

    C. Perform a 3-minute charge test.

    D. Replace the battery.

17. A battery has an open circuit voltage of 12.1 volts.

    Technician A says a battery capacity test should now be performed.

    Technician B says the battery should now be charged.

    Who is correct?

    A. A only   C. Both A and B
    B. B only   D. Neither A nor B

18. Battery test series is being discussed.

    Technician A says to perform the battery capacity test first.

    Technician B says the state of charge test is always the last test performed.

    Who is correct?

    A. A only   C. Both A and B
    B. B only   D. Neither A nor B

19. All of the following could cause a slow cranking condition **EXCEPT**:

    A. Overadvanced ignition timing.

    B. Shorted neutral safety switch.

    C. Misaligned starter mounting.

    D. Low-battery state of charge.

20. A vehicle owner states that occasionally they are unable to start their car: The engine starts cranking at normal speed and then all of a sudden they hears a "whee" sound and at that point the engine stops cranking. After four or five attempts, the engine finally starts.

    Technician A says the starter drive gear may be slipping.

    Technician B says there may be excessive voltage drop across the starter-solenoid contacts.

    Who is correct?

    A. A only   C. Both A and B

    B. B only   D. Neither A nor B

21. A vehicle with a no-crank condition is being tested. When the starter-solenoid battery and start terminals are connected with a jumper wire, the starter begins to crank the engine.

    Technician A says the starter solenoid may be faulty.

    Technician B says the ignition switch may be faulty.

    Who is correct?

    A. A only   C. Both A and B

    B. B only   D. Neither A nor B

22. A vehicle is being tested for a slow-crank condition. One lead of a voltmeter is connected to the positive battery post and the other lead is connected to the motor terminal of the starter-solenoid. With the ignition key on, the voltmeter indicates 12 volts; when the engine is cranked, the voltmeter indicates 0.2 volt.

    Technician A says the positive side of the starter-solenoid load circuit is OK.

    Technician B says the negative side of the starter circuit may have excessive voltage drop.

    Who is correct?

    A. A only   C. Both A and B

    B. B only   D. Neither A nor B

23. A technician performs an output test on an AC generator (alternator) rated at 100 amps with an internal regulator. During the test, it produces 30 amps. The cause of the low output may be:

    A. Shorted diode.

    B. Open sense circuit to the regulator.

    C. Worn brushes.

    D. Faulty capacitors.

24. A charging system voltage output test reveals the following information:

    1. Base voltage: 12.6 volts.
    2. At 1,500 rpm: 13.3 volts.
    3. At 2,000 rpm with loads on: 13.0 volts.

    Technician A says the alternator output voltage is lower than normal.

    Technician B says there may be excessive resistance on the insulated side of the charging circuit.

    Who is correct?

    A. A only   C. Both A and B

    B. B only   D. Neither A nor B

25. Charging system voltage and amperage on a vehicle equipped with an external voltage regulator is 12.1 volts and 0 amp at 1,500 rpm with all loads applied. When the alternator is full-fielded under the same conditions, the values rise to 13.8 volts and 95 amps. All of the following statements concerning these tests are correct **EXCEPT**:

    A. The alternator may have an open field circuit.

    B. The voltage regulator's voltage limiter contacts may have excessive resistance.

    C. The voltage limiter may have an open coil winding.

    D. The alternator is capable of charging correctly.

26. Alternator diode testing is being discussed.

    Technician A says a voltmeter set to AC voltage can be used to check for faulty diodes.

    Technician B says an ammeter can be used to check for faulty diodes.

    Who is correct?

    A. A only   C. Both A and B

    B. B only   D. Neither A nor B

27. *Technician A* says that a short to ground in the PCI bus network circuit will result in a total bus network failure.

    Technician B says an open PCI bus circuit wire to the BCM will result in total bus network failure.

Who is correct?

A. A only   C. Both A and B

B. B only   D. Neither A nor B

28. *Technician A* says that the normal at-rest voltage for the PCI bus is 0 volt.

Technician B says that the normal active voltage on the PCI bus is 12 volts.

Who is correct?

A. A only   C. Both A and B

B. B only   D. Neither A nor B

29. *Technician A* says that the location of an open CAN B bus circuit can be found by shorting the other circuit to ground while observing the module communications on the scan tool.

Technician B says if the fault code indicated that a CAN C circuit is shorted, this would cause a total failure of the CAN C bus.

Who is correct?

A. A only   C. Both A and B

B. B only   D. Neither A nor B

30. *Technician A* says the normal voltmeter reading on the CAN B(+) circuit with the ignition key in the RUN position is between 280 and 920 mV.

Technician B says normal CAN C bus termination is 60 ohms.

Who is correct?

A. A only   C. Both A and B

B. B only   D. Neither A nor B

31. How much total resistance is in a 12-volt circuit that is drawing 4 amps?

A. 0.333 ohm

B. 3 ohms

C. 4.8 ohms

D. 48 ohms

32. A replacement halogen bulb for a composite lamp was replaced but it lasted for a very short period of time; all of the other bulbs on the car are working fine. Which of the following could account for the premature failure of the bulb?

A. Excessive charging system voltage.

B. Fingerprints on the bulb.

C. Excessive voltage drop in the power feed circuit to the bulb.

D. Excessive resistance in the ground circuit of the bulb.

33. The turn signals of a vehicle are inoperative. The green indicator bulbs that are supposed to flash when the turn signal switch is moved to either the left- or the right-turn position do not turn on at all.

Technician A says the turn signal flasher contacts could have fused closed.

Technician B says the circuit from the turn signal flasher to the turn signal switch may be open.

Who is correct?

A. A only   C. Both A and B

B. B only   D. Neither A nor B

NOTE: Question 34 refers to the headlight switch on page 658.

34. Referring to the wiring diagram: the lighting circuit of this vehicle has the wire that is connected to the headlamp dimmer switch shorted to ground. Which of the following statements concerning this problem is true?

A. The 10-amp fuse will blow when the headlights are turned on.

B. The 15-amp fuse will blow immediately.

C. The 10-amp fuse will blow immediately.

D. The circuit breaker will open the circuit when the headlights are turned on.

NOTE: Question 35 refers to the multifunction switch on page 659.

35. Which of the following statements about the circuit is true?

A. The retract relay-up coil power feed comes from terminal 106 of the multifunction switch.

B. The power feed to the multifunction switch comes from the theft warning subcontrol amp.

C. The retract relay-up load circuit power feed comes from terminal 111 of the multifunction switch.

D. Terminal 106 of the multifunction switch is connected to ground.

36. *Technician A* says a voltmeter connected to the input wire of an IVR should indicate a fluctuating voltage.

Technician B says an open IVR on a single-gauge system could prevent the proper operation of the coolant temperature warning light.

Who is correct?

A. A only   C. Both A and B

B. B only   D. Neither A nor B

37. The fuel gauge of a vehicle equipped with an electromagnetic gauge cluster indicates a full tank, regardless of the actual amount of fuel in the tank. Which of the following could be the cause of this problem?

  A. Open IVR.

  B. Open fuel gauge sender unit.

  C. An open circuit between the gauge and the sending unit.

  D. A grounded circuit between the fuel gauge and the fuel gauge sender unit.

  NOTE: Question 38 refers to the charging system wiring diagram on page 660.

38. Referring to the wiring diagram, this vehicle's charge indicator light does not come on when the ignition switch is in the RUN position and the engine is off. Which of the following statements concerning this problem is true?

  A. The 10-ohm resistor inside the voltage regulator may be open.

  B. The field relay contacts may be stuck closed.

  C. The rotor winding may be shorted.

  D. The 50-ohm resistor may be open.

  NOTE: Question 39 refers to the two-speed wiper circuit on page 661.

39. The low-speed position of the windshield wiper system is inoperative; the high-speed position is working fine.
  Technician A says circuit 58 may be open.
  Technician B says circuit 63 may be open.
  Who is correct?

  A. A only   C. Both A and B

  B. B only   D. Neither A nor B

  NOTE: Question 40 refers to the blower motor circuit on page 662.

40. The medium-high speed of the blower motor circuit is inoperative; the rest of the blower speeds are fine.
  Technician A says circuit 752 may be open.
  Technician B says the middle resistor in the blower motor resistor assembly may be open.
  Who is correct?

  A. A only   C. Both A and B

  B. B only   D. Neither A nor B

  NOTE: Question 41 refers to the power window circuit on page 662.

41. The power window motor in the wiring diagram is completely inoperative. With the master window switch placed in the DOWN position, the following voltages are measured at each terminal:

Terminal Number	Voltage
1	12
2	0
3	12
4	0
5	12
6	12
7	0
8	12
9	0
10	12
11	0

Which of the following statements represents the cause of this problem?

  A. The master switch is faulty.

  B. The window switch is faulty.

  C. The motor is faulty.

  D. There is a poor ground in the circuit.

  NOTE: Question 42 refers to the body computer on page 663.

42. The "liftgate ajar" lamp remains on even after the liftgate is closed. Voltage at BCM terminal J2-4 is 0.0 volt.
  Technician A says that the BCM may be faulty.
  Technician B says that the liftgate switch may be stuck closed.
  Who is correct?

  A. A only   C. Both A and B

  B. B only   D. Neither A nor B

43. A Chrysler multiplex system is being discussed.
  Technician A says a voltmeter connected to the "Bus +" of any multiplexed component should indicate 12 volts with the ignition switch in the ON position.
  Technician B says a 12-volt test light can be used to test for power at various multiplexed components.
  Who is correct?

  A. A only   C. Both A and B

  B. B only   D. Neither A nor B

44. Sweep-testing a potentiometer with a voltmeter and an ohmmeter is being discussed.
  Referring to a voltmeter, *Technician A* says that the positive lead should be connected to the

feedback terminal and that the negative lead should be connected to the ground terminal. Referring to an ohmmeter, *Technician B* says that the positive lead should be connected to the feedback terminal and that the negative lead should be connected to the reference terminal. Who is correct?

A. A only
C. Both A and B
B. B only
D. Neither A nor B

NOTE: Question 45 refers to the diagnostic chart on page 664.

45. The diagnostic chart is being used to trouble-shoot a GM vehicle that has a current diagnostic trouble code 24 stored in memory. The ECC display panel indicates the correct vehicle speed. Which of the following represents the probable cause of the trouble code?

A. Open circuit 437.

B. Faulty speedometer gear.

C. Faulty vehicle speed sensor.

D. Possible ECM problem.

NOTE: Question 46 refers to the Ford-illuminated entry system on page 665.

46. The illuminated entry system is inoperative; neither the right nor the left outer door handle activates the system. The courtesy lights and each lock cylinder LED operates when the courtesy lamp switch is turned on. Which of the following represents the possible cause of this problem?

A. Open in circuit 54.

B. Short to ground in circuit 465.

C. Open in circuit 57.

D. Short to ground in circuit 54.

NOTE: Question 47 refers to the instrument cluster schematic on page 666.

47. The electronic instrument cluster is inoperative; none of the display segments illuminate.
Technician A says there may be an open in circuit 389.
Technician B says circuit 151 may be open.
Who is correct?

A. A only
C. Both A and B
B. B only
D. Neither A nor B

48. The ambient temperature sensor of a vehicle equipped with an SATC A/C system is being tested. The resistance of the sensor is about 1 kΩ.
Technician A says the sensor is out of range and needs to be replaced.
Technician B says the temperature of the sensor needs to be known before it can be accurately tested.
Who is correct?

A. A only

B. B only

C. Both A and B

D. Neither A nor B

49. Air bag module replacement is being discussed.
Technician A says the negative battery cable should be disconnected at the beginning of the repair.
Technician B says the reserve energy should be allowed to dissipate after removing battery power.
Who is correct?

A. A only

B. B only

C. Both A and B

D. Neither A nor B

NOTE: Question 50 refers to the GM electronic sunroof circuit on page 667.

50. The electric sunroof of the vehicle will close but not open. All of the following could be the cause of this problem **EXCEPT**:

A. Open wire between the "open" relay and the motor.

B. Faulty "open" relay load circuit contacts.

C. Faulty control switch.

D. Open wire between the control switch and the "open" relay.

Question 11 Horn circuit.

Rear Wiper/Washer

Question 12 Rear wiper/washer circuit.

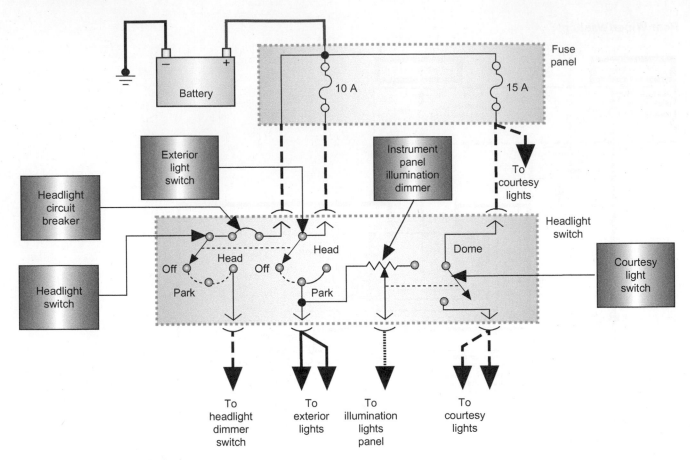

Question 34 Headlight switch circuit.

Question 35 Multifunction switch schematic.

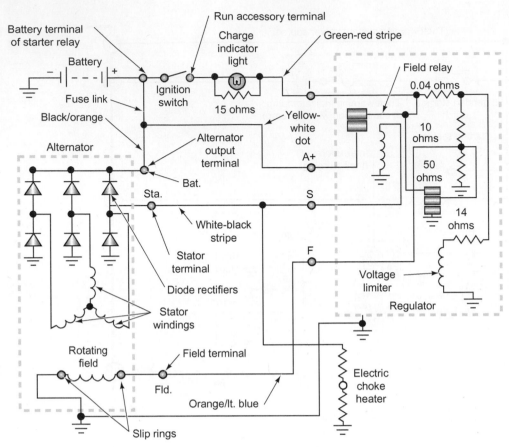

Question 38 Charging system wiring diagram.

Question 39 Two-speed wiper circuit.

Question 40 Blower motor circuit.

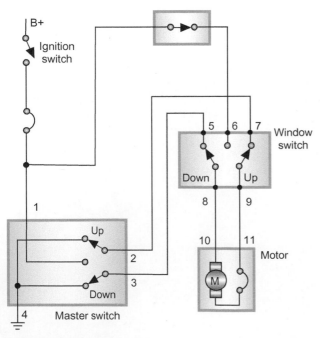

Question 41 Power window circuit.

Question 42 Body computer inputs and outputs.

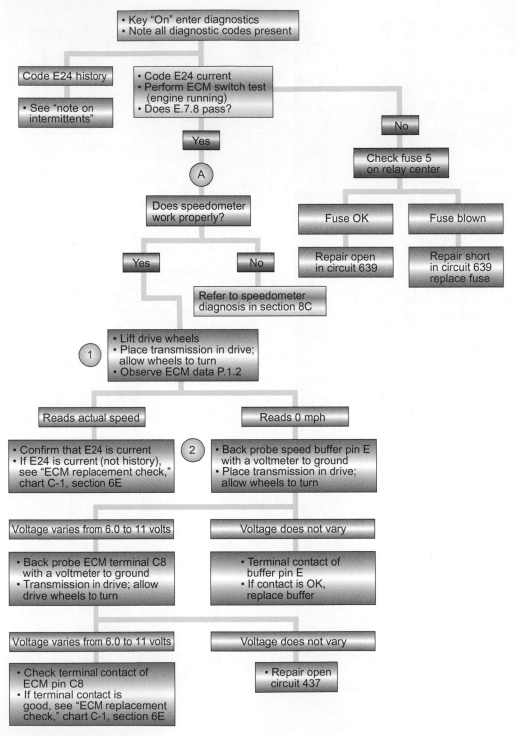

When all diagnoses and repairs are completed, clear codes and verify operation.

Question 45 Diagnostic chart.

Question 46 Ford-illuminated entry system diagram.

Question 47 Instrument cluster schematic.

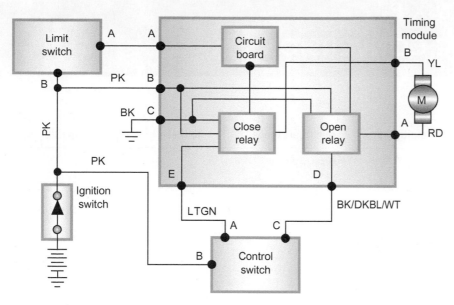

Question 50 GM electronic sunroof diagram.

APPENDIX B
METRIC CONVERSIONS

to convert these	to these	multiply by
TEMPERATURE		
Centigrade degrees	Fahrenheit degrees	1.8 then +32
Fahrenheit degrees	Centigrade degrees	0.556 after −32
LENGTH		
Millimeters	Inches	0.03937
Inches	Millimeters	25.4
Meters	Feet	3.28084
Feet	Meters	0.3048
Kilometers	Miles	0.62137
Miles	Kilometers	1.60935
AREA		
Square centimeters	Square inches	0.155
Square inches	Square centimeters	6.45159
VOLUME		
Cubic centimeters	Cubic inches	0.06103
Cubic inches	Cubic centimeters	16.38703
Cubic centimeters	Liters	0.001
Liters	Cubic centimeters	1,000
Liters	Cubic inches	61.025
Cubic inches	Liters	0.01639
Liters	Quarts	1.05672
Quarts	Liters	0.94633
Liters	Pints	2.11344
Pints	Liters	0.47317
Liters	Ounces	33.81497
Ounces	Liters	0.02957

to convert these	to these	multiply by
WEIGHT		
Grams	Ounces	0.03527
Ounces	Grams	28.34953
Kilograms	Pounds	2.20462
Pounds	Kilograms	0.45359
WORK		
Centimeter kilograms	Inch pounds	0.8676
Inch pounds	Centimeter kilograms	1.15262
Meter kilograms	Foot pounds	7.23301
Foot pounds	Newton meters	1.3558
PRESSURE		
Kilograms/sq. cm	Pounds/sq. inch	14.22334
Pounds/sq. inch	Kilograms/sq. cm	0.07031
Bar	Pounds/sq. inch	14.504
Pounds/sq. inch	Bar	0.06895

Alltest Instruments, Inc.
Farmingdale, NJ

Autel
Port Washington, NY

Baum Tools Unlimited, Inc.
Sarasota, FL

Bosch Automotive Service Solutions
Warren, MI

Carquest Corp.
Raleigh, NC

Fluke Corp.
Everett, WA

Kent-Moore, Div. of SPX Corp.
Charlotte, NC

Mac Tools
Westerville, OH

Matco Tools
Stow, OH

NAPA Hand/Service Tools
Atlanta, GA

OTC, Div. of Bosch Automotive Service Solutions
Warren, MI

Parts Plus, Division of Automotive Distribution Network
Germantown, TN

Pico Technology
Tyler, TX

Power Probe Group Inc.
Charlotte, NC

Snap-On Tools Corp.
Kenosha, WI

NEW TERM	NEW ACRONYMS/ ABBREVIATIONS	OLD ACRONYMS/ TERM	NEW TERM	NEW ACRONYMS/ ABBREVIATIONS	OLD ACRONYMS/ TERM
Accelerator pedal	AP	Accelerator	Diagnostic trouble code	DTC	Self-test code
Air cleaner	ACL	Air cleaner	Distributor ignition	DI	CBD
Air conditioning	A/C	A/C Air conditioning			DS TFI Closed bowl distributor Duraspark ignition Thick film ignition
Barometric pressure	BARO	BP Barometric pressure			
Battery positive voltage	B+	BATT+ Battery positive	Early fuel evaporation	EFE	EFE Early fuel evaporation
Camshaft position	CMP	Camshaft sensor	Electrically erasable programmable read-only memory	EEPROM	E2PROM Electrically erasable programmable read-only memory
Carburetor	CARB	CARB Carburetor			
Charge air cooler	CAC	After cooler Intercooler			
Closed loop	CL	EEC			
Closed throttle position	CTP	CTP Closed throttle position	Electronic ignition	EI	DIS EDIS Distributorless ignition system Electronic distributorless ignition system
Clutch pedal position	CPP	CES CIS Clutch engage switch Clutch interlock switch			
			Engine control module	ECM	ECM Engine control module
Continuous fuel injection	CFI	Continuous fuel injection	Engine coolant level	ECL	Engine coolant level
Continuous trap oxidizer	CTOX	CTO Continuous trap oxidizer	Engine coolant temperature	ECT	ECT Engine coolant temperature
Crankshaft position	CKP	CPS VRS Variable reluctance sensor	Engine speed	RPM	RPM Revolutions per minute
Data link connector	DLC	Self-test connector	Erasable programmable read-only memory	EPROM	EPROM Erasable programmable read-only memory
Diagnostic test mode	DTM	Self-test mode			

NEW TERM	NEW ACRONYMS/ ABBREVIATIONS	OLD ACRONYMS/ TERM	NEW TERM	NEW ACRONYMS/ ABBREVIATIONS	OLD ACRONYMS/ TERM
Evaporative emission	EVAP	EVP sensor EVR solenoid	Inertia fuel shutoff	IFS	Inertia switch
Exhaust gas recirculation	EGR	EGR Exhaust gas recirculation	Intake air temperature	IAT	ACT Air charge temperature
Fan control	FC	EDF Electro-drive fan	Knock sensor	KS	KS Knock sensor
Flash electrically erasable programmable read-only memory	FEEPROM	FEEPROM Flash electrically erasable programmable read-only memory	Malfunction indicator lamp	MIL	CEL "CHECK ENGINE" light "SERVICE ENGINE SOON" light
			Manifold absolute pressure	MAP	MAP Manifold absolute pressure
Flash-erasable programmable read-only memory	FEPROM	FEPROM Flash-erasable programmable read-only memory	Manifold differential pressure	MDP	MDP Manifold differential pressure
Flexible fuel	FF	FCS FFS FFV Fuel compensation sensor Flex fuel sensor	Manifold surface temperature	MST	MST Manifold surface temperature
Fourth gear	4GR	Fourth gear	Manifold vacuum zone	MVZ	MVZ Manifold vacuum zone
Fuel pump	FP	FP Fuel pump	Mass airflow	MAF	MAF Mass airflow
Generator	GEN	ALT Alternator	Mixture control	MC	Mixture control
Ground	GND	GND Ground	Multiport fuel injection	MFI	EFI Electronic fuel injection
Heated oxygen sensor	HO_2S	HEGO Heated exhaust gas oxygen sensor	Nonvolatile random access memory	NVRAM	NVM Nonvolatile memory
Idle air control	IAC	IAC Idle air bypass control	On-board diagnostic	OBD	Self-test On-board diagnostic
Idle speed control	ISC	Idle speed control	Open loop	OL	OL Open loop
Ignition control module	ICM	DIS module EDIS module TFI module	Oxidation catalytic converter	OC	COC Conventional oxidation catalyst
Indirect fuel injection	IFI	IDFI Indirect fuel injection	Oxygen sensor	O2S	EGO

NEW TERM	NEW ACRONYMS/ABBREVIATIONS	OLD ACRONYMS/TERM	NEW TERM	NEW ACRONYMS/ABBREVIATIONS	OLD ACRONYMS/TERM
PARK/NEUTRAL position	PNP	NDS NGS TSN Neutral drive switch Neutral gear switch Transmission select neutral	Secondary air injection	AIR	AM CT MTA Air management Conventional thermactor Managed thermactor air Thermactor
Periodic trap oxidizer	PTOX	PTOX Periodic trap oxidizer	Sequential multiport fuel injection	SFI	SEFI Sequential electronic fuel injection
Power steering pressure	PSP	PSPS Power steering pressure switch	Service reminder indicator	SRI	SRI Service reminder indicator
Powertrain control module	PCM	ECA ECM ECU EEC processor Engine control assembly Engine control module Engine control unit	Smoke puff limiter	SPL	SPL Smoke puff limiter
			Supercharger	SC	SC Supercharger
			Supercharger bypass	SCB	SCB Supercharger bypass
Programmable read-only memory	PROM	PROM Programmable read-only memory	System readiness test	SRT	—
Pulsed secondary air injection	PAIR	MPA PA Thermactor II Managed pulse air Pulse air	Thermal vacuum valve	TVV	Thermal vacuum switch
			Third gear	3GR	Third gear
Random access memory	RAM	RAM Random access memory	Three-way catalytic converter	TWC	TWC Three-way catalytic converter
Read-only memory	ROM	ROM Read-only memory	Three-way + oxidation catalytic converter	TWC + OC	TWC and COC Dual bed Three-way catalyst and conventional oxidation catalyst
Relay module	RM	RM Relay module	Throttle body	TB	TB Throttle body
Scan tool	ST	GST NGS Generic scan tool New-generation STAR tester Enhanced scan tool OBD II ST	Throttle body fuel injection	TBI	CFI Central fuel injection EFI
			Throttle position	TP	TP Throttle position

NEW TERM	NEW ACRONYMS/ ABBREVIATIONS	OLD ACRONYMS/ TERM	NEW TERM	NEW ACRONYMS/ ABBREVIATIONS	OLD ACRONYMS/ TERM
Torque converter clutch	TCC	CCC CCO MCCC Converter clutch control Converter clutch override Modulated converter clutch control	Voltage regulator	VR	VR Voltage regulator
			Volume airflow	VAF	VAF Volume airflow
Transmission control module	TCM	4EAT module	Warm-up oxidation catalytic converter	WU-OC	WU-OC Warm-up oxidation catalytic converter
Transmission range	TR	—	Warm-up three-way catalytic converter	WU-TWC	WU-TWC Warm-up three-way catalytic converter
Turbocharger	TC	TC Turbocharger	Wide-open throttle	WOT	Full throttle WOT Wide-open throttle
Vehicle speed sensor	VSS	VSS Vehicle speed sensor			

Note: **Terms are highlighted in bold,** followed by **Spanish translation in color.**

A circuit A generator circuit that uses an external grounded field circuit. The regulator is on the ground side of the field coil.

Circuito A Circuito regulador del generador que utiliza un circuito inductor externo puesto a tierra. En el circuito A, el regulador se encuentra en el lado a tierra de la bobina inductora.

Activation test A bidirectional scan tool function that allows the technician to command an actuator to test its operation.

Prueba de activación Una función de la herramienta de exploración bidireccional que le permite al técnico comandar un actuador para probar su operación.

Actuators Devices that perform the actual work commanded by the computer. They can be in the form of a motor, relay, switch, or solenoid.

Accionadores Dispositivos que realizan el trabajo efectivo que ordena la computadora. Dichos dispositivos pueden ser un motor, un relé, un conmutador o un solenoide.

Adaptive cruise control (ACC) An additional function of the cruise control system that uses laser radar or infrared transceiver sensors to determine the vehicle-to-vehicle distances and relational speeds. The system maintains both the vehicle speed and a set distance between vehicles.

Control de crucero adaptable (ACC) Una función adicional del sistema de control de crucero que utiliza un radar láser o sensores de transmisores infrarrojos para determinar las distancias entre vehículos y las velocidades de relación. El sistema mantiene tanto la velocidad del vehículo como una distancia fija entre vehículos.

Adaptive headlight system (AHS) A vehicle safety feature that is designed to enhance nighttime safety by providing drivers additional time to act or react to approaching road hazards by predicting upcoming corners or turns in the road.

Sistema de faros activos (AHS) Una característica de seguridad del vehículo diseñada para mejorar la seguridad nocturna, proporcionando a los conductores tiempo adicional para actuar o reaccionar ante los peligros de la carretera al acercarse a las esquinas o virajes próximos en la carretera.

Air bag module Composed of the air bag and inflator assembly that is packaged into a single module.

Unidad del Airbag Formada por el conjunto del Airbag y el inflador. Este conjunto se empaqueta en una sola unidad.

Air bag simulator load tool A tool that that is connected in place of air bag components to represent a known good circuit and resistances. The simulator allows for a safe test sequence that should identify any circuit problems.

Herramienta de carga del simulador de bolsa de aire Herramienta conectada en lugar de los componentes de la bolsa de aire para representar un circuito conocido bueno y las resistencias. El simulador permite realizar una secuencia de prueba segura para identificar cualquier problema del circuito.

Air bag system A supplemental restraint that deploys a bag out of the steering wheel or passenger-side dash panel to provide additional protection against head and face injuries during an accident.

Sistema de Airbag Resguardo complementario que expulsa una bolsa del volante o del panel de instrumentos del lado del pasajero para proveer protección adicional contra lesiones a la cabeza y a la cara en caso de un accidente.

Air gap The space between the trigger wheel teeth and the sensor.

Entrehierro El espacio entre los dientes de la rueda disparadora y el sensor.

Ambient temperature The temperature of the outside air.

Temperatura ambiente Temperatura del aire ambiente.

Ambient temperature sensor Thermistor used to measure the temperature of the air entering the vehicle.

Sensor de temperatura ambiente Termistor utilizado para medir la temperatura del aire que entra al vehículo.

Ammeter A test meter used to measure current draw.

Amperímetro Instrumento de prueba utilizado para medir la intensidad de una corriente.

Amperes *See* current.

Amperios *Véase* corriente.

Analog A voltage signal that is infinitely variable or can be changed within a given range.

Señal analógica Señal continua y variable que debe traducirse a valores numéricos discontinuos para poder ser tratada por una computadora.

Analog signal Varying voltage with infinite values within a defined range.

Señal análoga Voltaje variable con infinidad de valores dentro de un límite definido de velocidad.

Antilock brake system (ABS) A brake system that automatically pulsates the brakes to prevent wheel lock-up under panic stop and poor traction conditions.

Frenos antibloqueo Sistema de frenos que pulsa los frenos automáticamente para impedir el bloqueo de las ruedas en casos de emergencia y de tracción pobre.

Antitheft device A device or system that prevents illegal entry or driving of a vehicle. Most are designed to deter entry.

Dispositivo a prueba de hurto Un dispositivo o sistema quepreviene la entrada o conducción ilícita de un vehículo. Lamayoría se diseñan para detener la entrada.

A-pillar The pillar in front of the driver or passenger that supports the windshield.

Soporte A Soporte enfrente del conductor o del pasajero que sostiene el parabrisas.

Arming sensor A device that places an alarm system into "ready" to detect an illegal entry.

Sensor de armado Un dispositivo que pone "listo" un sistema dealarma para detectar una entrada ilícita.

Asbestos *Asbestos* is the name given to a set of six naturally occurring silicate minerals that are resistant to heat and corrosion. The mineral fibers include chrysotile, amosite, crocidolite, tremolite, anthophyllite, and actinolite.

Asbesto Asbestoes el nombre dado a un conjunto de seis minerales de silicato que son resistentes al calor y la corrosión de origen natural. Las fibras minerales incluyen; crisotilo, amosita, crocidolita, tremolita, antofilita, actinolita.

Aspirator Tubular device that uses a Venturi effect to draw air from the passenger compartment over the in-car sensor. Some manufacturers use a suction motor to draw the air over the sensor.

Aspirador Dispositivo tubular que utiliza un efecto venturi para extraer aire del compartimiento del pasajero sobre el sensor dentro del vehículo. Algunos fabricantes utilizan un motor de succion para extraer el aire sobre el sensor.

Audio system The sound system for a vehicle; can include radio cassette player, CD player, amplifier, and speakers.

Sistema de audio El sistema de sonido de un vehículo; puedeincluir el radio, el tocacaset, el toca discos compactos, el amplificador, y las bocinas.

Automatic door locks A system that automatically locks all doors by activating one switch.

Cerraduras automáticas de puerta Un sistema que cierra todas laspuertas automaticamente al activar un solo conmutador.

Automatic seat belt system Automatically puts the shoulder and/or lap belt around the driver or occupant. The automatic seat belt is operated by DC motors that move the belts by means of carriers on tracks.

Sistema automático de correas de asiento Funciona automáticamente, poniendo el cinturón de seguridad sobre el hombro y el pecho del chofer y el pasajero del automóvil. La correa automática del asiento trabaja por medio de motores DC, o sea corriente directa, que da movimiento a las correas transportadoras.

Automatic traction control A system that prevents slippage of one of the drive wheels. This is done by applying the brake at that wheel and/or decreasing the engine's power output.

Control automático de tracción Un sistema que previene el patinaje de una de las ruedas de mando. Esto se efectúa aplicando el freno en esa rueda y/o disminuyendo la salida de potencia del motor.

Average responding A method used to read AC voltage.

Respuesta media Un método que se emplea para leer la tensión decorriente alterna.

B circuit A generator regulator circuit that is internally grounded. In the B circuit, the voltage regulator controls the power side of the field circuit.

Circuito B Circuito regulador del generador puesto internamente a tierra. En el circuito B, el regulador de tensión controla el lado de potencia del circuito inductor.

Backprobe A term used to mean that a test is being performed on the circuit while the connector is still connected to the component. The test probes are inserted into the back of the wire connector.

Sonda exploradora de retorno Término utilizado para expresar que se está llevando a cabo una prueba del circuito mientras el conectador sigue conectado al componente. Las sondas de prueba se insertan a la parte posterior del conectador de corriente.

BAT Terminal of a generator, starter, or solenoid that has direct battery feed connected to it.

Forma de aproximación del balancín Terminal de un generador, encendedor o solenoide conectado directamente a la batería.

Battery ECU Used to monitor the condition of the HV battery assembly in a hybrid electric vehicle (HEV). The battery ECU determines the SOC of the HV battery by monitoring voltage, current, and temperature.

UCE de la batería Se utiliza para monitoreard del ensamblado de la batería del VH en un VHE. La UCE de la batería determina el EDC de la batería del VH al monitorear el voltaje, la corriente y la temperatura.

Battery holddowns Brackets that secure the battery to the chassis of the vehicle.

Portabatería Los sostenes que fijan la batería al chasis del vehículo.

Battery leakage test Used to determine if current is discharging across the top of the battery case.

Prueba de pérdida de corriente de la batería Prueba utilizada para determinar si se está descargando corriente a través de la parte superior de la caja de la batería.

Battery terminals Extend through the case of the battery to provide a means of connecting the battery plates to the vehicle's electrical system. All automotive batteries have two terminals.

Bornes de la batería Se extienden por fuera de la cubierta de la batería para permitir la conexión de las placas de la batería con el sistema eléctrico del vehículo. Todas las baterías de automóvil cuentan con dos bornes.

Battery terminal test Checks for poor electrical connections between the battery cables and terminals; uses a voltmeter to measure voltage drop across the cables and terminals.

Prueba del borne de la batería Verifica si existen conexiones eléctricas pobres entre los cables y los bornes de la batería. Utiliza un voltímetro para medir caídas de tensión entre los cables y los bornes.

B-codes DTCs that are assigned to the vehicle's body systems and control modules.

Códigos B Instrucciones de transmisión digital (DTC) que se asignan a los sistemas de la carrocería del vehículo y a los módulos de control.

Bench test A term used to indicate that the unit is to be removed from the vehicle and tested.

Prueba de banco Término utilizado para indicar que la unidad será removida del vehículo para ser examinada.

Bendix drive A type of starter drive that uses the inertia of the spinning starter motor armature to engage the drive gear to the gears of the flywheel. This type of starter drive was used on early models of vehicles and is rarely seen today.

Acoplamiento Bendix Un tipo del acoplamiento del motor de arranque que usa la inercia de la armadura del motor de arranque giratorio para endentar el engranaje de mando con los engranajes del volante. Este tipo de acoplamiento del motor de arranque se usaba en los modelos vehículos antiguos y se ven raramente.

Bezel The retaining trim around a component.

Bisel El resto del decorado alrededor de un componente.

Binary numbers Strings of zeroes and ones, or on and off, which represent a numeric value.

Números binarios Franjas que indican ceros (0) y unos (1), de prendido/apagado, representando el valor numérico.

B-Pillar The pillar located over the shoulder of the driver or passenger.

Soporte B Soporte ubicado sobre el hombro del conductor o del pasajero.

Breakout box A diagnostic tool that allows the technician to test circuits, sensors, and actuators by providing test points.

Caja de interruptores Herramienta de diagnóstico que proporciona al técnico puntos de prueba para verificar los circuitos, los sensores y los actuadores.

Brushes Electrically conductive sliding contacts, usually made of copper and carbon.

Escobillas Contactos deslizantes de conducción eléctrica, por lo general hechos de cobre y de carbono.

Bucking coil One of the coils in a three-coil gauge. It produces a magnetic field that bucks or opposes the low reading coil.

Bobina compensadora Una de las bobinas de un calibre de tres bobinas. Produce un campo magnético que es contrario o en oposición a la bobina de baja lectura.

Buffer A buffer cleans up a voltage signal. These are used with PM generator sensors to change the AC voltage to a digitalized signal.

Separador Un separador aguza una señal del tensión. Estos se usan con los sensores generadores PM para cambiar la tensión de corriente alterna a una señal digitalizado.

Bulkhead connector A large connector that is used when many wires pass through the bulkhead or firewall.

Conectador del tabique Un conectador que se usa al pasar muchos alambres por el tabique o mamparo de encendidos.

Bus bar A common electrical connection to which all of the fuses in the fuse box are attached. The bus bar is connected to battery voltage.

Barra colectora Conexión eléctrica común a la que se conectan todos los fusibles de la caja de fusibles. La barra colectora se conecta a la tensión de la batería.

Capacitance discharge sensor A form of piezo sensor that uses the discharge from a variable capacitor.

Sensor de descarga de capacitancia Un tipo de sensor que utiliza la descarga de un capacitor variable.

Capacitor An electrical device made from layers of conductors and dielectric used to store and release electrical energy; also used to absorb voltage changes in the circuit.

Capacitor Un dispositivo eléctrico que se fabrica de capas de conductores y material dieléctrico y se utiliza para almacenar y liberar energía eléctrica; también se utiliza para absorber los cambios de voltaje en el circuito.

Capacity test A dynamic test whereby a simulated load (in amperes or watts) is imposed on the battery system for a specified time.

Prueba de capacidad Prueba dinámica mediante la cual el sistema de baterías es sometido a una carga simulada (en amperes o vatios) durante un tiempo especificado.

Captured signal A signal stored in a DSO's memory.

Señal de captura La señal restauradora se mantiene en la memoria indicada con las letras DSO.

Carbon monoxide An odorless, colorless, and toxic gas that is produced as a result of combustion.

Monóxido de carbono Gas inodoro, incoloro y tóxico producido como resultado de la combustión.

Carbon tracking A condition where paths of carbon will allow current to flow to points that are not intended. This condition is most commonly found inside distributor caps.

Rastreo de carbón   Una condición en la cual las trayectorias del carbón permiten fluir el corriente a los puntos no indicados. Esta condición se encuentra comunmente dentro de las tapas del distribuidor.

Cartridge fuses   *See* maxi-fuse.

Fusibles cartucho   *Véase* maxifusible.

Cathode ray tube   Similar to a television picture tube. It contains a cathode that emits electrons and an anode that attracts them. The screen of the tube will glow at points that are hit by the electrons.

Tubo de rayos catódicos   Parecidos a un tubo de pantalla de televisor. Contiene un cátodo que emite los electrones y un ánodo que los atrae. La pantalla del tubo iluminará en los puntos en donde pegan los electrones.

Caustic   Chemicals that have the ability to destroy or eat through something and that are extremely corrosive.

Cáusticos   Químicos que tienen la habilidad de destruir o carcomer algo, y que son extremadamente corrosivos.

Cell element   The assembly of a positive and a negative plate in a battery.

Elemento de pila   La asamblea de una placa positiva y negativa en una batería.

Charge   To pass an electric current through a battery in an opposite direction than during discharge.

Cargar   Pasar una corriente eléctrica por la batería en una dirección opuesta a la usada durante la descarga.

Charge rate   The speed at which the battery can safely be recharged at a set amperage.

Indicador de carga eléctrica   La velocidad a la cual la batería puede ser recargada seguramente a un amperaje establecido.

Charging system requirement test   Diagnostic test used to determine the total electrical demand of the vehicle's electrical system.

Prueba del requisito del sistema de carga   Prueba diagnóstica utilizada para determinar la exigencia eléctrica total del sistema eléctrico del vehículo.

CHMSL   The abbreviation for center high-mounted stop light, often referred to as the third brake light.

CHMSL   La abreviación para el faro de parada montada alto en el centro que suele referirse como el faro de freno tercero.

Circuit   The path of electron flow consisting of the voltage source, conductors, load component, and return path to the voltage source.

Circuito   Trayectoria del flujo de electrones, compuesto de la fuente de tensión, los conductores, el componente de carga y la trayectoria de regreso a la fuente de tensión.

Clamping diode   A diode that is connected in parallel with a coil to prevent voltage spikes from the coil reaching other components in the circuit.

Diodo de bloqueo   Un diodo que se conecta en paralelo con una bobina para prevenir que los impulsos de tensión lleguen a otros componentes en el circuito.

Clockspring   Maintains a continuous electrical contact between the wiring harness and the air bag module.

Muelle de reloj   Mantiene un contacto eléctrico continuo entre el cableado preformado y la unidad del Airbag.

Closed circuit   A circuit that has no breaks in the path and allows current to flow.

Circuito cerrado   Circuito de trayectoria ininterrumpida que permite un flujo continuo de corriente.

Cold cranking amps (CCA)   Rating that indicates the battery's ability to deliver a specified amount of current to start an engine at low ambient temperatures.

Amperaje de arranque en frío (CCA, por su sigla en inglés)   Régimen que indica la capacidad de la batería para proporcionar una cantidad de corriente específica, capaz de hacer arrancar un motor a temperatura ambiente baja.

Cold soak   A key-off condition that allows the engine coolant temperature to equal with ambient temperature.

Amperios de arranque en frío   Tasa indicativa de la capacidad de la batería para producir una cantidad específica de corriente para arrancar un motor a bajas temperaturas ambiente.

Color codes   Used to assist in tracing the wires. In most color codes, the first group of letters designates the base color of the insulation and the second group of letters indicates the color of the tracer.

Códigos de colores   Utilizados para facilitar la identificación de los alambres. Típicamente, el primer alfabeto representa el color base del aislamiento y el segundo representa el color del indicador.

Common connector   A connector that is shared by more than one circuit and/or component.

Conector común   Un conector que se comparte entre más de un circuito y/o componente.

Commutator   A series of conducting segments located around one end of the armature.

Conmutador   Serie de segmentos conductores ubicados alrededor de un extremo de la armadura.

Component locator   Service manual used to find where a component is installed in the vehicle. The component locator uses both drawings and text to lead the technician to the desired component.

Manual para indicar los elementos componentes   Manual de servicio utilizado para localizar dónde se ha instalado un componente en el vehículo. En dicho manual figuran dibujos y texto para guiar al mecánico al componente deseado.

Composite bulb   A headlight assembly that has a replaceable bulb in its housing.

Bombilla compuesta   Una asamblea de faros cuyo cárter tiene una bombilla reemplazable.

Compound motor   A motor that has the characteristics of a series-wound and a shunt-wound motor.

Motor compuesta   Un motor que tiene las características de un motor exitado en serie y uno en derivación.

Computer   An electronic device that stores and processes data and is capable of operating other devices.

Computadora   Dispositivo electrónico que almacena y procesa datos y que es capaz de ordenar a otros dispositivos.

Conductance   A measurement of the battery's plate surface that is available for chemical reaction, determining how much power the battery can supply.

Conductancia   Medida de la superficie de la placa de la batería que está lista para la reacción química, determinando así cuánta potencia puede suplir la batería.

Conducted interference   An electromagnetic interference (EMI) caused by the physical contact of the conductors.

Interferencia conducida   Una interferencia electromagnética (EMI) provocada por el contacto físico de los conductores.

Conductor   A material in which electrons flow or move easily.

Conductor   Una material en la cual los electrones circulen o se mueven fácilmente.

Continuity   Refers to the circuit being continuous with no opens.

Continuidad   Se refiere al circuito ininterrumpido, sin aberturas.

Control assembly   Provides for driver input into the automatic temperature control microprocessor. The control assembly is also referred to as the control panel.

Montaje de control   Un microprocesador automático que le da al Chofer la información necesaria sobre el control de temperatura. El montaje de control es llamado también panel de control.

Converter   A DC/DC transformer that converts the voltage from 270 volts DC to 14 volts DC to recharge the auxiliary battery and to power 12-volt electrical components on an HEV.

Convertidor   Transformador CD-CD que convierte el voltaje de 270 voltios a 14 voltios de CD para recargar la batería auxiliar y para darles potencia a los componentes eléctricos de 12 voltios en un VHE.

Corona effect   A condition where high voltage leaks through a wire's insulation and produces a light or an illumination; worn insulation on spark plug wires causes this.

Efecto corona   Una condición en la cual la alta tensión se escapa por la insulación del alambre y produce una luz o una iluminación; esto se causa por la insulación desgastada en los alambres de las bujías.

Cowl   The top portion of the front of the automobile body that supports the windshield and dashboard.

Capucha   Esta es la parte principal de la carrocería en el frente del automóvil y es la que sostiene el parabrisas y el tablero de instrumentos.

Crash sensor   Normally open electrical switch designed to close when subjected to a predetermined amount of jolting or impact.

Sensor de impacto   Un conmutador normalmente abierto diseñado a cerrarse al someterse a un sacudo de una fuerza predeterminada o un impacto.

Crimping   The process of bending, or deforming by pinching, a connector so that the wire connection is securely held in place.

Engarzado   Proceso a través del cual se curva o deforma un conectador mediante un pellizco para que la conexión de alambre se mantenga firme en su lugar.

Crocus cloth   A sandpaper like material with a cloth backing used to polish metals. While polishing, the cloth removes very little metal.

Paño de color azafrán   Material similar a la lija, con un refuerzo de tela, usado para pulir metales. Al pulir, la tele remueve muy poco metal.

Cross-fire   The undesired firing of a spark plug that results from the firing of another spark plug. This is caused by electromagnetic induction.

Encendido transversal   El encendido no deseable de una bujía que resulta del encendido de otra bujía. Esto se causa por la inducción electromagnética.

CRT   The common acronym for a cathode ray tube.

CRT   La sigla común de un tubo de rayos catódicos.

Curb height   The height of the vehicle when it has no passengers or loads, and normal fluid levels and tire pressure.

Altura del contén   La altura del vehículo cuando no lleva pasajeros ni cargas, y los niveles de los fluidos y de la presión de las llantas son normales.

Current   The aggregate flow of electrons through a wire. One ampere represents the movement of 6.25 billion billion electrons (or one coulomb) past one point in a conductor in one second.

Corriente   Flujo combinado de electrones a través de un alambre. Un amperio representa el movimiento de 6,25 mil millones de mil millones de electrones (o un colombio) que sobrepasa un punto en un conductor en un segundo.

Current-draw test   Diagnostic test used to measure the amount of current that the starter draws when actuated. It determines the electrical and mechanical condition of the starting system.

Prueba de la intensidad de una corriente   Prueba diagnóstica utilizada para medir la cantidad de corriente que el arrancador tira cuando es accionado. Determina las condiciones eléctricas y mecánicas del sistema de arranque.

Current output test   Diagnostic test used to determine the maximum output of the AC generator.

Prueba de la salida de una corriente   Prueba diagnóstica utilizada para determinar la salida máxima del generador de corriente alterna.

Cycle   One set of changes in a signal that repeats itself several times.

Ciclo   Una serie de cambios en una señal que se repite varias veces.

Darlington   A special type of configuration usually consisting of two transistors fabricated on the same chip or mounted in the same package.

Darlington   Tipo especial de configuración que generalmente consiste en 2 transistores fabricados en el mismo chip o montados en el mismo paquete.

D'Arsonval gauge   A gauge design that uses the interaction of a permanent magnet and an electromagnet and the total field effect to cause needle movement.

Calibrador D'Arsonval   Calibrador diseñado para utilizar la interacción de un imán permanente y de un electroimán, y el efecto inductor total para generar el movimiento de la aguja.

D'Arsonval movement   A small coil of wire mounted in the center of a permanent horseshoe-type magnet. A pointer or needle is mounted to the coil.

Movimiento D'Arsonval   Se refiere a una pequeña bobina de cable montada en el centro de un permanente imán diseñado como la herradura de caballo. El puntero o la aguja está montada sobre la bobina.

Data link connector (DLC)   A multi-pin connector that houses the data communication lines to the vehicle's computers for on-board diagnostics (OBD) and gathering live data streams with a scan tool.

Conector de enlace de datos   Conector multi-pin en el que se encuentran las líneas de comunicación de datos con las computadoras del vehículo para el diagnóstico a bordo y la recopilación de datos en tiempo real con un instrumento de exploración.

Deep cycling   Discharging the battery completely before recharging it.

Operacion cíclica completa   La descarga completa de la batería previo al recargo.

Demonstration mode   SmartBeam™ function that allows operation of the automatic high beams and high-beam indicator while the vehicle is stationary and under any ambient lighting conditions.

Modo de demostración   Función del sistema SmartBeam™ que permite la operación de las luces altas automáticas y del indicador de luces altas mientras el vehículo no está en funcionamiento y en cualquier condición de iluminación ambiental.

Deployment Energy Reserve Module (DERM)   Provides an energy reserve of 36 volts to assure air bag deployment for a few seconds when vehicle voltage is low or lost. The DERM also maintains constant diagnostic monitoring of the air bags electrical system. It will store a code if a fault is found and provide driver notification by illuminating the warning light.

DERM   El significado de este término se refiere al diseño que mantiene una reserva de energía de 36 voltios la que asegura, por unos pocos segundos, la salida de la corriente cuando el voltaje del vehículo es bajo o se ha perdido. El DERM mantiene también un constante diagnóstico monitor del sistema eléctrico. Guarda también una clave en caso de que se encuentre alguna falla y le da al chofer una alerta iluminando las señales.

Diagnostic module   Part of an electronic control system that provides self-diagnostics and/or a testing interface.

Módulo de diagnóstico   Parte de un sistema controlado electronicamente que provee autodiagnóstico y/o una interfase de pruebas.

Diagnostic trouble code (DTC)   Output of the self-diagnostics program in the form of a alphanumeric code that indicates faulty circuits or components in a monitored system.

Códigos de falla de diagnóstico   Resultado del programa de autodiagnóstico en forma de código alfanumérico que marca los circuitos o componentes defectuosos del sistema monitoreado.

Diaphragm   A thin, flexible, circular plate that is held around its outer edge by the horn housing, allowing the middle to flex.

Diafragma   Es una fina placa circular flexible que es sostenida alrededor de su borde externo por el cuerno del embrague, permitiendo que el centro se doble.

Digital   A voltage signal is either on-off, yes-no, or high-low.

Digital   Una señal de tensión está Encendida-Apagada, es Sí-No o Alta-Baja.

Digital multimeter (DMM)   Displays values using liquid crystal displays instead of a swinging needle. They are basically computers that determine the measured value and display it for the technician.

Multímetro digital   Exhibe valores usando cristal líquido para desplegar, en cambio de una aguja giratoria. Son básicamente computadores que determinan el valor medido y lo muestran al técnico.

Digital signal   A voltage value that has two states. The states can be on-off or high-low.

Señal digital   El valor del voltaje que tiene dos ventajas. Una señal muestra prendido/apagado y la otra alta/baja.

Dimmer switch   A switch in the headlight circuit that provides the means for the driver to select either high-beam or low-beam operation, and to switch between the two. The dimmer switch is connected in series within the headlight circuit and controls the current path for high and low beams.

Conmutador reductor   Conmutador en el circuito para faros delanteros que le permite al conductor elegir la luz larga o la luz corta, y conmutar entre las dos. El conmutador reductor se conecta en serie dentro del circuito para faros delanteros y controla la trayectoria de la corriente para la luz larga y la luz corta.

Diode   An electrical one-way check valve that will allow current to flow only in one direction.

Diodo   Válvula eléctrica de retención, de una vía, que permite que la corriente fluya en una sola dirección.

Diode rectifier bridge   A series of diodes that are used to provide a reasonably constant DC voltage to the vehicle's electrical system and battery.

Puente rectificador de diodo   Serie de diodos utilizados para proveerles una tensión de corriente continua bastante constante al sistema eléctrico y a la batería del vehículo.

Diode trio   Used by some manufacturers to rectify the stator of an AC generator current so that it can be used to create the magnetic field in the field coil of the rotor.

Trío de diodos   Utilizado por algunos fabricantes para rectificar el estátor de la corriente de un generador de corriente alterna y poder así utilizarlo para crear el campo magnético en la bobina inductora del rotor.

Direct drive   A situation where the drive power is the same as the power exerted by the device that is driven.

Transmisión directa Una situación en la cual el poder de mando es lo mismo que la potencia empleada por el dispositivo arrastrado.

Discriminating sensors Part of the air bag circuitry; these sensors are calibrated to close with speed changes that are great enough to warrant air bag deployment. These sensors are also referred to as crash sensors.

Sensores discriminadores Una parte del conjunto de circuitos de Airbag; estos sensores se calibran para cerrar con los cambios de la velocidad que son bastante severas para justificar el despliegue del Airbag. Estos sensores también se llaman los sensores de impacto.

Drive coil A hollowed field coil used in a positive engagement starter to attract the movable pole shoe of the starter.

Bobina de excitación Una bobina inductora hueca empleada en un encendedor de acoplamiento directo para atraer la pieza polar móvil del encendedor.

DSO A common acronym for a digital storage oscilloscope.

DSO Una sigla común del osciloscopio de almacenamiento digital.

Dual ramping A sensor circuit that provides a more precise method of determining the temperature. This circuit starts with a high-voltage value when cold and decreases as temperature increases. When the voltage reaches a predetermined value, the PCM switches to lower resistance circuit causing the sense reading to go high again and provides a second set of inputs.

Voltaje de diente de sierra doble Un circuito de sensor que brinda un método más preciso de determinar la temperatura. Este circuito comienza con un valor de voltaje alto cuando está frío y desciende a medida que la temperatura se incrementa. Cuando el voltaje alcanza un valor predeterminado, el PCM pasa a un circuito de resistencia más baja, lo que hace que la lectura del sensor vuelva a subir y proporciona un segundo conjunto de información entrante.

Dump valve A safety switch that releases vacuum to the servo when the brake pedal is pressed.

Válvula de descarga Un interruptor de seguridad que facilita el vacío al servo cuando se presiona el pedal del freno.

Duty cycle The percentage of on time to total cycle time.

Ciclo de trabajo Porcentaje del trabajo efectivo a tiempo total del ciclo.

Dynamic sensor See sensor cluster.

Sensor dinámico Véase dispositivo de sensores.

Eddy currents Small induced currents.

Corriente de Foucault Pequeñas corrientes inducidas.

Electrical load The working device of the circuit.

Carga eléctrica Dispositivo de trabajo del circuito.

Electrochemical The chemical action of two dissimilar materials in a chemical solution.

Electroquímico Acción química de dos materiales distintos en una solución química.

Electrolysis The production of chemical changes by passing electrical current through an electrolyte.

Electrólisis La producción de los cambios químicos al pasar un corriente eléctrico por un electrolito.

Electrolyte A solution of 64% water and 36% sulfuric acid.

Electrolito Solucion de un 64 percent de agua y un 36 percent de ácido sulfúrico.

Electromagnetic gauge Gauge that produces needle movement by magnetic forces.

Calibrador electromagnético Calibrador que genera el movimiento de la aguja mediante fuerzas magnéticas.

Electromagnetic induction The production of voltage and current within a conductor as a result of relative motion within a magnetic field.

Inducción electrómagnética Producción de tension y de corriente dentro de un conductor como resultado del movimiento relativo dentro de un campo magnético.

Electromagnetic interference (EMI) An undesirable creation of electromagnetism whenever current is switched on and off.

Interferencia electromagnética Fenómeno de electromagnetismo no deseable que resulta cuando se conecta y se desconecta la corriente.

Electromagnetism A form of magnetism that occurs when current flows through a conductor.

Electromagnetismo Forma de magnetismo que ocurre cuando la corriente fluye a través de un conductor.

Electromechanical A device that uses electricity and magnetism to cause a mechanical action.

Electromecánico Un dispositivo que causa una acción mecánica por medio de la electricidad y el magnetismo.

Electromotive force (EMF) *See* voltage.

Fuerza electromotriz *Véase* tensión.

Electronic level controller (ELC) The computer that controls the automatic suspension system.

Controler de nivel electrónico Este es el computador que controla el sistema automático de suspensión.

Electrostatic discharge (ESD) strap Grounds your body to prevent static discharges that may damage electronic components.

Correas electroestáticas de descargue Estas aislan su cuerpo para prevenir descargas estáticas que puedan dañar los components electrónicos.

EMI Electromagnetic interference.

EMI La interferencia electromagnética.

Environmental data A saved snapshot of the conditions when the fault occurred.

Datos ambientales Una instantánea guardada de las condiciones en que se produjo el fallo.

Excitation current Current that magnetically excites the field circuit of the AC generator.

Corriente de excitación Corriente que excita magnéticamente al circuito inductor del generador de corriente alterna.

Face shields Clear plastic shields that protect the entire face.

Careta de soldador Caretas de plástico claro que protegen toda la cara.

Fail safe Computer substitution of a fixed input value if a sensor circuit should fail. This provides for system operation, but at a limited function; also referred to as the "limp-in" mode.

Falla Sustitución por la computadora de un valor fijo de entrada en caso de que ocurra una falla en el circuito de un sensor. Esto asegura el funcionamiento del sistema, pero a una capacidad limitada.

False wipes Unnecessary wipes that occur when the control knob of the multifunction switch is in one of the five automatic wiper sensitivity positions and no rain or moisture is apparent within the wipe pattern on the windshield glass.

Limpieza de parabrisas falsa Limpieza de parabrisas innecesaria que se produce cuando la perilla de control del interruptor multifunción se encuentra en una de las cinco posiciones de sensibilidad del limpiaparabrisas, y no hay lluvia ni humedad aparente dentro del patrón ni del cristal del parabrisas.

Fast charging Battery charging using a high amperage for a short period of time.

Carga rápida Carga de la batería que utiliza un amperaje máximo por un corto espacio de tiempo.

Feedback 1. Data concerning the effects of the computer's commands is fed back to the computer as an input signal; used to determine if the desired result has been achieved. 2. A condition that can occur when electricity seeks a path of lower resistance, but the alternate path operates another component than that intended. Feedback can be classified as a short.

Realimentación 1. Datos referentes a los efectos de las órdenes de la computadora se suministran a la misma como señal de entrada. La realimentación se utiliza para determinar si se ha logrado el resultado deseado. 2. Condición que puede ocurrir cuando la electricidad busca una trayectoria de menos resistencia, pero la trayectoria alterna opera otro componente que aquel deseado. La realimentación puede clasificarse como un cortocircuito.

Fiber optics A medium of transmitting light through polymethyl methacrylate plastic that keeps the light rays parallel even if there are extreme bends in the plastic.

Transmisión por fibra óptica Técnica de transmisión de luz por medio de un plástico de polimetacrilato de metilo que mantiene los rayos de luz paralelos aunque el plástico esté sumamente torcido.

Field current-draw test Diagnostic test that determines if there is current available to the field windings.

Prueba de la intensidad de una corriente inductora Prueba diagnóstica que determina si se está generando corriente a los devanados inductores.

Field relay   The relay that controls the amount of current going to the field windings of a generator. This is the main output control unit for a charging system.

Relé inductor   El relé que controla la cantidad del corriente a los devanados inductores de un generador. Es la unedad principal de potencia de salida de un sistema de carga.

Fire blanket   A safety device consisting of a sheet of fire retardant material that is thrown on top of a small fire in order to smother it.

Manta contra incendios   Un dispositivo de seguridad que consiste en una sábana de material retardante de fuego que se lanza sobre un pequeño incendio para reprimirlo.

Fire extinguisher   A portable apparatus that contains chemicals, water, foam, or special gas that can be discharged to extinguish a small fire.

Extinctor de incendios   Aparato portátil que contiene elementos químicos, agua, espuma o gas especial que pueden descargarse para extinguir un incendio pequeño.

Flammable   A substance that supports combustion.

Inflamable   Sustancia que promueve la combustión.

Flash   To remove the existing programming and overwrite it with new software.

Limpiar   Así se remueve el programa existente y lo reemplaza con programas nuevos.

Flash codes   Diagnostic trouble codes (DTCs) that are displayed by flashing a lamp or LED.

Códigos intermitentes   Códigos de fallas de diagnóstico (CFD) que se muestran en una lámpara intermitente o LED.

Flat rate   A pay system in which technicians are paid for the amount of work they do. Each job has a flat rate time. Pay is based on that time, regardless of how long it takes to complete the job.

Tarifa bloque   Sistema de pago en el que se le paga al técnico por cada hora de trabajo realizado. Cada trabajo tiene un tiempo de tarifa bloque. El pago se basa en ese tiempo sin importar cuánto tiempo se lleve para terminarlo.

Floating   The circuit isn't connected to either power or ground.

Flotación   El circuito no está conectado ni a la fuente de energía ni a tierra.

Floor jack   A portable hydraulic tool used to raise and lower a vehicle.

Gato de pie   Herramienta hidráulica portátil utilizada para levantar y bajar un vehículo.

Forward bias   A positive voltage that is applied to the P-type material and a negative voltage to the N-type material of a semiconductor.

Polarización directa   Tensión positiva aplicada al material P y tensión negativa aplicada al material N de un semiconductor.

Freeze frame   An OBD II requirement which specifies that engine conditions be recorded when an emissions-related fault is first detected.

Cuadro congelado   Un requisito OBD II que especifica que las condiciones del motor deben registrarse en cuanto se detecta una falla relacionada con las emisiones.

Freeze frame data   A recording of the driving condition when the malfunction occurred. This is useful for determining how the vehicle was operating at the time and for locating any input or output values that are out of range.

Datos de trama fija   Archivo de las condiciones de manejo cuando sucede una falla. Es conveniente para determinar cómo operaba el vehículo en ese momento y para localizar cualesquier valores de entrada y salida que estén fuera de banda.

Frequency   The number of complete oscillations that occur during a specific time, measured in hertz.

Frecuencia   El número de oscilaciones completas que ocurren durante un tiempo específico, medidas en Hertz.

Full field   Field windings that are constantly energized with full battery current. Full fielding will produce maximum AC generator output.

Campo completo   Devanados inductores que se excitan constantemente con corriente total de la batería. EL campo completo producirá la salida máxima de un generador de corriente alterna.

Full fielding   Supplies a constant full battery voltage to the field windings to produce maximum AC generator output.

Operación con excitación no regulada   Proporciona una tensión de batería completa a los devanados para obtener la máxima potencia de salida del generador de CA.

Full field test   Diagnostic test used to isolate if the detected problem lies in the AC generator or the regulator.

Prueba de campo completo   Prueba diagnóstica utilizada para determinar si el problema descubierto se encuentra en el generador de corriente alterna o en el regulador.

Functional test   Checks the operation of the system as the technician observes the different results.

Prueba funcional   Verifica la operación del sistema mientras el técnico observa los diferentes resultados.

Fuse   A replaceable circuit protection device that will melt should the current passing through it exceed its rating.

Fusible   Dispositivo reemplazable de protección del circuito que se fundirá si la corriente que fluye por el mismo excede su valor determinado.

Fuse box   A term used to indicate the central location of the fuses contained in a single holding fixture.

Caja de fusibles   Término utilizado para indicar la ubicación central de los fusibles contenidos en un solo elemento permanente.

Fusible link   A wire made of meltable material with a special heat-resistant insulation. When there is an overload in the circuit, the link melts and opens the circuit.

Cartucho de fusible   Alambre hecho de material fusible con aislamiento especial resistente al calor. Cuando ocurre una sobrecarga en el circuito, el cartucho se funde y abre el circuito.

Gassing   The conversion of a battery's electrolyte into hydrogen and oxygen gas.

Burbujeo   La conversión del electrolito de una bateria al gas de hidrógeno y oxígeno.

Gauge   1. A device that displays the measurement of a monitored system by the use of a needle or pointer that moves along a calibrated scale. 2. The number that is assigned to a wire to indicate its size. The larger the number, the smaller the diameter of the conductor.

Calibrador   1. Dispositivo que muestra la medida de un sistema regulado por medio de una aguja o indicador que se mueve a través de una escala calibrada. 2. El número asignado a un alambre indica su tamaño. Mientras mayor sea el número, más pequeño será el diámetro del conductor.

Gauss gauge   A meter that is sensitive to the magnetic field surrounding a wire conducting current. The gauge needle will fluctuate over the portion of the circuit that has current flowing through it. Once the ground has been passed, the needle will stop fluctuating.

Calibrador gauss   Instrumento sensible al campo magnético que rodea un alambre conductor de corriente. La aguja del calibrador se moverá sobre la parte del circuito a través del cual fluye la corriente. Una vez se pasa a tierra, la aguja dejará de moverse.

Glitches   Unwanted voltage spikes that are seen on a voltage trace. These are normally caused by intermittent opens or shorts.

Irregularidades espontáneos   Impulsos de tensión no deseables que se ven en una traza de tensión. Estos se causan normalmente por las aberturas o cortos intermitentes.

Grids   A series of horizontal, ceramic, silver-compounded lines that are baked into the surface of the window.

Rejillas   Una serie de líneas horizontales, de cerámica plateada, combinadas y endurecidas (horneadas) en la superficie de la ventana.

Ground   The common negative connection of the electrical system that is the point of lowest voltage.

Tierra   Conexión negativa común del sistema eléctrico. Es el punto de tensión más baja.

Ground circuit test   A diagnostic test performed to measure the voltage drop in the ground side of the circuit.

Prueba del circuito a tierra   Prueba diagnóstica llevada a cabo para medir la caída de tensión en el lado a tierra del circuito.

Ground side   The portion of the circuit that is from the load component to the negative side of the source.

Lado a tierra   Parte del circuito que va del componente de carga al lado negativo de la fuente.

Ground strap   A length of wire or small cable used to connect two components together so current can flow through both components; commonly used to provide a ground path from the engine block to the chassis.

Correa de tierra   Un trozo de alambre o un cable pequeño que se utiliza para conectar dos componentes entre sí para que la corriente pueda fluir entre ellos; se utiliza comúnmente para suministrar una puesta a tierra del bloque del motor al chasis.

Grounded circuit   An electrical defect that allows current to return to ground before it has reached the intended load component.

Circuito puesto a tierra   Falla eléctrica que permite el regreso de la corriente a tierra antes de alcanzar el componente de carga deseado.

Growler   Test equipment used to test starter armatures for shorts and grounds. It produces a very strong magnetic field that is capable of inducing a current flow and magnetism in a conductor.

Indicador de cortocircuitos   Equipo de prueba utilizado para localizar cortociruitos y tierra en armaduras de arranque. Genera un campo magnético sumamente fuerte, capaz de inducir flujo de corriente y magnetismo en un conductor.

Hall-effect switch   A sensor that operates on the principle that if a current is allowed to flow through a thin conducting material exposed to a magnetic field, another voltage is produced.

Conmutador de efecto Hall   Sensor que funciona basado en el principio de que si se permite el flujo de corriente a través de un material conductor delgado que ha sido expuesto a un campo magnético, se produce otra tensión.

Halogen   The term used to identify a group of chemically related non-metallic elements. These elements include chlorine, fluorine, and iodine.

Halógeno   Término utilizado para identificar un grupo de elementos no metálicos relacionados químicamente. Dichos elementos incluyen el cloro, el flúor y el yodo.

Hand tools   Tools that use only the force generated from the body to operate.

Herramientas manuales   Herramientas que sólo utilizan la fuerza que genera el cuerpo para manejarlas.

Hard codes   Failures that were detected the last time the BCM tested the circuit.

Códigos indicadores de dureza   Fallas que fueron detectadas la última vez que el funcionamiento indicado con las letras BCM probó el circuito.

Hard-shell connector   An electrical connector that has a hard plastic shell which holds the connecting terminals of separate wires.

Conectador de casco duro   Conectador eléctrico con casco duro de plástico que sostiene separados los bornes conectadores de alambres individuales.

Hazard Communication Standard   The original bases of the right-to-know laws.

Normalización de Comunicado sobre Riesgos   Las bases originales de las leyes de derecho de información.

Hazardous materials   Materials that can cause illness, injury, or death or that pollute water, air, or land.

Materiales peligrosos   Materiales que pueden causar enfermedades, daños, o la muerte, o que contaminan el agua o la tierra.

Hazardous waste   A waste that is on the Environmental Protection Agency (EPA) list of known and harmful materials.

Desecho peligroso   Desecho que está en la lista de materiales conocidos y peligrosos de la Agencia de Protección Ambiental (EPA).

Heat shrink tube   A hollow insulation material that shrinks to an airtight fit over a connection when exposed to heat.

Tubería contraída térmicamente   Material aislante hueco que se contrae para acomodarse herméticamente sobre una conexión cuando se encuentra expuesto al calor.

Hertz   A measurement of frequency.

Hertzios   Es una unidad de frecuencia.

High-efficiency particulate arresting (HEPA)   A special vacuum cleaner used to capture asbestos dust.

Detención de partículas de alto rendimiento (HEPA)   Una aspiradora especial que se utiliza para capturar polvo de asbesto.

High-voltage electronic control unit (HV ECU)   In an HEV, this unit controls the motor generators and the engine based on torque demand.

Unidad de control electrónico de alto voltaje (UCE AV)   En un VHE controla los generadores del motor y el motor basándose en la demanda del par motor.

High-voltage service plug   The EV and HEV is equipped with a high-voltage service plug that disconnects the HV battery from the system.

Bujía de servicio de alto voltaje   El VE y el VHE están equipados con una bujía de servicio de alto voltaje que desconecta la batería del VH del sistema.

History data   Data that is stored when a fault is set in an EV or HEV, which can be used to determine if a customer's concern is actually a problem with the system.

Historial de datos   Datos que se archivan cuando se produce una falla en un VE o VHE y que pueden usarse para determinar si la preocupación del cliente es en realidad un problema relativo al sistema.

Hoist   A lift that is used to raise an entire vehicle.

Elevador   Montacargas utilizado para elevar el vehículo en su totalidad.

Hold-in winding   A winding that holds the plunger of a solenoid in place after it moves to engage the starter drive.

Devanado de retención   Un devanado que posiciona el núcleo móvil de un solenoide después de que mueva para accionar el acoplamiento del motor de arranque.

Hydrometer   A test instrument used to check the specific gravity of the electrolyte to determine the battery's state of charge.

Hidrómetro   Instrumento de prueba utilizado para verificar la gravedad específica del electrolito y así determinar el estado de carga de la batería.

Hydrostatic lock   A condition where noncompressible liquid that is trapped in the combustion chamber of the engine prevents the piston from moving upward.

Bloqueo hidrostático   Una condición en la cual el líquido no comprimible que queda atrapado en la cámara de combustión del motor evita que el pistón se mueva hacia arriba.

Igniter   1. A combustible device that converts electrical energy into thermal energy to ignite the inflator propellant in an air bag system. 2. An electronic device used in an HID headlight to provide the initial 10,000 to 25,000 volts (V) to jump the gap of the electrodes.

Encendedor   1. Dispositivo combustible que convierte la energía eléctrica en energía térmica para encender el propelente inflador del sistema de bolsas de aire. 2. Dispositivo electrónico usado en un faro DAI para proporcionar los 10,000 a 25,000 V iniciales requeridos para saltar la separación de los electrodos.

Illuminated entry actuator   Contains a printed circuit and a relay.

Actuador iluminado de entradas   Contiene un circuito impreso y un relevador.

Immobilizer system   Designed to provide protection against unauthorized vehicle use by disabling engine starting if an invalid key is used to start the vehicle or if an attempt to hot-wire the ignition system is made.

Sistema de inmovilización   Sistema diseñado para proporcionar protección contra el uso no autorizado del vehículo que desactiva el motor si una llave no válida se utiliza para encender el motor o para hacer un puente con la intención de encender el motor sin la llave correspondiente.

Impedance   The combined opposition to current created by the resistance, capacitance, and inductance of a test meter or circuit.

Impedancia   Oposición combinada a la corriente generada por la resistencia, la capacitancia y la inductancia de un instrumento de prueba o de un circuito.

Inductive reactance   The result of current flowing through a conductor and the resultant magnetic field around the conductor that opposes the normal flow of current.

Reactancia inductiva   El resultado de un corriente que circule por un conductor y que resulta en un campo magnético alrededor del conductor que opone el flujo normal del corriente.

Information codes   EV and HEV codes that provide more information and freeze frame data concerning the DTCs. Information codes along with the DTC indicate more precisely the location of the fault.

Códigos de información   Códigos del VE y VHE que proporcionan más información y datos de trama fija relativos a los códigos de diagnóstico de fallas (CDF). Junto con los CDF, los códigos de información indican con más precisión la localización de la falla.

Instrument voltage regulator (IVR)   Provides a constant voltage to the gauge regardless of the voltage output of the charging system.

Instrumento regulador de tensión   Le provee tensión constante al calibrador, sin importar cual sea la salida de tensión del sistema de carga.

Insulated circuit resistance test   A voltage drop test that is used to locate high resistance in the starter circuit.

Prueba de la resistencia de un circuito aislado   Prueba de la caída de tensión utilizada para localizar alta resistencia en el circuito de arranque.

Insulated side   The portion of the circuit from the positive side of the source to the load component.

Lado aislado   Parte del circuito que va del lado positivo de la fuente al componente de carga.

Insulator   A material that does not allow electrons to flow easily through it.

Aislador   Una material que no permite circular fácilmente los electrones.

Integrated circuit (IC chip)   A complex circuit of thousands of transistors, diodes, resistors, capacitors, and other electronic devices that are formed into a small silicon chip. As many as 30,000 transistors can be placed on a chip that is 1/4 inch (6.35 mm) square.

Circuito integrado (Fragmento CI)   Circuito complejo de miles de transistores, diodos, resistores, condensadores, y otros dispositivos electrónicos formados en un fragmento pequeño de silicio. En un fragmento de 1/4 de pulgada (6,35 mm) cuadrada, pueden colocarse hasta 30.000 transistores.

Intermittent codes   DTCs for faults that have occurred in the past but were not present during the last BCM test of the circuit.

Códigos intermitentes   Códigos de diagnóstico de fallas pasadas que no se presentaron durante la última prueba BCM del circuito.

Inverter   Controls current flow between the motor generators and the HV battery in an EV or HEV; converts HV battery DC voltage into three-phase alternating current for the motor generators and also converts high-voltage AC from the motor generators to DC voltage to charge the HV battery.

Inversor   Controla el flujo de corriente entre los generadores del motor y la batería del VH en un VE o VHE; convierte el voltaje CC de la batería del VH en corriente alterna de tres fases para los generadores del motor, y también convierte la corriente alterna de alto voltaje de los generadores del motor en voltaje de corriente continua para cargar la batería VH.

J1962 breakout box (BOB)   A special tool that provides a passthrough test point which connects in series between the DLC and the scan tool. This provides easy testing of voltages and resistance of any of the DLC circuits without risk of damage to the DLC.

Caja de conexiones J1962   Herramienta especial que proporciona un paso mediante el punto de prueba que se conecta en series entre el circuito del control del enlace de datos (DLC) y el instrumento de exploración. Esto proporciona una prueba fácil de los voltajes y la resistencia en cualquiera de los circuitos del DLC sin riesgo de dañar el DLC.

Jump assist   A feature of the HEV to provide for the 12-volt auxiliary battery to recharge the HV battery.

Puente auxiliar   Una característica del HEV para que la batería auxiliar de 12 voltios recargue la batería HV.

Jumper wire   A wire used in diagnostics that is made up of a length of wire with a fuse or circuit breaker and has alligator clips on both ends.

Cable conector   Una alambre empleado en los diagnósticos que se comprende de un trozo de alambre con un fusible o un interruptor y que tiene una pinza de conexión en ambos lados.

Keyless entry   A lock system that allows for locking and unlocking of a vehicle with a touch keypad instead of a key.

Entrada sin llave   Un sistema de cerradura que permite cerrar y abrir un vehículo por medio de un teclado en vez de utilizar una llave.

Knock   The spontaneous auto-ignition of the remaining air-fuel mixture in the engine combustion chamber that occurs after normal combustion has started causing the formation of standing ultrasonic waves.

Autoencendido   El encendido automático espontáneo de la mezcla de combustible/aire restante en la cámara de combustión del motor que se produce después de que comienza la combustión normal, y que causa la formación de ondas ultrasónicas estacionarias.

Knock sensor   Measures engine knock, or vibration, and converts the vibration into a voltage signal.

Sensor de autoencendido   Mide el autoencendido del motor, o vibración, y convierte la vibración en una señal de voltaje.

KOEO   Key ON, Engine OFF.

KOEO   Llave puesta, motor apagado.

KOER   Key ON, Engine RUNNING.

KOER   Llave puesta, motor en marcha.

kV   Kilovolt or 1,000 volts.

kV   Kilovolito o 1000 voltios.

Lamination   The process of constructing something with layers of materials that are firmly connected.

Laminación   El proceso de construir algo de capas de materiales unidas con mucha fuerza.

Lamp   A device that produces light as a result of current flow through a filament. The filament is enclosed within a glass envelope and is a type of resistance wire that is generally made from tungsten.

Lámpara   Dispositivo que produce luz como resultado del flujo de corriente a través de un filamento. El filamento es un tipo de alambre de resistencia hecho por lo general de tungsteno, que es encerrado dentro de una bombilla.

Lamp sequence check   Used to determine problems by observing the operation of the warning lights under different conditions.

Lámpara verificadora de secuencias   Usada para determinar problemas al observar el funcionamiento de las luces de alerta bajo condiciones diferentes.

Light-emitting diode (LED)   A gallium-arsenide diode that converts the energy developed when holes and electrons collide during normal diode operation into light.

Diodo emisor de luz   Diodo semiconductor de galio y arseniuro que convierte en luz la energía producida por la colisión de agujeros y electrones durante el funcionamiento normal del diodo.

Limit switch   A switch used to open a circuit when a predetermined value is reached. Limit switches are normally responsive to a mechanical movement or temperature changes.

Disyuntor de seguridad   Un conmutador que se emplea para abrir un circuito al alcanzar un valor predeterminado. Los disyuntores de seguridad suelen ser responsivos a un movimiento mecánico o a los cambios de temperatura.

Liquid crystal display (LCD)   A display that sandwiches electrodes and polarized fluid between layers of glass. When voltage is applied to the electrodes, the light slots of the fluid are rearranged to allow light to pass through.

Visualizador de cristal líquido   Visualizador digital que consta de dos láminas de vidrio selladas, entre las cuales se encuentran los electrodos y el fluido polarizado. Cuando se aplica tensión a los electrodos, se rompe la disposición de las moléculas para permitir la formación de carácteres visibles.

Live module   An ABS module that has not been deployed.

Módulo activo   El módulo que no ha sido desplegado.

Load shedding   The process of reducing electrical power consumption by turning off non-essential electrical loads.

Desconexión de cargas   Proceso por el que se reduce el consumo de energía eléctrica desconectando cualquier carga eléctrica que no sea esencial.

Logic probe   A test instrument used to detect a pulsing signal.

Sonda lógica   Un instrumento de prueba que se emplea para detectar una señal pulsante.

Magnetic induction sensors   Sensors that use the principle of inducing a voltage into a winding by using a moving magnetic field.

Sensores de inducción magnética   Sensores que utilizan el principio de inducir un voltaje en un devanado mediante un campo magnético móvil.

Magnetic pulse generator   Sensor that uses the principle of magnetic induction to produce a voltage signal. Magnetic pulse generators are commonly used to send data concerning the speed of the monitored component to the computer.

Generador de impulsos magnéticos   Sensor que funciona según el principio de inducción magnética para producir una señal de tensión. Los generadores de impulsos magnéticos se utilizan comúnmente para transmitir datos a la computadora relacionados a la velocidad del componente regulado.

Magnetism   An energy form resulting from atoms aligning within certain materials, giving the materials the ability to attract other metals.

Magnetismo   Forma de energía que resulta de la alineación de átomos dentro de ciertos materiales y que le da a éstos la capacidad de atraer otros metales.

Magnetoresistive (MR) sensors   Consist of the magnetoresistive sensor element, a permanent magnet, and an integrated signal conditioning circuit to make use of the magnetoresistive effect.

Sensores de movimiento magneto resistivos (MR) Están compuestos por el elemento sensor de movimiento magneto resistivo, un imán permanente y un circuito de condicionamiento de señales integrado para hacer uso del efecto magneto resistivo.

Material safety data sheets (MSDS) Contain information about each hazardous material in the workplace.

Hojas de datos de seguridad del material (HDSM) Contienen información sobre cada material peligroso en el lugar de trabajo.

Maxi-fuse A circuit protection device that looks similar to blade-type fuses except they are larger and have a higher amperage capacity. Maxi-fuses are used because they are less likely to cause an underhood fire when there is an overload in the circuit. If the fusible link burns in two, it is possible that the hot side of the fuse could come into contact with the vehicle frame and the wire could catch on fire.

Maxifusible Dispositivo de protección del circuito parecido a un fusible de tipo de cuchilla, pero más grande y con mayor capacidad de amperaje. Se utilizan maxifusibles porque existen menos probabilidades de que ocasionen un incendio debajo de la capota cuando ocurra una sobrecarga en el circuito. Si el cartucho de fusible se quemase en dos partes, es posible que el lado "cargado" del fusible entre en contacto con el armazón del vehículo y que el alambre se encienda.

Metri-pack connector Special wire connectors used in some computer circuits. They seal the wire terminals from the atmosphere, thereby preventing corrosion and other damage.

Conector metri-pack Los conectores de alambres especiales que se emplean en algunos circuitos de computadoras. Impermealizan los bornes de los alambres, así previniendo la corrosión y otros daños.

Millisecond Equals 1/100th of a second.

Milesegundos Equivale a 1/100 de un segundo.

Molded connector An electrical connector that usually has one to four wires that are molded into a one-piece component.

Conectador moldeado Conectador eléctrico que por lo general tiene hasta un máximo cuatro alambres que se moldean en un componente de una sola pieza.

MSDS Material safety data sheet.

MSDS Hojas de Dato de Seguridad de los Materiales.

Multifunction switch Can have a combination of any of the following switches in a single unit: headlights, turn signal, hazard, dimmer, horn, and flash to pass.

Interruptor de múltiples funciones Este permite lograr una combinación de cualquiera de los siguientes interruptores: faroles, señal de doblaje, peligro, interruptor reducidor de luz, cuerno, señal de pasada a otro carril.

Multimeter A test instrument that measures more than one electrical property.

Multímetro Un instrumento diagnóstico que mide más de una propriedad eléctrica.

Negative temperature coefficient (NTC) Thermistors reduce their resistance as the temperature increases.

Coeficiente de temperatura negativa (NTC, por su sigla en inglés) Termistores cuya estructura de cristal disminuye el valor de la resistencia elecétrica a medida que aumenta la temperatura.

Neutral safety switch A switch used to prevent the starting of an engine unless the transmission is in PARK or NEUTRAL.

Disyuntor de seguridad en neutral Un conmutador que se emplea para prevenir que arranque un motor al menos de que la transmisión esté en posición PARK r NEUTRAL.

No crank A term used to mean that when the ignition switch is placed in the START position, the starter does not turn the engine.

Sin arranque Término utilizado para expresar que cuando el botón conmutador de encendido está en la posición START, el arrancador no enciende el motor.

No-crank test Diagnostic test performed to locate any opens in the starter or control circuits.

Prueba sin arranque Prueba diagnóstica llevada a cabo para localizar aberturas en los circuitos de arranque o de mando.

Noise An unwanted voltage signal that rides on a signal. Noise is usually the result of radio frequency interface (RFI) or electromagnetic induction (EMI).

Ruido Una señal indeseada del voltaje que aparece montada en una señal. El ruido es generalmente el resultado de la radio frecuencia de contacto (RFI) o de la inducción electromagnética (EMI).

Normally closed (NC) switch A switch designation denoting that the contacts are closed until acted upon by an outside force.

Conmutador normalmente cerrado Nombre aplicado a un conmutador cuyos contactos permanecerán cerrados hasta que sean accionados por una fuerza exterior.

Normally open (NO) switch A switch designation denoting that the contacts are open until acted upon by an outside force.

Conmutador normalmente abierto Nombre aplicado a un conmutador cuyos contactos permanecerán abiertos hasta que sean accionados por una fuerza exterior.

NTC Negative temperature coefficient.

NTC Las iniciales NTC representan la temperatura coeficiente negativa.

OBD II Stands for on-board diagnostics, second generation.

OBD II Las letras OBD II se refieren a los diagnósticos del tablero, segunda generación.

Occupational safety glasses Eye protection that is designed with special high-impact lens and frames, and side protection.

Gafas de protección para el trabajo Gafas diseñadas con cristales y monturas especiales resistentes y provistas de protección lateral.

OCS service kit An OCS part set that consists of the seat foam, the bladder, the pressure sensor, the occupant classification module (OCM), and the wiring. The service kit is calibrated as an assembly. The service kit OCM has a special identification data bit that is transmitted once it is connected.

Kit de servicio SCO Juego de partes SCO que consiste en la huleespuma del asiento, el depósito, el sensor de presión, el módulo de clasificación del ocupante (MCO) y el alambrado. El kit de servicio está calibrado como un ensamblado. El kit de servicio SCO tiene un bit de datos de identificación especial que se transmiten cuando se haya conectado.

OCS validation test Special procedure that confirms the OCS can properly classify the occupant. This task usually requires the use of special weights.

Prueba de revalidación del SCO Procedimiento especial que confirma que el SCO puede clasificar apropiadamente al ocupante. Esta tarea generalmente requiere el uso de pesas especiales.

Odometer A mechanical or digital counter in the speedometer unit that indicates total miles accumulated on the vehicle.

Odómetro Contador mecánico o digital del velocímetro que indica el total de millas recorridas por el vehículo.

Offset Placed off center. Refers to the number of degrees a timing light or meter should be set to provide accurate ignition timing readings.

Desviación Ubicado fuera de lo central. Se refiere al número de grados que se debe ajustar una luz de temporización o un medidor para proveer las lecturas exactas del tiempo de encendido.

Ohm Unit of measure for resistance. One ohm is the resistance of a conductor such that a constant current of one ampere in it produces a voltage of one volt between its ends.

Ohmio Unidad de resistencia eléctrica. Un ohmio es la resistencia de un conductor si una corriente constante de 1 amperio en el conductor produce una tensión de 1 voltio entre los dos extremos.

Ohmmeter A test meter used to measure resistance and continuity in a circuit.

Ohmiómetro Instrumento de prueba utilizado para medir la resistencia y la continuidad en un circuito.

Ohm's law Defines the relationship between current, voltage, and resistance.

Ley de Ohm Define la relación entre la corriente, la tensión y la resistencia.

One-hand rule A service procedure used on high-voltage EV and HEV vehicles that requires working with only one hand to prevent an electric shock from the high-voltage system passing through your body.

Regla de una mano Un procedimiento de servicio que se utiliza en vehículos EV y HEV de alto voltaje, el cual se opera con una sola mano para evitar que pase una descarga electrostática del sistema de alto voltaje a través del cuerpo.

Open An electrical term used to indicate that a circuit is not complete or is broken.

Abierto Es un término de electricidad, usado para indicar que el circuito no está completo o está roto.

Open circuit A term used to indicate that current flow is stopped. By opening the circuit, the path for electron flow is broken.

Circuito abierto Término utilizado para indicar que el flujo de corriente ha sido detenido. Al abrirse el circuito, se interrumpe la trayectoria para el flujo de electrones.

Open circuit testing Testing for voltage by disconnecting a connector and then measuring for applied voltage.

Prueba de circuito abierto Prueba de voltaje en donde se desconecta un conector y luego se mide el voltaje aplicado.

Open circuit voltage test Used to determine the battery's state of charge. It is used when a hydrometer is not available or cannot be used.

Prueba de la tensión en un circuito abierto Sirve para determinar el estado de carga de la batería. Esta prueba se lleva a cabo cuando no se dispone de un hidrómetro o cuando el mismo no puede utilizarse.

Optics test Test of the automatic high-beam system to test the ability to recognize ambient light through the lens and the windshield.

Prueba óptica Prueba del sistema automático de luces altas para averiguar la habilidad de reconocer las luces ambientales mediante la lente y el parabrisas.

Overload Excess current flow in a circuit.

Sobrecarga Flujo de corriente superior a la que tiene asignada un circuito.

Overrunning clutch A clutch assembly on a starter drive used to prevent the engine's flywheel from turning the armature of the starter motor.

Embrague de sobremarcha Una asamblea de embrague en un acoplamiento del motor de arranque que se emplea para prevenir que el volante del motor dé vueltas al armazón del motor de arranque.

Oxygen sensor A voltage-generating sensor that measures the amount of oxygen present in an engine's exhaust.

Sensor de oxígeno Un sensor generador de tensión que mide la cantidad del oxígeno presente en el gas de escape de un motor.

Parallel circuit A circuit that provides two or more paths for electricity to flow.

Circuito en paralelo Circuito que provee dos o más trayectorias para que circule la electricidad.

Parasitic drains Constant drains of the battery due to accessories that draw small amounts of current.

Drenaje parásita Los constantes drenajes en la batería son causados debido a que los accesorios atraen pequeñas cantidades de corriente.

Parasitic loads Electrical loads that are still present when the ignition switch is in the OFF position.

Cargas parásitas Cargas eléctricas que todavía se encuentran presentes cuando el botón conmutador de encendido está en la posición OFF.

Park assist system (PAS) A parking aid that alerts the driver to obstacles that are located in the path immediately behind the driver.

Sistema de ayuda al aparcamiento (PAS) Ayuda al aparcamiento que avisa al conductor de los obstáculos que se encuentran en la trayectoria inmediatamente detrás del conductor.

Park switch Contact points located inside the wiper motor assembly that supply current to the motor after the wiper control switch has been turned to the PARK position. This allows the motor to continue operating until the wipers have reached their PARK position.

Conmutador PARK Puntos de contacto ubicados dentro del conjunto del motor del frotador que le suministran corriente al motor después de que el conmutador para el control de los frotadores haya sido colocado en la posición PARK. Esto permite que el motor continue su funcionamiento hasta que los frotadores hayan alcanzado la posición original.

Passive seat belt system Seat belt operation that automatically puts the shoulder and/or lap belt around the driver or occupant. The automatic seat belt is moved by DC motors that move the belts by means of carriers on tracks.

Sistema pasivo de cinturones de seguridad Función de los cinturones de seguridad que automáticamente coloca el cinturón superior y/o inferior sobre el conductor o pasajero. Motores de corriente continua accionan los cinturones automáticos mediante el uso de portadores en pistas.

Peak The highest voltage value in one cycle of an AC voltage sine wave.

Máxima eficiencia El valor más alto del voltaje de un ciclo de corriente alterna (AC) en un voltaje de onda senoidal.

Peak-to-peak voltage The total voltage measured between the peaks of an AC sine wave.

Voltaje de máximas eficiencias La medida total del voltaje cuando los puntos son de máxima eficiencia en una onda senoidal de corriente alterna (AC).

Photocell A variable resistor that uses light to change resistance.

Fotocélula Resistor variable que utiliza luz para cambiar la resistencia.

Photocell resistance assembly A technician-made test tool that replaces the photocell to produce predictable results.

Montaje de resistencia fotocélula Una herramienta técnica de prueba que remplaza la fotocélula para producir resultados visibles.

Phototransistor A transistor that is sensitive to light.

Fototransistor Transistor sensible a la luz.

Photovoltaic diodes Diodes capable of producing a voltage when exposed to radiant energy.

Diodos fotovoltaicos Diodos capaces de generar una tensión cuando se encuentran expuestos a la energía de radiación.

Pickup coil The stationary component of the magnetic pulse generator consisting of a weak permanent magnet that is wound around by fine wire. As the timing disc rotates in front of it, the changes of magnetic lines of force generate a small voltage signal in the coil.

Bobina captadora Componente fijo del generador de impulsos magnéticos compuesta de un imán permanente débil devanado con alambre fino. Mientras gira el disco sincronizador enfrente de él, los cambios de las líneas de fuerza magnética generan una pequeña señal de tensión en la bobina.

Piezoelectric sensors Produce a proportional voltage output resulting from deformation of the element as pressure is applied.

Sensores piezoeléctricos Producen una salida de voltaje proporcional que es el resultado de una deformación del elemento a medida que se aplica presión.

Piezoelectric transducer A sensor that is capable of measuring the pressures associated with ultrasonic waves.

Transductor piezoeléctrico Un sensor capaz de medir las presiones relacionadas con las ondas ultrasónicas.

Piezoresistive Resistance that changes in value as the pressure applied to the sensing material changes.

Piezoresistencia Resistencia que cambia su valor a medida que varía la presión que se aplica al material de detección.

Piezoresistive sensor A sensor that is sensitive to pressure changes.

Sensor piezoresistivo Sensor susceptible a los cambios de presión.

PIN code A personal identification code used by the manufacturer to allow reprogramming and other services to the immobilizer system.

Código NIP Un código de identificación personal utilizado por el fabricante para permitir la reprogramación y otros servicios en el sistema inmovilizador.

Pinion gear A small gear; typically refers to the drive gear of a starter drive assembly or the small drive gear in a differential assembly.

Engranaje de piñon Un engranaje pequeño; tipicamente se refiere al engranaje de arranque de una asamblea de motor de arranque o al engranaje de mando pequeño de la asamblea del diferencial.

Pinpoint test A specific component test.

Prueba de precisión Una prueba hecha con un componente específico.

Plate straps Metal connectors used to connect the positive or negative plates in a battery.

Abrazaderas de la placa Los conectores metálicos que sirven para conectar las placas positivas o negativas de una batería.

Plates The basic structure of a battery cell; each cell has at least one positive plate and one negative plate.

Placas La estructura básica de una celula de batería; cada celula tiene al menos una placa positiva y una placa negativa.

PMGR An abbreviation for permanent magnet gear reduction.

PMGR Una abreviación de desmultiplicación del engranaje del imán permanente.

Pneumatic tools   Power tools that are powered by compressed air.

Herrimientas neumáticas   Herramientas mecánicas accionadas por aire comprimido.

Polarizers   Glass sheets that make light waves vibrate in only one direction. This converts light into polarized light.

Polarizadores   Las láminas de vidrio que hacen vibrar las ondas de luz en un sólo sentido. Esto convierte la luz en luz polarizada.

Polarizing   The process of light polarization or of setting one end of a field as a positive or negative point.

Polarizadora   El proceso de polarización de la luz o de establecer un lado de un campo como un punto positivo o negativo.

Polymer   Compound that consists of a plastic with carbon grains.

Polímero   Compuesto fabricado a partir de un plástico con granos de carbón.

Positive engagement starter   A type of starter that uses the magnetic field strength of a field winding to engage the starter drive into the flywheel.

Acoplamiento de arranque positivo   Un tipo de arrancador que utilisa la fuerza del campo magnético del devanado inductor para accionar el acoplamiento del arrancador en el volante.

Positive temperature coefficient (PTC)   Thermistors that increase their resistance as the temperature increases.

Coeficiente de temperatura positiva (PTC, por su sigla en inglés)   Termistores que aumentan su resistencia a medida que aumenta la temperatura.

Potential   The ability to do something; typically voltage is referred to as the potential. If you have voltage, you have the potential for electricity.

Potencial   La capacidad de efectuar el trabajo; típicamente se refiere a la tensión como el potencial. Si tiene tensión, tiene la potencial para la electricidad.

Potentiometer   A voltage divider circuit that is used to measure linear or rotary movement.

Potenciómetro   Un circuito divisor de voltaje que se utiliza para medir el movimiento lineal o rotativo.

Power formula   A formula used to calculate the amount of electrical power a component uses. The formula is P = I × E, where P stands for power (measured in watts), I stands for current, and E stands for voltage.

Formula de potencia   Una formula que se emplea para calcular la cantidad de potencia eléctrica utilizada por un componente. La formula es P = I × E, en el que el P quiere decir potencia (medida en wats), I representa el corriente y el E representa la tensión.

Power tools   Tools that use forces other than those generated from the body. They can use compressed air, electricity, or hydraulic pressure to generate and multiply force.

Herramientas mecánicas   Herramientas que utilizan fuerzas distintas a las generadas por el cuerpo. Dichas fuerzas pueden ser el aire comprimido, la electricidad, o la presión hidráulica para generar y multiplicar la fuerza.

Powertrain control module (PCM)   The computer that controls the engine operation.

Módulo de control de potencia   El computador que controla la operación del motor.

Pressure control solenoid   A solenoid used to control the pressure of a fluid, commonly found in electronically controlled transmissions.

Solenoide de control de la presión   Un solenoide que controla la presión de un fluido, suele encontrarse en las transmisiones controladas electronicamente.

Primary wiring   Conductors that carry low voltage and low current. The insulation of primary wires is usually thin.

Hilos primarios   Hilos conductores de tensión y corriente bajas. El aislamiento de hilos primarios es normalmente delgado.

Program   A set of instructions that the computer must follow to achieve desired results.

Programa   Conjunto de instrucciones que la computadora debe seguir para lograr los resultados deseados.

PROM (programmable read only memory)   Memory chip that contains specific data which pertains to the exact vehicle the computer is installed in. This information may be used to inform the CPU of the accessories that are equipped on the vehicle.

PROM (memoria de sólo lectura programable)   Fragmento de memoria que contiene datos específicos referentes al vehículo particular en el que se instala la computadora. Esta información puede utilizarse para informar a la UCP sobre los accesorios de los cuales el vehículo está dotado.

Protection device   Circuit protector that is designed to turn off the system that it protects. This is done by creating an open to prevent a complete circuit.

Dispositivo de protección   Protector de circuito diseñado para "desconectar" el sistema al que provee protección. Esto se hace abriendo el circuito para impedir un circuito completo.

Prove-out circuit   A function of a circuit that completes the warning light circuit to ground when the ignition switch is in the START position. The warning light is on during engine cranking to indicate to the driver that the bulb is working properly.

Circuito de prueba   Función de un circuito que completa el circuito de la luz de advertencia a tierra cuando el interruptor de encendido se encuentra en la posición ENCENDIDO. La luz de advertencia se encenderá durante el arranque del motor para avisar al conductor que la bombilla funciona correctamente.

PTC   Positive temperature coefficient.

PTC   Iniciales identificadoras de la temperatura positiva coeficiente.

Pull to seat   A method used to install the terminals into the connector.

Tire el asiento   Un método usado para instalar los terminales dentro del conectador.

Pulse width   The length of time in milliseconds that an actuator is energized.

Duración de impulsos   Espacio de tiempo en milisegundos en el que se excita un accionador.

Pulse width modulation   On-off cycling of a component. The period of time for each cycle does not change; only the amount of on time in each cycle changes.

Modulación de duración de impulsos   Modulación de impulsos de un componente. El espacio de tiempo de cada ciclo no varía; lo que varía es la cantidad de trabajo efectivo de cada ciclo.

Push to seat   A method used to install the terminals into the connector.

Empuje el asiento   Un método usado para instalar los terminales dentro del conectador.

Quartz swing needle   Displays that are similar in design to the air-core electromagnetic gauges used in conventional analog instrument panels.

Aguja de oscilación del cuarzo   Mostrarios que son similares en diseño al del núcleo de aire de los calibres electromagnéticos usados convencionalmente en los paneles de instrumentos análogos.

Quick test   Isolates the problem area and determines whether the starter motor, solenoid, or control circuit is at fault.

Prueba rápida   Aisla el área con problemas y determina si el arrancador del motor, el selenoid, y el circuito de control están fallando.

Radial grid   A type of battery grid that has its patterns branching out from a common center.

Rejilla radial   Un tipo de rejilla de bateria cuyos diseños extienden de un centro común.

Radiated noise   Signals that are generated from an electrical accessory and are detected by the antenna.

Ruido radiado   Señales que se generan desde un accesorio eléctrico y que son detectadas por la antena.

Radio choke   Absorbs voltage spikes and prevents static in the vehicle's radio.

Impedancia del radio   Absorba los impulsos de la tensión y previene la presencia del estático en el radio del vehículo.

Radio frequency interface (RFI)   Produced when electromagnetic radio waves of sufficient amplitude escape from a wire or connector.

Radio frecuencia de contacto (RFI)   Se produce cuando las ondas de radio eletromagnéticas y de suficiente amplitud se escapan de un cable o conector.

Radio interference   All undesirable signals that enter the radio and are heard as static or noise.

Interferencia de radio   Toda señal indeseable que entra a través de la radio y que se escucha en forma de estática o ruido.

Radio noise   An undesired interference or static noise that obstructs the normal sound of the radio station.

Ruido de radio   Interferencia o ruido estático indeseado que obstaculiza la emisión normal de las estaciones de radio.

Ratio   A mathematical relationship between two or more things.

Razón   Una relación matemática entre dos cosas o más.

Rectification   The conversion of AC current to DC current.

Rectificación   Proceso a través del cual la corriente alterna es transformada en una corriente continua.

Refractometer   A special meter used to measure the specific gravity of a liquid by the refraction of light.

Refractómetro   Herramienta especial que usa para medir la gravedad específica de un líquido por la refracción de la luz.

Relative compression test   A test method for determining if a cylinder has lower compression relative to other cylinders by using current draw during cranking.

Prueba de compresión relativa   Método de prueba para determinar si un cilindro tiene compresión más baja comparada a los otros cilindros al utilizar llamada de corriente durante la desencoladura.

Relay   A device that uses low current to control a high-current circuit. Low current is used to energize the electromagnetic coil, while high current is able to pass over the relay contacts.

Relé   Dispositivo que utiliza corriente baja para controlar un circuito de corriente alta. La corriente baja se utiliza para excitar la bobina electromagnética, mientras que la corriente alta puede transmitirse a través de los contactos del relé.

Remote keyless entry (RKE)   A radio frequency system that allows the vehicle to be unlocked or locked from a distance.

Entrada remota sin llave (RKE)   Un sistema de radiofrecuencia que permite el vehículo para ser abierta o cerrada desde una distancia.

Reserve-capacity rating   An indicator, in minutes, of how long a vehicle can be driven with the headlights on, if the charging system should fail. The reserve-capacity rating is determined by the length of time, in minutes, that a fully charged battery can be discharged at 25 amps before battery cell voltage drops below 1.75 volts per cell.

Clasificación de capacidad en reserva   Indicación, en minutos, de cuánto tiempo un vehículo puede continuar siendo conducido, con los faros delanteros encendidos, en caso de que ocurriese una falla en el sistema de carga. La clasificación de capacidad en reserva se determina por el espacio de tiempo, en minutos, en el que una batería completamente cargada puede descargarse a 25 amperios antes de que la tensión del acumulador de la batería disminuya a un nivel inferior de 1,75 amperios por acumulador.

Resistance   Opposition to current flow.

Resistencia   Oposición que presenta un conductor al paso de la corriente eléctrica.

Resistance wire   A special type of wire that has some resistance built into it. These typically are rated by ohms per foot.

Alambre de resistencia   Un tipo de alambre especial que por diseño tiene algo de resistencia. Estos tipicamente tienen un valor nominal de ohm por pie.

Resistive shorts   Shorts to ground that pass through a form of resistance first.

Cortocircuitos resistivos   Cortocircuitos a tierra que primero pasan por una forma de resistencia.

Resistor block   A series of resistors with different values.

Bloque resistor   Serie de resistores que tienen valores diferentes.

Resource Conservation and Recovery Act (RCRA)   Law that makes hazardous waste generators responsible for the waste from the time it becomes a waste material until proper waste disposal is completed.

Ley de la Conservación y Recuperación de los Recursos   Ley que hace responsables a los que generan residuos peligrosos por su desecho desde el momento en que se convierte en material de desecho hasta que se completa su destrucción apropiada de desechos.

Reversed bias   A positive voltage is applied to the N-type material and a negative voltage is applied to the P-type material of a semiconductor.

Polarización inversa   Tensión positiva aplicada al material N y tensión negativa aplicada al material P de un semiconductor.

RFI   Common acronym for radio frequency interference.

RFI   Una sigla común de la interferencia de radiofrecuencia.

Rheostat   A two-terminal variable resistor used to regulate the strength of an electrical current.

Reostático   Un resistor variable de dos terminales usado para regular la potencia de un circuito eléctrico.

Right-to-know laws   Laws concerning hazardous materials and wastes that protect every employee in a workplace by requiring employers to provide a safe working place as it relates to hazardous materials. The right-to-know laws state that employees have a right to know when the materials they use at work are hazardous.

Leyes de derecho de información   Leyes contra materiales peligrosos y desechos que protegen a cada empleado en un lugar de trabajo al requerir que los patrones proporcionen un lugar de trabajo segura contra los materiales peligrosos. Las leyes de derecho de información declaran que los empleados tienen derecho a saber cuando y qué materiales que usan en el trabajo son peligrosos.

RMS   Root mean square; a method for measuring AC voltage.

RMS   Raíz de la media de los cuadrados; un método para medir la tensión del corriente alterna.

Root mean square (RMS)   Meters convert AC signal to a comparable DC voltage signal.

Corriente efectiva (RMS)   Los medidores convierten la señal de corriente alterna (AC) a una señal de voltaje comparable (DC).

Rotor   The component of the AC generator that is rotated by the drive belt and creates the rotating magnetic field of the AC generator.

Rotor   Parte rotativa del generador de corriente alterna accionada por la correa de transmisión y que produce el campo magnético rotativo del generador de corriente alterna.

Safety goggles   Eye protection device that fits against the face and forehead to seal off the eyes from outside elements.

Gafas de seguridad   Dispositivo protector que se coloca delante de los ojos para preservarlos de elementos extraños.

Safety stands   Support devices used to hold the vehicle off the floor after it has been raised by a floor jack.

Soportes de seguridad   Dispositivos de soporte utilizados para sostener el vehículo sobre el suelo después de haber sido levantado con el gato de pie.

Scan tool   A microprocessor designed to communicate with the BCM. It accesses trouble codes and runs system operation, actuator, and sensor tests.

Herramienta analizadora   Un microprocesador diseñado para comunicarse usando la información de las iniciales BCM. Así se encuentran los códigos que indican problemas, cómo funciona el sistema de conducción, los actuadores y monitores de prueba.

Scanner   See scan tool.

Dispositivo de exploración   Véase herramienta analizadora.

Sealed-beam headlight   A self-contained glass unit that consists of a filament, an inner reflector, and an outer glass lens.

Faro delantero sellado   Unidad de vidrio que contiene un filamento, un reflector interior y una lente exterior de vidrio.

Secondary wiring   Conductors, such as battery cables and ignition spark plug wires, that are used to carry high voltage or high current. Secondary wires have extra-thick insulation.

Hilos secundarios   Conductores, tales como cables de batería e hilos de bujías del encendido, utilizados para transmitir tensión o corriente alta. Los hilos secundarios poseen un aislamiento sumamente grueso.

Self-test input (STI)   The single pigtail connector located next to the self-test connector.

Autoprueba de entrada (STI)   Simple cable flexible de conección localizado junto al conector de autoprueba.

Semiconductors   An element that is neither a conductor nor an insulator. Semiconductors are materials that conduct electric current under certain conditions, yet will not conduct under other conditions.

Semiconductores Elemento que no es ni conductor ni aislante. Los semiconductores son materiales que transmiten corriente eléctrica bajo ciertas circunstancias, pero no la transmiten bajo otras.

Sender unit The sensor for the gauge. It is a variable resistor that changes resistance values with changing monitored conditions.

Unidad emisora Sensor para el calibrador. Es un resistor variable que cambia los valores de resistencia según cambian las condiciones reguladas.

Sensitivity controls A potentiometer that allows the driver to adjust the sensitivity of the automatic dimmer system to surrounding ambient light conditions.

Controles de sensibilidad Un potenciómetro que permite que el conductor ajusta la sensibilidad del sistema de intensidad de iluminación automático a las condiciones de luz ambientales.

Sensor Any device that provides an input to the computer.

Sensor Cualquier dispositivo que le transmite información a la computadora.

Sensor cluster A sensor module containing both the lateral (side-to-side) and yaw (vehicle rotational) sensors.

Dispositivo de sensores Módulo que contiene los sensores laterales (lado a lado) y de rotación (vehicular).

Sensor dropout When an induction sensor stops producing an output voltage due to slow speeds of the target.

Desenganche del sensor Cuando un sensor de inducción deja de producir voltaje de salida debido a velocidades bajas del objetivo.

Series circuit A circuit that provides a single path for current flow from the electrical source through all the circuit's components, and back to the source.

Circuito en serie Circuito que provee una trayectoria única para el flujo de corriente de la fuente eléctrica a través de todos los componentes del circuito, y de nuevo hacia la fuente.

Series–parallel circuit A circuit that has some loads in series and some in parallel.

Circuito en series paralelas Circuito que tiene unas cargas en serie y otras en paralelo.

Series-wound motor A type of motor that has its field windings connected in series with the armature. This type of motor develops its maximum torque output at the time of initial start. Torque decreases as motor speed increases.

Motor con devanados en serie Un tipo de motor cuyos devanados inductores se conectan en serie con la armadura. Este tipo de motor desarrolla la salida máxima de par de torsión en el momento inicial de ponerse en marcha. El par de torsión disminuye al aumentar la velocidad del motor.

Servo Controls the position of the throttle by receiving a controlled amount of vacuum.

Control servo Este sirve para controlar la posición de la válvula reguladora al recibir una cantidad controlada de vacío.

Servomotor An electrical motor that produces rotation of less than a full turn. A feedback mechanism is used to position itself to the exact degree of rotation required.

Servomotor Motor eléctrico que genera rotación de menos de una revolución completa. Utiliza un mecanismo de realimentación para ubicarse al grado exacto de la rotación requerida.

Short An unwanted electrical path; sometimes this path goes directly to ground.

Corto Una trayectoria eléctrica no deseable; a veces este trayectoria viaja directamente a tierra.

Short to ground A condition that allows current to return to ground before it has reached the intended load component.

Corto a la tierra Su función permite que la corriente regrese a la tierra, antes de que haya llegado a un componente intencionalmente cargado.

Shorted circuit Allows current to bypass part of the normal path.

Circuito corto Este circuito permite que la corriente pase por una parte del recorrido normal.

Shunt-wound motor A type of motor whose field windings are wired in parallel to the armature. This type of motor does not decrease its torque as speed increases.

Motor con devanados en derivación Un tipo de motor cuyos devanados inductores se cablean paralelos a la armadura. Este tipo de motor no disminuya su par de torsión al aumentar la velocidad.

Shutter wheel A metal wheel consisting of a series of alternating windows and vanes. It creates a magnetic shunt that changes the strength of the magnetic field from the permanent magnet of the Hall-effect switch or magnetic pulse generator.

Rueda obturadora Rueda metálica compuesta de una serie de ventanas y aspas alternas. Genera una derivación magnética que cambia la potencia del campo magnético, del imán permanente del conmutador de efecto Hall o del generador de impulsos magnéticos.

Sine wave A waveform that shows voltage changing polarity.

Onda senoidal Una forma de onda que muestra un cambio de polaridad en la tensión.

Single-phase voltage The sine wave voltage induced in one conductor of the stator during one revolution of the rotor.

Tensión monofásica La tensión en forma de onda senoidal inducida en un conductor del estator durante una revolución del rotor.

Sinusoidal A waveform that is a true sine wave.

Senoidal Una forma de onda que es una onda senoidal verdadera.

Slow charging Battery charging rate between 3 and 15 amps for a long period of time.

Carga lenta Índice de carga de la batería de entre 3 y 15 amperios por un largo espacio de tiempo.

Slow crank A term used to mean that the starter drive engages the ring gear, but the engine turns too slow to start.

Arranque lento Término utilizado para expresar que el mecanismo de arranque engrana la corona pero que el motor arranca con mucha lentitud.

Soft codes Codes those that have occurred in the past, but were not present during the last BCM test of the circuit.

Códigos suaves Códigos que han ocurrido en el pasado, pero que no estaban presentes durante la última prueba BCM del circuito.

Soldering The process of using heat and solder (a mixture of lead and tin) to make a splice or connection.

Soldadura Proceso a través del cual se utiliza calor y soldadura (una mezcla de plomo y de estaño) para hacer un empalme o una conexión.

Solderless connectors Hollow metal tubes that are covered with insulating plastic. They can be butt connectors or terminal ends.

Conectadores sin soldadura Tubos huecos de metal cubiertos de plástico aislante. Pueden ser extremos de conectadores o de bornes.

Solenoid An electromagnetic device that uses movement of a plunger to exert a pulling or holding force.

Solenoide Dispositivo electromagnético que utiliza el movimiento de un pulsador para ejercer una fuerza de arrastre o de retención.

Solenoid circuit resistance test Diagnostic test used to determine the electrical condition of the solenoid and the control circuit of the starting system.

Prueba de la resistencia de un circuito solenoide Prueba diagnóstica utilizada para determinar la condición eléctrica del solenoide y del circuito de mando del sistema de arranque.

Speakers Turn electrical energy provided by the radio receiver amplifier into acoustical energy using a permanent magnet and electromagnet to move air to produce sound.

Altavoces Convierten la energía eléctrica que suministra el amplificador receptor de radio en energía acústica mediante el uso de un imán permanente y un electroimán, los cuales mueven el aire para producir sonido.

Specific gravity The weight of a given volume of a liquid divided by the weight of an equal volume of water.

Gravedad específica El peso de un volumen dado de líquido dividido por el peso de un volumen igual de agua.

Speedometer An instrument panel gauge that indicates the speed of the vehicle.

Velocímetro Calibrador en el panel de instrumentos que marca la velocidad del vehículo.

Splice The joining of single wire ends or the joining of two or more electrical conductors at a single point.

Empalme La unión de los extremos de un alambre o la unión de dos o más conductores eléctricos en un solo punto.

Splice clip   A special connector used along with solder to assure a good connection. The splice clip is different from solderless connectors in that it does not have insulation.

Grapa para empalme   Conectador especial utilizado junto con la soldadura para garantizar una conexión perfecta. La grapa para empalme se diferencia de los conectadores sin soldadura porque no está provista de aislamiento.

Square waves   Identified by having straight vertical sides and a flat top.

Ondas cuadradas   Para saber cuáles son, deben tener lados rectos verticales y una punta plana.

Stabilize   Removing the surface charge of the battery by placing a large load on the battery for 15 seconds.

Estabilizar   Para hacerlo hay que remover la superficie cargada de la batería, colocando en la batería, por l5 segundos, una carga grande.

Starter drive   The part of the starter motor that engages the armature to the engine flywheel ring gear.

Transmisión de arranque   Parte del motor de arranque que engrana la armadura a la corona del volante de la máquina.

State of charge   The condition of a battery's electrolyte and plate materials at any given time.

Estado de carga   Condición del electrolito y de los materiales de la placa de una batería en cualquier momento dado.

Stator   The stationary coil of the AC generator in which current is produced.

Estátor   Bobina fija del generador de corriente alterna donde se genera corriente.

Stator neutral junction   The common junction of Wye stator windings.

Unión de estátor neutral   La unión común de los devanados de un estátor Y.

Stepped resistor   A resistor that has two or more fixed resistor values.

Resistor de secciones escalonadas   Resistor que tiene dos o más valores de resistencia fija.

Stepper motor   An electrical motor that contains a permanent magnet armature with two or four field coils; can be used to move the controlled device to the desired location. By applying voltage pulses to selected coils of the motor, the armature turns a specific number of degrees. When the same voltage pulses are applied to the opposite coils, the armature rotates the same number of degrees in the opposite direction.

Motor paso a paso   Motor eléctrico que contiene una armadura magnética fija con dos o cuatro bobinas inductoras. Puede utilizarse para mover el dispositivo regulado a cualquier lugar deseado. Al aplicárseles impulsos de tensión a ciertas bobinas del motor, la armadura girará un número específico de grados. Cuando estos mismos impulsos de tensión se aplican a las bobinas opuestas, la armadura girará el mismo número de grados en la dirección opuesta.

Steering angle sensor (SAS)   A sensor used to determine steering wheel position and degrees of rotation.

Sensor de ángulo de dirección   Sensor usado para determinar la posición del volante de dirección y los grados de rotación.

System main relay (SMR)   Used to connect and disconnect the power source of the high-voltage system in an EV or HEV.

Relé principal del sistema (RPS)   Se usa para conectar y desconectar la fuente de energía del sistema de alto voltaje en un VE o VHE.

Tachometer   An instrument that measures the speed of the engine in revolutions per minute (rpm).

Tacómetro   Instrumento que mide la velocidad del motor en revoluciones por minuto (rpm).

Test light   Checks for electrical power in a circuit.

Luz de prueba   Verifica la potencia eléctrica en un circuito.

Thermistor   A solid-state variable resistor made from a semiconductor material that changes resistance as a function of temperature.

Termistor   Un resistor variable de estado sólido, fabricado a partir de material semiconductor que cambia la resistencia en función de la temperatura.

Timer control   A potentiometer that is part of the headlight switch in some systems. It controls the amount of time the headlights stay on after the ignition switch is turned off.

Control temporizador   Un potenciómetro que es parte del conmutador de los faros en algunos sistemas. Controla la cantidad del tiempo que quedan prendidos los faros después de apagarse la llave del encendido.

Tin   Process of applying solder to the tip of the soldering iron to provide better heating control.

Estaño   El proceso de aplicar soldadura en la punta de un hierro soldador para dar mejor control al calor.

Tire pressure monitoring (TPM) system   A safety system that notifies the driver if a tire is underinflated or overinflated.

Sistema de monitoreo de presión de las llantas (TPM, por su sigla en inglés)   Un sistema de seguridad que informa al conductor si una llanta está muy inflada o poco inflada.

Torque converter   A hydraulic device found on automatic transmissions. It is responsible for controlling the power flow from the engine to the transmission; works like a clutch to engage and disengage the engine's power to the drive line.

Convertidor de par   Un dispositivo hidraúlico en las transmisiones automáticas. Se encarga de controlar el flujo de la potencia del motor a la transmisión; funciona como un embrague para embragar y desembragar la potencia del motor con la flecha motríz.

Trimotor   A three-armature motor.

Trimotor   Es un motor de tres armaduras.

Troubleshooting   The diagnostic procedure of locating and identifying the cause of the fault. It is a step-by-step process of elimination using cause and effect.

Detección de fallas   Procedimiento diagnóstico a través del cual se localiza e identifica la falla. Es un proceso de eliminación que se lleva a cabo paso a paso por medio de causa y efecto.

TVRS   An abbreviation for television-radio-suppression cable.

TVRS   Una abreviación del cable de supresíon del televisión y radio.

Two-button diagnostics   The buttons on the instrument panel that, when pressed in the right combination, place the module into self-test mode.

Diagnósticos de dos botones   Estos son los botones que aparecen en el panel de instrumentos, los cuales al presionarlos en la combinación correcta, ponen el módulo en estado de autoprueba.

Two-coil gauge   A gauge design that uses the interaction of two electromagnets and the total field effect upon an armature to cause needle movement.

Calibrador de dos bobinas   Calibrador diseñado para utilizar la interacción de dos electroimanes y el efecto inductor total sobre una armadura para generar el movimiento de la aguja.

U-codes   Diagnostic trouble codes assigned to the vehicle communication network.

Códigos en U   Códigos de fallo de diagnóstico asignados a la red de comunicación del vehículo.

Vacuum distribution valve   A valve used in vacuum-controlled concealed headlight systems. It controls the direction of vacuum to various vacuum motors or to vent.

Válvula de distribución al vacío   Válvula utilizada en el sistema de faros delanteros ocultos controlado al vacío. Regula la dirección del vacío a varios motores al vacío o sirve para dar salida del sistema.

Vacuum fluorescent display (VFD)   A display type that uses anode segments coated with phosphor and bombarded with tungsten electrons to cause the segments to glow.

Visualización de fluorescencia al vacío   Tipo de visualización que utiliza segmentos ánodos cubiertos de fósforo y bombardeados de electrones de tungsteno para producir la luminiscencia de los segmentos.

Valve body   A unit that consists of many valves and hydraulic circuits. This unit is the central control point for gear shifting in an automatic transmission.

Cuerpo de la válvula   Una unedad que consiste de muchas válvulas y circuitos hidráulicos. Esta unedad es el punto central de mando para los cambios de velocidad en una transmisión automática.

Variable resistor   A resistor that provides for an infinite number of resistance values within a range.

Resistor variable   Resistor que provee un número infinito de valores de resistencia dentro de un margen.

Vehicle identification number (VIN)   A number that is assigned to a vehicle for identification purposes. The identification plate is usually located on the cowl, next to the left-upper instrument panel.

Número de identificación del vehículo Número asignado a cada vehículo para fines de identificación. Por lo general, la placa de identificación se ubica en la bóveda, al lado del panel de instrumentos superior de la izquierda.

Vehicle lift points The areas that the manufacturer recommends for safe vehicle lifting. They are the areas that are structurally strong enough to sustain the stress of lifting.

Puntos para elevar el vehículo Áreas específicas que el fabricante recomienda para sujetar el vehículo a fin de lograr una elevación segura. Son las áreas del vehículo con una estructura suficientemente fuerte para sostener la presión de la elevación.

Vehicle module scan A special function of the scan tools that queries all modules on the bus to respond and then lists those that did reply.

Explorador del módulo del vehículo Función especial de los instrumentos de exploración que interroga a todos los módulos en el bus para que respondan, y luego hace una lista de los que respondieron.

Volatile A substance that vaporizes or explodes easily.

Volátil Sustancia que se evapora o explota con facilidad.

Volatility The tendency of a fluid to evaporate quickly or pass off in the form of vapor.

Volatilidad La tendencia de un fluido a evaporarse rápidamente o disiparse en forma de vapor.

Volt The unit used to measure the amount of electrical force.

Voltio Unidad práctica de tensión para medir la cantidad de fuerza eléctrica.

Voltage The difference or potential that indicates an excess of electrons at the end of the circuit the farthest from the electromotive force. It is the electrical pressure that causes electrons to move through a circuit. One volt is the amount of pressure required to move one amp of current through one ohm of resistance.

Tensión Diferencia o potencial que indica un exceso de electrones al punto del circuito que se encuentra más alejado de la fuerza electromotriz. La presión eléctrica genera el movimiento de electrones a través de un circuito. Un voltio equivale a la cantidad de presión requerida para mover un amperio de corriente a través de un ohmio de resistencia.

Voltage drop A resistance in the circuit that reduces the electrical pressure available after the resistance. The resistance can be the load component, the conductors, any connections, or an unwanted resistance.

Caída de tensión Resistencia en el circuito que disminuye la presión eléctrica disponible después de la resistencia. La resistencia puede ser el componente de carga, los conductores, cualquier conexión o resistencia no deseada.

Voltage drop test Determines if the battery, regulator, and AC generator are all operating at the same potential.

Prueba de la caída de tensión en el voltaje Determina si la batería, el regulador, y el generador de corriente alterna AC están todos funcionando al mismo potencial.

Voltage limiter Connected through the resistor network of a voltage regulator. It determines whether the field will receive high, low, or no voltage. It controls the field voltage for the required amount of charging.

Limitador de tensión Conectado por el red de resistores de un regulador de tensión. Determina si el campo recibirá alta, baja o ninguna tensión. Controla la tensión de campo durante el tiempo indicado de carga.

Voltage output test Used to make a quick determination about whether the charging system is working properly. If the charging system is operating correctly, then check for battery drain.

Prueba de salida del voltaje Se usa para determinar en forma rápida si el sistema de carga está trabajando correctamente. Si el sistema de carga está funcionando correctamente, entonces revíselo todo para saber si hay algún drenaje en la batería.

Voltage regulator Used to control the output voltage of the AC generator, based on charging system demands, by controlling field current.

Regulador de tensión Dispositivo cuya función es mantener la tensión de salida del generador de corriente alterna, de acuerdo a las variaciones en la corriente de carga, controlando la corriente inductora.

Voltmeter A test meter used to read the pressure behind the flow of electrons.

Voltímetro Instrumento de prueba utilizado para medir la presión del flujo de electrones.

Warning light A lamp that illuminates to warn the driver of a possible problem or hazardous condition.

Luz de aviso Lámpara que se enciende para avisarle al conductor sobre posibles problemas o condiciones peligrosas.

Watt The unit of measure of electrical power, which is the equivalent of horsepower. One horsepower is equal to 746 watts.

Watio Unidad de potencia eléctrica, equivalente a un caballo de vapor. 746 watios equivalen a un caballo de vapor (CV).

Wattage A measure of the total electrical work being performed per unit of time.

Vataje Medida del trabajo eléctrico total realizado por unidad de tiempo.

Waveform The electronic trace that appears on a scope; it represents voltage over time.

Forma de onda La trayectoria electrónica que aparece en un osciloscopio; representa la tensión a través del tiempo.

Weather-pack connector An electrical connector that has rubber seals on the terminal ends and on the covers of the connector half to protect the circuit from corrosion.

Conectador resistente a la intemperie Conectador que tiene sellos de caucho en los extremos de los bornes y en las cubiertas de la parte del conectador para proteger el circuito contra la corrosión.

Wet fouling A condition of a spark plug in which it is wet with oil.

Engrase húmedo Una condición de la bujía en la cual se moja de aceite.

Wheatstone bridge A series–parallel arrangement of resistors between an input terminal and ground. Flexing of the disc on which the resistors are laid changes their value.

Puente Wheatstone Una disposición en serie paralela de resistores entre un terminal de entrada y la tierra. La flexión del disco en el que están colocados los resistores cambia su valor.

Wiring diagram An electrical schematic that shows a representation of actual electrical or electronic components and the wiring of the vehicle's electrical systems.

Esquema de conexiones Esquema en el que se muestran las conexiones internas de los componentes eléctricos o electronicos reales y las de los sistemas eléctricos del vehículo.

Wiring harness A group of wires enclosed in a conduit and routed to specific areas of the vehicle.

Cableado preformado Conjunto de alambres envueltos en un conducto y dirigidos hacia áreas específicas del vehículo.

Work order A legal document representing an agreement for a repair facility to provide services to a customer and give the shop the authorization to perform the service at the quoted estimate. The work order contains customer, vehicle, and shop information and describes issues customers experience with their vehicle. It is also used by the technician to track diagnostic and repair times.

Orden de trabajo Un documento legal que representa un acuerdo para que una instalación de reparación preste servicios a un cliente y le dé autorización al taller para realizar el servicio según la estimación cotizada. La orden de trabajo contiene información del cliente, vehículo y taller, y describe los problemas que experimentan los clientes con su vehículo. También la utiliza el técnico para rastrear los diagnósticos y tiempos de reparación.

Workplace hazardous materials information systems (WHMIS) Canadian equivalent to MSDS.

Sistemas de información acerca de los materiales peligrosos en el área de trabajo (SIMPAT o WHMIS) Equivalente canadiense a las hojas de datos de la seguridad de un material.

Worm gear A type of gear whose teeth wrap around the shaft. The action of the gear is much like that of a threaded bolt or screw.

Engranaje de tornillo sin fin Un tipo de engranaje cuyos dientes se envuelven alrededor del vástago. El movimiento del engranaje es muy parecido a un perno enroscado o una tuerca.

Wye connection A type of stator winding in which one end of the individual windings is connected at a common point. The structure resembles the letter *Y*.

Conexión Y Un tipo de devanado estátor en el cual una extremidad de los devanados individuales se conectan en un punto común. La estructura parece la letra "Y".

Index